Lecture Notes in Computer Science 3988

Commenced Publication in 1973
Founding and Former Series Editors:
Gerhard Goos, Juris Hartmanis, and Jan van Leeuwen

T0145081

Arnold Beckmann   Ulrich Berger
Benedikt Löwe   John V. Tucker (Eds.)

Logical Approaches to Computational Barriers

Second Conference on Computability in Europe, CiE 2006
Swansea, UK, June 30-July 5, 2006
Proceedings

 Springer

Volume Editors

Arnold Beckmann
Ulrich Berger
John V. Tucker
University of Wales Swansea
Department of Computer Science
Singleton Park, Swansea SA2 8PP, UK
E-mail: {a.beckmann,u.berger,J.V.Tucker}@swansea.ac.uk

Benedikt Löwe
Universiteit van Amsterdam
Institue for Logic, Language and Computation (ILLC)
Plantage Muidergracht 24, 1018 TV Amsterdam, The Netherlands
E-mail: bloewe@science.uva.nl

Library of Congress Control Number: 2006927797

CR Subject Classification (1998): F.1, F.2.1-2, F.4.1, G.1.0, I.2.6, J.3

LNCS Sublibrary: SL 1 – Theoretical Computer Science and General Issues

ISSN   0302-9743
ISBN-10   3-540-35466-2 Springer Berlin Heidelberg New York
ISBN-13   978-3-540-35466-6 Springer Berlin Heidelberg New York

Springer is a part of Springer Science+Business Media

springer.com

© Springer-Verlag Berlin Heidelberg 2006
Printed in Germany

Typesetting: Camera-ready by author, data conversion by Scientific Publishing Services, Chennai, India
Printed on acid-free paper   SPIN: 11780342   06/3142   5 4 3 2 1 0

Preface

CiE 2006: Logical Approaches to Computational Barriers
Swansea, Wales, June 30 - July 5, 2006

Computability in Europe (CiE) is an informal network of European scientists working on computability theory, including its foundations, technical development, and applications. Among the aims of the network is to advance our theoretical understanding of what can and cannot be computed, by *any* means of computation. Its scientific vision is broad: computations may be performed with discrete or continuous data by all kinds of algorithms, programs, and machines. Computations may be made by experimenting with any sort of physical system obeying the laws of a physical theory such as Newtonian mechanics, quantum theory or relativity. Computations may be very general, depending upon the foundations of set theory; or very specific, using the combinatorics of finite structures. CiE also works on subjects intimately related to computation, especially theories of data and information, and methods for formal reasoning about computations. The sources of new ideas and methods include practical developments in areas such as neural networks, quantum computation, natural computation, molecular computation, and computational learning. Applications are everywhere, especially, in algebra, analysis and geometry, or data types and programming.

This volume, *Logical Approaches to Computational Barriers*, is the proceedings of the second in a series of conferences of CiE that was held at the Department of Computer Science, Swansea University, 30 June - 5 July, 2006.

The first meeting of CiE was at the University of Amsterdam, June 8–12, 2005, and its proceedings, edited by S. Barry Cooper, Benedikt Löwe and Leen Torenvliet, was published as *Springer Lecture Notes in Computer Science*, Volume 3526. We are sure that all of the 200+ mathematicians and computer scientists attending that conference had their view of computability theory enlarged and transformed: they discovered that its foundations were deeper and more mysterious, its technical development more vigorous, its applications wider and more challenging than they had known. We believe the same is certainly true of the Swansea meeting.

CiE 2005 and CiE 2006 are at the start of a new conference series *Computability in Europe*. The series is coordinated by the CiE Steering Committee:

S. Barry Cooper (Leeds)
Benedikt Löwe (Amsterdam, Chair)
Elvira Mayordomo (Zaragoza)
Dag Normann (Oslo)
Andrea Sorbi (Siena)
Peter van Emde Boas (Amsterdam).

We will reconvene 2007 in Siena, 2008 in Athens, 2009 in Heidelberg, and 2010 in Lisbon.

Structure and Programme of the Conference

The conference was based on invited tutorials and lectures, and a set of special sessions on a range of subjects; there were also many contributed papers and informal presentations. This volume contains 30 of the invited lectures and 39.7% of the submitted contributed papers, all of which have been refereed. There will be a number of post-proceedings publications, including special issues of *Theoretical Computer Science*, *Theory of Computing Systems*, and *Journal of Logic and Computation*.

Tutorials

Samuel R. Buss (San Diego, CA), *Proof Complexity and Computational Hardness*
Julia Kempe (Paris), *Quantum Algorithms*

Invited Plenary Talks

Jan Bergstra (Amsterdam), *Elementary Algebraic Specifications of the Rational Function Field*
Luca Cardelli (Cambridge), *Biological Systems as Reactive Systems*
Martin Davis (New York), *The Church-Turing Thesis: Consensus and Opposition*
John W. Dawson (York, PA), *Gödel and the Origins of Computer Science*
Jan Krajíček (Prague), *Forcing with Random Variables and Proof Complexity*
Elvira Mayordomo (Zaragoza), *Two Open Problems on Effective Dimension*
István Németi (Budapest), *Can General Relativistic Computers Break the Turing Barrier?*
Helmut Schwichtenberg (Munich), *Inverting Monotone Continuous Functions in Constructive Analysis*
Andreas Weiermann (Utrecht), *Phase Transition Thresholds for Some Natural Subclasses of the Computable Functions*

Special Sessions

Proofs and Computation, organized by Alessandra Carbone and Thomas Strahm

Kai Brünnler (Bern), *Deep Inference and Its Normal Form of Derivations*
Roy Dyckhoff (St. Andrews), *LJQ: A Strongly Focused Calculus for Intuitionistic Logic*
Thomas Ehrhard (Marseille), *Böhm Trees, Krivine's Machine and the Taylor Expansion of Lambda-Terms*
Georges Gonthier (Cambridge), *Using Reflection to Prove the Four-Colour Theorem*

Computable Analysis, organized by Peter Hertling and Dirk Pattinson

Margarita Korovina (Aarhus), *Upper and Lower Bounds on Sizes of Finite Bisimulations of Pfaffian Hybrid Systems*
Paulo Oliva (London), *Understanding and Using Spector's Bar Recursive Interpretation of Classical Analysis*
Matthias Schröder (Edinburgh), *Admissible Representations in Computable Analysis*
Xizhong Zheng (Cottbus), *A Computability Theory of Real Numbers*

Challenges in Complexity, organized by Klaus Meer and Jacobo Torán

Johannes Köbler (Berlin), *Complexity of Graph Isomorphism for Restricted Graph Classes*
Sophie Laplante (Paris), *Lower Bounds Using Kolmogorov Complexity*
Johann A. Makowsky (Haifa), *From a Zoo to a Zoology: Descriptive Complexity for Graph Polynomials*
Mihai Prunescu (Freiburg), *Fast Quantifier Elimination Means P = NP*

Foundations of Programming, organized by Inge Bethke and Martín Escardó

Erika Ábrahám (Freiburg), *Heap-Abstraction for an Object-Oriented Calculus with Thread Classes*
Roland Backhouse (Nottingham), *Datatype-Generic Reasoning*
James Leifer (Le Chesnay), *Transactional Atomicity in Programming Languages*
Alban Ponse (Amsterdam), *An Introduction to Program and Thread Algebra*

Mathematical Models of Computers and Hypercomputers, organized by Joel D. Hamkins and Martin Ziegler

Jean-Charles Delvenne (Louvain-la-Neuve), *Turing Universality in Dynamical Systems*
Benedikt Löwe (Amsterdam), *Space Bounds for Infinitary Computation*
Klaus Meer (Odense), *Optimization and Approximation Problems Related to Polynomial System Solving*
Philip Welch (Bristol), *Non-Deterministic Halting Times for Hamkins-Kidder Turing Machines*

Gödel Centenary: Gödel's Legacy for Computability, organized by Matthias Baaz and John W. Dawson

Arnon Avron (Tel Aviv), *From Constructibility and Absoluteness to Computability and Domain Independence*

Torkel Franzén † (Luleå), *What Does the Incompleteness Theorem Add to the Unsolvability of the Halting Problem?*

Wilfried Sieg (Pittsburgh, PA), *Gödel's Conflicting Approaches to Effective Calculability*

Richard Zach (Calgary, AB), *Kurt Gödel and Computability Theory*

Organization and Acknowledgements

The CiE 2006 conference was organized by the logicians and theoretical computer scientists at Swansea: Arnold Beckmann, Ulrich Berger, Phil Grant, Oliver Kullmann, Faron Moller, Monika Seisenberger, Anton Setzer, John V. Tucker; and with the help of S. Barry Cooper (Leeds) and Benedikt Löwe (Amsterdam).

The Programme Committee was chaired by Arnold Beckmann and John V. Tucker and consisted of:

Samson Abramsky (Oxford)
Klaus Ambos-Spies (Heidelberg)
Arnold Beckmann (Swansea, Co-chair)
Ulrich Berger (Swansea)
Olivier Bournez (Nancy)
S. Barry Cooper (Leeds)
Laura Crosilla (Firenze)
Costas Dimitracopoulos (Athens)
Abbas Edalat (London)
Fernando Ferreira (Lisbon)
Ricard Gavaldà (Barcelona)
Giuseppe Longo (Paris)

Benedikt Löwe (Amsterdam)
Yuri Matiyasevich (St.Petersburg)
Dag Normann (Oslo)
Giovanni Sambin (Padova)
Uwe Schöning (Ulm)
Andrea Sorbi (Siena)
Ivan N. Soskov (Sofia)
Leen Torenvliet (Amsterdam)
John V. Tucker (Swansea, Co-chair)
Peter van Emde Boas (Amsterdam)
Klaus Weihrauch (Hagen)

We are delighted to acknowledge and thank the following for their essential financial support: the Department of Computer Science at Swansea, IT Wales, the Welsh Development Agency, the UK's Engineering and Physical Sciences Research Council, the British Logic Colloquium, the London Mathematical Society, and the Kurt Gödel Society of Vienna. Furthermore, we thank our sponsors the Association for Symbolic Logic, the European Association for Theoretical Computer Science, and the British Computer Society.

The high scientific quality of the conference was possible through the conscientious work of the Programme Committee, the special session organizers and the referees. We are grateful to all members of the Programme Committee for their efficient evaluations and extensive debates, which established the final programme. We also thank the following referees:

Klaus Aehlig
Jeremy Avigad
George Barmpalias
Andrej Bauer
A.P. Beltiukov
Jens Blanck
Stefan Bold
Ana Bove
Vasco Brattka
Robert Brijder
Andrei Bulatov
Wesley Calvert
Véronique Cortier
Ugo Dal Lago
Víctor Dalmau
K. Djemame
Mário Jorge Edmundo
Ioannis Emiris
Juan L. Esteban
Graham Farr
T. Flaminio
Hervé Fournier
Torkel Franzén †
Nicola Galesi
Nicola Gambino
Bert Gerards
Sergey S. Goncharov

Emmanuel Hainry
Neal Harman
Takis Hartonas
Peter Hertling
Pascal Hitzler
Jan Johannsen
Reinhard Kahle
Ker-I Ko
Margarita Korovina
N.K. Kosovsky
Sven Kosub
Oliver Kullmann
Antoni Lozano
Maria Emilia Maietti
Guillaume Malod
Simone Martini
Klaus Meer
Wolfgang Merkle
Christian Michaux
Joseph Miller
Faron Moller
Franco Montagna
Yiannis Moschovakis
Philippe Moser
Sara Negri
Stela Nikolova
Milad Niqui

A. Pastor
Dirk Pattinson
Hervé Perdry
Bruno Poizat
Chris Pollett
Diane Proudfoot
Michael Rathjen
Jan Reimann
Robert Rettinger
Fred Richman
Piet Rodenburg
Panos Rondogiannis
Anton Setzer
Cosma Shalizi
Dieter Spreen
Umberto Straccia
Sebastiaan Terwijn
Neil Thapen
Sergey Verlan
Jouko Väänänen
Philip Welch
Damien Woods
Y. Yang
Xizhong Zheng
Ning Zhong
Martin Ziegler
Jeff Zucker

Of course, the conference was primarily an event, and a rather complicated event at that. We are delighted to thank our colleagues on the Organizing Committee for their many contributions and our research students for practical help at the conference. We owe a special thanks to Beti Williams, Director of IT Wales, and her team for invaluable practical work. For example, IT Wales arranged a "business breakfast" where the wider computing community was addressed on why the research of CiE might be intriguing and of value to them in the years to come: a special thanks to Jan Bergstra for undertaking the task of briefing our captains of industry at break of day.

Finally, we thank Andrej Voronkov for his Easy Chair system which facilitated the work of the Programme Committee and the editors considerably.

Swansea and Amsterdam, April 2006

Arnold Beckmann
Ulrich Berger
Benedikt Löwe
John V. Tucker

After completing this volume, we heard the sad news that our invited Special Session speaker, Torkel Franzén, died on April 19, 2006. Torkel Franzén's work on the philosophy of logic and mathematics had gained more and more international recognition in recent years. His death is a huge loss for the scientific community and he will be very much missed at CiE 2006. Torkel Franzén did send us an abstract of his planned contribution to this conference which we have included in this volume.

The Editors

Table of Contents

Heap-Abstraction for an Object-Oriented Calculus with Thread Classes[*]

Erika Ábrahám[1], Andreas Grüner[2], and Martin Steffen[2]

[1] Albert-Ludwigs-University Freiburg, Germany
eab@informatik.uni-freiburg.de

[2] Christian-Albrechts-University Kiel, Germany
{ang, ms}@informatik.uni-kiel.de

Abstract. This paper formalizes an open semantics for a calculus featuring thread classes, where the environment, consisting in particular of an overapproximation of the heap topology, is abstractly represented.

We extend our prior work not only by adding thread classes, but also in that thread names may be *communicated*, which means that the semantics needs to account explicitly for the possible acquaintance of objects with threads. We show soundness of the abstraction.

Keywords: class-based oo languages, thread-based concurrency, open systems, formal semantics, heap abstraction, observable behavior.

1 Introduction

An *open* system is a program fragment or component interacting with its environment. The behavior of the component can be understood to consist of message traces at the interface, i.e., of sequences of component-environment interaction. Even if the environment is absent, it must be assured that the component together with the (abstracted) environment gives a well-formed program adhering to the syntactical and the context-sensitive restrictions of the language at hand. Technically, for an exact representation of the interface behavior, the semantics of the open program needs to be formulated under *assumptions* about the environment, capturing those restrictions. The resulting assumption-commitment framework gives insight to the semantical nature of the language. Furthermore, a characterization of the interface behavior with environment and component abstracted can be seen as a trace logic under the most general assumptions, namely conformance to the inherent restrictions of the language and its semantics.

With these goals in mind, we deal with languages supporting:

- *types and classes:* the languages are statically typed, and only well-typed programs are considered. For class-based languages, complications arise as classes play the role of types and additionally act as *generators* of objects.
- *concurrency:* the languages feature concurrency based on *threads* and *thread classes* (as opposed to processes or active objects).

---

[*] Part of this work has been financially supported by the NWO/DFG project Mobi-J (RO 1122/9-4) and by the DFG project AVACS (SFB/TR-14-AVACS).

A. Beckmann et al. (Eds.): CiE 2006, LNCS 3988, pp. 1–10, 2006.

– *references:* each object carries a unique *identity.* New objects are dynamically allocated on the heap as *instances of classes.*

The interface behavior is phrased in an assumption-commitment framework and is based on three orthogonal abstractions:

– a static abstraction, i.e., the type system;
– an abstraction of the stacks of recursive method invocations, representing the recursive and reentrant nature of method calls in a multi-threaded setting;
– an abstraction of the *heap topology,* approximating potential connectivity of objects and threads. The heap topology is dynamic due to object creation and tree-structured in that previously separate object groups may merge.

In [1,2] we showed that the need to represent the heap topology is a direct consequence of considering *classes* as a language concept. Their foremost role in object-oriented languages is to act as *"generators of state".* With *thread classes,* there is also a mechanism for *"generating new activity".* This extension makes cross-border activity generation a possible component-environment interaction, i.e., the component may create threads in the environment and vice versa.

Thus, the technical contribution of this paper is threefold. We extend the class-based calculus [1,2] with *thread classes* and allow to communicate thread names. This requires to consider cross-border *activity* generation and to incorporate the connectivity of objects *and* threads. Secondly, we characterize the potential traces of *any* component in an assumption-commitment framework in a novel derivation system: The branching nature of the heap abstraction—connected object groups may merge by communication—is reflected in the branching structure of the derivation system. Finally, we show soundness of the abstractions.

Overview. The paper is organized as follows. Section 2 defines syntax and semantics of the calculus. Section 3 characterizes the observable behavior of an open system and presents the soundness results. Related and future work is discussed in Section 4. See [3] for a full description of semantics and type system.

2   A Multi-threaded Calculus with Thread Classes

2.1   Syntax

The abstract syntax is given in Table 1. For names, we will generally use o and its syntactic variants for objects, c for classes (in particular c_t for thread classes), and n when being unspecific. A class $c[\![O]\!]$ with name c defines its methods and fields. A method $\varsigma(self{:}c).t_a$ provides the method body abstracted over the ς-bound "self" and the formal parameters. An object $o[c, F]$ of type c stores the current field values. We use l for fields, $l = f$ for field declaration, field access is written as $v.l$, and field update as $v.l := v'$. *Thread classes* $c_t\langle\!\langle t_a\rangle\!\rangle$ with name c_t carry their abstract code in t_a. A thread $n\langle t\rangle$ with name n is basically either a value or a sequence of expressions, notably method calls $v.l(\vec{v})$, object creation $new\ c$, and *thread instantiation* $spawn\ c_t(\vec{v})$.

Table 1. Abstract syntax

$$
\begin{aligned}
C &::= \mathbf{0} \mid C \parallel C \mid \nu(n{:}T).C \mid n[\![O]\!] \mid n[n, F] \mid n\langle t\rangle \mid n\langle\!\langle t_a \rangle\!\rangle && \text{program} \\
O &::= F, M && \text{object} \\
M &::= l = m, \ldots, l = m && \text{method suite} \\
F &::= l = f, \ldots, l = f && \text{fields} \\
m &::= \varsigma(n{:}T).t_a && \text{method} \\
f &::= \varsigma(n{:}T).\lambda().v \mid \varsigma(n{:}T).\lambda().\bot_c && \text{field} \\
t_a &::= \lambda(x{:}T, \ldots, x{:}T).t && \text{parameter abstraction} \\
t &::= v \mid stop \mid let\ x{:}T = e\ in\ t && \text{thread} \\
e &::= t \mid \text{if } v = v \text{ then } e \text{ else } e \mid \text{if } undef(v.l) \text{ then } e \text{ else } e && \text{expression} \\
&\quad \mid v.l(v, \ldots, v) \mid v.l := v \\
&\quad \mid currentthread \mid new\ n \mid spawn\ n(v, \ldots, v) \\
v &::= x \mid n && \text{values}
\end{aligned}
$$

As types we have *thread* for threads, class names c as object types, $T_1 \times \ldots \times T_k \to T$ as the type of methods and thread classes (in last case T equals *thread*), $[l_1{:}U_1, \ldots, l_k{:}U_k]$ for unnamed objects, and $[\![l_1{:}U_1, \ldots, l_k{:}U_k]\!]$ for classes.

2.2 Operational Semantics

For lack of space we concentrate on the interface behavior and omit the definitions of the component-internal steps like internal method calls [3].

The external steps define the interaction between component and environment in an assumption-commitment context. The static part of the context corresponds to the static type system [3] and takes care that, e.g., only well-typed values are received from the environment. The context, however, needs to contain also a *dynamic* part dealing with the potential *connectivity* of objects and thread names, corresponding to an abstraction of the heap topology.

The component-environment interaction is represented by labels a:

$$
\begin{aligned}
\gamma &::= n\langle call\ o.l(\vec{v})\rangle \mid n\langle return(v)\rangle \mid \langle spawn\ n\ of\ c(\vec{v})\rangle \mid \nu(n{:}T).\gamma \\
a &::= \gamma? \mid \gamma!
\end{aligned}
$$

For call and return, n is the active thread. For spawning, n is the new thread. There are no labels for object creation: Externally instantiated objects are created only when they are accessed for the first time (*"lazy instantiation"*). For labels $a = \nu(\Phi).\gamma?$ or $a = \nu(\Phi).\gamma!$ with Φ a sequence of $\nu(n{:}T)$ bindings and γ does not contain any binders, $\lfloor a \rfloor = \gamma$ is the *core* of the label a.

2.2.1 Connectivity Contexts

In the presence of cross-border instantiation, the semantics must contain a representation of the connectivity, which is formalized by a relation on names and which can be seen as an abstraction of the program's heap; see Eq. (2) and (3) below for the exact definition. The external semantics is formalized as labeled transitions between judgments of the form

$$\Delta, \Sigma; E_\Delta \vdash C : \Theta, \Sigma; E_\Theta , \tag{1}$$

where $\Delta, \Sigma; E_\Delta$ are the *assumptions* about the environment of C and $\Theta, \Sigma; E_\Theta$ the *commitments*. The assumptions consist of a part Δ, Σ concerning the existence (plus static typing information) of *named entities* in the environment. By convention, the contexts Σ (and their alphabetic variants) contain exactly all bindings for thread names. The semantics maintains as invariant that for all judgments $\Delta, \Sigma; E_\Delta \vdash C : \Theta, \Sigma; E_\Theta$ that Δ, Σ, and Θ are pairwise disjoint.

The semantics must book-keep which objects of the environment have been told which identities: It must take into account the *relation* of objects from the assumption context Δ amongst each other, and the knowledge of objects from Δ about thread names and names exported by the component, i.e., those from Θ. In analogy to the name contexts Δ and Θ, the connectivity context E_Δ expresses assumptions about the environment, and E_Θ commitments of the component:

$$E_\Delta \subseteq \Delta \times (\Delta + \Sigma + \Theta) \quad and \quad E_\Theta \subseteq \Theta \times (\Theta + \Sigma + \Delta) . \tag{2}$$

Since thread names may be communicated, we must include pairs from $\Delta \times \Sigma$ (resp. $\Theta \times \Sigma$) into the connectivity. We write $o \hookrightarrow n$ ("o may know n") for pairs from E_Δ and E_Θ. Without full information about the complete system, the component must make worst-case assumptions concerning the proliferation of knowledge, which are represented as the *reflexive*, *transitive*, and *symmetric* closure of the $\hookrightarrow$-pairs of *objects from* Δ. We write $\leftrightharpoons$ for this closure:

$$\leftrightharpoons \triangleq (\hookrightarrow\downarrow_\Delta \cup \hookleftarrow\downarrow_\Delta)^* \subseteq \Delta \times \Delta , \tag{3}$$

where $\hookrightarrow\downarrow_\Delta$ is the projection of $\hookrightarrow$ to Δ. We also need the union $\leftrightharpoons \cup \leftrightharpoons; \hookrightarrow \subseteq \Delta \times (\Delta + \Sigma + \Theta)$, where the semicolon denotes relational composition. We write $\leftrightharpoons\hookrightarrow$ for that union. As judgment, we use $\Delta, \Sigma; E_\Delta \vdash o_1 \leftrightharpoons o_2 : \Theta, \Sigma$, resp. $\Delta, \Sigma; E_\Delta \vdash o \leftrightharpoons\hookrightarrow n : \Theta, \Sigma$. For Θ, Σ, E_Θ, and Δ, Σ, the definitions are dual.

The relation $\leftrightharpoons$ partitions the objects from Δ (resp. Θ) into equivalence classes. We call a set of object names from Δ (or dually from Θ) such that for all objects o_1 and o_2 from that set, $\Delta, \Sigma; E_\Delta \vdash o_1 \leftrightharpoons o_2 : \Theta, \Sigma$, a *clique*, and if we speak of *the* clique of an object we mean the equivalence class.

If a thread is instantiated *without* connection to any object, like the initial thread, we need a syntactical representation $\odot_n$ for the clique the thread n starts in. If the single initial thread starts within the component, the contexts of the initial configuration $\Delta_0 \vdash C : \Theta_0$ assert $\Theta_0 \vdash \odot$. Otherwise, $\Delta_0 \vdash \odot$.

As for the relationship of communicated values, incoming and outgoing communication play dual roles: E_Θ over-approximates the actual connectivity of the component and is updated in incoming communications, while the assumption context E_Δ is consulted to exclude impossible incoming values, and is updated in outgoing communications. Incoming new names update both E_Θ and E_Δ.

2.2.2 Augmentation

We extend the syntax by two auxiliary expressions o_1 *blocks for* o_2 and o_2 *returns to* o_1 v, denoting a method body in o_1 waiting for a return from o_2, and dually for the return of v from o_2 to o_1. We augment the method definitions accordingly, such that each method call and spawn action is annotated by the caller. I.e., we write

$$\varsigma(self\!:\!c).\lambda(\vec{x}\!:\!\vec{T}).(\dots self\,x.l(\vec{y})\dots self\ spawn\ c_t(\vec{z})\dots)\ .$$

instead of $\varsigma(self\!:\!c).\lambda(\vec{x}\!:\!\vec{T}).(\dots x.l(\vec{y})\dots spawn\ c_t(\vec{z})\dots)$. Thread classes are augmented by $\odot$ instead of *self*. If a thread n is instantiated, $\odot$ is replaced by $\odot_n$. For a thread class of the form $c_t \langle\!\langle (\lambda(\vec{x}\!:\!\vec{T}).t)\rangle\!\rangle$, let $c_t(\vec{v})$ denote $t[\odot_n, \vec{v}/\odot, \vec{x}]$. The initial thread n_0, which is not instantiated but is given directly (in case it starts in the component), has $\odot_{n_0}$ as augmentation. We omit the adaptation of the internal semantics and the typing rules for the augmentation.

2.2.3 Use and Change of Contexts

Notation 1. *We abbreviate the triple of name contexts* Δ, Σ, Θ *as* Φ, *the context* $\Delta, \Sigma, \Theta, E_\Delta, E_\Theta$ *combining assumptions and commitments as* Ξ, *and write* $\Xi \vdash C$ *for* $\Delta, \Sigma; E_\Delta \vdash C : \Theta, \Sigma; E_\Theta$. *We use syntactical variants analogously.*

The operational semantics is formulated as transitions between typed judgments $\Xi \vdash C \xrightarrow{a} \acute{\Xi} \vdash \acute{C}$. The assumption context $\Delta, \Sigma; E_\Delta$ is an abstraction of the environment, as it represents the potential behavior of all possible environments. The check whether the current assumptions are met in an incoming communication step is given in Definition 1. Note that in case of an incoming *call* label, $fn(a)$, the free names in a, includes the receiver o_r and the thread name.

Definition 1 (Connectivity check). *An incoming core label a with sender o_s and receiver o_r is well-connected wrt.* $\acute{\Xi}$ *(written* $\acute{\Xi} \vdash o_s \xrightarrow{a} o_r$:*ok* *) if* $\acute{\Delta}, \acute{\Sigma}; \acute{E}_\Delta \vdash o_s \rightleftharpoons\hookrightarrow fn(a) : \acute{\Theta}, \acute{\Sigma}.$

Besides *checking* whether the connectivity assumptions are met before a transition, the contexts are *updated* by a step, reflecting the change of knowledge.

Definition 2 (Name context update: $\Phi + a$ **).** *The update* $\acute{\Phi} = \Phi + a$ *of an assumption-commitment context* Φ *wrt. an incoming label* $a = \nu(\Phi')\lfloor a\rfloor?$ *is defined as follows.*

1. $\acute{\Theta} = \Theta + \Theta'$. *For spawning,* $\acute{\Theta} = \Theta + (\Theta', \odot_n)$ *with n the spawned thread.*
2. $\acute{\Delta} = \Delta + (\odot_{\Sigma'}, \Delta')$. *For spawning of thread n,* $\odot_{\Sigma'\setminus n}$ *is used instead of* $\odot_{\Sigma'}$.
3. $\acute{\Sigma} = \Sigma + \Sigma'$.

The notation $\odot_{\Sigma'}$ *abbreviates* $\odot_n$ *for all thread identities from* Σ'. *The update for outgoing communication is defined dually (* $\odot_n$ *of a spawn label is added to* Δ *instead of* Θ, *and the* $\odot_{\Sigma'}$ *resp.* $\odot_{\Sigma'\setminus n}$ *are added to* Θ, *instead of* Δ).

Definition 3 (Connectivity context update). *The update* $(\acute{E}_\Delta, \acute{E}_\Theta) = (E_\Delta,$ $E_\Theta) + o_s \xrightarrow{a} o_r$ *of an assumption-commitment context* (E_Δ, E_Θ) *wrt. an incoming label* $a = \nu(\Phi')\lfloor a \rfloor$? *with sender* o_s *and receiver* o_r *is defined as follows.*

1. $\acute{E}_\Theta = E_\Theta + o_r \hookrightarrow fn(\lfloor a \rfloor)$.
2. $\acute{E}_\Delta = E_\Delta + o_s \hookrightarrow \Phi', \odot_{\Sigma'}$. *For spawning of* n, $\odot_{\Sigma' \setminus n}$ *is used instead of* $\odot_{\Sigma'}$.

Combining Definitions 2 and 3, we write $\acute{\Xi} = \Xi + o_s \xrightarrow{a} o_r$ when updating the name and the connectivity at the same time.

Besides the connectivity check of Definition 1, we must also check the *static* assumptions, i.e., whether the transmitted values are of the correct types. In slight abuse of notation, we write $\Phi \vdash o_s \xrightarrow{a} o_r : T$ for that check, where T is type of the expression in the program that gives rise to the label (see [3] for the definition). We combine the connectivity check of Definition 1 and the type check into a single judgment $\Xi \vdash o_s \xrightarrow{a} o_r : T$.

2.2.4 Operational Rules

Three CALLI-rules for incoming calls deal with three different situations: A call reentrant on the level of the component, a call of a thread whose name is already known by the component, and a call of a thread new to the component. For all three cases, the contexts are *updated* to $\acute{\Xi}$ to include the information concerning new objects, threads, and connectivity transmitted in that step. Furthermore, it is *checked* whether the label statically type-checks and that the step is possible according to the (updated) connectivity assumptions $\acute{\Xi}$. Remember that the update from Ξ to $\acute{\Xi}$ includes guessing of connectivity.

To deal with component entities (threads and objects) that are being created during the call, $C(\Theta', \Sigma')$ stands for $C(\Theta') \parallel C(\Sigma')$, where $C(\Theta')$ are the lazily instantiated objects mentioned in Θ'. Furthermore, for each thread name n' in Σ', a new component $n'\langle stop \rangle$ is included, written as $C(\Sigma')$.

For reentrant method calls in rule CALLI$_1$, the thread is blocked, i.e., it has left the component previously via an outgoing call. The object o_s that had been the target of the call is remembered as part of the augmented block syntax, and is used now to represent the sender's clique for the current incoming call.

In CALLI$_2$, the thread is not in the component, but the thread's name is already known. If $\Delta \vdash \odot_n$ and $n\langle stop \rangle$ is part of the component code, it is assured that the thread either has never actively entered the component before (and does so right now) or has left the component to the environment by some last outgoing return. In either case, the incoming call is possible now, and in both cases we can use $\odot_n$ as representative of the caller's identity.

In CALLI$_3$ a new thread n enters the component for the first time, as assured by $\Sigma' \vdash n : thread$. The new thread must be an instance of an environment thread class created by an environment clique, otherwise the cross-border instantiation would have been observed and the thread name would not be fresh. Since *any* existing environment clique is a candidate, the update to $\acute{\Xi}$ *non-deterministically*

Table 2. External steps

$$\acute{\Xi} = \Xi + o_s \xrightarrow{a} o_r \quad \acute{\Xi} \vdash o_s \xrightarrow{\lfloor a \rfloor} o_r : T$$
$$a = \nu(\Phi'). \; n\langle call \; o_r.l(\vec{v})\rangle? \quad t_{blocked} = let \; x':T' = o \; blocks \; for \; o_s \; in \; t$$
$$\rule{11cm}{0.4pt} \text{CALLI}_1$$
$$\Xi \vdash \nu(\Phi_1).(C \parallel n\langle t_{blocked}\rangle) \xrightarrow{a}$$
$$\acute{\Xi} \vdash \nu(\Phi_1).(C \parallel C(\Theta', \Sigma') \parallel n\langle let \; x:T = o_r.l(\vec{v}) \; in \; o_r \; returns \; to \; o_s \; x; t_{blocked}\rangle)$$

$$a = \nu(\Phi'). \; n\langle call \; o_r.l(\vec{v})\rangle? \quad \Delta \vdash \odot_n \quad \acute{\Xi} = \Xi + \odot_n \xrightarrow{a} o_r \quad \acute{\Xi} \vdash \odot_n \xrightarrow{\lfloor a \rfloor} o_r : T$$
$$\rule{11cm}{0.4pt} \text{CALLI}_2$$
$$\Xi \vdash C \parallel n\langle stop \rangle \xrightarrow{a} \acute{\Xi} \vdash C \parallel C(\Theta', \Sigma') \parallel n\langle let \; x:T = o_r.l(\vec{v}) \; in \; o_r \; returns \; to \; \odot_n \; x; stop \rangle$$

$$a = \nu(\Phi'). \; n\langle call \; o_r.l(\vec{v})\rangle? \quad \Delta \vdash o \quad \Sigma' \vdash n \quad \acute{\Xi} = \Xi + o \xrightarrow{a} o_r \quad \acute{\Xi} \vdash \odot_n \xrightarrow{\lfloor a \rfloor} o_r : T$$
$$\rule{11cm}{0.4pt} \text{CALLI}_3$$
$$\Xi \vdash C \xrightarrow{a} \acute{\Xi} \vdash C \parallel C(\Theta', \Sigma' \setminus n) \parallel n\langle let \; x:T = o_r.l(\vec{v}) \; in \; o_r \; returns \; to \; \odot_n \; x; stop \rangle$$

$$a = \nu(\Phi'). \; n\langle call \; o_r.l(\vec{v})\rangle! \quad \Phi' = fn(\lfloor a \rfloor) \cap \Phi_1 \quad \acute{\Phi}_1 = \Phi_1 \setminus \Phi' \quad \Delta \vdash o_r \quad \acute{\Xi} = \Xi + o_s \xrightarrow{a} o_r$$
$$\rule{11cm}{0.4pt} \text{CALLO}$$
$$\Xi \vdash \nu(\Phi_1).(C \parallel n\langle let \; x:T = o_s \; o_r.l(\vec{v}) \; in \; t\rangle) \xrightarrow{a}$$
$$\acute{\Xi} \vdash \nu(\acute{\Phi}_1).(C \parallel n\langle let \; x:T = o_s \; blocks \; for \; o_r \; in \; t\rangle)$$

$$\acute{\Xi} = \Xi + o_s \xrightarrow{a} \odot_n \quad \acute{\Xi} \vdash o_s \xrightarrow{\lfloor a \rfloor} \odot_n : thread$$
$$a = \nu(\Phi'). \langle spawn \; n \; of \; c_t(\vec{v})\rangle? \quad \acute{\Theta} \vdash \odot_n \quad \Delta \vdash o_s \quad \Theta \vdash c_t \quad \Sigma' \vdash n$$
$$\rule{11cm}{0.4pt} \text{SPAWNI}$$
$$\Xi \vdash C \xrightarrow{a} \acute{\Xi} \vdash C \parallel C(\Theta', \Sigma' \setminus n) \parallel n\langle c_t(\vec{v})\rangle$$

---

guesses to which environment clique the thread's origin $\odot_n$ belongs to. Note that $\odot_{\Sigma'}$ contains $\odot_n$ since $\Sigma' \vdash n$, which means $\Delta \vdash \odot_n$ after the call.

For incoming thread creation in SPAWNI the situation is similar to CALLI$_3$, in that the spawner needs to be guessed. The last rule deals with outgoing call and is simpler, as the "check-part" is omitted: With the code of the program present, the checks are guaranteed to be satisfied. In Table 2 we omitted the rules for outgoing spawning, for returns, and for the initial steps [3].

3   Legal Traces

Next we present an independent characterization of the possible interface behavior. "Half" of the work has been already done by the abstractly represented environment. For the legal traces, we analogously abstract away from the component, making the system completely symmetric.

3.1   A Branching Derivation System Characterizing Legal Traces

Instead of connectivity contexts, now the *tree structure* of the derivation represents the connectivity and its change. There are two variants of the derivation system, one from the perspective of the *component,* and one for the *environment.* Each derivation corresponds to a *forest,* with each tree representing a component resp. environment clique. In judgments $\Delta, \Sigma \vdash_\Theta r \triangleright s : trace \,\Theta, \Sigma$, r represents the history, and s the future interaction. We write $\vdash_\Theta$ to indicate that legality is checked from the perspective of the component. From that perspective, we maintain as invariant that on the commitment side, the context Θ represents one single clique. Thus the connectivity among objects of Θ needs no longer be remembered. What needs to be remembered still are the thread names known by Θ and the cross-border object connectivity, i.e., the acquaintance of the clique represented by Θ with objects of the environment. This is kept in Δ resp. Σ. Note that this corresponds to the environmental objects mentioned in $E_\Theta \subseteq \Theta \times (\Theta + \Delta + \Sigma)$, projected onto the component clique under consideration, in the linear system. The connectivity of the environment is *ignored* which implies that the system of Table 3 *cannot* assure that the environment behaves according to a possible connectivity. On the other hand, dualizing the rules checks whether the environment adheres to possible connectivity.

Table 3. Legal traces, branching on Θ (incoming call and skip)

$$\frac{\Phi = \bigoplus_\Theta \Phi_j \quad \Phi \vdash r \triangleright o_s \xrightarrow{a} o_r \quad \acute{\Phi} = \Phi_0, \Phi + a \quad \acute{\Phi} \vdash o_s \xrightarrow{\lfloor a \rfloor} o_r : ok}{\forall j.\; a_j = a \downarrow_{\Theta_j} \wedge \Theta_j \vdash \lfloor a \rfloor \quad a = \nu(\Phi').\, n\langle call\; o_r.l(\vec{v})\rangle? \quad r \neq \epsilon \quad \acute{\Phi} \vdash r\, a \triangleright s : trace}{\Phi_1 \vdash r \triangleright a_1\, s : trace \quad \ldots \quad \Phi_k \vdash r \triangleright a_k\, s : trace} \text{L-CALLI}$$

$$\frac{a = \gamma? \quad \Phi \vdash r \triangleright o_s \xrightarrow{a} o_r \quad \Theta \nvdash \lfloor a \rfloor, o_r \quad \Phi \vdash r a \triangleright s : trace \quad r \neq \epsilon}{\Phi \vdash r \triangleright s : trace} \text{L-SKIPI}$$

In L-CALLI of Table 3, the incoming call is possible only when the thread is input call enabled after the current history. This is checked by the premise $\Phi \vdash r \triangleright o_s \xrightarrow{a} o_r$, which also determines caller and callee. As from the perspective of the component, the connectivity of the environment is no longer represented as assumption, there are *no* premises checking connectivity! Interesting is the treatment of the commitment context: Incoming communication may *update* the component connectivity, in that new cliques may be created or existing cliques may merge. The merging of component cliques is now represented by a branching of the proof system. Leaves of the resulting tree (respectively forest) correspond to freshly created cliques. In L-CALLI, the context Θ in the premise corresponds to the merged clique, the Θ_i below the line to the still split cliques before the merge. The Θ_i's form a partitioning of the component objects before communication, Θ is the disjoint combination of the Θ_i's plus the lazily instantiated objects from Θ'. For the cross-border connectivity, i.e., the environmental objects known by the component cliques, the different component cliques Θ_i may of course share acquaintance; thus, the parts Δ_i and Σ_i are not merged disjointly, but by ordinary "set" union. These restrictions are covered by $\bigoplus \Xi_i$.

The skip-rules stipulate that an action a which does not belong to the component clique under consideration, is omitted from the component's "future" (interpreting the rule from bottom to top). We omit the remaining rules (see [3]).

Definition 4 (Legal traces, tree system). *We write* $\Delta \vdash_\Theta t : trace\,\Theta$, *if there exists a derivation forest using the rules of Table 3 with roots* $\Delta_i, \Sigma_i \vdash t \rhd \epsilon : trace\,\Theta_i, \Sigma_i$ *and a leaf justified by one of the initial rules* L-CALLI$_0$ *or* L-CALLO$_0$. *Using the dual rules, we write* $\vdash_\Delta$ *instead of* $\vdash_\Theta$. *We write* $\Delta \vdash_{\Delta\wedge\Theta} t : trace\,\Theta$, *if there exits a pair of derivations in the* $\vdash_\Delta$- *and the* $\vdash_\Theta$- *system with a consistent pair of root judgments.*

To accommodate for the simpler context structures, we adapt the notational conventions (cf. Notation 1) appropriately. The way a communication step updates the name context can be defined as simplification of the treatment in the operational semantics (cf. Definition 2). As before we write $\Phi \mid u$ for the update.

3.2   Soundness of the Abstractions

With E_Δ and E_Θ as part of the judgment, we must still clarify what it "means", i.e., when does $\Delta, \Sigma; E_\Delta \vdash C : \Theta, \Sigma; E_\Theta$ hold? The relation E_Θ asserts about the component C that the connectivity of the objects from the component is *not larger than* the connectivity entailed by E_Θ. Given a component C and two names o from Θ and n from $\Theta + \Delta + \Sigma$, we write $C \vdash o \hookrightarrow n$, if $C \equiv \nu(\Phi).(C' \parallel o[\ldots, f = n, \ldots])$ where o and n are not bound by Φ, i.e., o contains in one of its fields a reference to n. We can thus define:

Definition 5. *The judgment* $\Delta, \Sigma; E_\Delta \vdash C : \Theta, \Sigma; E_\Theta$ *holds, if* $\Delta, \Sigma \vdash C: \Theta, \Sigma$, *and if* $C \vdash n_1 \hookrightarrow n_2$, *then* $\Theta, \Sigma; E_\Theta \vdash n_1 \rightleftharpoons\hookrightarrow n_2 : \Delta, \Sigma$.

We simply write $\Delta, \Sigma; E_\Delta \vdash C : \Theta, \Sigma; E_\Theta$ to assert that the judgment is satisfied. Note that references mentioned in threads do not "count" as acquaintance.

Lemma 1 (Subject reduction). *Assume* $\Xi \vdash C \xRightarrow{s} \acute{\Xi} \vdash \acute{C}$. *Then*

1. $\acute{\Delta}, \acute{\Sigma} \vdash \acute{C} : \acute{\Theta}, \acute{\Sigma}$. *A fortiori: If* $\Delta, \Sigma, \Theta \vdash n : T$, *then* $\acute{\Delta}, \acute{\Sigma}, \acute{\Theta} \vdash n : T$.
2. $\acute{\Xi} \vdash \acute{C}$.

Definition 6 (Conservative extension). *Given 2 pairs* (Φ, E_Δ) *and* $(\acute{\Phi}, \acute{E}_\Delta)$ *of name context and connectivity context, i.e.,* $E_\Delta \subseteq \Phi \times \Phi$ *(and analogously for* $(\acute{\Phi}, \acute{E}_\Delta)$*), we write* $(\Phi, E_\Delta) \vdash (\acute{\Phi}, \acute{E}_\Delta)$ *if the following two conditions holds:*

1. $\acute{\Phi} \vdash \Phi$ *and*
2. $\acute{\Phi} \vdash n_1 \rightleftharpoons n_2$ *implies* $\Phi \vdash n_1 \rightleftharpoons n_2$, *for all* n_1, n_2 *with* $\Phi \vdash n_1, n_2$.

Lemma 2 (No surprise). *Let* $\Delta, \Sigma; E_\Delta \vdash C : \Theta, \Sigma; E_\Theta \xrightarrow{a} \acute{\Delta}, \acute{\Sigma}; \acute{E}_\Delta \vdash \acute{C} : \acute{\Theta}, \acute{\Sigma}; \acute{E}_\Theta$ *for some incoming label a. Then* $\Delta, \Sigma; E_\Delta \vdash \acute{\Delta}, \acute{\Sigma}; \acute{E}_\Delta$. *For outgoing steps, the situation is dual.*

Lemma 3 (Soundness of legal trace system). *If* $\Delta_0; \vdash C : \Theta_0;$ *and* $\Delta_0; \vdash C : \Theta_0; \xRightarrow{t}$, *then* $\Delta_0 \vdash t : trace\,\Theta_0$.

4   Conclusion

Related work [8] presents a fully abstract model for *Object-Z*, an object-oriented extension of the *Z* specification language. It is based on a refinement of the simple trace semantics called the complete-readiness model, which is related to the readiness model of Olderog and Hoare. In [9], full abstraction in an object calculus with subtyping is investigated. The setting is slightly different from the one here, as the paper does not compare a contextual semantics with a denotational one, but a semantics by translation with a direct one. The paper considers neither concurrency nor aliasing. Recently, Jeffrey and Rathke [7] extended their work [6] on trace-based semantics from an object-based setting to a core of *Java*, called *JavaJr*, including classes and subtyping. However, their semantics avoids object connectivity by using a notion of *package*. [5] tackles full abstraction and observable component behavior and connectivity in a UML-setting.

Future work. We plan to extend the language with further features to make it more resembling *Java* or $C^\#$. Besides *monitor synchronization* using object locks and wait and signal methods, as provided by *Java*, another interesting direction concerns *subtyping* and *inheritance*. This is challenging especially if the component may inherit from environment classes and vice versa. Another direction is to extend the semantics to a *compositional* one. Finally, we work on adapting the full abstraction proof of [1] to deal with thread classes. The results of Section 3.2 are covering the soundness-part of the full-abstraction result.

References

1. E. Ábrahám, M. M. Bonsangue, F. S. de Boer, and M. Steffen. Object connectivity and full abstraction for a concurrent calculus of classes. In Z. Li and K. Araki, editors, *ICTAC'04*, volume 3407 of *LNCS*, pages 37–51. Springer-Verlag, July 2004.
2. E. Ábrahám, F. S. de Boer, M. M. Bonsangue, A. Grüner, and M. Steffen. Observability, connectivity, and replay in a sequential calculus of classes. In Bosangue et al. [4], pages 296–316.
3. E. Ábrahám, A. Grüner, and M. Steffen. Dynamic heap-abstraction for open, object-oriented systems with thread classes. Technical Report 0601, Institut für Informatik und Praktische Mathematik, Christian-Albrechts-Universität zu Kiel, Jan. 2006.
4. M. Bosangue, F. S. de Boer, W.-P. de Roever, and S. Graf, editors. *Proceedings of FMCO 2004*, volume 3657 of *LNCS*. Springer-Verlag, 2005.
5. F. S. de Boer, M. Bonsangue, M. Steffen, and E. Ábrahám. A fully abstract trace semantics for UML components. In Bosangue et al. [4], pages 49–69.
6. A. Jeffrey and J. Rathke. A fully abstract may testing semantics for concurrent objects. In *Proceedings of LICS '02*. IEEE, Computer Society Press, July 2002.
7. A. Jeffrey and J. Rathke. Java Jr.: A fully abstract trace semantics for a core Java language. In M. Sagiv, editor, *Proceedings of ESOP 2005*, volume 3444 of *LNCS*, pages 423–438. Springer-Verlag, 2005.
8. G. P. Smith. *An Object-Oriented Approach to Formal Specification*. PhD thesis, Department of Computer Science, University of Queensland, Oct. 1992.
9. R. Viswanathan. Full abstraction for first-order objects with recursive types and subtyping. In *Proceedings of LICS '98*. IEEE, Computer Society Press, July 1998.

From Constructibility and Absoluteness to Computability and Domain Independence

Arnon Avron

School of Computer Science
Tel Aviv University, Tel Aviv 69978, Israel
aa@math.tau.ac.il

Abstract. Gödel's main contribution to set theory is his proof that GCH is consistent with ZFC (assuming that ZF is consistent). For this proof he has introduced the important ideas of constructibility of sets, and of absoluteness of formulas. In this paper we show how these two ideas of Gödel naturally lead to a simple unified framework for dealing with computability of functions and relations, domain independence of queries in relational databases, and predicative set theory.

1 Introduction: Absoluteness and Constructibility

Gödel classical work [6] on the constructible universe L is best known for its applications in pure set theory, especially consistency and independence proofs. Its relevance to computability theory was mostly ignored. Still, in this work Gödel introduced at least two ideas which are quite important from a computational point of view:

Computations with Sets. The notion of computation is usually connected with discrete structures, like the natural numbers, or strings of symbols from some alphabet. In this respect [6] is important, first of all, in being the first comprehensive research on (essentially) computability within a completely different framework (technically, the name Gödel used was "constructibility" rather than "computability", but the difference is not really significant). No less important (as we shall see) is the particularly important data structure for which computability issues were investigated in [6]: sets. Specifically, for characterizing the "constructible sets" Gödel identified operations on sets (which we may call "computable"), that may be used for "effectively" constructing new sets from given ones (in the process of creating the universe of "constructible" sets). Thus, binary union and intersection are "effective" in this sense, while the powerset operation is not. Gödel has even provided a finite list of basic set operations, from which all other "effective" constructions can be obtained through compositions.

Absoluteness. A formula in the language of set theory is absolute if its truth value in a transitive class M, for some assignment v of objects from M to its free variables, depends only on v, but not on M (i.e. the truth value is the

A. Beckmann et al. (Eds.): CiE 2006, LNCS 3988, pp. 11–20, 2006.

same in all structures M, in which v is legal). Absoluteness is a property of formulas which was crucial for Gödel consistency proof. However, it is not a decidable property. The following set Δ_0 of absolute formulas is therefore extensively used as a syntactically defined approximation:

- Every atomic formula is in Δ_0.
- If φ and ψ are in Δ_0, then so are $\neg\varphi$, $\varphi \vee \psi$, and $\varphi \wedge \psi$.
- If x and y are two different variables, and φ is in Δ_0, then so is $\exists x \in y\varphi$.

Now there is an obvious analogy between the roles in set theory of absolute formulas and of Δ_0 formulas, and the roles in formal arithmetic and computability theory of decidable formulas and of arithmetical Δ_0 formulas (i.e. Smullyan's "bounded" formulas). This analogy was noticed and exploited in the research on set theory. However, the reason for this analogy remains unclear, and beyond this analogy the importance and relevance of these two ideas of Gödel to other areas have not been noticed. As a result, strongly related ideas and theory have been redeveloped from scratch in relational database theory.

2   Domain Independence and Computability in Databases

From a logical point of view, a relational database DB of a scheme $\{P_1, \ldots, P_n\}$ is just a tuple $\langle \underline{P_1}, \ldots, \underline{P_n} \rangle$ of *finite* interpretations (called "tables") of the predicate symbols $P_1, \ldots, P_n$. DB can be turned into a structure S for a first-order language L with equality, the signature of which includes $\{P_1, \ldots, P_n\}$ and constants, by specifying a domain D, and an interpretation of the constants of L in it (different interpretations for different constants). The domain D should be at most countable (and usually it is finite), and should of course include the union of the domains of the tables in DB. A query for DB is simply a formula ψ of L. If ψ has free variables, then the answer to ψ in S is the set of tuples which satisfy it in S. If ψ is closed, then the answer to the query is either "yes" or "no", depending on whether ψ holds in S or not (The "yes" and "no" can be interpreted as $\{\emptyset\}$ and $\emptyset$, respectively). Now not every formula ψ of a L can serve as a query. Acceptable are only those the answer for which is a *computable function* of $\langle \underline{P_1}, \ldots, \underline{P_n} \rangle$ alone (and does not depend on the identity of the intended domain D. This in particular entails that the answer should be finite). Such queries are called *domain independent* ([8, 11, 1]). The exact definition is:

Definition 1. [1]*Let σ be a signature which includes $\overrightarrow{P} = \{P_1, \ldots, P_n\}$, and optionally constants and other predicate symbols (but no function symbols). A query $\varphi(x_1, \ldots, x_n)$ in σ is called $\overrightarrow{P}$–d.i. ($\overrightarrow{P}$–domain-independent), if whenever S_1 and S_2 are structures for σ, S_1 is a substructure of S_2, and the interpretations of $\{P_1, \ldots, P_n\}$ in S_1 and S_2 are identical, then for all $a_1 \in S_2, \ldots, a_n \in S_2$:*

$$S_2 \models \varphi(a_1, \ldots, a_n) \quad \leftrightarrow \quad a_1 \in S_1 \wedge \ldots \wedge a_n \in S_1 \wedge S_1 \models \varphi(a_1, \ldots, a_n)$$

---

[1] This is a slight generalization of the definition in [12], which in turn is a generalization of the usual one ([8, 11]). The latter applies only to free Herbrand structures which are generated by adding to σ some new set of constants.

Practical database query languages are designed so that only d.i. queries can be formulated in them. Unfortunately, it is undecidable which formulas are d.i. (or "safe" according to any other reasonable notion of safety of queries, like "finite and computable"). Therefore all commercial query languages (like SQL) allow to use as queries only formulas from some syntactically defined class of d.i. formulas. Many explicit proposals of decidable, syntactically defined classes of safe formulas have been made in the literature. The simplest among them (and the closer to what has actually been implemented) is perhaps the following class $SS(\overrightarrow{P})$ ("syntactically safe" formulas for a database scheme $\overrightarrow{P}$) from [11] (originally designed for languages with no function symbols) [2]:

1. $P_i(t_1, \ldots, t_{n_i}) \in SS(\overrightarrow{P})$ in case P_i (of arity n_i) is in $\overrightarrow{P}$.
2. $x = c$ and $c = x$ are in $SS(\overrightarrow{P})$ (where x is a variable and c is a constant).
3. $\varphi \vee \psi \in SS(\overrightarrow{P})$ if $\varphi \in SS(\overrightarrow{P})$, $\psi \in SS(\overrightarrow{P})$, and $Fv(\varphi) = Fv(\psi)$ (where $Fv(\varphi)$ denotes the set of free variables of φ).
4. $\exists x \varphi \in SS(\overrightarrow{P})$ if $\varphi \in SS(\overrightarrow{P})$.
5. If $\varphi = \varphi_1 \wedge \varphi_2 \wedge \ldots \wedge \varphi_k$, then $\varphi \in SS(\overrightarrow{P})$ if the following conditions are met:
   (a) For each $1 \leq i \leq k$, either φ_i is atomic, or φ_i is in $SS(\overrightarrow{P})$, or φ_i is a negation of a formula of either type.
   (b) Every free variable x of φ is limited in φ. This means that there exists $1 \leq i \leq k$ such that x is free in φ_i, and either $\varphi_i \in SS(\overrightarrow{P})$, or there exists y which is already limited in φ, and $\varphi_i \in \{x = y, y = x\}$.

It should be noted that there is one clause in this definition which is somewhat strange: the last one, which treats conjunction. The reason why this clause does not simply tell us (like in the case of disjunction) when a conjunction of *two* formulas is in $SS(\overrightarrow{P})$, is the desire to take into account the fact that once the value of y (say) is known, the formula $x = y$ becomes domain independent. In the unified framework described in the next section this problematic clause is replaced by a more concise one (which at the same time is more general).

A more important fact is that given $\{\underline{P_1}, \ldots, \underline{P_n}\}$, the set of relations which are answers to some query in $SS(\overrightarrow{P})$ is exactly the closure of $\{\underline{P_1}, \ldots, \underline{P_n}\}$ under a finite set of basic operations called "the relational algebra" ([1, 11]). This set is quite similar to set of basic operations used by Gödel in [6] for constructing the constructible universe.

3 Partial Domain Independence and Absoluteness

There is an obvious similarity between the concepts of d.i. in databases, and absoluteness in Set Theory. However, the two notions are not identical. Thus, the formula $x = x$ is not d.i., although it is clearly absolute. To exploit the similarity,

---

[2] What we present below is both a generalization and a simplification of Ullman's original definition.

the formula *property* of d.i. was turned in [2] into the following *relation* between a formula φ and finite subsets of $Fv(\varphi)$:

Definition 2. *Let σ be like in Definition 1. A formula $\varphi(x_1, \ldots, x_n, y_1, \ldots, y_k)$ in σ is $\vec{P}$-d.i. with respect to $\{x_1, \ldots, x_n\}$, if whenever S_1 and S_2 are structures as in Definition 1, then for all $a_1 \in S_2, \ldots, a_n \in S_2$ and $b_1 \in S_1, \ldots, b_k \in S_1$:*

$$S_2 \models \varphi(\vec{a}, \vec{b}) \quad \leftrightarrow \quad a_1 \in S_1 \wedge \ldots \wedge a_n \in S_1 \wedge S_1 \models \varphi(\vec{a}, \vec{b})$$

Note that φ is d.i. iff it is d.i. with respect to $Fv(\varphi)$. On the other hand the formula $x = y$ is only partially d.i.: it is d.i. with respect to $\{x\}$ and $\{y\}$, but not with respect to $\{x, y\}$. Note also that a formula φ is d.i. with respect to $\emptyset$ if whenever S_1 and S_2 are structures as in Definition 1 then for all $b_1, \ldots, b_k \in S_1$ $S_2 \models \varphi(\vec{b}) \quad \leftrightarrow \quad S_1 \models \varphi(\vec{b})$. Under not very different conditions concerning S_1 and S_2, this is precisely Gödel's idea of absoluteness. We'll return to this below.

Another important observation is that given a domain S for the database, if $\varphi(x_1, \ldots, x_n, y_1, \ldots, y_k)$ is $\vec{P}$-d.i. with respect to $\{x_1, \ldots, x_n\}$ then the function $\lambda y_1, \ldots, y_k.\{\langle x_1, \ldots, x_n \rangle \mid \varphi\}$ is a *computable* function from S^k to the set of finite subsets of S^n, the values of which depend only on the values of the arguments $y_1, \ldots, y_k$, but not on the identity of S. In case $n = 0$ the possible values of this function are $\{\langle\rangle\}$ and $\emptyset$, which can be taken as "true" and "false", respectively. Hence in this particular case what we get is a computable k-ary predicate on S. From this point of view k-ary predicates on a set S should be viewed as a special type of functions from S^k to the set of finite sets of S-tuples, rather than as a special type of functions from S^k to S, with arbitrary chosen two elements from S serving as the two classical truth values (while like in set theory, functions from S^k to S should be viewed as a special type of $(k+1)$-ary predicates on S).

Now it is easy to see that partial d.i. has the following properties (where $\varphi \succ X$ means that φ is $\vec{P}$-d.i. with respect to X):

0. If $\varphi \succ X$ and $Z \subseteq X$, then $\varphi \succ Z$.
1. $\varphi \succ Fv(\varphi)$ if φ is $p(t_1, \ldots, t_n)$ (where $p \in \vec{P}$).
2. $x \neq x \succ \{x\}$, $t = x \succ \{x\}$, and $x = t \succ \{x\}$ if $x \notin Fv(t)$.
3. $\neg\varphi \succ \emptyset$ if $\varphi \succ \emptyset$.
4. $\varphi \vee \psi \succ X$ if $\varphi \succ X$ and $\psi \succ X$.
5. $\varphi \wedge \psi \succ X \cup Y$ if $\varphi \succ X$, $\psi \succ Y$, and $Y \cap Fv(\varphi) = \emptyset$.
6. $\exists y \varphi \succ X - \{y\}$ if $y \in X$ and $\varphi \succ X$.

These properties can be used for defining a syntactic approximation $\succ_P$ of the semantic $\vec{P}$-d.i. relation. It can easily be checked that the set $\{\varphi \mid \varphi \succ_P Fv(\varphi)\}$ strictly extends $SS(\vec{P})$ (but note how the complicated last clause in the definition of $SS(\vec{P})$ is replaced here by a concise clause concerning conjunction!).

Note: For convenience, we are taking here $\wedge, \vee, \neg$ and $\exists$ as our primitives. Moreover: we take $\neg(\varphi \rightarrow \psi)$ as an abbreviation for $\varphi \wedge \neg\psi$, and $\forall x_1, \ldots, x_k \varphi$ as an abbreviation for $\neg \exists x_1, \ldots, x_k \neg \varphi$. This entails the following important property of "bounded quantification": If $\succ$ is a relation satisfying the above properties, and $\varphi \succ \{x_1, \ldots, x_n\}$, while $\psi \succ \emptyset$, then $\exists x_1 \ldots x_n(\varphi \wedge \psi) \succ \emptyset$ and $\forall x_1 \ldots x_n(\varphi \to \psi) \succ \emptyset$ (recall that $\varphi \succ \emptyset$ is our counterpart of absoluteness).

4 Partial Domain Independence in Set Theory

We return now to set theory, to see how the idea of partial d.i. applies there. In order to fully exploit it, we use a language with abstraction terms *for sets*. However, we allow only terms which are known to be d.i. in a sense we now explain. For simplicity of presentation, we assume the accumulative universe V of ZF, and formulate our definitions accordingly.

Definition 3. *Let $\mathcal{M}$ be a transitive class. Define the relativization to $\mathcal{M}$ of terms and formulas recursively as follows:*

- $t_{\mathcal{M}} = t$ *if t is a variable or a constant.*
- $\{x \mid \varphi\}_{\mathcal{M}} = \{x \mid x \in \mathcal{M} \wedge \varphi_{\mathcal{M}}\}$.
- $(t = s)_{\mathcal{M}} = (t_{\mathcal{M}} = s_{\mathcal{M}})$ $(t \in s)_{\mathcal{M}} = (t_{\mathcal{M}} \in s_{\mathcal{M}})$.
- $(\neg \varphi)_{\mathcal{M}} = \neg \varphi_{\mathcal{M}}$ $(\varphi \vee \psi)_{\mathcal{M}} = \varphi_{\mathcal{M}} \vee \psi_{\mathcal{M}}$. $(\varphi \wedge \psi)_{\mathcal{M}} = \varphi_{\mathcal{M}} \wedge \psi_{\mathcal{M}}$.
- $(\exists x \varphi)_{\mathcal{M}} = \exists x(x \in \mathcal{M} \wedge \varphi_{\mathcal{M}})$.

Definition 4. *Let T be a theory such that $V \models T$.*

1. *Let t be a term, and let $Fv(t) = \{y_1, \ldots, y_n\}$. We say that t is T-d.i., if the following is true (in V) for every transitive model $\mathcal{M}$ of T:*

$$\forall y_1 \ldots \forall y_n . y_1 \in \mathcal{M} \wedge \ldots \wedge y_n \in \mathcal{M} \to t_{\mathcal{M}} = t$$

2. *Let φ be a formula, and let $Fv(\varphi) = \{y_1, \ldots, y_n, x_1, \ldots, x_k\}$. We say that φ is T-d.i. for $\{x_1, \ldots, x_k\}$ if $\{\langle x_1, \ldots, x_k\rangle \mid \varphi\}$ is a set for all values of the parameters $y_1, \ldots, y_n$, and the following is true (in V) for every transitive model $\mathcal{M}$ of T:*

$$\forall y_1 \ldots \forall y_n . y_1 \in \mathcal{M} \wedge \ldots \wedge y_n \in \mathcal{M} \to [\varphi \leftrightarrow (x_1 \in \mathcal{M} \wedge \ldots \wedge x_k \in \mathcal{M} \wedge \varphi_{\mathcal{M}})]$$

Thus, a term is T-d.i. if it has the same interpretation in all transitive models of T which contain the values of its parameters, while a formula is T-d.i. for $\{x_1, \ldots, x_k\}$ if it has the same extension (which should be a set) in all transitive models of T which contain the values of its other parameters. In particular: φ is T-d.i. for $\emptyset$ iff it is absolute relative to T in the original sense of set theory, while φ is T-d.i. for $Fv(\varphi)$ iff it is domain-independent in the sense of database theory (see Definition 1) for transitive models of T.

The set-theoretical notion of d.i., we have just introduced, is again a semantic notion that one cannot characterize in a constructive manner, and so a syntactic approximation of it should be used in practice. The key observation for this is that the transitive classes are the structures for which the atomic formula $x \in y$ (where y is different from x) is d.i. with respect to $\{x\}$. Accordingly, an appropriate approximation is most naturally obtained by adapting the definition of $\succ_P$ above to the present language, taking into account this key observation:

Definition 5. *The relation* $\succ_{RST}$ *is inductively defined as follows:*

1. $\varphi \succ_{RST} \emptyset$ *if* φ *is atomic.*
2. $\varphi \succ_{RST} \{x\}$ *if* $\varphi \in \{x \neq x, x = t, t = x, x \in t\}$, *and* $x \notin Fv(t)$.
3. $\neg\varphi \succ_{RST} \emptyset$ *if* $\varphi \succ_{RST} \emptyset$.
4. $\varphi \vee \psi \succ_{RST} X$ *if* $\varphi \succ_{RST} X$ *and* $\psi \succ_{RST} X$.
5. $\varphi \wedge \psi \succ_{RST} X \cup Y$ *if* $\varphi \succ_{RST} X$, $\psi \succ_{RST} Y$, *and* $Y \cap Fv(\varphi) = \emptyset$.
6. $\exists y\varphi \succ_{RST} X - \{y\}$ *if* $y \in X$ *and* $\varphi \succ_{RST} X$.

Note: It can easily be proved by induction on the complexity of formulas that the clause 0 in the definition of $\succ_P$ is also satisfied by $\succ_{RST}$: if $\varphi \succ_{RST} X$ and $Z \subseteq X$, then $\varphi \succ_{RST} Z$.

A first (and perhaps the most important) use of $\succ_{RST}$ is for defining the set of legal terms of the corresponding system RST (Rudimentary Set Theory). Unlike the languages for databases (in which the only terms were variables and constants), the language of RST used here has a very extensive set of terms. It is inductively defined as follows:

- Every variable is a term.
- If x is a variable, and φ is a formula such that $\varphi \succ_{RST} \{x\}$, then $\{x \mid \varphi\}$ is a term (and $Fv(\{x \mid \varphi\}) = Fv(\varphi) - \{x\}$).

(Actually, the relation $\succ_{RST}$, the set of terms of RST, and the set of formulas of RST are defined together by a simultaneous induction).

A second use of $\succ_{RST}$ is that the set $\{\varphi \mid \varphi \succ_{RST} \emptyset\}$ is a natural extension of the set Δ_0 of bounded formulas. Moreover, we have:

Theorem 1. *Let RST be the theory consisting of the following axioms:*

Extensionality: $\forall y(y = \{x \mid x \in y\})$
Comprehension: $\forall x(x \in \{x \mid \varphi\} \leftrightarrow \varphi)$

Then given an extension T of RST, any valid term t of RST is T-d.i., and if $\varphi \succ_{RST} X$, then φ is T-d.i. for X.

The following theorem connects $\succ_{RST}$ with the class of rudimentary set functions (introduced independently by Gandy ([5]) and Jensen ([7]). See also [4]) — a refined version of Gödel basic set functions:

Theorem 2

1. *If F is an n-ary rudimentary function, then there exists a formula φ s. t.:*
   (a) $Fv(\varphi) = \{y, x_1, \ldots, x_n\}$
   (b) $\varphi \succ_{RST} \{y\}$
   (c) $F(x_1, \ldots, x_n) = \{y \mid \varphi\}$.
2. *If φ is a formula such that:*
   (a) $Fv(\varphi) = \{y_1, \ldots, y_k, x_1, \ldots, x_n\}$
   (b) $\varphi \succ_{RST} \{y_1, \ldots, y_k\}$
   then there exists a rudimentary function F such that:

$$F(x_1, \ldots, x_n) = \{\langle y_1, \ldots, y_k \rangle \mid \varphi\}$$

Corollary 1. *If $Fv(\varphi) = \{x_1, \ldots, x_n\}$, and $\varphi \succ_{RST} \emptyset$, then φ defines a rudimentary predicate P. Conversely, if P is a rudimentary predicate, then there is a formula φ such that $\varphi \succ_{RST} \emptyset$, and φ defines P.*

4.1   On Predicative Set Theory

In his writings Gödel expressed the view that his hierarchy of constructible sets codified the predicatively acceptable means of set construction, and that the only impredicative aspect of the constructible universe L is its being based on the full class On of ordinals. This seems to us to be only partially true. We think that indeed the predicatively acceptable instances of the comprehension schema are those which determine the collections they define in an absolute way, independently of any "surrounding universe". Therefore a formula ψ is predicative (with respect to x) if the collection $\{x \mid \psi(x, y_1, \ldots, y_n)\}$ is completely and uniquely determined by the identity of the parameters $y_1, \ldots, y_n$, and the identity of other objects referred to in the formula (all of which should be well-determined before). In other words: ψ is predicative (with respect to x) iff it is d.i. (with respect to x). It follows that all the operations used by Gödel are indeed predicatively acceptable, and even capture what is intuitively predicatively acceptable in the language of RST. However, we believe that one should go beyond first-order languages in order to capture all the predicatively acceptable means of set construction. In [3] we suggest that an adequate language for this is obtained by adding to the the language of RST an operation TC for transitive closure of binary relations, and then replacing $\succ_{RST}$ by the relation $\succ_{PZF}$, which is defined like $\succ_{RST}$, but with the following extra clause: $(TC_{x,y}\varphi)(x,y) \succ_{PZF} X$ if $\varphi \succ_{PZF} X$, and $\{x,y\} \cap X \neq \emptyset$. See [3] for more details.

5   Domain Independence: A General Framework

In this section we introduce a general abstract framework for studying domain independence and absoluteness (originally introduced in [2]).

Definition 6. *A d.i.-signature is a pair (σ, F), where σ is an ordinary first-order signature, and F is a function which assigns to every n-ary symbol s from σ (other than equality) a subset of $\mathcal{P}(\{1, \ldots, n\})$.*

Definition 7. *Let (σ, F) be a d.i.-signature. Let S_1 and S_2 be two structures for σ s.t. $S_1 \subseteq S_2$. S_2 is called a (σ, F)–extension of S_1 if the following conditions are satisfied:*

- *If $p \in \sigma$ is a predicate symbol of arity n, $I \in F(p)$, and $a_1, \ldots, a_n$ are elements of S_2 such that $a_i \in S_1$ in case $i \notin I$, then $S_2 \models p(a_1, \ldots, a_n)$ iff $a_i \in S_1$ for all i, and $S_1 \models p(a_1, \ldots, a_n)$.*
- *If $f \in \sigma$ is a function symbol of arity n, $a_1, \ldots, a_n \in S_1$, and b is the value of $f(a_1, \ldots, a_n)$ in S_2, then $b \in S_1$, and b is the value of $f(a_1, \ldots, a_n)$ in S_1. Moreover: if $I \in F(f)$, and $a_1, \ldots, a_n$ are elements of S_2 such that $a_i \in S_1$ in case $i \notin I$, then $S_2 \models b = f(a_1, \ldots, a_n)$ iff $a_i \in S_1$ for all i, and $S_1 \models b = f(a_1, \ldots, a_n)$.*

Definition 8. *Let (σ, F) be as in Definition 7. A formula φ of σ is called (σ, F)–d.i. w.r.t. X ($\varphi \succ^{di}_{(\sigma, F)} X$) if whenever S_2 is a (σ, F)–extension of S_1,* *and φ^* results from φ by substituting values from S_1 for the free variables of φ that are not in X, then the sets of tuples which satisfy φ^* in S_1 and in S_2 are identical.* [3] *A formula φ of σ is called (σ, F)–d.i. if $\varphi \succ_{(\sigma,F)}^{di} Fv(\varphi)$, and (σ, F)–absolute if $\varphi \succ_{(\sigma,F)}^{di} \emptyset$.*

Note: We assume that we are talking only about first-order languages with equality, and so we do not include the equality symbol in our first-order signatures. Had it been included then we would have defined $F(=) = \{\{1\}, \{2\}\}$ (meaning that $x_1 = x_2$ is d.i. w.r.t. both $\{x_1\}$ and $\{x_2\}$, but not w.r.t. $\{x_1, x_2\}$).

Examples

  – Let σ be a signature which includes $\vec{P} = \{P_1, \ldots, P_n\}$, and optionally constants and other predicate symbols (but no function symbols). Assume that the arity of P_i is n_i, and define $F(P_i) = \{\{1, \ldots, n_i\}\}$. Then φ is (σ, F)–d.i. w.r.t. X iff it is $\vec{P}$–d.i. w.r.t. X in the sense of Definition 2.
  – Let $\sigma_{ZF} = \{\in\}$ and let $F_{ZF}(\in) = \{\{1\}\}$. In this case the universe V is a (σ_{ZF}, F_{ZF})– extension of the transitive sets and classes. Therefore a formula is σ_{ZF}-absolute iff it is absolute in the usual sense of set theory.

Again the relation of (σ, F)–d.i. is a semantic notion that in practice should be replaced by a syntactic approximation. The following definition generalizes in a very natural way the relations $\succ_P$ and $\succ_{RST}$:

Definition 9. *The relation $\succ_{(\sigma,F)}$ is inductively defined as follows:*

0. *If $\varphi \succ_{(\sigma,F)} X$ and $Z \subseteq X$, then $\varphi \succ_{(\sigma,F)} Z$.*
1a. *If p is an n-ary predicate symbol of σ; $x_1, \ldots, x_n$ are n distinct variables, and $\{i_1, \ldots, i_k\}$ is in $F(p)$, then $p(x_1, \ldots, x_n) \succ_{(\sigma,F)} \{x_{i_1}, \ldots, x_{i_k}\}$.*
1b. *If f is an n-ary function symbol of σ; $y, x_1, \ldots, x_n$ are $n+1$ distinct variables, and $\{i_1, \ldots, i_k\} \in F(f)$, then $y = f(x_1, \ldots, x_n) \succ_{(\sigma,F)} \{x_{i_1}, \ldots, x_{i_k}\}$.*
2. *$\varphi \succ_{(\sigma,F)} \{x\}$ if $\varphi \in \{x \neq x, x = t, t = x\}$, and $x \notin Fv(t)$.*
3. *$\neg\varphi \succ_{(\sigma,F)} \emptyset$ if $\varphi \succ_{(\sigma,F)} \emptyset$.*
4. *$\varphi \vee \psi \succ_{(\sigma,F)} X$ if $\varphi \succ_{(\sigma,F)} X$ and $\psi \succ_{(\sigma,F)} X$.*
5. *$\varphi \wedge \psi \succ_{(\sigma,F)} X \cup Y$ if $\varphi \succ_{(\sigma,F)} X$, $\psi \succ_{(\sigma,F)} Y$, and $Y \cap Fv(\varphi) = \emptyset$.*
6. *$\exists y\varphi \succ_{(\sigma,F)} X - \{y\}$ if $y \in X$ and $\varphi \succ_{(\sigma,F)} X$.*

Again it is easy to see that if $\varphi \succ_{(\sigma,F)} X$, then $\varphi \succ_{(\sigma,F)}^{di} X$. The converse fails, of course. However, we suggest the following conjecture (that for reasons to become clear in the next section, may be viewed as a generalized Church Thesis):

Conjecture. Given a d.i. signature (σ, F), a formula is upward (σ, F)-absolute iff it is logically equivalent to a formula of the $\exists y_1, \ldots, y_n\psi$, where $\psi \succ_{(\sigma,F)} \emptyset$ ($\varphi(x_1, \ldots, x_n)$ is upward (σ, F)-absolute if whenever S_2 is a (σ, F)–extension of S_1, and $\models_{S_1} \varphi(a_1, \ldots, a_n)$, then $\models_{S_2} \varphi(a_1, \ldots, a_n)$).

---

[3] φ^* is a formula only in a generalized sense, but the intention should be clear.

6   Absoluteness and Computability in $\mathcal{N}$

Finally, we turn to the connections between the above ideas and computability in the Natural numbers.

Definition 10. *The d.i. signature $(\sigma_{\mathcal{N}}, F_{\mathcal{N}})$ is defined as follows:*

- $\sigma_{\mathcal{N}}$ *is the first-order signature which includes the constants 0 and 1, the binary predicate $<$, and the ternary relations P_+ and $P_\times$.*
- $F_{\mathcal{N}}(<) = \{\{1\}\}$, $F_{\mathcal{N}}(P_+) = F_{\mathcal{N}}(P_\times) = \{\emptyset\}$.

Definition 11. *The standard structure $\mathcal{N}$ for $\sigma_{\mathcal{N}}$ has the set of natural numbers as its domain, with the usual interpretations of 0, 1, and $<$, and the (graphs of the) operations $+$ and $\times$ on N (viewed as ternary relations on N) as the interpretations of P_+ and $P_\times$, respectively.*

It is easy now to see that $\mathcal{N}$ is a $(\sigma_{\mathcal{N}}, F_{\mathcal{N}})$-extension of a structure S for $\sigma_{\mathcal{N}}$ iff the domain of S is an initial segment of $\mathcal{N}$ (where the interpretations of the relation symbols are the corresponding reductions of the interpretations of those symbols in $\mathcal{N}$). Accordingly, if $\varphi \succ_{(\sigma_{\mathcal{N}}, F_{\mathcal{N}})} \emptyset$, then for any assignment in N it gets the same truth value in all initial segments of $\mathcal{N}$ (including $\mathcal{N}$ itself) which contain the values assigned to its free variables. Now the set of formulas φ such that $\varphi \succ_{(\sigma_{\mathcal{N}}, F_{\mathcal{N}})} \emptyset$ is a straightforward extension of Smullyan's set of bounded formulas ([10]). This set is defined of course using the relation $\succ_N = \succ_{(\sigma_{\mathcal{N}}, F_{\mathcal{N}})}$. From definitions 9 and 10 it easily follows that this relation can be characterized as follows (compare with Definition 5!):

1. $\varphi \succ_N \emptyset$ if φ is atomic.
2. $\varphi \succ_N \{x\}$ if $\varphi \in \{x \neq x, x = t, t = x, x < t\}$, and $x \notin Fv(t)$.
3. $\neg\varphi \succ_N \emptyset$ if $\varphi \succ_N \emptyset$.
4. $\varphi \vee \psi \succ_N X$ if $\varphi \succ_N X$ and $\psi \succ_N X$.
5. $\varphi \wedge \psi \succ_N X \cup Y$ if $\varphi \succ_N X$, $\psi \succ_N Y$, and $Y \cap Fv(\varphi) = \emptyset$.
6. $\exists y \varphi \succ_N X - \{y\}$ if $y \in X$ and $\varphi \succ_N X$.

Now the crucial connection between Gödel's work on absoluteness in set theory, and computability in the natural numbers, is given in the following Theorem:

Theorem 3. *The following conditions are equivalent for a relation R on N:*

1. *R is semi-decidable.*
2. *R is definable by a formula of the form $\exists y_1, \ldots, y_n \psi$, where $\psi \succ_N \emptyset$.*
3. *R is definable by a formula of the form $\exists y_1, \ldots, y_n \psi$, where the formula ψ is $(\sigma_{\mathcal{N}}, F_{\mathcal{N}})$-absolute.*

Proof. 2. follows from 1. by the Thesis of Church and Smullyan's characterization in [10] of the r.e. subsets of N using his set of bounded formulas (recall that if ψ is bounded, then $\psi \succ_N \emptyset$). That 3. follows from 2. is immediate from the fact that if $\psi \succ_N \emptyset$, then ψ is $(\sigma_{\mathcal{N}}, F_{\mathcal{N}})$-absolute. To show that 3. entails 1., assume that R is definable by a formula of the form $\exists y_1, \ldots, y_n \psi$, where the formula $\psi(x_1, \ldots, x_k, y_1, \ldots, y_n)$ is $(\sigma_\mathcal{N}, F_\mathcal{N})$-absolute. Given numbers $n_1, \ldots, n_k$ we search whether $R(n_1, \ldots, n_k)$ by examining all the finite initial segments of N that contain $n_1, \ldots, n_k$, and return "true" if we find in one of them numbers $m_1, \ldots, m_n$ such that $\psi(n_1, \ldots, n_k, m_1, \ldots, m_n)$ is true in it. From the fact that ψ is $(\sigma_\mathcal{N}, F_\mathcal{N})$-absolute, it easily follows that this procedure halts with the correct answer in case $R(n_1, \ldots, n_k)$, and never halt otherwise.

The last theorem shows a very close connection between (semi)-computability and (upward) absoluteness. However, further research is needed in order to understand the full connection between these notions. A key problem that one has to solve in order to provide a general computability theory based on d.i. relations and absoluteness, is what is so special about the standard interpretations in $\mathcal{N}$ of P_+ and $P_\times$ that makes the last theorem possible. We suspect that in order to provide a satisfactory answer (and develop the desired theory), one should go beyond first-order languages (most probably to first-order language with a transitive closure operation). We leave that for future investigations.

References

1. S. Abiteboul, R. Hull and V. Vianu, **Foundations of Databases**, Addison-Wesley, 1995.
2. A. Avron, *Safety Signatures for First-order Languages and Their Applications*, in **First-Order Logic revisited** (Hendricks et al, eds.), 37-58, Logos Verlag Berlin, 2004.
3. A. Avron, *A New Approach to Predicative Set Theory*, to appear.
4. K. J. Devlin, **Constructibility**, Perspectives in Mathematical Logic, Springer-Verlag, 1984.
5. Gandy, R. O., em Set-theoretic functions for elementary syntax, In **Axiomatic set theory, Part 2**, AMS, Providence, Rhode Island, 1974, 103-126.
6. K. Gödel, **The Consistency of the Continuum Hypothesis**, Annals of Mathematical Studies, No. 3, Princeton University Press, Princeton, N.J., 1940.
7. R. B. Jensen, *The Fine Structure of the Constructible Hierarchy*, Annals of Mathematical Logic 4 (1971), 229-308, AMS, Providence, Rhode Island, 1974, 143-176.
8. M. Kiffer, *On Safety Domain independence and capturability of database queries*, Proc. International Conference on database and knowledge bases, 405-414, Jerusalem 1988.
9. K. Kunen, **Set Theory, An Introduction to Independence Proofs**, North-Holland, 1980.
10. R. M. Smullyan, **The Incompleteness Theorems**, Oxford University Press, 1992.
11. J.D. Ullman, **Principles of database and knowledge-base systems**, Computer Science Press, 1988.
12. D. Suciu, *Domain-independent queries on databases with external functions*, Theoretical Computer Science 190, 279-315, 1998.

Datatype-Generic Reasoning

Roland Backhouse

School of Computer Science and Information Technology, University of Nottingham,
Nottingham NG8 1BB, England
rcb@cs.nott.ac.uk

Abstract. Datatype-generic programs are programs that are parameterised by a datatype. Designing datatype-generic programs brings new challenges and new opportunities. We review the allegorical foundations of a methodology of designing datatype-generic programs. The effectiveness of the methodology is demonstrated by an extraordinarily concise proof of the well-foundedness of a datatype-generic occurs-in relation.

Keywords: Datatype, generic programming, relation algebra, allegory, programming methodology.

1 Introduction

The central issue of computing science is the development of practical programming methodologies. Characteristic of a programming methodology is that it involves a *discipline* designed to maximise confidence in the reliability of the end product. The discipline constrains the construction methods to those that are demonstrably simple and easy to use, whilst still allowing sufficient flexibility that the creative process of program construction is not impeded. For example, an insight that played an important role in the development of a methodology for sequential programs is that it is possible to restrict attention —without loss of generality— to just the class of **while** programs. It is neither necessary nor desirable to consider arbitrary **goto** programs.

The systematic use of induction on the structure of datatypes is another such discipline; defining and exploiting application-specific datatypes is sound practice, as is well-known, particularly among functional programmers. This has led to the development of a new programming concept, called *(datatype-)generic* programming [1, 2, 3, 4]. Datatype-generic programs are programs that are parameterised by a data structure. For example, the compression of data can be much more effective if the specific structure of the data is known in advance — the compression of *eg* computer programs can exploit the specific syntactic structure of the programs to achieve a higher compression ratio [3].

The idea of making data structure a parameter opens up new challenges and new opportunities. A major new insight is to consider the algebraic structure of data structures — how complex data structures are built from simpler components. In this paper, we review the theoretical foundations of reasoning

A. Beckmann et al. (Eds.): CiE 2006, LNCS 3988, pp. 21–34, 2006.

about datatype-generic programs. We review the notion of "F-reductivity", introduced by Doornbos [5, 6, 7], and show its application to establishing the well-foundedness of the occurs-in relation in a dataype-generic unification algorithm [8, 9].

2   Relation Algebra

2.1   Basic Definitions

Although much recent work on datatype-generic programming has been conducted within the paradigm of *functional* programming, there are far-reaching arguments for adopting a relational framework. Two directly relevant to the current paper are: specifications are typically nondeterministic (i.e. relations, not functions) and termination arguments are almost always conducted within the framework of well-founded relations. So, for us, a program is an input-output relation. The convention we use when defining relations is that the input is on the right and the output on the left (as in functional programming). Formally, a (binary) relation is a triple consisting of a pair of types I and J, say, and a subset of the cartesian product $I \times J$. We write $R :: I \leftarrow J$ (read "R has type I from J"), the left-pointing arrow indicating that we view I as the set of possible outputs and J as the set of possible inputs. I is called the *target* and J the *source* of the relation R, and $I \leftarrow J$ is called its *type*. We use a raised infix dot to denote relational composition. Thus $R \cdot S$ denotes the *composition* of relations R and S. The *converse* of relation R is denoted by $R^\cup$. Relations of the same type are ordered by set inclusion denoted in the conventional way by the infix $\subseteq$ operator.

For each set I, there is an identity relation which we denote by id_I. Thus $\mathsf{id}_I :: I \leftarrow I$. Relations of type $I \leftarrow I$ contained in id_I will be called *coreflexives*. By convention, we use R, S, T to denote arbitrary relations and A, B and C to denote coreflexives. Clearly, the coreflexives of type $I \leftarrow I$ are in one-to-one correspondence with the subsets of I; we exploit this correspondence by identifying subsets of I with the coreflexives of type $I \leftarrow I$.

Functions are "single-valued" relations; a relation R is *single-valued* if $R \cdot R^\cup \subseteq \mathsf{id}_I$ where I is the target of R. We use an infix dot to denote function application. Thus $f.x$ denotes application of function f to argument x. Dual to the notion of single-valued is the notion of injectivity. A relation R with source J is *injective* if $R^\cup \cdot R \subseteq \mathsf{id}_J$. Which of the properties $R \cdot R^\cup \subseteq \mathsf{id}_I$ or $R^\cup \cdot R \subseteq \mathsf{id}_J$ one calls "single-valued" and which "injective" is a matter of interpretation. The choice here fits in with the convention that input is on the right and output on the left. More importantly, it fits with the convention of writing $f.x$ rather than say x^f (that is the function to the left of its argument). A sensible consequence is that type arrows point from right to left.

2.2   Domains and Division Operators

The *left domain* of a relation R is, informally, the set of output values that are related by R to at least one input value. Formally, the right domain $R_>$ of a relation R of type $I \leftarrow J$ is a coreflexive of type $I \leftarrow I$ satisfying the property that

$$\langle \forall A : A \subseteq \mathrm{id}_I : A \cdot R = R \equiv R < \subseteq A \rangle \ . \tag{1}$$

Given a coreflexive A, $A \subseteq \mathrm{id}_I$, the relation $A \cdot R$ can be viewed as the relation R restricted to outputs in the set A. Thus, in words, the left domain of R is the least coreflexive A that maintains R when R is restricted to outputs in the set A. The *right domain* $R>$ is defined symetrically by reversing the composition $R \cdot A$. The left/right domain should not be confused with the target/source of the relation.

In general, for relations R of type $I \leftarrow J$ and T of type $I \leftarrow K$ there is a relation $R \backslash T$ of type $J \leftarrow K$ satisfying the Galois connection, for all relations S,

$$R \cdot S \subseteq T \equiv S \subseteq R \backslash T \ .$$

The operator $\backslash$ is called a *division* operator (because of the similarity of the above rule to the rule of division in ordinary arithmetic). The relation $R \backslash T$ is called a *residual* or a *factor* of the relation T. Interpreting relations as specifications, the above Galois connection defines $R \backslash T$ to be the "weakest" specification of a program S such that executing R after S satisfies specification T. With this interpretation, $R \backslash T$ has been called a *weakest prespecification* [10].

The *weakest liberal precondition* operator will be denoted here by the symbol "$\backslash$". Formally, if R is a relation of type $I \leftarrow J$ and A is a coreflexive of type $I \leftarrow I$ then $R \backslash A$ is a coreflexive of type $J \leftarrow J$ characterised by the property that, for all coreflexives B of type $J \leftarrow J$,

$$(R \cdot B) < \ \subseteq A \equiv B \subseteq R \backslash A \ . \tag{2}$$

Again, we use a division-like notation, rather than "wlp", to emphasise the similarity with division in normal arithmetic.

3 Allegories and Relators

We assume that the reader is familiar with the most basic notions of category theory, namely objects, arrows, functors, natural transformations and (initial) algebras We use *Fun* to denote the category with sets as objects and functions between sets as arrows. We use *Rel* to denote the category with sets as objects and binary relations as arrows. We also assume familiarity with the relevance of these concepts to functional programming: functors correspond to type constructors and natural transformations correspond to polymorphic functions.

The categorical notion of functor is too weak to describe type constructors in the context of a relational theory of datatypes. The notion of an "allegory" [11] extends the notion of a category in order to better capture the essential properties of relations, and the notion of a "relator" [12, 13, 14] extends the notion of a functor in order to better capture the relational properties of datatype constructors.

Formally, an *allegory* is a category such that, for each pair of objects A and B, the class of arrows of type $A\leftarrow B$ forms an ordered set. In addition there is a converse operation on arrows and a meet (intersection) operation on pairs of arrows of the same type. These are the minimum requirements. For practical purposes, more is needed. A *locally-complete, tabulated, unitary, division allegory* is an allegory such that, for each pair of objects A and B, the partial ordering on the set of arrows of type $A\leftarrow B$ is complete ("locally-complete"), the division operators introduced in section 2.2 are well-defined ("division allegory"), the allegory has a unit (which is a relational extension of the categorical notion of a unit — "unitary") and, finally, the allegory is "tabulated". "Tabulated" captures the fact that relations are subsets of the cartesian product of a pair of sets [15]. (Tabularity is vital because it provides the link between categorical properties and their extensions to relations.)

A suitable extension to the notion of functor is the notion of a "relator" [12]. A *relator* is a functor whose source and target are both allegories, and is monotonic with respect to the subset ordering on relations of the same type, and commutes with converse. Thus, a *relator* F is a function to the objects of an allegory $\mathcal{C}$ from the objects of an allegory $\mathcal{D}$ together with a mapping to the arrows (relations) of $\mathcal{C}$ from the arrows of $\mathcal{D}$ satisfying the following properties:

$$F.R \text{ has type } F.I \xleftarrow{\mathcal{C}} F.J \text{ whenever } R \text{ has type } I \xleftarrow{\mathcal{D}} J. \tag{3}$$

$$F.R \cdot F.S = F.(R \cdot S) \quad \text{for each } R \text{ and } S \text{ of composable type,} \tag{4}$$

$$F.\mathrm{id}_A = \mathrm{id}_{F.A} \quad \text{for each object } A, \tag{5}$$

$$F.R \subseteq F.S \ \Leftarrow \ R \subseteq S \quad \text{for each } R \text{ and } S \text{ of the same type,} \tag{6}$$

$$(F.R)^\cup = F.(R^\cup) \quad \text{for each } R. \tag{7}$$

For example, List is a unary relator, and product is a binary relator. If R is a relation of type $I \leftarrow J$ then List.R relates a list of Is to a list of Js whenever the two lists have the same length and corresponding elements are related by R. The relation $R \times S$ relates two pairs if the first components are related by R and the second components are related by S. List is an example of an inductively-defined datatype; in [16] it was observed that all inductively-defined datatypes are relators.

A design requirement, that dictates the above definition of a relator, is that a relator should extend the notion of a functor but in such a way that it coincides with the latter notion when restricted to functions. Formally, relation R of type $I \leftarrow J$ is *total* iff $\mathrm{id}_J \subseteq R^\cup \cdot R$. A *function* is a relation that is both total and single-valued. It is easy to verify that total relations are closed under composition, as are single-valued relations. Hence, functions are closed under composition too. In other words, the functions form a sub-category. For an allegory $\mathcal{A}$, we denote the sub-category of functions by $Map(\mathcal{A})$. In particular, $Map(Rel)$ is the category Fun. Now, the desired property of relators is that relator F of type $\mathcal{A} \leftarrow \mathcal{B}$ is a functor of type $Map(\mathcal{A}) \leftarrow Map(\mathcal{B})$. It is easily shown that our definition of relator guarantees this property.

(Bird and De Moor [15] omit (7) and define a relator to be a monotonic functor. However, their theorem 5.1, which purports to justify the omission, is false.)

Polymorphic functions play a major role in functional programming. An insight that has helped to increase the understanding of the relevance of category theory to functional programming is that polymorphic functions, like the flatten function on lists, are natural transformations [17, 18]. However, caution is needed when extending the categorical notion of natural transformation to allegories. In the latter context, the term *lax natural transformation* is sometimes used. The collection of lax natural transformations to relator F from G is denoted by $F \hookleftarrow G$ and defined by

$$\alpha :: F \hookleftarrow G \quad \equiv \quad (F.R \cdot \alpha_J \supseteq \alpha_I \cdot G.R \quad \text{for each } R :: I \leftarrow J) \ . \quad (8)$$

A relationship between naturality in the allegorical sense and in the categorical sense is the following [19]. Recall that relators respect functions, i.e. relators are functors on the sub-category Map. Then, in the case that all elements of the collection α are *functions*,

$$\alpha :: F \hookleftarrow G \quad \text{in } \mathcal{A} \quad \equiv \quad \alpha :: F \leftarrow G \quad \text{in } Map(\mathcal{A})$$

where by "in X" we mean that all quantifications in the definition of the type of natural transformation range over the objects and arrows of X. This means that the notion of "lax" natural transformation is the more appropriate allegorical extension of the categorical notion of natural transformation rather than being a natural transformation in the underlying category. Thus we shall not use the qualifier "lax". For us, a natural transformation is as defined by (8).

4 A Programming Paradigm

4.1 Hylo Programs

Characteristic of a programming methodology is that it involves a *discipline* designed to maximise confidence in the reliability of the end product. The discipline constrains the construction methods to those that are demonstrably simple and easy to use, whilst still allowing sufficient flexibility that the creative process of program construction is not impeded.

In standard treatments of the discipline of sequential programming, the class of programs considered is the class of **while** programs; it has long been accepted that arbitrary **goto** programs are undesirable. But, whilst theoretically expressive enough, **while** programs are inadequate to express many elegant and well-known *recursive* programs, like quicksort. On the other hand, arbitrary recursion is also undesirable. Restriction to a more limited class of recursive programs is desirable for a sound discipline of datatype-generic programming.

The programs in the class on which our discipline is based are called *hylomorphisms*. The fact that many recursively defined functional programs are hylomorphisms was identified by Fokkinga, Meijer and Paterson [20], the name having been coined by Meijer [21]. Unlike [20], however, the current paper is not restricted to functional programs.

Definition 1 (Hylos). Let R and S be relations and F a relator. An equation in X of the form $X = R \cdot F.X \cdot S$ is said to be a *hylo equation* or *hylo program*. □

Space does not allow us to give detailed examples of hylo programs here. Briefly, the hylo recursion scheme offers substantial freedom in designing programs because the solution strategy is a parameter of the scheme. The solution strategy is encapsulated in the relator, F. For instance relator $\langle X :: I + X \rangle$ encapsulates repetition, $\langle X :: I + X \times X \rangle$ encapsulates a divide and conquer strategy, and $\langle X :: F.(I \times X) \rangle$ encapsulates primitive recursion. A first step in the design of hylo programs is the choice of the relator [5]. Extending hylo programs to allow relations as components is also a significant advance on the functional paradigm. Relations on strings, like the prefix, suffix, subsequence and segment relations are easy to express as hylo equations, as can quite complex problems like context-free language recognition (even in the most general case) [22].

Crucial to developing a discipline of hylo programming is that the meaning of a hylo equation is well-understood, both as a specification of a relation, and operationally as a program that can be executed. The operational meaning demands an understanding of how hylo equations are executed, including when they are guaranteed to terminate. This is discussed in section 4.2. The specificational meaning can be understood in several ways. One is to extrapolate from the now well-understood notion of a catamorphism on an initial F-algebra. This is captured by theorem 1, below. The definition of a "relational initial F-algebra" is needed first.

Definition 2. Assume that F is an endorelator. Then (I, in) is a *relational initial F-algebra* iff in has type $I \leftarrow F.I$ (and thus is an F-algebra), and there is a mapping $(\![\_]\!)$ defined on all F-algebras such that

$$(\![R]\!) :: A \leftarrow I \quad \text{if } R \text{ has type } A \leftarrow F.A \ , \tag{9}$$

$$(\![\text{in}]\!) = \text{id}_I \ , \text{ and} \tag{10}$$

$$(\![R]\!) \cdot (\![S]\!)^\cup = \langle \mu X :: R \cdot F.X \cdot S^\cup \rangle \ . \tag{11}$$

That is, $(\![R]\!) \cdot (\![S]\!)^\cup$ is the smallest solution of the equation in X, $R \cdot F.X \cdot S^\cup \subseteq X$. □

Definition 2 makes use of the "banana brackets", $(\![\_]\!)$, introduced by Malcolm [23, 24] to denote a functional/relational catamorphism. In categorical terms, catamorphisms are the unique arrows from the initial object in the category of F-algebras; in programming terms, catamorphisms are programs defined by structural induction on a datatype. The definition extends the categorical notion of an initial F-algebra to allegories in a way that is made precise by the hylo theorem below. Recall that $Map(\mathcal{A})$ denotes the sub-category of functions in the allegory $\mathcal{A}$. For clarity, we distinguish between the endorelator F and the corresponding endofunctor, F', defined on $Map(\mathcal{A})$.

Theorem 1 (Hylo Theorem [25]). Suppose F is an endorelator on a locally-complete, tabular allegory $\mathcal{A}$. Let F' denote the endofunctor obtained by restricting F to the objects and arrows of $Map(\mathcal{A})$. Then, in is an initial F'-algebra equivales it is a relational initial F-algebra.   □

Note that the hylo theorem states an equivalence between two definitions. Considering first the implication (loosely speaking, an initial F-algebra is a relational initial F-algebra), property (11) is the property that is most often understood as the "hylo theorem". Property (9) is a necessary prerequisite; essentially it states that catamorphisms are well-defined on relations given that they are well-defined on functions. Property (10) is the key to proving Lambek's lemma that an initial F-algebra is an isomorphism between its source and its target. A consequence of the opposite implication (a relational initial F-algebra is an initial F-algebra) is that catamorphisms on functions are the unique solutions of their defining equations.

4.2 Reductivity

A discipline of programming should always provide the programmer with straightforward-to-use techniques for guaranteeing termination of programs. For datatype-generic programs this is provided by the theory of so-called "reductivity" [5,7] . The major innovatory aspect of this concept is that it is parameterised by a relator, making it possible to explore how properties of termination are induced by properties of datatypes and (natural) transformations between datatypes.

A hylo program, $X = R \cdot F.X \cdot S$, is executed by first unfolding the equation and then computing the argument for the recursive call by executing S. This procedure is repeated until a base case is reached and no further unfoldings are necessary. Then the output is computed by executing R as often as the equation was unfolded. Assuming R and S are both guaranteed to terminate, termination of the recursion is thus dependent only on S, and not on R. Furthermore, if S is nondeterministic, a demonic semantics demands termination irrespective of which output from the unfoldings of S is chosen. This is the familiar execution scheme applied by the implementations of imperative and functional languages. Because of this execution scheme, the computed input-output relation is the least solution of the hylo program.

Suppose that execution begins in a state described by the coreflexive A, and suppose B describes the "safe set" of the hylo program: the maximal set of states from which execution is guaranteed to terminate. Then, execution of S must guarantee that recursive calls begin from a state in B. That is, $(S \cdot A)^< \subseteq F.B$, or, equally, $A \subseteq S \backslash F.B$. Since, B is the maximal set of such states, A, and since the semantics defines the input-ouput relation to be the least solution of the hylo equation, the safe set of program $X = R \cdot F.X \cdot S$ is the coreflexive $\langle \mu A :: S \backslash F.A \rangle$. Termination is guaranteed if this is the identity relation on the domain of S. Hence, the definition of reductivity:

Definition 3 (*F*-reductivity). Relation S of type $F.I \leftarrow I$ is said to be *F-reductive* if and only if $\langle \mu A :: S \backslash F.A \rangle = \text{id}_I$. □

Let us now check that the notion of F-reductivity is compatible with more familiar accounts of program termination.

A programmer proves termination by using well-founded relations: they prove that the argument of every recursive call is "smaller" than the original argument. For program $X = R \cdot F.X \cdot S$ this means that all values stored in an output F-structure of S have to be smaller than the corresponding input of S. More formally, with $x\langle\text{mem}\rangle y$ standing for "x is a member of F-structure y" (or, x is a value stored in F-structure y"), we need for all x and z

$$\langle \forall y :: x\langle\text{mem}\rangle y \wedge y\langle S\rangle z \Rightarrow x \prec z \rangle \quad ,$$

for some well-founded ordering $\prec$. That is, a relation S is F-reductive if and only if there is a well-founded relation $\prec$ such that whenever an F-structure is related by S to some y, it is the case that every value stored in the F-structure is related to y by $\prec$.

To make this statement precise we need to formalise the concept of "values stored in an F-structure". Hoogendijk and De Moor [26, 19] have shown that this is possible for so-called "container types". For the relators from this class, one can define a membership relation, say mem. For example, for the list relator this relation holds between a point of the universe and a list precisely when the point is in the list. For product, the relation holds between x and (x,y) and also between y and (x,y).

A precise characterisation of the membership relation of a relator is the following:

Definition 4 (Membership). Relation mem $\quad :: \quad I \leftarrow F.I$ is a membership relation of relator F if and only if $F.A = \text{mem} \backslash A$ for all coreflexives A, $A \subseteq I$. □

Using this definition of membership we get a precise relationship between reductivity and well-foundedness. Indeed, for coalgebra S with carrier I and coreflexive A below I, we have:

$$
\begin{aligned}
& S \backslash F.A \\
= \quad & \{ \quad \text{definition 4} \quad \} \\
& S \backslash (\text{mem} \backslash A) \\
= \quad & \{ \quad \text{factors (2)} \quad \} \\
& (\text{mem} \cdot S) \backslash A \quad .
\end{aligned}
$$

Now, well-foundedness of relation R of type $I \leftarrow I$ is the condition that the least prefix point of the function $\langle A :: R \backslash A \rangle$ is I [27], whereas reductivity of $S ::$ $F.I \leftarrow I$ is the condition that the least prefix point of the function $\langle A :: S \backslash F.A \rangle$

is I. So, for coalgebra $S :: F.I \leftarrow I$, the statement that S is F-reductive is equivalent to the statement that $\text{mem} \cdot S$ is well-founded. Formally,

$$S \text{ is } F\text{-reductive} \equiv \text{mem} \cdot S \text{ is well-founded} .$$

Conversely,

$$R \text{ is well-founded} \equiv \text{mem} \backslash R \text{ is } F\text{-reductive} .$$

Summarising, we have:

Theorem 2. Suppose mem is the membership relation for relator F. Then the functions $\langle S :: \text{mem} \cdot S \rangle$ and $\langle R :: \text{mem} \backslash R \rangle$ form a Galois connection between the F-reductive relations, S, and the well-founded relations, R.   □

Bird and De Moor [15, chapter 6] avoid the introduction of the notion of reductivity by always requiring that $\text{mem} \cdot S$ is well-founded whenever F-reductivity of S is required. The main advantage of defining termination in terms of reductivity instead of well-foundedness and membership is that it is possible to formulate theorems relating reductivity of one type to reductivity of another type. The rules presented in section 5 are of this nature.

5   A Calculus of Reductive Relations

Theorem 3. The converse of an initial F-algebra is F-reductive.

Proof. Let in $:: I \leftarrow F.I$ be an initial F-algebra and Λ an arbitrary coreflexive of type $I \leftarrow I$. We must show that

$$I \subseteq A \quad \Leftarrow \quad \text{in}^\cup \backslash F.A \subseteq A .$$

We start with the antecedent and derive the consequent:

$\quad \text{in}^\cup \backslash F.A \subseteq A$

$=\qquad \{\qquad$ for function f and coreflexive B, $f \backslash B = f^\cup \cdot B \cdot f$,

$\qquad\qquad \text{in}^\cup$ is a function and $F.A$ is a coreflexive $\quad\}$

$\quad \text{in} \cdot F.A \cdot \text{in}^\cup \subseteq A$

$\Rightarrow\qquad \{\qquad$ hylo theorem $\quad\}$

$\quad (\![\text{in}]\!) \cdot (\![\text{in}]\!)^\cup \subseteq A$

$=\qquad \{\qquad$ identity rule: (10), in $:: I \leftarrow F.I$ is an initial F-algebra $\quad\}$

$\quad I \subseteq A .$ □

Theorem 4. Let Q be G-reductive and S be a natural transformation of type $F \leftarrow \text{Id}$, where Id denotes the identity relator. Then $F.Q \cdot S$ is $(F \circ G)$-reductive.

Proof. We prove the stronger:

$$\langle \mu A :: Q \backslash G.A \rangle \subseteq \langle \mu A :: (F.Q \cdot S) \backslash F.(G.A) \rangle .$$

First, we observe a general fact about natural transformations α of type $F \hookleftarrow H$, namely, for all objects I and all coreflexives A such that $A \subseteq I$,

$$H.A \subseteq \alpha_I \backslash F.A \quad , \tag{12}$$

since

$$H.A \subseteq \alpha_I \backslash F.A$$

$\qquad = \qquad \{ \qquad$ factors: (2) $\quad \}$

$$(\alpha_I \cdot H.A)^< \subseteq F.A$$

$\qquad = \qquad \{ \qquad$ domains: (1) $\quad \}$

$$F.A \cdot \alpha_I \cdot H.A = \alpha_I \cdot G.A$$

$\qquad = \qquad \{ \qquad \alpha$ has type $F \hookleftarrow H$. Thus, $F.A \cdot \alpha_I \supseteq \alpha_I \cdot H.A$.

$\qquad\qquad\qquad\qquad A$ is a coreflexive, so $H.A \cdot H.A = H.A \quad \}$

$$F.A \cdot \alpha_I \cdot H.A \subseteq \alpha_I \cdot H.A$$

$\qquad = \qquad \{ \qquad F.A \subseteq \mathrm{id}_{F.I} \quad \}$

true .

The theorem follows, by monotonicity of the fixpoint operator μ, from the fact that, for all A,

$$(F.Q \cdot S) \backslash F.(G.A)$$

$\qquad = \qquad \{ \qquad$ factors: (2) $\quad \}$

$$S \backslash (F.Q \backslash F.(G.A))$$

$\qquad \supseteq \qquad \{ \qquad$ factors: (2) $\quad \}$

$$S \backslash F.(Q \backslash G.A)$$

$\qquad \supseteq \qquad \{ \qquad S$ has type $F \hookleftarrow \mathsf{Id}$, (12) $\quad \}$

$$Q \backslash G.A \quad . \qquad\qquad\qquad\qquad\qquad\qquad\qquad\qquad\qquad \square$$

6   Generic Unification

In this section, we apply the notion of F-reductivity to a key lemma in the proof of correctness of a generic unification algorithm. Such an algorithm was first formulated by Jeuring and Jansson [28] and is further elaborated in [9]. The algorithm is "generic" in the sense that it is parameterised by a relator F that specifies the structure of expressions to be unified.

Here, we show that the "occurs-properly-in" relation on expressions is well-founded. Particularly remarkable about our proof is that it is very simple. This is a result of its not requiring the definition of a size function on expressions in any way, the key to the proof being instead the fact that the converse of an initial F-algebra is F-reductive.

(The reader is invited to compare the proof presented here with the one given in [9]. Although the one presented here was the first to be developed, it was considered expedient at the time not to burden the reader of [9] with too many new ideas, and to present a more conventional proof instead.)

In its generic form, unification is expressed as follows. A parameter is a relator F. A second parameter is a type V, elements of which are called *variables*. Given these two, we may define a relator F_V which maps relation X to $F.X + \mathrm{id}_V$. Then we assume that in is an initial F_V-algebra with carrier F^*V. That is,

$$\mathsf{in} \ :: \ F^*V \leftarrow F.F^*V + V \ .$$

The relator F^* (together with appropriately defined unit and multiplier) is a monad which, as the Kleene-star-like notation suggests, is obtained by repeated application of the relator F. Elements of F^*V are called *expressions*; the parameter F limits the way that new expressions are built up out of subexpressions. Substitution of an expression for a variable can now be defined in such a way that the composition of substitutions is Kleisli composition in the monad. The ordering "more general than" on substitutions is defined in the usual way. Generic unification is then the problem of finding a substitution that unifies two expressions and is more general than any other unifier.

A fundamental lemma in a proof of correctness of unification is to show that if a variable occurs in an expression then the variable and expression are not unifiable. The way to do this is to define an "occurs-properly-in" relation between expressions, show that this relation is well-founded (and thus is irreflexive) and finally show that it is preserved by substitution. Here we will just show the first two of these steps as an illustration of the reductivity calculus.

Suppose mem is the membership relation of the relator F. Let $\mathsf{inl}_{A,B}$ denote the injection function of type $A+B \leftarrow A$. (We will drop subscripts from now on for simplicity.) Then we can define the relation occurs_properly_in of type $F^*V \leftarrow F^*V$ by

$$\mathsf{occurs\_properly\_in} = (\mathsf{mem} \cdot (\mathsf{in} \cdot \mathsf{inl})^{\cup})^{+} \ .$$

Informally, the relation $(\mathsf{in} \cdot \mathsf{inl})^{\cup}$ (which has type $F.(F^*V) \leftarrow F^*V$) destructs an element of F^*V into an F-structure and then mem identifies the data stored in that F-structure. Thus $\mathsf{mem} \cdot (\mathsf{in} \cdot \mathsf{inl})^{\cup}$ destructs an element of F^*V into a number of immediate subcomponents. Application of the transitive-closure operation repeats this process thus breaking the structure down into all its subcomponents.

The occurs_properly_in relation has a very simple structure. We ought to be able to see that it is well-founded almost directly just from that structure. Indeed this is what the reductivity calculus allows us to do. The lemma and its proof follow. The first step involves a well-known property of well-founded relations. Otherwise, every non-trivial step uses the reductivity calculus.

Theorem 5.   The relation occurs_properly_in is well-founded.

Proof

occurs_properly_in is well-founded

=   {   definition of occurs_properly_in,

R is well-founded $\equiv$ R^+ is well-founded }

mem $\cdot$ (in$\cdot$inl)$^\cup$ is well-founded

$\Leftarrow$ {   mem $\cdot R$ is well-founded $\equiv$ R is F-reductive }

(in$\cdot$inl)$^\cup$ is F-reductive

=   {   (in$\cdot$inl)$^\cup$ = inl$^\cup \cdot$ in$^\cup$, }

inl$^\cup \cdot$ in$^\cup$ is F-reductive

$\Leftarrow$ {   theorem 4 }

in$^\cup$ is F_V-reductive $\wedge$ inl$^\cup$ :: $F \hookleftarrow F_V$

$\Leftarrow$ {   theorem 3, definition of $\hookleftarrow$ }

true $\wedge$ $\langle \forall R :: F.R \cdot \text{inl}^\cup{}_J \supseteq \text{inl}^\cup{}_I \cdot F_V.R \rangle$

=   { $F_V.A = F.A + \text{id}_V$, converse and defn. of inl }

true . $\square$

Note that the proof is entirely algebraic and does not involve any notion of the "size" of expressions. Many well-foundedness arguments are based on defining a variant function with range the natural numbers and exploiting their well-foundedness. The above proof is based on the basic reductivity theorem that the converse of an initial F-algebra is F-reductive, a consequence of which theorem is that the natural numbers are well-founded. Introducing the natural numbers into the proof would be introducing unnecessary detail.

Acknowledgements

This work was supported by EPSRC grant GR/S27085/01, Data-type generic programming.

References

1. Jeuring, J., Jansson, P.: Polytypic programming. In Launchbury, J., Meijer, E., Sheard, T., eds.: Proceedings of the Second International Summer School on Advanced Functional Programming Techniques, Springer-Verlag (1996) 68–114 LNCS 1129.
2. Hinze, R.: Polytypic values possess polykinded types. Science of Computer Programming **43**(2-3) (2002) 129–159

3. Hinze, R., Jeuring, J., Löh, A.: Type-indexed data types. Science of Computer Programming **51**(1-2) (2004) 117–151
4. Löh, A., Clarke, D., Jeuring, J.: Dependency-style Generic Haskell. In Shivers, O., ed.: Proceedings of the International Conference, ICFP'03, ACM Press (2003) 141–152
5. Doornbos, H., Backhouse, R.: Induction and recursion on datatypes. In Möller, B., ed.: Mathematics of Program Construction, 3rd International Conference. Volume 947 of LNCS., Springer-Verlag (1995) 242–256
6. Doornbos, H.: Reductivity arguments and program construction. PhD thesis, Eindhoven University of Technology, Department of Mathematics and Computing Science (1996)
7. Doornbos, H., Backhouse, R.: Reductivity. Science of Computer Programming **26**(1–3) (1996) 217–236
8. Jansson, P., Jeuring, J.: Functional pearl: Polytypic unification. Journal of Functional Programming (1998)
9. Backhouse, R., Jansson, P., Jeuring, J., Meertens, L.: Generic programming. An introduction. In Swierstra, S., ed.: 3rd International Summer School on Advanced Functional Programming, Braga, Portugal, 12th-19th September, 1998. Volume LNCS 1608., Springer Verlag (1999) 28–115
10. Hoare, C., He, J.: The weakest prespecification. Fundamenta Informaticae **9** (1986) 51–84, 217–252
11. Freyd, P., Ščedrov, A.: Categories, Allegories. North-Holland (1990)
12. Backhouse, R.: Naturality of homomorphisms. Lecture notes, International Summer School on Constructive Algorithmics, vol. 3, 1989 (1989)
13. Backhouse, R., Bruin, P.d., Malcolm, G., Voermans, T., Woude, J.v.d.: Relational catamorphisms. In B., M., ed.: Proceedings of the IFIP TC2/WG2.1 Working Conference on Constructing Programs from Specifications, Elsevier Science Publishers B.V. (1991) 287–318
14. Backhouse, R., Woude, J.v.d.: Demonic operators and monotype factors. Mathematical Structures in Computer Science **3**(4) (1993) 417–433
15. Bird, R.S., de Moor, O.: Algebra of Programming. Prentice-Hall International (1996)
16. Backhouse, R., Bruin, P.d., Hoogendijk, P., Malcolm, G., Voermans, T., Woude, J.v.d.: Polynomial relators. In Nivat, M., Rattray, C., Rus, T., Scollo, G., eds.: Proceedings of the 2nd Conference on Algebraic Methodology and Software Technology, AMAST'91, Springer-Verlag, Workshops in Computing (1992) 303–326
17. Reynolds, J.: Types, abstraction and parametric polymorphism. In Mason, R., ed.: IFIP '83. Elsevier Science Publishers (1983) 513–523
18. Wadler, P.: Theorems for free! In: 4'th Symposium on Functional Programming Languages and Computer Architecture, ACM, London. (1989)
19. Hoogendijk, P.: A Generic Theory of Datatypes. PhD thesis, Department of Mathematics and Computing Science, Eindhoven University of Technology (1997)
20. Meijer, E., Fokkinga, M., Paterson, R.: Functional programming with bananas, lenses, envelopes and barbed wire. In: FPCA '91: Functional Programming Languages and Computer Architecture. Number 523 in LNCS, Springer-Verlag (1991) 124–144
21. Meijer, E.: Calculating Compilers. PhD thesis, University of Nijmegen (1992)
22. Backhouse, R., Doornbos, H.: Mathematics of recursive program construction. Internet publication available from http://www.cs.nott.ac.uk/~rcb/MPC/papers (2001)

23. Malcolm, G.: Algebraic data types and program transformation. PhD thesis, Groningen University (1990)
24. Malcolm, G.: Data structures and program transformation. Science of Computer Programming 14(2–3) (1990) 255–280
25. Backhouse, R., Hoogendijk, P.: Final dialgebras: From categories to allegories. Theoretical Informatics and Applications 33(4/5) (1999) 401–426
26. Hoogendijk, P., de Moor, O.: Container types categorically. Journal of Functional Programming 10(2) (2000) 191–225
27. Doornbos, H., Backhouse, R., van der Woude, J.: A calculation approach to mathematical induction. Theoretical Computer Science 179 (1997) 103–135
28. Jansson, P., Jeuring, J.: PolyP - a polytypic programming language extension. In: POPL '97: The 24th ACM SIGPLAN-SIGACT Symposium on Principles of Programming Languages, ACM Press (1997) 470–482

The Logical Strength of the Uniform Continuity Theorem

Josef Berger

University of Canterbury
Department of Mathematics and Statistics
Private Bag 4800
Christchurch, New Zealand
Josef.Berger@canterbury.ac.nz

Abstract. We introduce a notion of complexity for sets of finite binary sequences such that the corresponding fan theorem is constructively equivalent to the uniform continuity theorem. This settles an open question.

Keywords: Constructive reverse mathematics, uniform continuity theorem.

It is well known that the uniform continuity theorem implies the fan theorem for detachable bars [5]. Furthermore, it is easy to see that the fan theorem for Π_1^0–bars implies the uniform continuity theorem.[1] Here we present a very natural version of the fan theorem which exactly hits the logical strength of the uniform continuity theorem in the sense of constructive reverse mathematics [6, 7].

We use the framework **BISH** of Bishop's constructive mathematics [3, 4, 5]. We regard **BISH** as simply mathematics with intuitionistic, rather than classical, logic, together with some suitable foundation such as CZF [1].

Since the most complex objects we deal with are integer-valued functions on the Cantor space, our results can easily be carried over to intuitionistic finite-type arithmetic **HA**$^\omega$; see [9, 11] for an introduction to this system. We keep the formalisation aspect in view; but for the sake of better readability, we do not carry it out rigorously. Whenever we use a version of the axiom of choice, we mention this explicitly.

Let $\mathbb{N} = \{1, 2, 3, \dots\}$ denote the set of all natural numbers m, n, N. Let $\{0, 1\}^{\mathbb{N}}$ denote the set of all infinite binary sequences α, β, γ. Let $\overline{\alpha}n$ denote the restriction of α to the first n components, where $n \in \mathbb{N} \cup \{0\}$. Thus $\overline{\alpha}0$ is the empty sequence ( ). Under the compact[2] metric

$$d(\alpha, \beta) = \inf \left\{ 2^{-n} \mid \overline{\alpha}n = \overline{\beta}n \right\}$$

on $\{0, 1\}^{\mathbb{N}}$, pointwise continuity of functions $F : \{0, 1\}^{\mathbb{N}} \to \mathbb{N}$ reads as

$$\forall \alpha \, \exists n \, \forall \beta \left(\overline{\alpha}n = \overline{\beta}n \to F(\alpha) = F(\beta) \right)$$

---

[1] We do not give a proof of this fact since it follows from Proposition 1.

[2] A metric space is *compact* if it is complete and totally bounded.

A. Beckmann et al. (Eds.): CiE 2006, LNCS 3988, pp. 35–39, 2006.

whereas uniform continuity reads as

$$\exists n \, \forall \alpha, \beta \, (\overline{\alpha}n = \overline{\beta}n \rightarrow F(\alpha) = F(\beta)) \, .$$

The uniform continuity theorem is the following principle:

UC Every pointwise continuous function $F \, : \, \{0,1\}^{\mathbb{N}} \, \rightarrow \, \mathbb{N}$ is uniformly continuous.

Let $\{0,1\}^*$ denote the set of all finite binary sequences u, w. Concatenation of u and v is denoted by $u * w$. For $F : \{0,1\}^{\mathbb{N}} \rightarrow \mathbb{N}$ and u we set

$$F(u) = F(u * 0 * 0 * 0 * \dots) \, .$$

This leads to a characterisation of uniform continuity:

Lemma 1. *Let* $F : \{0,1\}^{\mathbb{N}} \rightarrow \mathbb{N}$ *be pointwise continuous. Then we have*

$$\forall \alpha \, \exists n \, \forall w \, (F(\overline{\alpha}n) = F(\overline{\alpha}n * w)) \, .$$

Furthermore, F *is uniformly continuous if and only if*

$$\exists n \, \forall \alpha \, \forall w \, (F(\overline{\alpha}n) = F(\overline{\alpha}n * w)) \, . \tag{1}$$

These results can be proved in **BISH** *as well as in* **HA**$^{\omega}$.

Proof. We show only the most interesting part of the lemma. Suppose that F is pointwise continuous and that (1) holds. We have to show that F is uniformly continuous. By (1) there is N such that

$$\forall \alpha \, \forall w \, (F(\overline{\alpha}N) = F(\overline{\alpha}N * w)) \, .$$

Fix α, β with $\overline{\alpha}N = \overline{\beta}N$. We claim that $F(\alpha) = F(\beta)$. By the pointwise continuity of F there is $m \geq N$ such that

$$\forall \gamma \, (\overline{\alpha}m = \overline{\gamma}m \rightarrow F(\alpha) = F(\gamma)) \, .$$

and

$$\forall \gamma \, (\overline{\beta}m = \overline{\gamma}m \rightarrow F(\beta) = F(\gamma)) \, .$$

Putting the pieces together yields

$$F(\alpha) = F(\overline{\alpha}m) = F(\overline{\alpha}N) = F(\overline{\beta}m) = F(\beta) \, .$$

A subset Y of a set X is *detachable* from X if

$$\forall x \in X \, (x \in Y \vee x \notin Y) \, .$$

A subset C of $\{0,1\}^*$ is a c–set if there is a detachable subset D of $\{0,1\}^*$ such that

$$\forall u \, (u \in C \leftrightarrow \forall w \, (u * w \in D)) \, .$$

Thus a sequence u belongs to C if every extension of u belongs to D. The letter c in the expression c–set should indicate that this notion of complexity is related to continuity. A subset B of $\{0,1\}^*$ is a bar if

$$\forall \alpha \, \exists n \, (\overline{\alpha} n \in B)$$

and a uniform bar if

$$\exists N \, \forall \alpha \, \exists n \leq N \, (\overline{\alpha} n \in B).$$

Now we can introduce the fan theorem for c–sets:

c–FT every bar which is a c–set is uniform

Every c–set C is closed under extensions; that means

$$\forall u, w \, (u \in C \rightarrow u * w \in C);$$

therefore a c–set C is a uniform bar if and only if

$$\exists N \, \forall \alpha \, (\overline{\alpha} N \in C).$$

Proposition 1. *c–FT implies* **UC** *over both* **BISH** *and* **HA**$^\omega$.

Proof. Assume **c–FT** and fix a pointwise continuous function $F : \{0,1\}^{\mathbb{N}} \rightarrow \mathbb{N}$. We define

$$D = \{u \mid F(u) = F(u * 1)\}, \ B = \{u \mid \forall w \, (u * w \in D)\}.$$

Then D is detachable from $\{0,1\}^*$ and B is a c–set. From

$$\forall \alpha \, \exists n \, \forall w \, (F(\overline{\alpha} n) = F(\overline{\alpha} n * w))$$

we obtain

$$\forall \alpha \, \exists n \, \forall w \, (F(\overline{\alpha} n * w) = F(\overline{\alpha} n * w * 1))$$

and thus

$$\forall \alpha \, \exists n \, \forall w \, (\overline{\alpha} n * w \in D).$$

Hence B is a bar and therefore, by assumption, a uniform bar. We thus can find N such that

$$\forall \alpha \, (\overline{\alpha} N \in B),$$

which reads as

$$\forall \alpha \, \forall w \, (F(\overline{\alpha} N * w) = F(\overline{\alpha} N * w * 1)).$$

We can deduce

$$\forall \alpha \, \forall w \, (F(\overline{\alpha} N) = F(\overline{\alpha} N * w)),$$

which is just the uniform continuity of F. We thus have shown **UC**.

The proof of the converse requires a version of unique choice.

AC* Let X be a detachable subset of $\{0,1\}^{\mathbb{N}} \times \mathbb{N}$ such that $\forall \alpha \exists! n \, (\alpha, n) \in X$. Then there exists a function $F : \{0,1\}^{\mathbb{N}} \to \mathbb{N}$ such that $\forall \alpha \, (\alpha, F(\alpha) \in X$.

A similar axiom is called $\mathbf{AC}_{1,0}!$ in [9]. Note that unique choice is admissible in **BISH** but not in $\mathbf{HA}^{\omega}$.

Proposition 2. *UC implies c–FT over BISH. Furthermore, we have*

$$\mathbf{HA}^{\omega} + \mathbf{AC}^* \vdash \mathbf{UC} \to \mathbf{c\text{-}FT}.$$

Proof. Assume **UC**. Let D be a detachable subset of $\{0,1\}^*$ such that

$$B = \{u \mid \forall w \, (u * w \in D)\}$$

is a bar. For every α define

$$D_{\alpha} = \{n \mid \overline{\alpha}n \notin D\} \cup \{1\}.$$

Since B is a bar, D_{α} is bounded; by **AC*** there is a function $F : \{0,1\}^{\mathbb{N}} \to \mathbb{N}$ such that

$$\forall \alpha \, (F(\alpha) = \max D_{\alpha}).$$

Fix α. There is n such that $\overline{\alpha}n \in B$. Now, for β with $\overline{\alpha}n = \overline{\beta}n$, we have $D_{\alpha} = D_{\beta}$; this implies that $F(\alpha) = F(\beta)$. Thus F is pointwise and therefore uniformly continuous. By Corollary 4.3 in Chapter 4 of [4], F is bounded, thus we can find N such that

$$\forall \alpha \, (F(\alpha) < N).$$

We now can conclude that

$$\forall \alpha \forall w \, (\overline{\alpha}N * w \in D);$$

therefore B is a uniform bar and we have shown **c–FT**.

The fan theorem for detachable bars reads as:

Δ–FT every detachable bar is uniform

A subset B of $\{0,1\}^*$ is a Π_1^0-set if there is a detachable set

$$D \subseteq \{0,1\}^* \times \{0,1\}^*$$

such that

$$\forall u \, (u \in B \leftrightarrow \forall w \, (u, w) \in D).$$

The fan theorem for Π_1^0-bars reads as:

Π_1^0–FT every bar which is a Π_1^0-set is uniform

It is easy to see that

$$\Pi^0_1\text{-}\mathbf{FT} \Rightarrow \mathbf{c\text{-}FT} \Rightarrow \Delta\text{-}\mathbf{FT}.$$

It remains to investigate which of these implications are strict. To this end we will have to place even more emphasis on formalisation. Some authors prove the equivalence of $\mathbf{UC}$ and $\Delta\text{-}\mathbf{FT}$, with the help of additional assumptions. This is continuous choice in the case of [5]. In [2, 8, 12] the authors work with pointwise continuous functions which possess a modulus of pointwise continuity, whereas in [7, 10] the authors work with so called neighborhood functions. We did not make use of any such hypothesis.

We presume that further propositions in analysis are equivalent to $\mathbf{c\text{-}FT}$.

Acknowledgments. The author benefits from many fruitful discussions about uniform continuity. Special thanks go to Douglas Bridges, Hajime Ishihara, Ulrich Kohlenbach, Peter Schuster, Thomas Streicher, and Wim Veldman.

References

1. Peter Aczel and Michael Rathjen. *Notes on constructive set theory.* Technical Report 40, Institut Mittag-Leffler, The Royal Swedish Academy of Sciences (2001)
2. Josef Berger. *The fan theorem and uniform continuity.* In: S. Barry Cooper, Benedikt Löwe, Leen Torenvliet (eds.), *New Computational Paradigms. First Conference on Computability in Europe, CiE 2005, Amsterdam, The Netherlands, June 8–12, 2005, Proceedings*, Lecture Notes in Computer Science 3526. Springer–Verlag (2005) pages 18–22
3. Errett Bishop. *Foundations of Constructive Analysis.* McGraw–Hill, New York (1967)
4. Errett Bishop and Douglas Bridges. *Constructive Analysis.* Grundlehren der mathematischen Wissenschaften 279. Springer (1985)
5. Douglas Bridges and Fred Richman. *Varieties of Constructive Mathematics.* London Mathematical Society Lecture Note Series 97. Cambridge University Press (1987)
6. Hajime Ishihara. *Constructive Reverse Mathematics: Compactness Properties.* In: L. Crosilla and P. Schuster (eds). *From Sets and Types to Topology and Analysis.* Oxford Logic Guides 48. Oxford University Press (2005) pages 245–267
7. Hajime Ishihara. *Reverse mathematics in Bishop's constructive mathematics.* To appear in Philosophia Scientiae.
8. Iris Loeb. *Equivalents of the (Weak) Fan Theorem.* Annals of Pure and Applied Logic, Volume 132, Issue 1 (2005) pages 51–66
9. Anne S. Troelstra (ed). *Metamathematical Investigation of Intuitionistic Arithmetic and Analysis.* Lecture Notes in Mathematics 344. Springer (1973)
10. Anne S. Troelstra and Dirk van Dalen. *Constructivism in Mathematics. An Introduction. Vol I.* Studies in Logic and the Foundation of Mathematics, Vol. 121, North–Holland (1988)
11. Anne S. Troelstra and Dirk van Dalen. *Constructivism in Mathematics. An Introduction. Vol II.* Studies in Logic and the Foundation of Mathematics, Vol. 123, North–Holland (1988)
12. Wim Veldman. *Brouwer's fan theorem as an axiom and as a contrast to Kleene's alternative.* Preprint, Radboud University, Nijmegen (2005)

Elementary Algebraic Specifications of the Rational Function Field

J.A. Bergstra

University of Amsterdam, Programming Research Group
Kruislaan 403, 1098 SJ Amsterdam, Netherlands
janb@science.uva.nl
and
Utrecht University, Applied Logic Group
Heidelberglaan 8, 3584 CS Utrecht, Netherlands
janb@phil.uu.nl

Abstract. The elementary algebraic specifications form a small subset of the range of techniques available for algebraic specifications and are based on equational specifications with hidden functions and sorts and initial algebra semantics. General methods exist to show that all semicomputable and computable algebras can be characterised up to isomorphism by such specifications. Here we consider these specification methods for specific computable rational number arithmetics. In particular, we give an elementary equational specification of the 0-totalised rational function field $\mathbb{Q}_0(X)$ with its degree operator as an auxiliary function.

1 Introduction

Between 1979 and 1995 in cooperation with J V Tucker we wrote a series of papers that classified the computable, semicomputable and cosemicomputable data types using algebraic specifications (see Bergstra and Tucker [2, 3, 4, 5]). Work has continued on this subject, refining notions such as finality (e.g., including Meseguer and Goguen [13] and Moss, Meseguer and Goguen [26]), and on open questions (e.g., by Marongiu and Tulipani [22] and by Khoussainov [20, 21]).

Recently, we have returned to the foundations of the subject in [7, 8], tackling the specification of basic data types such as the rational numbers, and we continue here. We will use the *elementary algebraic specifications*, as proposed in [8] which are close to the basic techniques of the ADJ Group of the 1970s.

The set $\mathbb{Q}$ of rational numbers is a number system designed to denote measurements. Rationals are used to define the real and complex numbers via approximation. The rationals are the numbers with which we make finite computations with full precision. Algebras made by equipping $\mathbb{Q}$ with some constants and operations we call *rational arithmetics*. We usually calculate with the algebra $(\mathbb{Q}|0, 1, +, -, \cdot, {}^{-1})$ which is called the *field* of rational numbers where the operations satisfy certain standard axioms.

A. Beckmann et al. (Eds.): CiE 2006, LNCS 3988, pp. 40–54, 2006.

In addition to rational arithmetics, of particular interest are field extensions of the rational number field. One important field extension is the field of rational functions, based on the set

$$\mathbb{Q}(X) = \{p(X)/q(X)|p, q \in \mathbb{Q}[X]\}.$$

This has special operations such as the degree operator $d\colon \mathbb{Q}(X) \to \mathbb{N}$.

The algebras of rational numbers, such as the field and its extensions by real and complex numbers, are among the truly fundamental data types. Despite the fact they have been known and used for over two millennia, they are neglected in the modern theory of data types. After over 30 years of data type theory, many questions about the specification of rational arithmetics and their extensions are open. There is an obvious technical obstacle: the axioms concerning division are not equations and, indeed, it is known that the class of fields cannot be defined by a set of equations.

Now the common rational arithmetics and field extensions are all computable algebras. Indeed, in the *theory of computable rings and fields* there is a wealth of constructions of computable algebras that start with the rationals and the finite fields: see the introduction and survey Stoltenberg-Hansen and Tucker [28]. Therefore, according to the general theory of algebraic specifications of computable data types they have various equational specifications under initial and final algebra semantics. Computable algebras also have equational specifications that are complete term rewriting systems ([5]). However, these general specification theorems for computable data types involve (binary) hidden functions and are based on equationally definable enumerations of the data type.

Recently Moss found in [25] that there exists an equational specification of the ring of rationals (i.e., without division or inverse) with just *one* unary hidden function. He used a remarkable enumeration theorem for the rationals in Calkin and Wilf [9]. He also gave specifications of other rational arithmetics and asked if hidden functions were necessary. In [7] we proved that there exists a finite equational specification under initial algebra semantics, *without* further hidden functions, but making use of an inverse operation, of the field of rational numbers. Here we will continue this line of work and in particular we prove:

Theorem 1. *There exists a finite equational specification under initial algebra semantics, without hidden functions, of the algebra*

$$Q_0(X, d) = (\mathbb{Q}(X)|0, 1, X, +, -, \cdot, ^{-1}, d)$$

of rational functions with field and degree operations that are all total.

The structure of the paper is this. In Section 2, we discuss the basics of specification theory and define the elementary algebraic specifications. In Section 3, we describe the algebras and the axioms we will use to specify them. In Section 4 we prove the main theorem. Finally, in Section 5 we discuss some open problems.

We thank J V Tucker for many valuable discussions on the results of this paper.

2   Elementary Algebraic Specifications (EAS)

2.1   What Are the Elementary Algebraic Specifications?

Algebraic specification starts with the idea of modelling - e.g., data, processes, syntax, hardware, etc. - using sets and functions. Wherever there are sets and functions there are algebras! For example, the sets X, Y and function $f : X \to Y$ are combined to form the many sorted algebra $(X, Y | f)$. A particular algebra A is a mathematical model of a specific concrete representation of the system equipped with concrete operations. The need to understand the system, its representations and the extent to which they are unique leads to the concepts of (i) axiomatic theories for the chosen operators, and (ii) homomorphisms and isomorphisms for the comparison of algebras. The simplest axioms are equations. The simplest deductions are are those of equational logic based on the rewriting of terms.

The line of thought that focussed on the role of initial algebras in semantic modelling and specification was expounded in [16]. Joseph Goguen, Jim Thatcher and Eric Wagner, writing as the ADJ Group, provided a mathematical basis for modelling and specifying abstract data types, starting in [17]. The ADJ Group established most of the basic concepts by combining the technical ideas of many sorted algebras, equations, conditional equations, hidden functions and sorts, term rewriting and initial algebras. In Kamin [19], a rapidly growing literature was organised and problems identified and clearly stated, such as when were hidden functions and sorts necessary? Goguen and Wagner have reflected on the ADJ Group in [11] and [30], respectively.

The theory of computable data types demonstrates that any computable system can be modelled in this way. Therefore, we define the basic elements of EAS as follows.

Definition 1. *An algebraic specification* (Σ', E') *of a* Σ *algebra* A *is elementary if it involves only*

1. *A many sorted signature* Σ' *that is non-void. A signature is non-void if there is a closed term of every sort.*
2. *A set* E' *of equations or conditional equations.*
3. *An initial algebra semantics such that* $I(\Sigma', E')|_\Sigma \cong A$.

In particular, the elementary specifications *require* total functions, *allow* hidden functions and sorts, and may or may not be complete term rewriting systems. Clearly, there are plenty of restrictions in force: see subsection 2.2 below.

Definition 2. *The specification problem is this: Given a* Σ *algebra* A, *can one find an algebraic specification* (Σ', E') *such that* $I(\Sigma', E')|_\Sigma \cong A$.

An EAS is "better" if it is finite rather than infinite, contains equations rather than conditional equations, or features nice term rewriting properties such as confluency and termination. A standard way of validating an elementary specification is to check these properties:

Definition 3. *An algebraic specification* (Σ', E') *of a* Σ *algebra* A *satisfies Goguen's conditions if it the following are true:*

No Junk or Minimality. *The algebra* A *is* Σ*-minimal.*
No Confusion or Completeness. *For all closed* Σ *terms* t, t', *we have*

$$A \models t = t' \text{ if, and only if, } E' \vdash t = t'.$$

In particular, the Goguen conditions imply that

$$I(\Sigma', E')|_{\Sigma} \cong A.$$

What makes these features *elementary*? The purpose of developing a specification is to model, analyse and understand. In simple terms, these algebraic tools are fundamental for any modelling using sets and functions: they are used to abstract and analyse the properties models of an idea, component, or system. One chooses a set of operators and postulates a set of laws they satisfy; the laws are expressed as equations or conditional equations. The terms express all possible operations that can be derived by combining operations, and the equational identities express the consequent facts about the model. The term rewriting is a completely basic mechanism for both abstract reasoning and computation. This view suggests the elementary character of the equations and that we cannot make do with less. There is also an argument that they need extension in special circumstances.

Now, the whole modelling and specification process for elementary specifications is mathematically *robust* in the sense that the syntax and semantics have virtually no special conditions, neither subtle nor obvious.

In modelling using an elementary algebraic specification one simply starts playing with operators, the equations and rewrites. There are no side conditions, side effects, and semantic errors to beware. The elementary algebraic specifications work simply in all cases. The only mistakes possible are mistakes in understanding what one is trying to model.

Technically, all computable algebras can be specified with hidden functions, and all semicomputable algebras can be specified with hidden sorts and functions. In general this is the best possible. For computable algebras specifications may take the from of complete term rewriting systems following [5]. One such result is in [2]:

Theorem 2. *A* Σ *algebra* A *with* n *sorts is computable if, and only if, it possesses an elementary equational specification* (Σ', E') *containing just* $2(n + 1)$ *equations and* $3(n + 1)$ *hidden functions that defines* A *under both initial and final algebra semantics.*

This result can be adapted by means of a very tedious proof to yield algebraic specifications involving exclusively unary hidden functions for each computable data type (though with less efficient bounds on their numbers). Providing an attractive proof of that fact which merits publication is still an open issue. Interestingly in the specification theory for fields we find that a single unary hidden function does the job in various cases. Moss used a unary hidden function in [25]. We gave a simplified EAS for the rational number field using the modulus function as a unary auxiliary operator in [7]. We used the complex conjugate operator as an auxiliary one for giving an EAS for the 0-totalised complex rational numbers in [8], and below we will make use of the degree operator as an auxiliary operator in the context of the rational function field.

2.2   What Are the Non-elementary Algebraic Specifications?

Since the first examples of algebraic specifications of data types in the 1970s, there has been a steady growth in the features that one may add to the basic techniques to be found the early ADJ papers such as [17]. The new techniques have been introduced for a number of obvious reasons: they have been found to be natural, or useful, or necessary to solve problems, or they have been used to extend or explore simpler techniques. The development of languages and tools (such as OBJ, ASF-SDF, Maude, CASL, etc.) for algebraic specification has increased the number and complexity of features in use.

What features have we excluded from the Definition 1 and hence have "declared" to be *not* elementary, and why?

We have excluded *final algebra semantics* because final algebras of equational specifications do not always exist and there are different interpretations possible (see Moss, Meseguer and Goguen [26]).

We have excluded *loose semantics* because we are focussed on specifying algebras up to isomorphism rather than classes of possible models.

The *multi-equations* studied by Adamek et. al. [1] are a convincing generalization of equations, but non-elementary by being less well-known. Further *Priority rewriting* and *innermost rewriting* are considered non-elementary due to their semantic subtleties.

Partial functions Partiality is an essential aspect of computation, but logics of partial functions are quite sophisticated and by no means elementary.

Subsorts occur naturally and help with modelling subtyping, errors, etc. However, there are different theories none of which are as obvious as EAS: see, for example, the survey [15].

In addition EAS excludes features such as: *empty sorts* (see Goguen and Meseguer [14]) because of logical difficulties; *errors and exceptions; modularity; and parameterization.* Each of these invite a proliferation of semantical foundations vastly exceeding the basics of EAS.

2.3   Totalisation of Algebras

Informally totalised algebras emerge by making algebras total which are usually considered to contain partial operators. Unavoidably totalisation introduces an element of arbitrariness which may be considered artificial because values are added which do not belong to the primary intuitions at hand. If a data type starts its life as a partial algebra an EAS treatment of it will involve totalisation at some stage.

Totalisation is not without problems when specifying a stack, as we have seen in our [6]. Totalisation is a matter of costs and benefits and in some cases the theory of a totalised data type, even when specified by means of a convincing EAS, may be harder to swallow than some of its non-elementary expositions, even including the required meta-theory for those non-elementary features. Stacks are a candidate of such a data type. But in the case of fields we have found totalisation a convincing technique. For that we have four arguments:

(1) Totalisation of fields leads to a specification which itself has a larger model class consisting of the so-called meadows having remarkably natural properties in particular in the finite case,

(2) The EAS specification theory of totalised fields is attractive.

(3) EAS provides a fundamental decoupling of syntax and semantics. All simple answers on the question why 0^{-1} fails to exist depend on the observation that this piece of syntax should not have been written down in the first place because it carries no intended meaning. The partial inverse operator cannot be syntactically decoupled from its meaning. Exactly this interplay between syntax and semantics is completely removed in the setting of EAS and totalised fields.

(4) The costs of totalisation, due to the introduction of a 'fake' value for 0^{-1} and its impact on the theory of numbers are already compensated by the gains mentioned in (2) and (3) above.

2.4   Technical Preliminaries on Algebraic Specifications

We assume the reader is familiar with using equations and conditional equations and initial algebra semantics to specify data types. Some accounts of this are: ADJ [17], Meseguer and Goguen [24], or Wirsing [32]. The theory of algebraic specifications is based on theories of universal algebras (e.g., Wechler [31], Meinke and Tucker [23]), computable and semicomputable algebras (Stoltenberg-Hansen and Tucker [27]), and term rewriting (Terese [29]).

We use standard notations: typically, we let Σ be a many sorted signature and A a total Σ algebra. The class of all total Σ algebras is $Alg(\Sigma)$ and the class of all total Σ-algebras satisfying all the axioms in a theory T is $Alg(\Sigma, T)$. The word 'algebra' will mean total algebra.

3   Specifications for Rational Arithmetics

We will build our specifications in stages. The primary signature Σ is simply that of the *field* of rational numbers:

signature Σ
sorts *field*
operations
0: $\rightarrow$ *field*;
1: $\rightarrow$ *field*;
+: *field* $\times$ *field* $\rightarrow$ *field*;

$-:\ field \rightarrow field;$
$\cdot:\ field \times field \rightarrow field;$
$^{-1}:\ field \rightarrow field$
end

3.1 Commutative Rings

The first set of axioms is that of a *commutative ring with* 1, which establishes the standard properties of $+$, $-$, and $\cdot$. We will refer to these axioms by $CR1, \ldots, CR8$ etc.

equations CR

$$(x + y) + z = x + (y + z) \tag{1}$$
$$x + y = y + x \tag{2}$$
$$x + 0 = x \tag{3}$$
$$x + (-x) = 0 \tag{4}$$
$$(x \cdot y) \cdot z = x \cdot (y \cdot z) \tag{5}$$
$$x \cdot y = y \cdot x \tag{6}$$
$$x \cdot 1 = x \tag{7}$$
$$x \cdot (y + z) = x \cdot y + x \cdot z \tag{8}$$

end

These axioms generate a wealth of properties of $+, -, \cdot$ which we will assume the reader is familar.

3.2 Totalised Fields

In working with the rational numbers the usual axiom for division $^{-1}$ is that found among the axioms of fields. The axioms of a field simply add to CR the following the *general inverse law* (Gil) for division:

$$x \neq 0 \implies x \cdot x^{-1} = 1$$

and the *axiom of separation* (Sep) for the constants:

$$0 \neq 1.$$

Neither axioms are equations. In field theory the value of 0^{-1} is either left undefined, or left unspecified. However, in working with elementary specifications operations are total.

Let (Σ, T_{field}) be the axiomatic specification of fields, where

$$T_{field} = CR + Gil + Sep.$$

The class $Alg(\Sigma, T_{field})$ is the class of *total* algebras satisfying the axioms in T_{field}. For emphasis, we refer to these algebras as *totalised fields*.

Now, for all totalised fields $A \in Alg(\Sigma, T_{field})$ and all $x \in A$, the inverse x^{-1} is defined. In particular, 0_A^{-1} is defined. The actual value $0_A^{-1} = a$ can be anything but it is convenient to set $0^{-1} = 0$ (see [7], and compare, e.g., Hodges [18], p. 695). Later we will use a specification which forces $0^{-1} = 0$ (Lemma 1). A field with $0^{-1} = 0$ is called 0-totalised.

The main Σ-algebras we are interested in are these: first,

$$Q_0 = (\mathbb{Q}|0, 1, +, -, \cdot, ^{-1})$$

where the inverse is total

$$\begin{aligned} x^{-1} &= 1/x && \text{if } x \neq 0; \\ &= 0 && \text{if } x = 0 \end{aligned}$$

This total algebra satisfies the axioms of a field T_{field} and is a totalised field of rationals. Next, we are interested in the totalised field extension

$$Q_0(X) = (\mathbb{Q}|X, 0, 1, +, -, \cdot, ^{-1})$$

and its expansion by the degree operator

$$Q_0(X, d) = (\mathbb{Q}|X, 0, 1, +, -, \cdot, ^{-1}, d)$$

Our first objective is to replace the axioms Gil and Sep by equations, which requires an investigation of divison.

3.3   Strong Inverse Properties

Our first set SIP of axioms for $^{-1}$ contain the following three equations, which we call the *strong inverse properties* following [7]. They are "strong" because they are equations in involving $^{-1}$ *without any guards*, such as $x \neq 0$:

equations SIP

$$\begin{aligned} (-x)^{-1} &= -(x^{-1}) && (9) \\ (x \cdot y)^{-1} &= x^{-1} \cdot y^{-1} && (10) \\ (x^{-1})^{-1} &= x && (11) \end{aligned}$$

end

Our specification $CR \cup SIP$ draws attention to division by zero. From [7] we find:

Lemma 1. *The following equations are provable from $CR \cup SIP$:*

$$\begin{aligned} 0^{-1} &= 0 && (12) \\ 0 \cdot x &= 0 && (13) \end{aligned}$$

Thus, $0 \cdot 0^{-1} = 0$. In dealing with division it is helpful to have functions such as

$$Z(x) = 1 - x \cdot x^{-1}$$

Clearly, $Z(x) = 0 \Leftrightarrow x \cdot x^{-1} = 1$.

In particular, in our [7] (Theorem 3.5) we add an axiom L, based on Lagrange's Theorem to give an equational specification of the the rationals. Lagrange's Theorem states that every natural number can be represented as the sum of four squares. We define a special equation L (for Lagrange):

$$Z(1 + x^2 + y^2 + z^2 + u^2) = 0.$$

L expresses that for a large collection of numbers, in particular those q which can be written as 1 plus the sum of four squares, $q \cdot q^{-1}$ equals 1. The following result is then found.

Theorem 3. *There exists a finite elementary equational specification* $(\Sigma, CR + SIP + L)$, *without hidden functions and under initial algebra semantics, of* Q_0.

3.4   Meadows, *Ril* and *Mil*

In [7] we also add to $CR + SIP$ the *restricted inverse law* (*Ril*): $x \cdot (x \cdot x^{-1}) = x$, which, using commutativity and associativity, expresses that $x \cdot x^{-1}$ is 1 in the presence of x. We note that:

Lemma 2. $Ril \vdash x \cdot x^{-1} = 0 \Longrightarrow x = 0$

Proof. Assuming $x \cdot x^{-1} = 0$ one obtains $x \cdot x^{-1} \cdot x = 0 \cdot x$ by multiplication with x on both sides. Thus, $x = 0$ by applying Ril to the LHS and Lemma 1 to the RHS.

Whilst the initial algebra of CR is the ring of integers, we found in [7] that

Lemma 3. *The initial algebra of CR+SIP+Ril is a computable algebra but it is not an integral domain.*

The models of $CR + SIP + Ril$ are algebras with reasonable properties, in spite of not being fields nor even integral domains. We propose to name this theory ENA for elementary number algebra (or equational number algebra if one prefers that explanation). 'Number algebra' then represents the (EAS styled) study of ENA and its extensions, and our use of this phrase corresponds to our use of the phrases 'process algebra', 'program algebra', 'thread algebra' and 'module algebra' in other work. For models of ENA the following convention is taken from [7].

Definition 4. *A model of* $ENA(= CR + SIP + Ril)$ *is called a* meadow.

All fields are clearly meadows but not conversely. That the initial algebra of $CR + SIP + Ril$ is not a field follows from the fact that $(1 + 1) \cdot (1 + 1)^{-1} = 1$

cannot be derivable because it fails to hold in the prime field of characteristic 2 which is a model of these equations as well.

Lemma 4. Let A be a meadow. The following are equivalent:

1. A is a field.
2. For all $x \in A$, $x \neq 0 \implies Z(x) = 0$.
3. For all $x \in A$, $Z(x) = 0$ or $Z(x) = 1$.

Proof. That (1) implies (2) is easy: if $x \neq 0$ then $x \cdot x^{-1} = 1$, by the field axioms, and hence $Z(x) = 1 - x \cdot x^{-1} = 0$. That (2) implies (3) is also clear. For any $x \in A$ there are two cases: if $x = 0$ then $Z(x) = 1 - x \cdot x^{-1} = 1 - 0 \cdot 0^{-1} = 1 - 0 = 1$; and if $x \neq 0$ then $Z(x) = 0$ by (2). Finally, suppose that (3) is true. Then assume $x \neq 0$ and $x \cdot x^{-1} \neq 1$. Now $Z(x) \neq 0$ and therefore $Z(x) = 1$ and $x \cdot x^{-1} = 0$. Using Ril $x = x \cdot x \cdot x^{-1} = x \cdot 0 = 0$ contradicting the assumption on x.

In fact, consider the axiom the *minimal inverse law*, Mil: $x \neq 0 \implies x \cdot x^{-1} \neq 0$. Mil is true of fields and of meadows. However, here exist structures that satisfy Mil but which are not meadows and, in particular, Mil is weaker than Ril. For example consider $A = (\mathbb{Q} | 0, 1, +, -, \cdot, ^{-1})$ where the inverse is redefined as $x^{-1} = x$. Then this algebra A satisfies Mil and yet is not a meadow as it does not satisfy Ril: for $x \in \mathbb{Q}$, if $x \neq 0$ then $x \cdot (x \cdot x^{-1}) = x \cdot x \cdot x \neq x$.

Theorem 4. For any closed terms $t, t' \in T(\Sigma)$, the following are equivalent

1. $t = t'$ is true in all 0-totalised fields.
2. $t = t'$ is true in all meadows.

This was shown in [7]. If $t = t'$ holds in all fields then in all totalised fields and in all 0-totalised fields and therefore in all meadows. Here is an example of a non-trivial equation equation true of all meadows: $2 \cdot Z(2 - \frac{x}{x}) = 0$. Working modulo 2 and taking $x = 0$, $2 - \frac{0}{0} = 1$ from which it follows that $Z(2 - \frac{0}{0})$ is not 0 in all meadows and in particular not in the initial meadow. Stated differently: in the initial meadow $\underline{2}$ is a zero divisor, which is not the case in any homomorphic image constituting a field.

3.5   Finite and Minimal Meadows

Writing $\underline{0}$ for 0 and $\underline{n+1}$ for $\underline{n}+1$ and given a positive natural number k we can define M_k for the initial algebra of $CR + SIP + Ril + Z_k$ with Z_k the equation $\underline{k} = 0$. It is easily seen that for k a prime number M_k is the 0-totalised prime field of characteristic k. Moreover if k is a product of different primes (no factor twice) M_k has exactly k elements. In this case we call k the characteristic of M_k and M_k the minimal meadow of characteristic k. If k and l have the same prime factors then $M_k \cong M_l$. If k is a divisor of l then M_k is a homomorphic image of M_l and all finite and minimal meadows are of the form M_k for some positive natural number k. If its non-zero characteristic is not a prime a finite meadow has proper zero-divisors and fails to be an integral domain.

4 An EAS of the Rational Function Field

We add to the field signature Σ two items:

(i) the degree operation which, given that the integers are contained in the rationals, we treat as a function $d\colon field \to field$;

(ii) the indeterminate X which we treat as a constant $X\colon \to field$;

which together forms the signature $\Sigma_{X,d}$ of the 0-totalised field of rational polynomial functions with degree operator. According to [28] this is a computable algebra which implies the existence of an initial algebra specification with hidden functions. We will establish a specification without auxiliary functions. However, we will rather consider the degree operator a unary hidden function, used to specify the 0-totalised field of rational functions.

First, define $N(x) = 1 - Z(x) = x \cdot x^{-1}$, now consider these equations over the signature $\Sigma_{X,d}$:

equations DG

$$d(0) = 0 \tag{14}$$
$$d(1) = 0 \tag{15}$$
$$d(X) = 1 \tag{16}$$
$$d(X + 1) = 1 \tag{17}$$
$$d(-x) = d(x) \tag{18}$$
$$d(x^{-1}) = -d(x) \tag{19}$$
$$d(d(x)) = 0 \tag{20}$$
$$N(y) \cdot d(x) + N(x) \cdot d(y) = d(x \cdot y) \tag{21}$$
$$Z(d(y + 1) - d(y)) \cdot Z(d(x) - 1 - d(y)) \cdot (d(x + 1) - d(x)) = 0 \tag{22}$$
$$N(d(x)) \cdot Z(x) = 0 \tag{23}$$

end

Theorem 5. *There exists a finite elementary equational specification* $(\Sigma_{X,d}, E)$, *without hidden functions, of the algebra* $Q_0(X, d)$ *of rational polynomial functions with field and degree operations that are all total, under initial algebra semantics. That is,*

$$I(\Sigma_{X,d}, E) \cong Q_0(X, d)$$

where $E = ENA + L + DG$.

Proof. First we must verify that our specification is true of the algebra.

Lemma 5. $Q_0(X, d) \models ENA + L + DG$

Proof. Most of the axioms are straightforward to check. Those of ENA are easy. For the axiom L one can argue that the 0-totalised rational function field is isomorphic to the rationals expanded with a positive (and real) transcendental number (e.g. π). It follows that it can be totally ordered with the effect that 1 + the sum of four squares is always positive and therefore non-zero. This fact remains when forgetting the ordering. Next we consider the axioms of DG. Only the last three axioms need attention.

(a) Consider: $d(x \cdot y) = N(y) \cdot d(x) + N(x) \cdot d(y)$.
If x, y are non-zero then $N(x) = N(y) = 1$ and we have $d(x \cdot y) = d(x) + d(y)$. Supposing $x = 0$ the the LHS is $d(x \cdot y) = d(0 \cdot y) = d(0) = 0$ and the RHS is $N(y) \cdot d(0) + N(0) \cdot d(y) = N(y) \cdot 0 + 0 \cdot d(y) = 0$. Hence the two sides agree.

(b) Consider: $Z(d(y + 1) - d(y)) \cdot Z(d(x) - 1 - d(y)) \cdot (d(x + 1) - d(x)) = 0$
Here notice that $d(y)$ is always an integer and that $d(y+1) - d(y) = 0$ for all and only y with a non-negative degree. (Notice: $d(X^{-1} + 1) = d(X^{-1} + X \cdot X^{-1}) = d((X+1) \cdot X^{-1}) - 1 + (-1) - 0 \neq d(X^{-1}) = 1$.) Now the equation has the form $Z(r(y)) \cdot Z(p(x, y)) \cdot q(x) = 0$. This can be read as: if $r = 0$ and $p = 0$ (which implies $Z(r) = 1$ and $Z(p) = 1$) then $q = 0$. So assume $d(y + 1) - d(y) = 0$ and $d(x) - 1 - d(y) = 0$ then $d(y)$ is nonnegative and $d(x) = 1 + d(y)$ and so $d(x)$ is also a nonnegative integer (within the rationals). For such x we have $d(x) - d(x + 1) = 0$.

(c) Consider: $N(d(x)) \cdot Z(x) = 0$.
If $N(d(x)) = 0$ then $N(d(x)) \cdot Z(x) = 0$. If $N(d(x)) \neq 0$ then $d(x) \cdot d(x)^{-1} \neq 0$ and also $d(x) \neq 0$. Hence, $x \neq 0$ and $Z(x) = 0$ which implies $N(d(x)) \cdot Z(x) = 0$.

By Lemma 5, there is a $\Sigma_{X,d}$-homomorphsim $\phi \colon I(\Sigma_{X,d}, E) \rightarrow Q_0(X, d)$. Since the algebra $Q_0(X, d)$ is $\Sigma_{X,d}$-minimal the map ϕ is surjective. Thus, to complete the theorem we must prove that ϕ is injective. We introduce a notation for an equation between closed terms t, r in $T(\Sigma_{X,d})$: $t \sharp r \Leftrightarrow Z(t - r) = 0 \Leftrightarrow \frac{t-r}{t-r} = 1$. $E \vdash t \sharp r$ expresses that t is provably different from r. Suppose for a contradiction that ϕ is not injective. Then there are closed terms s, s' in $T(\Sigma_{X,d})$ such that $[s] \neq [s']$ in $I(\Sigma, E)$ and $\phi([s]) = \phi([s'])$ in $Q_0(X, d)$. We need the following fact:

Lemma 6. *For all closed terms t, r in $T(\Sigma_{X,d})$, either $E \vdash t = r$ or $E \vdash t \sharp r$.*

Applying Lemma 6 to the pair s, s' above, which contradicts the injectivity of ϕ, we find that $E \vdash s \sharp s'$ because otherwise $E \vdash s = s'$ against the assumptions. So we find:

$$\phi(Z([s] - [s'])) = \quad \phi(0) \qquad \qquad \text{applying the map}$$
$$Z(\phi([s] - [s'])) = \quad 0 \qquad \qquad \phi \text{ is homomorphism}$$
$$Z(\phi([s]) - \phi([s']))) = \quad 0 \qquad \qquad \phi \text{ is homomorphism}$$
$$Z(0) = \quad 0 \qquad \text{by assumption that } \phi \text{ is not injective}$$

but this is a contradiction since $Z(0) = 1$, which demonstrates the injectivity of ϕ. It remains to prove Lemma 6.

Proof. It suffices to show that for all t, either $E \vdash t = 0$ or $E \vdash t \sharp 0$. For suppose that it is not the case that $E \vdash t = r$ then neither $E \vdash t - r = 0$. Then $E \vdash (t - r) \sharp 0$ and thus $E \vdash t \sharp r$.

In the proof of Theorem 3 as given in [7] we found that for all terms s of the form $\pm m \cdot n^{-1}$, where m and n are non-zero, we have $CR + SIP + L \vdash s \sharp 0$.

We write $INT = \{-\underline{n} | n > 0\} \cup \{\underline{n} | n > 0\} \cup \{\underline{0}\}$.

Note $d(X) = \underline{1}$ and $d(0) = d(1) = \underline{0}$. In fact all the integers appear in this range of values of the form $d(t)$: with induction one proves: $d(X^k) = \underline{k}$, and $d(X^{-k}) = -\underline{k}$. The induction step is thus (for $k > 0$):
$$d(X \cdot X^k) = N(X) \cdot d(X^k) + N(X^k) \cdot d(X) = d(X^k) + N(X^k) \text{ because } d(X) = 1$$
and thus $N(d(X)) = 1$ which using equation (22) implies $Z(X) = 0$ and $N(X) = 1$.

As induction hypothesis suppose that $d(X^k) = \underline{k}$ for $k > 0$. Since $k > 0$, $N(\underline{k}) = 1$ and thus $N(d(X^k)) = 1$ which (using equation 21) implies $Z(X^k) = 0$ and $N(X^k) = 1$. Then $d(X \cdot X^{k-1}) = d(X^{k-1}) + d(X) = \underline{k-1} + \underline{1} = \underline{k}$. A similar argument works for negative powers of X.

Thus, for $r \in INT$, there is a term t such that $d(t) = r$. So $d(r) = d(d(t)) = 0$ by axiom $d(d(x)) = 0$ in DG. This implies that the degree of all integers is 0. Using the axioms $d(x^{-1}) = -d(x)$ and $d(x \cdot y) = N(y) \cdot d(x) + N(x) \cdot d(y)$ we can show that the degree of all rationals is provably zero too.

The next stage is to prove that a polynomial $p = p_k X^k + \cdots + p_1 X + p_0$ has degree k provided $p_k \sharp 0$. As a consequence it is non-zero and $Z(p) = 0$, i.e. $p \sharp 0$.

This is done by induction on k. The basis is clear: for $k = 0$, p has degree 0. Suppose it is true for polynomials of degree $k = n$ and consider case $k = n + 1$. Consider $p = p_{n+1} X^{n+1} + \cdots + p_1 X + p_0$. There are two cases: $p_0 = 0$ and $p_0 \sharp 0$.

(i) Case $p_0 = 0$: Write $p = p_{n+1} X^{n+1} + \cdots + p_1 X = q \cdot X$. We know by induction that $E \vdash d(q) = \underline{n}$ and $q \sharp 0$. Thus:
$$d(p) = d(q \cdot X) = N(q) \cdot d(X) + N(X) \cdot d(q) = 1 \cdot d(X) + 1 \cdot d(q) = \underline{n+1}.$$
(This is because $d(X) = 1$ and $N(d(X)) = 1$; by axiom $N(d(x)) \cdot Z(x) = 0$ in DG we have $Z(X) = 0$ which implies $N(X) = 1$.)

(ii) Case $p_0 \sharp 0$: Write $p = p_{n+1} X^{n+1} + \cdots + p_1 X + p_0 = q \cdot X + p_0 = p_0(\frac{1}{p_0} q \cdot X + 1)$.

Now $d(\frac{1}{p_0} q \cdot X) = d(q \cdot X) = \underline{n+1}$ as $p_0 \sharp 0$. We apply $Z(d(y+1) - d(y)) \cdot Z(d(x) - 1 - d(y)) \cdot (d(x+1) - d(x)) = 0$ with $x = \frac{1}{p_0} q \cdot X$ and $y = X^n$. We obtain $d(y+1) - d(y) = 0$ because $n+1$ is nonnegative $d(x) - 1 - d(y) = 0$ via simple calculation and as a consequence $Z(d(y+1) - d(y)) \cdot Z(d(x) - 1 - d(y)) = 1$ and therefore, $d(x+1) - d(x) = 0$, which gives $d(\frac{1}{p_0} q \cdot X + 1) = d(\frac{1}{p_0} q \cdot X) = \underline{n+1}$ and thus $d(p) = \underline{n+1}$. Together with axiom (22) this also proves that $p \sharp 0$.

At this stage we have shown that all for polynomials of positive degree p, $d(p)$ equals a positive integer and $p \sharp 0$, i.e., $\frac{p}{p} = 1$. This allows us to write each closed term as the quotient of polynomials p and q or 0. In fact this matter takes an induction argument over all terms but it poses no difficulty.

Now let $E \vdash t = \frac{p}{q}$ where we may assume that $d(q) > 0$. If $d(p) \neq d(q)$ then $d(t) \sharp 0$, whence $t \sharp 0$ (by axiom 22). If $d(p) = d(q) > 1$ then $N(p) = N(q) = 1$ and $N(\frac{p}{q}) = \frac{p}{q} / \frac{p}{q} = \frac{p}{p} \cdot \frac{q}{q} = 1 \cdot 1 =$ and thus $t \sharp 0$.

Now all cases have been dealt with and we have shown lemma 6.

This also completes the proof of Theorem 5.

5   Concluding Remarks

We have demonstrated that totalisation is an effective strategy for obtaining a specification theory of computable fields and meadows and this has been illustrated on the field of rational functions. In addition to questions listed in [7, 8] the following two problems arise from this work:

Problem 1. Is there a finite elementary equational specification of the totalised field $Q_0(X)$ of rational functions, without the use of the degree function d as a hidden function, and under initial algebra semantics?

Problem 2. Is there a finite elementary equational specification of either of the algebras $Q_0(X, d)$ or $Q_0(X)$ of rational functions, without hidden functions, and under initial algebra semantics, which constitutes a complete term rewriting system?

References

1. J. ADAMEK, M. HEBERT AND J. ROSICKY On abstract data types presented by multiequations *Theoretical Computer Science* 275 (2002) 427 - 462.
2. J A BERGSTRA AND J V TUCKER, The completeness of the algebraic specification methods for data types, *Information and Control*, 54 (1982) 186-200.
3. J A BERGSTRA AND J V TUCKER, Initial and final algebra semantics for data type specifications: two characterisation theorems, *SIAM Journal on Computing*, 12 (1983) 366-387.
4. J A BERGSTRA AND J V TUCKER, Algebraic specifications of computable and semicomputable data types, *Theoretical Computer Science*, 50 (1987) 137-181.
5. J A BERGSTRA AND J V TUCKER, Equational specifications, complete term rewriting systems, and computable and semicomputable algebras, *Journal of ACM*, 42 (1995) 1194-1230.
6. J A BERGSTRA AND J V TUCKER, The data type variety of stack algebras, *Annals of Pure and Applied Logic*, 73 (1995) 11-36.
7. J A BERGSTRA AND J V TUCKER, The rational numbers as an abstract data type, submitted.
8. J A BERGSTRA AND J V TUCKER, Elementary algebraic specifications of the rational complex numbers, submitted.
9. N CALKIN AND H S WILF, Recounting the rationals, *American Mathematical Monthly*, 107 (2000) 360-363.
10. E CONTEJEAN, C MARCHE AND L RABEHASAINA, Rewrite systems for natural, integral, and rational arithmetic, in *Rewriting Techniques and Applications 1997*, Springer Lecture Notes in Computer Science 1232, 98-112, Springer, Berlin,1997.
11. J A GOGUEN, Memories of ADJ, *Bulletin of the European Association for Theoretical Computer Science*, 36 (October 1989), pp 96-102.
12. J A GOGUEN, Tossing algebraic flowers down the great divide, in C S Calude (ed.), *People and ideas in theoretical computer science*, Springer, Singapore, 1999, pp 93-129.
13. J MESEGUER AND J A GOGUEN, Initiality, induction, and computability, In M Nivat (editors) *Algebraic methods in semantics*, Cambridge University Press,1986 pp 459 - 541.

14. J MESEGUER AND J A GOGUEN, Remarks on remarks on many-sorted algebras with possibly emtpy carrier sets, *Bulletin of the EATCS*, 30 (1986) 66-73.
15. J A GOGUEN AND R DIACONESCU An Oxford Survey of Order Sorted Algebra *Mathematical Structures in Computer Science* 4 (1994) 363-392.
16. J A GOGUEN, J W THATCHER, E G WAGNER AND J B WRIGHT, Initial algebra semantics and continuous algebras, *Journal of ACM*, 24 (1977), 68-95.
17. J A GOGUEN, J W THATCHER AND E G WAGNER, An initial algebra approach to the specification, correctness and implementation of abstract data types, in R.T Yeh (ed.) *Current trends in programming methodology. IV. Data structuring*, Prentice-Hall, Engelwood Cliffs, New Jersey, 1978, pp 80-149.
18. W HODGES, *Model Theory*, Cambridge University Press, Cambridge, 1993.
19. S KAMIN, Some definitions for algebraic data type specifications, SIGLAN Notices 14 (3) (1979), 28.
20. B KHOUSSAINOV, Randomness, computability, and algebraic specifications, *Annals of Pure and Applied Logic*, (1998) 1-15.
21. B KHOUSSAINOV, On algebraic specifications of abstract data types, in *Computer Science Logic: 17th International Workshop*, Lecture Notes in Computer Science, Volume 2803, 299 313, 2003.
22. G MARONGIU AND S TULIPANI, On a conjecture of Bergstra and Tucker, *Theoretical Computer Science*, 67 (1989).
23. K MEINKE AND J V TUCKER, Universal algebra, in S. Abramsky, D. Gabbay and T Maibaum (eds.) *Handbook of Logic in Computer Science. Volume I: Mathematical Structures*, Oxford University Press, 1992, pp.189-411.
24. J MESEGUER AND J A GOGUEN, Initiality, induction and computability, in M Nivat and J Reynolds (eds.), *Algebraic methods in semantics*, Cambridge University Press, Cambridge, 1985, pp.459-541.
25. L MOSS, Simple equational specifications of rational arithmetic, *Discrete Mathematics and Theoretical Computer Science*, 4 (2001) 291-300.
26. L MOSS, J MESEGUER AND J A GOGUEN, Final algebras, cosemicomputable algebras, and degrees of unsolvability, *Theoretical Computer Science*, 100 (1992) 267-302.
27. V STOLTENBERG-HANSEN AND J V TUCKER, Effective algebras, in S Abramsky, D Gabbay and T Maibaum (eds.) *Handbook of Logic in Computer Science. Volume IV: Semantic Modelling* , Oxford University Press, 1995, pp.357-526.
28. V STOLTENBERG-HANSEN AND J V TUCKER, Computable rings and fields, in E Griffor (ed.), *Handbook of Computability Theory*, Elsevier, 1999, pp.363-447.
29. TERESE, *Term Rewriting Systems*, Cambridge Tracts in Theoretical Computer Science 55, Cambridge University Press, Cambridge, 2003.
30. E WAGNER, Algebraic specifications: some old history and new thoughts, *Nordic Journal of Computing*, 9 (2002), 373 - 404.
31. W WECHLER, *Universal algebra for computer scientists*, EATCS Monographs in Computer Science, Springer, 1992.
32. M WIRSING, Algebraic specifications, in J van Leeuwen (ed.), *Handbook of Theoretical Computer Science. Volume B: Formal models and semantics*, North-Holland, 1990, pp 675-788.

Random Closed Sets[*]

Paul Brodhead, Douglas Cenzer, and Seyyed Dashti

Department of Mathematics, University of Florida,
P.O. Box 118105, Gainesville, Florida 32611
Fax: 352-392-8357
cenzer@math.ufl.edu

Abstract. We investigate notions of randomness in the space $\mathcal{C}[2^{\mathbb{N}}]$ of nonempty closed subsets of $\{0,1\}^{\mathbb{N}}$. A probability measure is given and a version of the Martin-Löf Test for randomness is defined. Π_2^0 random closed sets exist but there are no random Π_1^0 closed sets. It is shown that a random closed set is perfect, has measure 0, and has no computable elements. A closed subset of $2^{\mathbb{N}}$ may be defined as the set of infinite paths through a tree and so the problem of compressibility of trees is explored. This leads to some results on a Chaitin-style notion of randomness for closed sets.

Keywords: Computability, Randomness, Π_1^0 Classes.

1 Introduction

The study of algorithmic randomness has been of great interest in recent years. The basic problem is to quantify the randomness of a single real number and here we will extend this problem to the randomness of a finite-branching tree. Early in the last century, von Mises [10] suggested that a random real should obey reasonable statistical tests, such as having a roughly equal number of zeroes and ones of the first n bits, in the limit. Thus a random real would be *stochastic* in modern parlance. If one considers only *computable* tests, then there are countably many and one can construct a real satisfying all tests.

Martin-Löf [8] observed that stochastic properties could be viewed as special kinds of meaure zero sets and defined a random real as one which avoids certain effectively presented measure 0 sets. That is, a real $x \in 2^{\mathbb{N}}$ is Martin-Löf random if for any effective sequence $S_1, S_2, \ldots$ of c.e. open sets with $\mu(S_n) \leq 2^{-n}$, $x \notin \cap_n S_n$.

At the same time Kolmogorov [6] defined a notion of randomness for finite strings based on the concept of *incompressibility*. For infinite words, the stronger notion of prefix-free complexity developed by Levin [7], Gács [5] and Chaitin [3] is needed. Schnorr later proved that the notions of Martin-Löf randomness and Chaitin randomness are equivalent.

In this paper we want to consider algorithmic randomness on the space $\mathcal{C}$ of nonempty closed subsets P of $2^{\mathbb{N}}$. For a finite string $\sigma \in \{0,1\}^n$, let $|\sigma| = n$. For

---

[*] Research was partially supported by the National Science Foundation.

A. Beckmann et al. (Eds.): CiE 2006, LNCS 3988, pp. 55–64, 2006.
© Springer-Verlag Berlin Heidelberg 2006

two string σ, τ, say that τ extends σ and write $\sigma \sqsubseteq \tau$ if $|\sigma| \leq |\tau|$ and $\sigma(i) = \tau(i)$ for $i < |\sigma|$. Similarly $\sigma \sqsubset x$ for $x \in 2^{\mathbb{N}}$ means that $\sigma(i) = x(i)$ for $i < |\sigma|$. Let $\sigma^{\frown}\tau$ denote the concatenation of σ and τ and let $\sigma^{\frown}i$ denote $\sigma^{\frown}(i)$ for $i = 0, 1$. Let $x\lceil n = (x(0), \ldots, x(n-1))$. Now a nonempty closed set P may be identified with a tree $T_P \subseteq \{0, 1\}^*$ as follows. For a finite string σ, let $I(\sigma)$ denote $\{x \in 2^{\mathbb{N}} : \sigma \sqsubset x\}$. Then $T_P = \{\sigma : P \cap I(\sigma) \neq \emptyset\}$. Note that T_P has no dead ends, that is if $\sigma \in T_P$ then either $\sigma^{\frown}0 \in T_P$ or $\sigma^{\frown}1 \in T_P$.

For an arbitrary tree $T \subseteq \{0, 1\}^*$, let $[T]$ denote the set of infinite paths through T, that is,

$$x \in [T] \iff (\forall n) x\lceil n \in T.$$

It is well-known that $P \subseteq 2^{\mathbb{N}}$ is a closed set if and only if $P = [T]$ for some tree T. P is a Π_1^0 class, or effectively closed set, if $P = [T]$ for some computable tree T. P is a strong Π_2^0 class, or Π_2^0 closed set, if $P = [T]$ for some Δ_2^0 tree. The complement of a Π_1^0 class is sometimes called a c.e. open set. We remark that if P is a Π_1^0 class, then T_P is a Π_1^0 set, but not in general computable. There is a natural effective enumeration $P_0, P_1, \ldots$ of the Π_1^0 classes and thus an enumeration of the c.e. open sets. Thus we can say that a sequence $S_0, S_1, \ldots$ of c.e. open sets is *effective* if there is a computable function, f, such that $S_n = 2^{\mathbb{N}} - P_{f(n)}$ for all n. For a detailed development of Π_1^0 classes, see [1] or [2].

To define Martin-Löf randomness for closed sets, we give an effective homeomorphism with the space $\{0, 1, 2\}^{\mathbb{N}}$ and simply carry over the notion of randomness from that space.

Chaitin randomness for reals is defined as follows. Let M be a prefix-free function with domain $\subset \{0, 1\}^*$. For any finite string τ, let $K_M(\tau) = min\{|\sigma| : M(\sigma) = \tau\}$. There is a *universal* prefix-free function U such that, for any prefix-free M, there is a constant c such that for all τ

$$K_U(\tau) \leq K_M(\tau) + c.$$

We let $K(\sigma) = K_U(\sigma)$. Then x is said to be *Chaitin random* if there is a constant c such that $K(x\lceil n) \geq n - c$ for all n. This means that the initial segments of x are not *compressible*.

For a tree T, we want to consider the compressibility of $T_n = T \cap \{0, 1\}^n$. This has a natural representation of length 2^n since there are 2^n possible nodes of length n. We will show that any tree T_P can be compressed, that is, $K(T_n) \geq 2^n - c$ is impossible for a tree with no dead ends.

2   Martin-Löf Randomness of Closed Sets

In this section, we define a measure on the space $\mathcal{C}$ of nonempty closed subsets of $2^{\mathbb{N}}$ and use this to define the notion of randomness for closed sets. We then obtain several properties of random closed sets.

An effective one-to-one correspondence between the space $\mathcal{C}$ and the space $3^{\mathbb{N}}$ is defined as follows. Let a closed set Q be given and let $T = T_Q$ be the tree without dead ends such that $Q = [T]$.

Then define the code $x = x_Q \in \{0, 1, 2\}N$ for Q as follows. Let $\sigma_0 = \emptyset, \sigma_1, \ldots$ enumerate the elements of T in order, first by length and then lexicographically. We now define $x = x_Q = x_T$ by recursion as follows. For each n, $x(n) = 2$ if $\sigma_n^\frown 0$ and $\sigma_n^\frown 1$ are both in T, $x(n) = 1$ if $\sigma_n^\frown 0 \notin T$ and $\sigma_n^\frown 1 \in T$ and $x(n) = 0$ if $\sigma_n^\frown 0 \in T$ and $\sigma_n^\frown 1 \notin T$.

Then define the measure μ^* on $\mathcal{C}$ by

$$\mu^*(\mathcal{X}) = \mu(\{x_Q : Q \in \mathcal{X}\}).$$

Informally this means that given $\sigma \in T_Q$, there is probability $\frac{1}{3}$ that both $\sigma^\frown 0 \in T_Q$ and $\sigma^\frown 1 \in T_Q$ and, for $i = 0, 1$, there is probability $\frac{1}{3}$ that only $\sigma^\frown i \in T_Q$. In particular, this means that $Q \cap I(\sigma) \neq \emptyset$ implies that for $i = 0, 1$, $Q \cap I(\sigma^\frown i) \neq \emptyset$ with probability $\frac{2}{3}$.

Let us comment briefly on why some other natural representations were rejected. Suppose first that we simply enumerated all strings in $\{0, 1\}^*$ as $\sigma_0, \sigma_1, \ldots$ and then represent T by its characteristic function so that $x_T(n) = 1 \iff \sigma_n \in T$. Then in general a code x might not represent a tree. That is, once we have $(01) \notin T$ we cannot later decide that $(011) \in T$. Suppose then that we allow the empty closed set by using codes $x \in \{0, 1, 2, 3\}^*$ and modifying our original definition as follows. Let $x(n) = i$ give the same definition as above for $i \leq 2$ but let $x(n) = 3$ mean that neither $\sigma_n^\frown 0$ nor $\sigma^\frown 1$ is in T. Informally, this would mean that for $i = 0, 1$, $\sigma \in T$ implies that $\sigma^\frown i \in T$ with probability $\frac{1}{2}$. The advantage here is that we can now represent all trees. But this is also a disadvantage, since for a given closed set P, there are many different trees T with $P = [T]$. The second problem with this approach is that we would have $[T] = \emptyset$ with probability 1. It then follows that $[T]$ would have to be empty for a random tree, that is the only random closed set would be the empty set.

Now we will say that a closed set Q is (Martin-Löf) random if the code x_Q is Martin-Löf random. Since random reals exists, it follows that random closed sets exists. Furthermore, there are Δ_2^0 reals, so we have the following.

Theorem 1. *There exists a random Π_2^0 closed set.* □

Next we obtain some properties of random closed sets.

Proposition 1. *P is a random closed set if and only if, for every $\sigma \in T_P$, $P \cap I(\sigma)$ is a random closed set.*

Proof. One direction is immediate. For the other direction, suppose that $P \cap I(\sigma)$ is not random and let $S_0, S_1, \ldots$ be an effective sequence of c. e. open sets in $\mathcal{C}$ with $\mu^*(S_n) < 2^{-n}$, such that $P \cap I(\sigma) \in \cap_n S_n$. Let $S_n' = \{Q : Q \cap I(\sigma) \in S_n\}$. Then $P \in \cap_n S_n'$ and $\mu^*(S_n') \leq \mu^*(S_n) < 2^{-n}$ for all n. □

This implies that a random closed set P must be nowhere dense, since certainly no interval $I(\sigma)$ is random and hence $P \cap I(\sigma) \neq I(\sigma)$ for any $\sigma \in \{0, 1\}^*$.

Theorem 2. *If Q is a random closed set, then Q has no computable members.*

Proof. Suppose that Q is random and let y be a computable real. For each n, let

$$S_n = \{P : P \cap I(y\lceil n) \neq \emptyset\}.$$

Then the $S_0, S_1, \ldots$ is an effective sequence of clopen sets in $\mathcal{C}$, and an easy induction shows that $\mu^*(S_n) = (2/3)^n$. This is a Martin-Löf test and it follows that $Q \notin S_n$ for some n, so that $y \notin Q$. □

Theorem 3. *If Q is a random closed set, then Q has no isolated elements.*

Proof. Let $Q = [T]$ and suppose by way of contradiction that Q contains an isolated path x. Then there is some node $\sigma \in T$ such that $Q \cap I(\sigma) = \{x\}$. For each n, let

$$S_n = \{P \in \mathcal{C} : card(\{\tau \in \{0,1\}^n : P \cap I(\sigma^\frown \tau) \neq \emptyset\}) = 1\}.$$

That is, $P \in S_n$ if and only if the tree T_P has exactly one extension of σ of length $n + |\sigma|$. It follows that

$$card(P \cap I(\sigma)) = 1 \iff (\forall n)P \in S_n$$

Now for each n, S_n is a clopen set in $\mathcal{C}$ and again by induction, S_n has measure $(2/3)^n$. Thus the sequence $S_0, S_1, \ldots$ is a Martin-Löf test. It follows that for some n, $Q \notin S_n$. Thus there are at least two extensions in T_Q of σ of length $n + |\sigma|$, contradicting the assumption that x was the unique element of $Q \cap I(\sigma)$. □

Corollary 1. *If Q is a random closed set, then Q is perfect and hence has continuum many elements.* □

Theorem 4. *If Q is a random closed set, then $\mu(Q) = 0$.*

Proof. We will show that in the space $\mathcal{C}$, $\mu(P) = 0$ with probability 1. This is proved by showing that for each m, $\mu(P) \geq 2^{-m}$ with probability 0. For each m, let

$$S_m = \{P : \mu(P) \geq 2^{-m}\}.$$

We claim that for each m, $\mu^*(S_m) = 0$. The proof is by induction on m.

For $m = 0$, we have $\mu(P) \geq 1$ if and only if $P = 2^{\mathbb{N}}$, which is if and only if $x_P = (2, 2, \ldots)$, so that S_0 is a singleton and thus has measure 0.

Now assume by induction that S_m has measure 0. Then the probability that a closed set $P = [T]$ has measure $\geq 2^{-m-1}$ can be calculated in two parts.

(i) If T does not branch at the first level, say $T_0 = \{(0)\}$ without loss of generality. Now consider the closed set $P_0 = \{y : 0^\frown y \in P\}$. Then $\mu(P) \geq 2^{-m-1}$ if and only if $\mu(P_0) \geq 2^{-m}$, which has probability 0 by induction, so we can discount this case.

(ii) If T does branch at the first level, let $P_i = \{y : i^\frown y \in P\}$ for $i = 0, 1$. Then $\mu(P) = \frac{1}{2}(\mu(P_0) + \mu(P_1))$, so that $\mu(P) \geq 2^{-m-1}$ implies that at least one of $\mu(P_i) \geq 2^{-m-1}$. Let $p = \mu^*(S_{m+1})$. The observations above imply that

$$p \leq \frac{1}{3}(1 - (1-p)^2) = \frac{2}{3}\,p - \frac{1}{3}p^2,$$

and therefore $p = 0$.

To see that a random closed set Q must have measure 0, fix m and let $S = S_m$. Then S is the intersection of an effective sequence of clopen sets V_ℓ, where for $P = [T]$,

$$P \in V_\ell \iff \mu([T_\ell]) \geq 2^{-m}.$$

Since these sets are uniformly clopen, the sequence $m_\ell = \mu^*(V_\ell)$ is computable. Since $\lim_\ell m_\ell = 0$, it follows that this is a Martin-Löf Test and therefore no random set Q belongs to $\cap_\ell V_\ell$. Then in general, no random set can have measure $\geq 2^{-m}$ for any m.   □

No computable real can be random and it follows that no decidable Π^0_1 class can be random, where P is decidable if $P = [T]$ for some computable tree T with no dead ends. We can now extend this to arbitrary Π^0_1 classes.

Lemma 1. *For any closed set Q, $\mu^*(\{P : P \subseteq Q\}) \leq \mu(Q)$.*

Proof. Let $\mathcal{P}_C(Q)$ denote $\{P : P \subseteq Q\}$. We first prove the result for (nonempty) clopen sets U by the following induction. Suppose $U = \cup_{\sigma \in S} I(\sigma)$, where $S \subseteq \{0, 1\}^n$. For $n = 1$, either $\mu(U) = 1 = \mu^*(\mathcal{P}_C(U))$ or $\mu(U) = \frac{1}{2}$ and $\mu^*(\mathcal{P}_C(Q)) = \frac{1}{3}$. For the induction step, let $S_i = \{\sigma : i^\frown\sigma \in S$, let $U_i = \cup_{\sigma \in S_i} I(\sigma)$, let $m_i = \mu(U_i)$ and let $v_i = \mu^*(\mathcal{P}_C(U_i))$, for $i = 0, 1$. Then considering the three cases in which S includes both initial branches or just one, we calculate that

$$\mu^*(\mathcal{P}_C(U)) = \frac{1}{3}(v_0 + v_1 + v_0 v_1).$$

Thus by induction we have

$$\mu^*(\mathcal{P}_C(U)) \leq \frac{1}{3}(m_0 + m_1 + m_0 m_1).$$

Now

$$2m_0 m_1 \leq m_0^2 + m_1^2 \leq m_0 + m_1,$$

and therefore

$$\mu^*(\mathcal{P}_C(U)) \leq \frac{1}{3}(m_0 + m_1 + m_0 m_1) \leq \frac{1}{2}(m_0 + m_1) = \mu(U).$$

For a closed set Q, let $Q = \cap_n U_n$, with $U_{n+1} \subseteq U_n$ for all n. Then $P \subset Q$ if and only if $P \subseteq U_n$ for all n. Thus

$$\mathcal{P}_C(Q) = \cap_n \mathcal{P}_C(U_n),$$

so that

$$\mu^*(\mathcal{P}_C(Q)) = \lim_{n\to\infty} \mu^*(\mathcal{P}_C(U_n)) \leq \lim_{n\to\infty} \mu(U_n) = \mu(Q).$$

This completes the proof of the lemma.   □

Theorem 5. *Let Q be a Π^0_1 class with measure 0. Then no subset of Q is random.*

Proof. Let T be a computable tree (possibly with dead ends) and let $Q = [T]$. Then $Q = \cap_n U_n$, where $U_n = [T_n]$. Since $\mu(Q) = 0$, it follows from the lemma that $lim_n \mu^*(\mathcal{P}_C(U_n)) = 0$. But $\mathcal{P}_C(U_n)$ is a computable sequence of clopen sets in $\mathcal{C}$ and $\mu^*(\mathcal{P}_C(U_n))$ is a computable sequence of rationals with limit 0. Thus $\mathcal{P}_C(U_n)$ is a Martin-Löf Test, so that no random closed set can be a member. □

Since any random class has measure 0, we have the following immediate corollary.

Corollary 2. *No Π_1^0 class can be random.* □

3   Chaitin Complexity of Closed Sets

In this section, we consider randomness for closed sets and trees in terms of incompressibility.

Of course, Schnorr's theorem tells us that P is random if and only if the code $x_P \in \{0, 1, 2\}^*$ for P is Chaitin random, that is, $K_3(x\lceil n) \geq n - O(1)$. Here we write K_3 to indicate that we would be using a universal prefix-free function $U : \{0, 1, 2\}^* \to \{0, 1, 2\}^*$.

However, many properties of trees and closed sets depend on the levels $T_n = T \cap \{0, 1\}^n$ of the tree. For example, if $[T_n] = \cup\{I(\sigma) : \sigma \in T_n\}$, then $[T] = \cap_n [T_n]$ and $\mu([T]) = lim_{n \to \infty} \mu([T_n])$.

So we want to consider the compressibility of a tree in terms of $K(T_n)$. Now there is a natural representation of T_n as a subset of $\{0, 1\}^n$ which has length 2^n. That is, list $\{0, 1\}^\ell$ in lexicographic order as $\sigma_1, \ldots, \sigma_{2^\ell}$ and represent T_ℓ by the string $e_1, \ldots, e_{2^\ell}$ where $e_i = 1$ if $\sigma \in T_\ell$ and $e_i = 0$ otherwise. Henceforth we identify T_ℓ with this natural representation. It is interesting to note that there is no immediate connection between T_n and $x_T\lceil n$. For example, if x is the code for the full tree $\{0, 1\}^*$, then $x = (2, 2, \ldots)$ and the code for T_n is a string of $2^n - 1$ 2's. On the other hand, if $[T] = \{y\}$ is a singleton, then $x = y$ and the code for T_n is $x\lceil n$. For the remainder of this section, we will focus on the natural representation, rather than the code.

The natural question here is whether there is a formulation of randomness in terms of the incompressibility of T_n. In this section, we will explore this question and give some partial answers.

At first it seems plausible that P is random if and only if there is a constant c such that $K(T_\ell) \geq 2^\ell - c$ for all ℓ.

As usual, let U be a universal prefix-free Turing machine and let $K(T_\ell) = min\{|\sigma| : U(\sigma) = T_\ell\}$. Now suppose that $P = [T]$ is a random closed set with code X. Since X is 1-random, we know that $K(X\lceil n) \geq n - c$ for some constant c. In general, we can compute $X\lceil n$ from T_n and hence

$$K(T_n) \geq K(X\lceil n) - b.$$

for some constant b.

That is, define the (not necessarily prefix-free) machine M so that $M(T_n) = X_T\lceil n$ and then let

$$V(\sigma) = M(U(\sigma)).$$

Then $K_V(X_T \lceil n) \leq K(T_n)$, so that for some constant b, $K(X_T \lceil n) \leq K(T_n) + b$ and hence

$$K(T_n) \geq K(X \lceil n) - b \geq n - b - c.$$

Thus we have shown

Proposition 2. *If P is a random closed set and $T = T_P$, then there is a constant c such that $K(T_n) \geq n - c$ for all n.* □

But this is a very weak condition.

Going in the other direction, we can compute T_ℓ uniformly from $X \lceil 2^\ell$, so that as above, $K(X \lceil 2^\ell) \geq K(T_\ell) - b$ for some b. Thus in order to conclude that X is random, we would need to know that $K(T_\ell) \geq 2^\ell - c$ for some c. Our first observation is that this is not possible, since trees are naturally compressible.

Theorem 6. *For any tree $T \subseteq \{0,1\}^*$, there are constants $k > 0$ and c such that $K(T_\ell) \leq 2^\ell - 2^{\ell-k} + c$ for all ℓ.*

Proof. For the full tree $\{0,1\}^*$, this is clear so suppose that $\sigma \notin T$ for some $\sigma \in \{0,1\}^m$. Then for any level $\ell > m$, there are $2^{\ell-m}$ possible nodes for T which extend σ and T_ℓ may be uniformly computed from σ and from the characteristic function of T_ℓ restricted to the remaining set of nodes. That is, fix σ of length m and define a prefix-free computer M which computes only on input of the form $0^\ell 1\tau$ where $|\tau| = 2^\ell - 2^{\ell-m}$ and outputs the standard representation of a tree T_ℓ where no extension of σ is in T_ℓ and where τ tells us whether strings not extending σ are in T_ℓ. M is clearly prefix-free and we have $K_M(T_\ell) = \ell + 1 + 2^\ell - 2^{\ell-m}$. Thus $K((T_\ell) \leq \ell + 1 + 2^\ell - 2^{\ell-m} + c$ for some constant c. Now $\ell + 1 < 2^{\ell-m-1}$ for sufficiently large ℓ and thus by adjusting the constant c, we can obtain c' so that

$$K(T_\ell) \leq 2^\ell - 2^{\ell-m-1} + c'.$$ □

We might next conjecture that $K(T_\ell) > 2^{\ell-c}$ is the right notion of Chaitin randomness. However, classes with small measure are more compressible.

Theorem 7. *If $\mu([T]) < 2^{-k}$, then there exists c such that, for all ℓ,*
   $K(T_\ell) \leq 2^{\ell-k} + c$.

Proof. Suppose that $\mu([T]) < 2^{-k}$. Then for some level n, T_n has $< 2^{n-k}$ nodes $\sigma_1, \ldots, \sigma_t$. Now for any $\ell > n$, T_ℓ can be computed from the list $\sigma_1, \ldots, \sigma_t$ and the list of nodes of T_ℓ taken from the at most $2^{\ell-k}$ extensions of $\sigma_1, \ldots, \sigma_t$. It follows as in the proof of Theorem 6 above that for some constant c,
   $K(T_\ell) \leq 2^{\ell-k} + c$. □

Note that if $\mu([T]) = 0$, then for any k, there is a constant c such that $K(T_\ell) \leq 2^{\ell-k} + c$. But by Theorem 4, random closed sets have measure zero. Thus if P is random, then it is not the case that $K(T_n) \geq 2^{n-k}$.

Next we try to directly construct trees with not too much compressibility. The standard example of a random real [3] is a so-called c.e. real and therefore Δ_2^0.

Thus there exists a Δ_2^0 random tree T and by our observations above $K(T_\ell) \geq \ell - c$ for some c. Our first result here shows that we can get a Π_1^0 tree with this property.

Theorem 8. *There is a Π_1^0 class $P = [T]$ such that $K(T_n) \geq n$ for all n.*

Proof. Recall the universal prefix-free machine U and let $S = \{\sigma \in Dom(U) : |U(\sigma)| \geq 2^{|\sigma|}\}$. Then S is a c.e. set and can be enumerated as $\sigma_1, \sigma_2, \ldots$.

The tree $T = \cap_s T^s$ where T^s is defined at stage s. Initially we have $T^0 = \{0,1\}^*$.

σ_t *requires action* at stage $s \geq t$ when $\tau = U(\sigma_t) = T_n^s$ for some n (so that $|\tau| = 2^n$ and $n \geq |\sigma|$). Action is taken by selecting some path ρ_t of length n and define T^{s+1} to contain all nodes of T^s which do not extend ρ_t. Then $\tau \neq T_n^{s+1}$ and furthermore $\tau \neq T_n^r$ for any $r \geq s+1$ since future action will only remove more nodes from T_n.

At stage $s+1$, look for the least $t \leq s+1$ such that σ_t requires action and take the action described if there is such a t. Otherwise, let $T^{s+1} = T^s$.

Recall that $\sum_t 2^{-|\sigma_t|} < 1$. Since $|\rho_t| \geq |\sigma_t|$, it follows that $\sum_{t \in A} 2^{-|\rho_t|} < 1$ as well. Now $\mu([T]) = 1 - \sum_t 2^{-|\rho_t|} > 0$ and

$$\mu([T^s]) = 1 - \sum_{t \in A, t \leq s} 2^{-|\rho_t|} > 2^{-|\rho_{s+1}|}.$$

Thus when $U(\sigma_{s+1}) = \tau = T_n^s$ with $n \geq |\sigma_{s+1}|$, there will always be at least 2 nodes of length n in T_s so that we can remove one of them without killing the tree.

Let $T = \cap T^s$ and observe that by the construction each T^s is nonempty and therefore T is nonempty by compactness.

It follows from the construction that for each t, action is taken for σ_t at most once.

Now suppose by way of contradiction that $U(\sigma) = T_n$ for some σ_t with $|\sigma| \leq n$. There must be some stage $r \geq t$ such that for all $s \geq r$, $T_n^s = T_n$ and such that action is never taken on any $t' < t$ after stage r. Then σ_t will require action at stage $r+1$ which makes $T_n^{r+1} \neq T_n^r$, a contradiction. $\qquad\square$

Next we construct a closed Π_2^0 class with a stronger incompressibility property.

Theorem 9. *There is a Π_2^0 class $P = [T]$ such that $K(T_\ell) \geq 2^{\sqrt{\ell}}$ for all ℓ.*

Proof. We will construct a tree T such that T_{n^2} can not be computed from fewer than 2^n bits. We will assume that $U(\emptyset) \uparrow$ to take care of the case $n = 0$. At stage s, we will define the (nonempty) level T_{s^2} of T, using an oracle for $\mathbf{0}'$.

We begin with $T_0 = \{\emptyset\}^*$.

At stage $s > 0$, we consider

$$D_s = \{\sigma \in Dom(U) : |\sigma| < 2^s\}.$$

Since U is prefix-free, $card(D_s) < 2^{2^s}$. Now there are at least $2^{2^{2s-1}}$ trees of height s^2 which extend $T_{(s-1)^2}$ and we can use the oracle to choose some finite extension $T' = T_{s^2}$ of $T_{(s-1)^2}$ such that, for any $\sigma \in D_s$, $U(\sigma) \neq T'$ and furthermore, $U(\sigma) \neq T_r$ for any possible extension T_r with $s^2 \leq r$. That is, since there are $< 2^{2^s}$ finite trees which equal $U(\sigma)$ for some $\sigma \in D_s$, there is some extension T' of $T_{(s-1)^2}$ which differs from all of these at level s^2. We observe that the oracle for $\mathbf{0}'$ is used to determine the set D_s.

At stage s, we have ensured that for any extension $T \subseteq \{0,1\}^*$ of T_{s^2}, any σ with $|\sigma| \leq 2^{s^2}$ and any $n \geq s^2$, $U(\sigma) \neq T_n$. It is immediate that $K(T_n) \geq 2^{\sqrt{n}}$. $\qquad\square$

4 Conclusions and Future Research

In this paper we have proposed a notion of randomness for closed sets and derived several interesting properties of random closed sets. Random strong Π_2^0 classes exist but no Π_1^0 class is random. A random closed set has measure zero and contains no computable elements. A random closed set is perfect and hence uncountable. There are many other properties of closed sets and also of effectively closed sets which can be studied for random closed sets. For example, does a random closed set contain a c.e. real, or a real of low degree. We have obtained a Π_2^0 random closed set, whereas there is no Π_1^0 random closed set. We conjecture that every element of a random closed set is a random real and in fact the members are mutually random.

A real x is said to be K-trivial if $K(x\lceil n) \leq K(n) + c$ for some c. Much interesting work has been done on the K-trivial reals. Chaitin showed that if A is K-trivial, then $A \leq_T \mathbf{0}'$. Solovay constructed a noncomputable K-trivial real. Downey, Hirschfeldt, Nies and Stephan [4] showed that no K-trivial real is c.e. complete. It should be interesting to consider K-trivial closed sets.

We have examined the notion of compressibility for trees based on the Chaitin complexity of the nth level T_n of a tree. We constructed a Π_1^0 class $P = [T]$ such that $K(T_n) \geq n$ for all n and also a Π_2^0 class $Q = [T]$ such that $K(T_n) \geq 2^{\sqrt{n}}$ for all n. Much remains to be done here. It should not be difficult to improve the result of Proposition 2 to at least $K(T_n) \geq n^2$ or perhaps even $2^{\sqrt{n}}$. For the other direction, we need some level of incompressibility which implies randomness. We conjecture that $K(T_n) \geq 2^{n/c}$ should imply that $[T]$ is random for any constant $c > 0$. We would like to explore the notion that Π_1^0 classes more compressible than arbitrary closed sets.

For many mathematical problems, the set of solutions can be viewed as a closed set in $\{0,1\}^{\mathbb{N}}$. This includes combinatorial problems such as graph-coloring and matching, problems from logic and algebra such as finding a complete consistent extension of a consistent theory, and problems from analysis such as finding a zero or a fixed point of a continuous function. See [2] for a detailed discussion of many such problems. Of course the notion of a random graph is well-known. The connection between a randomly posed problem and the randomness of the set of solutions should be of interest.

References

1. D. Cenzer, Effectively Closed Sets, ASL Lecture Notes in Logic, to appear.
2. D. Cenzer and J. B. Remmel, Π_1^0 classes, in Handbook of Recursive Mathematics, Vol. 2: Recursive Algebra, Analysis and Combinatorics, editors Y. Ersov, S. Goncharov, V. Marek, A. Nerode, J. Remmel, Elsevier Studies in Logic and the Foundations of Mathematics, Vol. 139 (1998) 623-821.
3. G. Chaitin, Information-theoretical characterizations of recursive infinite strings, Theor. Comp. Sci. 2 (1976), 45-48.
4. R. Downey, D. Hirschfeldt, A. Nies and F. Stephan, Trivial reals, in Proc. 7th and 8th Asian Logic Conference, World Scientific Press, Singapore (2003), 101-131.
5. P. Gács, On the symmetry of algorithmic information, Soviet Mat. Dokl. 15 (1974), 1477-1480.
6. A. N. Kolmogorov, Three approaches to the quantitative defintion of information, in Problems of Information Transmission, Vol. 1 (1965), 1-7.
7. L. Levin, On the notion of a random sequence, Soviet Mat. Dokl. 14 (1973), 1413-1416.
8. P. Martin-Löf, The definition of random sequences, Information and Control 9 (1966), 602-619.
9. M. van Lambalgen, Random Sequences, Ph.D. Dissertation, University of Amsterdam (1987).
10. R. von Mises, Grundlagen der Wahrscheinlichkeitsrechnung, Math. Zeitschrift 5 (1919), 52-99.

Deep Inference and Its Normal Form of Derivations

Kai Brünnler

Institut für angewandte Mathematik und Informatik
Neubrückstr. 10, CH – 3012 Bern, Switzerland
kai@iam.unibe.ch
www.iam.unibe.ch/~kai/

Abstract. We see a notion of normal derivation for the calculus of structures, which is based on a factorisation of derivations and which is more general than the traditional notion of cut-free proof in this formalism.

1 Introduction

An inference rule in a traditional proof theoretical formalism like the sequent calculus or natural deduction only has access to the main connective of a formula. It does not have the feature of *deep inference*, which is the ability to access subformulas at arbitrary depth. Proof theoretical systems which do make use of this feature can be found as early as in Schütte [21], or, for a recent example, in Pym [19]. The *calculus of structures* is a formalism due to Guglielmi [11] which is centered around deep inference. Thanks to deep inference it drops the distinction between logical and structural connectives, a feature which already Schütte desired [20]. It also drops the tree-shape of derivations to expose a vertical symmetry which is in some sense new. One motivation of the calculus of structures is to find cut-free systems for logics which lack cut-free sequent systems. There are plenty of examples of such logics, and many are relevant to computer science: important modal logics like S5, many temporal and also intermediate logics. The logic that gave rise to the calculus of structures is the substructural logic BV which has connectives that resemble those of a process algebra and which can not be expressed without deep inference [26]. Systems in the calculus of structures so far have been studied for linear logic [24], non-commutative variants of linear logic [13, 7], classical logic [2] and several modal logics [22].

In this paper we ask the question what the right notion of *cut-free, normal* or *analytic* proof should be in the calculus of structures, and we see one such notion which is a factorisation of derivations and which generalises the notion that is used in the works cited above. This factorisation has independently been discovered by McKinley in [18]. The existence of normal derivations follows easily from translations between sequent calculus and calculus of structures. Here we consider systems for classical predicate logic, i.e. system LK [9] and system SKSgq [2] as examples, but it is a safe conjecture that this factorisation applies to any logic which has a cut-free sequent calculus.

A. Beckmann et al. (Eds.): CiE 2006, LNCS 3988, pp. 65–74, 2006.

After recalling a system for classical predicate logic in the calculus of structures, as well as the traditional notion of cut admissibility for this system, we see a more general notion based on factorisation as well as a proof that each derivation can be factored in this way. An outlook on some current research topics around the calculus of structures concludes this paper.

2   A Proof System for Classical Logic

The *formulas* for classical predicate logic are generated by the grammar

$$A ::= \mathsf{f} \mid \mathsf{t} \mid a \mid [A, A] \mid (A, A) \mid \exists x A \mid \forall x A \quad,$$

where f and t are the units *false* and *true*, a is an *atom*, which is a predicate symbol applied to some terms, possibly negated, $[A, B]$ is a *disjunction* and (A, B) is a *conjunction*. Atoms are denoted by a, b, c, formulas are denoted by A, B, C, D. We define $\bar{A}$, the *negation* of the formula A, as usual by the De Morgan laws. There is a *syntactic equivalence relation* on formulas, which is the smallest congruence relation induced by commutativity and associativity of conjunction and disjunction, the capture-avoiding renaming of bound variables as well as the following equations:

$$[A, \mathsf{f}] = A \qquad [\mathsf{t}, \mathsf{t}] = \mathsf{t}$$
$$(A, \mathsf{t}) = A \qquad (\mathsf{f}, \mathsf{f}) = \mathsf{f} \qquad \forall x A = A = \exists x A \quad \text{if } x \text{ is not free in } A \quad.$$

Thanks to associativity, we write $[A, B, C]$ instead of $[A, [B, C]]$, for example.

The *inference rules* in the calculus of structures are just rewrite rules known from term rewriting that work on formulas modulo the equivalence given above. There is the notational difference that here the context $S\{ \ \}$, in which the rule is applied, is made explicit. Here are two examples of inference rules:

$$\mathsf{i}\!\downarrow \frac{S\{\mathsf{t}\}}{S[A, \bar{A}]} \qquad \text{and} \qquad \mathsf{i}\!\uparrow \frac{S(A, \bar{A})}{S\{\mathsf{f}\}} \quad.$$

The name of the rule on the left is $\mathsf{i}\!\downarrow$ (read *i–down* or *identity*), and seen from top to bottom or from premise to conclusion it says that wherever the constant t occurs inside a formula, it can be replaced by the formula $[A, \bar{A}]$ where A is an arbitrary formula. The rule on the right (read *i–up* or *co-identity* or *cut*), also seen from top to bottom, says that anywhere inside a formula the formula $(A, \bar{A})$ can be replaced by the constant f. The two rules are *dual* meaning that one is obtained from the other by exchanging premise and conclusion and replacing each connective by its De Morgan dual. Here is another example of an inference rule, which is called *switch* and which happens to be its own dual:

$$\mathsf{s} \frac{S([A, B], C)}{S[(A, C), B]} \quad.$$

$$\mathsf{i}{\downarrow}\ \frac{S\{t\}}{S[A,\bar{A}]} \qquad\qquad \mathsf{i}{\uparrow}\ \frac{S(A,\bar{A})}{S\{f\}}$$

$$\mathsf{s}\ \frac{S([A,B],C)}{S[(A,C),B]}$$

$$\mathsf{u}{\downarrow}\ \frac{S\{\forall x[A,B]\}}{S[\forall xA,\exists xB]} \qquad\qquad \mathsf{u}{\uparrow}\ \frac{S(\forall xA,\exists xB)}{S\{\exists x(A,B)\}}$$

$$\mathsf{w}{\downarrow}\ \frac{S\{f\}}{S\{A\}} \qquad\qquad \mathsf{w}{\uparrow}\ \frac{S\{A\}}{S\{t\}}$$

$$\mathsf{c}{\downarrow}\ \frac{S[A,A]}{S\{A\}} \qquad\qquad \mathsf{c}{\uparrow}\ \frac{S\{A\}}{S(A,A)}$$

$$\mathsf{n}{\downarrow}\ \frac{S\{A[x/t]\}}{S\{\exists xA\}} \qquad\qquad \mathsf{n}{\uparrow}\ \frac{S\{\forall xA\}}{S\{A[x/t]\}}$$

Fig. 1. Predicate logic in the calculus of structures

A *derivation* is a finite sequence of instances of inference rules. For example

$$\mathsf{s}\ \frac{\mathsf{s}\ \dfrac{([A,C],[B,D])}{[A,(C,[B,D])]}}{[A,B,(C,D)]}\ .$$

The topmost formula in a derivation is its *premise* of the derivation, and the formula at the bottom is its *conclusion*. A *proof* is a derivation with the premise t. Dually, a *refutation* is a derivation with the conclusion f.

Figure 1 shows system SKSgq from [2]: a system for classical predicate logic. It is *symmetric* in the sense that for each rule in the system, the dual rule is also in the system. Like all systems in the calculus of structures it consists of two dual fragments: an up- and a down-fragment. The *down-fragment* is the system $\{\mathsf{i}{\downarrow},\mathsf{s},\mathsf{w}{\downarrow},\mathsf{c}{\downarrow},\mathsf{u}{\downarrow},\mathsf{n}{\downarrow}\}$ and the *up-fragment* is the system $\{\mathsf{i}{\uparrow},\mathsf{s},\mathsf{w}{\uparrow},\mathsf{c}{\uparrow},\mathsf{u}{\uparrow},\mathsf{n}{\uparrow}\}$. We also denote these two systems respectively by $\downarrow$ and $\uparrow$ and their union, the symmetric system, by $\updownarrow$. The letters w, c, u, n are respectively for *weakening, contraction, universal* and *instantiation*. It is proved in [2] that the down-fragment is complete in the sense that it has a proof for each valid formula, the up-fragment is complete in the sense that it has a refutation for each unsatisfiable formula and their union is complete also in the sense that for each valid implication it has a derivation from the premise to the conclusion of this implication.

3 Cut Elimination

The importance of cut-free proofs in the sequent calculus comes from the fact that they have the subformula property. Now, clearly the subformula property does not make sense for the calculus of structures in the same way as a "subsequent property" does not make sense for the sequent calculus. So the question for the calculus of structures is: what is is a cut-free proof?

Definition 1 (Cut-free Proof). A proof in the calculus of structures is *cut-free* if it does not contain any up-rules.

The cut elimination theorem takes the following form:

Theorem 2 (Up-fragment Admissibility). For each proof in the symmetric system there is a proof in the down-fragment with the same conclusion.

The above notion seems reasonable for our system for classical predicate logic, since it gives us the usual immediate consequences of Gentzen's Hauptsatz such as consistency and Herbrand's Theorem [3]. Craig Interpolation also follows, but it would be a bit of a stretch to call it an immediate consequence. It requires some work because rules are less restricted in the calculus of structures than in the sequent calculus.

Since for classical predicate logic there is a cut-free sequent system, Theorem 2 can be proved easily: we first translate derivations from the calculus of structures into this sequent system, using the cut in the sequent system to cope with the deep applicability of rules. Then we apply the cut elimination theorem for the sequent system. Finally we translate back the cut-free proof into the calculus of structures, which does not introduce any up-rules. Details are in [2]. To give an idea of how derivations in the sequent calculus are translated into the calculus of structures and to justify why the i↑-rule is also named *cut*, we see the translation of the cut rule:

$$
\text{Cut} \frac{\Phi \vdash A, \Psi \quad \Phi', A \vdash \Psi'}{\Phi, \Phi' \vdash \Psi, \Psi'} \qquad \text{translates into} \qquad
\text{s} \cfrac{
\text{s} \cfrac{
\text{i↑} \cfrac{
= \cfrac{
([\bar{\Phi}, A, \Psi], [\bar{\Phi}', \bar{A}, \Psi'])
}{[\bar{\Phi}, \Psi, (A, [\bar{\Phi}', \Psi', \bar{A}])]}
}{[\bar{\Phi}, \bar{\Phi}', \Psi, \Psi', (A, \bar{A})]}
}{[\bar{\Phi}, \bar{\Phi}', \Psi, \Psi', \mathsf{f}]}
}{[\bar{\Phi}, \bar{\Phi}', \Psi, \Psi']}
$$

A natural question here is whether there is an *internal* cut elimination procedure, i.e. one which does not require a detour via the sequent calculus. Such a procedure was nontrivial to find, since the deep applicability of rules renders the techniques of the sequent calculus useless. It has been given in [2, 1] for the propositional fragment and has been extended to predicate logic in [3].

Now we see a more general notion of cut-free or normal derivation. It is not characterised by the absence of certain inference rules, but by the the way in which the inference rules are composed:

Definition 3 (Normal Derivation). A derivation in the calculus of structures is *normal* if no up-rule occurs below a down-rule. To put it differently, a normal derivation has the form

$$
\begin{array}{c}
A \\
\Big\| \uparrow \\
B \\
\Big\| \downarrow \\
C
\end{array} \quad .
$$

This definition subsumes the definition of a cut-free proof: a proof is cut-free if and only if it is normal. Consider a proof, i.e. a derivation with premise syntactically equivalent to t, of the form given in the definition above. Since the conclusion of all rules in the up-fragment is equivalent to t if their premise is equivalent to t, then B has to be equivalent to t. We thus have a proof of C in the down-fragment. So the following theorem subsumes the admissibility of the up-fragment:

Theorem 4 (Normalisation). For each derivation in the symmetric system there is a normal derivation with the same premise and conclusion.

We will see a proof of this theorem shortly, but let us first have a look at an immediate consequence. Since no rule in the up-fragment introduces new predicate symbols going down and no rule in the down-fragment introduces new predicate symbols going up, the formula that connects the derivation in the up- with the derivation in the down-fragment is an interpolant:

Corollary 5 (Craig Interpolation). For each two formulas A, C such that A implies C there is a formula B such that A implies B, B implies C and all the predicate symbols that occur in B occur in both A and C.

To prove Theorem 4 we go the easy route just as for Theorem 2, we use cut elimination for LK and translations that we see in the two lemmas that follow. However, there is a crucial difference between the translations used to obtain Theorem 2 and the translations that we are going to see now: while the former just squeeze a tree into a sequence by glueing together the branches, the latter will rotate the proof by ninety degrees. We use a version of LK, which works on multisets of formulas, has multiplicative rules and which is restricted to formulas in negation normal form. LK is cut-free, we denote the system with the cut rule as LK + Cut. It is easy to check that we preserve cut admissibility when we replace the negation rules by two additional axioms:

$$
A, \bar{A} \vdash \qquad \text{and} \qquad \vdash A, \bar{A} \quad .
$$

Formulas of the calculus of structures and sequents of the sequent calculus are easily translated into one another: for a sequent

$$
\varPhi \vdash \varPsi \quad = \quad A_1, \ldots, A_m \vdash B_1, \ldots, B_n
$$

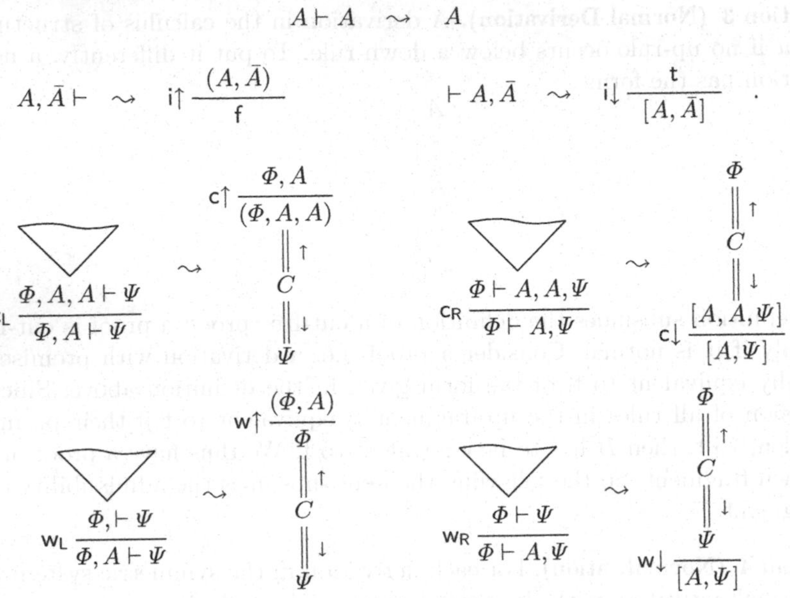

$$A \vdash A \quad \leadsto \quad A$$

$$A, \bar{A} \vdash \quad \leadsto \quad \mathsf{i}{\uparrow} \frac{(A, \bar{A})}{\mathsf{f}} \qquad\qquad \vdash A, \bar{A} \quad \leadsto \quad \mathsf{i}{\downarrow} \frac{\mathsf{t}}{[A, \bar{A]}} \quad .$$

$$\mathsf{cL} \frac{\Phi, A, A \vdash \Psi}{\Phi, A \vdash \Psi} \quad \leadsto \quad \mathsf{c}{\uparrow} \frac{\Phi, A}{(\Phi, A, A)} \\ \parallel {\uparrow} \\ C \\ \parallel {\downarrow} \\ \Psi \qquad\qquad \mathsf{cR} \frac{\Phi \vdash A, A, \Psi}{\Phi \vdash A, \Psi} \quad \leadsto \quad \begin{array}{c} \Phi \\ \parallel {\uparrow} \\ C \\ \parallel {\downarrow} \\ \mathsf{c}{\downarrow} \frac{[A, A, \Psi]}{[A, \Psi]} \end{array}$$

$$\mathsf{wL} \frac{\Phi, \vdash \Psi}{\Phi, A \vdash \Psi} \quad \leadsto \quad \mathsf{w}{\uparrow} \frac{(\Phi, A)}{\Phi} \\ \parallel {\uparrow} \\ C \\ \parallel {\downarrow} \\ \Psi \qquad\qquad \mathsf{wR} \frac{\Phi \vdash \Psi}{\Phi \vdash A, \Psi} \quad \leadsto \quad \begin{array}{c} \Phi \\ \parallel {\uparrow} \\ C \\ \parallel {\downarrow} \\ \mathsf{w}{\downarrow} \frac{\Psi}{[A, \Psi]} \end{array}$$

Fig. 2. Axioms and structural rules

we obtain two formulas that we denote by Φ and Ψ as well: $(A_1, \ldots, A_m)$ and $[B_1, \ldots, B_n]$. We identify an empty conjunction with t and an empty disjunction with f.

Lemma 6 (SKS to LK). For each derivation from A to B in SKSgq there is a proof of $A \vdash B$ in LK + Cut.

Proof. By induction on the length of the derivation, where we count applications of the equivalence as inference rules. The base case gives an axiom in the sequent calculus. The inductive case looks as follows:

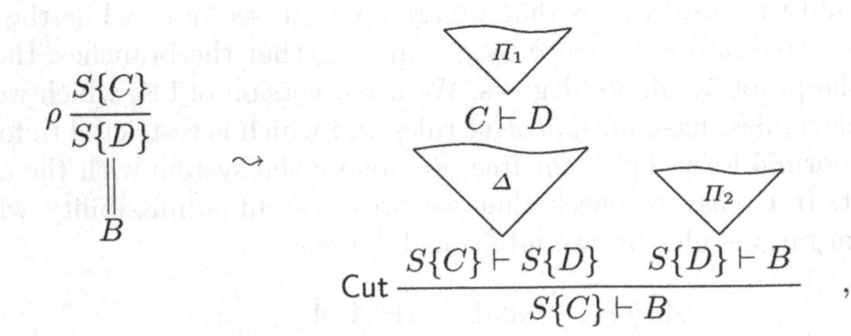

$$\rho \frac{S\{C\}}{S\{D\}} \\ \parallel \\ B \qquad \leadsto \qquad \mathsf{Cut} \frac{\overset{\displaystyle \Delta}{\overline{S\{C\} \vdash S\{D\}}} \quad \overset{\displaystyle \Pi_2}{S\{D\} \vdash B}}{S\{C\} \vdash B} \,,$$

where Π_2 exists by induction hypothesis, the existence of the derivation Δ can be easily shown for arbitrary formulas C, D and the existence of the proof Π_1 can be easily shown for each rule $\rho \in$ SKSgq and for the equations which generate the syntactic equivalence. $\square$

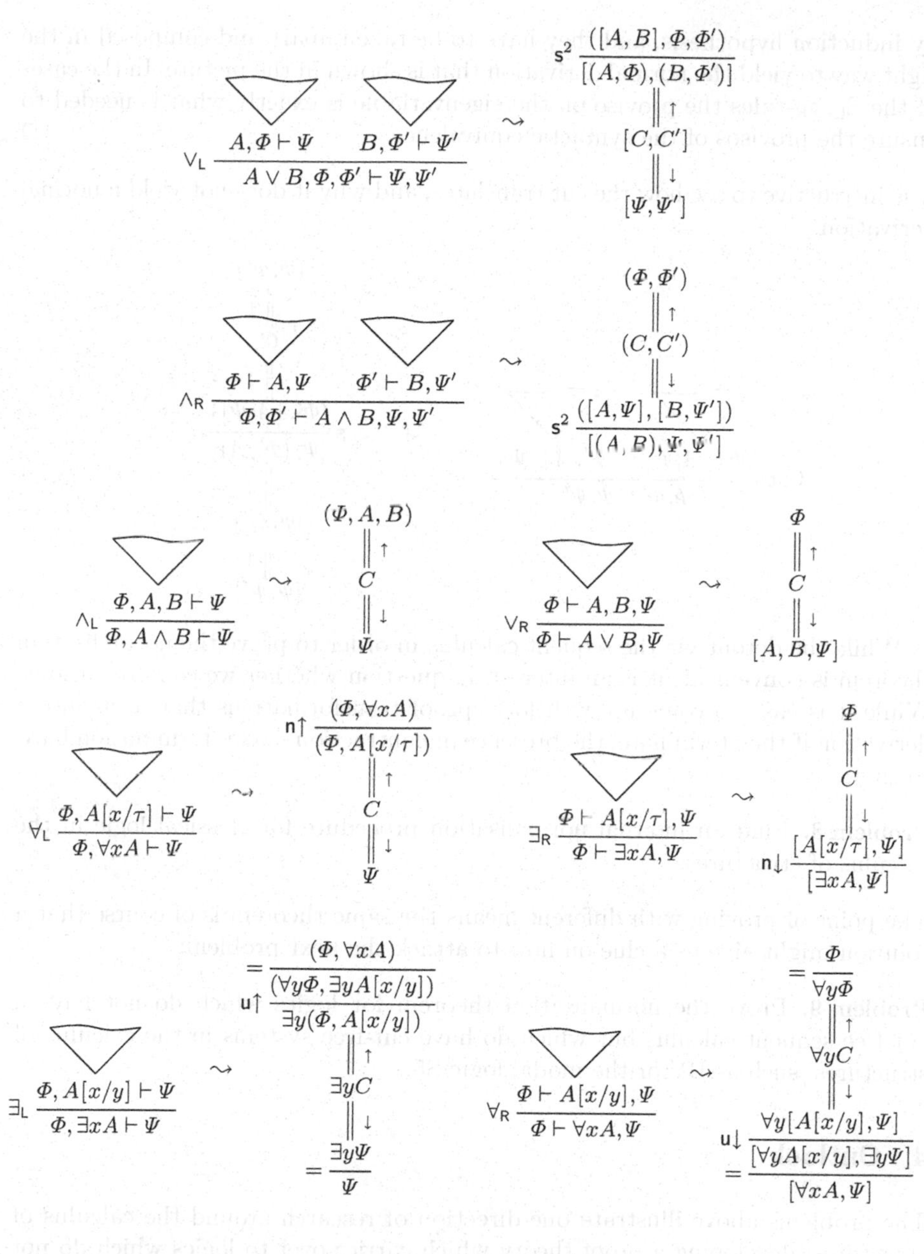

$$\mathsf{V_L}\ \frac{A,\Phi \vdash \Psi \qquad B,\Phi' \vdash \Psi'}{A \vee B,\Phi,\Phi' \vdash \Psi,\Psi'} \quad\leadsto\quad \mathsf{s}^2\ \frac{([A,B],\Phi,\Phi')}{[(A,\Phi),(B,\Phi')]}$$
$$\|\uparrow$$
$$[C,C']$$
$$\|\downarrow$$
$$[\Psi,\Psi']$$

$$\mathsf{\wedge_R}\ \frac{\Phi \vdash A,\Psi \qquad \Phi' \vdash B,\Psi'}{\Phi,\Phi' \vdash A \wedge B,\Psi,\Psi'} \quad\leadsto\quad \frac{(\Phi,\Phi')}{}$$
$$\|\uparrow$$
$$(C,C')$$
$$\|\downarrow$$
$$\mathsf{s}^2\ \frac{([A,\Psi],[B,\Psi'])}{[(A,B),\Psi,\Psi']}$$

$$\mathsf{\wedge_L}\ \frac{\Phi,A,B \vdash \Psi}{\Phi,A \wedge B \vdash \Psi} \quad\leadsto\quad \frac{(\Phi,A,B)}{}$$
$$\|\uparrow$$
$$C$$
$$\|\downarrow$$
$$\Psi$$

$$\mathsf{V_R}\ \frac{\Phi \vdash A,B,\Psi}{\Phi \vdash A \vee B,\Psi} \quad\leadsto\quad \frac{\Phi}{}$$
$$\|\uparrow$$
$$C$$
$$\|\downarrow$$
$$[A,B,\Psi]$$

$$\mathsf{\forall_L}\ \frac{\Phi,A[x/\tau] \vdash \Psi}{\Phi,\forall x A \vdash \Psi} \quad\leadsto\quad \mathsf{n}\uparrow\ \frac{(\Phi,\forall x A)}{(\Phi,A[x/\tau])}$$
$$\|\uparrow$$
$$C$$
$$\|\downarrow$$
$$\Psi$$

$$\mathsf{\exists_R}\ \frac{\Phi \vdash A[x/\tau],\Psi}{\Phi \vdash \exists x A,\Psi} \quad\leadsto\quad \frac{\Phi}{}$$
$$\|\uparrow$$
$$C$$
$$\|\downarrow$$
$$\mathsf{n}\downarrow\ \frac{[A[x/\tau],\Psi]}{[\exists x A,\Psi]}$$

$$\mathsf{\exists_L}\ \frac{\Phi,A[x/y] \vdash \Psi}{\Phi,\exists x A \vdash \Psi} \quad\leadsto\quad = \frac{(\Phi,\forall x A)}{}$$
$$\mathsf{u}\uparrow\ \frac{(\forall y \Phi,\exists y A[x/y])}{\exists y(\Phi,A[x/y])}$$
$$\|\uparrow$$
$$\exists y C$$
$$\|\downarrow$$
$$= \frac{\exists y \Psi}{\Psi}$$

$$\mathsf{\forall_R}\ \frac{\Phi \vdash A[x/y],\Psi}{\Phi \vdash \forall x A,\Psi} \quad\leadsto\quad = \frac{\Phi}{\forall y \Phi}$$
$$\|\uparrow$$
$$\forall y C$$
$$\|\downarrow$$
$$\mathsf{u}\downarrow\ \frac{\forall y[A[x/y],\Psi]}{= \frac{[\forall y A[x/y],\exists y \Psi]}{[\forall x A,\Psi]}}$$

Fig. 3. Logical rules

Lemma 7 (LK to SKS). *For each proof of $\Phi \vdash \Psi$ in* LK *there is a normal derivation from Φ to Ψ in* SKSgq.

Proof. By induction on the depth of the proof tree. All cases are shown in Figure 2 and Figure 3. In the cases of the $\mathsf{V_L}$, $\mathsf{\wedge_R}$-rules we get two normal derivations by induction hypothesis, and they have to be taken apart and composed in the right way to yield the normal derivation that is shown in the picture. In the cases of the $\exists_L, \forall_R$-rules the proviso on the eigenvariable is exactly what is needed to ensure the provisos of the syntactic equivalence. $\square$

It is instructive to see how the cut translates, and why it does not yield a normal derivation:

$$
\text{Cut} \; \frac{\Phi \vdash A, \Psi \qquad \Phi', A \vdash \Psi'}{\Phi, \Phi' \vdash \Psi, \Psi'}
\qquad \rightsquigarrow \qquad
\text{s} \; \begin{array}{c}
(\Phi, \Phi') \\
\| \uparrow \\
C \\
\| \downarrow \\
\dfrac{(\Phi', [A, \Psi])}{[\Psi, (\Phi', A)]} \\
\| \uparrow \\
\left[\Psi, C'\right] \\
\| \downarrow \\
\left[\Psi, \Psi'\right]
\end{array}
$$

While the detour via the sequent calculus in order to prove the normalisation theorem is convenient, it is an interesting question whether we can do without. While it is easy to come up with local proof transformations that normalise a derivation if they terminate, the presence of contraction makes termination hard to show.

Problem 8. Find an internal normalisation procedure for classical logic in the calculus of structures.

The point of proving with different means the same theorem is of course that a solution might give us a clue on how to attack the next problem:

Problem 9. Prove the normalisation theorem for logics which do not have a cut-free sequent calculus but which do have cut-free systems in the calculus of structures, such as BV or the modal logic S5.

4   Outlook

The problems above illustrate one direction of research around the calculus of structures: developing a proof theory which carries over to logics which do not have cut-free sequent systems. Examples are modal logics which can not be captured in the (plain vanilla) sequent calculus, like S5. Hein, Stewart and Stouppa are working on the project of obtaining modular proof systems for modal logic in the calculus of structures [15, 22, 23].

Another research thread is that of proof semantics. There is still the question of the right categorical axiomatisation of classical proofs. For predicate logic there is an approach by McKinley [18] which is derived from the concept of *classical category* by Führmann and Pym [8] and which is partly inspired by the shape of inference rules in the calculus of structures. A second approach is based on the notion of a *boolean category* by Lamarche and Straßburger [17, 25]. It is also involves concepts from the calculus of structures, in particular the fact that contraction can be reduced to atomic form and the so-called *medial* rule [5], which achieves that reduction.

The proof complexity of systems in the calculus of structures is also a topic of current research. The cut-free calculus of structures allows for an exponential speedup over the cut-free sequent calculus, as Bruscoli and Guglielmi [6] show using Statman's tautologies. Among the many open questions is whether there are short proofs for the pigeonhole principle. Short cut-free proofs in the calculus of structures of course come with a price: there is much more choice in applying rules than in the sequent calculus, which is an obstacle to implementation and applications. Work by Kahramanoğullari [16] is attacking this issue.

Finally there is a war against bureaucracy, which is also known as the quest for *deductive proof nets*, due to Guglielmi [12]. We say that a formalism contains bureaucracy if it allows to form two different derivations that differ only due to trivial rule permutations and are thus morally identical. Proof nets, for example, do not contain bureaucracy, while the sequent calculus and the calculus of structures do. Deductive proof nets, which still do not exist, should not contain bureaucracy (and thus be like proof nets and unlike sequent calculus), but should also have a locally and/or easily checkable correctness criterion (and thus be like sequent calculus and unlike proof nets). Approaches to the identification and possibly elimination of bureaucracy can be found in Guiraud [14] and Brünnler and Lengrand [4].

This has been a subjective and incomplete outlook, but more open problems and conjectures can be found on the calculus of structures website [10].

References

1. Kai Brünnler. Atomic cut elimination for classical logic. In M. Baaz and J. A. Makowsky, editors, *CSL 2003*, volume 2803 of *Lecture Notes in Computer Science*, pages 86–97. Springer-Verlag, 2003.
2. Kai Brünnler. *Deep Inference and Symmetry in Classical Proofs*. PhD thesis, Technische Universität Dresden, September 2003.
3. Kai Brünnler. Cut elimination inside a deep inference system for classical predicate logic. *Studia Logica*, 82(1):51–71, 2006.
4. Kai Brünnler and Stéphane Lengrand. On two forms of bureaucracy in derivations. In Paola Bruscoli, François Lamarche, and Charles Stewart, editors, *Structures and Deduction*, pages 69–80. Technische Universität Dresden, 2005.
5. Kai Brünnler and Alwen Fernanto Tiu. A local system for classical logic. In R. Nieuwenhuis and A. Voronkov, editors, *LPAR 2001*, volume 2250 of *Lecture Notes in Artificial Intelligence*, pages 347–361. Springer-Verlag, 2001.
6. Paola Bruscoli and Alessio Guglielmi. On the proof complexity of deep inference. In preparation, 2006.
7. Pietro Di Gianantonio. Structures for multiplicative cyclic linear logic: Deepness vs cyclicity. In J. Marcinkowski and A. Tarlecki, editors, *CSL 2004*, volume 3210 of *Lecture Notes in Computer Science*, pages 130–144. Springer-Verlag, 2004.

8. Carsten Fürmann and David Pym. Order-enriched categorical models of the classical sequent calculus. submitted, 2004.

9. Gerhard Gentzen. Investigations into logical deduction. In M. E. Szabo, editor, *The Collected Papers of Gerhard Gentzen*, pages 68–131. North-Holland Publishing Co., Amsterdam, 1969.

10. Alessio Guglielmi. The calculus of structures website. Available from http://alessio.guglielmi.name/res/cos/index.html.

11. Alessio Guglielmi. A system of interaction and structure. Technical Report WV-02-10, Technische Universität Dresden, 2002. To appear in ACM Transactions on Computational Logic.

12. Alessio Guglielmi. The problem of bureaucracy and identity of proofs from the perspective of deep inference. In Paola Bruscoli, François Lamarche, and Charles Stewart, editors, *Structures and Deduction*, pages 53–68. Technische Universität Dresden, 2005.

13. Alessio Guglielmi and Lutz Straßburger. Non-commutativity and MELL in the calculus of structures. In L. Fribourg, editor, *CSL 2001*, volume 2142 of *Lecture Notes in Computer Science*, pages 54–68. Springer-Verlag, 2001.

14. Yves Guiraud. The three dimensions of proofs. to appear in Annals of pure and applied logic, 2005.

15. Robert Hein and Charles Stewart. Purity through unravelling. In Paola Bruscoli, François Lamarche, and Charles Stewart, editors, *Structures and Deduction*, pages 126–143. Technische Universität Dresden, 2005.

16. Ozan Kahramanoğullari. Reducing nondeterminism in the calculus of structures. Manuscript, 2005.

17. François Lamarche and Lutz Straßburger. Naming proofs in classical propositional logic. In Paweł Urzyczyn, editor, *Typed Lambda Calculi and Applications, TLCA 2005*, volume 3461 of *Lecture Notes in Computer Science*, pages 246–261. Springer-Verlag, 2005.

18. Richard McKinley. Categorical models of first-order classical proofs. submitted for the degree of Doctor of Philosophy of the University of Bath, 2005.

19. D.J. Pym. *The Semantics and Proof Theory of the Logic of Bunched Implications*, volume 26 of *Applied Logic Series*. Kluwer Academic Publishers, 2002.

20. Kurt Schütte. Schlussweisen-Kalküle der Prädikatenlogik. *Mathematische Annalen*, 122:47–65, 1950.

21. Kurt Schütte. *Proof Theory*. Springer-Verlag, 1977.

22. Charles Stewart and Phiniki Stouppa. A systematic proof theory for several modal logics. In Renate Schmidt, Ian Pratt-Hartmann, Mark Reynolds, and Heinrich Wansing, editors, *Advances in Modal Logic*, volume 5 of *King's College Publications*, pages 309–333, 2005.

23. Phiniki Stouppa. A deep inference system for the modal logic S5. To appear in Studia Logica, 2006.

24. Lutz Straßburger. *Linear Logic and Noncommutativity in the Calculus of Structures*. PhD thesis, Technische Universität Dresden, 2003.

25. Lutz Straßburger. On the axiomatisation of boolean categories with and without medial. Manuscript, 2005.

26. Alwen Fernanto Tiu. Properties of a Logical System in the Calculus of Structures. Master's thesis, Technische Universität Dresden, 2001.

Logspace Complexity of Functions and Structures*

Douglas Cenzer[1] and Zia Uddin[2]

[1] Department of Mathematics, University of Florida,
P.O. Box 118105, Gainesville, Florida 32611
Fax: 352-392-8357
cenzer@math.ufl.edu
[2] Department of Mathematics, 421 Robinson Hall
Lock Haven University of Pennsylvania, Lock Haven, PA 17745
zuddin@lhup.edu

Abstract. Logspace complexity of functions and structures is based on
the notion of a Turing machine with input and output as in Papadmitriou
[16]. For any $k > 1$, we construct a logspace isomorphism between $\{0,1\}^*$
and $\{0, 1, \ldots, k\}^*$. We improve results of Cenzer and Remmel [5] by characterizing the sets which are logspace isomorphic to $\{1\}^*$. We generalize
Proposition 8.2 of [16] by giving upper bounds on the space complexity of
compositions and use this to obtain the complexity of isomorphic copies
of structures with different universes. Finally, we construct logspace models with standard universe $\{0,1\}^*$ of various additive groups, including
$Z(p^\infty)$ and the rationals.

Keywords: Computability, Complexity, Computable Model Theory.

1 Introduction

Complexity theory has been a central theme of computer science and related
areas of mathematics since the middle of the last century. Much work has been
done on the time complexity of sets, functions and structures. The practical goal
is to find efficient algorithms for computing functions and solving problems. In his
encyclopedic book [12], Knuth examines in detail the quest for fast multiplication
and also considers the problem of radix conversion of numbers between binary
and decimal representation. Recent striking advances include the proof that
primality is polynomial-time decidable [1] and the result that division of integers
can be computed in logspace [10]. The latter result will be used below in our
construction of a logspace model for the additive group of rationals.

Complexity theoretic model theory and algebra was developed by Nerode,
Remmel and Cenzer [13, 14, 15, 4, 5]; see the handbook article [9] for details. One
of the basic questions which have been studied is the existence problem, that is,
whether a given computable structure is isomorphic, or computably isomorphic,
to a resource-bounded (typically polynomial time) structure. For example, it was

---

* Research was partially supported by the National Science Foundation.

A. Beckmann et al. (Eds.): CiE 2006, LNCS 3988, pp. 75–84, 2006.

shown in [4] that every computable relational structure is computably isomorphic to a polynomial time structure and that the standard model of arithmetic $(\mathbb{N}, +, -, \cdot, <, 2^x)$ has a polynomial time model, where 2^x indicates the unary exponential function. The fundamental effective completeness theorem states that any decidable theory has a decidable model and it follows that any decidable theory has a polynomial time model. However, there is a fundamental difference between computable structures and complexity theoretic structures. Any two infinite computable sets of integers are computably isomorphic and therefore any computable structure may be taken to have a standard universe $\mathbb{N}$. However, it is not the case that any two infinite polynomial time sets are polynomial time isomorphic. Thus the refined existence question is whether a given computable structure has a polynomial time model with a standard universe, meaning either $Bin(\mathbb{N})$ (the set of binary representations of natural numbers) or $Tal(\mathbb{N})$ (the set of tally representations). It was shown in [5] that there is a family of Abelian p-groups, including the computably categorical p-groups of [17] which are computably isomorphic to polynomial time groups with a standard universe. At the same time, Abelian p-groups were constructed in [5] which are not computably isomorphic to polynomial time groups with a standard universe. The question of uniqueness of representation, that is, categoricity, was studied further in [6, 8].

In the present paper, we consider the efficient use of space. It was established by Hopcroft and Ullman [11] that an appropriate model for function calculation is the machine with read-only input and write-only output. The motivation for the input/output approach is that simple functions such as addition can be performed in logarithmic space (in fact in zero space) whereas including the input and/or output would automatically require at least space n.

In particular, addition of integers can be computed with zero space and multiplication can be computed in logspace. Recent work of Chiu et al [10] has shown that division can also be computed in logspace. It then follows from [2] that powering and iterated multiplication can also be computed in logspace. On the other hand, the best upper bound for radix conversion seems to require space $log\ n\ log\ log\ n$. (see [2]). We show that, nevertheless, for each k, there is a logspace isomorphism between the binary and k-ary representations of natural numbers.

We improve some results of Cenzer and Remmel [5] by characterizing the sets of natural numbers which are logspace isomorphic to $\{1\}^*$ and by giving various lemmas which ensure that a given sum or product of logspace sets is logspace isomorphic to $Tal(\mathbb{N})$ or to $Bin(\mathbb{N})$.

The family of logspace functions is closed under composition and therefore this notion of logspace computation is robust. We give a generalization of this result which gives upper bounds for the complexity of the composition of functions of arbitrary space complexity.

All of these results come together in the construction of logspace models for certain standard Abelian groups, such as the additive groups $\mathbb{Q}_p$ of p-adic rationals and $\mathbb{Z}(p^\infty)$ of p-adic rationals modulo 1, where p is a prime. We also construct logspace models for the additive groups $\mathbb{Q}\ mod\ 1$ and $\mathbb{Q}$.

We conclude this section with some definitions.

Our model of computation is the multi-tape Turing machine of Papadimitriou [16]. The cursor of each tape can move independently of the cursors of other tapes. Our Turing machines are both *read-only* (input tape symbols are never overwritten) and *write-only* (the output-string cursor never moves left).

Let $\mathbb{N}$ denote the set $\{0, 1, 2, \ldots\}$ of natural numbers and $\mathbb{N}^+ = \mathbb{N} - \{0\}$. A function $F(x) : \mathbb{N}^+ \to \mathbb{N}^+$ is a *proper complexity function* if F is nondecreasing and furthermore, there is a Turing machine M with input and output which, on any input x, computes the string $1^{F(|x|)}$ in $\leq \mathcal{O}(|x| + F(|x|))$ steps and uses space $\leq \mathcal{O}(F(|x|))$. Some examples are constant functions, $k \log x$, $(\log x)^k$, kx, x^k, $2^{(\log x)^k}$, 2^{kx}, 2^{x^k}, or $2^{2^{kx}}$. (We use $\log x$ as an abbreviation for $\log_2 x$.)

Fix a finite alphabet Σ and a proper complexity function G. Then a function $f : (\Sigma^*)^k \to \Sigma^*$ is computable in SPACE(G) if there is a Turing machine M with input and output which computes $f(x_1, \ldots, x_k)$ using space $\leq G(|x|)$; f is computable in TIME(G) if there is a Turing machine M with input and output which computes $f(x_1, \ldots, x_k)$ using time $\leq G(|x|)$. For time complexity, the restriction on input and output does not change the capability of the Turing machine, by Proposition 2.2 of [16].

We are primarily interested in the following families

$$LOG = LOGSPACE = \cup_{c \in \mathbb{N}} SPACE(c \log n);$$
$$PLOGSPACE = \cup_{c \in \mathbb{N}} SPACE((\log n)^c);$$
$$P = PTIME = \cup_{c \in \mathbb{N}} TIME(n^c).$$

A function mapping Σ^* to Σ^* is sometimes said to be *FLOG* computable, or simply *FLOG* if it is in *LOG*. The following is part of Theorem 7.4 of [16].

Lemma 1. *For any proper complexity function G:*

(a) $TIME(G) \subseteq SPACE(G)$;
(b) $SPACE(G) \subseteq TIME(k^{\log n + G(n)})$ for some k. $\square$

This implies in particular that $LOG \subseteq P$ and hence the following fact.

Lemma 2. *For any function f in $FLOG$, there is a constant k such that $|f(x)| \leq |x|^k$ for all inputs x.*

The standard universes for computation are the following. Let $Tal(0) = 0$ and for $n > 0$, let $Tal(n) = 1^n$. Then $Tal(\mathbb{N}) = \{0\} \cup \{1\}^* = \{Tal(n) : n \in \mathbb{N}\}$. For each $n \in \mathbb{N}$ and each $k > 1$, let $B_k(n) = b_0 b_1 \ldots b_r \in \{0, 1, \ldots, k-1\}^*$ be the standard (reverse order) k-ary representation, so that $b_r > 0$ and $n = b_0 + b_1 k + \ldots + b_r k^r$. Then

$$B_k(\mathbb{N}) = \{B_k(n) : n \in \mathbb{N}\} = \{b_0 \ldots b_r \in \{0, 1, \ldots, k-1\}^* : b_r \neq 0\}.$$

In particular, let $Bin(n) = B_2(n)$ and $Bin(\mathbb{N}) = B_2(\mathbb{N})$.

2   Composition

In this section, we consider the space complexity of composite functions. We give a general result which provides an upper bound on the complexity of a composition of functions and some specific corollaries which we will need for the study of resource-bounded structures.

Theorem 1. *Let F and G be nonconstant proper complexity functions and let g be a unary function in $SPACE(G)$ and f an n-ary function in $SPACE(F)$. Then the composition $g \circ f$ can be computed in $SPACE \leq G(2^{kF})$ for some constant k.*

Proof. The proof is a generalization of the standard proof that the composition of two $LOGSPACE$ functions is in $LOGSPACE$. In particular, note that by Lemma 1 for $x = (x_1, \ldots, x_n)$, $f(x)$ can be computed in time $c|x|^n 2^{kF(|x|)}$ for some constants c and k which bounds the length of $f(x)$. The logspace algorithm uses a binary counter to keep track of the location of the pointer for the g work tape and recreates the bits of $f(x)$ as needed.

Corollary 1. *(a) $LOGSPACE \circ LINSPACE = LINSPACE$;*
(b) $PLOGSPACE \circ PLOGSPACE = PLOGSPACE$;

3   Logspace Sets and Radix Representation

In this section, we establish a few lemmas about logspace isomorphisms of sets which will be needed for the discussion of logspace structures.

The first lemma characterizes sets isomorphic to $Tal(\mathbb{N})$ and is similar to Lemma 2.4 of [5].

Lemma 3. *Let A be a $LOGSPACE$ subset of $Tal(\mathbb{N})$, and list the elements $a_0, a_1, a_2 \ldots$ of A in the standard ordering. Then the following are equivalent:*

(a) A is $LOGSPACE$ set-isomorphic to $Tal(\mathbb{N})$.
(b) For some k and all $n \geqslant 2$, we have $|a_n| \leq n^k$.
(c) The canonical bijection between $Tal(\mathbb{N})$ and A that associates 1^n with a_n, $n \geq 0$, is in $LOGSPACE$.

Proof. The map taking a_n to 1^n is $FLOG$ even without assumption (b). That is, given tally input $a = a_n$, one proceeds as follows. First convert a to binary b and write this on a worktape. Now a second tape will begin with $Bin(0)$ and increment at stage $t + 1$ from $Bin(t)$ to $Bin(t + 1)$ as long as $Bin(t) \leq b$. The output tape will begin with 0. Then at stage t, we will simulate testing whether $Tal(t) \in A$ as follows. Use the standard $LINSPACE$ conversion $Bin(t)$ into $Tal(t)$ and then the LOG test of whether $Tal(T) \in A$. It follows from Corollary 1 that this computation is $LINSPACE$ in the input $Bin(t)$ and since $Tal(t) \leq a_n$, the computation can be done in $LOGSPACE$ with respect to input a_n. If the test is positive, then a "1" is appended to the output tape.

For the map taking 1^n to $a_n = Tal(m)$, assume (b) and use the following procedure. As above, at stage $t \leq n$, we will have $Bin(t)$ on one work tape and test whether $Tal(t) \in A$. If the test is positive, then we move the cursor on the input tape to the right and otherwise not. Once the end of the input tape is reached, we will have $Bin(m)$ on the work tape. The final step is to convert this to $a_n = Tal(m)$. Since $a_n \leq n^k$, it follows that $|Bin(m)| \leq log(n^k)$, so that the computation can be done in $LOGSPACE$.

The next lemma is crucial for building structures with a standard universe.

Lemma 4. *For each $k \geq 2$, the following sets are $LOGSPACE$ isomorphic:*

(a) $Bin(\mathbb{N})$;
(b) $B_k(\mathbb{N})$;
(c) $\{0, 1, \ldots, k-1\}^*$.

Furthermore, there exists a $LOGSPACE$ bijection $f : Bin(\mathbb{N}) \to B_k(\mathbb{N})$ and constants $c_1, c_2 > 0$ such that, for every $n \in \mathbb{N}$:

(i) $|f(Bin(n))| \leq c_1 |Bin(n)|$ and
(ii) $|f^{-1}(B_k(n))| \leq c_2 |B_k(n)|$.

Proof. It is easy to see that $Bin(\mathbb{N})$ is logspace isomorphic to $\{0, 1\}^*$. For any $k > 2$, $\{0, 1, \ldots, k-1\}^*$ is logspace isomorphic to $\{0, 1\}^*$ by the function f defined as follows.

First define $g : \{0, 1, \ldots, k-1\} \longrightarrow \{0, 1\}^*$ by $g(0) = 0^{k-1}$ and $g(i) = 0^{i-1}1$ for $1 \leq i < k$. Then let $f(\emptyset) = \emptyset$, $f(0^n) = 0^n$, $f(\sigma^\frown 0^n) = f(\sigma)^\frown 0^n$ when σ is a string with at least one non-0 symbol, and $f(a_0 a_1 \ldots a_{n-1}) = g(a_0)^\frown g(a_1)^\frown \ldots ^\frown g(a_{n-1})$ where each $a_i \in \{0, 1, \ldots, k-1\}$ and at least one $a_i \neq 0$.

Observe that $B_k(\mathbb{N}) - \{0\}$ is the set of strings in $\{0, 1, \ldots, k-1\}^* - \{\emptyset\}$, beginning with $i = 1, 2, \ldots, k-1$ and is therefore logspace isomorphic to $k-1$ copies of $\{0, 1 \ldots, k-1\}^*$. On the other hand, $\{0, 1, \ldots, k-1\}^*$ is naturally isomorphic to k copies of itself. We will show that $\{0, 1 \ldots, k-1\}^* - \{\emptyset\}$ is logspace isomorphic to $k-1$ copies of itself and thus $B_k(\mathbb{N})$ is logspace isomorphic to $\{0, 1, \ldots, k-1\}^*$. Elements of $\{0, 1 \ldots, k-1\}^*$ are denoted below by σ and elements of the $k-1$ fold disjoint union by $\langle j, \sigma \rangle$, and arbitrary elements of $\{0, 1 \ldots, k-1\}^*$ are denoted by τ. The mapping is defined by the following sets of rules. For strings not beginning with 0 or 1, we have:

$$2 \to \langle 1, 0 \rangle \qquad\qquad 2^\frown 0^n \to \langle 1, 0^{n+1} \rangle \qquad\qquad 2^\frown \sigma \to \langle 1, \sigma \rangle$$
$$3 \to \langle 2, 0 \rangle \qquad\qquad 3^\frown 0^n \to \langle 2, 0^{n+1} \rangle \qquad\qquad 3^\frown \sigma \to \langle 2, \sigma \rangle$$
$$\vdots \qquad\qquad\qquad \vdots \qquad\qquad\qquad\qquad \vdots$$
$$k-2 \to \langle k-3, 0 \rangle \qquad (k-2)^\frown 0^n \to \langle k-3, 0^{n+1} \rangle \qquad (k-2)^\frown \sigma \to \langle k-3, \sigma \rangle$$
$$k-1 \to \langle k-2, 0 \rangle \qquad (k-1)^\frown 0^n \to \langle k-2, 0^{n+1} \rangle \qquad (k-1)^\frown \sigma \to \langle k-2, \sigma \rangle$$

For strings beginning with 1, we have:

$$1 \to \langle 0, (k-1)^\frown 0 \rangle \qquad 1^\frown 0^n \to \langle 0, (k-1)^\frown 0^{n+1} \rangle \qquad 1^\frown \sigma \to \langle 0, (k-1)^\frown \sigma \rangle$$

For strings beginning with 0, we have:

$$0^n \;\to\; \langle 0, 0^n \rangle$$

$$0^\frown(k-1)^\frown\tau \;\to\; \langle 0, (k-2)^\frown\tau \rangle \qquad 0^{n+1\frown}(k-1)^\frown\tau \;\to\; \langle 0, 0^{n\frown}(k-2)^\frown\tau \rangle$$

$$0^\frown(k-2)^\frown\tau \;\to\; \langle 0, (k-3)^\frown\tau \rangle \qquad 0^{n+1\frown}(k-2)^\frown\tau \;\to\; \langle 0, 0^{n\frown}(k-3)^\frown\tau \rangle$$

$$\vdots \qquad\qquad\qquad\qquad\qquad\qquad \vdots$$

$$02^\frown\tau \;\to\; \langle 0, 1^\frown\tau \rangle \qquad\qquad 0^{n+1\frown}2^\frown\tau \;\to\; \langle 0, 0^{n\frown}1^\frown\tau \rangle$$

$$01^\frown\tau \;\to\; \langle 0, 0^\frown(k-1)^\frown\tau \rangle \qquad 0^{n+1\frown}1^\frown\tau \;\to\; \langle 0, 0^{n+1\frown}(k-1)^\frown\tau \rangle$$

It is not hard to see that this defines a bijection f and that both f and f^{-1} are logspace computable, in fact, can be computed without using any space.

Lemma 5. *Let A be a nonempty LOGSPACE subset of $Tal(\mathbb{N})$. Then*

(a) *The set $A \oplus Tal(\mathbb{N})$ is LOGSPACE isomorphic to $Tal(\mathbb{N})$ and the set $A \oplus Bin(\mathbb{N})$ is LOGSPACE set-isomorphic to $Bin(\mathbb{N})$.*

(b) *The set $A \times Tal(\mathbb{N})$ is LOGSPACE isomorphic to $Tal(\mathbb{N})$ and the set $A \times Bin(\mathbb{N})$ is LOGSPACE set-isomorphic to $Bin(\mathbb{N})$.*

(c) *Both $Bin(\mathbb{N}) \oplus Bin(\mathbb{N})$ and $Bin(\mathbb{N}) \times Bin(\mathbb{N})$ are LOGSPACE isomorphic to $Bin(\mathbb{N})$.*

(d) *If B is a nonempty finite subset of $Bin(\mathbb{N})$, then both $B \oplus Bin(\mathbb{N})$ and $B \times Bin(\mathbb{N})$ are LOGSPACE isomorphic to $Bin(\mathbb{N})$.*

Proof. The tally cases of parts (a) and (b) follow from Lemma 3. That is, for example, $A \oplus Tal(\mathbb{N})$ contains all odd numbers and therefore the nth element is certainly $\leq 2n + 1$.

For the binary cases of (a) and (b), first observe that $Bin(\mathbb{N}) - Tal(\mathbb{N})$ is logspace isomorphic to $Bin(\mathbb{N})$ via the map $f(x) = x + 1 - |x|$ and $Tal(\mathbb{N}) \otimes Bin(\mathbb{N})$ is logspace isomorphic to $Bin(N)$ via the map g defined as follows:

$g(\langle 1^m, 0 \rangle) = 0^{m\frown}1$ and $g(\langle 1^m, Bin(n) \rangle) = 0^{m\frown}1^\frown Bin(n)$ for $n \neq 0$;
$g(\langle 1, Bin(n) \rangle) = 1^\frown Bin(n)$;
$g(\langle 0, 0 \rangle] = 0$ and $g(\langle 0, 1 \rangle) = 1$.

Then $A \oplus Bin(\mathbb{N})$ is logspace isomorphic to $A \oplus Tal(\mathbb{N}) \oplus (Bin(\mathbb{N}) - Tal(\mathbb{N}))$, which is logspace isomorphic to $Tal(\mathbb{N}) \oplus Bin(\mathbb{N}) - Tal(\mathbb{N})$ by the tally case and thus logspace isomorphic to $Bin(\mathbb{N})$. Finally, $A \otimes B$ is logspace isomorphic to $A \otimes Tal(\mathbb{N}) \otimes Bin(\mathbb{N})$, which is logspace isomorphic to $Tal(\mathbb{N}) \otimes Bin(\mathbb{N})$ by the tally case and thus logspace isomorphic to $Bin(\mathbb{N})$.

For part (c), partition $\mathbb{N} \times \mathbb{N}$ into an infinite disjoint union as follows. For each $n \geqslant 1$, define

$$
\begin{aligned}
A_n &= \{0, 1, \ldots, 2^n - 1\} \times \{2^n, 2^n + 1, \ldots, 2^{n+1} - 1\}, \\
B_n &= \{2^n, 2^n + 1, \ldots, 2^{n+1} - 1\} \times \{0, 1, \ldots, 2^n - 1\}, \\
C_n &= \{2^n, 2^n + 1, \ldots, 2^{n+1} - 1\} \times \{2^n, 2^n + 1, \ldots, 2^{n+1} - 1\}.
\end{aligned}
$$

Define the map f from $\mathbb{N} \times \mathbb{N}$ to $\mathbb{N}$ by $f(0,0) = 0$, $f(1,0) = 2$, $f(1,1) = 3$ and for each $n \geq 1$,

$$
\begin{aligned}
(x,y) &\mapsto 2^n x + y + 2^{2n} - 2^n && \text{if } (x,y) \in A_n, \\
(x,y) &\mapsto 2^n x + y + 2^{2n} && \text{if } (x,y) \in B_n, \\
(x,y) &\mapsto 2^n x + y + 2^{2n+1} - 2^n && \text{if } (x,y) \in C_n.
\end{aligned}
$$

Then it can be shown that the corresponding map from $Bin(\mathbb{N}) \times Bin(\mathbb{N})$ to $Bin(\mathbb{N})$ is a logspace isomorphism.

Part (d) is not difficult.

4   Logspace Structures

This section contains the main construction of resource-bounded models, with standard universe, of certain basic groups, including the p-groups $Z(p^\infty)$ and the additive group $\mathbb{Q}$ of rationals. We begin with some necessary lemmas.

Lemma 6. *Let A be a LOGSPACE structure and let ϕ be a LOGSPACE bijection from A (the universe of A) onto a set B. Then B is a LOGSPACE structure, where the functions and relations on its universe B are defined to make ϕ an isomorphism of the structures.*

Lemma 7. *Let M be a structure with universe $M \subseteq \mathbb{N}$, and let $A = Tal(M)$ and $B = B_k(M)$, where $k \geq 2$. Then we have*

(a) If $B \in LOGSPACE$, then $A \in PLOGSPACE$.

(b) If $B \in LINSPACE$ and for all functions f^B, $|f^B(m_1, \ldots, m_n)| \leqslant c(|m_1| + \ldots + |m_n|)$ for some fixed constant c and all but finitely many n-tuples, then $A \in LOGSPACE$.

The direct sum, or external weak product, of a sequence $A_i = (A_i, +_i, -_i, e_i)$ of groups is defined as usual to have elements $(a_0, a_1, \ldots)$ where, for all but finitely many i, $a_i = e_i$ and the operations are coordinatewise. The sequence is *fully uniformly LOGSPACE* over B (either $Bin(\mathbb{N})$ or $Tal(\mathbb{N})$) if the following hold.
(i) The set $\{\langle B(n), a \rangle : a \in A_n\}$ is a LOGSPACE subset of $B \otimes B$, where $B(n) = Tal(n)$ if $B = Tal(\mathbb{N})$ and $B(n) = Bin(n)$ if $B = Bin(\mathbb{N})$.
(ii) The functions F and G, defined by $F(B(n), a, b) = a +_n b$ and $G(B(n), a, b) = a -_n b$, are *LOGSPACE* computable.
(iii) The function $e : Tal(\mathbb{N}) \to B$, defined by $e(Tal(i)) = e_i$, is in $LOGSPACE$.

Lemma 8. *Let B be either $Tal(\mathbb{N})$ or $Bin(\mathbb{N})$. Suppose that the sequence $\{A_i\}_{i \in \mathbb{N}}$ of groups is fully uniformly LOGSPACE over B. Then*

(a) The direct sum $\oplus_i A_i$ is recursively isomorphic to a LOGSPACE group with universe contained in $Bin(\mathbb{N})$.

(b) If A_i is a subgroup of A_{i+1} for all i, and if there is a LOGSPACE function $f : \{0,1\}^* \to B$ such that for all $a \in \bigcup_i A_i$, we have $a \in A_{f(a)}$, then the union $\bigcup_i A_i$ is a LOGSPACE group with universe contained in B.

(c) If the sequence is finite, one of the components has universe B, and the remaining components have universes that are $LOGSPACE$ subsets of $Tal(\mathbb{N})$, then the direct sum is recursively isomorphic to a $LOGSPACE$ group with universe B.

(d) If the sequence is infinite and if each component has universe $Bin(\mathbb{N})$, then the direct sum is recursively isomorphic to a $LOGSPACE$ group with universe $Bin(\mathbb{N})$.

(e) If each component has universe $Tal(\mathbb{N})$ and there is a uniform constant c such that for each i and any $a, b \in A_i$, we have both $|a +_i b| \leqslant c(|a| +_i |b|)$ and $|a -_i b| \leqslant c(|a| +_i |b|)$, then the direct sum is recursively isomorphic to a $LOGSPACE$ group with universe $Tal(\mathbb{N})$.

Proof. We just sketch the proof of (d) to give the idea. Elements of the sum may be viewed as finite sequences from $Bin(\mathbb{N})$ with coordinatewise operations and therefore $LOGSPACE$. For the isomorphism, sequences of length $n > 0$ may be mapped to $\langle n, m \rangle \in Bin(\mathbb{N}) \oplus Bin(\mathbb{N})$ since the set of sequences of length n is a finite sum of $n - 1$ copies of $Bin(\mathbb{N})$ with one copy of $Bin(\mathbb{N}) - \{0\}$ and hence isomorphic to $Bin(\mathbb{N})$ by Lemma 5. This will give a logspace isomorphism of the universe which then leads to a group isomorphism by Lemma 6.

For a given prime p, let $\mathbb{Q}_p$ denote the additive group of all p-adic rationals

Theorem 2. *Let $k > 1$ be in $\mathbb{N}$ and let p be a prime. Each of the groups $\mathbb{Z}$, $\bigoplus_\omega \mathbb{Z}_k$, $\mathbb{Z}(p^\infty)$, and $\mathbb{Q}_p$ are computably isomorphic to $LOGSPACE$ groups $\mathcal{A}$ with universe $Bin(\mathbb{N})$, and $\mathcal{B}$ with universe $Tal(\mathbb{N})$.*

Proof. The standard structure for $\mathbb{Z}$ is clearly logspace and can be made to have universe $Bin(\mathbb{N})$ or $Tal(\mathbb{N})$ by mapping n to $2n$ for $n \geq 0$ and mapping $-n$ to $2n + 1$ for $n > 0$. For any k, $\bigoplus_\omega \mathbb{Z}_k$ is easily seen to be in logspace by Lemma 4.

For a fixed prime number p, the group $\mathbb{Z}(p^\infty)$ consists of rational numbers of the form a/p^i where $a, i \in \mathbb{N}$, $0 \leq a < p^i$ and $i > 0$ with addition modulo 1. For our logspace model $\mathcal{G}(p^\infty)$, we let the string $e_0 e_1 \ldots e_{n-1} \in B_p(\mathbb{N})$ represent the p-adic rational

$$\frac{e_0}{p} + \frac{e_1}{p^2} + \cdots + \frac{e_{n-1}}{p^n}.$$

It can be verified that the addition operation on these strings is indeed $FLOG$ computable so that $(\mathcal{G}(p^\infty), +^G)$ is a logspace model of $\mathbb{Z}(p^\infty)$ with universe $B_p(\mathbb{N})$. Note that in $Z(p^\infty)$, the sum $x +^G y$ of two rationals either equals $x + y$ (if $x + y < 1$) or equals $x + y - 1$ (if $x + y \geq 1$), and these cases can be determined in logspace. Now Lemma 7 implies that there is a logspace model with universe $Bin(\mathbb{N})$. Furthermore, $|a \oplus b| \leqslant \max(|a|, |b|)$, so that by Lemmas 4 and 7, there is a logspace model with universe $Tal(\mathbb{N})$.

The group $\mathbb{Q}_p$ is almost the direct sum of the groups $\mathbb{Z}$ and $Z(p^\infty)$. That is, the universe of $\mathbb{Q}_p$ is the product of the universes of the two groups, but for the addition, we have to check as in the remarks above, whether the elements of $Z(p^\infty)$, viewed as rational numbers, have a sum less than 1, or not.

Now let $(\mathcal{B}_1, +_1)$ be our logspace model of $\mathbb{Z}$ and let $(\mathcal{B}_2, +_2)$ be our logspace model of $Z(p^\infty)$ and let 1^B denote the element of B_1 corresponding to the integer 1. The desired model of $\mathbb{Q}_p$ will have elements $\langle b_1, b_2 \rangle$ with $b_1 \in B_1$ and $b_2 \in B_2$. To compute $(b_1, b_2) + (c_1, c_2)$, first compute $b_1 +_1 c_1$ and $b_2 +_2 c_2$. Note from the remarks above that we can also decide in logspace whether the $b_2 +_2 c_2 = b_2 + c_2$ or equals $b_2 + c_2 - 1$. In the former case, $(b_1, b_2) + (c_1, c_2) = (b_1 +_1 c_1, b_2 +_2 c_2)$ and in the latter case, $(b_1, b_2) + (c_1, c_2) = (b_1 +_1 c_1 +_1 1, b_2 +_2 c_2)$. This construction will carry over to the models with binary and tally universes.

Theorem 3. *The additive group $\mathbb{Q}$ of rationals and also the additive groups $\mathbb{Q}$ mod 1, are computably isomorphic to LOGSPACE groups with universe $Bin(\mathbb{N})$, and to LOGSPACE groups with universe $Tal(\mathbb{N})$.*

Proof. The group $\mathbb{Q}$ mod 1 can be represented as the infinite sum of the groups $Z(p^\infty)$, for prime p, and Lemma 8 implies that there are logspace models with universe $Bin(\mathbb{N})$ and with universe $Tal(\mathbb{N})$. We will briefly explain how this sum can be obtained in a fully uniformly $LOGSPACE$ fashion. Let $\mathcal{A}_p$ be a $LOGSPACE$ group with universe $B = Tal(\mathbb{N})$ and define $\mathcal{C}_p$ to be a copy of $\mathcal{A}_p$ with the element a replaced by $\langle Tal(p), a \rangle$. Given $x = \langle Tal(n), \langle Tal(p), Tal(a) \rangle \rangle$, $x \in \mathcal{C}_{p_n}$ if and only if $p = p_n$, the nth prime. Since the set of primes is polynomial time in Binary, it is $LOGSPACE$ in Tally and therefore we can check whether $p = p_n$ in $LOGSPACE$. The second clause in the definition of uniformly $LOGSPACE$ follows from the uniformity of the proof of Theorem 2. Part (e) of Lemma 8 now gives a group with universe $Tal(\mathbb{N})$. Omitting the first component C_2 from the sequence, we get a group with universe $Tal(\mathbb{N})$ which can then by combined with a binary copy of $\mathbb{Z}(2^\infty)$ to obtain a copy of $\mathbb{Q}$ mod 1 with universe $Bin(\mathbb{N})$, by Lemma 5.

For the group $\mathbb{Q}$, we proceed as in the proof of Theorem 2. That is, the universe of $\mathbb{Q}$ is the product of the universes of models for $\mathbb{Z}$ and for $\mathbb{Q}$ mod 1 and thus by Lemma 8 may be taken to be $Bin(\mathbb{N})$ or $Tal(\mathbb{N})$ as desired. However, for the addition, we have to add the elements from $\mathbb{Q}$ mod 1 as rationals and then carry the integer part over. Now in our model of $\mathbb{Q}$ mod 1, a finite sequence of strings $\sigma^1, \ldots, \sigma^n$ where each $\sigma^i = (e_0^i, e_2^i, \ldots, e_{k_i-1}^i) \in \mathcal{B}_{p_i}^{k_i}$ represents the p_i-adic rational $\frac{e_0^i}{p_i} + \cdots + \frac{e_{k_i-1}}{p_i^{k_i}}$. To compute the sum $\sigma_1 + \ldots + \sigma_n$ requires taking a common denominator $p_1^{k_1} \cdot p_2^{k_2} \ldots p_n^{k_n}$ and using iterated multiplication and addition to obtain the numerator and finally division to obtain desired carry value c to be added to the integer sum. The results of [2, 10] imply that this can be done in logspace.

5   Conclusion and Further Research

In this paper we have begun to examine the role of logspace computability (and other notions of space complexity) in complexity theoretic model theory and algebra. We have established basic results on the standard universes $Bin(\mathbb{N})$ and $Tal(\mathbb{N})$ and on the composition of functions. Finally, we have constructed logspace models with standard universes of certain computable Abelian groups.

For future research, we will investigate space complexity of other structures, including in particular torsion-free Abelian groups and equivalence structures (see [3]). The general notion of categoricity is of particular interest. Research of Cenzer and Remmel [6, 7, 8] found very few positive categoricty results for time complexity, but we believe that space complexity holds more promise.

References

1. M. Agrawal, N. Kayhal and N. Saxena, *PRIMES is in P*, Annals of Mathematics 160 (2004), 781-793.
2. P. W. Beame, S. A. Cook and H. J. Hoover, *Log depth circuits for division and related problems*, SIAM J. Computing 15 (1986), 994-1003.
3. W. Calvert, D. Cenzer, V. Harizanov and A. Morozov, Δ_2^0 *categoricity of equivalence structures*, Ann. Pure Appl. Logic, to appear.
4. D. Cenzer and J. B. Remmel, *Polynomial-time versus recursive models*, Ann. Pure Appl. Logic 54 (1991), 17-58.
5. D. Cenzer and J.B. Remmel, *Polynomial-time Abelian groups*, Ann. Pure Appl. Logic 56 (1992), 313-363.
6. D. Cenzer and J.B. Remmel, *Feasibly categorical Abelian groups*, in Feasible Mathematics II, ed. P. Clote and J. Remmel, Prog. in Comp. Science and Appl. Logic 13, Birkhäuser (1995), 91-154.
7. D. Cenzer and J.B. Remmel, *Feasibly categorical models*, in Logic and Computational Complexity, ed. D. Leivant, Springer Lecture Notes in Computer Science 960 (1995), 300-312.
8. D. Cenzer and J.B. Remmel, *Complexity and categoricity*, Information and Computation 140 (1998), 2-25.
9. D. Cenzer and J.B. Remmel, *Complexity-theoretic model theory and algebra*, in Handbook of Recursive Mathematics (Vol. I), ed. Y. Ershov, S.S. Goncharov, A. Nerode and J.B. Remmel, Elsevier Studies in Logic and Found. Math. 138 (1998), 381-513.
10. A. Chiu, G. Davida and B. Litow, *Division in logspace-uniform NC^1*, Theor. Inform. Appl. 35 (2001), 259-275.
11. J.E. Hopcroft and J.D. Ullman, *Formal Languages and their Relation to Automata*, Addison-Wesley (1969).
12. *The Art of Computer Programming, Volume 2: Seminumerical Algorithms*, Addison-Wesley, 1998.
13. A. Nerode and J.B. Remmel, *Complexity-theoretic algebra II: Boolean algebras*, Ann. Pure Appl. Logic 44 (1989), 71-79.
14. A. Nerode and J.B. Remmel, *Polynomial time equivalence types*, in Logic and Computation, ed. W. Sieg, Contemp. Math. 106 (1990), 221-249.
15. A. Nerode and J.B. Remmel, *Polynomially isolated sets*, in Recursion Theory Week (Oberwolfach 1989), ed. K. Ambos-Spies, G.H. Muller and G.E. Sacks, Springer Lecture Notes in Math. 1432 (1990), 323-362.
16. C. H. Papadimitriou, *Computational Complexity*, Addison-Wesley (1995).
17. R. Smith, *Two theorems on autostability in p-groups*, Logic Year 1979-80 (Storrs, CT), Lecture Notes in Math. 859, Springer-Verlag, Berlin (1981), 302-311.

Prefix-Like Complexities and Computability in the Limit[*]

Alexey Chernov[1] and Jürgen Schmidhuber[1,2]

[1] IDSIA, Galleria 2, 6928 Manno, Switzerland
[2] TU Munich, Boltzmannstr. 3, 85748 Garching, München, Germany

Abstract. Computability in the limit represents the non-plus-ultra of constructive describability. It is well known that the limit computable functions on naturals are exactly those computable with the oracle for the halting problem. However, prefix (Kolmogorov) complexities defined with respect to these two models may differ. We introduce and compare several natural variations of prefix complexity definitions based on generalized Turing machines embodying the idea of limit computability, as well as complexities based on oracle machines, for both finite and infinite sequences.

Keywords: Kolmogorov complexity, limit computability, generalized Turing machine, non-halting computation.

1 Introduction

Limit computable functions are functions representable as a limit of a computable function over an extra argument. They are a well-known extension of the standard notion of computability, and appear in many contexts, e. g. [1,6,14]. It was argued that many human activities (such as program debugging) produce the final result only in the limit, and that limit computability is the non-plus-ultra of constructive describability—even more powerful models of computation cannot be classified as constructive any more, e. g. [11,13]. Several papers discuss the possibility of infinite computations in the physical world, e. g. [5,10,19].

Several authors considered variants of Kolmogorov complexity based on limit computations and computations with the halting problem oracle $0'$, e. g. [2,3,9]. Limit computable functions are exactly functions computable with the oracle $0'$ by the well-known Shoenfield limit lemma [21]. In algorithmic information theory, however, we cannot simply apply the Shoenfield lemma to replace limit computability by $0'$-computability. The reason is that the lemma is proven for functions on naturals, whereas definitions of prefix and monotone complexity require functions on sequences satisfying some kind of prefix property—see [7,12,16,17].

In the present paper, we prove equalities and inequalities for prefix complexity with the oracle $0'$ and several natural variations of prefix complexity based on

---

[*] This work was sponsored by SNF grant 200020-107590/1.

A. Beckmann et al. (Eds.): CiE 2006, LNCS 3988, pp. 85–93, 2006.

generalized Turing machines (GTMs), one of the natural models for limit computability [20] (also compare [4]). GTMs never halt and are allowed to rewrite their previous output, with the only requirement being that each output bit eventually stabilizes forever. We prove that depending on the subtleties of the definition, the corresponding complexities may differ up to a logarithmic term.

Originally, Kolmogorov [15] defined the complexity of an object as the minimal size of its description with respect to some effective specifying method (mode of description), i. e. a mapping from the set of descriptions to the set of objects. The specifying methods may be implemented on various computing devices (such as ordinary TMs, possibly non-halting TMs, possibly supplied with an oracle, etc.). All yield different complexity variants. Restricting oneself to values defined up to a bounded additive term, one can speak about complexity with respect to a certain class of machines (containing a universal one).

Even for a given machine, however, we get different complexity variants defining the machine input and the input size in different ways. Let us consider a generic machine with a single one-way infinite input tape containing only zeros and ones, reading the input tape bit by bit, and generating some output object. Researchers used (sometimes implicitly) at least three variants of "input mode":

Strong Prefix Mode. The machine has to separate explicitly the description ("a significant part of the input") from the rest of the input. More formally, the description is the initial part of the input actually read during generating the output object. The size of the description is its length; the set of possible descriptions is prefix-free: no description is a prefix of another one.

Weak Prefix Mode. The description is a finite sequence such that the machine generates the object if the input tape contains any prolongation of this sequence; the size of the description is its length. The set of descriptions is not prefix-free, but if a description is a prefix of another one, they describe the same object. Every strong prefix description is also a weak one, but the converse does not hold. In the weak prefix case, the set of the "shortest descriptions" (those that are not prolongations of other descriptions) is prefix-free, but in general this set cannot be enumerated effectively, unlike the strong prefix case.

For machines with halting computations, the weak prefix mode can be interpreted with the help of an "interactive" input model. Instead of reading the input off an input tape, the machine obtains its finite or infinite input sequence bit by bit from the user (or some physical or computational process), who decides when the next bit is provided. The result of any computation may not depend on the timing of the input bits, but depends on the input sequence only. Clearly, if the machine generates some object on input x, the machine will generate the same object on all prolongations of x (since the user may provide x at the beginning and the rest once the machine has halted). On the other hand, one may assume the following property: if the machine halts and generates some object on all prolongations of x, then the machine halts and generates the same object also on x. (It is sufficient to note that one can enumerate all such x: Consider the set of y such that the machine halts on xy, but does not halt on any prefix of xy. This set contains a prefix of any infinite prolongation of x. This set is finite, otherwise the machine does not halt on some infinite prolongation of x.) Clearly, the input sequences of a machine with this property are exactly the weak prefix descriptions.

Probabilistic Mode. In this case, the input tape is interpreted as a source of random bits, and the probability of generating some object serves to measure its complexity (complex objects are unlikely). More formally, a description is any set of infinite sequences such that the machine generates the object when the input tape contains an element of this set. The size of the description is the negative logarithm of its uniform measure. If x is a weak prefix description of size n, then the set of all prolongations of x is a probabilistic description of size n. On the other hand, for any collection of non-overlapping probabilistic descriptions there is a prefix-free set of finite sequences of the same sizes (the set of strong prefix descriptions), but in general one cannot find it effectively.

For any machine model, one may consider these three input modes and get three complexity types. The strong prefix mode complexity is the largest, the probabilistic mode complexity the smallest. In fact, two important results of algorithmic complexity theory can be interpreted as comparing these input modes for specific machine models. These results concern prefix and monotone complexity, and provide examples of machine models where the three kinds of complexity coincide and where they are different. In both cases, the standard TM is used, and the difference is in the definition of the computational result (the computed object) only.

For prefix complexity, the machine is said to generate an object if the machine prints the object and halts (thus, objects are identifiable with finite sequences). Levin's Coding Theorem [16] (see also [7]) implies that in this case all three input modes lead to the same complexity (up to an additive constant). Informally speaking, the theorem says that the probability of guessing any program for the given data is essentially equal to the probability of guessing its shortest program. Technically, the Levin theorem simplifies many proofs allowing us to switch at any moment to the most suitable of the three definitions (see [23] for an extensive discussion of the strong and weak prefix modes for prefix complexity).

For monotone complexity, the objects are finite and infinite sequences, the TM prints its output bit by bit and does not necessarily halt. We say a finite sequence is generated by the machine if this sequence appears on the output tape at some point (subsequently the machine may prolong the output); an infinite sequence is generated if all its finite prefixes appear during the computation. In this case, the probabilistic mode gives the value known as the logarithm of the a priori semimeasure on binary sequences. The weak prefix mode is used for the main definition of monotone complexity by Gács in [12] (which is referred to as type 2 monotone complexity in [17, pp. 312–313]). The strong prefix mode is used for definition of monotone complexity in the main text of [17] (where it also referred to as type 3 monotone complexity, pp. 312–313). All three values coincide up to a logarithmic additive term. Gács [12] proved that the difference between the monotone complexity (under his definition) and the logarithm of the a priori semimeasure (between the probabilistic and weak prefix modes in our terms) is unbounded on finite sequences. It is unknown whether the two monotone complexities (the weak and strong prefix modes) coincide; all known theorems hold for both.

Here the three modes are studied for finite and infinite output sequences computed on GTMs. Informally speaking, it turns out that the strong prefix-mode complexity differs from the weak prefix-mode complexity by a logarithmic term, for both finite and infinite sequences. For finite sequences, the weak prefix-mode complexity coincides with the probabilistic-mode complexity up to a constant. For infinite sequences, they coincide up to a logarithm. It remains an open question whether this bound is tight. A diagram in Sect. 6 displays the results including relations to complexities with the halting problem oracle $0'$. The rest of the paper is organized as follows: Sect. 2 contains definitions of GTM and complexities; Sect. 3 provides technical lemmas connecting GTMs and oracle machines; the main results are given in Sect. 4 for finite sequences and in Sect. 5 for infinite sequences. For the full proofs, see technical report [8].

2   Definition of GTMs and Complexities

Denote by $\mathbb{B}^*$ the space of finite sequences over the binary alphabet $\mathbb{B} = \{0, 1\}$ and by $\mathbb{B}^\infty$ the space of infinite sequences. Denote by $\ell(x)$ the length of $x \in \mathbb{B}^*$, and put $\ell(x) = \infty$ for $x \in \mathbb{B}^\infty$. For $x \in \mathbb{B}^* \cup \mathbb{B}^\infty$ and $n \in \mathbb{N}$, let x_n be the n-th bit of x (0 or 1) if $n \leq \ell(x)$ and a special symbol "blank" otherwise.

A *generalized Turing machine* (GTM) is a machine with one read-only input tape, several work tapes, and one output tape; all tapes are infinite in one direction. A GTM never halts; it reads the input tape bit by bit from left to right; it can print on the output tape in any order, i.e. can print or erase symbols in any cell many times. For a machine T and an input sequence $p \in \mathbb{B}^\infty$, denote by $T_t(p)$ the finite binary sequence[1] on the output tape at the moment t. We say that a GTM T on an input $p \in \mathbb{B}^\infty$ *converges* to $x \in \mathbb{B}^* \cup \mathbb{B}^\infty$ (write $T(p) \rightsquigarrow x$) if $\forall n \, \exists t_n \, \forall t > t_n \, [T_t(p)]_n = x_n$ (informally speaking, each bit of the output stabilizes eventually). The sequence x is called the *output* of T on p, and p is called a *program* for x.

We say that a GTM T on an input $p \in \mathbb{B}^*$ *strongly converges* to $x \in \mathbb{B}^* \cup \mathbb{B}^\infty$ (write $T(p) \rightrightarrows x$) if $T(p0^\infty) \rightsquigarrow x$ and T reads exactly p during the computation. We say that a GTM T on an input $p \in \mathbb{B}^*$ *weakly converges* to $x \in \mathbb{B}^* \cup \mathbb{B}^\infty$ (write $T(p) \rightarrowtail x$) if $T(pq) \rightsquigarrow x$ for any $q \in \mathbb{B}^\infty$. These two kinds of convergence reflect the strong and weak prefix modes. Clearly, if $T(p) \rightrightarrows x$, then $T(p) \rightarrowtail x$.

Recall that for the weak prefix mode we had two equivalent models in the case of halting computations. For non-halting computations, there are several (non-equivalent) ways of defining some analogue of the "interactive" machine

---

[1] For technical convenience, we assume that the content of the output tape is always a finite sequence of zeros and ones followed by blanks (without blanks inside). This assumption is not restrictive: for any T one can consider T' that emulates T but postpones printing a bit to the output tape if the requirement is violated; clearly, the results of converging computations are not affected.

(where the user sometimes provides a new bit). The following variant is chosen for conveniently relating GTMs to oracle machines below. We say that a GTM T on an input $p \in \mathbb{B}^*$ *uniformly weakly converges* to $x \in \mathbb{B}^* \cup \mathbb{B}^\infty$ (write $T(p) \rightarrowtail x$) if $\forall n \exists t_n \forall t > t_n \forall q \in \mathbb{B}^\infty [T_t(pq)]_n = x_n$. The last formula differs from the definition of weak convergence *only by the order of quantifiers* $(T(p) \rightarrowtail x$ iff $\forall q \in \mathbb{B}^\infty \forall n \exists t_n \forall t > t_n [T_t(pq)]_n = x_n)$. Informally speaking, in the uniform case, the moment of stabilization of a certain output bit is determined by some finite part of the input. It is easy to see that uniform weak convergence can be implemented by some kind of "interactive" machine. In the non-uniform case, however, for any initial part of the input sequence there may be a prolongation where this bit will change. One can show that strong, weak, and uniform weak convergence yield different classes of computable functions.

By the standard argument, there is a universal GTM U. For $x \subset \mathbb{B}^* \cup \mathbb{D}^\infty$, we define complexities corresponding to the strong and weak prefix modes:

$$K_{\Rightarrow}^G(x) = \min\{\ell(p) \mid U(p) \rightrightarrows x\},$$
$$K_{\rightharpoonup}^G(x) = \min\{\ell(p) \mid U(p) \rightarrowtail x\},$$
$$K_{\rightarrow}^G(x) = \min\{\ell(p) \mid U(p) \twoheadrightarrow x\}.$$

The idea of probabilistic mode is reflected by the a priori GTM-probability

$$P^G(x) = \lambda\{p \mid U(p) \rightsquigarrow x\},$$

where λ is the uniform measure on $\mathbb{B}^\infty$; we do not introduce a special sign for the corresponding complexity $-\log_2 P^G(x)$. These complexity measures are well-defined in the sense that if U is replaced by any other GTM, then $K_{\Rightarrow}^G(x)$, $K_{\rightharpoonup}^G(x)$, $K_{\rightarrow}^G(x)$, and $-\log_2 P^G(x)$ can decrease at most by a constant, the same for all x. Clearly, for $x \in \mathbb{B}^* \cup \mathbb{B}^\infty$,

$$-\log_2 P^G(x) \leq K_{\rightharpoonup}^G(x) \leq K_{\rightarrow}^G(p) \leq K_{\Rightarrow}^G(x). \tag{1}$$

As usual in complexity theory, many relations hold up to a bounded additive term, which is denoted by $\overset{+}{=}$, $\overset{+}{\leq}$ in the sequel.

The complexity $K_{\Rightarrow}^G(x)$ coincides with $K^G(x)$ originally defined in [20]. Poland [18] suggested a definition of complexity for enumerable output machines similar to $K_{\rightharpoonup}^G(x)$ in our case, and proved that for enumerable output machines, his complexity is equal to the logarithm of the a priori measure up to a constant (for GTMs such an equality was not known even with logarithmic accuracy [20]).

3   Oracle Machines and GTMs

Recall that an oracle Turing machine is a Turing machine with one additional operation: for any number n, the machine can check whether n belongs to a fixed set called an oracle (or, equivalently, the machine can get any bit of a certain infinite sequence). The oracle is not a part of the machine: the machine can work with different oracles but the result depends on the oracle used.

Denote by $0'$ the oracle for the halting problem (see [22]). By the Shoenfield limit lemma [21], $0'$-computable functions are exactly the limit computable functions. The GTM also embodies the idea of computability in the limit: it tries various answers and eventually (in the limit) gives the correct answer. But if the input is provided without delimiters, a difference arises. In the strong prefix mode case, an oracle machine can use the oracle to detect the input end (and to stop reading in time), while a GTM does not differ from an ordinary TM in this respect. In the probabilistic mode case, a GTM has an advantage since it does not halt and may use the entire (infinite) input sequence.

It turns out that uniform weak convergence for GTMs is equivalent to weak convergence for machines with the oracle for the halting problem[2].

Lemma 1. *1. For any GTM T there exists an oracle machine $\tilde{T}$ with two input tapes[3] such that: For any $p \in \mathbb{B}^*$, if $T(p) \twoheadrightarrow x$, then $\forall q \in \mathbb{B}^\infty \ \forall n \ \tilde{T}^{0'}(pq, n)$ halts and prints $x_{1:n}$.*
2. For any oracle machine T with two input tapes there exists a GTM $\tilde{T}$ with the following properties. For any $p \in \mathbb{B}^\infty$, if $T^{0'}(p, n)$ halts and prints $x_{1:n}$ for all n, then $\tilde{T}(p) \rightsquigarrow x$. If $T^{0'}(pq, n)$ halts and prints $x_{1:n}$ for some $p \in \mathbb{B}^*$, for all $q \in \mathbb{B}^\infty$ and for all n, then $\tilde{T}(p) \twoheadrightarrow x$.

Note that one cannot replace uniform weak convergence by weak convergence in the first statement of Lemma 1, because the behavior of the GTM may always depend on the *unread* part of the input, which is unacceptable for halting machines. Actually, there are functions computable on GTMs in the sense of weak convergence, but not on machines with the oracle $0'$. For example, let $f(n)$ be 0 if the n-th *oracle* machine with the oracle $0'$ halts on *all* inputs, and let $f(n)$ be undefined otherwise (compare [11]).

The next lemma relates probabilistic GTM-descriptions to $0'$-machines. For any halting machine (such as traditional prefix and monotone machines), it is easy to show that the probability of generating a certain object is enumerable from below, since the pre-image of any object is a countable union of cylinder sets (sets of all infinite sequences with a fixed prefix q), see [17]. In contrast, Example 7 in [18] shows that the set $\{p \in \mathbb{B}^\infty \mid \forall n \exists t_n \forall t > t_n \ [T_t(p)]_n = x_n\}$ may contain no cylinder set for some GTM T. Nevertheless, GTM-probabilities turn out to be $0'$-enumerable from below.

Lemma 2. *For any GTM T, the value*

$$R(x, m) = \lambda(\{p \in \mathbb{B}^\infty \mid \forall n \leq m \ \exists t_n \forall t > t_n \ [T_t(p)]_n = x_n\})$$

is $0'$-enumerable from below for any $x \in \mathbb{B}^* \cup \mathbb{B}^\infty$ and $m \in \mathbb{N}$.

---

[2] It was mentioned in [20] that the oracle complexity $K^{0'}$ equals the GTM-complexity $K^G (\overset{+}{=} K_{\rightrightarrows}^G)$, but without proof and without specifying accuracy. Surprisingly, our present refinement of the connection between oracle machines and GTMs is enough to refute (for $x \in \mathbb{B}^*$) Conjecture 5.3 from [20], namely, that $P^G(x) = O(2^{-K_{\rightrightarrows}^G(x)})$.

[3] The first tape contains the GTM input, the second tape provides the required length of the output (to deal with infinite GTM outputs).

4 Finite Sequences

The prefix complexity of $x \in \mathbb{B}^*$ is $K(x) = \min\{\ell(p) \mid V(p) = x\}$, where V is a universal prefix machine. The universal enumerable semimeasure on naturals (encoded by finite binary sequences) is $m(x) = \lambda\{pq \mid V(p) = x, p \in \mathbb{B}^*, q \in \mathbb{B}^\infty\}$ (see [17] for details). For any other enumerable semimeasure on naturals μ there is a constant c such that $m(x) \geq c\mu(x)$ for all x. By the Levin theorem [16], $K(x) \stackrel{+}{=} -\log_2 m(x)$. Relativizing w.r.t. the oracle $0'$, we get $K^{0'}(x) = \min\{\ell(p) \mid V^{0'}(p) = x\}$, the universal $0'$-enumerable semimeasure $m^{0'}(x)$, and $K^{0'}(x) \stackrel{+}{=} -\log_2 m^{0'}(x)$. The following two theorems provide a complete description of relations between GTM- and $0'$-complexities for finite sequences.

Theorem 1. *For $x \in \mathbb{B}^*$, it holds*

$$\log_2 m^{0'}(x) \stackrel{+}{=} -\log_2 P^G(x) \stackrel{\pm}{=} K_\rightleftarrows^G(x) \stackrel{\pm}{=} K_\rightarrow^G(x) \stackrel{\pm}{=} K^{0'}(x).$$

Theorem 2. *For $x \in \mathbb{B}^*$, it holds $K^{0'}(x) \stackrel{<}{\sim} K_\rightarrow^G(x) \stackrel{<}{\sim} K^{0'}(x) + K(K^{0'}(x))$. Both bounds are almost tight; namely, for some constant C and for infinitely many x it holds that $K^{0'}(x) \geq K_\rightarrow^G(x) - 2\log_2 \log_2 K^{0'}(x) - C$; and for infinitely many x it holds that $K_\rightarrow^G(x) \geq K^{0'}(x) + K(K^{0'}(x)) - 2\log_2 K(K^{0'}(x)) - C$.*

Note. In the tightness statements, $\log_2 K(K^{0'}(x))$ and $\log_2 \log_2 K^{0'}$ can be replaced by expressions with any number of logarithms.

5 Infinite Sequences

The complexity of infinite sequences can be defined with the help of halting machines having two inputs and generating the initial segment of the infinite output sequence (as in Lemma 1). It follows easily, however, that this approach will lead to the usual monotone complexity but restricted to infinite sequences.

Monotone machines are non-halting machines that print their output (finite or infinite) bit by bit (see [17] for details). Let W be a universal monotone machine. For $p \in \mathbb{B}^\infty$, by $W(p)$ denote the (complete) output of W on the input sequence p; for $p \in \mathbb{B}^*$ by $W(p)$ denote the output of W printed after reading just p and nothing else. In the book [17], the monotone complexity of $x \in \mathbb{B}^* \cup \mathbb{B}^\infty$ is defined as $Km(x) = \min\{\ell(p) \mid p \in \mathbb{B}^*, W(p) = xy, y \in \mathbb{B}^* \cup \mathbb{B}^\infty\}$ (which corresponds to the strong prefix mode in our terms). Gács [12] used another definition, $Km_I(x) = \min\{\ell(p) \mid p \in \mathbb{B}^*, \forall q \in \mathbb{B}^\infty W(pq) = xy, y \in \mathbb{B}^* \cup \mathbb{B}^\infty\}$ (corresponding to the weak prefix mode). The universal (a priori) probability of $x \in \mathbb{B}^* \cup \mathbb{B}^\infty$ is $M(x) = \lambda\{p \in \mathbb{B}^\infty \mid W(p) = xy, y \in \mathbb{B}^* \cup \mathbb{B}^\infty\}$. Gács [12] showed that $-\log_2 M(x) \stackrel{<}{\sim} Km_I(x) \stackrel{<}{\sim} Km(x) \stackrel{<}{\sim} -\log_2 M(x) + \log_2 Km(x)$ and the difference between $-\log_2 M(x)$ and $Km_I(x)$ is unbounded for $x \in \mathbb{B}^*$ (unlike the prefix complexity case). His proof does not bound the difference between $-\log_2 M(x)$ and $Km_I(x)$ for $x \in \mathbb{B}^\infty$, and the question about coincidence of $-\log_2 M(x)$, $Km_I(x)$, and $Km(x)$ for $x \in \mathbb{B}^\infty$ remains open. After relativization w.r.t. the oracle $0'$, we get $Km^{0'}$ and $M^{0'}$. Note that for $x \in \mathbb{B}^\infty$, $Km^{0'}(x)$ and $-\log_2 M^{0'}(x)$ are finite iff x is $0'$-computable.

Theorem 3. *For $x \in \mathbb{B}^\infty$, it holds $-\log_2 P^G(x) \overset{+}{=} -\log_2 M^{0'}(x)$.*

Theorem 4. *For $x \in \mathbb{B}^\infty$, it holds $K^G_\rightarrow(x) \overset{+}{=} Km^{0'}_I(x)$.*

Theorem 5. *For $x \in \mathbb{B}^\infty$, it holds $Km^{0'}(x) \overset{+}{\le} K^G_\Rightarrow(x) \overset{+}{\le} Km^{0'}(x) + K(Km^{0'}(x))$. The upper bound is almost tight: for some constant C and for infinitely many x it holds $K^G_\Rightarrow(x) \ge Km^{0'}(x) + K(Km^{0'}(x)) - 2\log_2 K(Km^{0'}(x)) - C$.*

6   Conclusion

Generalized Turing machines (GTMs) are a natural model for computability in the limit, and hence are closely related to machines with the oracle $0'$. It turns out, however, that there is no single obvious way of formally specifying a prefix complexity based on GTMs. Instead there are several closely related but slightly different ways that all seem natural. This paper introduced and studied them, exhibiting several relations between them, and also between them and $0'$-complexities, as summarized by the following diagram:

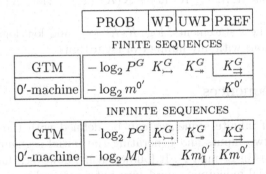

PROB	WP	UWP	PREF

FINITE SEQUENCES

GTM	$-\log_2 P^G$	$K^G_\rightharpoonup$	$K^G_\rightarrow$	$K^G_\Rightarrow$
$0'$-machine	$-\log_2 m^{0'}$			$K^{0'}$

INFINITE SEQUENCES

GTM	$-\log_2 P^G$	$K^G_\rightharpoonup$	$K^G_\rightarrow$	$K^G_\Rightarrow$
$0'$-machine	$-\log_2 M^{0'}$		$Km^{0'}_I$	$Km^{0'}$

The columns correspond to probabilistic, weak prefix, uniform weak prefix, and strong prefix descriptions, respectively. The borders between cells describe relation between the corresponding values: no border means that the values are equal up to a constant, the solid line separates values that differ by a logarithmic term, the dashed line shows that the relation is unknown.

The main **open question** is whether weak GTM-complexities ($K^G_\rightarrow$, $K^G_\rightharpoonup$, and $-\log_2 P^G$) coincide on infinite sequences. A closely related (and probably, difficult) question is if $Km^{0'}(x)$ and $Km^{0'}_I(x)$ coincide with $-\log_2 M^{0'}(x)$ for $x \in \mathbb{B}^\infty$, and if this holds for non-relativized $Km(x)$, $Km_I(x)$, and $-\log_2 M(x)$. If they do coincide, this would form a surprising contrast to the famous result of Gács [12] on the monotone complexity of finite sequences.

Acknowledgments. The authors are grateful to Marcus Hutter, Jan Poland, and Shane Legg for useful discussions and proofreading.

References

1. E. Asarin and P. Collins. Noisy Turing Machines. *Proc. of 32nd ICALP*, v. 3580 LNCS, pp. 1031–1042, Springer, 2005.
2. V. Becher and S. Figueira. Kolmogorov Complexity for Possibly Infinite Computations. *J. Logic, Language and Information*, 14(2):133–148, 2005.
3. V. Becher, S. Figueira, A. Nies, and S. Picchi. Program Size Complexity for Possibly Infinite Computations. *Notre Dame J. Formal Logic*, 46(1):51–64, 2005.
4. M. S. Burgin. Inductive Turing Machines. *Soviet Math., Doklady*, 27(3):730–734, 1983.
5. C. S. Calude and B. Pavlov. Coins, Quantum Measurements, and Turing's Barrier. *Quantum Information Processing*, 1(1-2):107–127, 2002.
6. J. Case, S. Jain, and A. Sharma. On Learning Limiting Programs. *Proc. of COLT'92*, pages 193–202, ACM Press, 1992.
7. G. J. Chaitin. A Theory of Program Size Formally Identical to Information Theory. *Journal of the ACM*, 22:329–340, 1975.
8. A. Chernov, J. Schmidhuber. Prefix-like Complexities of Finite and Infinite Sequences on Generalized Turing Machines. Technical Report IDSIA-11-05, IDSIA, Manno (Lugano), Switzerland, 2005.
9. B. Durand, A. Shen, and N. Vereshchagin. Descriptive Complexity of Computable Sequences. *Theoretical Computer Science*, 271(1-2):47–58, 2002.
10. G. Etesi and I. Nemeti. Non-Turing Computations via Malament-Hogarth Space-Times. *International Journal of Theoretical Physics*, 41:341, 2002.
11. R. V. Freivald. Functions Computable in the Limit by Probabilistic Machines. *Proc. of the 3rd Symposium on Mathematical Foundations of Computer Science*, pages 77–87, Springer, 1975.
12. P. Gács. On the Relation between Descriptional Complexity and Algorithmic Probability. *Theoretical Computer Science*, 22(1-2):71–93, 1983.
13. E. M. Gold. Limiting Recursion. *J. Symbolic Logic*, 30(1):28–46, 1965.
14. S. Hayashi and M. Nakata. Towards Limit Computable Mathematics. *Selected papers of TYPES'2000*, v. 2277 LNCS, pp. 125–144, Springer, 2002.
15. A. N. Kolmogorov. Three Approaches to the Quantitative Definition of Information. *Problems of Information Transmission*, 1(1):1–11, 1965.
16. L. A. Levin. Laws of Information (Nongrowth) and Aspects of the Foundation of Probability Theory. *Problems of Information Transmission*, 10(3):206–210, 1974.
17. M. Li and P. M. B. Vitányi. *An Introduction to Kolmogorov Complexity and its Applications (2nd edition)*. Springer, 1997.
18. J. Poland. A Coding Theorem for Enumerating Output Machines. *Information Processing Letters*, 91(4):157–161, 2004.
19. J. Schmidhuber. Algorithmic Theories of Everything. Technical Report IDSIA-20-00, quant-ph/0011122, IDSIA, Manno (Lugano), Switzerland, 2000.
20. J. Schmidhuber. Hierarchies of Generalized Kolmogorov Complexities and Nonenumerable Universal Measures Computable in the Limit. *International J. Foundations of Computer Science*, 13(4):587–612, 2002.
21. J. R. Shoenfield. On Degrees of Unsolvability. *Annals of Mathematics*, 69:644–653, 1959.
22. S. G. Simpson. Degrees of Unsolvability: A Survey of Results. In J. Barwise, editor, *Handbook of Mathematical Logic*, pp. 631–652. North-Holland, Amsterdam, 1977.
23. V. A. Uspensky, N. K. Vereshchagin, and A. Shen. Lecture Notes on Kolmogorov Complexity. unpublished, http://lpcs.math.msu.su/~ver/kolm-book.

Partial Continuous Functions and Admissible Domain Representations

Fredrik Dahlgren*

Department of mathematics, University of Uppsala,
P.O. Box 480, 751 06 Uppsala, Sweden
fredrik.dahlgren@math.uu.se

Abstract. It is well known that to be able to represent continuous functions between domain representable spaces it is critical that the domain representations
of the spaces we consider are dense. In this article we show how to develop a representation theory over a category of domains with morphisms partial continuous
functions. The raison d'être for introducing partial continuous functions is that by
passing to partial maps, we are free to consider totalities which are not dense. We
show that there is a natural subcategory of the category of representable spaces
with morphisms representable maps which is Cartesian closed. Finally, we consider the question of effectivity.

Keywords: Domain theory, domain representations, computability theory, computable analysis.

1   Introduction

One way of studying computability on uncountable spaces is through effective domain
representations. A domain representation of a space X is a domain D together with a
continuous function $\delta : D^R \to X$ onto X where D^R is some nonempty subset of D.
When D is an effective domain the computability theory on D lifts to a δ-computability
theory on the space X. If (E, E^R, ε) is a domain representation of the space Y we
say that $f : X \to Y$ is representable if there is a continuous function $\overline{f} : D \to E$
which takes δ-names of $x \in X$ to ε-names of $f(x)$ for each $x \in X$. If every continuous
function from X to Y is representable we may construct a domain representation of the
space of continuous function from X to Y over $[D \to E]$.

It thus becomes interesting to find necessary and sufficient conditions on the domain representations (D, D^R, δ) and (E, E^R, ε) to ensure that every continuous function from X to Y is representable. This problem has been studied by (among others)
Stoltenberg-Hansen, Blanck and Hamrin (c.f. [SH01], [Bla00] and [Ham05]). It turns
out that it is often important that the representation (D, D^R, δ) is dense. That is, that
the set D^R of δ-names is dense in D with respect to the Scott-topology on D. However, if (D, D^R, δ) is not dense there is no general effective construction which given
(D, D^R, δ) yields a dense and effective representation of X. This is perhaps not so
problematic as long as we are interested in building type structures over $\mathbb{R}^n$ or $\mathbb{C}^n$, but

---

* I would like to thank Professor Viggo Stoltenberg-Hansen for his support of my work.

A. Beckmann et al. (Eds.): CiE 2006, LNCS 3988, pp. 94–104, 2006.

if we would like to study computability on more complex topological spaces such as the space $C^\infty(\mathbb{R})$ of smooth functions from $\mathbb{R}$ to $\mathbb{C}$, or the space $\mathscr{D}$ of smooth functions with compact support which are considered in distribution theory, the requirement of denseness becomes a rather daunting exercise in computability theory. Indeed, it is still not known if there is an effective dense domain representation of the space $\mathscr{D}$ of smooth functions with compact support. The natural candidate for an effective domain representation of $\mathscr{D}$ is not dense, and so the standard arguments showing that every distribution is representable fail.

One way to effectively circumvent the problem of finding dense representations of the spaces under consideration is to represent continuous functions from X to Y by partial continuous functions from D to E. Here, a partial continuous function from D to E is a pair (S, f) where S is a subobject of D (in the sense of category theory) and f is a (total) continuous function from S to E.

To make sure that enough continuous functions are representable, and to be able to lift the order theoretic characterisations of continuity to partial continuous functions, we need to place some restrictions on the domain $S \subseteq D$ of a partial continuous function f from D to E. As we shall see, by a careful analysis of which properties we require of S we get a category of domains with morphisms partial continuous functions which is well suited for representation theory in general, and an effective theory of distributions in particular.

2 Preliminaries from Domain Theory

A *Scott-Ershov domain* (or simply domain) is a consistently complete algebraic cpo. Let D be a domain. Then D_c denotes the set of compact elements in D. Given $x \in D$ we write approx(x) for the set $\{a \in D_c; \ a \sqsubseteq x\}$. Since D is algebraic, approx(x) is directed and $\bigsqcup \text{approx}(x) = x$ for each $x \in D$.

D is *effective* if there is a surjective numbering[1] $\alpha : \mathbb{N} \to D_c$ of the set D_c such that $(D_c, \sqsubseteq, \sqcup, \text{cons}, \bot)$ is a computable structure with respect to α. When (D, α) and (E, β) are effective domains then $x \in D$ is α-*computable* if approx(x) is α-semidecidable, and if $f : D \to E$ is continuous then f is *effective* if the relation $\beta(n) \sqsubseteq_E f(\alpha(m))$ is r.e. We usually leave out the prefixes α, β and (α, β), if the numberings α and β are either clear from the context or not important.

Let X be a topological space. A *domain representation* of X is a triple (D, D^R, δ) where D is a domain, D^R a nonempty subset of D, and $\delta : D^R \longrightarrow X$ a continuous function from D^R onto X. We assume throughout that $\bot \notin D^R$. For $r \in D$ and $x \in X$ we write $r \prec x$ if $x \in \delta[\uparrow r \cap D^R]$. Thus, $r \prec x$ if and only if there is some $s \in D^R$ such that $r \sqsubseteq s$ and $\delta(s) = x$.

When D is effective then (D, D^R, δ) is an effective domain representation of X. Suppose (D, D^R, δ) and (E, E^R, ε) are effective domain representations of X and Y respectively. We say that $x \in X$ is δ-*computable* (or simply computable if the representation δ is clear from the context) if there is some computable $r \in D$ such that

---

[1] We will write $\alpha : \mathbb{N} \to D_c$ even though the domain of α may be a proper (decidable) subset of the natural numbers. For an introduction to the theory of numberings c.f. [Ers73, Ers75].

$\delta(r) = x$. A continuous function $f : X \to Y$ is *representable* if there is a continuous function $\overline{f} : D \to E$ such that $\overline{f}[D^R] \subseteq E^R$ and $f(\delta(r)) = \varepsilon(\overline{f}(r))$ for each $r \in D^R$. f is (δ, ε)-*computable* (or simply computable) if $\overline{f}$ is effective. A domain representation (D, D^R, δ) of a space X is *dense* if the set D^R of representing elements is dense in D with respect to the Scott-topology on D.

Definition 1. *Let (E, E^R, ε) be a domain representation of Y. (E, E^R, ε) is called admissible[2] if for each triple (D, D^R, δ) where D is a countably based domain, $D^R \subseteq D$ is dense in D, and $\delta : D^R \to Y$ is a continuous function from D^R to Y, there is a continuous function $\overline{\delta} : D \to E$ such that $\overline{\delta}[D^R] \subseteq E^R$ and $\delta(r) = \varepsilon(\overline{\delta}(r))$ for each $r \in D^R$.*

The following simple observation indicates why admissibility is interesting from a purely representation theoretic point of view.

Theorem 1. *Suppose (D, D^R, δ) is a countably based dense representation of X and (E, E^R, ε) is an admissible representation of Y. Then every sequentially continuous function from X to Y is representable.*

(For a proof of Theorem 1 see [Ham05].)

Theorem 1 can be used as a tool in constructing a representation of the space of continuous functions from X to Y over the domain $[D \to E]$ of continuous functions from D to E. However, if the representation (D, D^R, δ) is not dense Theorem 1 does not tell us anything.

3   Partial Continuous Functions

In the case when (D, D^R, δ) is a countably based and dense representation of the space X and (E, E^R, ε) is an admissible representation of Y Theorem 1 tells that every sequentially continuous function $f : X \to Y$ from X to Y lifts to a continuous function $\overline{f} : D \to E$ such that $\varepsilon(\overline{f}(x)) = f(\delta(x))$ for each $x \in D^R$. In the case when the representation (D, D^R, δ) of X is not dense there is a standard construction in domain theory which constructs a dense representation of X from (D, D^R, δ): Let

$$D_c^D = \{a \in D_c;\ a \sqsubseteq x \text{ for some } x \in D^R\}$$

and

$$D^D = \{\bigsqcup A;\ A \subseteq D_c^D \text{ is directed}\}.$$

D^D is sometimes called the *dense part* of D. It is not difficult to show that $D^D = (D^D, \sqsubseteq_D, \bot_D)$ is a domain and that $D^R \subseteq D^D$. In fact, D^R is a subspace of D^D. Thus, (D^D, D^R, δ) is a domain representation of X and D^R is dense in D^D by construction.

---

[2] This notion of an admissible representation corresponds to that of an ω-admissible representation found in [Ham05].

We may now apply Theorem 1 to show that every sequentially continuous function $f : X \to Y$ from X to Y has a continuous representation $\overline{f} : D^D \to E$ from D^D to E. However, there is no a priori reason why the relation $(\exists x \in D^R)(a \sqsubseteq x)$ on D_c should be decidable, even when (D, D^R, δ) is effective, and thus (D^D, D^R, δ) is noneffective in general.

3.1   Partial Continuous Functions with Scott-Closed Domains

A reasonable alternative to working over D^D would be to view $(D^D, \overline{f})$ as a partial function (in the sense of category theory) from D to E. Partial continuous functions in categories of domains have been studied before by (among many others) Plotkin and Fiore (c.f. [Plo85] and [Fio94]). Here a partial continuous function from D to E is a pair (U, f) where $U \subseteq D$ is a Scott-open subset of D and $f : U \to E$ is a continuous function from U to E. However, this notion of a partial continuous function is not the appropriate one for our purposes for essentially two different reasons:

  i. D^D is not an open subset of D in general. (In fact, it is easy to see that D^D is open if and only if $D^D = D$.)
  ii. Every partial continuous function (U, f) from D to E with open domain U extends to a total continuous function $f : D \to E$. Thus nothing is gained from a representation theoretic point of view when going to partial continuous functions with open domains.

Thus, as a first step we would like to distinguish a class of subobjects $\mathcal{M}$ in the category of domains which contains $D^D \hookrightarrow D$ for each representation (D, D^R, δ) and which is admissible[3] in the sense of [RR88]. (A collection of subobjects $\mathcal{M}$ from a category $\mathbb{C}$ is admissible in the sense of [RR88] if the objects in $\mathbb{C}$ together with the collection of partial morphisms (S, f) where $S \in \mathcal{M}$ form a category.) The following three observations are crucial.

Lemma 1. *Let D be a domain and let S be a nonempty subset of D. Then the closure $\overline{S}$ of S satisfies*

$$\overline{S} = \{\bigsqcup A; \ A \subseteq \, \downarrow S \text{ is directed}\}.$$

Thus in particular, D^D is closed in D since D_c^D is downwards closed. If D is a domain and $s : S \rightarrowtail D$ is a subobject of D then $s : S \rightarrowtail D$ is *Scott-closed*[4] if S is isomorphic via s to a Scott-closed subset of D. Now we have

Lemma 2. *Let D be a domain and suppose $S \subseteq D$. Then S is a Scott-closed subobject of D if and only if S is a nonempty closed subset of D.*

By Lemma 2 D^D is a Scott-closed subobject of D. Finally, we note that

---

[3] The term admissible is unfortunate here since it clashes with the notion of an admissible domain representation defined earlier. However, it is only used here to provide an inspiration for the definition of a partial continuous function below.

[4] We leave it to the interested reader to convince him or herself that this definition is independent of how we choose to represent $s : S \to D$.

Lemma 3. *The class of Scott-closed subobjects is admissible (in the sense of [RR88]) in the category of domains with morphisms strict continuous functions.*

Inspired by this we make the following definition:

Definition 2. *Let D and E be domains. A partial continuous function from D to E is a pair (S, f) where $S \subseteq D$ is a nonempty closed subset of D and f is a strict continuous function from S to E.*

We write $f : D \rightharpoonup E$ if $(\mathrm{dom}(f), f)$ is a partial continuous function from D to E and we denote by $[D \rightharpoonup E]$ the set of partial continuous functions from D to E. We will write $[D \to_{\perp} E]$ for the domain of strict continuous functions from D to E.

As a motivation for Definition 2 we give a new characterisation of admissibility in terms of partial continuous functions.

Theorem 2. *Let (E, E^R, ε) be a domain representation of Y. Then (E, E^R, ε) is admissible if and only if for every triple (D, D^R, δ) where D is a countably based domain, $D^R \subseteq D$, and $\delta : D^R \to Y$ is a continuous function from D^R to Y, there is a partial continuous function $\overline{\delta} : D \rightharpoonup E$ such that*

i. $D^R \subseteq \mathrm{dom}(\overline{\delta})$.
ii. $\overline{\delta}[D^R] \subseteq E^R$.
iii. $\delta(x) = \varepsilon(\overline{\delta}(x))$ for each $x \in D^R$.

Theorem 2 suggests representing continuous functions in the category of topological spaces using partial continuous functions on domains. We now make this idea more precise. If (D, D^R, δ) is a domain representation X and (E, E^R, ε) a domain representation of Y we say that $\overline{f} : D \rightharpoonup E$ *represents* the continuous function $f : X \to Y$ if

i. $D^R \subseteq \mathrm{dom}(\overline{f})$.
ii. $\overline{f}[D^R] \subseteq E^R$.
iii. $f(\delta(x)) = \varepsilon(\overline{f}(x))$ for each $x \in D^R$.

$f : X \to Y$ is called *representable* if there is a partial continuous function $\overline{f} : D \rightharpoonup E$ satisfying the conditions i – iii above. (If we would like to distinguish between partial representations and total representations we say that f is *partially representable* if the continuous function $\overline{f}$ representing f is a partial continuous function and f is *totally representable* if $\overline{f}$ is a total function.)

Now Theorem 2 immediately yields a version of Theorem 1 for partial continuous functions.

Theorem 3. *Let (D, D^R, δ) be a countably based domain representation of X and let (E, E^R, ε) be an admissible representation of Y. Then every sequentially continuous function $f : X \to Y$ is representable by a partial continuous function $\overline{f} : D \rightharpoonup E$.*

Just as in the case when the representation (D, D^R, δ) is dense we have the following characterisation of the sequentially continuous functions from X to Y. (For a thorough study of the dense and total case, see [Ham05].)

Theorem 4. *Let (D, D^R, δ) be a countably based admissible domain representation of X and let (E, E^R, ε) be a domain representation of Y. If $f : X \to Y$ is representable then f is sequentially continuous.*

Corollary 1. *Let (D, D^R, δ) be a countably based admissible domain representation of X and let (E, E^R, ε) be an admissible domain representation of Y. Then $f : X \to Y$ is representable if and only if f is sequentially continuous.*

3.2 The Domain of Partial Continuous Functions

It is well known that the category of domains with morphisms total continuous functions forms a Cartesian closed category (c.f. [SLG94]). This convenient circumstance is employed in representation theory to build domain representable type structures over domain representable spaces. It is only natural to expect that much of this categorical structure will be lost when going to partial continuous functions. However, as we shall see not all is lost.

As a first step we show that the Scott-closed subobjects of a domain D have a natural domain structure. Let $\operatorname{cl}(D)$ be the collection of all nonempty Scott-closed subsets of D. We order $\operatorname{cl}(D)$ by

$$S \sqsubseteq T \iff S \subseteq T.$$

Then $\sqsubseteq$ is a partial order on D with least element $\{\bot\}$. More is true however.

Theorem 5. *Let D be a domain. Then $\operatorname{cl}(D) = (\operatorname{cl}(D), \sqsubseteq, \{\bot\})$ is a domain and $S \in \operatorname{cl}(D)$ is compact if and only if $S = (\downarrow a_0) \cup (\downarrow a_1) \cup \ldots \cup (\downarrow a_n)$ for some $n \in \mathbb{N}$ and compact $a_0, a_1, \ldots a_n$ in D.*

Remark 1. Note that when D is countably based, so is $\operatorname{cl}(D)$. Furthermore, if $S = (\downarrow a_0) \cup (\downarrow a_1) \cup \ldots \cup (\downarrow a_m)$ and $T = (\downarrow b_0) \cup (\downarrow b_1) \cup \ldots \cup (\downarrow b_n)$ are compact in $\operatorname{cl}(D)$, then $S \sqsubseteq T \iff$ for each a_i there is some b_j such that $a_i \sqsubseteq_D b_j$. It follows that $\operatorname{cl}(D)$ is isomorphic to the Hoare power domain over D.

Now, let D and E be domains. To show that $[D \rightharpoonup E]$ admits a natural domain structure we define a partial order on $[D \rightharpoonup E]$ by

$$f \sqsubseteq g \iff \operatorname{dom}(f) \subseteq \operatorname{dom}(g) \text{ and } f(x) \sqsubseteq_E g(x) \text{ for each } x \in \operatorname{dom}(f).$$

$\sqsubseteq$ is a partial order on $[D \rightharpoonup E]$ with least element $\bot_{[D \rightharpoonup E]} = (\{\bot_D\}, \lambda x.\bot_E)$.

Theorem 6. *Let D and E be domains. Then $[D \rightharpoonup E] = ([D \rightharpoonup E], \sqsubseteq, \bot_{[D \rightharpoonup E]})$ is a domain and $g : D \rightharpoonup E$ is compact in $[D \rightharpoonup E]$ if and only if $\operatorname{dom}(g)$ is compact in $\operatorname{cl}(D)$ and g is compact in $[\operatorname{dom}(g) \rightharpoonup_\bot E]$.*

Remark 2. It follows immediately by the characterisation of $[D \rightharpoonup E]_c$ that $[D \rightharpoonup E]$ is countably based whenever D and E are countably based.

We now apply Theorems 3 and 6 to construct a countably based and admissible domain representation of the space of sequentially continuous functions from X to Y, given countably based and admissible representations of the spaces X and Y.

Let X and Y be topological spaces. We write $[X \to_\omega Y]$ for the space of sequentially continuous functions from X to Y. If $A \subseteq X$ and $B \subseteq Y$ we let $M(A, B) = \{f \in [X \to_\omega Y]; f[A] \subseteq B\}$. The collection of all sets $M(\{x_n\}_n \cup \{x\}, U)$ where $(x_n)_n \longrightarrow x$ in X and U is an open subset of Y form a subbasis for a topology τ_ω on $[X \to_\omega Y]$. The topology τ_ω is a natural generalisation of the topology of pointwise convergence on $[X \to_\omega Y]$, the latter being generated by the collection of all finite intersection of sets $M(\{x\}, U)$ where $x \in X$ and U is open in Y. The following lemma characterises the convergence relation on $([X \to_\omega Y], \tau_\omega)$.

Lemma 4. *Let X and Y be topological spaces. $(f_n)_n \longrightarrow f$ in $([X \to_\omega Y], \tau_\omega)$ if and only if $(f_n(x_n))_n \longrightarrow f(x)$ in Y for each convergent sequence $(x_n)_n \longrightarrow x$ in X.*

Now, suppose (D, D^R, δ) is a countably based admissible domain representation of X and (E, E^R, ε) is a countably based admissible representation of Y. By Theorem 3, every sequentially continuous function $f : X \to Y$ from X to Y is representable by a partial continuous function $\overline{f} : D \to E$ from D to E. Let $[D \to E]^R$ be the set $\{\overline{f} : D \to E; \overline{f}$ represents some sequentially continuous function $f : X \to Y\}$ and define $[\delta \to \varepsilon] : [D \to E]^R \to [X \to_\omega Y]$ by

$$[\delta \to \varepsilon](\overline{f}) = f \iff \overline{f} \text{ represents } f.$$

Then $[\delta \to \varepsilon]$ is well defined and $[\delta \to \varepsilon]$ is surjective by Theorem 3.

Theorem 7. *Let (D, D^R, δ) be a countably based admissible domain representation of X and let (E, E^R, ε) be a countably based admissible domain representation Y. Then $([D \to E], [D \to E]^R, [\delta \to \varepsilon])$ is a countably based admissible domain representation of $([X \to_\omega Y], \tau_\omega)$.*

Proof. To show that $[\delta \to \varepsilon]$ is continuous it is enough to show that $[\delta \to \varepsilon]$ is sequentially continuous since $[D \to E]$ is countably based. That $[\delta \to \varepsilon]$ is sequentially continuous follows by an application of Lemma 4.

That $([D \to E], [D \to E]^R, [\delta \to \varepsilon])$ is admissible follows since the standard representation $([D^D \to E], [D^D \to E]^R, [\delta \to \varepsilon])$ of $([X \to_\omega Y], \tau_\omega)$ is admissible. $\square$

Before we go on to study evaluation and type conversion we take note of the following fact. (For a proof, see [Ham05].)

Fact 1. *Let (D, D^R, δ) and (E, E^R, ε) be countably based and admissible domain representations of the spaces X and Y. Then $(D \times E, D^R \times E^R, \delta \times \varepsilon)$ is a countably based and admissible domain representation of $X \times Y$. Moreover, the projections $\pi_1 : X \times Y \to X$ and $\pi_2 : X \times Y \to Y$ are sequentially continuous and thus representable by Theorem 3.*

Since evaluation $(f, x) \mapsto f(x)$ is a sequentially continuous by Lemma 4, and $([D \to E], [D \to E]^R, [\delta \to \varepsilon])$ is admissible, it follows immediately by Fact 1 and Theorem 3 that

Proposition 1. *$(f, x) \mapsto f(x)$ is representable.*

Let X, Y, and Z be topological spaces. If $f : X \times Y \to Z$ is sequentially continuous then $y \mapsto f(x, y)$ is sequentially continuous for each $x \in X$. We write $f^* : X \to [Y \to_\omega Z]$ for the map $x \mapsto f(x, \cdot)$.

Proposition 2. *Let (D, D^R, δ), (E, E^R, ε), and (F, F^R, φ) be countably based and admissible domain representations of the spaces X, Y and Z and suppose $f : X \times Y \to Z$ is sequentially continuous. Then $f^* : X \to [Y \to_\omega Z]$ is representable.*

Proof. The result follows since f^* is sequentially continuous using Lemma 4.   □

We may summarise the results of this section in the following way: Let **ADM** be the category with objects ordered pairs (D, X) where D is a countably based and admissible domain representation of the space X, and morphisms $f : (D, X) \to (E, Y)$ sequentially continuous functions from X to Y which are representable by some partial continuous function $\bar{f} : D \to E$. We now have the following theorem:

Theorem 8. **ADM** *is Cartesian closed.*

3.3   Effectivity

In this section we will show that the constructions from the previous section are effective. We first consider the domain of Scott-closed subsets of an effective domain D.

Let (D, α) be an effective domain. We define $\mathrm{cl}(\alpha) : \mathbb{N} \to \mathrm{cl}(D)$ by $\mathrm{cl}(\alpha)(k) = (\downarrow a_1) \cup (\downarrow a_2) \cup \ldots \cup (\downarrow a_n) \iff k = \langle k_1, k_2, \ldots k_n \rangle$, where $k_i \in \mathrm{dom}(\alpha)$ and $\alpha(k_i) = a_i$ for each $1 \leq i \leq n$. It is clear that $\mathrm{cl}(\alpha)$ is surjective.

Theorem 9. *Let (D, α) be an effective domain. Then $(\mathrm{cl}(D), \mathrm{cl}(\alpha))$ is effective.*

Remark 3. It follows from the proof of Theorem 9 that an index for $(\mathrm{cl}(D), \mathrm{cl}(\alpha))$ can be obtained uniformly from an index for (D, α).

If S is compact in $\mathrm{cl}(D)$ we define $\alpha\!\restriction_S : \mathbb{N} \to S$ by $\mathrm{dom}(\alpha\!\restriction_S) = \{n \in \mathrm{dom}(\alpha); \alpha(n) \in S\}$ and $\alpha\!\restriction_S(n) = \alpha(n)$ for each $n \in \mathrm{dom}(\alpha\!\restriction_S)$. We write $\alpha\!\restriction_k$ for the numbering $\alpha\!\restriction_{\mathrm{cl}(\alpha)(k)} : \mathbb{N} \to \mathrm{cl}(\alpha)(k)$.

Let (D, α) and (E, β) be effective domains and let $[\alpha \to_\perp \beta] : \mathbb{N} \to [D \to_\perp E]_c$ be the standard numbering of $[D \to_\perp E]_c$. We define $[\alpha \to \beta] : \mathbb{N} \to [D \to E]_c$ by $k \in \mathrm{dom}([\alpha \to \beta])$ if and only if $k = \langle l, m \rangle$ where $l \in \mathrm{dom}(\mathrm{cl}(\alpha))$ and $m \in \mathrm{dom}([\alpha\!\restriction_l \to_\perp \beta])$ and then $[\alpha \to \beta](k) = (\mathrm{cl}(\alpha)(l), [\alpha\!\restriction_l \to_\perp \beta](m))$.

Theorem 10. *Let (D, α) and (E, β) be effective domains. Then $([D \to E], [\alpha \to \beta])$ is an effective domain.*

Remark 4. As before, it actually follows from the proof of Theorem 10 that an index for $([D \to E], [\alpha \to \beta])$ can be obtained uniformly from indices for (D, α) and (E, β).

To be able to analyse the effective content of Propositions 1 and 2 we now introduce a notion of an effective partial continuous function.

Definition 3. *Let* (D, α) *and* (E, β) *be effective domains and let* $f : D \rightharpoonup E$ *be a partial continuous function from* D *to* E. $f : D \rightharpoonup E$ *is called* (α, β)-*effective if there is an r.e. relation* $R_f \subseteq \mathbb{N} \times \mathbb{N}$ *such that if* $m \in \mathrm{dom}(\alpha)$, $n \in \mathrm{dom}(\beta)$ *and* $\alpha(m) \in \mathrm{dom}(f)$, *then*

$$R_f(m, n) \iff \beta(n) \sqsubseteq_E f(\alpha(m)).$$

An r.e. index for R_f *is called an* index *for* $f : D \rightharpoonup E$.

If the numberings α and β are clear from the context we will economise on the notation and simply say that f is effective rather than (α, β)-effective. The following elementary lemma describes some basic but important properties of effective partial continuous functions.

Lemma 5. *Let* (D, α), (E, β), *and* (F, γ) *be effective domains, and suppose* $f : D \rightharpoonup E$ *and* $g : E \rightharpoonup F$ *are effective. Then*

  i. $g \circ f : D \rightharpoonup F$ *is effective and an index for* $g \circ f$ *is obtained uniformly from indices for* f *and* g.
  ii. *If* $x \in \mathrm{dom}(f)$ *is computable in* D *then* $f(x)$ *is computable in* E.

Remark 5. If (D, α) is an effective domain then $\mathrm{id}_D : D \rightarrow D$ is effective. It follows by Lemma 5 that the class of effective domains with morphisms effective partial continuous functions form a category.

We now have two notions of computability for partial continuous functions from (D, α) to (E, β). A partial continuous function $f : D \rightharpoonup E$ from D to E may either be effective in the sense of Definition 3, or $f : D \rightharpoonup E$ may be computable as an element of $([D \rightharpoonup E], [\alpha \rightharpoonup \beta])$. The next proposition relates these two notions of computability to each other.

Proposition 3. *Let* (D, α) *and* (E, β) *be effective domains and suppose* $f : D \rightharpoonup E$ *is a partial continuous function from* D *to* E. *Then* $f : D \rightharpoonup E$ *is computable if and only if* $f : D \rightarrow E$ *is effective and* $\mathrm{dom}(f)$ *is computable.*

Let (D, D^R, δ) and (E, E^R, ε) be effective domain representations of the spaces X and Y and suppose that $f : X \rightharpoonup Y$ is a continuous function from X to Y. Then $f : X \rightharpoonup Y$ is called (δ, ε)-*effective* when there is an effective partial continuous function $\bar{f} : D \rightharpoonup E$ from D to E which represents f. A (δ, ε)-*index* for f is an index for the effective partial continuous function $\bar{f} : D \rightharpoonup E$.

It is easy to see that the identity on X is (δ, δ)-effective, and if (F, F^R, φ) is an effective domain representation of Z and $g : Y \rightharpoonup Z$ is an (ε, φ)-effective continuous function, then $g \circ f : X \rightharpoonup Z$ is (δ, φ)-effective by Lemma 5. If the representations δ, ε and φ are clear from the context we will drop the prefixes and simply say that f, g and $g \circ f$ are effective.

We note that evaluation $(f, x) \mapsto f(x)$ is effective.

Proposition 4. *Let* (D, D^R, δ) *and* (E, E^R, ε) *be effective admissible domain representations of* X *and* Y. *Then* eval $: [X \rightharpoonup_\omega Y] \times X \rightharpoonup Y$ *is effective.*

Next, we consider type conversion in the category **ADM**. We will show that for a restricted class of effectively representable sequentially continuous functions $f : X \times Y \to Z$ in **ADM**, type conversion is effective and yields a new effective sequentially continuous function $f^* : X \to [Y \to_\omega Z]$.

We begin with some definitions: Let (D, α), (E, β) and (F, γ) be effective domains and suppose $f : D \times E \rightharpoonup F$. f is called *right-computable* if f is effective and $\text{dom}(f) = S \times T$ for some Scott-closed set $S \subseteq D$ and some computable Scott-closed set $T \subseteq E$. Left-computability for a partial continuous function $f : D \times E \rightharpoonup F$ is defined analogously. By Proposition 3, if $f : D \times E \rightharpoonup F$ and $\text{dom}(f) = S \times T$ then $f : D \times E \rightharpoonup F$ is computable if and only if f is both left- and right-computable.

Proposition 5. *Let (D, D^R, δ), (E, E^R, ε), and (F, F^R, φ) be effective admissible domain representations of the spaces X, Y and Z and suppose $f : X \times Y \to Z$ is sequentially continuous. If f is effectively representable by some right-computable partial continuous function then $f^* : X \to [Y \to_\omega Z]$ is effective.*

In many cases of interest, E^D is a computable Scott-closed subset of E. It turns out that this is enough to ensure that any effectively representable sequentially continuous function $f : X \times Y \to Z$ has a right-computable representation.

References

[AJ94] ABRAMSKY, S. AND JUNG, A. 1994, Domain Theory, *Handbook for Logic in Comp. Sci. 3*, Clarendon Press (Oxford).

[Bla00] BLANCK, J. 2000, Domain representations of topological spaces, *Theor. Comp. Sci. 247*, pp. 229-255.

[BST98] BLANCK, J. STOLTENBERG-HANSEN V. AND TUCKER, J. 1998, Streams, Stream Transformers and Domain Representations, *Prospects for Hardware Foundations*, Springer LNCS vol. 1546.

[BST02] BLANCK, J. STOLTENBERG-HANSEN V. AND TUCKER, J. 2002, Domain Representations of Partial Functions, with Applications to Spatial Objects and Constructive Volume Geometry, *Theor. Comp. Sci. 28*, pp. 207-240.

[Ers73] ERSHOV, Y. 1973, Theorie der Numerierungen I, *Zeitschrift für Mathematische Logik und Grundlagen der Mathematik 19*, pp. 289-388.

[Ers75] ERSHOV, Y. 1975, Theorie der Numerierungen II, *Zeitschrift für Mathematische Logik und Grundlagen der Mathematik 21*, pp. 473-584.

[Fio94] FIORE, M. 1994, Axiomatic Domain Theory in Categories of Partial Maps, *Ph.D. Thesis, University of Edinburgh*.

[Ham05] HAMRIN, G. 2005, Admissible Domain Representations of Topological Spaces, *U.U.D.M. report 2005:16*.

[Plo85] PLOTKIN, G. 1985 *Denotational Semantics with Partial Functions*, Lecture Notes from the C.S.L.I. Summer School in 1985.

[RR88] ROBINSON, E. AND ROSOLINI, G. 1988, Categories of Partial Maps, *Inf. and Comp. 79*, pp. 95-130.

[Sch02] SCHRÖDER, M. 2002, Extended Admissibility, *Theor. Comp. Sci. 284*, pp. 519-538.

[SH01] STOLTENBERG-HANSEN, V. 2001, *Notes on Domain Theory*, Lecture notes from the summer school on Proof and System Reliability in Marktoberdorf, 2001.

[SHT88]   STOLTENBERG-HANSEN, V. AND TUCKER J. 1988, Complete Local Rings as Domains, *Journal of Symbolic Logic 53*, pp. 603-624.
[SHT95]   STOLTENBERG-HANSEN, V. AND TUCKER J. 1995, Effective Algebra, *Handbook of Logic in Comp. Sci. IV*, Oxford University Press, pp. 357-526.
[SLG94]   STOLTENBERG-HANSEN, V., LINDSTRÖM, I. AND GRIFFOR, E. 1994, *Mathematical Theory of Domains*, Cambridge Tracts in Comp. Sci..

An Invariant Cost Model for
the Lambda Calculus

Ugo Dal Lago and Simone Martini

Dipartimento di Scienze dell'Informazione, Università di Bologna
via Mura Anteo Zamboni 7, 40127 Bologna, Italy
{dallago,martini}@cs.unibo.it

Abstract. We define a new cost model for the call-by-value lambda-calculus satisfying the invariance thesis. That is, under the proposed cost model, Turing machines and the call-by-value lambda-calculus can simulate each other within a polynomial time overhead. The model only relies on combinatorial properties of usual beta-reduction, without any reference to a specific machine or evaluator. In particular, the cost of a single beta reduction is proportional to the difference between the size of the redex and the size of the reduct. In this way, the total cost of normalizing a lambda term will take into account the size of all intermediate results (as well as the number of steps to normal form).

1   Introduction

Any computer science student knows that all computational models are extensionally equivalent, each of them characterizing the same class of computable functions. However, the definition of *complexity classes* by means of computational models must take into account several differences between these models, in order to rule out unrealistic assumptions about the cost of respective computation steps. It is then usual to consider only *reasonable* models, in such a way that the definition of complexity classes remain invariant when given with reference to any such reasonable model. If polynomial time is the main concern, this reasonableness requirement take the form of the *invariance thesis* [13]:

> *Reasonable machines can simulate each other within a polynomially-bounded overhead in time and a constant-factor overhead in space.*

Once we agree that Turing machines are reasonable, then many other machines satisfy the invariance thesis. Preliminary to the proof of polynomiality of the simulation on a given machine, is the definition of a *cost model*, stipulating when and how much one should account for time and/or space during the computation. For some machines (e.g., Turing machines) this cost model is obvious; for others it is much less so. An example of the latter kind is the type-free lambda-calculus, where there is not a clear notion of *constant time* computational step, and it is even less clear how one should count for consumed space.

The idea of counting the number of beta-reductions [6] is just too naïve, because beta-reduction is inherently too complex to be considered as an atomic

A. Beckmann et al. (Eds.): CiE 2006, LNCS 3988, pp. 105–114, 2006.

operation, at least if we stick to explicit representations of lambda terms. Indeed, in a beta step

$$(\lambda x.M)N \to M\{x/N\},$$

there can be as many as $|M|$ occurrences of x inside M. As a consequence, $M\{x/N\}$ can be as big as $|M||N|$. As an example, consider the term $\underline{n}\ 2$, where $\underline{n} \equiv \lambda x.\lambda y.x^n y$ is the Church numeral for n. Under innermost reduction this term reduces to normal form in $3n - 1$ beta steps, but there is an exponential gap between this quantity and the time needed to write the normal form, that is 2^n. Under outermost reduction, however, the normal form is reached in an exponential number of beta steps. This simple example shows that taking the number of beta steps to normal form as the cost of normalization is at least problematic. Which strategy should we choose[1]? How do we account for the size of intermediate (and final) results?

Clearly, a viable option consists in defining the cost of reduction as the time needed to normalize a term by another reasonable abstract machine, e.g. a Turing machine. However, in this way we cannot compute the cost of reduction from the structure of the term, and, as a result, it is very difficult to compute the cost of normalization for particular terms or for classes of terms. Another invariant cost model is given by the actual cost of outermost (normal order) evaluation, naively implemented [9]. Despite its invariance, it is a too generous cost model (and in its essence not much different from the one that counts the numbers of steps needed to normalize a term on a Turing machine). What is needed is a machine-independent, *parsimonious*, and invariant cost model. Despite some attempts [7,9,10] (which we will discuss shortly), a cost model of this kind has not appeared yet.

To simplify things, we attack in this paper the problem for the call-by-value lambda-calculus, where we do not reduce under an abstraction and we always fully evaluate an argument before firing a beta redex. Although simple, it is a calculus of paramount importance, since it is the reduction model of any call-by-value functional programming language. For this calculus we define a new, machine-independent cost model and we prove that it satisfies the invariance thesis for time. The proposed cost model only relies on combinatorial properties of usual beta-reduction, without any reference to a specific machine or evaluator. The basic idea is to let the cost of performing a beta-reduction step depend on the size of the involved terms. In particular, the cost of $M \to N$ will be related to the *difference* $|N| - |M|$. In this way, the total cost of normalizing a lambda term will take into account the size of all intermediate results (as well as the number of steps to normal form). The last section of the paper will apply this cost model to the combinatory algebra of closed lambda-terms, to establish some results needed in [3]. We remark that in this algebra the universal function (which maps two terms M and N to the normal form of MN) adds only a *constant* overhead to the time needed to normalize MN. This result, which is almost

---

[1] Observe that we cannot take the length of the longest reduction sequence, both because in several cases this would involve too much useless work, and because for some normalizing term there is *not* a longest reduction sequence.

obvious when viewed from the perspective of lambda-calculus, is something that cannot be obtained in the realm of Turing machines.

An extended version including all proofs is available [4].

1.1 Previous Work

The two main attempts to define a parsimonious cost model share the reference to optimal lambda reduction à la Lévy [11], a parallel strategy minimizing the number of (parallel) beta steps (see [2]).

Frandsen and Sturtivant [7] propose a cost model essentially based on the number of parallel beta steps to normal form. Their aim is to propose a measure of efficiency for functional programming language implementations. They show how to simulate Turing machines in the lambda calculus with a polynomial overhead. However, the paper does not present any evidence on the existence of a polynomial simulation in the other direction. As a consequence, it is not known whether their proposal is invariant.

More interesting contributions come from the literature of the nineties on optimal lambda reduction. Lamping [8] was the first to operationally present this strategy as a graph rewriting procedure. The interest of this technique for our problem stems from the fact that a single beta step is decomposed into several elementary steps, allowing for the duplication of the argument, the computation of the levels of nesting inside abstractions, and additional bookkeeping work. Since any such elementary step is realizable on a conventional machine in constant time, Lamping's algorithm provides a theoretical basis for the study of complexity of a single beta step. Lawall and Mairson [9] give results on the efficiency of optimal reduction algorithms, highlighting the so-called bookkeeping to be the bottleneck from the point of view of complexity. A consequence of Lawall and Mairson's work is evidence on the inadequacy of the cost models proposed by Frandsen and Sturtivant and by Asperti [1], at least from the point of view of the invariance thesis. In subsequent work [10], Lawall and Mairson proposed a cost model for the lambda calculus based on Lévy's labels. They further proved that Lamping's *abstract* algorithm satisfies the proposed cost model. This, however, does not imply by itself the existence of an algorithm normalizing *any* lambda term with a polynomial overhead (on the proposed cost). Moreover, studying the dynamic behaviour of Lévy labels is clearly more difficult than dealing directly with the number of beta-reduction steps.

2 Syntax

The language we study is the pure untyped lambda-calculus endowed with weak reduction (that is, we never reduce under an abstraction) and call-by-value reduction.

Definition 1. *The following definitions are standard:*
- *Terms are defined as follows:*

$$M ::= x \mid \lambda x.M \mid MM$$

Λ *denotes the set of all lambda terms.*

- *Values are defined as follows:*

$$V ::= x \mid \lambda x.M$$

Ξ *denotes the set of all closed values.*
- *Call-by-value reduction is denoted by $\rightarrow$ and is obtained by closing the rule $(\lambda x.M)V \rightarrow M\{V/x\}$ under all applicative contexts. Here M ranges over terms, while V ranges over values.*
- *The length $|M|$ of M is the number of symbols in M.*

Following [12] we consider this system as a complete calculus and not as a mere strategy for the usual lambda-calculus. Indeed, respective sets of normal forms are different. Moreover, the relation $\rightarrow$ is not deterministic although, as we are going to see, this non-determinism is completely harmless.

The way we have defined beta-reduction implies a strong correspondence between values and closed normal forms:

Lemma 1. *Every value is a normal form and every closed normal form is a value.*

The prohibition to reduce under abstraction enforces a strong notion of confluence, the so-called one-step diamond property, which instead fails in the usual lambda-calculus.

Proposition 1 (Diamond Property). *If $M \rightarrow N$ and $M \rightarrow L$ then either $N \equiv L$ or there is P such that $N \rightarrow P$ and $L \rightarrow P$.*

As an easy corollary of Proposition 1 we get an equivalence between all normalization strategies— once again a property which does not hold in the ordinary lambda-calculus.

Corollary 1 (Strategy Equivalence). *M has a normal form iff M is strongly normalizing.*

But we can go even further: in this setting, the number of beta-steps to the normal form is invariant from the evaluation strategy:

Proposition 2. *For every term M, there are at most one normal form N and one integer n such that $M \rightarrow^n N$.*

3   An Abstract Time Measure

We can now define an abstract time measure and prove a diamond property for it. Intuitively, every beta-step will be endowed with a positive integer cost bounding the difference (in size) between the reduct and the redex.

Definition 2. • *Concatenation of $\alpha, \beta \in \mathbb{N}^*$ is simply denoted as $\alpha\beta$.*

- $\twoheadrightarrow$ *will denote a subset of* $\Lambda \times \mathbb{N}^* \times \Lambda$. *In the following, we will write* $M \xrightarrow{\alpha} N$ *standing for* $(M, \alpha, N) \in \twoheadrightarrow$. *The definition of* $\twoheadrightarrow$ *(in SOS-style) is the following:*

$$\frac{}{M \xrightarrow{\varepsilon} M} \qquad \frac{M \to N \quad n = \max\{1, |N| - |M|\}}{M \xrightarrow{(n)} N} \qquad \frac{M \xrightarrow{\alpha} N \quad N \xrightarrow{\beta} L}{M \xrightarrow{\alpha\beta} L}$$

Observe we charge $\max\{1, |N| - |M|\}$ *for every step* $M \to N$. *In this way, the cost of a beta-step will always be positive.*
- *Given* $\alpha = (n_1, \ldots, n_m) \in \mathbb{N}^*$, *define* $||\alpha|| = \sum_{i=1}^{m} n_i$.

The result of Proposition 2 can be lifted to this new notion on weighted reduction.

Proposition 3. *For every term* M, *there are at most one normal form* N *and one integer* n *such that* $M \xrightarrow{\alpha} N$ *and* $||\alpha|| = n$.

We are now ready to define the abstract time measure which is the core of the paper.

Definition 3 (Difference cost model). *If* $M \xrightarrow{\alpha} N$, *where* N *is a normal form, then* $Time(M)$ *is* $||\alpha|| + |M|$. *If* M *diverges, then* $Time(M)$ *is infinite.*

Observe that this is a good definition, in view of Proposition 3. In other words, showing $M \xrightarrow{\alpha} N$ suffices to prove $Time(M) = ||\alpha|| + |M|$. This will be particularly useful in the following section.

As an example, consider again the term $\underline{n}\ 2$ we discussed in the introduction. It reduces to normal form in one step, because we do not reduce under the abstraction. To force reduction, consider $E \equiv \underline{n}\ 2\ c$, where c is a free variable; E reduces to

$$\lambda y_n.(\lambda y_{n-1} \ldots (\lambda y_2.(\lambda y_1.c^2 y_1)^2 y_2)^2 \ldots) y_n$$

in $\Theta(n)$ beta steps. However, $Time(E) = \Theta(2^n)$, since at any step the size of the term is duplicated.

4   Simulating Turing Machines

In this and the following section we will show that the difference cost model satisfies the polynomial invariance thesis. The present section shows how to encode Turing machines into the lambda calculus.

We denote by H the term MM, where $M \equiv \lambda x.\lambda f.f(\lambda z.xxfz)$. H is a call-by-value fixed-point operator: for every N, there is α such that

$$HN \xrightarrow{\alpha} N(\lambda z.HNz)$$
$$||\alpha|| = O(|N|)$$

The lambda term H provides the necessary computational expressive power to encode the whole class of computable functions.

The simplest objects we need to encode in the lambda-calculus are finite sets. Elements of any finite set $A = \{a_1, \ldots, a_n\}$ can be encoded as follows:

$$\ulcorner a_i \urcorner^A \equiv \lambda x_1.\ldots.\lambda x_n.x_i$$

Notice that the above encoding induces a total order on A such that $a_i \leq a_j$ iff $i \leq j$.

Other useful objects are finite strings over an arbitrary alphabet, which will be encoded using a scheme attributed to Scott [14]. Let $\Sigma = \{a_1, \ldots, a_n\}$ be a finite alphabet. A string in $s \in \Sigma^*$ can be represented by a value $\ulcorner s \urcorner^{\Sigma^*}$ as follows, by induction on s:

$$\ulcorner \varepsilon \urcorner^{\Sigma^*} \equiv \lambda x_1.\ldots.\lambda x_n.\lambda y.y$$
$$\ulcorner a_i u \urcorner^{\Sigma^*} \equiv \lambda x_1.\ldots.\lambda x_n \lambda y.x_i \ulcorner u \urcorner^{\Sigma^*}$$

Observe that representations of symbols in Σ and strings in Σ^* depend on the cardinality of Σ. In other words, if $u \in \Sigma^*$ and $\Sigma \subset \Delta$, $\ulcorner u \urcorner^{\Sigma^*} \neq \ulcorner u \urcorner^{\Delta^*}$. Besides data, we want to be able to encode functions between them. In particular, the way we have defined numerals lets us concatenate two strings in linear time in the underlying lambda calculus. The encoding of a string depends on the underlying alphabet. As a consequence, we also need to be able to convert representations for strings in one alphabet to corresponding representations in a bigger alphabet. This can be done efficiently in the lambda-calculus. A deterministic Turing machine $\mathcal{M}$ is a tuple $(\Sigma, a_{blank}, Q, q_{initial}, q_{final}, \delta)$ consisting of:

- A finite alphabet $\Sigma = \{a_1, \ldots, a_n\}$;
- A distinguished symbol $a_{blank} \in \Sigma$, called the *blank symbol*;
- A finite set $Q = \{q_1, \ldots, q_m\}$ of *states*;
- A distinguished state $q_{initial} \in Q$, called the *initial state*;
- A distinguished state $q_{final} \in Q$, called the *final state*;
- A partial *transition function* $\delta : Q \times \Sigma \to Q \times \Sigma \times \{\leftarrow, \to, \downarrow\}$ such that $\delta(q_i, a_j)$ is defined iff $q_i \neq q_{final}$.

A configuration for $\mathcal{M}$ is a quadruple in $\Sigma^* \times \Sigma \times \Sigma^* \times Q$. For example, if $\delta(q_i, a_j) = (q_l, a_k, \leftarrow)$, then $\mathcal{M}$ evolves from (ua_p, a_j, v, q_i) to $(u, a_p, a_k v, q_l)$ (and from $(\varepsilon, a_j, v, q_i)$ to $(\varepsilon, a_{blank}, a_k v, q_l)$). A configuration like (u, a_i, v, q_{final}) is *final* and cannot evolve. Given a string $u \in \Sigma^*$, the *initial configuration* for u is $(\varepsilon, a, u, q_{initial})$ if $u = av$ and $(\varepsilon, a_{blank}, \varepsilon, q_{initial})$ if $u = \varepsilon$. The string corresponding to the final configuration (u, a_i, v, q_{final}) is $ua_i v$.

A Turing machine $(\Sigma, a_{blank}, Q, q_{initial}, q_{final}, \delta)$ computes the function $f : \Delta^* \to \Delta^*$ (where $\Delta \subseteq \Sigma$) in time $g : \mathbb{N} \to \mathbb{N}$ iff for every $u \in \Delta^*$, the initial configuration for u evolves to a final configuration for $f(u)$ in $g(|u|)$ steps.

A configuration (s, a, t, q) of a machine $\mathcal{M} = (\Sigma, a_{blank}, Q, q_{initial}, q_{final}, \delta)$ is represented by the term

$$\ulcorner (u, a, v, q) \urcorner^{\mathcal{M}} \equiv \lambda x.x \ulcorner u^r \urcorner^{\Sigma^*} \ulcorner a \urcorner^{\Sigma} \ulcorner v \urcorner^{\Sigma^*} \ulcorner q \urcorner^{Q}$$

We now encode a Turing machine $\mathcal{M} = (\Sigma, a_{blank}, Q, q_{initial}, q_{final}, \delta)$ in the lambda-calculus. Suppose $\Sigma = \{a_1, \ldots, a_{|\Sigma|}\}$ and $Q = \{q_1, \ldots, q_{|Q|}\}$ We proceed by building up three lambda terms:

- First of all, we need to be able to build the initial configuration for u from u itself. This can be done in linear time.
- Then, we need to extract a string from a final configuration for the string. This can be done in linear time, too.
- Most importantly, we need to be able to simulate the transition function of $\mathcal{M}$, i.e. compute a final configuration from an initial configuration (if it exists). This can be done with cost proportional to the number of steps $\mathcal{M}$ takes on the input.

At this point, we can give the main simulation result:

Theorem 1. *If* $f : \Delta^* \to \Delta^*$ *is computed by a Turing machine* $\mathcal{M}$ *in time* g, *then there is a term* $U(\mathcal{M}, \Delta)$ *such that for every* $u \in \Delta^*$ *there is* α *with* $U(\mathcal{M}, \Delta)\ulcorner u \urcorner^{\Delta^*} \overset{\alpha}{\twoheadrightarrow} \ulcorner f(u) \urcorner^{\Delta^*}$ *and* $||\alpha|| = O(g(|u|))$

Noticeably, the just described simulation induces a linear overhead: every step of $\mathcal{M}$ corresponds to a constant cost in the simulation, the constant cost not depending on the input but only on $\mathcal{M}$ itself.

5 Evaluating with Turing Machines

We informally describe a Turing machine $\mathcal{R}$ computing the normal form of a given input term, if it exists, and diverging otherwise. If M is the input term, $\mathcal{R}$ takes time $O((Time(M))^4)$.

First of all, let us observe that the usual notation for terms does not take into account the complexity of handling variables, and substitutions. We introduce a notation in the style of deBruijn [5], with binary strings representing occurrences of variables. In this way, terms can be denoted by finite strings in a finite alphabet.

Definition 4. • *The alphabet* Θ *is* $\{\lambda, @, 0, 1, \blacktriangleright\}$.
- *To each lambda term* M *we can associate a string* $M^\# \in \Theta^+$ *in the standard deBruijn way, writing* @ *for (prefix) application. For example, if* $M \equiv (\lambda x.xy)(\lambda x.\lambda y.\lambda z.x)$, *then* $M^\#$ *is* @λ@$\blacktriangleright$0$\blacktriangleright \lambda\lambda\lambda \blacktriangleright$10. *In other words, free occurrences of variables are translated into* $\blacktriangleright$, *while bounded occurrences of variables are translated into* $\blacktriangleright$ *s, where* s *is the binary representation of the deBruijn index for that occurrence.*
- *The* true length $||M||$ *of a term* M *is the length of* $M^\#$.

Observe that $||M||$ grows more than linearly on $|M|$:

Lemma 2. *For every term* M, $||M|| = O(|M|\log|M|)$. *There is a sequence* $\{M_n\}_{n\in\mathbb{N}}$ *such that* $|M_n| = \Theta(n)$, *while* $||M_n|| = \Theta(|M_n|\log|M_n|)$.

$\mathcal{R}$ has nine tapes, expects its input to be in the first tape and writes the output on the same tape. The tapes will be referred to as *Current* (the first one), *Preredex*, *Functional*, *Argument*, *Postredex*, *Reduct*, *StackTerm*, *StackRedex*, *Counter*. $\mathcal{R}$ operates by iteratively performing the following four steps:

1. First of all, $\mathcal{R}$ looks for redexes in the term stored in *Current* (call it M), by scanning it. The functional part of the redex will be put in *Functional* while its argument is copied into *Argument*. Everything appearing before (respectively, after) the redex is copied into *Preredex* (respectively, in *Postredex*). If there is no redex in M, then $\mathcal{R}$ halts. For example, consider the term $(\lambda x.\lambda y.xyy)(\lambda z.z)(\lambda w.w)$ which becomes $@@\lambda\lambda@@\blacktriangleright 1\blacktriangleright 0\blacktriangleright 0\lambda\blacktriangleright 0\lambda\blacktriangleright 0$ in deBruijn notation. Table 1 summarizes the status of some tapes after this initial step.

Table 1. The status of some tapes after step 1

Preredex	$@@$
Functional	$\lambda\lambda@@\blacktriangleright 1\blacktriangleright 0\blacktriangleright 0$
Argument	$\lambda\blacktriangleright 0$
Postredex	$\lambda\blacktriangleright 0$

2. Then, $\mathcal{R}$ copies the content of *Functional* into *Reduct*, erasing the first occurrence of λ and replacing every occurrence of the bounded variable by the content of *Argument*. In the example, *Reduct* becomes $\lambda@@\lambda\blacktriangleright 0\blacktriangleright 0\blacktriangleright 0$.
3. $\mathcal{R}$ replaces the content of *Current* with the concatenation of *Preredex*, *Reduct* and *Postredex* in this particular order. In the example, *Current* becomes $@\lambda@@\lambda\blacktriangleright 0\blacktriangleright 0\blacktriangleright 0\lambda\blacktriangleright 0$, which correctly correspond to $(\lambda y.(\lambda z.z)yy)(\lambda w.w)$.
4. Finally, the content of every tape except *Current* is erased.

Every time the sequence of steps from 1 to 4 is performed, the term M in *Current* is replaced by another term which is obtained from M by performing a normalization step. So, $\mathcal{R}$ halts on M if and only if M is normalizing and the output will be the normal form of M.

Tapes *StackTerm* and *StackRedex* are managed in the same way. They help keeping track of the structure of a term as it is scanned. The two tapes can only contain symbols A_λ, $F_@$ and $S_@$. In particular:

- The symbol A_λ stands for the argument of an abstraction;
- the symbol $F_@$ stands for the first argument of an application;
- the symbol $S_@$ stands for the second argument of an application;

StackTerm and *StackRedex* can only be modified by the usual stack operations, i.e. by pushing and popping symbols from the top of the stack. Anytime a new symbol is scanned, the underlying stack can possibly be modified:

- If $@$ is read, then $F_@$ must be pushed on the top of the stack.
- If λ is read, then A_λ must be pushed on the top of the stack.
- If $\blacktriangleright$ is read, then symbols $S_@$ and A_λ must be popped from the stack, until we find an occurrence of $F_@$ (which must be popped and replaced by $S_@$) or the stack is empty.

Now, consider an arbitrary iteration step, where M is reduced to N. We claim that the steps 1 to 4 can all be performed in $O((\|M\| + \|N\|)^2)$.

Lemma 3. *If $M \to^n N$, then $n \leq Time(M)$ and $|N| \leq Time(M)$.*

Proof. Clear from the definition of $Time(M)$.

Theorem 2. $\mathcal{R}$ *computes the normal form of the term M in $O((Time(M))^4)$ steps.*

6   Closed Values as a Partial Combinatory Algebra

If U and V are closed values and UV has a normal form W (which must be a closed value), then we will denote W by $\{U\}(V)$. In this way, we can give $\varXi$ the status of a partial applicative structure, which turns out to be a partial combinatory algebra. The abstract time measure induces a finer structure on $\varXi$, which we are going to sketch in this section. In particular, we will be able to show the existence of certain elements of $\varXi$ having both usual combinatorial properties as well as bounded behaviour. These properties are exploited in [3], where elements of $\varXi$ serves as (bounded) realizers in a semantic framework.

In the following, $Time(\{U\}(V))$ is simply $Time(UV)$ (if it exists). Moreover, $\langle V, U \rangle$ will denote the term $\lambda x.xVU$.

First of all, we observe the identity and basic operations on couples take constant time. For example, there is a term M_{swap} such that $\{M_{swap}\}(\langle V, U \rangle) = \langle U, V \rangle$ and $Time(\{M_{swap}\}(\langle V, U \rangle)) = 5$. There is a term in $\varXi$ which takes as input a pair of terms $\langle V, U \rangle$ and computes the composition of the functions computed by V and U. The overhead is constant, i.e. do not depend on the intermediate result.   We need to represent functions which go beyond the realm of linear logic. In particular, terms can be duplicated, but linear time is needed to do it: there is a term M_{cont} such that $\{M_{cont}\}(V) = \langle V, V \rangle$ and $Time(\{M_{cont}\}(V)) = O(|V|)$. From a complexity viewpoint, what is most interesting is the possibility to perform higher-order computation with constant overhead. In particular, the universal function is realized by a term M_{eval} such that $\{M_{eval}\}(\langle V, U \rangle) = \{V\}(U)$ and $Time(\{M_{eval}\}(\langle V, U \rangle)) = 4 + Time(\{U\}(V))$. The fact that a "universal" combinator with a constant cost can be defined is quite remarkable. It is a consequence of the inherent higher-order of the lambda-calculus. Indeed, this property does not hold in the context of Turing machines.

7   Conclusions

We have introduced and studied the difference cost model for the pure, untyped, call-by-value lambda-calculus. The difference cost model satisfies the invariance thesis, at least in its weak version [13]. We have given sharp complexity bounds on the simulations establishing the invariance and giving evidence that the difference cost model is a parsimonious one. We do not claim this model is the definite word on the subject. More work should be done, especially on lambda-calculi based on other evaluation models.

The availability of this cost model allows to reason on the complexity of call-by-value reduction by arguing on the structure of lambda-terms, instead of using complicated arguments on the details of some implementation mechanism. In this way, we could obtain results for eager functional programs without having to resort to, e.g., a SECD machine implementation.

We have not treated space. Indeed, the very definition of space complexity for lambda-calculus—at least in a less crude way than just "the maximum ink used [9]"—is an elusive subject which deserves better and deeper study.

References

1. Andrea Asperti. On the complexity of beta-reduction. In *Proc. 23rd ACM SIG-PLAN Symposium on Principles of Programming Languages*, pages 110–118, 1996.
2. Andrea Asperti and Stefano Guerrini. *The Optimal Implementation of Functional Programming Languages*, volume 45 of *Cambridge Tracts in Theoretical Computer Science*. Cambridge University Press, 1998.
3. Ugo Dal Lago and Martin Hofmann. Quantitative models and implicit complexity. In *Proc. Foundations of Software Technology and Theoretical Computer Science*, volume 3821 of *LNCS*, pages 189–200. Springer, 2005.
4. Ugo Dal Lago and Simone Martini. An invariant cost model for the lambda calculus. Extended Version. Available at http://arxiv.org/cs.LO/0511045, 2005.
5. Nicolaas Govert de Bruijn. Lambda calculus with nameless dummies, a tool for automatic formula manipulation, with application to the church-rosser theorem. *Indagationes Mathematicae*, 34(5):381–392, 1972.
6. Mariangiola Dezani-Ciancaglini, Simona Ronchi della Rocca, and Lorenza Saitta. Complexity of lambda-terms reductions. *RAIRO Informatique Theorique*, 13(3):257–287, 1979.
7. Gudmund Skovbjerg Frandsen and Carl Sturtivant. What is an efficient implementation of the lambda-calculus? In *Proc. 5th ACM Conference on Functional Programming Languages and Computer Architecture*, pages 289–312, 1991.
8. John Lamping. An algorithm for optimal lambda calculus reduction. In *Proc. 17th ACM SIGPLAN Symposium on Principles of Programming Languages*, pages 16–30, 1990.
9. Julia L. Lawall and Harry G. Mairson. Optimality and inefficiency: What isn't a cost model of the lambda calculus? In *Proc. 1996 ACM SIGPLAN International Conference on Functional Programming*, pages 92–101, 1996.
10. Julia L. Lawall and Harry G. Mairson. On global dynamics of optimal graph reduction. In *Proc. 1997 ACM SIGPLAN International Conference on Functional Programming*, pages 188–195, 1997.
11. Jean-Jacques Lévy. Réductions corrected et optimales dans le lambda-calcul. Université Paris 7, Thèses d'Etat, 1978.
12. Simona Ronchi Della Rocca and Luca Paolini. *The parametric lambda-calculus*. Texts in Theoretical Computer Science: An EATCS Series. Springer-Verlag, 2004.
13. Peter van Emde Boas. Machine models and simulation. In *Handbook of Theoretical Computer Science, Volume A: Algorithms and Complexity (A)*, pages 1–66. MIT Press, 1990.
14. Christopher Wadsworth. Some unusual λ-calculus numeral systems. In J.P. Seldin and J.R. Hindley, editors, *To H.B. Curry: Essays on Combinatory Logic, Lambda Calculus and Formalism*. Academic Press, 1980.

On the Complexity of the Sperner Lemma*

Stefan Dantchev

Department of Computer Science, Durham University
Science Labs, South Road, Durham, DH1 3LE, UK
s.s.dantchev@durham.ac.uk

Abstract. We present a reduction from the Pigeon-Hole Principle to the classical Sperner Lemma. The reduction is used
  1. to show that the Sperner Lemma does not have a short constant depth Frege proof, and
  2. to prove lower bounds on the Query Complexity of the Sperner Lemma in the Black-Box model of Computation.

Keywords: Propositional Proof Complexity, Constant-Depth Frege, Search Problems, Query Complexity, Sperner Lemma, Pigeon-Hole Principle.

1 Introduction

The classical Sperner Lemma, which was introduced and proven by E. Sperner in [11], is a combinatorial statement about a vertex-coloured regular triangulation of an equilateral triangle. It is of great importance in Topology since various fixed-point theorems can be easily derived from the lemma.

The Sperner Lemma is of interest in Computational Complexity, too. Indeed, it is one of the problems considered by Papadimitriou in [10] whose motivation was to classify the total search problems, i.e. computational problems whose solution is guaranteed to exists by some well known combinatorial principle. In was proven there that the Sperner Lemma belongs to one of the important complexity classes of total-function search problems, the so-called *PPAD*. A three-dimensional variant of the lemma was in fact proven to be complete for *PPAD*. Another line of research studied the Query Complexity of Search Problems in the Black-Box Model of Computation. A number of algorithms and, more importantly, lower bounds concerning the Sperner Lemma were proven in [7, 8].

As far as we are aware, though, the *Proof Complexity of the Sperner Lemma* has not been studied so far. Thus the main motivation of our work was to show a hardness result in this setting. The main theorem of the paper is that the *Sperner Lemma is hard for Constant-Depth Frege proof systems*, i.e. every such proof is exponential in the size of the propositional encoding of the lemma. We prove this result through a reduction from another well-known combinatorial principle, the Pigeon-Hole Principle. Our reduction is inspired from and very

* This research was funded by EPSRC under grant EP/C526120/1.

A. Beckmann et al. (Eds.): CiE 2006, LNCS 3988, pp. 115–124, 2006.

similar to a reduction shown by S. Buss in [5]. Our reduction can be used to reprove the optimal deterministic query lower bound from [7], and we conjecture that it can be used to achieve the randomised query lower bound from as well.

The rest of the paper is organised as follows. We first give the necessary background. We then explain the reduction from the Pigeon-Hole Principle to the Sperner Lemma, and prove that it preserves constant-depth Frege proofs, thus showing that the Sperner Lemma is hard for constant-depth Frege. Finally, we give a simple argument that proves the optimal deterministic query lower bound for the Pigeon-Hole Principle, and show that it translates via the reduction into an optimal lower bound for the Sperner Lemma.

2   Preliminaries

The (Classical) Sperner lemma is the following well-known combinatorial principle: An equilateral triangle with side length n is regularly triangulated into unit triangles. Every vertex of the triangulation is coloured in black, white or grey, and so that each colour is forbidden for a different side of the big triangle (see figure 1(a) for an example of legitimate Sperner colouring where grey is forbidden for the horizontal side, black is forbidden for the right side, and white is forbidden for the left side; note also that this enforces unique colouring of the vertices of the big triangle). The Sperner lemma asserts that, for every Sperner colouring, there exists a trichromatic unit triangle (in fact there are an odd number of such triangles).

A classical proof, which also gives an algorithm for finding a trichromatic triangle, is the so-called path-following argument, depicted on figure 1(b). Let us call "door" an edge of the triangulation that has a white and a black end, and let us assume that the doors are one-way, i.e. one can go through a door so that the white vertex remains on the right. Let us now observe that every "room" (a small triangle) can have no doors, 1 door or 2 doors. Moreover, if a room has two doors, one of them is an entrance and the other is an exit, i.e. we can only go through such a room. If a room has a single door, then it is a trichromatic triangle. Now observe that the horizontal side of the big triangle contains a sequence of alternating entrances and exits, which starts and finishes with an

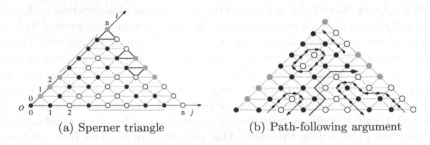

(a) Sperner triangle   (b) Path-following argument

Fig. 1. Two-dimensional Sperner lemma

entrance. This means that when one tries all entrances by walking through and then continuing as far as possible, from all of them but one, one will leave the triangle through an exit. The entrance, which is not matched by any exit, though, will lead to a room that can only be entered, i.e. a trichromatic triangle (see figure figure 1(b)) .

We shall be interested in formalising this or any other proof in a certain propositional proof system. Thus we first need to encode the Sperner lemma in propositional logic. Let us assume a numbering of the rows and the (non-vertical) columns of the triangulation as shown on figure 1(a), and let us introduce propositional variables s_{ij}^p for $0 \leq i, j \leq n$, $i + j \leq n$, $p \in \{1, 2, 3\}$ to mean "the point in row i and column j has colour p (assuming 1 is black, 2 is white, and 3 is grey)". It is not hard to see that the following set of clause, which we shall refer to as $Sperner_n$, correctly encodes the negation of the Sperner lemma,

$$s_{ij}^1 \vee s_{ij}^2 \vee s_{ij}^3 \quad 0 \leq i, j < n, \, i + j \leq n$$
$$\neg s_{ij}^p \vee \neg s_{ij}^q \quad 0 \leq i, j < n, \, i + j \leq n, \, 1 \leq p < q \leq 3$$
$$\neg s_{0j}^3 \quad 0 \leq j \leq n$$
$$\neg s_{i0}^2 \quad 0 \leq i \leq n$$
$$\neg s_{i\,n-i}^1 \quad 0 \leq i \leq n$$
$$\neg s_{ij}^p \vee \neg s_{i\,j+1}^q \vee \neg s_{i+1\,j}^r \quad 0 \leq i, j < n, \, 0 \leq i + j < n, \, \{p, q, r\} = \{1, 2, 3\}$$
$$\neg s_{ij}^p \vee \neg s_{i\,j-1}^q \vee \neg s_{i-1\,j}^r \quad 0 < i, j \leq n, \, 0 < i + j \leq n, \, \{p, q, r\} = \{1, 2, 3\}.$$

The first two lines say that every vertex is coloured in exactly one colour, the next three lines impose the restrictions on the sides of the big triangle, and the las two lines claim that no small triangle is trichromatic.

The (Bijective) Pigeon-Hole Principle. Simply says that there is no bijection between a (finite) set of $n + 1$ pigeons and a (finite) set of n holes. Its negation, which we call PHP_n^{n+1}, can be encoded as the following set of clauses:

$$\bigvee_{j=1}^{n} p_{ij} \quad 1 \leq i \leq n + 1$$
$$\neg p_{ij} \vee \neg p_{ik} \quad 1 \leq i \leq n + 1, \, 1 \leq j < k \leq n$$
$$\bigvee_{i=1}^{n+1} p_{ij} \quad 1 \leq j \leq n$$
$$\neg p_{ij} \vee \neg p_{kj} \quad 1 \leq j \leq n, \, 1 \leq i < k \leq n + 1$$

(obviously p_{ij} stands for "pigeon i goes to hole j").

Frege and Constant-Depth Frege Propositional Proof Systems. Frege proof systems are the usual "text-book", Hilbert-style, proof systems for propositional logic based on modus ponens [6]. Proof lines of a Frege proof are propositional formulae built upon finite set of variables as well as the logical connectives $\neg$,

$\vee$, $\wedge$ and $\rightarrow$. There are a finite number of axiom schemata (propositional tautologies, such as $\varphi \rightarrow \varphi \vee \psi$, where φ and ψ could be arbitrary propositional formulae), and w.l.o.g. there is a single derivation rule, the modus ponens:

$$\frac{\varphi \quad \varphi \rightarrow \psi}{\psi}.$$

It is well known that Frege systems are sound and complete.

The propositional proof systems, which we consider in the paper, is a special restricted variant of Frege systems called *Constant-Depth Frege* (or *Bounded-Depth Frege*) and denoted further by $cd\mathcal{F}$. We restrict the proof lines to formulae with *negation on variables only*, and with a *constant number of alternation of the connectives* $\vee$ and $\wedge$ ($\rightarrow$ is not permitted).

Additionally, we consider *refutations rather than proofs*. That is, instead of the formula we want to prove, we take its negation, usually encoded as a set (conjunction) of simpler formulae (usually clauses). We then use these formulae as axioms in a Frege (or $cd\mathcal{F}$) proof whose goal is now to derive the empty formula (constant $false$), thus refuting the negation of the original formula.

A well-known result from Propositional Proof Complexity, which we shall use, is that *the Pigeon-Hole Principle is hard for constant-depth Frege*, i.e. every proof of PHP_n^{n+1} in depth d Frege is of size $2^{n^{1/2\Omega(d)}}$: in a great breakthrough, Ajtaj first proved a super-polynomial lower bound in [1], which was later improved in [3] to the exponential lower bound we have just mentioned.

Search Problems and Black-Box Complexity. are two concepts that arise in connection with propositional contradictions [9]. The Search Problem for an unsatisfiable set of clauses is as follows: given an assignment, find a clause which is violated by the assignment. In a different context, Search problems were first defined and studied by Papadimitriou [10] whose motivation was classifying computational problems that are guaranteed to have a solution (as opposed to decision problems commonly studied in Structural Complexity). Since then, a number of reductions and separations have been obtained among different (classes of) search problems [2, 4]. As the latter paper shows, there is a strong link between Search problems and Propositional Proof Complexity. In the present paper, though, we are interested in the Query Complexity of the classical Sperner Lemma in the so-called Black-Box Model of Computation: given the propositional encoding and an assignment of the variables, what is the minimal number of variables that an algorithm has to query in order to find a contradiction, i.e. a clause which is falsified under the assignment? The query complexity of the Sperner Lemma was previously studied in [7] and [8].

3　Complexity Results

We first show a

Reduction from the Pigeon-Hole Principle to the Sperner Lemma. It is depicted on figure 2(a). The pigeons, $P_1, P_2 \ldots P_n, P_{n+1}$ and the holes $H_1, H_2 \ldots H_n$ are

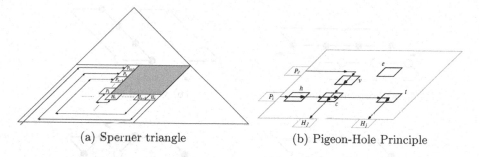

(a) Sperner triangle   (b) Pigeon-Hole Principle

Fig. 2. Reduction from PHP_n^{n+1} to $Sperner_{12n+6}$

represented by small 3×3 rhombs. The arrow represent paths, i.e. they cross a sequence of edges whose left end is black and whose right end is white (here left and right are relative to the direction of the arrow). There is a single "door" on the horizontal side of the big triangle (this means that the horizontal side consist of a sequence of (few) black vertices followed by a (long) sequence of white vertices). The walk through the door leads to the $n + 1$st pigeon. There are long paths starting from a hole i and ending into a pigeon i for $1 \le i \le n$. There are no paths inside the two triangles, the topmost and the rightmost ones (this means that they contain mostly grey vertices).

The essence of the reduction, though, is concentrated into the rhomb in the middle (filled in grey on figure 2(a)), which is enlarged on figure 2(b). If pigeon i goes into hole j in the Pigeon-Hole formula, there is a path in the Sperner formula starting from left side of the rhomb on row i, going horizontally as far as column j, then turning left and thus going down straight to the bottom side of the rhomb (in column j). We could imagine that a path is represented by means of several different types of 3×3 rhombs that we shall call *tiles*. Clearly, we need horizontal and vertical tiles denoted by h and v, respectively, as well as an empty tile e and a turn tile t (see figure 2(b)). The last type of tile, which we need, is a "cross" tile c: if there are pigeons $i < i'$ and holes $j > j'$ such that pigeon i goes to hole j while pigeon i' goes to hole j', the two corresponding paths cross each other at position (i, j') - the intersection of row j and column j'. Note that the tile of type c does not actually make two paths to cross but to switch over; however only the positions of incoming and outgoing edges are important, so we might as well think that the paths cross each other. The implementation of these tiles by 3×3 rhombs is depicted on figure 3 (the empty tile e, which is not shown there, has all vertices coloured in grey).

It is obvious that a (fictional) violation of the Pigeon-Hole Principle (i.e. a satisfying assignment of the clauses PHP_n^{n+1} translates via our reduction into a violation of the Sperner Lemma (i.e. a satisfying assignment of the clauses $Sperner_{12n+6}$). We are now ready to explain why

The Sperner Lemma is Hard for Constant-Depth Frege. We shall show how to transform a Constant-Depth Frege (*cdF*) refutation of $Sperner_{12n+6}$ of size S

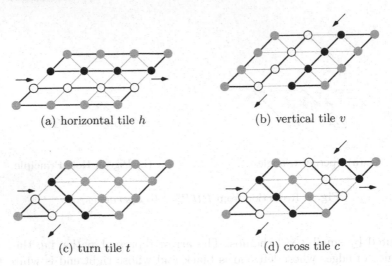

(a) horizontal tile h (b) vertical tile v

(c) turn tile t (d) cross tile c

Fig. 3. Different kinds of tiles

into a $cd\mathcal{F}$ refutation of PHP_n^{n+1} of size at most $\text{poly}\,(n, S)$. The fact that PHP_n^{n+1} is hard for Constant-Depth Frege would then give the desired result.

Given a $cd\mathcal{F}$ refutation of $Sperner_{12n+6}$, we need to show

1. how to replace each Sperner variable by a small-depth and small-size formula over the Pigeon-Hole variables, and
2. that this substitution preserves the correctness of the refutation.

First of all, going back to the smaller picture of the reduction (figure 2(b)), we note that the presence of a particular kind of tile on a particular position can be expressed as a formula over the Pigeon-Hole variables as follows.

$$t_{i\,j} \equiv p_{i\,j}$$

$$e_{i\,j} \equiv \bigwedge_{l=j}^{n} \neg p_{i\,l} \wedge \bigwedge_{k=i}^{n+1} \neg p_{k\,j}$$

$$h_{i\,j} \equiv \bigwedge_{l=1}^{j} \neg p_{i\,l} \wedge \bigwedge_{k=i}^{n+1} \neg p_{k\,j}$$

$$v_{i\,j} \equiv \bigwedge_{l=j}^{n} \neg p_{i\,l} \wedge \bigwedge_{k=1}^{i} \neg p_{k\,j}$$

$$c_{i\,j} \equiv \bigwedge_{l=1}^{j} \neg p_{i\,l} \wedge \bigwedge_{k=1}^{i} \neg p_{k\,j}.$$

Indeed, the case of a turn tile $t_{i\,j}$ is the easiest one: if such a tile is present, the pigeon i has to go hole j. An empty tile on position (i, j) prevents pigeon

i from going into the jth hole as well as any hole which is to the right, i.e. all holes whose number is greater than or equal to j; similarly, hole j cannot be occupied by any pigeon whose number is greater than or equal to i. The other three cases, those of tile types h, v and c, can be easily verified using the same way of reasoning.

We can now explain how the Sperner variables can be substituted by formulae over the Pigeon-Hole variables. Looking back at the bigger picture of the reduction (figure 2(a)), we note that all Sperner variables that are outside the grey rhomb in the middle are to be substituted by constants $true$ or $false$. As far as the other variables are concerned, we first observe that the tile (3×3 rhomb) at position (i, j) corresponding to the ith pigeon and the jth hole in PHP_n^{n+1} has its bottom left corner with coordinates $(3n - 3 + 3i, 3n + 3j)$. Every variable $s_{k\,l}^p$ within a tile (i, j) can be expressed in terms of variables $t_{i\,j}$, $e_{i\,j}$, $h_{i\,j}$, $v_{i\,j}$ and $c_{i\,j}$. We shall give several typical cases as examples:

1. Looking at figure 3, we easily observe that the corners of the tiles are always grey, i.e.

$$s_{3n-3+3i\ 3n+3j}^3 \equiv true$$
$$s_{3n-3i\ 3n+3j}^1 \equiv false$$
$$s_{3n-3i\ 3n+3j}^2 \equiv false$$

   for $1 \leq i \leq n + 2$, $1 \leq j \leq n + 1$.
2. If the point $(2, 1)$ within a tile (assuming the bottom left corner is $(0, 0)$) is coloured in black, then the tile has to be a horizontal tile, i.e.

$$s_{3n-1+3i\ 3n+1+3j}^1 \equiv v_{i\,j}$$

   for $1 \leq i \leq n + 1$, $1 \leq j \leq n$.
3. If the point $(1, 1)$ within a tile is coloured in white, then the tile has to be of type either v or h, i.e.

$$s_{3n-2+3i\ 3n+1+3j}^2 \equiv v_{i\,j} \vee h_{i\,j}$$

   for $1 \leq i \leq n + 1$, $1 \leq j \leq n$.
4. If the point $(2, 2)$ within a tile is coloured in grey, then the tile may be of type e, t or c only, i.e.

$$s_{3n-2+3i\ 3n+1+3j}^3 \equiv e_{i\,j} \vee t_{i\,j} \vee c_{i\,j}$$

   for $1 \leq i \leq n + 1$, $1 \leq j \leq n$.

All the remaining substitutions can be easily resolved in the same manner, and thus all Sperner variables can be equivalently expressed in terms of Pigeon-Hole variables (recall that the variables $t_{i\,j}$, $e_{i\,j}$, $h_{i\,j}$, $v_{i\,j}$ and $c_{i\,j}$ are just short-hand notation for certain formulae over the Pigeon-Hole variables $p_{i\,j}$).

The final step is to show that the new refutation obtained in this way is a correct $cd\mathcal{F}$ refutation, and that it is not much bigger than the original one. All axiom schemata as well as all derivation rules in Frege systems are closed under substitution. Thus, the only potential problems are the clauses of $Sperner_n$. It is straightforward, though rather tedious, task to verify that these clauses have been transformed by the substitution into formulae that have trivial $cd\mathcal{F}$ proofs from the PHP_n^{n+1} clauses. Thus, the new refutation is indeed a correct $cd\mathcal{F}$ refutation of the negation of the Pigeon-Hole principle. Moreover, every variable of the original refutation have been substituted by a formula of at most $O(n)$ Pigeon-Hole variables. Thus, the size of the new refutation is at most nS where S is the size of the original refutation. This completes the proof that our reduction preserves $cd\mathcal{F}$ refutations, and, together with the well-known result that every $cd\mathcal{F}$ refutation of PHP_n^{n+1} is of size exponential in n, gives the main result of the paper.

Proposition 1. *Every $cd\mathcal{F}$ refutation of $Sperner_n$ is of size exponential in n.*

Finally, we turn our attention to

The Query Complexity of the Sperner Lemma. We shall reprove the optimal deterministic query complexity lower bound from [7], and conjecture that the same randomised lower bound holds.

Our proof is via the reduction from the Pigeon-Hole Principle: it is clear that an $\Omega(q(n))$ lower bound for PHP_n^{n+1} translates into an $\Omega(q(n)/n)$ lower bound for $Sperner_n$ (as each Sperner variable is represented by a formula of at most $O((n))$ Pigeon-Hole variables).

We first prove an optimal Pigeon-Hole lower bound.

Lemma 1. *Every deterministic algorithm that solves the search problem for PHP_n^{n+1} has to make at least $n^2/8$ queries.*

Proof. We shall present an adversary argument. The idea behind it is that the adversary keeps pigeons/holes *free*, i.e. unassigned to a hole/pigeon, as long as possible, and can survive as long as there are "few" *busy*, i.e. committed to a hole/pigeon, pigeons/holes.

The adversary's strategy is simple: for each pigeon or hole, the adversary answers "no" to the first $n/2$ different queries regarding this pigeon/hole and a free hole/pigeon. After the last such query the adversary makes the pigeon/hole in question busy by assigning it to a free hole/pigeon. We claim that this is always possible provided that the number of the busy pigeons and holes altogether is smaller than $n/2$. Indeed, assume w.l.o.g. that the item in question is a pigeon. There are exactly $n/2$ holes forbidden for that pigeon by the negative answers plus fewer that $n/2$ holes forbidden because they have already been made busy. Thus there is at least one hole that can be assigned to the pigeon in question.

To conclude the argument, assume the adversary gives up when the number of busy items, pigeons and holes, reaches $n/2$. The algorithm has made at least $n/2$ queries for each of these items, and in the worst case every "final" query (i.e. the one that makes an item busy) makes two items busy at the same time - both the pigeon and the hole mentioned by the query, so that at this stage, at least $n^2/8$ queries have been made in total. $\square$

Combined with the initial observation, this results gives the following corollary.

Corollary 1. *Every deterministic algorithm that solves the search problem for* $Sperner_n$ *has to make* $\Omega(n)$ *queries.*

Under the plausible conjecture that the randomised query complexity of Pigeon-Hole contradiction can be lower bounded by $\Omega(n^2)$, we can conjecture that the randomised and the deterministic query complexities of the Sperner Lemma are the same.

Conjecture 1. Every randomised algorithm that solves the search problem for $Sperner_n$ has to make $\Omega(n)$ queries.

As far as the quantum query complexity is concerned, we do not believe that a lower bound can be achieved through our reduction, the reason being that it is very likely that the quantum query complexity of the Pigeon-Hole Principle is as low as $O(n)$ (it is believable that it is even $o(n)$).

4   Conclusion

We have shown a reduction from the Pigeon-Hole Principle to the classical Sperner Lemma. The reduction proves that the Sperner Lemma is hard for Constant-Depth Frege. It also gives an optimal deterministic query lower bound for the corresponding search problem. We have conjectured that an optimal randomised query lower bound can be proven in the same way: the open question, which we would need to resolve, is to prove a *quadratic randomised query lower bound for the Pigeon-Hole* contradiction.

References

1. M. Ajtai. The complexity of the pigeonhole principle. *Combinatorica*, 14(4):417–433, 1994. A preleminary version appeared at the 29th FOCS in 1998.
2. P. Beame, S. Cook, J. Edmonds, R. Impagliazzo, and T. Pitassi. The relative complexity of $\mathcal{NP}$ search problems. *J. Comput. System Sci.*, 57(1):3 – 19, 1998.
3. P. Beame, R. Impagliazzo, J. Krajicek, T. Pitassi, P. Pudlak, and A. Woods. Exponential lower bounds for the pigeonhole principle. In *Proceedings of the 24th Annual ACM Symposium on Theory Of Computing*, pages 200–220. ACM, May 1992.
4. J. Buresh-Oppenheim and T. Morioka. Relativized $\mathcal{NP}$ search problems and propositional proof systems. In *The 19th Conference on Computational Complexity*, pages 54 – 67, 2004.
5. S. R. Buss. Polynomial-size frege and resolution proofs of st-connectivity and hex tautologies. 2003.
6. S. Cook and R. Reckhow. The relative efficiency of propositional proof systems. *Journal of Symbolic Logic*, 44(1):36–50, March 1979.
7. P. Crescenzi and R. Silvestri. Sperner's lemma and robust machines. *Comput. Complexity*, 7(2):163 – 173, 1998.

8. K. Firedhl, G. Ivanyos, M. Santha, and Y.F. Verhoeven. On the black-box complexity of sperner's lemma. In *The 14th Internat. Symp. on Fund. of Comput. Theory, FCT'05*, pages 245 – 257, 2005.
9. J. Krajíček. *Bounded Arithmetic, Propositional Logic, and Complexity Theory*. Cambridge University Press, 1995.
10. C. Papadimitriou. On the complexity of the parity argument and other inefficient proofs of existence. *J. Comput. System Sci.*, 48(3):498 – 532, 1994.
11. E. Sperner. Neuer Beweis für die Invarianz der Dimensionzahl und des Gebietes. In *Abh. Math. Sem. Hamburg Univ.*, volume 6, pages 265 – 272. 1928.

The Church-Turing Thesis
Consensus and Opposition

Martin Davis

Mathematics Dept., University of California, Berkeley, CA 94720, USA
martin@eipye.com

Many years ago, I wrote [7]:

> It is truly remarkable (Gödel ... speaks of a kind of miracle) that it has
> proved possible to give a precise mathematical characterization of the
> class of processes that can be carried out by purely mechanical means.
> It is in fact the possibility of such a characterization that underlies the
> ubiquitous applicability of digital computers. In addition it has made it
> possible to prove the algorithmic unsolvability of important problems,
> has provided a key tool in mathematical logic, has made available an
> array of fundamental models in theoretical computer science, and has
> been the basis of a rich new branch of mathemtics.

A few years later I wrote [8]:

> The subject ... is Alan Turing's discovery of the universal (or all-purpose)
> digital computer as a mathematical abstraction. ... We will try to show
> how this very abstract work helped to lead Turing and John von Neu-
> mann to the modern concept of the electronic computer.

In the 1980s when those words were written, the notion that the work by
the logicians Church, Post, and Turing had a significant relationship with the
coming of the modern computer was by no means generally accepted. What was
innovative about the novel vacuum tube computers being built in the late 1940s
was still generally thought to be captured in the phrase "the stored program
concept". Much easier to think of this revolutionary paradigm shift in terms
of the use of a piece of hardware than to credit Turing's abstract pencil-and-
paper "universal" machines as playing the key role. However by the late 1990s
a consensus had developed that the Church-Turing Thesis is indeed the basis
of modern computing practice. Statements could be found characterizing it as a
natural law, without proper care to distinguish between the infinitary nature of
the work of the logicians and the necessarily finite character of physical computers. Even the weekly news magazine *Time* in its celebration of the outstanding
thinkers of the twentieth century proclaimed in their March 29, 1999 issue:

> ... the fact remains that everyone who taps at a keyboard, opening a
> spreadsheet or a word-processing program, is working on an incarnation
> of a Turing machine...

A. Beckmann et al. (Eds.): CiE 2006, LNCS 3988, pp. 125–132, 2006.

Virtually all computers today from $10 million supercomputers to the tiny chips that power cell phones and Furbies, have one thing in common: they are all "von Neumann machines," variations on the basic computer architecture that John von Neumann, building on the work of Alan Turing, laid out in the 1940s.

Despite all of this, computer scientists have had to struggle with the all-too-evident fact that from a practical point of view, Turing computability does not suffice. Von Neumann's awareness from the very beginning of not only the significance of Turing universality but also of the crucial need for attention in computer design to limitations of space and time comes out clearly in the report [3]:

> It is easy to see by formal-logical methods that there exist codes that are *in abstracto* adequate to control and cause the execution of any sequence of operations which are individually available in the machine and which are, in their entirety, conceivable by the problem planner. The really decisive considerations from the present point of view, in selecting a code, are of a more practical nature: simplicity of the equipment demanded by the code, and the clarity of its application to the actually important problems together with the speed of its handling those problems.

Steve Cook's ground-breaking work of 1971 establishing the NP-completeness of the satisfiability problem, and the independent discovery of the same phenomenon by Leonid Levin, opened a Pandora's box of NP-complete problems for which no generally feasible algorithms are known, and for which, it is believed, none exist. With these problems Turing computability doesn't help because, in each case, the number of steps required by the best algorithms available grows exponentially with the length of the input, making their use in practice problematical. How strange that despite this clear evidence that computbility alone does not suffice for practical purposes, a movement has developed under the banner of "hypercomputation" proposing the practicality of computing the non-computable. In a related direction, it has been proposed that in our very skulls resides the ability to transcend the computable. It is in this context, that this talk will survey the history of and evidence for Turing computability as a theoretical and practical upper limit to what can be computed.

1   The Birth of Computability Theory

This is a fascinating story of how various researchers approaching from different directions all arrived at the same destination. Emil Post working in isolation and battling his bipolar demons arrived at his notion of normal set already in the 1920s. After Alonzo Church's ambitious logical system was proved inconsistent by his students Kleene and Rosser, he saw how to extract from it a consistent subsystem, the λ-calculus. From this, Church and Kleene arrived at their notion of λ-definability, which Church daringly proposed as a precise equivalent of the intuitive notion of calculability. Kurt Gödel in lectures 1n 1934 suggested that this same intuitive notion would be captured by permitting functions to be specified by recursive definitions of the most general sort and even suggested one way this could be realized. Finally, Alan Turing in England, knowing none of this, came up with his own formulation of computability in terms of abstract machines limited to the most elemental operations but permitted unlimited space. Remarkably all these notions turned out to be equivalent.[1]

2   Computability and Computers

I quote from what I have written elsewhere [9]:

> ...there is no doubt that, from the beginning the logicians developing the theoretical foundations of computing were thinking also in terms of physical mechanism. Thus, as early as 1937, Alonzo Church reviewing Turing's classic paper wrote [4]:
>
>> [Turing] proposes as a criterion that an infinite sequence of digits 0 and 1 be 'computable' that it shall be possible to devise a computing machine, occupying a finite space and with working parts of finite size, which will write down the sequence to any desired number of terms if allowed to run for a sufficiently long time. As a matter of convenience, certain further restrictions are imposed on the character of the machine, but these are of such a nature as obviously to cause no loss of generality ...
>
> Turing himself speaking to the London Mathematical Society in 1947 said [24]:
>
>> Some years ago I was researching what now may be described as an investigation of the theoretical possibilities and limitations of digital computing machines. I considered a type of machine which had a central mechanism, and an infinite memory which was contained on an infinite tape. This type of machine seemed to be sufficiently general. One of my conclusions was that the idea of 'rule of thumb' process and 'machine process' were synonymous.
>
> Referring to the machine he had designed for the British National Physics Laboratory, Turing went on to say:
>
>> Machines such as the ACE (Automatic Computing Engine) may be regarded as practical versions of this same type of machine.

Of course one should not forget that the infinite memory of Turing's model can not be realized in the physical world we inhabit. It is certainly impressive to observe the enormous increase in storage capability in readily available computers over the years making more and more of the promise of universality enjoyed by Turing's abstract devices available to all of us. But nevertheless it all remains

---

[1] I tell the story in some detail in my [7] and provide additional references. My account is not entirely fair to Church; for a better account of his contribution see [21].

finite. Elsewhere I've emphasized the paradigm shift in our understanding of computation already implicit in Turing's theoretical work [11, 12]:

> Before Turing the ... supposition was that ... the three categories, ma-
> chine, program, and data, were entirely separate entities. The machine
> was a physical object ... hardware. The program was the plan for doing
> a computation ... The data was the numerical input. Turing's universal
> machine showed that the distinctness of these three categories is an illu-
> sion. A Turing machine is initially envisioned as a machine ..., *hardware*.
> But its code ... functions as a *program*, detailing the instructions to the
> universal machine ... Finally, the universal machine in its step-by-step
> actions sees the ... machine code as just more *data* to be worked on.
> This fluidity ... is fundamental to contemporary computer practice. A
> *program* ... is *data* to the ... compiler.

One can see this interplay manifested in the recent quite non-theoretical book [14], for example in pp. 6–11.

3   Trial and Error Computability as Hypercomputation

Consider a computation which produces output from time to time and which is guaranteed to *eventually* produce the correct desired output, but with no bound on the time required for this to occur. However once the correct output has been produced any subsequent output will simply repeat this correct result. Someone who wishes to know the correct answer would have no way to know at any given time whether the latest output is the correct output. This situation was analyzed by E.M. Gold and by Hilary Putnam [15, 19]. If the computation is to determine whether or not a natural number n as input belongs to some set S, then it turns out that sets for which such "trial and error" computation is available are exactly those in the Δ_2^0 class in the arithmetic hierarchy. These are exactly the sets that are computable relative to a $0'$ oracle, that is to an oracle that provides correct answers to queries concerning whether a given Turing machine (say with an initially empty tape) will eventually halt.

All of this has been well understood for decades. But now Mark Burgin in his [2] proposes that this kind of computation should be regarded as a "super-recursive algorithm". This book is in the series *Monographs in Computer Science* with distinguished editors. Yet it is hard to make any sense of the author's claims that these Δ_2^0 sets should be regarded as computable in some extended sense. It is generally understood that in order for a computational result to be useful one must be able to at least recognize that it is indeed the result sought. Indeed Burgin goes much further, ascending the full arithmetic hierarchy and insisting that for all the sets in that hierarchy, "super-recursive algorithms" are available.[2]

---

[2] Burgin's book discusses a very large number of abstract models of computation and their interrelationships. The present criticism is not about the mathematical discussion of these matters but only about the misleading claims regarding physical systems of the present and future.

4   "Hypercomputation" Via Real Numbers as Oracle

The term "hypercomputation" was coined by Jack Copeland[3] with particular reference to Turing's notion of computation with an oracle. He seems to believe (or have believed) that Turing intended this abstract theoretical discussion as an implied proposal to construct an actual non-Turing computable oracle, declaring that the search was on for such. His hint at the physical form such an oracle might take suggested that the information would be presented as the successive digits of an infinite-precision real number. In another direction Hava Siegelmann proposed computation by a model of computation using neural nets with infinite-precision real numbers again playing a key role. In both cases (as I pointed out in detail in my [9]), the claimed non-computability was nothing more than that of the real numbers provided at the beginning

5   The Human Mind as Hypercomputer

How nice to think that we can have a hypercomputer without having to work out how to make one. Each and every one of us has one between our own ears! In their book of over 300 pages [1], Selmer Bringsjord and Michael Zenzen make exactly this claim. How finite brains are to manifest the necessarily infinite capability implied by the word "hypercomputer" is never made clear.

Without making such outlandish claims, the illustrious mathematician and physicist Roger Penrose has joined those who have attempted to harness Gödel's incompleteness theorem to demonstrate that the human mind transcends the capability of any conceivable computer [17, 18]. The argument is that any computer programmed to generate theorems will be subject to the incompleteness theorem, so that there will be some proposition (indeed of a rather simple form) left undecided by that computer. Since, it is claimed, we can "see" that that very proposition is true, we cannot be equivalent to such a computer. The fallacy (in essence pointed out long ago by Turing) is that all we can really "see" is that if the computer program is consistent, that is will never generate two mutually contradictory assertions, then the proposition in question is true. Moreover, the computer program will generate this same implication, being quite able to prove that if it itself is consistent then that proposition is indeed true. So any claim that we do better than the computer boils down to claiming that we can "see" the consistency of the systems with which we work. But this claim is highly dubious. The list of logicians who have seriously proposed systems of logic that subsequently turned out to be inconsistent reads like an honor role.[4] Indeed the distinguished mathematician and computer scientist Jack Schwartz has recently even proposed as a serious possibility that the Peano postulates themselves might be inconsistent [20].

---

[3] For references to the writings of Copeland, Siegelmann, and Turing relevant to this section, see [9].

[4] It includes Frege, Church, Rosser, and Quine.

6   "Hypercomputation" Via Quantum Mechanics

Finally, a few words about the effort of Tien Kieu to use the Quantum Adiabatic Theorem, a well-known and important result in quantum mechanics, to provide a computational solution fo a problem known to be Turing non-computable. The problem in question is Hilbert's tenth problem which may be expressed as follows:

> Find an algorithm which given a polynomial equation in any number of unknowns with integer coefficients will determine whether or not that equation has a solution in positive integers.

Kieu's proposed solution [16] has been thoroughly criticized by people who understand quantum mechanics much better than I do,[5] and I don't propose to say a great deal about it. But there is one point that I do want to make because it illustrates how blithely the hypercomputationists manage to be blind to the significance of infinity. Kieu suggests that one way his method could be used to determine whether the given equation has a solution is to observe, when the process concludes, a tuple of positive integers, the "occupation numbers". These are then to be substituted into the given equation. Either they satisfy the equation, in which case we know there is a solution, or they do not, in which case Kieu assures us, the equation has no solutions. Now, evidently there are equations that have positive integer solutions, but for which the least such solution is enormous, for example so large, that to write the numbers in decimal notation would require a space larger than the diameter of our galaxy! In what sense could such numbers be read off a piece of equipment occupying a small part of our small planet? And how can we suppose that it would be feasible to substitute numbers of such magnitude into an equation and to carry out the arithmetic needed to determine whether they satisfy the equation?

7   A Hypercomputational Physics?

Despite all of the above, it would be foolhardy to claim that no future device will be able to compute the noncomputable. Indeed, it hardly needed the current "hypercomputation" movement to call this to our attention. In 1958 I wrote [5]:

> For how can we can we ever exclude the possibility of our being presented, some day ... with a ... device or "oracle" that "computes" a noncomputable function?

However, on what basis could it be claimed that some device is indeed a "hypercomputer"? What would be required is an appropriate physical theory, a theory that can be certified as being absolutely correct, unlike any existing theory, which physicists see as only an approximation to reality. Furthermore the theory would have to predict the value of some dimensionless uncomputable real

---

[5] See for example [22].

number to infinite precision. Finally, the device would have to follow exactly the requirements of the supposed theory. Needless to say, nothing like this is even remotely on the horizon.

Elsewhere I wrote [10]:

> The two pillars of contemporary physics are quantum mechanics and relativity theory. So it was inevitable that relativity theory would be brought to bear on solving the unsolvable. In [13] Etcsi and Nemeti argue that conditions in the vicinity of certain kinds of black holes in the context of the equations of general relativity indeed permit an infinite time span to occur that will appear as finite to a suitable observer. Assuming that such an observer can feed problems to a device subject to this compression of an infinite time span, such a device could indeed solve the unsolvable without recourse to Kieu's miracle of an infinite computation in a finite time period. Of course, even assuming that all this really does correspond to the actual universe in which we live, there is still the question of whether an actual device to take advantage of this phenomenon is possible. But the theoretical question is certainly of interest.

References

1. Bringsjord, Selmer and Michael Zenzen: *Superminds: People Harness Hypercomputation, and More.* Studies in Cognitive Systems **29** Kluwer 2003.
2. Burgin, Mark: *Super-Recursive Algorithms.* Springer 2005.
3. Burkes, A.W., H.H. Goldstine, and J. von Neumann: *Preliminary Discussion of the Logical Design of an Electronic Computing Instrument.* Institute for Advanced Study 1946. Reprinted: J. von Neumann, *Collected Works* A. H. Taub ed., vol. 5, Pergamon Press 1963.
4. Church, Alonzo: Review of [23]. J. Symbolic Logic. **2** (1937) 42–43
5. Davis, Martin: *Computability and Unsolvability* McGraw Hill 1958. Reprinted Dover 1983.
6. Davis, Martin, ed.: *The Undecidable.* Raven Press 1965. Reprinted: Dover 2004.
7. Davis, Martin: Why Gödel Didn't Have Church's Thesis. Information and Control **54** (1982) 3–24.
8. Davis, Martin: Mathematical Logic and the Origin of Modern Computers. *Studies in the History of Mathematics.* Mathematical Association of America, 1987, 137–165. Reprinted in: *The Universal Turing Machine - A Half-Century Survey,* Rolf Herken, editor. Verlag Kemmerer & Unverzagt, Hamburg, Berlin 1988; Oxford University Press, 1988, 149–174.
9. Davis, Martin: The Myth of Hypercomputation. *Alan Turing: Life and Legacy of a Great Thinker,* Christof Teuscher, ed. Springer 2004, 195–212.
10. Davis, Martin: Why There Is No Such Subject As Hypercomputation. Applied Matheamtics and Computation. Special issue on hypercomputation, Jos Flix Costa and Francisco Doria, eds. To appear 2006.
11. Davis, Martin: *The Universal Computer: The Road from Leibniz to Turing.* W.W. Norton 2000.

12. Davis, Martin: *Engines of Logic: Mathematicians and the Origin of the Computer.* W.W. Norton 2001 (paperback edition of [11]).

13. Etesi, Gabor and Istvan Nemeti, Non-Turing Computations via Malament-Hogarth Spacetimes. *International Journal of Theoretical Physics.* **41,2** 2002, 341-370.

14. Fisher, Joseph, Paolo Faraboschi, and Cliff Young: *Embedded Computing: A VLIW Approach to Architecture, Compilers and Tools.* Elsevier 2005.

15. Gold, E.M.: Limiting Recursion. Journal of Symbolic Logic **30** (1965) 28-46

16. Kieu, Tien: Computing the noncomputable. Contemporary Physics **44** (2003) 51-77.

17. Penrose, Roger: *The Emperor's New Mind.* Oxford 1989.

18. Penrose, Roger: *Shadows of the Mind: A Search for the Missing Science of Consciousness.* Oxford 1994.

19. Putnam, Hilary: Trial and Error Predicates and the Solution to a Problem of Mostowski. Journal of Symbolic Logic **30** (1965) 49-57.

20. Schwartz, Jack T.: Do the integers exists? The unknowability of arithmetic consistency. Commun. Pure Appl. Math. **58** (2005) 1280-1286.

21. Sieg, Wilfried: Step by Recursive Step: Church's Analysis of Effective Calculability. Bulletin of Symbolic Logic **3** (1997) 154-180.

22. Smith, Warren D.: Three Counterexamples Refuting Kieu's Plan for "Quantum Adiabatic Hypercomputation" and Some Uncomputable Quantum Mechanical Tasks. Applied Matheamtics and Computation. Special issue on hypercomputation, Jos Flix Costa and Francisco Doria, eds. To appear 2006.

23. Turing, Alan: On Computable Numbers, with an Application to the Entscheidungsproblem. Proc. London Math. Soc. **42** (1937) 230-265. Correction: Ibid. **43**, 544-546. Reprinted in [6] 155-222, [26] 18-56.

24. Turing, Alan: Lecture to the London Mathematical Society on 20 February 1947. *Alan Turing's ACE Report of 1946 and Other Papers.* B.E. Carpenter and R.W. Doran, eds. MIT Press 1986 106-124. Reprinted in [25] 87-105.

25. Turing, Alan: *Collected Works: Mechanical Intelligence.* D.C. Ince, ed. North-Holland 1992.

26. Turing, Alan: *Collected Works: Mathematical Logic.* R.O. Gandy and C.E.M. Yates, eds. North-Holland 2001.

Gödel and the Origins of Computer Science

John W. Dawson Jr.

Penn State York
1031 Edgecomb Avenue
York PA 17403, U.S.A.
jwd7@psu.edu

Abstract. The centenary of Kurt Gödel (1906–78) is an appropriate occasion on which to assess his profound, yet indirect, influence on the development of computer science. His contributions to and attitudes toward that field are discussed, and are compared with those of other pioneer figures such as Alonzo Church, Emil Post, Alan Turing, and John von Neumann, in order better to understand why Gödel's role was no greater than it was.

Kurt Gödel's impact on the development of computer science was at once seminal and indirect. His (first) incompleteness theorem, published in 1931 and later recast by Alan Turing in the guise of the Halting Problem, established bounds on what is and is not computable (and more recently, has been invoked to demonstrate that there can be no perfect virus checker [5]). In proving that theorem, Gödel gave a precise definition of the class of functions now called primitive recursive, which he employed in a way that, as Martin Davis ([4], 120) has remarked, "looks very much like a computer program" and anticipated "many of the issues that those designing [and using] programming languages" would later face. Furthermore, he introduced the fundamental technique of arithmetization of syntax ("Gödel-numbering") — the first explicit instance in which one mathematical data type (that of syntax) was represented by another (that of numbers) for the purpose of computation.

During the years 1932–33, before the enunciation of Church's Thesis, Gödel published two papers on decision problems for formulas of the predicate calculus, in one of which he established both that the validity of prenex formulas in one prefix class is a decidable question, and that for an arbitrary formula, the decision problem for validity is reducible to that of formulas in another such prefix class. Then in 1934, in a series of lectures at the Institute for Advanced Study, he went on to define the notion of *general* recursive function — one of several definitions that were later proved to be equivalent and adduced as evidence for Church's Thesis.[1] Nevertheless, Gödel was a bystander in the further development of recursion theory. He came to accept Church's Thesis only in the wake of Turing's characterization of abstract computing machines, and despite

---

[1] Gödel's definition was based on a suggestion of Jacques Herbrand, which Herbrand communicated to Gödel shortly before his own untimely death in a mountaineering accident. See [15] for a detailed analysis of that correspondence.

A. Beckmann et al. (Eds.): CiE 2006, LNCS 3988, pp. 133–136, 2006.

his presence at the Institute for Advanced Study — where von Neumann and others developed one of the earliest digital computing machines — he was never involved in the physical realization of computers (nor, as Turing and von Neumann were, with their applications to problems in ballistics, cryptography, or the development of nuclear weapons).

Similarly, in 1936, in his short note [7], Gödel stated an instance of what, thirty years afterward, would be called a "speed-up" theorem — a topic that subsequently became a major focus of research in computational complexity theory. But he gave no proof of the result he stated there and did not pursue the idea further. In 1956, however, he raised a related issue in a letter to von Neumann[2] (then terminally ill and unable to reply) — a question closely related to the $P = NP$ problem that is today the central problem in theoretical computer science.[3]

During his early years in Princeton Gödel was in contact with several of the pioneers of recursion theory, including Stephen C. Kleene, J. Barkley Rosser, and Alonzo Church, and he also had correspondence with Emil Post, who had come close to anticipating the incompleteness theorem twenty years before Gödel and whose approach to recursion theory foreshadowed the later development of automata theory. Regrettably, however, after Gödel's emigration in 1940 there is little documentary evidence to indicate how extensive his interaction was with those, such as Church and von Neumann, who lived nearby.

Turing, too, was in Princeton during the years 1936–38, working on his doctorate. But Gödel was away then, incapacitated by an episode of depression, and by the time he recovered Turing had returned to Britain and become involved with the ultra-secret cryptographic work at Bletchley Park. That, and Turing's untimely death in 1954, is no doubt the reason the two never met. Surprisingly, however — especially in view of their shared interest in the question whether the capabilities of the human mind exceed those of any machine — they seem never to have corresponded either.

In that regard, the contrast between Gödel's views and those of Post and Turing is stark. For whereas Post sought to establish the existence of absolutely unsolvable problems, in his Gibbs Lecture to the American Mathematical Society in 1951 Gödel maintained that the incompleteness theorems imply **either** that *"the human mind ... infinitely surpasses the powers of any finite machine"* **or** that *"there exist absolutely unsolvable Diophantine problems"*. (He did not fall into the error, first advanced by J.R. Lucas and subsequently by Roger Penrose, of asserting that the incompleteness theorems definitively *refute* mechanism; but it is clear that he believed the first alternative in the above disjunction to be the correct one.) And though, in a note he added in 1963 to the English translation of his incompleteness paper, Gödel conceded that Turing's work had provided "a precise and unquestionably adequate definition of the general notion of formal

---

[2] Published in [13], 372–375.

[3] See [14]. The claims Gödel made in his 1936 paper and in his letter to von Neumann attracted notice only posthumously. They were first verified by Samuel Buss in the 1990s. See in particular [1].

system" — systems in which "reasoning ..., in principle, can be completely replaced by mechanical devices" — he later maintained[4] that Turing had erred in arguing that "mental procedures cannot go beyond mechanical procedures." Specifically, Gödel claimed that Turing had overlooked that the human mind *"in its use, is not static but constantly developing."* He argued that though at each stage of its development "the number and precision of the abstract terms at our disposal may be *finite"*, in the course of time both might *"converge toward infinity"*, whence so might the "number of *distinguishable states of mind"* that Turing had invoked in his analysis.

That criticism of Turing, however, seems quite unfounded. For between the years 1948 and 1952 Turing had written and lectured on several occasions[5] on the question whether machines can be said to think. He had proposed a criterion — the so-called "Turing test" — for judging whether or not they do, had cogently rebutted many of the arguments adduced to show that they cannot (including those based on the incompleteness theorems), and had stressed the possibility of designing machines that would, like the human mind, be capable of *learning.*

Overall, then, how ought Gödel's role in the development of computer science to be assessed? Certainly, he introduced many crucial theoretical ideas; and his prescience, concerning such matters as the representation of data types and the significance of questions related to the $P = NP$ problem, is truly remarkable.[6] He did not contribute to the further development of many of those ideas, but that is in accord with his other mathematical work (in set theory, for example): He was a path-breaker, eager to move on to new conquests while leaving follow-up work to others; and his orientation was predominantly philosophical rather than practical.

Indeed, in many respects Gödel was quite otherworldly, and some of his pronouncements seem remarkably naive — such as his statement, in the Gibbs Lecture, that if there are absolutely undecidable propositions, then mathematics must not be a creation of the human mind, because a "creator necessarily knows all properties of his creatures, [which] can't have any others than those he has given them". (To be sure, in the very next paragraph he admitted that "we build machines and [yet] cannot predict their behavior in every detail." But he claimed that that was because "we don't create ... machines out of nothing, but build them out of some given material." [??]). One is reminded of an objection to the idea of machine intelligence that Turing considered in his 1948 report "Intelligent Machinery" ([3], pp. 411–412): To claim that "in so far as a machine can show intelligence, [that must] be regarded as nothing but a reflection of the intelligence of its creator", is, Turing says, like claiming that "credit

---

[4] In a note ([9]) first published posthumously in vol. III of his *Collected Works*.

[5] See chapters 9–14 in [3].

[6] To appreciate just *how* remarkable, compare Gödel's idea of arithmetizing syntax with the statement by computer pioneer Howard Aiken (in 1956!) that "If it should turn out that the basic logic of a machine designed for the numerical solution of differential equations coincides with [that] of a machine intended to make bills for a department store, [that would be] the most amazing coincidence I have ever encountered." (Quoted in [2], p. 43.)

for the discoveries of a pupil should be given to his teacher". "In such a case", he counters, "the teacher [sh]ould be pleased with the success of his methods of education" but "[sh]ould not claim the results themselves unless he ... actually communicated them to his pupil."

Gödel died in 1978, on the eve of the microcomputer revolution, so we are left to wonder what he might have thought of that development. No doubt he would have welcomed computers as tools for mathematical experimentation, which might lead to the formulation of new conjectures and extend our mathematical intuition. But I doubt he would have wavered in his belief that the power of the human mind surpasses that of any machine, and that we possess the ability to solve any mathematical problem we are capable of posing — a belief he shared with David Hilbert, despite his own incompleteness results, which had shown Hilbert's formalist program for securing the foundations of mathematics to be unrealizable.

References

1. Buss, S.S.: On Gödel's Theorems on Lengths of Proofs II: Lower Bounds for Recognizing k Symbol Provability. In: Clote, P., Remmel, J.B. (eds.): Feasible Mathematics II. Progress in Computer Science and Applied Logic, Vol. 13. Birkhäuser, Boston (1995) 57–90
2. Ceruzzi, P. E.: Reckoners, the Prehistory of the Digital Computer, from Relays to the Stored Program Concept, 1933–1945. Greenwood Press, Westport, Conn. (1983)
3. Copeland, J. (ed.): The Essential Turing. Oxford University Press, Oxford (2004).
4. Davis, M.: The Universal Computer. The Road from Leibniz to Turing. W.W. Norton and Company, New York London (2004)
5. Dowling, W.F.: There Are No Safe Virus Tests. American Mathematical Monthly **96** (1989) 835–6
6. Gödel, K.: Über formal unentscheidbare Sätze der *Principia Mathematica* und verwandter Systeme I. Monatshefte für Mathematik und Physik **38** (1931) 173–198. Reprinted with English translation in [10] 144–195
7. Gödel, K.: Über die Länge von Beweisen. Ergebnisse eines mathematischen Kolloquiums **7** (1936) 23–24. Reprinted with English translation in [10] 396–399
8. Gödel, K.: Some Basic Theorems on the Foundations of Mathematics and Their Philosophical Implications. In: [12] 304–23
9. Gödel, K.: Some Remarks on the Undecidability Results. In [11] 305–306
10. Feferman, S., et alii (eds.): Kurt Gödel Collected Works, vol. I. Oxford University Press, New York Oxford (1986)
11. Feferman, S., et alii (eds.): Kurt Gödel Collected Works, vol. II. Oxford University Press, New York Oxford (1990)
12. Feferman, S., et alii (eds.): Kurt Gödel Collected Works, vol. III. Oxford University Press, New York Oxford (1995)
13. Feferman, S., et alii (eds.): Kurt Gödel Collected Works, vol. V. Oxford University Press, New York Oxford (2003)
14. Hartmanis, J.: Gödel, von Neumann, and the $P = NP$ Problem. Bulletin of the European Association for Computer Science **38** (1989) 101–107
15. Sieg, W.: Only Two Letters: The Correspondence Between Herbrand and Gödel. The Bulletin of Symbolic Logic **11** (2005) 172–184

The Role of Algebraic Models and Type-2 Theory of Effectivity in Special Purpose Processor Design

Gregorio de Miguel Casado and Juan Manuel García Chamizo

University of Alicante, 03690, San Vicente del Raspeig, Alicante, Spain
{demiguel, juanma}@dtic.ua.es
http://www.ua.es/i2rc/index2-eng.html

Abstract. A theoretical approach to a novel design method for special purpose processors for computable integral transforms and related operations is presented. The method is based on algebraic processor models and Type-2 Theory of Effectivity and aims for specification formalization and calculation reliability together with implementation feasibility. The convolution operation is presented as a case of study.

Keywords: Theoretical VLSI design, Algebraic specification, Type-2 Theory of Effectivity, Online-arithmetic, Integral transforms, Convolution.

1 Introduction

Advances in Theoretical Computer Science provide new opportunities to deal with problems in other Computer Science fields. Nevertheless, when trying to apply them some feasibility barriers can appear. Scientific and engineering computing raise increasing reliability demands according to complexity growth of mathematical and physical modeling [11]. In this context, Computable Analysis deals with the computability and complexity issues so that to guarantee software development reliability [5]. Research work concerned with correctness in VLSI designs [9][10] has produced remarkable formal approaches for specifying and verifying processors using algebraic models [3][4][6]. In the field of Computer Arithmetic, the precision features of IEEE 754 floating point arithmetic limit the hardware support possibilities for Scientific Computing applications [8]. In this context, we recall the advances achieved in online-arithmetic for variable precision numerical calculations [2]. With respect to technology trends, we remark the sustained on-chip memory increase and the recent development of hybrid chips. This approach consists of a conventional CPU with reconfigurable hardware, allowing ad-hoc implementation of end user operations with the low cost advantages stemming from high scale manufacturing processes [1].

From the point of view of Computer Architecture, correctness and computability criteria are conventionally considered hardly applicable issues due to the limited utility of mathematical processor models and the inherent limited nature

A. Beckmann et al. (Eds.): CiE 2006, LNCS 3988, pp. 137–146, 2006.

of physical hardware resources, respectively. Nevertheless, we consider that the current technology achievements provide new opportunities to introduce paradigms of correctness and computability in hardware support design for high performance scientific calculation tasks.

This research is focused on hardware support design for numerical integral transforms and related calculations in $L_p(\mathbb{R})$. We focus our interest in the convolution operation as it appears in a wide range of scientific calculations. In this paper we present a novel method for special purpose processor design. A specification of a processor for $L_1(\mathbb{R})$ function convolution is developed as a case of study. The method embeds two paradigms: The Fox-Harman-Tucker algebraic processor model [4][6] and Type Two Theory of Effectivity (TTE) [11]. The conceptual sketch of the method (Figure 1) gathers the main decision processes (ovals), outputs and inputs (squares), transitions (solid lines) and feedbacks (dot line arrows).

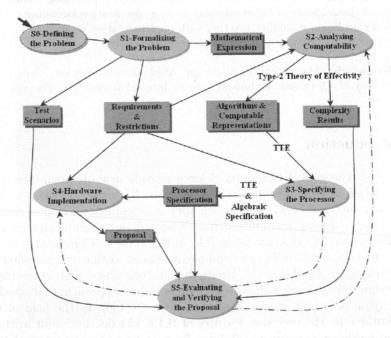

Fig. 1. Application of algebraic models and TTE in the design process of a special purpose processor

An informal description of the calculation problem (S0) feeds the problem formalization stage (S1), which provides a mathematical expression of the calculation, the requirements and restrictions for the hardware support and a set of test scenarios. The computability analysis process (S2) is input with the mathematical expression and the precision requirements. It outputs algorithms based on computable representations and complexity results. The processor specification stage (S3) is fed up by the requirements and restrictions produced in S1 and by the computable algorithms obtained in S2. This stage outputs a high level processor specification for a TTE-computable calculation. The hardware implementation process (S4) takes as inputs the global restrictions and requirements from S1 and the algebraic specification from S4 so that to produce a hardware prototype. The final stage (S5) is devoted to the evaluation and verification of the processor specification and the prototype. The verification can be carried out using specific tools for each of the design levels introduced. Finally, the performance of the prototype can be evaluated according to the test scenarios proposed. A feedback to stages S2, S3 and S4 can be obtained by evaluating the prototype.

This method aims for obtaining the benefits which stem from the specification of the processor using formal methods (the algebraic model) and from the calculation reliability introduced by a TTE approach, which is suitable for defining variable precision schemes. The paper is organized in the following sections: after this introduction, the problem formalization for the design of the convolution processor is developed in Section 2; the computability analysis is carried out in Section 3 and the processor algebraic specification is developed in Section 4. Finally, Section 5 summarizes the conclusions drawn from this paper. The complexity analysis of the convolution algorithm is omitted due space constraints. The algebraic verification and the hardware implementation evaluation are out of the scope of this paper.

2   Formalizing the Problem

The formalization of the specialized processor for function convolution calculation begins with a precise definition of the operations and mathematical objects involved. Next, a list of requirements and restrictions to be considered along the design process is developed. Finally, the mathematical expression of the calculation together with the system requirements and restrictions would help in describing the test scenarios for evaluating the prototype.

2.1   The Convolution Operation

The convolution operation between two functions outputs a function which accumulates the product of one of the input functions with a flipped and shifted version of the other input function. We define it as in [12]:

$$h\left(x\right) = f * g\left(x\right) = \int_{\mathbb{R}^d} f\left(x - y\right) g\left(y\right) dy = \int_{\mathbb{R}^d} f\left(y\right) g\left(x - y\right) dy. \qquad (1)$$

Particularly, we consider the convolution in the separable Banach space $L_1\left(\mathbb{R}\right)$ of Lebesgue-measurable integrable functions on $\mathbb{R}$ with values in $\mathbb{C}$ with the norm $\|f\|$ equal to $\int |f\left(x\right)| d\lambda\left(x\right)$ and λ denoting the Lebesgue measure.

2.2 Requirements and Restrictions

We propose the following requirements and restrictions to be considered throughout the design process of the processor: heterogeneous data sources such as those obtained from symbolic calculation programs and real world data series; scalability features for calculation parallelization; variable precision capabilities to support a wide range of calculation precision requirements; finally, calculation time restrictions and result quality management.

3 Analyzing Computability Issues

We recall the interest of introducing computability criteria so that to preserve reliability in numerical calculations. Among the whole approaches to Computable Analysis, TTE resembles to be one of the most accepted theories [11]. We remark two motivating facts: first, the TTE computability concept is based on tangible representations for computable objects via concrete naming systems and realizations; second, data features and data flow design are considered some of the keystones in computer architecture specification.

The process of analyzing computability given a mathematical expression provides a collection of algorithms based on effective representations. The level of abstraction of the representation objects and arithmetic operations involved has to be suitable for hardware implementation. Furthermore, depending on the representation features, a problem complexity analysis could be carried out.

3.1 Computable Representations for L_1 ($\mathbb{R}$) Spaces

Previous work in TTE computability on L_p ($\mathbb{R}$) spaces, distributions and convolution operation can be found in [7][11][12]. In these research works the TTE-computability is introduced by effective and computable metric spaces or limit spaces and achieved by approximations by infinite fast converging sequences of rational step functions or truncated polynomials.

Due to the nature of source functions and considering arithmetic implementation feasibility, rational step functions (or simple functions) are chosen as the basis for our representation. Let $RSF := \{\sum_{i=0}^{n} \chi_{(a_i,b_i)} c_i : a_i \leq b_i |\ n \in \mathbb{N},\ a_i, b_i \in \mathbb{Q},\ c_i \in \mathbb{Q}^2\}$ be the set of step functions in $\mathbb{R}$ with rational values defined by characteristic functions $\chi_{(a_i,b_i)}$ over intervals with rational endpoints a_i and b_i. The set RSF is a countable dense subset of the space L_1 ($\mathbb{R}$) as every integrable measurable function can be approximated by measurable step functions in the norm $|\cdot|$ and every measurable subset of $\mathbb{R}$ can be approximated from above by open sets with respect to the Lebesgue measure [7].

Rational Step Function Representation. Let $\Sigma = \{\bar{1}, 0, 1\}$ be a finite set. Let Σ^* and Σ^w denote the sets of finite and infinite sequences over Σ, respectively. We introduce the normalized representation ν_{nsd} to codify names of rational numbers with exponent $e \in \mathbb{Z}$ and mantissa $m \in \mathbb{Q}$ in ν_{sd} [11, Def. 3.1.2. and 7.2.4]. This aims for simplifying the arithmetic operators.

$$\nu_{sd}^{\exp} :\subset \Sigma^* \longrightarrow \mathbb{Z},$$
$$dom\,(\nu_{sd}^{\exp}) := \begin{cases} \text{all } a_n \ldots a_0 \in \Sigma^* \text{ for } n \geq 0, \\ a_i \in \Sigma \text{ for } i \leq n, \\ a_n \neq 0, \text{ if } n \geq 0 \text{ and } a_n a_{n-1} \notin \{1\bar{1}, \bar{1}1\}, \text{ if } n \geq 1, \end{cases} \quad (2)$$
$$\nu_{sd}^{\exp}\,(a_n \ldots a_0) := \rho_{sd}\,(a_n \ldots a_0 0^w).$$
$$\nu_{sd}^{man} :\subset \Sigma^* \longrightarrow \mathbb{Q},$$
$$dom\,(\nu_{sd}^{man}) := \{u_\bullet v \mid u = 0, v \in \Sigma^*, \ u_\bullet v 0^w \in dom\,(\rho_{sd})\},$$
$$\nu_{sd}^{man}\,(u_\bullet v) := \rho_{sd}\,(u_\bullet v 0^w).$$

$$\nu_{nsd} :\subset \Sigma^* \longrightarrow \mathbb{Q},$$
$$dom\,(\nu_{nsd}) := \{\exp man \mid \exp \in dom\,(\nu_{sd}^{\exp}), \ man \in dom\,(\nu_{sd}^{man})\}, \quad (3)$$
$$\nu_{nsd}\,(\exp man) := \nu_{sd}\,(\exp \cdot man).$$

The following representation for rational step functions is proposed:

$$\alpha_{RSF}^{\nu_{nsd}} :\subset \Sigma^* \longrightarrow RSF,$$
$$dom\,(\alpha_{RSF}^{\nu_{nsd}}) := \{nrlh\,\{(ab)\,(c_R c_I)_1 \ldots (ab)\,(c_R c_I)_n\} \mid$$
$$n \in dom\,(\nu_\mathbb{N}) \text{ and } r, l, h, a, b, c_R, c_I \in dom\,(\nu_{nsd})\}, \quad (4)$$
$$\alpha_{RSF}^{\nu_{nsd}}\,(nrlhs) = \langle nrlh\,\{s_1 \ldots s_n\}\rangle =$$
$$\iota\,(n)\,\iota\,(r)\,\iota\,(l)\,\iota\,(h)\,\iota\,(\iota\,\{\iota\,(s_1) \ldots \iota\,(s_n)\}), \text{ with } s_i = (ab)\,(c_R c_I)_1,$$

where $\nu_\mathbb{N}$ is a standard TTE binary notation for natural numbers [11, Def. 3.1.2]. The representation codifies names of rational step functions by the following word pairing: n is the number of steps of a given rational step function; r is the partition size for the function range related to the Lebesgue integral, l and h are the lower an upper rational numbers which define the domain of the rational step function and, finally, (ab) is the step domain interval and its corresponding step value $(c_R, c_I) \in \mathbb{Q}^2$.

$L_1\,(\mathbb{R})$ Function Representation.

Define the representation $\alpha_{L_1(\mathbb{R})}$ of the functions $f \in L_1\,(\mathbb{R})$ as a sequence of rational step functions:

$$\alpha_{L_1(\mathbb{R})} :\subset \Sigma^w \longrightarrow L_1\,(\mathbb{R})$$
$$dom\,(\alpha_{L_1(\mathbb{R})}) := \{p_0 p_1 \ldots p_i \ldots \in \Sigma^w \mid p_i \in dom\,(\alpha_{RSF}^{\nu_{nsd}}), \ i \in \mathbb{N}\}. \quad (5)$$

Let $p \in dom\,(\alpha_{L_1(\mathbb{R})})$ with $p = p_0 p_1 \ldots,$ $p_i \in dom\,(\alpha_{RSF}^{\nu_{nsd}}), i = 0, 1, \ldots.$ The sequence $p = p_0 p_1 \ldots$ must fulfil the convergence condition $\|f - s_i\|_{L_1(\mathbb{R})} \leq 2^{-i},$ where $f = \alpha_{L_1(\mathbb{R})}(p)$ and $s_i = \alpha_{RSF}^{\nu_{nsd}}(p_i)$ (see [11, Ex. 8.1.8]).

The next algorithm for $\alpha_{L_1(\mathbb{R})}$ representation is proposed:

1. Let $f_1, f_2 \in L_1\,(\mathbb{R})$ and $p1, p2 = \lambda$ (initialization to empty word).
2. Choose an initial partition size for the function range step $r \in \mathbb{Q}$. Then, obtain the corresponding function domain partitions.
3. Construct rational step functions $s1, s2$ for f_1, f_2 according to $\alpha_{RSF}^{\nu_{nsd}}$ features. An approximation in $\mathbb{Q}^2$ with k significative digits is obtained for each step.
4. Rearrange the rational step functions $s1, s2$ achieving a "reverse compatible partition scheme" for f_1, f_2 in order to ease the calculation of the convolution.

5. Check the convergence condition for the sequence of rational step functions $p1s1$ and $p2s2$. If the condition is fulfilled, update $p1$ and $p2$ by concatenating $s1$ and $s2$ to the corresponding sequence. If not, perform new interval splitting into halves with $r = \frac{r}{2}$ together with an increment $k = k + \triangle k$, according to the precision requirements established. If r and k exceed the values established, then stop; else, go to step 3.

Representations for discrete functions can be obtained in a similar way, considering value interpolation where necessary.

All the operations involved in the algorithm proposed for building the source functions are TTE computable: minimum, rational number $k-$digit truncating and rational number division and addition.

Convolution Between Rational Step Functions. The resulting sequence of rational step functions obtained from the convolution operation has convergence features due to the nature of the source representations. A TTE effective representation $p = p_0, p_1, \dots \in dom\left(\alpha_{L_1(\mathbb{R})}\right)$ will be achieved by checking the convergence condition $||f - s_i||_{L_1(\mathbb{R})} \leq 2^{-i}$, with $f = \alpha_{L_1(\mathbb{R})}(p)$ and $s_i = \alpha_{RSF}^{\nu_{nsd}}(p_i)$.
The convolution algorithm:

$$
\begin{aligned}
&f = f_1 f_2, \dots \in \Sigma^w, \; g = g_1 g_2, \dots \in \Sigma^w, \; f_i, g_i \in dom\left(\alpha_{RSF}^{\nu_{nsd}}\right) \\
&\forall f_i \wedge g_i, \; f_i \in f, g_i \in g, \; i = 1, 2, \dots \\
&h_i = f_i * g_i.
\end{aligned}
\tag{6}
$$

The following inner operations are codified and concatenated for each rational step function convolution result:

$$
h[i] = \begin{cases}
h_i \bullet n = f_i \bullet n + g_i \bullet n - 1, \\
h_i \bullet r = f_i \bullet r, \\
h_i \bullet l = f_i \bullet l, \; h_i \bullet h = (f_i \bullet h - f_i \bullet l) + (g_i \bullet h - g_i \bullet l), \\
\qquad \begin{cases} h_i \bullet (a)_j = f_i \bullet (a)_k, \\ h_i \bullet (b)_j = f_i \bullet (b)_k, \\ h_i \bullet (c_R c_I)_j = \sum_{k=0}^{j} f_i \bullet (c_R c_I)_k \cdot g_i \bullet (c_R c_I)_{[j-k]}, \end{cases} \\
j = 1, \dots, h_i \bullet n.
\end{cases}
\tag{7}
$$

The abstractions embedded in the function names, which are, namely the number of steps n, range step size r, lower interval limit l, upper interval limit h and step values a, b, c_R, c_I are accessed with the operator ".". This operator, which is not formally developed due to space constraints, extracts the corresponding abstractions within the RSF name. Remark that the arithmetic operations "+", "−" and "·" are computable with respect to the ν_{sd} representation.

4   Specifying the Convolution Processor

The first part of this section introduces a functional specification of the processor. Next, an algebraic specification at programmer's level is presented. The last part deals with a proposal for filling the implementation gap between TTE and the algebraic processor model by means of a set of memory mapping functions.

4.1   Functional Specification

The processor consists of five main modules: the system bus interface, the control unit, the processor memory and the RSF arithmetic unit. The system bus interface provides external control and scalability management for multiple processing units. The control unit has an instruction decoder which feeds a microinstruction encoder. This module translates the high abstraction level processor instructions into simpler machine instructions, which are associated to the datapath and the control signals of the bus. The control unit manages the arithmetic logic unit (ALU) and the register banks. It also interfaces the memory module an the RSF arithmetic unit. The memory module is split in two memory areas: the program area and the data memory area. The RSF arithmetic unit provides support for online arithmetic operations for parts of the computable RSF representations, attending to RSF word function name stride, pipes, precision, interval step and available calculation time provided during the processor configuration stage.

The instruction set of the processor consists of five instructions conceived for external interfacing throughout the processor program memory:

- $Status\_request()$: $Idle \lor Busy \lor Error(1)$. Provides the status of the processor.
- $Configuration\_request()$: $(pNum, wStride, nPipes, lPrec, hPrec, tAvail, tOP, addH, addF, addG) \lor Error$ $(2,..,11)$. Provides information about the configuration of the processor: $pNum$ is the processor number; $wStride$ is the RSF stride used for coordinating a calculation carried out by multiple processing units; $nPipes$ is the number of arithmetic pipes ;$lPrec$ and $hPrec$ are the lower and higher precision bounds; $tAvail$ is the available calculation time; tOP is the single operation time for a given precision; $addH$, $addF$ and $addG$ keep the memory base address for the source functions f and g and the result function h. Returns the number of operations done when halted.
- $Configuration\_set(pNum, wStride, nPipes, lPrec, hPrec, tAvail, tOP, addH, addF, addG)$: $Ack \lor Error$ $(12,\ldots,21)$. Processor configuration instruction.
- $Halt()$: $Ack \lor OPs \lor Error(22)$. This is an instruction with priority for stopping the processor. It returns the number of operations done.
- $Convolution()$: $Ack \lor OPs \lor Error$ $(23,\ldots,55)$. Returns the number of operations done when the operation is halted.

According to our estimations based on an algorithmic approach, a program counter PC and four banks of registers are needed:

- Configuration Registers (CR): PN (processor number), WS (word stride), NP (arithmetic pipes), LP and HP (precision bounds), TA (calculation time available) and TO (operation time).
- Base-Address Registers (BA): $AH \leftarrow addH$, $AF \leftarrow addF$, $AG \leftarrow addG$.
- Status Registers (SR): CI for the current instruction, OR for result storage, RA result storage address, OC for operation counter.
- Arithmetic Registers (AR). Provide arithmetic support (32 registers).

4.2   Algebraic Specification. Programmer's Level

By introducing an algebraic model for the processor specification formal behavior methods over time and data representation as well as operations can be isolated. In this section, an algebraic specification of the processor at programmer's level is developed following [6]. The instruction delay is chosen as abstract system clock T. The next subsection develops the state and the next-state algebras.

The State and Next-State Algebras. The processor state consists of the program counter PC, the register banks CR, BA, SR, AR and the memory system. The program memory PM and the data memory DM are modeled as a mapping of a subset of the natural numbers into the binary numbers: $PM, DM :\subseteq N \times B2$, $N \subseteq dom\,(\nu_{\mathbb{N}})$ and $B2 \subseteq dom\,(\nu_{b,2})$. PM also holds the instructions execution results. The state of the machine is defined by $Cc = PC \times CR \times BA \times SR \times AR \times Mem$, with $Mem = PC \longrightarrow PM \cup DM$.

There are 5 inputs with the corresponding outputs:

$$IIn = \{Stat\_req(),\ Conf\_req(),\ Conf\_set(nPs, ...),\ Halt(),\ Conv()\}$$
$$IOut = \{Idle \vee Busy \vee Error(1),\ ...,\ Ack \vee OPs \vee Error(23, ..., 55)\} \quad (8)$$

Therefore, input is defined as $In = IIn \times PC$ and output as $Out = IOut \times SR.CI \times Mem$. The state algebra can be expressed as:

Algebra Convolution Processor State
Sets $T, Cc, In, Out, [T \longrightarrow In]$
Operations $CC : T \times Cc \times [T \longrightarrow In] \longrightarrow Cc \times Out$
End Algebra

CC is defined as:

$$CC_1\,(0, g, i) = g,$$
$$CC_1\,(t + 1, g, i) = cc\,(CC_1\,(t, g, i)\,, i\,(t))\,, \quad (9)$$
$$CC_2\,(0, g, i) = out\,(CC_1\,(t, g, i))\,,$$

where $cc : Cc \times In \longrightarrow Cc$ is the next-state function, and $out : Cc \longrightarrow Out$ is the output function. Hence, the next-state algebra is defined as:

Algebra Convolution Processor Next-State
Sets $T, Cc, In, Out, [T \longrightarrow In]$
Constants $0 : T$
Operations $t + 1 : T \longrightarrow T$
$\qquad\qquad cc : Cc \times In \longrightarrow Cc$
$\qquad\qquad out : Cc \longrightarrow Out$
$\qquad\qquad eval : T \times [T \longrightarrow In] \longrightarrow In$
End Algebra

The stream evaluation function $eval : T \times [T \longrightarrow In] \longrightarrow In$, is defined by primitive recursive functions over the next-state algebra.

4.3   Data Memory Organization

The implementation in memory of source and result functions can be done by mapping every abstraction of a function name $p \in dom\,(\alpha_{L_1(\mathbb{R})})$ into a memory designed as an address space $M :\subseteq N \times B2$, $N \subseteq dom\,(\nu_{\mathbb{N}})$ and $B2 \subseteq dom\,(\nu_{b,2})$.

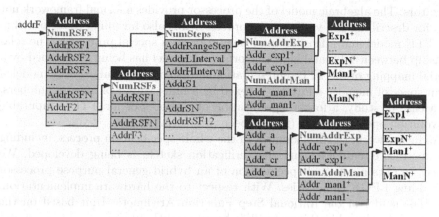

Fig. 2. Data memory mapping of a RSF representation

A realistic memory organization with flexibility criteria suggests the introduction of indirect addressing in the mapping process (see Fig. 2). This way, we introduce a set of mapping functions according to the different abstractions within the function name $p \in dom\left(\alpha_{L_1(\mathbb{R})}\right)$. Their formal definition is omitted due to space constraints.

$p_{headName}$: maps the starting function representation address.

p_{addrF} : maps the amount of RSFs and their corresponding addresses.

$p_{addrRSF}$: maps the number of steps and the address for the step size, the lower and upper intervals of a RSF and the step addresses.

$p_{headStep}$: maps the addresses for the step values.

$p_{addrRangeStep}, p_{addrLInt}, p_{addrHInt}$: map the number of addresses and initial addresses for exponent and mantissa of range step and interval bounds.

$p_{addrA}, p_{addrB}, p_{addrCr}, p_{addrCi}$: map the number of addresses and initial addresses for exponent and mantissa of step data (domain and value in $\mathbb{Q}^2$).

$p_{RangeStep}, p_{lInterval}, p_{hInterval}, p_a, p_b, p_{Cr}, p_{Ci}$: map the values of the step size, lower interval, upper interval and the step using a conversion type from the normalized signed digit representation into positive and negative string parts in $\nu_{\mathbb{N}}$ so that to use online arithmetic operators (see [2, Ch. 9]).

5   Conclusions

A novel theoretical approach for special purpose processor design based on Type Two Theory of Effectivity and the Fox-Harman-Tucker algebraic processor specification model has been presented. The TTE approach, which is based on a Type-2 Turing Machine computability model, provides criteria about the management of the calculation precision according to realistic limited memory resources, as the computable representations involved in the calculation embed precision features. In addition, the representation is based on signed digit values which are suitable for specific hardware support based on online arithmetic operators. The algebraic model of the processor provides a formal framework not only for describing and verifying the processor but also for filling the gap within the TTE model and a feasible implementation of a special processor. The relationship between TTE and algebraic processor model has being established by a partial mapping of the TTE representation into a conventional memory modeled as an space of memory addresses indexed by a finite subset of natural numbers. As a case of study, a formal specification of a processor for TTE computable convolution operation in L_1 ($\mathbb{R}$) has been proposed.

As future work, the development of the whole specification process, including the abstract circuit level and the verification sketch, is being developed. We expect to obtain a formal specification of an hybrid general purpose processor embedding FPGA capabilities. With respect to the hardware implementation, a VHDL model of the Rational Step Function Arithmetic Unit based on the datatypes developed in this research is being carried out.

References

1. Andrews, D.; Niehaus, D.; Jidin, R.; Finley, M.; Peck, W.; Frisbie, M.; Ortiz, J.; Komp, E. and Ashenden, P.: Programming Models for Hybrid FPGA-CPU Computational Components: A Missing Link, IEEE Micro, Vol. 24, (2004) 42–53
2. Ercegovac, M.D. and Lang, T.: Digital Arithmetic, M. Kaufmann, (2004)
3. Fox, A.C.J.: Formal specification and verification of ARM6. LNCS, Vol. 2758, (2003) 25–40
4. Fox, A.C.J. and Harman, N.A.: Algebraic Models of Correctness for Abstract Pipelines, The Journal of Algebraic and Logic Programming, Vol. 57, (2003) 71–107
5. Gowland, P. and Lester, D.: A Survey of Exact Arithmetic Implementations, LNCS, Vol. 2064, (2001) 30–47
6. Harman, N.A. and Tucker, J.V.: Algebraic models of microprocessors: the verification of a simple computer. Proc. 2nd IMA Conference, (1997) 135–170
7. Klunkle, D.: Type-2 computability on spaces of integrable functions, Math. Log. Quart., Vol. 50, (2004) 417–430
8. Lynch, T. and Schulte, M.: A High Radix On-line Arithmetic for Credible and Accurate Computing. Journal of UCS, Vol. 1 (1995) 439–453
9. McEvoy, K. and Tucker, J.V.: Theoretical Foundations of VLSI Design, Cambridge Tracts in Theoretical Computer Science, Vol. 10, (1990)
10. Möller, B. and Tucker, J. V. (editors): Prospects for hardware foundations. LNCS, Vol. 1546, (1998)
11. Weihrauch, K.: Computable Analysis. Springer-Verlag, (2000)
12. Zhong, N. and Weihrauch, K.: Computability Theory of Generalized Functions, Journal of the ACM, Vol. 50, (2003) 469–505

Turing Universality in Dynamical Systems

Jean-Charles Delvenne

Université catholique de Louvain,
Department of Mathematical Engineering,
Avenue Georges Lemaître 4, B-1348 Louvain-la-Neuve, Belgium
delvenne@inma.ucl.ac.be

Abstract. A computer is classically formalized as a universal Turing machine. However over the years a lot of research has focused on the computational properties of dynamical systems other than Turing machines, such cellular automata, artificial neural networks, mirrors systems, etc.
   In this talk we review some of the definitions that have been proposed for Turing universality of various systems, and the attempts to understand the relation between dynamical and computational properties of a system.

1   What Is a Dynamical System?

A dynamical system is intuitively anything that evolves in time. In this talk we mainly consider deterministic, discrete-time systems, given by an evolution map $f : X \to X$, where X is the *state space*, or *configuration space*. A state x is transformed into $f(x)$, then $f(f(x))$, and so on.

Examples include Turing machines, cellular automata, subshifts, piecewise affine maps, piecewise polynomial maps and neural networks.

We may also be interested in continuous-time systems, usually defined by a differential equation on (a part of) $\mathbb{R}^n$.

Here we do not consider quantum universal systems; see for instance [7].

2   What Is Universality?

We are interested to solve decision problems on integers, e.g., primality, by means of a dynamical system.

Informally, a universal dynamical system is a system with the same computing power as a universal Turing machine. Thus, a universal system can be used to (semi-)solve the same decision problems as a universal Turing machine.

Note that in this talk we are only interested in solving decision problems on *integers*, while computable analysis deals with computable functions and decision problems on the reals (e.g., checking invertibility of a real-valued matrix). See for instance [23, 3, 19, 17] on computable analysis.

Also, we do not consider systems with super-Turing capabilities, as done for instance in [20, 4].

A. Beckmann et al. (Eds.): CiE 2006, LNCS 3988, pp. 147–152, 2006.

Note that here we do not define universality as the ability to 'simulate any other system'. See for instance [18] for such a notion of universality in the case of cellular automata.

3   Point-to-Point and Point-to-Set Universality

The most remarkable feature of a universal Turing machine is r.e.-completeness of its halting problem. In fact, Davis [5] considered this property as the very definition of universality for a Turing machine.

Following that idea, we can say that a system $f : X \to X$ is universal if its point-to-point or point-to-set reachability problem is r.e.-complete. The *reachability problem* goes as follows: we are given two points x and y ('point-to-point') or a point x and a set Y ('point-to-set'), and the question is whether there is a t such that $f^t(x) = y$ or $f^t(x) \in Y$.

Such a definition is meaningful only if we restrict ourselves to countable families of points x, y and sets Y.

In cellular automata, point-to-point reachability with almost periodic configurations (made of a same pattern indefinitely repeated except for finitely many cells) is usually considered. For instance the automaton 110 and the Game of Life are universal according to this definition. Why almost periodic configurations and not a wider, or smaller, countable family of points? This is discussed in [21].

For systems in $\mathbb{R}^n$, points with rational coordinates and sets defined by polynomial inequalities with rational coefficients (e.g., polyhedra or euclidian balls) are usually considered. The choice of rational numbers seems to be guided by simplicity only.

Let us give some examples of universal systems according to this definition.

- A piecewise-affine continuous map in dimension 2 [9]. This map is defined by a finite set of affine maps of rational coefficients, on domains delimited by lines with rational coefficients.
- Artificial neural networks for several kinds of saturation functions [20].
- A closed-form analytic map in dimension 1 [10].

We can define in a very similar way universal systems in continuous time. Examples of such systems are:

- A piecewise-constant derivative system in dimension 3 [2]. The state space is partitioned on finitely domains delimited by hyperplanes with rational coefficients, and the vector field is constant with a rational value on every domain.
- A ray of light between a set of mirrors [14].

A variant to the choice of rational points is proposed in [16]: the system is endowed with a special point, call it 0, and two functions f_0 and f_1. Then the initial condition $f_0 f_0 f_1 f_0(0)$, say, encodes the binary word 0010. So instead of choosing a countable family of initial conditions, we can choose an initial point and two functions.

3.1 Set-to-Set Universality

Following step by step Turing's original argumentation to build his machines, another definition of universality is elaborated in [6], that holds for symbolic system. *Symbolic systems* are those whose states are sequences of symbols, such as Turing machines, cellular automata and subshifts.

The basic ingredient is to consider a set-to-set reachability problem. Sets are chosen as cylinders, i.e., sets of configurations characterized by the value of finitely many cells. The system is said to be universal if the existence of a trajectory going from one given cylinder to another is r.e.-complete to decide. Actually, variants of reachability are considered as well, such as the existence of a trajectory from given cylinder to another cylinder, then reaching a third given cylinder, or more generally any property that can be observed by a finite automaton.

This definition avoids the somewhat arbitrary choice of a countable family of initial states.

3.2 Robust Computation

Is universality preserved when the dynamics is perturbed by some noise?

In the point-to-point or point-to-set universality, the least uncertainty on the initial condition can *a priori* completely destroy the computation. In fact, ensuring that a physical system is, e.g., in a rational state is obviously an impossible task in practice. The set-to-set universality is less sensitive to perturbation on the initial state. In any case, the question remains if a noise is applied that would slightly perturb every step of the computation.

It has been shown that many reachability problems become decidable when perturbation is added to the dynamics; see for instance [1, 12, 8].

4 A Universal System: What Does It Look Like?

What is the link between the dynamics of a system and its computational capabilities?

Wolfram proposed a loose classification of 1-D cellular automata based on the patterns present in the space-time diagram of the automaton; see [24]. He then conjectured that the universal automata are in the so-called 'fourth class', associated to the most complex patterns.

Langton [11] advocated the idea of the 'edge of chaos', according to which a universal cellular automaton is likely to be neither globally stable (all points converging to one single configuration) nor chaotic. See also [13] for a discussion. Other authors argue that a universal system may be chaotic; see [20].

However it seems difficult to prove any non-trivial result of this kind with the point-to-point definition of universality. Moreover a countable set of points can be 'hidden' in a very small part of the state space (nowhere dense, with zero measure for instance), so the link between this set and the global dynamics is unclear in general.

The set-to-set definition of universality for symbolic systems is more treatable from this respect. In particular, it can be proved [6] that a universal system according to this definition has at least one proper closed subsystem, must have a sensitive point and can be Devaney-chaotic.

5   Decidability vs. Universality

System theorists are often interested by another kind of problem. We consider a countable family of systems. One must check whether a given system of the family has some fixed dynamical property. Note the difference with universality, where we consider a family of points/subsets in a single dynamical system.

For instance, it was proved [22] that global stability (all trajectories uniformly converge to 0) is undecidable for saturated linear systems with rational entries in dimension 4. A saturated linear system is a map $x \mapsto \sigma Ax$, where A is a linear map (with rational entries) and σ is the saturation function applied componentwise: $\sigma(x) = x$ if $x \in [-1, 1]$, $\sigma(x) = 1$ if $x > 1$ and $\sigma(x) = -1$ if $x < -1$.

6   Conclusion: What Is a Computer?

The search for universal computation in dynamical systems has lead to a great diversity of results associated to a great variety of definitions. The emergence of natural computing and quantum computing makes it only more crucial to understand what exactly we mean by computer.

A desirable achievement would thus be to relate definitions and results in a common general framework. This could perhaps lead to a precise understanding of how computational capabilities emerge from the purely dynamical properties of a system.

A motivation can be the characterization of physical systems able of computation. For instance, we can ask whether a gravitational N-body system [14], or a fluid governed by Navier-Stokes equations [15], is universal.

Acknowledgements

This abstract presents research results of the Belgian Programme on Interuniversity Attraction Poles, initiated by the Belgian Federal Science Policy Office. It has been also supported by the ARC (Concerted Research Action) "Large Graphs and Networks", of the French Community of Belgium. The scientific responsibility rests with its authors. The author is holding a FNRS fellowship (Belgian Fund for Scientific Research).

References

1. E. Asarin and A. Bouajjani. Perturbed turing machines and hybrid systems. In *Proceedings of the 16th Annual IEEE Symposium on Logic in Computer Science (LICS-01)*, pages 269–278, Los Alamitos, CA, June 16–19 2001. IEEE Computer Society.

2. E. Asarin, O. Maler, and A. Pnueli. Reachability analysis of dynamical systems having piecewise-constant derivatives. *Theoretical Computer Science*, 138(1):35–65, 1995.

3. L. Blum, M. Shub, and S. Smale. On a theory of computation and complexity over the real numbers: NP-completeness, recursive functions and universal machines. *Bulletin of the American Mathematical Society*, 21:1–46, 1989.

4. O. Bournez and M. Cosnard. On the computational power of dynamical systems and hybrid systems. *Theoretical Computer Science*, 168:417–450, 1990.

5. M. D. Davis. A note on universal Turing machines. In C.E. Shannon and J. McCarthy, editors, *Automata Studies*, pages 167–175. Princeton University Press, 1956.

6. J.-Ch. Delvenne, P. Kůrka, and V. D. Blondel. Computational universality in symbolic dynamical systems. *Fundamenta Informaticae*, 71:1–28, 2006.

7. D. Deutsch. Quantum theory, the Church-Turing principle and the universal quantum computer. *Proceedings of the Royal Society of London Ser. A*, A400:97–117, 1985.

8. Peter Gács. Reliable cellular automata with self-organization. In *38th Annual Symposium on Foundations of Computer Science*, pages 90–99, Miami Beach, Florida, 20–22 October 1997. IEEE.

9. P. Koiran, M. Cosnard, and M. Garzon. Computability with low-dimensional dynamical systems. *Theoretical Computer Science*, 132(1-2):113–128, 1994.

10. P. Koiran and Cr. Moore. Closed-form analytic maps in one and two dimensions can simulate universal Turing machines. *Theoretical Computer Science*, 210(1):217–223, 1999.

11. C. G. Langton. Computation at the edge of chaos. *Physica D*, 42:12–37, 1990.

12. W. Maass and P. Orponen. On the effect of analog noise in discrete-time analog computations. *Neural Computation*, 10(5):1071–1095, 1998.

13. M. Mitchell, P. T. Hraber, and J. P. Crutchfield. Dynamic computation, and the "edge of chaos": A re-examination. In G. Cowan, D. Pines, and D. Melzner, editors, *Complexity: Metaphors, Models, and Reality*, Santa Fe Institute Proceedings, Volume 19, pages 497–513. Addison-Wesley, 1994. Santa Fe Institute Working Paper 93-06-040.

14. Cr. Moore. Unpredictability and undecidability in dynamical systems. *Physical Review Letters*, 64(20):2354–2357, 1990.

15. Cr. Moore. Generalized shifts: Unpredictability and undecidability in dynamical systems. *Nonlinearity*, 4:199–230, 1991.

16. Cr. Moore. Dynamical recognizers: real-time language recognition by analog computers. *Theoretical Computer Science*, 201:99–136, 1998.

17. Cristopher Moore. Recursion theory on the reals and continuous-time computation. *Theoretical Computer Science*, 162(1):23–44, 1996.

18. N. Ollinger. The intrinsic universality problem of one-dimensional cellular automata. In H. Alt and M. Habib, editors, *Symposium on Theoretical Aspects of Computer Science (Berlin, Germany, 2003)*, volume 2607 of *Lecture Notes in Computer Science*, pages 632–641. Springer, Berlin, 2003.

19. Marian Boykan Pour-El and J. Ian Richards. Computability and noncomputability in classical analysis. *Transactions of the American Mathematical Society*, 275:539–560, 1983.
20. H. T. Siegelmann. *Neural Networks and Analog Computation: Beyond the Turing Limit*. Progress in Theoretical Computer Science. Springer-Verlag, 1999.
21. K. Sutner. Almost periodic configurations on linear cellular automata. *Fundamenta Informaticae*, 58(3–4):223–240, 2003.
22. P. Koiran V. Blondel, O. Bournez and J. Tsitsiklis. The stability of saturated linear dynamical systems is undecidable. journal of computer and system sciences. *Journal of Computer and System Sciences*, 62:442–462, 2001.
23. K. Weihrauch. *Computable Analysis*. Springer-Verlag, 2000.
24. S. Wolfram. *A new kind of science*. Wolfram Media, Inc., Champaign, IL, 2002.

Every Sequence Is Decompressible from a Random One*

David Doty

Department of Computer Science, Iowa State University, Ames, IA 50011, USA
ddoty@iastate.edu

Abstract. Kučera and Gács independently showed that every infinite sequence is Turing reducible to a Martin-Löf random sequence. We extend this result to show that every infinite sequence S is Turing reducible to a Martin-Löf random sequence R such that the asymptotic number of bits of R needed to compute n bits of S, divided by n, is precisely the constructive dimension of S. We show that this is the optimal ratio of query bits to computed bits achievable with Turing reductions. As an application of this result, we give a new characterization of constructive dimension in terms of Turing reduction compression ratios.

Keywords: Constructive dimension, Kolmogorov complexity, Turing reduction, compression, martingale, random sequence.

1 Introduction

An (infinite, binary) sequence S is Turing reducible to a sequence R, written $S \leq_T R$, if there is an algorithm M that can compute S, given oracle access to R. Any computable sequence is trivially Turing reducible to any other sequence. Thus, if $S \leq_T R$, then intuitively we can consider R to contain the uncomputable information that M needs to compute S.

Informally, a sequence is Martin-Löf random [Mar66] if it has no structure that can be detected by any algorithm. Kučera [Kuč85, Kuč89] and Gács [Gác86] independently obtained the surprising result that *every* sequence is Turing reducible to a Martin-Löf random sequence. Thus, it is possible to store information about an arbitrary sequence S into another sequence R, while ensuring that the storage of this information imparts no detectable structure on R. In the words of Gács, "it permits us to view even very pathological sequences as the result of the combination of two relatively well-understood processes: the completely chaotic outcome of coin-tossing, and a transducer algorithm."

Gács additionally demonstrated that his reduction does not "waste bits of R." Viewing R as a compressed representation of S, the asymptotic number of bits of R needed to compute n bits of S, divided by n, is essentially the compression

---

* This research was funded in part by grant number 9972653 from the National Science Foundation as part of their Integrative Graduate Education and Research Traineeship (IGERT) program.

A. Beckmann et al. (Eds.): CiE 2006, LNCS 3988, pp. 153–162, 2006.

ratio of the reduction. Gács showed that his reduction achieves a compression ratio of 1; only $n + o(n)$ bits of R are required to compute n bits of S. Merkle and Mihailović [MM04] have provided a simpler proof of this result using martingales, which are strategies for gambling on successive bits of a sequence.

Lutz [Lut03b] defined the *(constructive) dimension* $\dim(S)$ of a sequence S as an effective version of Hausdorff dimension (the most widely-used fractal dimension; see [Hau19, Fal90]). Constructive dimension is a measure of the "density of computably enumerable information" in a sequence. Lutz defined dimension in terms of constructive *gales*, a generalization of martingales. Mayordomo [May02] proved that for all sequences S, $\dim(S) = \liminf_{n \to \infty} \frac{K(S \restriction n)}{n}$, where $K(S \restriction n)$ is the Kolmogorov complexity of the nth prefix of S.

Athreya et. al. [AHLM04], also using gales, defined the *(constructive) strong dimension* $\dim(S)$ of a sequence S as an effective version of packing dimension (see [Tri82, Sul84, Fal90]), another type of fractal dimension and a dual of Hausdorff dimension. They proved the analogous characterization $\text{Dim}(S) = \limsup_{n \to \infty} \frac{K(S \restriction n)}{n}$. Since Kolmogorov complexity is a lower bound on the algorithmic compression of a finite string, $\dim(S)$ and $\text{Dim}(S)$ can respectively be considered to measure the best- and worst-case compression ratios achievable on finite prefixes of S.

Consider the following example. It is well known that K, the characteristic sequence of the halting language, has dimension and strong dimension 0 [Bar68]. The binary representation of Chaitin's halting probability $\Omega = \sum_{M \text{ halts}} 2^{-|M|}$ (where M ranges over all halting programs and $|M|$ is M's description length) is an algorithmically random sequence [Cha75]. It is known that $K \leq_{\text{T}} \Omega$ (see [LV97]). Furthermore, only the first n bits of Ω are required to compute the first 2^n bits of K, so the asymptotic compression ratio of this reduction is 0. Ω can be considered an optimally compressed representation of K, and it is no coincidence that the compression ratio of 0 achieved by the reduction is precisely the dimension of K.

We generalize this phenomenon to arbitrary sequences, extending the result of Kučera and Gács by pushing the compression ratio of the reduction down to its optimal lower bound. Compression can be measured by considering both the best- and worst-case limits of compression, corresponding respectively to measuring the limit inferior and the limit superior of the compression ratio on longer and longer prefixes of S. We show that, for every sequence S, there is a sequence R such that $S \leq_{\text{T}} R$, where the best-case compression ratio of the reduction is the dimension of S, and the worst-case compression ratio is the strong dimension of S. Furthermore, we show that the sequence R can be chosen to be Martin-Löf random, although this part is achieved easily by simply invoking the construction of Gács in a black-box fashion. The condition that R is random is introduced chiefly to show that our main result is a strictly stronger statement than the result of Kučera and Gács, but the compression is the primary result. Our result also extends a compression result of Ryabko [Rya86], discussed in section 3, although it is not a strict improvement, since Ryabko considered two-way reductions (Turing equivalence) rather than one-way reductions.

One application of this result is a new characterization of constructive dimension as the optimal compression ratio achievable on a sequence with Turing reductions. This compression characterization differs from Mayordomo's Kolmogorov complexity characterization in that the compressed version of a prefix of S does not change drastically from one prefix to the next, as it would in the case of Kolmogorov complexity. While the theory of Kolmogorov complexity assigns to each finite string an optimally compact representation of that string – its shortest program – this does not easily allow us to compactly represent an infinite sequence with another infinite sequence. This contrasts, for example, the notions of finite-state compression [Huf59] or Lempel-Ziv compression [ZL78], which are *monotonic*: for all strings x and y, $x \sqsubseteq y$ (x is a prefix of y) implies that $C(x) \sqsubseteq C(y)$, where $C(x)$ is the compressed version of x. Monotonicity enables these compression algorithms to encode and decode an infinite sequence – or in the real world, a data stream of unknown length – online, without needing to reach the end of the data before starting. However, if we let $\pi(x)$ and $\pi(y)$ respectively be shortest programs for x and y, then $x \sqsubseteq y$ does not imply that $\pi(x) \sqsubseteq \pi(y)$. In fact, it may be the case that $\pi(x)$ is longer than $\pi(y)$, or that $\pi(x)$ and $\pi(y)$ do not even share any prefixes in common.

Our characterization of sequence compression via Turing reductions, coupled with the fact that the optimal compression ratio is always achievable by a single oracle sequence and reduction machine, gives a way to associate with each sequence S another sequence R that is an optimally compressed representation of S. As in the case of Kolmogorov complexity, the compression direction is in general uncomputable; it is not always the case that $R \leq_{\mathrm{T}} S$.

2   Preliminaries

2.1   Notation

All logarithms are base 2. We write $\mathbb{R}$, $\mathbb{Q}$, $\mathbb{Z}$, and $\mathbb{N}$ for the set of all real numbers, rational numbers, integers, and non-negative integers, respectively. For $A \subseteq \mathbb{R}$, A^+ denotes $A \cap (0, \infty)$.

$\{0,1\}^*$ is the set of all finite, binary *strings*. The length of a string $x \in \{0,1\}^*$ is denoted by $|x|$. λ denotes the empty string. Let $s_0, s_1, s_2, \ldots \in \{0,1\}^*$ denote the standard enumeration of binary strings $s_0 = \lambda, s_1 = 0, s_2 = 1, s_3 = 00, \ldots$. For $k \in \mathbb{N}$, $\{0,1\}^k$ denotes the set of all strings $x \in \{0,1\}^*$ such that $|x| = k$. The Cantor space $\mathbf{C} = \{0,1\}^\infty$ is the set of all infinite, binary *sequences*. For $x \in \{0,1\}^*$ and $y \in \{0,1\}^* \cup \mathbf{C}$, xy denotes the concatenation of x and y, and $x \sqsubseteq y$ denotes that x is a prefix of y. For $S \in \{0,1\}^* \cup \mathbf{C}$ and $i, j \in \mathbb{N}$, we write $S[i]$ to denote the ith bit of S, with $S[0]$ being the leftmost bit, we write $S[i..j]$ to denote the substring consisting of the ith through j^{th} bits of S (inclusive), with $S[i..j] = \lambda$ if $i > j$, and we write $S \upharpoonright i$ to denote $S[0..i-1]$.

2.2 Reductions and Compression

Let M be a Turing machine and $S \in \mathbf{C}$. We say M *computes* S if, on input $n \in \mathbb{N}$, M outputs the string $S \upharpoonright n$.

We define an *oracle Turing machine* (OTM) to be a Turing machine M that can make constant-time queries to an oracle sequence, and we let OTM denote the set of all oracle Turing machines. For $R \in \mathbf{C}$, we say M *operates with oracle* R if, whenever M makes a query to index $n \in \mathbb{N}$, the bit $R[n]$ is returned.

Let $S, R \in \mathbf{C}$ and $M \in$ OTM. We say S *is Turing reducible to* R *via* M, and we write $S \leq_T R$ *via* M, if M computes S with oracle R. In this case, define $M(R) = S$. We say S *is Turing reducible to* R, and we write $S \leq_T R$, if there exists $M \in$ OTM such that $S \leq_T R$ via M.

Since we do not consider space or time bounds with Turing reductions, we may assume without loss of generality that an oracle Turing machine queries each bit of the oracle sequence at most once, caching the bit for potential future queries.

In order to view Turing reductions as decompression algorithms, we must define how to measure the amount of compression achieved. Let $S, R \in \mathbf{C}$ and $M \in$ OTM such that $S \leq_T R$ via M. Define $\#_S^R(M, n)$ to be the *query usage of* M *on* $S \upharpoonright n$ *with oracle* R, the number of bits of R queried by M when computing $S \upharpoonright n$. Let $S, R \in \mathbf{C}$ and $M \in$ OTM such that $S \leq_T R$ via M. Define

$$\rho_M^-(S, R) = \liminf_{n \to \infty} \frac{\#_S^R(M, n)}{n},$$

$$\rho_M^+(S, R) = \limsup_{n \to \infty} \frac{\#_S^R(M, n)}{n}.$$

$\rho_M^-(S, R)$ and $\rho_M^+(S, R)$ are respectively the best- and worst-case compression ratios as M decompresses R into S. Note that $0 \leq \rho_M^-(S, R) \leq \rho_M^+(S, R) \leq \infty$. Let $S \in \mathbf{C}$. The *lower and upper compression ratios of* S are respectively defined

$$\rho^-(S) = \min_{\substack{R \in \mathbf{C} \\ M \in \text{OTM}}} \left\{ \rho_M^-(S, R) \mid S \leq_T R \text{ via } M \right\},$$

$$\rho^+(S) = \min_{\substack{R \in \mathbf{C} \\ M \in \text{OTM}}} \left\{ \rho_M^+(S, R) \mid S \leq_T R \text{ via } M \right\}.$$

Note that $0 \leq \rho^-(S) \leq \rho^+(S) \leq 1$. As we will see, by Lemma 4.1 and Theorem 4.2, the two minima above exist. In fact, there is a single OTM M that achieves the minimum compression ratio in each case.

2.3 Constructive Dimension

See [Lut03a, Lut03b, AHLM04, Lut05] for a more comprehensive account of the theory of constructive dimension and other effective dimensions.

1. An *s-gale* is a function $d : \{0, 1\}^* \to [0, \infty)$ such that, for all $w \in \{0, 1\}^*$,

$$d(w) = 2^{-s}[d(w0) + d(w1)].$$

2. A *martingale* is a 1-gale.

Intuitively, a martingale is a strategy for gambling in the following game. The gambler starts with some initial amount of *capital* (money) $d(\lambda)$, and it reads an infinite sequence S of bits. $d(w)$ represents the capital the gambler has after reading the prefix $w \sqsubseteq S$. Based on w, the gambler bets some fraction of its capital that the next bit will be 0 and the remainder of its capital that the next bit will be 1. The capital bet on the bit that appears next is doubled, and the remaining capital is lost. The condition $d(w) = \frac{d(w0)+d(w1)}{2}$ ensures *fairness*: the martingale's expected capital after seeing the next bit, given that it has already seen the string w, is equal to its current capital. The fairness condition and an easy induction lead to the following observation.

Observation 2.1. *Let $k \in \mathbb{N}$ and let $d : \{0,1\}^* \to [0,\infty)$ be a martingale. Then*

$$\sum_{u \in \{0,1\}^k} d(u) = 2^k d(\lambda).$$

An *s-gale* is a martingale in which the capital bet on the bit that occurred is multiplied by 2^s, as opposed to simply 2, after each bit. The parameter s may be regarded as the *unfairness of the betting environment*; the lower the value of s, the faster money is taken away from the gambler. Let $d : \{0,1\}^* \to [0,\infty)$ be a martingale and let $s \in [0,\infty)$. Define the *s-gale induced by d*, denoted $d^{(s)}$, for all $w \in \{0,1\}^*$ by

$$d^{(s)}(w) = 2^{(s-1)|w|}d(w).$$

If a gambler's martingale is given by d, then, for all $s \in [0,\infty)$, its s-gale is $d^{(s)}$.

Let $S \in \mathbf{C}$, $s \in [0,\infty)$, and let $d : \{0,1\}^* \to [0,\infty)$ be an s-gale. d *succeeds on* S, and we write $S \in S^\infty[d]$, if

$$\limsup_{n \to \infty} d(S \restriction n) = \infty.$$

d *strongly succeeds on* S, and we write $S \in S^\infty_{\mathrm{str}}[d]$, if

$$\liminf_{n \to \infty} d(S \restriction n) = \infty.$$

An s-gale succeeds on S if, for every amount of capital C, it eventually makes capital at least C. An s-gale strongly succeeds on S if, for every amount of capital C, it eventually makes capital at least C and stays above C forever.

Let $d : \{0,1\}^* \to [0,\infty)$ be an s-gale. We say that d is *constructive (a.k.a. lower semicomputable, subcomputable)* if there is a computable function $\widehat{d} : \{0,1\}^* \times \mathbb{N} \to \mathbb{Q}$ such that, for all $w \in \{0,1\}^*$ and $t \in \mathbb{N}$,

1. $\widehat{d}(w,t) \le \widehat{d}(w,t+1) < d(w)$, and
2. $\lim_{t \to \infty} \widehat{d}(w,t) = d(w)$.

Let $R \in \mathbf{C}$. We say that R is *Martin-Löf random*, and we write $R \in \mathrm{RAND}$, if there is no constructive martingale d such that $R \in S^\infty[d]$. This definition of Martin-Löf randomness, due to Schnorr [Sch71], is equivalent to Martin-Löf's traditional definition (see [Mar66, LV97]).

The following well-known theorem (see [MM04]) says that there is a *single* constructive martingale that *strongly* succeeds on every $S \notin \mathrm{RAND}$.

Theorem 2.2. [MM04] *There is a constructive martingale* **d** *such that* $S_{\text{str}}^{\infty}[\mathbf{d}] =$ RAND^c.

Let $\hat{\mathbf{d}}:\{0,1\}^* \times \mathbb{N} \to \mathbb{Q}$ be the computable function testifying that **d** is constructive.

The following theorem, due independently to Hitchcock and Fenner, states that $\mathbf{d}^{(s)}$ is "optimal" for the class of constructive t-gales whenever $s > t$.

Theorem 2.3. [Hit03, Fen02] *Let* $s > t \in \mathbb{R}^+$, *and let* d *be a constructive* t-gale. *Then* $S^{\infty}[d] \subseteq S^{\infty}[\mathbf{d}^{(s)}]$ *and* $S_{\text{str}}^{\infty}[d] \subseteq S_{\text{str}}^{\infty}[\mathbf{d}^{(s)}]$.

By Theorem 2.3, the following definition of constructive dimension is equivalent to the definitions given in [Lut03b, AHLM04]. Let $X \subseteq \mathbf{C}$. The *constructive dimension* and the *constructive strong dimension* of X are respectively defined

$$\text{cdim}(X) = \inf\{s \in [0,\infty) \mid X \subseteq S^{\infty}[\mathbf{d}^{(s)}]\},$$
$$\text{cDim}(X) = \inf\{s \in [0,\infty) \mid X \subseteq S_{\text{str}}^{\infty}[\mathbf{d}^{(s)}]\}.$$

Let $S \in \mathbf{C}$. The *dimension* and the *strong dimension* of S are respectively defined

$$\dim(S) = \text{cdim}(\{S\}),$$
$$\text{Dim}(S) = \text{cDim}(\{S\}).$$

Intuitively, the (strong) dimension of S is the *most unfair betting environment* s in which the optimal constructive gambler **d** (strongly) succeeds on S. The following theorem – the first part due to Mayordomo and the second to Athreya et. al. – gives a useful characterization of the dimension of a sequence in terms of Kolmogorov complexity, and it justifies the intuition that dimension measures the *density of computably enumerable information* in a sequence.

Theorem 2.4. [May02, AHLM04] *For all* $S \in \mathbf{C}$,

$$\dim(S) = \liminf_{n\to\infty} \frac{K(S \upharpoonright n)}{n}, \ \text{ and } \ \text{Dim}(S) = \limsup_{n\to\infty} \frac{K(S \upharpoonright n)}{n}.$$

One of the most important properties of constructive dimension is that of *absolute stability*, shown by Lutz [Lut03b], which allows us to reason equivalently about the constructive dimension of individual sequences and sets of sequences:

Theorem 2.5. [Lut03b] *For all* $X \subseteq \mathbf{C}$,

$$\text{cdim}(X) = \sup_{S \in X} \dim(S), \ \text{ and } \ \text{cDim}(X) = \sup_{S \in X} \text{Dim}(S).$$

3   Previous Work

The next theorem says that every sequence is Turing reducible to a random sequence. Part 1 is due independently to Kučera and Gács, and part 2 is due to Gács.

Theorem 3.1. [Kuč85, Kuč89, Gác86] *There is an OTM M such that, for all $S \in \mathbf{C}$, there is a sequence $R \in \mathrm{RAND}$ such that*

1. *$S \leq_\mathrm{T} R$ via M.*
2. *$\rho_M^+(S, R) = 1$.*

Let $X \subseteq \mathbf{C}$. Define the *code cost* of X by

$$c_\mathrm{T}(X) = \inf_{M_e, M_d \in \mathrm{OTM}} \left\{ \sup_{S \in X} \rho_{M_d}^-(S, M_e(S)) \,\middle|\, (\forall S \in X)\, M_d(M_e(S)) = S \right\}.$$

$c_\mathrm{T}(X)$ is the optimal lower compression ratio achievable with *reversible* Turing reductions on sequences in X. The next theorem is due to Ryabko [Rya86].

Theorem 3.2. [Rya86] *For every $X \subseteq \mathbf{C}$, $c_\mathrm{T}(X) = \mathrm{cdim}(X)$.*

Theorem 3.2 achieves weaker compression results than the main results of this paper, Theorems 4.2 and 4.3. Theorem 3.2 does not include ρ^+ or cDim, and it requires optimizing over all OTMs. However, unlike Theorem 4.2, in which only the decompression is computable, the compression achieved in Theorem 3.2 is computable, by the definition of c_T.

4   Results

We now state the new results.

An OTM that computes a sequence S, together with a finite number of oracle bits that it queries, is a program to produce a prefix of S. Thus, the query usage of the Turing machine on that prefix cannot be far below the Kolmogorov complexity of the prefix. This is formalized in the following lemma, which bounds the compression ratio below by dimension.

Lemma 4.1. *Let $S, R \in \mathbf{C}$ and $M \in \mathrm{OTM}$ such that $S \leq_\mathrm{T} R$ via M. Then*

$$\rho_M^-(S, R) \geq \dim(S), \text{ and } \rho_M^+(S, R) \geq \mathrm{Dim}(S).$$

The next theorem is the main result of this paper. It shows that the compression lower bounds of Lemma 4.1 are achievable, and that a single OTM M suffices to carry out the reduction, no matter which sequence S is being computed. Furthermore, the oracle sequence R to which S reduces can be made Martin-Löf random.

Theorem 4.2. *There is an OTM M such that, for all $S \in \mathbf{C}$, there is a sequence $R \in \mathrm{RAND}$ such that*

1. *$S \leq_\mathrm{T} R$ via M.*
2. *$\rho_M^-(S, R) = \dim(S)$.*
3. *$\rho_M^+(S, R) = \mathrm{Dim}(S)$.*

Finally, these results give a new characterization of constructive dimension.

Theorem 4.3. *For every sequence $S \in \mathbf{C}$,*

$$\dim(S) = \rho^-(S), \text{ and } \text{Dim}(S) = \rho^+(S),$$

and, for all $X \subseteq \mathbf{C}$,

$$\text{cdim}(X) = \sup_{S \in X} \rho^-(S), \text{ and } \text{cDim}(X) = \sup_{S \in X} \rho^+(S).$$

Proof. Immediate from Lemma 4.1 and Theorems 4.2 and 2.5.   □

5   Conclusion

We have shown that every infinite sequence is Turing reducible to a Martin-Löf random infinite sequence with the optimal compression ratio possible. Since this optimal ratio is the constructive dimension of the sequence, this gives a new characterization of constructive dimension in terms of Turing reduction compression ratios.

The Turing reductions of Theorems 3.1, 3.2, and 4.2 satisfy the stronger properties of the *weak truth-table reduction* (see [Soa87]), which is a Turing reduction in which the query usage of the reduction machine M on input n is bounded by a computable function of n. For example, $2n + O(1)$ suffices. Thus, constructive dimension could also be defined in terms of weak truth-table reductions.

As noted in the introduction, for the sequences S and R in Theorems 3.1 and 4.2, it is not necessarily the case that $R \leq_T S$. In other words, though the decompression is computable, it is not computably reversible in all cases. For instance, if S is computable, then $R \not\leq_T S$, since no sequence $R \in \text{RAND}$ is computable. For this reason, Theorem 4.2 does not imply Theorem 3.2, which allows for the reduction to be computably reversed, subject to the trade-off that the compression requirements are weakened.

The compression of Theorem 4.2 may not be computable even if we drop the requirement that the oracle sequence be random. If the sequence S in Theorem 4.2 satisfies $\dim(S) > 0$ and $\text{Dim}(S) > 0$, then for all $P \in \mathbf{C}$ (not necessarily random) and $M \in \text{OTM}$ satisfying $S \leq_T P$ via M, $\rho_M^-(S, P) = \dim(S)$, and $\rho_M^+(S, P) = \text{Dim}(S)$, it follows that $\dim(P) = \text{Dim}(P) = 1$. This implies that the reversibility of decompression – whether $P \leq_T S$ – is related to an open question posed by Miller and Nies when considering Reimann and Terwijn's question concerning the ability to compute a random sequence from a sequence of positive dimension. Question 10.2 of [MN05] asks whether it is always possible, using an oracle sequence S of positive dimension, to compute a sequence P with dimension greater than that of S.

Acknowledgments. I am grateful to Philippe Moser and Xiaoyang Gu for their insightful discussions, to Jack Lutz and Jim Lathrop for their helpful advice in preparing this article, and to John Hitchcock for making useful corrections in an earlier draft. I would also like to thank anonymous referees for helpful suggestions.

References

[AHLM04] K. B. Athreya, J. M. Hitchcock, J. H. Lutz, and E. Mayordomo. Effective strong dimension, algorithmic information, and computational complexity. *SIAM Journal on Computing*, 2004. To appear. Preliminary version appeared in *Proceedings of the 21st International Symposium on Theoretical Aspects of Computer Science*, pages 632-643.

[Bar68] Y. M. Barzdin'. Complexity of programs to determine whether natural numbers not greater than n belong to a recursively enumerable set. *Soviet Mathematics Doklady*, 9:1251-1254, 1968.

[Cha75] G. J. Chaitin. A theory of program size formally identical to information theory. *Journal of the Association for Computing Machinery*, 22:329-340, 1975.

[Edg04] G. A. Edgar. *Classics on Fractals* Westview Press, Oxford, U.K., 2004.

[Fal90] K. Falconer. *Fractal Geometry: Mathematical Foundations and Applications*. John Wiley & Sons, 1990.

[Fen02] S. A. Fenner. Gales and supergales are equivalent for defining constructive Hausdorff dimension. Technical Report cs.CC/0208044, Computing Research Repository, 2002.

[Gác86] P. Gács. Every sequence is reducible to a random one. *Information and Control*, 70:186-192, 1986.

[Hau19] F. Hausdorff. Dimension und äusseres Mass. *Mathematische Annalen*, 79:157-179, 1919. English version appears in [Edg04], pp. 75-99.

[Hit03] J. M. Hitchcock. Gales suffice for constructive dimension. *Information Processing Letters*, 86(1):9-12, 2003.

[Huf59] D. A. Huffman. Canonical forms for information-lossless finite-state logical machines. *IRE Trans. Circuit Theory CT-6 (Special Supplement)*, pages 41-59, 1959. Also available in E.F. Moore (ed.), Sequential Machine: Selected Papers, Addison-Wesley, 1964, pages 866-871.

[Kuč85] A. Kučera. Measure, Π_1^0-classes and complete extensions of PA. *Recursion Theory Week, Lecture Notes in Mathematics*, 1141:245-259, 1985.

[Kuč89] A. Kučera. On the use of diagonally nonrecursive functions. In *Studies in Logic and the Foundations of Mathematics*, volume 129, pages 219-239. North-Holland, 1989.

[Lut03a] J. H. Lutz. Dimension in complexity classes. *SIAM Journal on Computing*, 32:1236-1259, 2003. Preliminary version appeared in *Proceedings of the Fifteenth Annual IEEE Conference on Computational Complexity*, pages 158-169, 2000.

[Lut03b] J. H. Lutz. The dimensions of individual strings and sequences. *Information and Computation*, 187:49-79, 2003. Preliminary version appeared in *Proceedings of the 27th International Colloquium on Automata, Languages, and Programming*, pages 902-913, 2000.

[Lut05] J. H. Lutz. Effective fractal dimensions. *Mathematical Logic Quarterly*, 51:62-72, 2005. (Invited lecture at the International Conference on Computability and Complexity in Analysis, Cincinnati, OH, August 2003.).

[LV97] M. Li and P. M. B. Vitányi. *An Introduction to Kolmogorov Complexity and its Applications*. Springer-Verlag, Berlin, 1997. Second Edition.

[Mar66] P. Martin-Löf. The definition of random sequences. *Information and Control*, 9:602-619, 1966.

[May02]   E. Mayordomo. A Kolmogorov complexity characterization of constructive Hausdorff dimension. *Information Processing Letters*, 84(1):1–3, 2002.

[MM04]   W. Merkle and N. Mihailović. On the construction of effective random sets. *Journal of Symbolic Logic*, pages 862–878, 2004.

[MN05]   J. S. Miller and A. Nies. Randomness and computability: Open questions. Technical report, University of Auckland, 2005.

[Rya86]   B. Ya. Ryabko. Noiseless coding of combinatorial sources. *Problems of Information Transmission*, 22:170–179, 1986.

[Sch71]   C. P. Schnorr. A unified approach to the definition of random sequences. *Mathematical Systems Theory*, 5:246–258, 1971.

[Soa87]   R. I. Soare. *Recursively Enumerable Sets and Degrees*. Springer-Verlag, Berlin, 1987.

[Sul84]   D. Sullivan. Entropy, Hausdorff measures old and new, and limit sets of geometrically finite Kleinian groups. *Acta Mathematica*, 153:259–277, 1984.

[Tri82]   C. Tricot. Two definitions of fractional dimension. *Mathematical Proceedings of the Cambridge Philosophical Society*, 91:57–74, 1982.

[ZL78]   J. Ziv and A. Lempel. Compression of individual sequences via variable-rate coding. *IEEE Transaction on Information Theory*, 24:530–536, 1978.

Reversible Conservative Rational Abstract Geometrical Computation Is Turing-Universal

Jérôme Durand-Lose

Laboratoire d'Informatique Fondamentale d'Orléans, Université d'Orléans,
B.P. 6759, F-45067 ORLÉANS Cedex 2
Jerome.Durand-Lose@univ-orleans.fr
http://www.univ-orleans.fr/lifo/Members/Jerome.Durand-Lose

Abstract. In *Abstract geometrical computation for black hole computation (MCU '04, LNCS 3354)*, the author provides a setting based on rational numbers, abstract geometrical computation, with super-Turing capability. In the present paper, we prove the Turing computing capability of reversible conservative abstract geometrical computation. Reversibility allows backtracking as well as saving energy; it corresponds here to the local reversibility of collisions. Conservativeness corresponds to the preservation of another energy measure ensuring that the number of signals remains bounded. We first consider 2-counter automata enhanced with a stack to keep track of the computation. Then we built a simulation by reversible conservative rational signal machines.

Keywords: Abstract geometrical computation, Conservativeness, Rational numbers, Reversibility, Turing-computability, 2-counter automata.

1 Introduction

Reversible computing is a very important issue because it allows on the one hand to backtrack a phenomenon to its source and on the other hand to save energy. Let us note that quantum computing relies on reversible operations. There are general studies on reversible computation [LTV98] as well as many model dependent results: on Turing machines [Lec63, Ben73, Ben88] and 2-counter machines [Mor96] (we do not use these), on logic gates [FT82, Tof80], and last but not least, on reversible Cellular automata on both finite configurations [Mor92, Mor95, Dub95] and infinite ones [DL95, DL97, DL02, Kar96, Tof77] as well as decidability results [Čul87, Kar90, Kar94] and a survey [TM90].

Abstract geometrical computation (AGC) [DL05b, DL05c] comes from the common use in the literature on *cellular automata* (CA) of Euclidean lines to model discrete lines in space-time diagrams of CA (*i.e.* colorings of $\mathbb{Z} \times \mathbb{N}$ with states as on the left of Fig. 1) to access dynamics or to design. The main characteristics of CA, as well as abstract geometrical computation, are: *parallelism, synchrony, uniformity* and *locality* of updating. Discrete lines are often observed and idealized as on Fig. 1. They can be the keys to understanding the dynamics like in [Ila01, pp. 87–94] or [BNR91, JSS02]. They can also be the tool to design

A. Beckmann et al. (Eds.): CiE 2006, LNCS 3988, pp. 163–172, 2006.

CA for precise purposes like Turing machine simulation [LN90]. These discrete line systems have also been studied on their own [MT99, DM02].

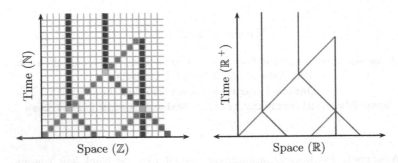

Fig. 1. Space-time diagram of a cellular automaton and its signal machine counterpart

Abstract geometrical computation considers Euclidean lines. The support of space and time is $\mathbb{R}$. Computations are produced by *signal machines* which are defined by finite sets of *meta-signals* and of *collision rules*. Signals are atomic information that correspond to meta-signals and move at constant speed thus generating Euclidean line segments on space-time diagrams. Collision rules are pairs *(incoming meta-signals, outgoing meta-signals)* that define a mapping over sets of meta-signals. They define what happens when signals meet. A configuration is a mapping from $\mathbb{R}$ to meta-signals, collision rules and two special values: void (*i.e.* nothing there) and accumulations (amounting for black holes). The time scale being $\mathbb{R}^+$, there is no such thing as a "next configuration". The following configurations are defined by the uniform movement of signals. In the configurations following a collision, incoming signals are replaced by outgoing signals according to a collision rule.

Zeno like acceleration and accumulation can be brought out as on Fig. 2 of Sect. 2. Accumulations can lead to an unbounded burst of signals producing infinitely many signals in finite time (as in the right of Fig. 2). To avoid this, a *conservativeness* condition is imposed: a positive energy is defined for every meta-signal, the sum of the energies must be preserved by each rule. Thus no energy creation is possible; the number of signals is bounded.

To our knowledge, AGC is the only computing model that is a dynamical system with continuous time and space but finitely many local values. The closest model we know of is the Mondrian automata of Jacopini and Sontacchi [JS90] which is also reversible. Their space-time diagrams are mappings from $\mathbb{R}^n$ to a finite set of colors representing bounded finite polyhedra. Another close model is the piecewise-constant derivative system [AM95, Bou99]: $\mathbb{R}^n$ is partitioned into finitely many polygonal regions, trajectories are defined by a constant derivative on each region and form sequences of (Euclidean) line segments.

In this paper, space and time are restricted to rational numbers. This is possible since all the operations used preserve rationality. All quantifiers and intervals are over $\mathbb{Q}$, not $\mathbb{R}$.

The Turing computing capability of conservative signal machines is proved in [DL05a, DL05b] by simulating any 2-counter automaton since Turing machines and 2-counter automata compute exactly the same functions. To build a reversible version of the simulation, we have to cope with the inherent irreversibility of counter automaton (the result of Morita [Mor96] is not used here because it does not fit well with our approach): branching (there is no way to guess the previous instruction) and subtracting (0 comes from both 0 and 1).

To cope with this we add stacks to store information for reversibility. A stack over an alphabet $\{1, 2, \ldots, l\}$ is encoded by a rational number σ such that: $\frac{1}{l+2}$ encodes the empty stack, otherwise $\frac{1}{l+1} < \sigma < 1$. After pushing a value v on a stack σ the new stack is $\frac{v+\sigma}{l+1}$. The top of the stack is $\lfloor (l+1)\sigma \rfloor$ and $(l+1)\sigma - \lfloor (l+1)\sigma \rfloor$ encodes the rest of the stack. This more or less corresponds to a base $l+1$ decimal number manipulation as can be found in e.g. [Bou99].

This simple memory scheme can be implemented inside a reversible conservative rational signal machine. We then finish the simulation by adapting the construction from [DL05b, DL05a] by adding the storing of current line number before passing to the next and of the previous value of a counter whenever it reaches 0.

The definition of signal machines can be found in Sect. 2. Section 3 deals with 2-counter automata and their enhancement with stacks. Section 4 shows how the stacks are implemented. In Sect. 5, the different pieces are gathered in order to achieve the simulation. Section 6 gives a short conclusion.

2   Abstract Geometrical Computations

Abstract geometrical computations are defined by the following machines:

Definition 1. *A* rational signal machine *is defined by* (M, S, R) *where* M *(meta-signals) is a finite set, S (speeds) a mapping from M to $\mathbb{Q}$ and R (collision rules) a partial mapping from the subsets of M of cardinality at least 2 into the subsets of M (speeds must differ in both domain and range).*

Each instance of a meta-signal is a *signal*. The mapping S assigns rational *speeds* to meta-signals. They correspond the slopes of the segments in space-time diagrams. The *collision rules*, denoted $\rho^- \rightarrow \rho^+$, define what happens when two or more signals meet.

The *extended value set*, V, is the union of M and R plus two symbols: one for void, $\oslash$, and one for an accumulation (or black hole) $*$. A *configuration*, c, is a total mapping from $\mathbb{Q}$ to V such that the set $\{x \in \mathbb{Q} \mid c(x) \neq \oslash\}$ is finite.

A signal corresponding to a meta-signal μ at a position x, i.e. $c(x) = \mu$, is moving uniformly with constant speed $S(\mu)$. A signal must start (resp. end) in the initial (resp. final) configuration or in a collision. These correspond to condition 2 in Def. 2. At a $\rho^- \rightarrow \rho^+$ collision signals corresponding to the meta-signals in ρ^- (resp. ρ^+) must end (resp. start); no other signal should be present (condition 3). A black hole corresponds to an accumulation of collisions and disappears without a trace (condition 4).

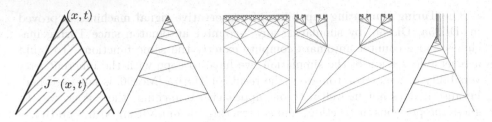

Fig. 2. Light-cone, a simple accumulation and three unwanted phenomena

Let S_{min} and S_{max} be the minimal and maximal speeds. The *causal past*, or *light-cone*, arriving at position x and time t, $J^-(x,t)$, is defined by all the positions that might influence the information at (x,t) through signals, formally:
$$J^-(x,t) = \{ (x',t') \mid x - S_{max}(t-t') \le x' \le x - S_{min}(t-t') \} .$$

Definition 2. *The* space-time diagram *issued from an initial configuration c_0 and lasting for T, is a mapping c from $[0,T]$ to configurations (i.e. a mapping from $\mathbb{Q} \times [0,T]$ to V) such that, $\forall(x,t) \in \mathbb{Q} \times [0,T]$:*

1. *$\forall t \in [0,T]$, $\{ x \in \mathbb{Q} \mid c_t(x) \ne \oslash \}$ is finite,*
2. *if $c_t(x)=\mu$ then $\exists t_i, t_f \in [0,T]$ with $t_i < t < t_f$ or $0=t_i=t<t_f$ or $t_i<t=t_f=T$ s.t.:*
   - *$\forall t' \in (t_i, t_f)$, $c_{t'}(x + S(\mu)(t' - t)) = \mu$,*
   - *$t_i = 0$ or $c_{t_i}(x_i) \in R$ and $\mu \in (c_{t_i}(x_i))^+$ where $x_i = x + S(\mu)(t_i - t)$,*
   - *$t_f = T$ or $c_{t_f}(x_f) \in R$ and $\mu \in (c_{t_f}(x_f))^-$ where $x_f = x + S(\mu)(t_f - t)$;*
3. *if $c_t(x)=\rho^- \to \rho^+ \in R$ then $\exists \varepsilon$, $0<\varepsilon$, $\forall t' \in [t-\varepsilon, t+\varepsilon] \cap [0,T]$, $\forall x' \in [x-\varepsilon, x+\varepsilon]$,*
   - *$c_{t'}(x') \in \rho^- \cup \rho^+ \cup \{\oslash\}$,*
   - *$\forall \mu \in M$, $c_{t'}(x')=\mu \Rightarrow \bigvee \begin{cases} \mu \in \rho^- \text{ and } t' < t \text{ and } x' = x + S(\mu)(t' - t)), \\ \mu \in \rho^+ \text{ and } t < t' \text{ and } x' = x + S(\mu)(t' - t)); \end{cases}$*
4. *if $c_t(x) = *$ then*
   - *$\exists \varepsilon > 0$, $\forall(x',t') \notin J^-(x,t)$, $(|x-x'|<\varepsilon$ and $|t-t'|<\varepsilon)\Rightarrow c_{t'}(x) = \oslash$,*
   - *$\forall \varepsilon > 0$, $\{ (x',t') \in J^-(x,t) \mid t-\varepsilon<t'<t \wedge c_{t'}(x') \in R \}$ is infinite.*

In the illustrations of space-time diagrams, time increases upwards.

2.1 Conservativeness

The three space-time diagrams on the right of Fig. 2 provide examples uncompatible with Def. 2 at the time of accumulation. In each case, the number of signals is bursting to infinity and black holes are not isolated. To prevent this, the following restriction is imposed.

Definition 3. *A signal machine is* conservative *iff there exists an energy from meta-signals to positive integers, $E : M \to \mathbb{N}^*$, such that the total energy of the system, i.e. the sum of the energies of all present signals, is constant.*

One can check easily that the total energy is constant iff for each rule the sum of the energies of incoming meta-signals and the sum of outgoing ones are equal. It follows automatically that given a conservative signal machine and an initial configuration, the number of signals in any following configuration, as well as the number of accumulations, is bounded (by the total energy divided by the least atomic energy). A simple sub-case of conservativeness is when all the meta-signals have the same energy and the numbers of in and out meta-signals are always equal.

2.2   Reversibility

A dynamical systems is said to be *reversible* when from any configuration it is possible to generate all the previous configurations. Moreover, the inverse dynamical system should be of the same nature.

Concerning signal machines, the "inversion" of an isolated signal is the same signal with opposite speed. Regarding collision, one has to guess its position and the in-coming signals from the out-going ones. Collisions resulting in nothing are impossible to guess going backward, and they cannot be conservative. Collisions resulting in only one signal are also impossible to predict. Reversibility holds if and only if the (partial mapping) M is one-to-one and always yields 2 or more out-going meta-signals. The inverse signal machine is the same with the rules reversed.

Definition 4. *A signal machine is* reversible *if and only if R is one-to-one and maps only on sets of cardinality at least 2.*

Let us point out this is true as long as there is no accumulation! The way an accumulation disappears is like a (second order) collision resulting in nothing. Moreover if its location could be guessed, there are infinitely many way to scale it since there is no absolute scale for space nor time.

3   2-Counter Automaton with Stack

A *2-counter automaton* is a finite automaton coupled with two counters, A and B. The possible actions on any counter are *add/subtract* 1 and *branch if non-zero*. Such an automaton can be described with a six-operations assembly language with branching labels as on the left part of Fig. 6 (see [Min67] for more on 2-counter automata). The configuration of a 2-counter automaton is defined by (n, a, b) (the line number and the values of the counters).

Two-counter automaton are intrinsically irreversible: subtracting 1 yields 0 for both values 0 and 1 and before a labeled instruction the instruction can be the one on the previous line or a branching to this label.

To achieve reversibility two stacks are added: one, Σ_i, records the instruction number (*i.e.* it records the values of the instruction counter) and another one, Σ_z, record the previous value (0 or 1) of any counter that holds zero after a subtraction. We write $x.\Sigma$ to indicate pushing x on S or that x is the top of the stack. The dynamics is described on Fig. 3 for instructions on A. The ones for B are similar. Discarding the stacks, one gets the usual dynamics. The inverse dynamics is automatic (as long as the sequence was generated legally, otherwise things like undoing adding 1 from a zero counter might happen).

Instruction at line n	Associated action
A ++	$(n, \quad a, b, \Sigma_i, \Sigma_z) \vdash (n+1, a+1, b, n.\Sigma_i, \quad \Sigma_z)$
A --	$(n, a+2, b, \Sigma_i, \Sigma_z) \vdash (n+1, a+1, b, n.\Sigma_i, \quad \Sigma_z)$
A --	$(n, \quad 1, b, \Sigma_i, \Sigma_z) \vdash (n+1, \quad 0, b, n.\Sigma_i, 1.\Sigma_z)$
A --	$(n, \quad 0, b, \Sigma_i, \Sigma_z) \vdash (n+1, \quad 0, b, n.\Sigma_i, 0.\Sigma_z)$
IF A !=0 m	$(n, \quad 0, b, \Sigma_i, \Sigma_z) \vdash (n+1, \quad 0, b, n.\Sigma_i, \quad \Sigma_z)$
IF A !=0 m	$(n, a+1, b, \Sigma_i, \Sigma_z) \vdash (\quad m, a+1, b, n.\Sigma_i, \quad \Sigma_z)$

Fig. 3. Dynamics of a 2-counter automata with memory stacks

4   Stack Implementation

Since we are dealing with rational numbers, it is very easy to implement an unbounded stack of natural numbers 1 to l in the following way: $\frac{1}{l+2}$ encodes the empty stack. Let σ be a rational number encoding of a stack ($0 < \sigma < 1$). After pushing a value v on top of a stack σ, the new stack is $\frac{v+\sigma}{l+1}$. This ensures that, as soon as the stack is not empty, $\frac{1}{l+1} < \sigma < 1$ and distinct from $\frac{i}{l+1}$, $i \in \{1, 2, \dots, l\}$. The top of the stack $\lfloor (l+1)\sigma \rfloor$ and rest of the stack is $(l+1)\sigma - \lfloor (l+1)\sigma \rfloor$.

To implement this in a signal machine, a scale is defined (because the space is continuous and scaleless) and then the push operation is implemented. We are not interested in the pop and test of emptiness because our 2-counter simulation only pushes values. The pop corresponds to the inverse of push and is thus implicitly built.

The rational σ is encoded with a zero-speed signal mem. The scale is defined with zero-speed signals mark_0, $\text{mark}_1, \dots$, mark_l. They are regularly positioned, never move and defined positions $0, 1, \dots, l$. Thus the normal position of mem is between mark_0 and mark_1. To push v, mem is translated by v (lower part of Fig. 4). Then this position is scaled by $\frac{1}{l+1}$ (upper part of Fig. 4) to get $\frac{\sigma+v}{l+1}$. The process starts at the arrival of $\overleftarrow{\text{store}_v}$.

Translating is very easy: $\overrightarrow{\text{mem}}$ and $\overrightarrow{\text{store}_v}$ are parallel. Their distance encodes σ. Their movement stops when the first one, $\overrightarrow{\text{store}_v}$, reaches mark_v. The signal catch is then issued to stop $\overrightarrow{\text{mem}}$ as in the middle of Fig. 4. This collision is distance v away from the original position of mem. This is ensured by the definition of speeds (we leave to the reader to verify this linear equation system based on the speeds given in Fig. 4). When this point is reached, a scaling remains to be done. Scaling by $\frac{1}{l+1}$, to go from $\sigma+v$ to $\frac{\sigma+v}{l+1}$, fix has to travel $(\sigma+v) - \frac{\sigma+v}{l+1} = (\sigma+v)\frac{l}{l+1}$ units, and $\overleftarrow{\text{mem}}$ and $\overrightarrow{\text{mem}}$ have to travel $(\sigma+v) + \frac{\sigma+v}{l+1} = (\sigma+v)\frac{l+2}{l+1}$ units. Thus if the speeds of $\overleftarrow{\text{mem}}$ and $\overrightarrow{\text{mem}}$ have the same absolute value then the speed of fix must be $\frac{l}{l+2}$ times the one of $\overleftarrow{\text{mem}}$ (notice in Fig. 4 that $-2 = -3\frac{4}{4+2}$).

If the stack is empty, then backward collision between $\overleftarrow{\text{mem}}$ and fix happens between mark_0 and mark_1. So that there is a (backward) collision between $\overrightarrow{\text{mem}}$ (regenerated from mark_0 and $\overleftarrow{\text{mem}}$) and catch. There no such a rule, it can be defined to have, for example, mem fixed and catch exiting on the left.

It is easy to check that the rules are invertible.

Let us note that, with slight modifications, $\overleftarrow{\text{store}_v}$ and ack could come from/be sent left or right. It is also possible to carry extra information (like a next line

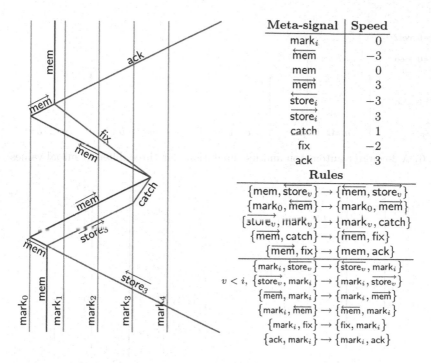

Meta-signal	Speed
mark_i	0
$\overleftarrow{\text{mem}}$	-3
mem	0
$\overrightarrow{\text{mem}}$	3
$\overleftarrow{\text{store}_i}$	-3
$\overrightarrow{\text{store}_i}$	3
catch	1
fix	-2
ack	3

Rules

$$\{\text{mem}, \overleftarrow{\text{store}_v}\} \rightarrow \{\overleftarrow{\text{mem}}, \overrightarrow{\text{store}_v}\}$$
$$\{\text{mark}_0, \overleftarrow{\text{mem}}\} \rightarrow \{\text{mark}_0, \overrightarrow{\text{mem}}\}$$
$$\{\overrightarrow{\text{store}_v}, \text{mark}_v\} \rightarrow \{\text{mark}_v, \text{catch}\}$$
$$\{\overrightarrow{\text{mem}}, \text{catch}\} \rightarrow \{\overleftarrow{\text{mem}}, \text{fix}\}$$
$$\{\overleftarrow{\text{mem}}, \text{fix}\} \rightarrow \{\text{mem}, \text{ack}\}$$

$$v < i, \quad \begin{aligned} \{\text{mark}_i, \overleftarrow{\text{store}_v}\} &\rightarrow \{\overleftarrow{\text{store}_v}, \text{mark}_i\} \\ \{\overrightarrow{\text{store}_v}, \text{mark}_i\} &\rightarrow \{\text{mark}_i, \overrightarrow{\text{store}_v}\} \\ \{\overrightarrow{\text{mem}}, \text{mark}_i\} &\rightarrow \{\text{mark}_i, \overrightarrow{\text{mem}}\} \\ \{\text{mark}_i, \overleftarrow{\text{mem}}\} &\rightarrow \{\overleftarrow{\text{mem}}, \text{mark}_i\} \\ \{\text{mark}_i, \text{fix}\} &\rightarrow \{\text{fix}, \text{mark}_i\} \\ \{\text{ack}, \text{mark}_i\} &\rightarrow \{\text{mark}_i, \text{ack}\} \end{aligned}$$

Fig. 4. Implementation the stack for $l = 4$ and push(3)

number) through the storing process. This is done by having a special set of $\overleftarrow{\text{store}_i}$, $\overrightarrow{\text{store}_i}$, catch, fix, and ack for each possible piece of information.

5 Reversible Computation

The idea is to use the construction provided in [DL05b] to simulate any 2-counter automaton with a conserving signal machine (which cannot be detailed here). Figure 5 shows how the counters are encoded using two fixed signals zero and one as a scale. A signal amounting for the current line zigzags between these signals. Figure 6 presents the code of a simple 2-counter automaton and some simulations. When a simulation stops, a signal stop appears and bounces between zero and one. In the reversible version, it exits the configuration on the right.

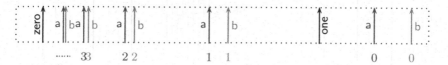

Fig. 5. Encoding positions of counters

Figure 7 show how everything is interconnected. Before going to the next configuration, the number of the just carried out instruction is stored on the

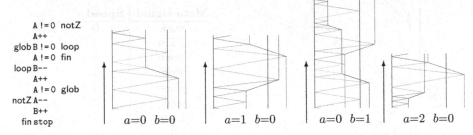

Fig. 6. A 2-counter automaton and its simulations for three different initial values

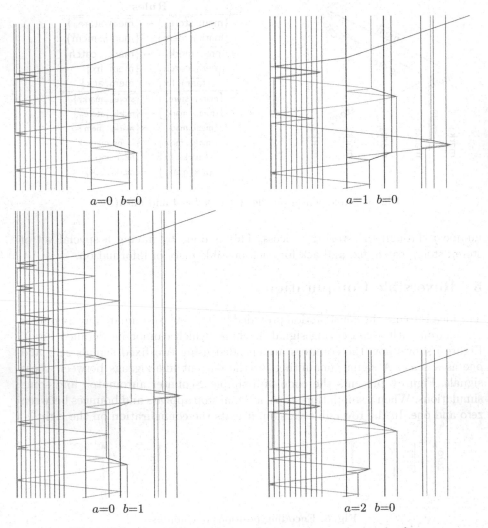

Fig. 7. The reversible simulations for same automaton and values

left while recording the next instruction number. Each time there is a subtract 1 generating a 0, the previous value of the counter is stored on the right (cases $(1, 0)$ and $(0, 1)$) whatever counter is concerned.

6   Conclusion

As far as there is no accumulation, reversible conservative rational abstract geo metrical computation has exactly the same computing capability as Turing machines (because rational numbers can be implemented exactly).

Two-counter automaton are exponentially slower than Turing machines. This is not important since we are only interested in computational issues. Nevertheless, it is easy to simulate reversibly (but without conservativeness) Turing machine in such a way that the number of collisions io propoi tional to the number of TM iterations. It would be interesting to do so to study the complexity.

It is possible to apply the iterated shrinking construction of [DL05b, DL05a] which preserves reversibility so that the black hole models can now be embedded in a reversible setting. We do not do it here for lack of room on one side and on the other side this would provoke an accumulation which is clearly not a reversible phenomena.

References

[Ada02] A. Adamatzky, ed. *Collision based computing*. Springer, 2002.

[AM95] E. Asarin and O. Maler. Achilles and the Tortoise climbing up the arithmetical hierarchy. In *FSTTCS '95*, number 1026 in LNCS, pp. 471–483, 1995.

[Ben73] C. H. Bennett. Logical reversibility of computation. *IBM J. Res. Dev.*, 6:525–532, 1973.

[Ben88] C. H. Bennett. Notes on the history of reversible computation. *IBM J. Res. Dev.*, 32(1):16–23, 1988.

[BNR91] N. Boccara, J. Nasser, and M. Roger. Particle-like structures and interactions in spatio-temporal patterns generated by one-dimensional deterministic cellular automaton rules. *Phys. Rev. A*, 44(2):866–875, 1991.

[Bou99] O. Bournez. Achilles and the Tortoise climbing up the hyper-arithmetical hierarchy. *Theoret. Comp. Sci.*, 210(1):21–71, 1999.

[Čul87] K. Čulik II. On invertible cellular automata. *Complex Systems*, 1:1035–1044, 1987.

[DL95] J. Durand-Lose. Reversible cellular automaton able to simulate any other reversible one using partitioning automata. In *LATIN '95*, number 911 in LNCS, pp. 230–244. Springer, 1995.

[DL97] J. Durand-Lose. Intrinsic universality of a 1-dimensional reversible cellular automaton. In *STACS '97*, number 1200 in LNCS, pp. 439–450. Springer, 1997.

[DL02] J. Durand-Lose. Computing inside the billiard ball model. In A. Adamatzky, ed., *Collision-based computing*, pp. 135–160. Springer, 2002.

[DL05a] J. Durand-Lose. Abstract geometrical computation 1: embedding black hole computations with rational numbers. Research Report RR-2005-05, LIFO, U. D'Orléans, France, 2005. http://www.univ-orleans.fr/lifo/prodsci/rapports.

[DL05b]   J. Durand-Lose. Abstract geometrical computation for black hole computation. In M. Margenstern, ed., *Machines, Computations, and Universality (MCU '04)*, number 3354 in LNCS, pp. 175–186. Springer, 2005.

[DL05c]   J. Durand-Lose. Abstract geometrical computation: Turing-computing ability and undecidability. In B. S. Cooper, B. Löwe, and L. Torenvliet, eds, *New Computational Paradigms, 1st Conference on Computability in Europe (CiE '04)*, number 3526 in LNCS, pp. 106–116. Springer, 2005.

[DM02]   M. Delorme and J. Mazoyer. Signals on cellular automata. in [Ada02], pp. 234–275, 2002.

[Dub95]   J.-C. Dubacq. How to simulate Turing machines by invertible 1d cellular automata. *International Journal of Foundations of Computer Science*, 6(4):395–402, 1995.

[FT82]   E. Fredkin and T. Toffoli. Conservative logic. *International Journal of Theoretical Physics*, 21, 3/4:219–253, 1982.

[Ila01]   A. Ilachinski. *Cellular automata –a discrete universe*. World Scientific, 2001.

[JS90]   G. Jacopini and G. Sontacchi. Reversible parallel computation: an evolving space-model. *Theoret. Comp. Sci.*, 73(1):1–46, 1990.

[JSS02]   M. H. Jakubowsky, K. Steiglitz, and R. Squier. Computing with solitons: a review and prospectus. in [Ada02], pp. 277–297, 2002.

[Kar90]   J. Kari. Reversibility of 2D cellular automata is undecidable. *Phys. D*, 45:379–385, 1990.

[Kar94]   J. Kari. Reversibility and surjectivity problems of cellular automata. *J. Comput. System Sci.*, 48(1):149–182, 1994.

[Kar96]   J. Kari. Representation of reversible cellular automata with block permutations. *Math. System Theory*, 29:47–61, 1996.

[Lec63]   Y. Lecerf. Machines de Turing réversibles. Récursive insolubilité en $n \in \mathbb{N}$ de l'équation $u = \theta^n u$, où θ est un isomorphisme de codes. *Comptes rendus des séances de l'académie des sciences*, 257:2597–2600, 1963.

[LN90]   K. Lindgren and M. G. Nordahl. Universal computation in simple one-dimensional cellular automata. *Complex Systems*, 4:299–318, 1990.

[LTV98]   M. Li, J. Tromp, and P. Vitanyi. Reversible simulation of irreversible computation. *Physica D*, 120:168–176, 1998.

[Min67]   M. Minsky. *Finite and infinite machines*. Prentice Hall, 1967.

[Mor92]   K. Morita. Computation-universality of one-dimensional one-way reversible cellular automata. *Inform. Process. Lett.*, 42:325–329, 1992.

[Mor95]   K. Morita. Reversible simulation of one-dimensional irreversible cellular automata. *Theoret. Comp. Sci.*, 148:157–163, 1995.

[Mor96]   K. Morita. Universality of a reversible two-counter machine. *Theoret. Comp. Sci.*, 168(2):303–320, 1996.

[MT99]   J. Mazoyer and V. Terrier. Signals in one-dimensional cellular automata. *Theoret. Comp. Sci.*, 217(1):53–80, 1999.

[TM90]   T. Toffoli and N. Margolus. Invertible cellular automata: a review. *Phys. D*, 45:229–253, 1990.

[Tof77]   T. Toffoli. Computation and construction universality of reversible cellular automata. *J. Comput. System Sci.*, 15:213–231, 1977.

[Tof80]   T. Toffoli. Reversible computing. In *ICALP*, volume 85 of *LNCS*, pp. 632–644. Springer, 1980.

LJQ: A Strongly Focused Calculus for Intuitionistic Logic*

Roy Dyckhoff and Stéphane Lengrand

School of Computer Science, University of St Andrews, St Andrews, Fife,
KY16 9SX, Scotland
{rd, sl}@dcs.st-and.ac.uk

Abstract. LJQ is a focused sequent calculus for intuitionistic logic, with
a simple restriction on the first premiss of the usual left introduction
rule for implication. We discuss its history (going back to about 1950,
or beyond), present the underlying theory and its applications both to
terminating proof-search calculi and to call-by-value reduction in lambda
calculus.

Keywords: Sequent calculus, purification, call-by-value semantics,
focused, depth-bounded, guarded logic.

1 Introduction

Proof systems for intuitionistic logic close to natural deduction are well-known
to be related to computation. For example, ordinary typed λ-calculus, with beta-
reduction, is the classic model of computation for typed functional programmes
with call-by-name (CBN) semantics; likewise, a system of uniform proofs for
Horn logic is a coherent explanation of proof search in pure Prolog, as argued
by (e.g.) [23]. The focused calculus **LJT** of Herbelin [17] (with antecedents in
work [20], [3] by Joinet et al) is an intuitionistic sequent calculus equivalent
to natural deduction (in the sense that its cut-free proofs are in a natural 1-1
correspondence with normal deductions); it also has a well-developed theory of
proof-reduction with strong normalisation [17], [10], [12]. It can thus be seen to
fulfil both these important roles, in being a basis both for proof search (where
the proofs are of interest in themselves [9]) and for functional program evaluation
with CBN-semantics. Work by the second author [21], [22] is developing the first
of these ideas for a wide range of type theories.

The purpose of the present paper is to consider a different focused calculus **LJQ**, as named by Herbelin [16] and with similar antecedents [20], [3]. We
present some aspects of its history and its applications both in structural proof
theory and in λ-calculus, with connections in the first instance to automated

---

* Thanks are due to James McKinna and Christian Urban for useful suggestions, to
  Hugo Herbelin and Sara Negri for stimulating questions, to José Carlos Espírito
  Santo for his unpublished [14] and to Jörg Hudelmaier for a copy of his thesis [18].

A. Beckmann et al. (Eds.): CiE 2006, LNCS 3988, pp. 173–185, 2006.

reasoning and in the second to call-by-value programming language semantics. Fuller details will appear elsewhere.

Vorob'ev [34] (detailing ideas published in 1952) showed (Theorem 3) that, in a minor variant of Gentzen's calculus **LJ** for intuitionistic logic, one may, without losing completeness, restrict instances of the left rule $L\supset$ for implication to those in which, if the antecedent A of the principal formula $A\supset B$ is atomic, then the first premiss is an axiom. Independently, Hudelmaier [18] showed that one could further restrict this rule to those instances where the first premiss was either an axiom or the conclusion of a right rule; the result was proved in his [18] and described in [19] as folklore. The same result is mentioned by Herbelin in [16] as the completeness of a certain calculus **LJQ**, described simply as **LJ** with the last-mentioned restriction.

It is convenient to formalise such restrictions in terms of a sequent calculus **LJQ'** with two forms of sequent; letting Γ range over multisets of formulae, we have the *ordinary* sequent $\Gamma \Rightarrow A$ to express the deducibility of the formula A from assumptions Γ, and the *focused* sequent $\Gamma \to A$ to impose the restriction that the last step in the deduction is by an axiom or a right rule (i.e. with the succedent formula principal). A natural deduction interpretation is straightforward. Note that the focused sequent $p \vee q \to p \vee q$ is not derivable; the last step of its derivation can only be a right-introduction step.

The rules of the calculus are then as presented below, in Sect. 2. We use the name **LJQ'** rather than **LJQ** both to indicate the explicit focusing (use of two kinds of sequent) and the extra focusing (in the premisses of right rules for $\vee$ and $\wedge$). In later sections, when we consider a term calculus to represent derivations, we revert to the generic name **LJQ** for this kind of calculus.

For example, the rule $L\supset'$ has, as conclusion and second premiss, ordinary sequents, but as first premiss a focused sequent, capturing the restriction on proofs discussed by Hudelmaier (given that the focused sequents are exactly the axioms or the conclusions of right introduction rules). However, further restrictions are allowed: our right rules for disjunction and conjunction also have focused premisses. This represents a strengthening of Hudelmaier's folklore result.

LJQ as described in [16] originates in linear logic, in work by Danos *et al* [3] without mention of disjunction and conjunction. This in turn goes back to the thesis [20] of Joinet. Focusing itself is a technique pioneered by Andreoli [1] (but one of the points of our paper is a demonstration of its origins in much earlier work).

Such calculi are of interest not just because of the restricted proof search imposed by the focusing but because the completeness of **LJQ** (or of **LJQ'**) has as an easy corollary the completeness of more specialised "depth-bounded" calculi (as devised e.g. by [34], [18], [5]) in which proof search has limited (e.g. linear) depth (a.k.a. "height"); [33] gives a convenient account of the **G4ip** calculus, as it is there called.

These focused calculi are complementary to other focused calculi like Herbelin's **LJT**, as studied in [16], [17], [24], [32], [33], [9], [10], [12].

The present extended abstract outlines the theory (Sect. 2), presents some variations (Sect. 3), summarises some applications (Sect. 4), presents the calculus with term annotations (Sect. 5) (including a strongly normalising reduction system for **LJQ** and a preservation theorem relating **LJQ** to Moggi's calculus λ_C) and summarises some related work (joint with Delia Kesner: Sect. 6).

2 LJQ'

Basic syntactic conventions are as in [33]; in particular, P is a metavariable for "proposition variables" and Γ indicates a multiset of formulae. The symbols p and q are distinct proposition variables. The rules of **LJQ'** are as given below in Fig. 1.

$$\frac{}{\Gamma, \bot \Rightarrow A} L\bot \qquad \frac{\Gamma \to A}{\Gamma \Rightarrow A} Der \qquad \frac{}{\Gamma, P \to P} Ax$$

$$\frac{\Gamma, A{\supset}B \to A \quad \Gamma, B \Rightarrow C}{\Gamma, A{\supset}B \Rightarrow C} L{\supset}' \qquad \frac{A, \Gamma \Rightarrow B}{\Gamma \to A{\supset}B} R{\supset}'$$

$$\frac{\Gamma, A \Rightarrow C \quad \Gamma, B \Rightarrow C}{\Gamma, A \vee B \Rightarrow C} L\vee' \qquad \frac{\Gamma \to A_i}{\Gamma \to A_0 \vee A_1} R\vee'$$

$$\frac{\Gamma, A, B \Rightarrow C}{\Gamma, A \wedge B \Rightarrow C} L\wedge' \qquad \frac{\Gamma \to A \quad \Gamma \to B}{\Gamma \to A \wedge B} R\wedge'$$

Fig. 1. Rules of **LJQ'**

Expressed in terms of our notation, the right rule for conjunction in the calculus **LJQ** of [16] would be

$$\frac{\Gamma \Rightarrow A \quad \Gamma \Rightarrow B}{\Gamma \to A \wedge B}$$

and similarly for disjunction; the definition of "pure" derivations in [18] could be expressed in similar terms. The rule Der is named after the dereliction rule in linear logic; the latter rule has (used from conclusion to premiss) a similar effect, enabling a transition between a sequent where a certain formula is optional to one where it is required. The restriction to proposition variables P in Ax has the effect that the natural deduction interpretations of derivations are in long normal form. Use of arbitrary axioms $\Gamma, A \to A$ would give a different notion of derivability, e.g. $p \vee q \to p \vee q$ would be derivable. A formula is *irreducible* when it is of the form P or $B{\supset}C$. To save space, proofs omit treatment of absurdity, conjunction and disjunction; details will appear in the full paper. Results as stated apply to the full calculus.

Lemma 1. *All sequents of the following form are derivable:*

1. *$\Gamma, A, A{\supset}B \Rightarrow B$;*
2. *$\Gamma, A \to A$ for irreducible A;*
3. *$\Gamma, A \Rightarrow A$.*

Proof: The three parts are handled by a simultaneous induction on the sizes of $A \supset B$, A and A respectively. Each part is allowed to depend on its predecessor (up to the same size) and on itself and its successors (at smaller sizes).   □

The condition "for irreducible A" is needed once absurdity, disjunction and conjunction are included in the language; if we omit them all, then the condition can be omitted. *Weakening* rules, some *Inversion* rules and *Contraction* rules are routinely shown to be admissible.

Theorem 1. *The following* Cut *rules are admissible:*

$$\frac{\Gamma \to A \quad A, \Gamma' \to B}{\Gamma, \Gamma' \to B} C_1 \qquad \frac{\Gamma \to A \quad A, \Gamma' \Rightarrow B}{\Gamma, \Gamma' \Rightarrow B} C_2 \qquad \frac{\Gamma \Rightarrow A \quad A, \Gamma' \Rightarrow B}{\Gamma, \Gamma' \Rightarrow B} C_3$$

Proof: Simultaneous induction on *cut rank* (size of cut formula A, height of first derivation d_1, height of second derivation d_2), with case analysis.   □

Note that $p \to p$ and $p, p \supset q \Rightarrow q$ and $q \to q$ and $q \Rightarrow q$ are all derivable but that $p, p \supset q \to q$ is not derivable, hence the rules

$$\frac{\Gamma \to A \quad A, \Gamma' \Rightarrow B}{\Gamma, \Gamma' \to B} \qquad \frac{\Gamma \Rightarrow A \quad A, \Gamma' \to B}{\Gamma, \Gamma' \to B} \qquad \frac{\Gamma \Rightarrow A \quad A, \Gamma' \Rightarrow B}{\Gamma, \Gamma' \to B}$$

are not admissible.

Corollary 1. *The following rules are admissible:*

$$\frac{\Gamma \Rightarrow A \quad A, \Gamma' \to B}{\Gamma, \Gamma' \Rightarrow B} C_4 \qquad \frac{\Gamma \to A \quad A, \Gamma' \to B}{\Gamma, \Gamma' \Rightarrow B} C_5$$

Proof: Using *Der*.   □

Corollary 2. *The following rules are admissible:*

$$\frac{\Gamma, A \Rightarrow B}{\Gamma \Rightarrow A \supset B} R \supset \qquad \frac{\Gamma, A \supset B \Rightarrow A \quad \Gamma, B \Rightarrow C}{\Gamma, A \supset B \Rightarrow C} L \supset$$

Proof: The first is derivable using $R \supset'$ and *Der*. The second can be achieved, using Lemma 1 for the premiss $A, A \supset B \Rightarrow B$, as

$$\frac{\Gamma, A \supset B \Rightarrow A \quad \dfrac{\dfrac{\cdots}{A, A \supset B \Rightarrow B} \quad \Gamma, B \Rightarrow C}{A, A \supset B, \Gamma \Rightarrow C} C_3}{\dfrac{\Gamma, A \supset B, \Gamma, A \supset B \Rightarrow C}{\Gamma, A \supset B \Rightarrow C} Contr} C_3 \qquad □$$

It follows from Corollary 2 and Lemma 1 (3) that this calculus **LJQ'** is as strong as **G3ip**. Since a derivation therein becomes a **G3ip** derivation if we ignore the distinction between the two kinds of sequent (and remove instances of Der), the two calculi are equivalent.

3   Variations

Several variations (and combinations of the variations) on the above are possible.

The first is to include the principal formula $A \supset B$ in the antecedent of the second premiss of $L\supset'$. This is preferable when we come to consider a term calculus, in Sect. 5 below; from the point of view of derivability it makes no difference.

The second removes the focusing from the premisses of the rules $R\wedge$ and $R\vee$ (this gives us the calculus **LJQ** of Herbelin [16]). Completeness of the calculus so modified is an immediate corollary of the completeness of **LJQ'**, since the focused versions (as we have presented them) are derivable using the unfocused versions and the Der rule.

The third is a multi-succedent version **LJQ*** (a variant of this appears in [16], page 78). We use two kinds of sequent as before; but this time, because of the need for a multiple succedent, we have a semi-colon to separate the focused formula (the *stoup*) from the rest of the succedent. The rules of **LJQ*** are as in Fig. 2 ($-$ indicates an empty multi-set).

$$\frac{}{\Gamma, \bot \Rightarrow \Delta} L\bot^* \qquad \frac{\Gamma \to A; \Delta}{\Gamma \Rightarrow A, \Delta} Der^* \qquad \frac{}{\Gamma, P \to P; \Delta} Ax^*$$

$$\frac{\Gamma, A\supset B \to A; - \quad \Gamma, B \Rightarrow \Delta}{\Gamma, A\supset B \Rightarrow \Delta} L\supset^* \qquad \frac{A, \Gamma \Rightarrow B}{\Gamma \to A\supset B; \Delta} R\supset^*$$

$$\frac{\Gamma, A \Rightarrow \Delta \quad \Gamma, B \Rightarrow \Delta}{\Gamma, A \vee B \Rightarrow \Delta} L\vee^* \qquad \frac{\Gamma \Rightarrow A, B, \Delta}{\Gamma \to A \vee B; \Delta} R\vee^*$$

$$\frac{\Gamma, A, B \Rightarrow \Delta}{\Gamma, A \wedge B \Rightarrow \Delta} L\wedge^* \qquad \frac{\Gamma \to A; \Delta \quad \Gamma \to B; \Delta}{\Gamma \to A \wedge B; \Delta} R\wedge^*$$

Fig. 2. Rules of the multi-succedent calculus **LJQ***

The crucial *Cut* rules are

$$\frac{\Gamma \to A; \Delta \quad A, \Gamma' \to B; \Delta'}{\Gamma, \Gamma' \to B; \Delta, \Delta'} C_1 \qquad \frac{\Gamma \to A; \Delta \quad A, \Gamma' \Rightarrow \Delta'}{\Gamma, \Gamma' \Rightarrow \Delta, \Delta'} C_2$$

$$\frac{\Gamma \Rightarrow \Delta, A \quad A, \Gamma' \Rightarrow \Delta'}{\Gamma, \Gamma' \Rightarrow \Delta, \Delta'} C_3$$

and these are admissible by a routine argument similar to that already given.

4 Applications

4.1 Completeness of G4ip

The calculus **G4ip** was introduced by Hudelmaier [18], who gives a completeness proof that is essentially the following. For brevity we omit consideration of absurdity, conjunction and disjunction. Formulae are *weighted* as follows: $w(P) = 0$ and $w(A \supset B) = 1 + w(A) + w(B)$. Sequents $\Gamma \Rightarrow A$ are then ordered using the multi-set ordering (on the multiset Γ, A); in effect, this allows us to refer to the *weight* of a sequent. The rules are as follows.

$$\frac{}{P, \Gamma \Rightarrow P} \; Ax. \qquad\qquad \frac{A, \Gamma \Rightarrow B}{\Gamma \Rightarrow A \supset B} \; R\supset$$

$$\frac{P, B, \Gamma \Rightarrow E}{P, P\supset B, \Gamma \Rightarrow E} \; L0\supset \qquad \frac{D\supset B, C, \Gamma \Rightarrow D \quad B, \Gamma \Rightarrow E}{(C\supset D)\supset B, \Gamma \Rightarrow E} \; L\supset\supset$$

Note that every inference has as its conclusion a sequent with greater weight than each premiss; so root-first proof search is terminating, in a depth bounded by the weight of the sequent being proved.

Proposition 1 (Completeness of G4ip)

1. *If $\Gamma \to E$ is derivable in* **LJQ′**, *then $\Gamma \Rightarrow E$ is derivable in* **G4ip**.
2. *If $\Gamma \Rightarrow E$ is derivable in* **LJQ′**, *then $\Gamma \Rightarrow E$ is derivable in* **G4ip**.

Proof: By simultaneous induction on the sequent weight, using case analysis on the last step of the derivation. For (1), the last step is either an axiom (in which case we are done) or an $R\supset'$ inference, where the inductive hypothesis (2) can be used. For (2), the last step is either a dereliction, in which case (1) (for the same weight) is used, or an $L\supset'$ inference with principal formula $A\supset B$. In the latter case, if A is an atom P, then the first premiss is an axiom, with P in Γ; the inductive hypothesis (2) applied to the second premiss followed by an $L0\supset$-inference provides the required **G4ip** derivation. Otherwise, with $A = C\supset D$ and $\Gamma = (C\supset D)\supset B, \Gamma'$, the premisses are $\Gamma', (C\supset D)\supset B \to C\supset D$ and $\Gamma', B \Rightarrow E$. The inductive hypothesis (2) provides a **G4ip** derivation of $\Gamma', B \Rightarrow E$. The first premiss must be the conclusion from $\Gamma', (C\supset D)\supset B, C \Rightarrow D$, whose derivability in **LJQ′** easily implies that of the less weighty sequent $\Gamma', D\supset B, C \Rightarrow D$. The induction hypothesis (2) provides a **G4ip** derivation of this, and an $L\supset\supset$ inference provides a **G4ip** derivation of $\Gamma', (C\supset D)\supset B \Rightarrow E$. □

An almost identical argument, using **LJQ*** from Section 3, demonstrates the completeness of the multi-succedent version of **G4ip** in [5].

4.2 Completeness of Dragalin's GHPC

Dragalin [4] presented a multi-succedent sequent calculus **GHPC** for intuitionistic predicate logic, with the feature that the first premiss of the left rule for implication was single-succedent. This feature also appears in **LJQ*** in Section 3. An easy argument based on the completeness of the calculus **LJQ*** shows the completeness of **GHPC**; every inference (except by *Der*) in **LJQ*** becomes an inference in **GHPC**, and *Der* can be simulated by *Weakening* in **GHPC**.

4.3 Calculi for (Intuitionistic) Guarded Logic

Guarded first-order classical logic is of interest for its ability to interpret modal logics. The "guarded" restriction on formulae is that universal quantifiers are allowed only in the form $\forall \mathbf{x}(P{\supset}A)$, where $\mathbf{x}$ is a list of variables, P is an atom, A is a formula with $FV(A) \subseteq FV(P)$ and all the variables bound by the quantifier are free in P, i.e. *guarded* by the atom P; there is a similar restriction on existential quantifiers. No function symbols are allowed. In such a situation, the free variables in P (and hence in A) are a combination of those in $\mathbf{x}$ and possibly others. So, we indicate by $P\mathbf{xy}$ (resp. $A\mathbf{xy}$) an atom (resp. formula) all of whose free variables are in $\mathbf{x}, \mathbf{y}$ and by $P\mathbf{zy}$ (resp. $A\mathbf{zy}$) the result of substituting $\mathbf{z}$ for $\mathbf{x}$ therein. The notation $(\forall \mathbf{x}{:}P\mathbf{xy})A\mathbf{xy}$ then abbreviates $\forall \mathbf{x}(P\mathbf{xy}{\supset}A\mathbf{xy})$.

It is of interest to see whether this specialised form of quantification leads to a specialised inference rule. We treat this in the intuitionistic case; full details of this (treating also the existential quantifier) and the classical case are given in [11], including cut-admissibility proofs. The relevant inference rules are

$$\frac{P\mathbf{zy}, \Gamma, (\forall \mathbf{x}{:}P\mathbf{xy})A\mathbf{xy}, A\mathbf{zy} \Rightarrow B}{P\mathbf{zy}, \Gamma, (\forall \mathbf{x}{:}P\mathbf{xy})A\mathbf{xy} \Rightarrow B} L\forall' \qquad \frac{\Gamma, P\mathbf{zy} \Rightarrow A\mathbf{zy}}{\Gamma \Rightarrow (\forall \mathbf{x}{:}P\mathbf{xy})A\mathbf{xy}} R\forall'$$

where the variables $\mathbf{z}$ are *fresh* (i.e. disjoint from the free variables of $\Gamma, A\mathbf{xy}$) in the $R\forall'$ rule. We regard the atom $P\mathbf{zy}$ in the conclusion of the $L\forall'$ rule as a *key* that *unlocks* the guard on $A\mathbf{xy}$.

The first of these may be considered to be the composition of a standard $L\forall$ rule and a $L{\supset}$ rule, as in

$$\frac{\dfrac{}{P\mathbf{zy}, \Gamma, (\forall \mathbf{x}{:}P\mathbf{xy})A\mathbf{xy}, P\mathbf{zy}{\supset}A\mathbf{zy} \Rightarrow P\mathbf{zy}} Ax \qquad P\mathbf{zy}, \Gamma, (\forall \mathbf{x}{:}P\mathbf{xy})A\mathbf{xy}, A\mathbf{zy} \Rightarrow B}{\dfrac{P\mathbf{zy}, \Gamma, (\forall \mathbf{x}{:}P\mathbf{xy})A\mathbf{xy}, P\mathbf{zy}{\supset}A\mathbf{zy} \Rightarrow B}{P\mathbf{zy}, \Gamma, (\forall \mathbf{x}{:}P\mathbf{xy})A\mathbf{xy} \Rightarrow B} L\forall} L{\supset}$$

with the same restriction as in **LJQ** that the first premiss of the $L{\supset}$ inference have its succedent principal. Since this succedent is (by the guarded restriction) an atom $P\mathbf{zy}$, that means it must occur in the antecedent, as indicated. Thus, the **LJQ** restriction occurs also in this context, of intuitionistic guarded logic.

4.4 Negri's Conservativity Theorem

Negri [26] showed conservativity of the intuitionistic propositional theory of apartness over the theory of equality defined as the negation of apartness. The first complete proof used the calculus **G3ip** as basic; this was simplified in [27] once the completeness of the calculus **G4ip** (extended with rules for apartness) was demonstrated (in [8]). The use of **G4ip** was explained in [27] as "allowing a better control on derivations". In retrospect, it appears[1] that the use of the **LJQ** calculus would have sufficed.

---

[1] Personal communication from Sara Negri (Summer 2004).

5   LJQ with Terms

In this section we describe **LJQ** as the typing system of a term syntax, which we then use to establish a connection between **LJQ** and the call-by-value λ-calculus λ_C of Moggi [25]. For brevity, we consider in this section implication only, and the main distinction between **LJQ** and **LJQ**' can therefore be ignored; so, hereafter we just use the name **LJQ**.

5.1   A Term Calculus for LJQ

This term syntax is described as follows:

$$V, V' \quad ::= x \mid \lambda x.M \mid C_1(V, x.V')$$
$$M, N, P ::= \uparrow V \mid x(V, y.N) \mid C_2(V, x.N) \mid C_3(M, x.N)$$

The terms $C_i(-, -.-)$ are explicit substitutions, to be distinguished from the meta-notation $M\{x = N\}$ standing for "M with x replaced by N". Binding occurrences of variables are those immediately followed by ".". A term without any occurrence of a C_i is said to be *cut-free*. *Values* are cut-free terms of the form V.

The typing rules, shown in Fig. 3, are naturally derived from Fig. 1. Note that the rules Ax and $R\supset'$ with focused conclusions are those that type values. There are three changes, all more appropriate for the consideration of proof-terms.

The first change allows Ax to have an arbitrary formula as principal; by Lemma 1 this is acceptable in the implicational case.

The second change is that $L\supset'$ now allows the use of the formula $A\supset B$ in the proof of its second premiss, thus widening the space of proofs(-terms), as in Sect. 3. For instance, when establishing a connection with λ-calculus, this enables the proper representation of Church numerals; otherwise they would all (except 0) be mapped to the same proof-term.

$$\frac{}{\Gamma, x : A \to x : A} Ax \qquad\qquad \frac{\Gamma \to V : A}{\Gamma \Rightarrow \uparrow V : A} Der$$

$$\frac{\Gamma, x : A \Rightarrow M : B}{\Gamma \to \lambda x.M : A\supset B} R\supset' \qquad \frac{\Gamma, x : A\supset B \to V : A \quad \Gamma, x : A\supset B, y : B \Rightarrow N : C}{\Gamma, x : A\supset B \Rightarrow x(V, y.N) : C} L\supset'$$

$$\frac{\Gamma \to V : A \quad \Gamma, x : A \to V' : B}{\Gamma \to C_1(V, x.V') : B} C_1 \qquad \frac{\Gamma \to V : A \quad \Gamma, x : A \Rightarrow N : B}{\Gamma \Rightarrow C_2(V, x.N) : B} C_2$$

$$\frac{\Gamma \Rightarrow M : A \quad \Gamma, x : A \Rightarrow N : B}{\Gamma \Rightarrow C_3(M, x.N) : B} C_3$$

Fig. 3. LJQ with terms

The third change is that we include the cut rules as primitive; in contrast to those earlier, they are context-sharing (i.e. additive) rather than context-splitting (i.e. multiplicative or context-independent). This removes the need to formulate the admissibility of *Contraction* separately from the admissibility of cuts, the former being an easy sub-case of the latter.

(B) $C_3(\uparrow \lambda x.M, y.y(V, z.P))$	$\longrightarrow$	$C_3(C_3(\uparrow V, x.M), z.P)$
		if $y \notin FV(V) \cup FV(P)$
$C_3(\uparrow x, y.N)$	$\longrightarrow$	$N\{y = x\}$
$C_3(M, y.\uparrow y)$	$\longrightarrow$	M
$C_3(z(V, y.P), x.N)$	$\longrightarrow$	$z(V, y.C_3(P, x.N))$
$C_3(C_3(\uparrow V', y.y(V, z.P)), x.N)$	$\longrightarrow$	$C_3(\uparrow V', y.y(V, z.C_3(P, x.N)))$
		If $y \notin FV(V) \cup FV(P)$
$C_3(C_3(M, y.P), x.N)$	$\longrightarrow$	$C_3(M, y.C_3(P, x.N))$
		if the redex is not one of the previous rule
$C_3(\uparrow \lambda y.M, x.N)$	$\longrightarrow$	$C_2(\lambda y.M, x.N)$
		if N is not an x-covalue (see below)
$C_1(V, x.x)$	$\longrightarrow$	V
$C_1(V, x.y)$	$\longrightarrow$	y
$C_1(V, x.\lambda y.M)$	$\longrightarrow$	$\lambda y.C_2(V, x.M)$
$C_2(V, x.\uparrow V')$	$\longrightarrow$	$\uparrow C_1(V, x.V')$
$C_2(V, x.x(V', z.P))$	$\longrightarrow$	$C_3(\uparrow V, x.x(C_1(V, x.V'), z.C_2(V, x.P)))$
$C_2(V, x.x'(V', z.P))$	$\longrightarrow$	$x'(C_1(V', x.V), z.C_2(V, x.P))$
$C_2(V, x.(C_3(M, y.P)))$	$\longrightarrow$	$C_3(C_2(V, x.M), y.C_2(V, x.P))$
η $\lambda x.y(x, z.z)$	$\longrightarrow$	y

N is an x-*covalue* iff $N = \uparrow x$ or N is of the form $x(V, z.P)$ with $x \notin FV(V) \cup FV(P)$

Fig. 4. LJQ-reductions

The reduction rules for the calculus are shown in Fig. 4. This reduction system has the following properties:

1. It reduces any term that is not cut-free;
2. It satisfies the *Subject Reduction* property;
3. It is confluent;
4. It is *Strongly Normalising*;
5. A fortiori, it is *Weakly Normalising*.

As a corollary of 1, 2 and 5, we have the admissibility of *Cut*. It is interesting to see in the proof of *Subject Reduction* how these reductions transform the proof derivations and to compare them to those used in the proof of Theorem 1—details will be in the full paper. Apart from the differences between the inference rules already mentioned, there are also differences between the proof-transformations. The reduction system here is more subtle, because we are now interested not only in its weak normalisation but also in its strong normalisation and its connection with call-by-value λ-calculus.

The main reduction rule (B), breaking a cut on an implication into cuts on its direct sub-formulae, is now done with C_3 rather than with C_2. The reason is that we use C_3 to encode each β-redex of λ-calculus and C_2 to simulate the evaluation of its substitutions. Just as in λ-calculus, where substitutions can be pushed through β-redexes, so may C_2 be pushed through C_3, by use of the penultimate rule of Fig. 4 (which is not needed if the only concern is cut-admissibility).

Similarly, the last rule, (η), which has nothing to do with cut-elimination, is needed to account for η-conversion in (call-by-value) λ-calculus. It is interesting to see its meaning in proof theory: it generates an axiom on an implication, given a proof on the same sequent built from axioms on each of its direct sub-formulae, and then left and right introductions of the implication. In fact, recursive application of the reverse transform is precisely what is used to prove that one can safely restrict **LJQ** (in the implicational case) to atomic axioms.

5.2   Connection with Call-by-Value λ-Calculus

We will now be precise about what we call *CBV λ-calculus*. In [30], Plotkin introduces λ_V, a calculus whose terms are exactly those of Church's λ-calculus and whose reduction rule, called β_V, is merely β-reduction restricted to the case where the argument is a *value*, i.e. a variable (typed by an axiom) or an abstraction (typed by implication introduction).

However, the equational theory produced by β_V-conversion is shown [30] to be incomplete with respect to some canonical call-by-value semantics called *Continuation Passing Style*. Therefore, λ_V was later extended [25] to λ_C with a let . = ... in ...-construct (like our *Cut*-constructs for **LJQ**) and additional reduction rules; [31] shows, in effect, that the equational theory matches the CBV-semantics. Terms of λ_C are defined as follows:

$$M, N, P ::= x \mid \lambda x.M \mid M\ N \mid \text{let } x = M \text{ in } N$$

We use V as a meta-variable ranging only over values. The reduction rules of λ_C are as follows:

$(\lambda x.M)\ V$	$\longrightarrow M\{x = V\}$	
let $x = V$ in M	$\longrightarrow M\{x = V\}$	
$M\ N$	$\longrightarrow$ let $x = M$ in $(x\ N)$	(M not a value)
$V\ N$	$\longrightarrow$ let $y = N$ in $(V\ y)$	(N not a value)
let $x = M$ in x	$\longrightarrow M$	
let $y = (\text{let } x = M \text{ in } N)$ in P	$\longrightarrow$ let $x = M$ in (let $y = N$ in P)	

The reduction η_V can usefully be added: $\lambda x.(V\ x) \longrightarrow_{\eta_V} V$ if $x \notin FV(V)$ In the presence of β_V, the following rule has the same effect:

$$\lambda x.(y\ x) \longrightarrow_{\eta_V} y \qquad \text{if } x \neq y$$

We define the translation $.^\flat$ from **LJQ**-terms to λ_C by induction on the structure of terms:

$$
\begin{aligned}
x^\flat &= x \\
(\lambda x.M)^\flat &= \lambda x.M^\flat \\
(\uparrow V)^\flat &= V^\flat \\
(x(V, y.M))^\flat &= \text{let } y = x\, V^\flat \text{ in } M^\flat \\
(C_3(N, x.M))^\flat &= \text{let } x = N^\flat \text{ in } M^\flat \\
(C_2(V, x.M))^\flat &= M^\flat\{x = V^\flat\} \\
(C_1(V, x.V'))^\flat &= V'^\flat\{x = V^\flat\}
\end{aligned}
$$

We define the translation $.^\sharp$ from λ_C to **LJQ** by a similar induction, using an auxiliary translation $.^\natural$ from values to values (a measure shows that the definitions are well-founded).

$$
\begin{aligned}
x^\natural &= x \\
(\lambda x.M)^\natural &= \lambda x.M^\sharp \\
V^\sharp &= \uparrow V^\natural \\
(\text{let } y = x\, V \text{ in } P)^\sharp &= x(V^\natural, y.P^\sharp) \\
(\text{let } y = (\lambda x.M)\, V \text{ in } P)^\sharp &= C_3(\lambda x.M^\sharp, z.z(V^\natural, y.P^\sharp)) \\
(\text{let } z = V\, N \text{ in } P)^\sharp &= (\text{let } y = N \text{ in } (\text{let } z = V\, y \text{ in } P))^\sharp \\
& \qquad \text{if } N \text{ is not a value} \\
(\text{let } z = M\, N \text{ in } P)^\sharp &= (\text{let } x = M \text{ in } (\text{let } z = x\, N \text{ in } P))^\sharp \\
& \qquad \text{if } M \text{ is not a value} \\
(\text{let } z = (\text{let } x = M \text{ in } N) \text{ in } P)^\sharp &= (\text{let } x = M \text{ in } (\text{let } z = N \text{ in } P))^\sharp \\
(\text{let } y = V \text{ in } P)^\sharp &= C_3(V^\natural, y.P^\sharp) \\
(M\, N)^\sharp &= (\text{let } y = M\, N \text{ in } y)^\sharp
\end{aligned}
$$

Notice that if M is a C_1/C_2-free term of **LJQ**, $M^{\flat\sharp} = M$ and that for any term M of λ_C, $M \longleftrightarrow^* M^{\sharp\flat}$. Now we can state (using $\longrightarrow^*$ for the reflexive transitive closure of $\longrightarrow$, etc) the following:

Theorem 2 (Preservation Theorem)

1. *For any terms M and N of λ_C, if $M \longrightarrow N$ then $M^\sharp \longrightarrow^* N^\sharp$.*
2. *For any terms M and N of **LJQ**, $M \longleftrightarrow^* N$ iff $M^\flat \longleftrightarrow^* N^\flat$.*

Hence, if a term M of **LJQ** is given the CBV-semantics of $M^\flat$, **LJQ** inherits from λ_C a semantics that captures exactly its equational theory.

Ongoing work includes refining the connection above and generalising it to a framework that would also account for the call-by-name discipline, by using a calculus introduced by Espírito Santo [14].

6   G4ip with Terms

Bringing some of the above ideas together, we can regard **G4ip** itself as the typing system for a term calculus. The associated reduction system for cuts also has the strong normalisation property. Details are in [6]. The main point of interest is the avoidance of auxiliary operations (corresponding to admissibility lemmas) in favour of uses of instances of the explicit substitution operation.

7 Conclusion

We have presented, proved complete and shown some applications of a strongly focused calculus **LJQ'**, incorporating and extending the restrictions on derivations explicit in the calculus **LJQ** of Herbelin [16], implicit in the work on "purification" in Hudelmaier [18] and with early traces in the work of Vorob'ev [34]. These applications range from sequent calculi for automated proof search to CBV-semantics of λ-calculus.

References

1. J.-M. Andreoli. *Logic programming with focusing proofs in linear logic*, J. Logic & Computation **2**, 297–347, 1992.
2. B. Cooper, J. Truss. "Sets and Proofs", (Proceedings of Logic Colloquium 97), Cambridge University Press, 1999.
3. V. Danos, J.-B. Joinet, H. Schellinx. *LKQ and LKT: sequent calculi for second order logic based upon dual linear decompositions of classical implication*, in [15], pp 211–224.
4. A. G. Dragalin. "Mathematical Intuitionism", Translations of Mathematical Monographs, **67**, Amer. Math. Soc., Providence, Rhode Island, 1988.
5. R. Dyckhoff. *Contraction-free sequent calculi for intuitionistic logic*, J. Symbolic Logic **57**, 795–807, 1992.
6. R. Dyckhoff, S. Lengrand, D. Kesner. *Strong cut-elimination systems for Hudelmaiers depth-bounded sequent calculus for implicational logic*, submitted, 2006.
7. R. Dyckhoff, S. Negri. *Admissibility of structural rules for contraction-free systems of intuitionistic logic*, J. Symbolic Logic **65**, 1499–1518, 2000.
8. R. Dyckhoff, S. Negri. *Admissibility of structural rules for extensions of contraction-free sequent calculi*, Logic J. of the IGPL **9**, 573–580, 2001
9. R. Dyckhoff, L. Pinto. *Proof search in constructive logics*, in [2], 53–65.
10. R. Dyckhoff, L. Pinto. *Cut-elimination and a permutation-free sequent calculus for intuitionistic logic*, Studia Logica **60**, 107–118, 1998.
11. R. Dyckhoff, A. Simpson. *Proof theory of guarded logics*, MS, 2004.
12. R. Dyckhoff, C. Urban. *Strong normalization of Herbelin's explicit substitution calculus with substitution propagation*, J. Logic & Comput. **13**, 689–706, 2003.
13. R. Dyckhoff. *Variations on a theme of Hudelmaier*, MS, 2006.
14. J. Espírito Santo. *Unity in structural proof theory and structural extensions of the λ-calculus*, MS available from http://www.math.uminho.pt/~jes/Publications.htm , Jul 2005.
15. J.-Y. Girard, Y. Lafont, L. Regnier (eds). "Advances in Linear Logic", LMS Lecture Note Series **222**, Cambridge University Press, 1995.
16. H. Herbelin. *Séquents qu'on calcule*, Thèse de Doctorat, Université Paris 7, 1995.
17. H. Herbelin. *A λ-calculus structure isomorphic to sequent calculus structure*, in [29], 67–75.

18. J. Hudelmaier. *Bounds for cut-elimination in intuitionistic propositional logic*, PhD thesis, Tübingen, 1989.
19. J. Hudelmaier. *An O(n log(n)) decision procedure for intuitionistic propositional logic*, J. of Logic & Computation **3**, 63–75, 1993.
20. J.-B. Joinet. *Étude de la normalisation du calcul des séquents classique à travers la logique linéaire*, Thèse de Doctorat, Université Paris 7, 1993.
21. S. Lengrand. *Normalisation and equivalence in proof theory and type theory*, Draft PhD thesis (Paris 7 & St Andrews), 2006.
22. S. Lengrand, R. Dyckhoff. *Type theory in sequent calculus*, submitted, 2006.
23. D. Miller, G. Nadathur, F. Pfenning, A. Scedrov. *Uniform proofs as a foundation for logic programming*, Ann. Pure Appl. Log. **91**, 125–157, 1991.
24. G. Mints. *Normal forms for sequent derivations*, in [28], 469–492.
25. E. Moggi. *Computational lambda-calculus and monads*, Report ECS-LFCS-88-66, University of Edinburgh, 1988.
26. S. Negri. *Sequent calculus proof theory of intuitionistic apartness and order relations*, Arch. Math. Logic **38**, 521–547, 1999.
27. S. Negri. *Conservativity of apartness over equality, revisited*, Research Report CS/99/4, University of St Andrews, 6 pp, 1999.
28. P. Odifreddi (ed). "Kreiseliana", A. K. Peters, Wellesley, Mass., 1996.
29. L. Pacholski, J. Tiuryn (eds). "Proceedings of Computer Science Logic 1994", LNCS **933**, Springer, 1995.
30. G. Plotkin. *Call-by-name, call-by-value and the lambda-calculus*, Theoret. Comput. Sci. **1**, 125–159, 1975.
31. A. Sabry, M. Felleisen. *Reasoning about programs in continuation-passing style*, Lisp and Symb. Comput. **6**, 289–360, 1993.
32. A. S. Troelstra, *Marginalia on sequent calculi*, Studia Logica **62**, 291–303, 1999.
33. A. S. Troelstra, H. Schwichtenberg. "Basic Proof Theory", Cambridge University Press, 2nd ed., 2000.
34. N. N. Vorob'ev. *A new algorithm for derivability in the constructive propositional calculus*, Amer. Math. Soc. Translations, ser. 2, vol. 94, 37–71, 1970.

Böhm Trees, Krivine's Machine and the Taylor Expansion of Lambda-Terms

Thomas Ehrhard[1] and Laurent Regnier[2,⋆]

[1] Preuves, Programmes et Systèmes (UMR 7126)
Thomas.Ehrhard@pps.jussieu.fr
[2] Institut de Mathématiques de Luminy (UMR 6206)
Laurent.Regnier@iml.univ-mrs.fr

Abstract. We introduce and study a version of Krivine's machine which provides a precise information about how much of its argument is needed for performing a computation. This information is expressed as a term of a resource lambda-calculus introduced by the authors in a recent article; this calculus can be seen as a fragment of the differential lambda-calculus. We use this machine to show that Taylor expansion of lambda-terms (an operation mapping lambda-terms to generally infinite linear combinations of resource lambda-terms) commutes with Böhm tree computation.

1 Introduction

After having introduced the differential lambda-calculus in [1], we studied in [2] a subsystem of the differential lambda-calculus which turns out to be very similar to resource oriented versions of the lambda-calculus previously introduced and studied by various authors [3,4,5]: the *resource lambda-calculus*. It is a finitary calculus in the sense that it enjoys strong normalization, even in the untyped case.

Resource lambda-calculus as the target language of the complete Taylor expansion of lambda-terms. Our viewpoint on this resource lambda-calculus is that it is the sublanguage of the differential lambda-calculus where the *complete*[1]Taylor expansions of ordinary lambda-terms can be written.

Indeed, the only notion of application available in this resource calculus consists in taking a term s (of type $A \to B$ if the calculus is typed) and a finite number of terms $s_1, \ldots, s_n$ (of type A) and applying s to the multiset consisting of the terms s_i, written multiplicatively $s_1 \ldots s_n$. This application is written $\langle s \rangle (s_1 \ldots s_n)$. In differential calculus, this operation would correspond to taking the nth derivative of s at 0, which is a symmetric n-linear map, and applying this derivative to the tuple $(s_1, \ldots, s_n)$.

Defining a beta-reduction in this calculus (as in the original differential lambda-calculus) requires the possibility of adding terms, because the analogue

---

⋆ This work has been supported by the ACI project GEOCAL.
[1] By complete, we mean that all applications in the lambda-terms are expanded.

A. Beckmann et al. (Eds.): CiE 2006, LNCS 3988, pp. 186–197, 2006.

of substitution is a notion of *formal partial derivative* whose inductive definition is based on Leibniz rule[2], and the expression $\langle s \rangle (s_1 \ldots s_n)$ is linear in s, $s_1, \ldots, s_n$; the connection between algebraic linearity and this syntactical notion of linearity is discussed in the introduction of [1]. The logical significance of this derivative, and the linear logic analogue of this resource lambda-calculus are discussed in [6], where *differential interaction nets* are introduced. The striking fact is that this new structure appears in this linear logic setting as new operations associated to the exponentials, completely dual to the traditional *structural* operations (weakening, contraction), and to dereliction.

In constrast, the usual lambda-calculus has a notion of application which is linear in the function but not in the argument, for which we used the notation $(M) N$ (parenthesis around the function, not around the argument). The connection between these two applications is given by the Taylor formula.

Taylor expansion and normalization. In [2], we explained how to Taylor expand arbitrary lambda-terms (of the usual lambda-calculus) as (generally infinite) linear combinations of resource lambda-terms with rational coefficients. We showed moreover that, when normalizing the resource terms which occur in such a Taylor expansion, one gets – generally infinitely many – finite linear combinations of normal resource terms (with positive integers as coefficients) which *do not overlap*; so it makes sense to sum up all these linear combinations. Moreover, the numerical coefficients "behave well" during the reduction, in a sense which is made precise in the corresponding statement, recalled as Theorem 1 in the present paper.

1.1 Overview

We show that this sum s of normal resource terms obtained by normalizing the Taylor expansion of a lambda-term M is simply the Taylor expansion of the Böhm tree of M (the extension of Taylor expansion to Böhm trees is straightforward). Thanks to the results obtained in [2], this reduces to showing that a normal resource term appears in s with a nonzero coefficient iff it appears with a nonzero coefficient in the Taylor expansion of the Böhm tree of M. The "only if" part of this equivalence is fairly straightforward, whereas the "if" part requires the introduction of a version of Krivine's machine which also provides an appealing computational interpretation of the result.

Krivine's machine. Usually, Krivine's machine [7] is described as an abstract environment machine which performs the weak linear head reduction on lambda-terms: given a term M which is beta-equivalent to a variable x, starting from the state $(M, \emptyset, \emptyset)$ (empty environment and empty stack[3]), after a certain number of steps, the machine will produce the result $(x, E, \emptyset)$ where the resulting variable x is not bound by the environment E.

---

[2] In [6], it is shown that Leibniz rule is more pecisely related to the interaction between derivation and contraction.

[3] The stack is there as usual for pushing the arguments of applications.

This computation can be understood as a special kind of reduction of lambda-terms (mini-reduction, aka. linear head reduction [8,9]) which cannot be described exactly as a beta-reduction because, at each reduction step, only the *leftmost occurrence* of variable in the term is substituted. As an example, consider the term $M_0 = (\lambda x\,(x)\,(x)\,y)\,\lambda z\,z$. After one step of linear head reduction, we get $M_1 = (\lambda x\,(\lambda z\,z)\,(x)\,y)\,\lambda z\,z$. Observe that the argument and the lambda of the main redex are still there, and that the function still contains an occurrence of the variable x. Now the leftmost variable occurrence is z and the term M_1 reduces to $M_2 = (\lambda x\,(\lambda z\,(x)\,y)\,(x)\,y)\,\lambda z\,z$. The leftmost occurrence of variable is x again and we get $M_3 = (\lambda x\,(\lambda z\,(\lambda z\,z)\,y)\,(x)\,y)\,\lambda z\,z$ which reduces to $M_4 = (\lambda x\,(\lambda z\,(\lambda z\,y)\,y)\,(x)\,y)\,\lambda z\,z$. We arrive to a term M_4 whose redexes are all K-redexes[4] and reduces to the variable y.

This is exactly this kind of computation that Krivine's machine performs, with the restriction that one does not reduce under the lambda's, in some sense (whence the word "weak"). We extend Krivine's machine in two directions[5].

- First, we accept to reduce under lambda's.
- Second, when Krivine's machine arrives to a state (x, E, Π) where the environment E does not bind x and Π is a non-empty stack, it classically stops with an error. Here instead we continue the computation by running the machine on each element of Π. This corresponds, in the linear head reduction process, to reducing within the arguments of the head variable when a head normal form has been reached.

We call K this extended machine. When fed with a triple $(M, E, \emptyset)$ where E does not bind the free variables of M, this machine produces the Böhm tree of M (all finite approximations being obtained in a finite number of steps).

A more informative version of the machine. Then we define a version $\widehat{\mathsf{K}}$ of that machine where a "tracing mechanism" is added. The idea is to count precisely how many times the various parts of the term M have been used, starting from the state $(M, \emptyset, \emptyset)$, for reaching the state $(x, E', \emptyset)$ (when one knows that M is equivalent to the variable x). This information is summarized as a resource term which has the same shape as M (or, equivalently, appears in the Taylor expansion of M with a nonzero coefficient). For example, in the example of M_0, the corresponding resource term is $\langle \lambda x\,\langle x \rangle\,\langle x \rangle\,y \rangle\,(\lambda z\,z)^2$, which appears with coefficient $\frac{1}{2}$ in the Taylor expansion of M_0.

But there is no reason for limiting our attention to lambda-terms equivalent to a variable: when M reduces to a Böhm tree B, we just add a parameter to our Krivine's machine, which is a resource term u occurring in the Taylor expansion of B. Then $\widehat{\mathsf{K}}(M, \emptyset, \emptyset, u)$ produces a resource term s which appears in the Taylor expansion of M and, in some sense, counts how much of M the machine uses for

---

[4] A K-redex is a redex $(\lambda x\,M)\,N$ such that x does not occur free in M. In M_4, the outermost redex is not a K-redex, but becomes a K-redex after reduction of the internal K-redexes.

[5] These extensions are fairly standard and are part of the folklore.

producing u. This resource term s will depend on M and on u: the larger will be u, the larger will be s.

This machine also gives us a proof for the "if" part of our main result (see the beginning of this "Overview" section), because u appears with a nonzero coefficient in the normal form of the resource term s produced by the machine.

2 Ordinary Notions

We use the word "ordinary" for qualifying the usual lambda-calculus (as opposed to the resource lambda-calculus to be introduced later), and we adopt Krivine's notations: application of M to N is written $(M)N$, and $(M)N_1 \ldots N_p$ stands for $(((M)N_1)\ldots)N_p$.

Böhm trees. An *elementary Böhm tree* (EBT) is a normal term in the lambda-calculus extended with the constant Ω subject to the following equations: $(\Omega)M = \Omega$ and $\lambda x\,\Omega = \Omega$. In other words: Ω is an elementary Böhm tree and if $x, x_1, \ldots, x_n$ are variables and $B_1, \ldots, B_k$ are elementary Böhm trees, then $\lambda x_1 \ldots x_n\,(x)\,B_1 \ldots B_k$ is an elementary Böhm tree. The following clauses define an order relation on EBTs:

- $\Omega \leq B$ for all EBT B;
- $\lambda x_1 \ldots x_n\,(x)\,B_1 \ldots B_k \leq C$ if $C = \lambda x_1 \ldots x_n\,(x)\,C_1 \ldots C_k$ with $B_j \leq C_j$ for all j.

A (general) Böhm tree is now defined as an ideal of elementary Böhm trees, in other word, it is a set B of EBTs which is downwards closed and directed (and hence non-empty). We define now a family of functions from lambda-terms to EBTs.

- $\mathsf{BT}_0(M) = \Omega$;
- $\mathsf{BT}_{n+1}(\lambda x_1 \ldots x_p\,(x)\,M_1 \ldots M_k) = \lambda x_1 \ldots x_p\,(x)\,\mathsf{BT}_n(M_1) \ldots \mathsf{BT}_n(M_k)$;
- $\mathsf{BT}_{n+1}(\lambda x_1 \ldots x_p\,((\lambda y\,P)\,Q)\,M_1 \ldots M_k)$
$$= \mathsf{BT}_n(\lambda x_1 \ldots x_p\,(P\,[Q/y])\,M_1 \ldots M_k)$$

It is straightforward to check that $\mathsf{BT}_n(M)$ is a non decreasing sequence of EBTs. Then the Böhm tree of M is the downwards closure of the set $\{\mathsf{BT}_n(M) \mid n \in \mathbb{N}\}$, which is an ideal of EBTs.

Krivine's abstract machine. If $f : S \to S'$ is a partial function, $a \in S$ and $b \in S'$, we denote by $f_{a \mapsto b}$ the partial function $g : S \to S'$ which is defined like f but for a, where it is defined and takes the value b.

By simultaneous induction, we define the two following concepts: a *closure* is a pair $\Gamma = (M, E)$ where M is a lambda-term and E is an environment such that $\mathrm{FV}(M) \subseteq \mathrm{Dom}\,E$ and an *environment* is a finite partial function on variables, taking closures or the distinguished symbol free as value. We use $\mathrm{Dom}_c\,E$ for the subset of $\mathrm{Dom}\,E$ whose elements are not mapped to free. We need also an auxiliary concept: a *stack* is a finite list Π of closures.

We define a sequence of functions from states to EBTs.

- $K_0(\Gamma, \Pi) = \Omega$;
- $K_{n+1}(x, E, \Pi) = K_n(E(x), \Pi)$ if $x \in \text{Dom}_c(E)$;
- $K_{n+1}(x, E, \Pi) = (x) K_n(\Gamma_1, \emptyset) \ldots K_n(\Gamma_k, \emptyset)$ where $\Pi = (\Gamma_1, \ldots, \Gamma_n)$, if $E(x) = \text{free}$;
- $K_{n+1}(\lambda x\, M, E, \emptyset) = \lambda x\, K_n(M, E_{x \mapsto \text{free}}, \emptyset)$ (assuming that $x \notin \text{Dom}(E)$ and that x does not appear free in any of the terms mentioned in E);
- $K_{n+1}(\lambda x\, M, E, \Gamma :: \Pi) = K_n(M, E_{x \mapsto \Gamma}, \Pi)$ (with similar assumptions for x);
- $K_{n+1}((M)\, N, E, \Pi) = K_n(M, E, (N, E) :: \Pi)$.

Observe that the definition is correct in the sense that, in all "recursive calls" of the function K, the closures are well formed (the domain of their environment contains the free variables of their term).

Let $S = (\Gamma, \Pi)$ be a state. One checks easily that $(K_n(S))_{n \in \mathbb{N}}$ is a non decreasing sequence of EBTs. We define $K(S)$ as the downwards closure of the set $\{K_n(S) \mid n \in \mathbb{N}\}$; this set is a Böhm tree.

We define another total function T, from closures to lambda-terms. Given a closure $\Gamma = (M, E)$, we set $T(\Gamma) = M\, [T(E(x))/x]_{\dot{x} \in \text{Dom}_c\, E}$. This is a definition by induction on the height of closures, seen as finitely branching trees. We extend this mapping to states: $T(\Gamma, (\Gamma_1, \ldots, \Gamma_n)) = (T(\Gamma))\, T(\Gamma_1) \ldots T(\Gamma_n)$.

The main, standard, property of Krivine's machine is that $K(S) = \text{BT}(T(S))$ for any state S. This "soundness" result shows in particular that Krivine's machine computes the Böhm tree of lambda-terms: $\text{BT}(M) = K(M, E, \emptyset)$, where E is any environment mapping all the free variables of M to the value free.

3 Resource Notions

Notations. Let E be a set. A multiset on E is a function $m : E \to \mathbb{N}$. The support $\text{supp}(m)$ of m is the set of all $a \in E$ such that $m(a) \neq 0$. The multiset m is finite if $\text{supp}(m)$ is finite. The number $m(a)$ is the multiplicity of a in m. We denote by $\mathcal{M}_{\text{fin}}(E)$ the set of all finite multisets on E.

3.1 The Resource Lambda-Calculus

We give a short account of the resource lambda-calculus, as developed in [2]. We recall the syntax and terminology of [2]. As usual we are given a countable set of variables.

Simple terms and poly-terms

- If x is a variable, then x is a simple term.
- If x is a variable and t is a simple term, then $\lambda x\, t$ is a simple term.
- If t is a simple term and T is a simple poly-term, then $\langle t \rangle\, T$ is a simple term.
- A simple poly-term is a multiset of simple terms. We use multiplicative notations for these multisets: 1 denotes the empty poly-term, if t is a simple term, we use also t for denoting the simple poly-term whose only element is t, and if S and T are simple poly-terms, we use ST for the multiset union (sum) of S and T.

We use the greek letters $\sigma, \tau \ldots$ for simple terms or poly-terms when we do not want to be specific. We use Δ for the set of all simple terms, $\Delta^!$ for the set of all simple poly-terms and $\Delta^{(!)}$ for one of these two sets when we don't want to be specific.

Linear combinations and reduction. We use $\mathbb{Q}^+$ (the rig of non-negative rational numbers) as set of scalars. If A is a set, we use $\mathbb{Q}^+\langle A \rangle$ for the free $\mathbb{Q}^+$-module generated by A. If $\alpha \in \mathbb{Q}^+\langle A \rangle$, we use $\mathrm{Supp}(\alpha)$ for the set of all $a \in A$ such that $\alpha_a \neq 0$. We use $\mathbb{N}\langle A \rangle$ for the elements of $\mathbb{Q}^+\langle A \rangle$ whose coefficients are integers.

A redex is a simple term of the shape $r = \langle \lambda x\, s \rangle\, S$. It reduces to $0 \in \mathbb{N}\langle \Delta \rangle$ if the cardinality of the multiset S is distinct from the number of free occurrences of x in s, and otherwise reduces to

$$\partial_x(s, S) = \sum_{f \in \mathfrak{S}_d} s\left[s_1, \ldots, s_d / x_{f(1)}, \ldots, x_{f(d)} \right] \in \mathbb{N}\langle \Delta \rangle$$

where $S = s_1 \ldots s_d$ and $x_1, \ldots, x_d$ are the d free occurrences of x in s. In this expression, $\mathfrak{S}_d$ stands for the group of all permutations on the set $\{1, \ldots, d\}$.

This notion of reduction extends to all simple (poly-)terms, using the fact that all constructions of the syntax are linear. For instance, if $s_1, \ldots, s_n \in \Delta$ and for each i, s_i reduces to $s_i' \in \mathbb{N}\langle \Delta \rangle$, then the simple poly-term $s_1 \ldots s_n$ reduces to $\prod_{i=1}^n s_i' \in \mathbb{N}\langle \Delta^! \rangle$.

This notion of reduction is a relation $\rightsquigarrow$ from $\Delta^{(!)}$ to $\mathbb{N}\langle \Delta^{(!)} \rangle$; it is extended to a relation from $\mathbb{Q}^+\langle \Delta^{(!)} \rangle$ to itself by linearity (the linear span of $\rightsquigarrow$ in the product space $\mathbb{Q}^+\langle \Delta^{(!)} \rangle \times \mathbb{Q}^+\langle \Delta^{(!)} \rangle$). This relation is confluent, and strongly normalizing if we only consider integer coefficients. We use Δ_0 for the set of all normal simple terms, and NF for the normalization map $\mathbb{N}\langle \Delta^{(!)} \rangle \to \mathbb{N}\langle \Delta_0^{(!)} \rangle$, which is linear.

Taylor expansion of ordinary lambda-terms. Let us give an intuition of the resource lambda-calculus, explaining why it is related to the idea of Taylor expansion. Usually, when f is a sufficiently regular function from a vector space E to a vector space F (finite dimensional spaces, or Banach spaces, typically), at all point $x \in E$, f has nth derivatives for all $n \in \mathbb{N}$, and these derivatives are maps $f^{(n)} : E \times E^n \to F$ with the same regularity as f and such that $f^{(n)}(x, u_1, \ldots, u_n) = f^{(n)}(x) \cdot (u_1, \ldots, u_n)$ is n-linear and symmetric in $u_1, \ldots, u_n$. When one is lucky, and usually locally only, the Taylor formula holds. Around 0, it reads

$$f(x) = \sum_{n=0}^{\infty} \frac{1}{n!} f^{(n)}(0) \cdot (u, \ldots, u)$$

If we want to Taylor-expand lambda-terms, which after all are functions, we need to extend the language with explicit differentials, or more precisely a construction of *differential application* of a term M to n terms $N_1, \ldots, N_n$, as we did in [1] (a simplified version of that calculus is now available in [10]). The idea is that if M represents a function f from E to F and if $N_1, \ldots, N_n$ represent n vectors

$u_1, \ldots, u_n \in E$, then this new construction $D^n M \cdot (N_1, \ldots, N_n)$ will represent the function from E to F which maps x to $f^{(n)}(x) \cdot (u_1, \ldots, u_n)$, and therefore this construction is linear and symmetric in the N_i's.

The Taylor expansion of a single lambda-calculus application $(M) N$ would then read

$$\sum_{n=0}^{\infty} \frac{1}{n!} (D^n M \cdot (N, \ldots, N)) 0$$

If we want now to Taylor expand *all* the applications occurring in a lambda-term, we see that the usual lambda-calculus application in its generality will become useless: only application to 0 is needed. This is exactly the purpose of the construction $\langle s \rangle s_1 \ldots s_n$ of the resource lambda-calculus; with the notations of the differential lambda-calculus, the expression $\langle s \rangle s_1 \ldots s_n$ stands for $(D^n s \cdot (s_1, \ldots, s_n)) 0$.

So the resource lambda-calculus is a "target language" for completely Taylor expanding ordinary lambda-terms. The expansion of a term M will be an infinite linear combination of resource terms, with rational coefficients (actually, inverses of positive integers). Let us use M^* for the complete Taylor expansion of M. By what we said, this operation should obey $(M) N^* = \sum_{n=0}^{\infty} \frac{1}{n!} \langle M^* \rangle (N^*)^n$ as well as $x^* = x$ and $(\lambda x\, M)^* = \lambda x\, M^*$. From these equations, we obtain, applying the multinomial formula, that

$$M^* = \sum_{s \in \mathcal{T}(M)} \frac{1}{\mathrm{m}(s)} s$$

where $\mathcal{T}(M) \subseteq \Delta$ (the set of resource terms which have "the same shape" as M) is defined inductively by $\mathcal{T}(x) = \{x\}$, $\mathcal{T}(\lambda x\, M) = \{\lambda x\, s \mid s \in \mathcal{T}(M)\}$ and $\mathcal{T}((M) N) = \{\langle s \rangle S \mid s \in \mathcal{T}(M) \text{ and } S \in \mathcal{M}_{\mathrm{fin}}(\mathcal{T}(N))\}$. The positive number $\mathrm{m}(\sigma)$ associated to each (poly-)term σ is called its *multiplicity coefficient*; see the definition and properties of these numbers in [2]. We can recall now the main result proven in that paper.

Theorem 1. *Let M be an ordinary lambda-term.*

1. *If $s, s' \in \mathcal{T}(M)$ and s and s' are not α-equivalent, then $\mathrm{Supp}(\mathrm{NF}(s)) \cap \mathrm{Supp}(\mathrm{NF}(s')) = \emptyset$.*
2. *If $s \in \mathcal{T}(M)$ and $u \in \mathrm{Supp}(\mathrm{NF}(s))$, then the coefficient $\mathrm{NF}(s)_u$ of u in $\mathrm{NF}(s)$ (remember that this coefficient must be a positive integer) is equal to $\mathrm{m}(s)/\mathrm{m}(u)$.*

Proving this result involved a coherence relation on simple terms for the first part, and some considerations on groups of permutations of simple term variables for the second part.

Given an ordinary lambda-term M, it makes sense therefore to apply NF to each of the simple terms occurring in its Taylor expansion, defining $\mathrm{NF}(M^*) = \sum_{s \in \mathcal{T}(M)} \frac{1}{\mathrm{m}(s)} \mathrm{NF}(s)$. Indeed by Theorem 1, if u is a normal simple term, there is at most one $s \in \mathcal{T}(M)$ such that $\mathsf{NF}(s)_u \neq 0$.Moreover, if such a simple term s exists, the coefficient of u in the sum above is

$$* \quad \mathsf{NF}(M^*)_u = \frac{1}{\mathsf{m}(s)} \, \mathsf{NF}(s)_u = \frac{1}{\mathsf{m}(u)}$$

by Theorem 1 again.

We want to prove that this sum is equal to $\mathsf{BT}(M)^*$, the Taylor expansion of the Böhm tree of M. To give a meaning to this notion, we need first to define $\mathcal{T}(B)$ when B is an EBT: the definition is the same as for ordinary lambda-terms, with the additional clause that $\mathcal{T}(\Omega) = \emptyset$. For instance $\mathcal{T}((x)\,\Omega) = \{\langle x \rangle\,1\}$. Observe that $B \leq C \Rightarrow \mathcal{T}(B) \subseteq \mathcal{T}(C)$.

We generalize this notion to arbitrary Böhm trees: $\mathcal{T}(\mathsf{B}) = \bigcup_{R \leq \mathsf{B}} \mathcal{T}(R)$ (this is a directed union since B is an ideal). Of course, all these resource terms are normal. Given a Böhm tree B, it makes sense finally to define its Taylor expansion, as we did for ordinary lambda-terms: $\mathsf{B}^* = \sum_{b \in \mathcal{T}(\mathsf{B})} (1/\mathsf{m}(b))b$.

3.2   Resource Closures and Resource Stacks

We adapt now the concepts of closure and stack to the framework of the resource lambda-calculus, introducing multi-set based versions thereof. We stick to our multiplicative conventions for denoting multi-sets.

- A *resource environment* is a *total* function e on variables, taking resource closures or the symbol free as values. We extend pointwise the multi-set notations to resource environments, e.g. $(ee')(x) = e(x)e'(x)$ (equal to free when one of these two values is equal to free). For an environment e, we require moreover $e(x) = 1$ for almost all variables x, where 1 is the unit resource closure (see below the definition of resource closures). If x is a variable and c is a resource closure, we denote by $[x \mapsto c]$ the resource environment which takes the value 1 for all variables but for x, for which it takes the value c. If e is a resource environment, $e \setminus x$ denotes the resource environment which takes the same values as e but for x where it takes the value free. We use $\mathrm{Dom}_c\, e$ for the (co-finite) set of all variables where e does not take the value free.

- A *resource closure* is a pair $c = (T, e)$ where T is a simple resource poly-term and e is a resource environment, or is the special *unit closure* 1. Intuitively, this special closure is "equal" to any closure of the shape $(1, e)$ where e maps all variables to free, to the unit closure 1 or to any closure of the shape we are now describing.

  Poly-term multiplication is extended to closures in the obvious way: the unit closure 1 is neutral, and $(T, e)(T', e') = (TT', ee')$.

  A resource closure (T, e) will be said to be *elementary* if the multi-set T has exactly one element. All resource closures are product (in many different ways, usually) of elementary resource closures. We use the letters $c, c', \ldots$ for general resource closures and $\gamma, \gamma' \ldots$ for elementary resource closures.

Finally, a *resource stack* π is a finite sequence of resource closures.

A *resource state* is a triple (t, e, π) where t is a simple resource term, e is a resource environment and π is a resource stack. In such a resource state, the pair (t, e) will be considered as an elementary resource closure.

By mutual induction, we define $\mathcal{T}(E)$ and $\mathcal{T}(\Gamma)$, the set of all resource environments and resource closures of shape E (ordinary environment) and Γ (ordinary closure) respectively:

- $\mathcal{T}(E)$ is the set of all resource environments e such that
  - if $E(x) = \mathsf{free}$, then $e(x) = \mathsf{free}$;
  - otherwise and if $E(x)$ is defined, then $e(x) \in \mathcal{T}(E(x))$;
  - if $E(x)$ is undefined, then $e(x) = 1$.
- If $\Gamma = (M, E)$, then $\mathcal{T}(\Gamma) = (\mathcal{M}_{\mathrm{fin}}(\mathcal{T}(M)) \times \mathcal{T}(E)) \cup \{1\}$.

This extends to standard stacks and resource stacks in the obvious way, defining $\pi \in \mathcal{T}(\Pi)$. Last we set $\mathcal{T}(\Gamma, \Pi) = \mathcal{T}(\Gamma) \times \mathcal{T}(\Pi)$.

As we did for the ordinary lambda-calculus, we associate to each resource closure c a (generally not simple) resource poly-term $\mathsf{T_D}(c) \in \mathbb{N}\langle \Delta^! \rangle$ by the following inductive definition

$$
\mathsf{T_D}(c) = \begin{cases} 1 & \text{if } c = 1 \\ \partial_{x_1, \ldots, x_n}(T, \mathsf{T_D}(e(x_1)), \ldots, \mathsf{T_D}(e(x_n))) & \text{if } c = (T, e) \end{cases}
$$

where $x_1, \ldots, x_n$ is any repetition-free sequence of variables which contains all the variables of $\mathrm{Dom}_c\, e$ which are free in T or satisfy $e(x) \neq 1$ (in particular, this expression is equal to 0 if there exists a variable x not free in T and such that $e(x) \neq 1$).

Due to the basic properties of partial derivatives explained in [2], the expression above of $\mathsf{T_D}(c)$ does not depend on the choice of the sequence of variables $x_1, \ldots, x_n$.

Observe that when c is elementary, $\mathsf{T_D}(c)$ can be seen as a resource term.

Last, we extend this definition to resource states (γ, π) where $\pi = (c_1, \ldots, c_k)$ is a resource stack (γ and the c_i's are therefore resource closures, and we know moreover that γ is elementary), setting

$$
\mathsf{T_D}(\gamma, \pi) = \langle \cdots \langle \mathsf{T_D}(\gamma) \rangle\, \mathsf{T_D}(c_1) \cdots \rangle\, \mathsf{T_D}(c_k) \in \mathbb{N}\langle \Delta \rangle
$$

4   A Resource Driven Krivine's Machine

We define a new version $\widehat{\mathsf{K}}$ of Krivine's machine which, fed with an ordinary closure Γ, an ordinary stack Π and a *normal* resource term u, will return a pair $(\gamma, \pi) \in \mathcal{T}(\Gamma, \Pi)$ where γ is an elementary resource closure, or will be undefined.

We use the symbol "$\uparrow$" for the result of the function when it is undefined. As before, we define by induction on n an increasing sequence of partial functions $\widehat{\mathsf{K}}_n$ and we set $\widehat{\mathsf{K}} = \bigcup_{n=0}^{\infty} \widehat{\mathsf{K}}_n$.

The base case is trivial: $\widehat{\mathsf{K}}_0(\Gamma, \Pi, t) = \uparrow$, always.

The inductive step is by case on the shape of the first element of the closure $\Gamma = (M, E)$ (remember that we assume that $\mathrm{FV}(M) \subseteq \mathrm{Dom}\, E$).

- If $M = x$ is a variable, we have two subcases.
  - Assume first that $x \in \mathrm{Dom}_c(E)$. If $\widehat{\mathsf{K}}_n(E(x), \Pi, u) = \uparrow$,
    then $\widehat{\mathsf{K}}_{n+1}(\Gamma, \Pi, u) = \uparrow$ and otherwise, let $(\gamma, \pi) = \widehat{\mathsf{K}}_n(E(x), \Pi, u)$, then

$$\widehat{\mathsf{K}}_{n+1}(M, E, \Pi, u) = (x, e, \pi) \quad \text{where} \quad e(y) = \begin{cases} \gamma & \text{if } y = x \\ \text{free} & \text{if } E(y) = \text{free} \\ 1 & \text{otherwise.} \end{cases}$$

  - Otherwise, we have $x \in \mathrm{Dom}(E)$ and $E(x) = \text{free}$. The stack Π is a sequence $(\Gamma_1, \ldots, \Gamma_k)$ of ordinary closures.
    * If $u = \langle \cdots \langle x \rangle V_1 \cdots \rangle V_k$ and for each $j = 1, \ldots, k$ and $v \in \mathrm{supp}(V_j)$, there exists an elementary resource closure $\gamma_j(v)$ such that $\widehat{\mathsf{K}}_n(\Gamma_j, \emptyset, v) = (\gamma_j(v), \emptyset)$, then

$$\overset{\circ}{\mathsf{K}}_{n+1}(M, E, \Pi, u) = (x, e, \pi) \quad \text{where} \quad e(y) = \begin{cases} \text{free} & \text{if } E(y) = \text{free} \\ 1 & \text{otherwise.} \end{cases}$$

    and where $\pi = (c_1, \ldots, c_k)$ with $c_j = \prod_{v \in \mathrm{supp}(V_j)} \gamma_j(v)^{V_j(v)}$ (this product has to be understood as a product of resource closures, in the sense defined above — remember that $V_j(v)$ is a positive integer, the multiplicity of v in the multiset V_j).
    * Otherwise, $\widehat{\mathsf{K}}_{n+1}(M, E, \Pi, u) = \uparrow$.
- Assume now that $M = \lambda x\, N$. Without loss of generality, we can assume that $E(x) = \uparrow$. Again, we have two subcases.
  - Assume first that $\Pi = \emptyset$ is the empty stack.
    If $u = \lambda x\, v$ and $\widehat{\mathsf{K}}_n(N, E_{x \mapsto \text{free}}, \emptyset, v) = (t, e, \emptyset)$ with $e(x) = \text{free}$, then

$$\widehat{\mathsf{K}}_{n+1}(M, E, \emptyset, u) = (\lambda x\, t, e_{x \mapsto 1}, \emptyset)$$

    and otherwise, $\widehat{\mathsf{K}}_{n+1}(M, E, \emptyset, u) = \uparrow$.
  - Assume next that $\Pi = \Gamma :: \Pi'$. If $\widehat{\mathsf{K}}_n(N, E_{x \mapsto \Gamma}, \Pi', u) = (t, e, \pi')$ with $e(x) \neq \text{free}$, then

$$\widehat{\mathsf{K}}_{n+1}(M, E, \Pi, u) = (\lambda x\, t, e_{x \mapsto 1}, e(x) :: \pi')$$

    and otherwise, $\widehat{\mathsf{K}}_{n+1}(M, E, \emptyset, u) = \uparrow$.
- Last assume that $M = (P)\, Q$. If $\widehat{\mathsf{K}}_n(P, E, (Q, E) :: \Pi, u) = (t, e, (T, e') :: \pi)$, then

$$\widehat{\mathsf{K}}_{n+1}(M, E, \Pi, u) = (\langle t \rangle\, T, ee', \pi)$$

and otherwise, $\widehat{\mathsf{K}}_{n+1}(M, E, \emptyset, u) = \uparrow$.

The following lemmas summarize the main properties of this machine.

Lemma 1. *Let Γ be an ordinary closure, Π be an ordinary stack and u be a simple resource term.*
If $\widehat{\mathsf{K}}(\Gamma, \Pi, u)$ is defined, then u is normal and $\widehat{\mathsf{K}}(\Gamma, \Pi, u)$ is a resource state (γ, π) which belongs to $\mathcal{T}(\Gamma, \Pi)$.

Lemma 2. *Let Γ be an ordinary closure, Π be an ordinary stack and u be a normal simple resource term.*

For each $n \in \mathbb{N}$, we have the following equivalence:

$$u \in \mathcal{T}(\mathsf{K}_n(\Gamma, \Pi)) \quad \text{iff} \quad \widehat{\mathsf{K}}_n(\Gamma, \Pi, u) \text{ is defined}$$

Lemma 3. *Let Γ be an odinary closure, Π be an ordinary stack and u be a normal simple resource term.*

Let $n \in \mathbb{N}$. If $\widehat{\mathsf{K}}_n(\Gamma, \Pi, u) = (\gamma, \pi)$, then $u \in \mathrm{Supp}(\mathrm{NF}(\mathsf{T}_\mathsf{D}(\gamma, \pi)))$

5   Normal Form of the Taylor Expansion

Using the lemmas proven so far and some natural properties relating substitution in ordinary and resource lambda-calculi, we can prove the main theorem of the paper.

Theorem 2. *Let M be an ordinary lambda-term and let u be a normal simple resource term. Then $u \in \mathcal{T}(\mathsf{BT}(M))$ if and only if there exists $s \in \mathcal{T}(M)$ such that $u \in \mathrm{Supp}(\mathrm{NF}(s))$. Moreover, when this simple term s exists, it is unique.*

From this result and from Theorem 1, we can derive the announced commutation property.

Corollary 1. *Let M be an ordinary lambda-term. One has*

$$\mathsf{BT}(M)^* = \mathrm{NF}(M^*) = \sum_{s \in \mathcal{T}(M)} \frac{1}{\mathrm{m}(s)} \mathrm{NF}(s)$$

6   Concluding Remarks

By Theorem 2, there exists a partial function $\mathsf{E} : \Lambda \times \Delta_0 \to \Delta$ such that $\mathsf{E}(M, u)$ is defined if and only if $u \in \mathcal{T}(\mathsf{BT}(M))$ and then takes as value the unique simple term $s \in \mathcal{T}(M)$ such that $u \in \mathrm{Supp}(\mathrm{NF}(s))$. In the proof of that theorem, one sees how this function E can be defined, using a modified version of Krivine's machine (an implementation of that machine is available at `http://iml.univ-mrs.fr/~regnier/taylor/`).

When $\mathsf{BT}(M)$ is a variable $\star$, the situation is particularly simple: we have $\mathcal{T}(\mathsf{BT}(M)) = \{\star\}$ and $\mathsf{E}(M, \star)$ is the unique $s \in \mathcal{T}(M)$ which has a non-zero normal form, and the normal form of s must be $\mathrm{m}(s)\star$. In that particular case, it is interesting to observe that the "size" of s (easy to define by a simple induction on s) is the number of steps in the reduction of M to $\star$ by Krivine's machine, which seems to be a sensible measure of the complexity of the reduction of M.

The map $\mathsf{S} \circ \mathsf{E} : \Lambda \times \Delta_0 \to \mathbb{N}$ seems therefore to provide more generally a way of measuring the complexity of the reduction of lambda-terms. The interesting point is that this measure is associated to the algebraic property stated by Theorems 2 and 1.

References

1. Ehrhard, T., Regnier, L.: The differential lambda-calculus. Theoretical Computer Science **309**(1-3) (2003) 1–41
2. Ehrhard, T., Regnier, L.: Uniformity and the Taylor expansion of ordinary lambda-terms. Technical report, Institut de mathématiques de Luminy (2005) submitted to Theoretical Computer Science
3. Boudol, G.: The lambda calculus with multiplicities. Technical Report 2025, INRIA Sophia-Antipolis (1993)
4. Boudol, G., Curien, P.L., Lavatelli, C.: A semantics for lambda calculi with resource. Mathematical Structures in Computer Science **9**(4) (1999) 437–482
5. Kfoury, A.J.: A linearization of the lambda-calculus. Journal of Logic and Computation **10**(3) (2000) 411–436
6. Ehrhard, T., Regnier, L.: Differential interaction nets In: Proceedings of WoLLIC'04. Volume 103 of Electronic Notes in Theoretical Computer Science., Elsevier Science (2004) 35–74
7. Krivine, J.L.: A call-by-name lambda-calculus machine. Higher-Order and Symbolic Computation (2005) To appear
8. De Bruijn, N.: Generalizing Automath by means of a lambda-typed lambda calculus. In Kueker, D., Lopez-Escobar, E., Smith, C., eds.: Mathematical Logic and Theoretical Computer Science. Lecture Notes in Pure and Applied Mathematics, Marcel Dekker (1987) 71–92 Reprinted in: Selected papers on Automath, Studies in Logic, volume 133, pages 313-337, North-Holland, 1994
9. Danos, V., Regnier, L.: Reversible, irreversible and optimal lambda-machines. Theoretical Computer Science **227**(1-2) (1999) 273–291
10. Vaux, L.: The differential lambda-mu calculus. Technical report, Institut de Mathématiques de Luminy (2005) Submitted for publication

What Does the Incompleteness Theorem Add to the Unsolvability of the Halting Problem?

Torkel Franzén[†]

Systemteknik, LTU, SE-971 87 Luleå, Sweden
torkel@sm.luth.se

Turing's paper including a proof of the unsolvability of the halting problem appeared five years after Gödel's 1931 paper, and ever since there have been many and close ties between uncomputability and incompleteness. The purpose of my talk is to discuss the specific role and contribution of the "Gödelian" approach to incompleteness.

[†] Torkel Franzén died on April 19, 2006.

A. Beckmann et al. (Eds.): CiE 2006, LNCS 3988, p. 198, 2006.
© Springer-Verlag Berlin Heidelberg 2006

An Analysis of the Lemmas of Urysohn and Urysohn-Tietze According to Effective Borel Measurability

Guido Gherardi

Dipartimento di Scienze Matematiche e Informatiche
"R. Magari", Università di Siena
Pian dei Mantellini 44 - 53100 Siena Italy
gherardi3@unisi.it

Abstract. In [6], K. Weihrauch studied the computational properties of the Urysohn Lemma and of the Urysohn-Tietze Lemma within the framework of the TTE-theory of computation. He proved that with respect to negative information both lemmas cannot in general define computable single valued mappings. In this paper we reconsider the same problem with respect to positive information. We show that in the case of positive information neither the Urysohn Lemma nor the Dieudonné version of Urysohn-Tietze Lemma define computable functions. We analyze the degree of the incomputability of such functions (or more precisely, of the incomputability of some of their realizations in the Baire space) according to the theory of effective Borel measurability. In particular, we show that with respect to positive information both the Urysohn function and the Dieudonné function are Σ_2^0-computable and in some cases even Σ_2^0-complete.

Keywords: *Computable Analysis, Borel Measurability, Urysohn Lemma, Urysohn-Tietze Lemma.*

1 Preliminaries

We assume that the reader is familiar with the basic concepts of computable analysis as outlined in [5] and of effective Borel measurability as introduced in [1] and [2]. For the reader's convenience, we recall however in the following some basic concepts and definitions. Notations and terminology that are not standard will be explicitly introduced and carefully explained. As to the theory of representations, we use the approach of [5], except that the Cantor space is replaced by the Baire space, as in [1].

Definition 1. $\mathbb{N}^*$ *($\mathbb{N}^{\mathbb{N}}$) is the set of all finite (infinite) sequences of natural numbers.*
For $y \in \mathbb{N}^* \cup \mathbb{N}^{\mathbb{N}}$ and $n \in \mathbb{N}$, the expression "$n \in y$" means that y lists n, thus there are $y_0 \in \mathbb{N}^*$ and $y_1 \in \mathbb{N}^* \cup \mathbb{N}^{\mathbb{N}}$ such that $y = y_0 n y_1$.
Given any $p \in \mathbb{N}^{\mathbb{N}}$, "$p[n]$" denotes the initial segment of p of length $n \in \mathbb{N}$.

A. Beckmann et al. (Eds.): CiE 2006, LNCS 3988, pp. 199–208, 2006.

For any given word $w \in \mathbb{N}^*$, *let "$w\mathbb{N}^{\mathbb{N}}$" denote the set* $\{p \in \mathbb{N}^{\mathbb{N}} : w = p[m]\}$, *where* $m = \text{length}(w)$. *Recall that such set is an open ball in the Baire space. Since any* $p \in \mathbb{N}^{\mathbb{N}}$ *is a function* $p : \mathbb{N} \to \mathbb{N}$, *the symbol "$p(n)$" denotes the n-th number listed by* p.

Definition 2 (Naming systems). *Given a set* S, *a notation (a representation)* γ *of* S *is a surjective function* $\gamma :\subseteq \mathbb{N} \to S$ $(\gamma :\subseteq \mathbb{N}^{\mathbb{N}} \to S)$.
A represented set is a pair (S, γ) *where* S *is a set and* γ *is a representation of* S.

For the sake of clarity, given a represented set (S, δ) and an element $x \in S$, we may use x itself to label all its δ-names. Thus, $p_x \in \mathbb{N}^{\mathbb{N}}$ is a δ-name of x, where δ is clear from the context.

From now on the definitions and results refer to a generic computable complete metric space $\mathbf{M} = (M, d, Q, \nu_Q)$ (see [5]), unless otherwise specified. For the sake of generality, we assume that $M \neq \emptyset$ and $\text{dom}(\nu_Q) = \mathbb{N}$.
$\mathcal{A}$ will be the class of all the closed sets in the topology generated by d.
As a particular case, $\mathbb{R}$ is the computable metric space $(\mathbb{R}, d', \mathbb{Q}, \nu_{\mathbb{Q}})$, where d' is the Euclidean metric and $\nu_{\mathbb{Q}}$ is a notation for the set of the rational numbers which is given by some recursive enumeration of $\mathbb{Q}$ itself.

Definition 3. *Let a computable metric space* $\mathbf{M} = (M, d, Q, \nu_Q)$ *be given. The set of all balls* $B(q, \alpha)$ *with* $q \in Q$ *and* $\alpha \in \mathbb{Q}^+$ *is a base for the topology generated by* d. *We call the elements of this set noted open balls of* $\mathbf{M}$. *Each noted open ball is uniquely determined by its center and its radius, thus one can define through a pairing function a notation* $\nu_{\mathbf{M}}$ *for this base. Let* $\text{dom}(\nu_{\mathbf{M}})$ *be the (whole) set* $\mathbb{N}^+$, *so that 0 denotes no ball (in this context 0 means "no information"). The ball* $\nu_{\mathbf{M}}(n)$, *for* $n > 0$, *will be denoted by "$I_n^{\mathbf{M}}$".*

Definition 4 (Standard representation). $\delta_{\mathbf{M}}$ *is the* standard representation *of* M *associated with* $\mathbf{M}$: *for* $p \in \mathbb{N}^{\mathbb{N}}$, *let*

$$\delta_{\mathbf{M}}(p) = x \in M \Leftrightarrow \{n > 0 : n \in p\} = \{n : x \in I_n^{\mathbf{M}}\}.$$

As usual, ρ denotes the standard representation associated with $\mathbb{R}$, and in the sequel $\mathbb{N}^{\mathbb{N}}$ will always be represented by the standard representation $\delta_{\mathbb{B}}$ associated with the Baire computable metric space $\mathbb{B}$.

Definition 5 (Effective Borel measurability). *A function* $F :\subseteq \mathbb{N}^{\mathbb{N}} \to \mathbb{N}^{\mathbb{N}}$ *is called* Σ_k^0-computable, *for* $k \geq 1$, *if there is a computable function* $G :\subseteq \mathbb{N}^{\mathbb{N}} \to \mathbb{N}^{\mathbb{N}}$ *mapping each* $\delta_{\Sigma_1^0(\mathbb{B})}$-*name of any* $O \in \Sigma_1^0(\mathbb{B})$ *to some* $\delta_{\Sigma_k^0(\mathbb{B})}$-*name of a set* $V \in \Sigma_k^0(\mathbb{B})$ *such that* $F^{-1}(O) = V \cap \text{dom}(F)$.[1]
Let represented sets (S_i, δ_i), *for* $i = 0, 1$, *and a function* $f :\subseteq S_1 \to S_0$ *be given. Any function* $F :\subseteq \mathbb{N}^{\mathbb{N}} \to \mathbb{N}^{\mathbb{N}}$ *is said to be a* (δ_1, δ_0)-realization *of* f *when* $f\delta_1(p) = \delta_0 F(p)$ *for all* $p \in \text{dom}(f\delta_1)$. *The function* f *is said to be* Σ_k^0-computable *with respect to representations* (δ_1, δ_0) *(written "Σ_k^0-computable w.r.t. (δ_1, δ_0)"), for* $k \geq 1$, *if it has a* Σ_k^0-computable (δ_1, δ_0)-realization. *Notice that for k=1 the function* f *is* (δ_1, δ_0)-computable *as defined in [5].*

---

[1] For the representations $\delta_{\Sigma_1^0(\mathbb{B})}, \delta_{\Sigma_k^0(\mathbb{B})}$ see [1].

To prove that the function f is Σ_k^0-computable with respect to the representations δ_1, δ_0 one can show, by Corollary 3.9 in [1] and Definition 5, that f has some (δ_1, δ_0)-realization F such that $F = F_1 \circ \ldots \circ F_n$ for some $n \in \mathbb{N}$, $F_i :\subseteq \mathbb{N}^{\mathbb{N}} \to \mathbb{N}^{\mathbb{N}}$ is $\Sigma_{k_i}^0$-computable for $1 \leq i \leq n$ and $k = k_1 + \ldots + k_n - n + 1$.

Definition 6 (Reducibility). *Let $F :\subseteq \mathbb{N}^{\mathbb{N}} \to \mathbb{N}^{\mathbb{N}}$ and $G :\subseteq \mathbb{N}^{\mathbb{N}} \to \mathbb{N}^{\mathbb{N}}$ be two given functions. F is (computably) reducible to G (written "$F \leq_c G$") if there are two computable functions $\mathbf{A} :\subseteq \mathbb{N}^{\mathbb{N}} \to \mathbb{N}^{\mathbb{N}}$ and $\mathbf{B} :\subseteq \mathbb{N}^{\mathbb{N}} \to \mathbb{N}^{\mathbb{N}}$ such that*

$$F(p) = \mathbf{A}(p, G \circ \mathbf{B}(p))$$

for all $p \in \mathrm{dom}(F)$.

Let represented sets (S_i, δ_i), for $1 \leq i \leq 4$, be given and let functions $f :\subseteq S_1 \to S_2$, $g :\subseteq S_3 \to S_4$ be given. The function f is said to be (computably) reducible to g with respect to representations $(\delta_1, \delta_2, \delta_3, \delta_4)$ (written "$f \leq_c g$ w.r.t $(\delta_1, \delta_2, \delta_3, \delta_4)$") if there are a $(\delta_1, \delta_4, \delta_2)$-computable function $\mathbf{a} :\subseteq S_1 \times S_4 \to S_2$ and a (δ_1, δ_3)-computable function $\mathbf{b} :\subseteq S_1 \to S_3$ such that

$$f(x) = \mathbf{a}(x, g \circ \mathbf{b}(x))$$

for all $x \in \mathrm{dom}(f)$. Recall that for $S_1 = S_2 = \mathbb{N}^{\mathbb{N}}$ we let $\delta_1 = \delta_2 = \delta_{\mathbb{B}}$ and so we omit the reference to the representations δ_1, δ_2. Thus, we speak simply of reducibility w.r.t. (δ_3, δ_4), because, by the first notion of reducibility, f can be identified with some of its $(\delta_{\mathbb{B}}, \delta_{\mathbb{B}})$-realizations. The same holds of $S_3 = S_4 = \mathbb{N}^{\mathbb{N}}$.

By applying Proposition 5.2 of [1] to Definition 6, we deduce that if $f \leq_c g$ w.r.t. $(\delta_1, \delta_2, \delta_3, \delta_4)$ and g is Σ_{k+1}^0-computable w.r.t. (δ_3, δ_4), then f is Σ_{k+1}^0-computable w.r.t. (δ_1, δ_2).

Definition 7 (Completeness). *For any $k \in \mathbb{N}$ let $C_k : \mathbb{N}^{\mathbb{N}} \to \mathbb{N}^{\mathbb{N}}$ be the function:*

$$C_k(p)(n) = \begin{cases} 0 & \text{if } \exists n_k \forall n_{k-1} \exists n_{k-2} \ldots Q n_1 : p\langle n, n_k, n_{k-1} \ldots, n_1 \rangle \neq 0 \\ 1 & \text{otherwise} \end{cases}$$

where $Q n_1 = \exists n_1$ if k is odd, and $Q n_1 = \forall n_1$ else.

Given represented sets (S_i, δ_i), for $i = 0, 1$, a function $f :\subseteq S_1 \to S_0$ is Σ_{k+1}^0-complete w.r.t. (δ_1, δ_0) if it is Σ_{k+1}^0-computable w.r.t. (δ_1, δ_0) and $C_k \leq_c f$ w.r.t. (δ_1, δ_0).

The rationale behind Definition 7 is the following fact, which is a consequence of the application of Theorem 5.5 of [1] to Definition 6: given any function $g :\subseteq S_3 \to S_4$, one has that $g \leq_c C_k$ w.r.t. (δ_3, δ_4) if and only if g is Σ_{k+1}^0-computable w.r.t. (δ_3, δ_4). Moreover, by Theorem 5.5 and Proposition 8.5 of [1], for any $k \in \mathbb{N}$, the function C_k is Σ_{k+1}^0-computable but not Σ_k^0-computable.

Definition 8. *Consider the set:*

$$F^{\omega\omega} = \{F :\subseteq \mathbb{N}^{\mathbb{N}} \to \mathbb{N}^{\mathbb{N}} : F \text{ is } (\mathbb{B}, \mathbb{B})\text{-continuous and } \mathrm{dom}(F) \text{ is a } G_\delta\text{-set}\}.$$

In the following, let $\eta :\subseteq \mathbb{N}^\mathbb{N} \to F^{\omega\omega}$ be any standard representation of $F^{\omega\omega}$ (see [5]) satisfying the universal Turing machine (utm-) property, the parameters (smn-) property, and the fact that any computable function $F \in F^{\omega\omega}$ has some computable η-name.

As usual, we write "η_p" as an abbreviation for "$\eta(p)$".[2]

2   Urysohn Lemma

We show that with respect to positive information there is a Σ_2^0-computable Urysohn function u mapping any pair of disjoint closed sets in a given metric space to some continuous real function satisfying certain properties. The function u is defined in terms of a map associating any given closed set A with some continuous function o such that $A = o^{-1}[\{0\}]$. We adapt proofs contained in [5] and [6], using techniques from the effective Borel measurability theory. By these techniques we also achieve some completeness results.

For closed sets, positive information consists in the following representation:

Definition 9. ψ_+ is the representation of the class $\mathcal{A}$ of the closed subsets of M defined in the following way: for $p \in \mathbb{N}^\mathbb{N}$ let

$$\psi_+(p) = A \in \mathcal{A} \Leftrightarrow \{n > 0 : n \in p\} = \{n : A \cap I_n^M \neq \emptyset\}.$$

The representation ψ_+ is largely used in the literature, although denoted by different symbols (e.g. by "$\delta_<$" in [3], while for the computable metric spaces $\mathbb{R}$ it coincides with "$\psi_<$" in [5]). The next representation is very used as well, and it is well-defined by the Main Theorem of [5]:

Definition 10. Let $\delta_{\mathrm{M}\mathbb{R}}$ be the representation of the set $C(M)$ of all total continuous real functions $f : M \to \mathbb{R}$ defined as follows for all $p \in \mathrm{dom}(\eta)$:

$$\delta_{\mathrm{M}\mathbb{R}}(p) = f \in C(M) \Leftrightarrow \eta_p \in F^{\omega\omega} \text{ is a } (\delta_\mathrm{M}, \rho)\text{-realization of } f.$$

Lemma 1. There is a Σ_2^0-computable function $o : \mathcal{A} \to C(M)$ w.r.t. $(\psi_+, \delta_{\mathrm{M}\mathbb{R}})$ mapping any closed set $A \subseteq M$ to some continuous function $o_A : M \to \mathbb{R}$ such that $A = o_A^{-1}[\{0\}]$.

Proof. For any non-empty closed set A let d_A be the distance function of A: $d_A(x) = \inf\{d(x,y) : y \in A\}$ for all $x \in M$. Let:

$$o_A(x) = \begin{cases} \min\{1, d_A(x)\} & \text{if } A \neq \emptyset \\ 1 & \text{otherwise.} \end{cases}$$

One can define a computable function $H :\subseteq \mathbb{N}^\mathbb{N} \times \mathbb{N}^\mathbb{N} \to \mathbb{N}^\mathbb{N}$ which outputs a list (of all the $\nu_\mathbb{Q}$-names) of the rational upper bounds of $o_A(x)$, given as input

---

[2] By using a pairing function in $\mathbb{N}^\mathbb{N}$, $F^{\omega\omega}$ can be considered as a set of functions with (finitely) many arguments. The utm- and the smn-property of η immediately extend to this case.

any ψ_+-name of the set A and any δ_M-name of a point $x \in M$. To accomplish this, just consider the following property which holds for all $\epsilon \in \mathbb{Q}^+$:

$$o_A(x) < \epsilon \Leftrightarrow \epsilon > 1 \vee (\exists c \in Q, \alpha \in \mathbb{Q}^+ : B(c, \alpha) \cap A \neq \emptyset \wedge d(x, c) + \alpha < \epsilon).$$

Similarly, it is easy to define a computable function $G :\subseteq \mathbb{N}^{\mathbb{N}} \times \mathbb{N}^{\mathbb{N}} \to \mathbb{N}^{\mathbb{N}}$ which outputs all rational lower bounds of $o_A(x)$, given as input a list (of all the ν_M-names) of the noted open balls not intersecting A and a δ_M-name of x. To do this consider that for any $\epsilon \in \mathbb{Q}^+$:

$$\epsilon < o_A(x) \Leftrightarrow \epsilon < 1 \wedge (\exists c \in Q, \alpha \in \mathbb{Q}^+ : B(c, \alpha) \cap A = \emptyset \wedge \alpha - d(x, c) > \epsilon).$$

There is a Σ_2^0-computable function L mapping any given ψ_+-name p_A of A to some enumeration $L(p_A) \subset \mathbb{N}^{\mathbb{N}}$ of the set $\{n : I_n^M \cap A = \emptyset\}$ (possibly empty). Put $L(p_A) = \lim_B (L_j(p_A))_{j \in \mathbb{N}}$, where:

$$L_j(p_A)(n) = \begin{cases} n \text{ if } n \notin p_A[j] \\ 0 \text{ otherwise.} \end{cases} \tag{1}$$

The function $\lim_B$ is Σ_2^0-computable by Proposition 9.1 of [1] and $\lambda p.(L_j(p))_{j \in \mathbb{N}}$ is computable. Therefore by Corollary 3.9 of [1], L is Σ_2^0-computable.

Hence, there is a computable function $\xi(p_1, p_2, p_3)$ which outputs a ρ-name of $o_A(x)$ on input $(p_A, L(p_A), r_x)$, where r_x is any δ_M-name of x. By the smn-property, there is a computable function S such that

$$\eta_{S(p_A, L(p_A))}(r_x) = \xi(p_A, L(p_A), r_x).$$

$S(p_A, L(p_A))$ is a δ_{MR}-name of o_A and the mapping $\lambda p_A.S(p_A, L(p_A))$ is Σ_2^0-computable by Corollary 3.9 of [1]. $\square$

In general, a map o satisfying Lemma 1 is not necessarily Σ_2^0-complete. For example, for $M = \{x\}$ (the singleton metric space), even if o cannot be (ψ_+, δ_{MR})-continuous, it has a (ψ_+, δ_{MR})-realization mapping computable objects to computable ones. This means, by the Invariance Theorem of [1], that $C_1 \not\leq_c o$. Nevertheless, the next proposition shows that in some cases Σ_2^0-completeness obtains:

Proposition 1. *For $M = \mathbb{R}$, it obtains $C_1 \leq_c o$ w.r.t. $(\psi_+, \delta_{\mathbb{R}\mathbb{R}})$.*

Proof. We give the proof for $\mathbb{R}$, but the method can be applied to other computable metric spaces with similar features. Let $p \in \mathbb{N}^{\mathbb{N}}$ be given. Then consider the closed set $A \subseteq \mathbb{R}$:

$$A = \{n : \exists m \, (p\langle n, m \rangle \neq 0)\}.$$

Let $G : \mathbb{N}^{\mathbb{N}} \to \mathbb{N}^{\mathbb{N}}$ be any computable function such that $\psi_+(G(p)) = A$. For any $n \in \mathbb{N}$, one can directly check if there is an m such that $p\langle n, m \rangle \neq 0$, and if so, $n \in A$, whence $o_A(n) = 0$. Otherwise, if such an m does not exist, then

$n \notin A$ and $o_A(n) \neq 0$. It is sufficient in this case to compute a ρ-name of $o_A(n)$ and verify that this is not a ρ-name of 0. More formally, if O is a $(\psi_+, \delta_{\mathbb{IRR}})$-realization of o, $OG(p)$ is a $\delta_{\mathbb{IRR}}$-name of o_A; given any δ_M-name r_n of n, we check computably (by the utm-property) whether $\rho(\eta_{o_{G(p)}}(r_n)) \neq 0$ (observe that there is a computable sequence $\{r_n\}_{n \in \mathbb{N}}$ of ρ-names such that $\rho(r_n) = n$ for all $n \in \mathbb{N}$). □

Theorem 1. *There is a Σ_2^0-computable function $u :\subseteq \mathcal{A} \times \mathcal{A} \to C(M)$ w.r.t. $(\psi_+, \psi_+, \delta_{\mathrm{MR}})$, mapping every disjoint pair of closed subsets of M to some total continuous function $u_{A,B} : M \to \mathbb{R}$ such that $u_{A,B}(x) = 0$ for $x \in A$, $u_{A,B}(x) = 1$ for $x \in B$, and $0 < u_{A,B}(x) < 1$ otherwise.*

Proof. For any closed set A let o_A be defined as in the proof of Lemma 1. Consider the function $u_{A,B}$ for $A, B \in \mathcal{A}$:

$$u_{A,B} = \frac{o_A}{o_A + o_B}.$$

Lemma 1 together with Corollary 3.9 and Proposition 3.8(3) of [1] give then the result for $u : (A, B) \mapsto u_{A,B}$. □

Considering again the singleton metric space, one can show that in general the function u of Theorem 1 is not Σ_2^0-complete. Nevertheless, the following proposition shows a case in which u is Σ_2^0-complete:

Proposition 2. *For $\mathbf{M} = \mathbb{R}$, it obtains $C_1 \leq_c u$ w.r.t. $(\psi_+, \psi_+, \delta_{\mathbb{IRR}})$.*

Proof. Again, the proof is valid for other cases of metric spaces similar to $\mathbb{R}$. Let $p \in \mathbb{N}^{\mathbb{N}}$ be given. For any $n \in \mathbb{N}$ define the closed sets $A_n, B_n \subseteq \mathbb{R}$:

$$A_n = \overline{\{n + 2^{-(k+2)} : \exists m \leq k \, (p\langle n, m \rangle \neq 0)\}},$$
$$B_n = \overline{\{n - 2^{-(k+2)} : \forall m \leq k \, (p\langle n, m \rangle = 0)\}}.$$

Then put $A = \bigcup_{n \in \mathbb{N}} A_n$ and $B = \bigcup_{n \in \mathbb{N}} B_n$. Let G, G' be two computable functions such that $\psi_+(G(p)) = A$, $\psi_+(G'(p)) = B$. Let $u_{A,B} \in C(M)$ be such that $u_{A,B}[A] = \{0\}$, $u_{A,B}[B] = \{1\}$. Given $n \in \mathbb{N}$, if there is an m for which $p\langle n, m \rangle \neq 0$ then $n \in A$, whence $u_{A,B}(n) = 0$. But if such an m does not exist, then $n \in B$ and $u_{A,B}(n) = 1$. □

3   Dieudonné's Function for the Urysohn-Tietze Lemma

In [6], Weihrauch observes that Dieudonné's approach to Urysohn-Tietze Lemma is not computable with respect to negative information.

We analyze again Dieudonné's solution, but for the case of positive information, and by the tools of effective Borel measurability.

First, we give the concept of positive information for continuous partial real functions according to [5] (the corresponding notion with respect to negative information is fundamental in [6]):

Definition 11. *Let $C^p(M)$ be the set of all continuous partial real functions $f :\subseteq M \to \mathbb{R}$ with closed domain and let δ^p_{MR} be the representation of such set defined in the following way:*

$$\delta^p_{\mathrm{MR}}\langle p, q \rangle = f \Leftrightarrow \eta_p \text{ is a } (\delta_{\mathrm{M}}, \rho)\text{-realization of } f \text{ and } \mathrm{dom}(f) = \psi_+(q).$$

Lemma 2. *Let $f : M \times A \to \mathbb{R}$ be some continuous bounded real function, with $A \subseteq M$ closed, such that the function $h = \lambda x.\inf\{f(x,y) : y \in A\}$ is continuous. Then there is a function mapping f to h which is Σ^0_2-computable w.r.t. $(\delta^p_{\mathrm{MMR}}, \delta_{\mathrm{MR}})$.*

Proof. By the sake of simplicity, consider $\langle p, q_A \rangle$ as a δ^p_{MMR}-name of f when η_p is a $(\delta_{\mathrm{M}}, \delta_{\mathrm{M}}, \rho)$-realization of f and $\psi_+(q_A) = A$. Let r_x and r_y be any δ_{M}-names of points $x \in M$, $y \in A$, respectively. Through (p, r_x, r_y) it is possible to compute a ρ-name of $f(x,y)$ using the utm-property. Obviously, $\alpha \in \mathbb{Q}$ is bigger than $h(x)$ if and only if there exists some $y \in A$ such that $f(x,y) < \alpha$. Given then a list s of all rational numbers bigger than $h(x)$, a list t of all rational numbers smaller than $h(x)$ is obtained using $\lim_{\mathbb{B}}$ similarly to (1), except that we need suitable adjustment in order to take account of the case in which $h(x) \in \mathbb{Q}$. Since the set A may be uncountable and the function $\lim_{\mathbb{B}}$ is Σ^0_2-computable, t may not be computable on the given input[3]. Therefore the smn-property is bound to depend also on such argument. But the information coded in t depends on x, whereas by applying the smn-property we want to find a δ_{MR}-name of h which depends only on $\langle p, q_A \rangle$. Hence, what we actually do is that we compute an "oracle" $K\langle p, q_A \rangle$ for the function h which is defined on suitable initial segments of δ_{M}-names, and such that $\xi(K\langle p, q_A \rangle, r_x)$ is a ρ-name of $h(x)$, for some computable function $\xi(p_1, p_2)$.

By definition of computable metric space (see [5]), the set $\{u \in \mathbb{N}^* : u\mathbb{N}^{\mathbb{N}} \cap \mathrm{dom}(\delta_{\mathrm{M}}) \neq \emptyset\}$ is r.e.: let $\{u_0, u_1, u_2, ...\}$ be a 1-1 computable enumeration of it. Similarly, let $\{w_0^{q_A}, w_1^{q_A}, w_2^{q_A}, ...\}$ be an enumeration, through q_A, of the set $\{w \in \mathbb{N}^* : \delta_{\mathrm{M}}[w\mathbb{N}^{\mathbb{N}}] \cap A \neq \emptyset\}$.

For $j \in \mathbb{N}$ let $\mathcal{M}$ be a machine which computes a function K_j such that $K_j\langle p, q_A \rangle \langle n, m \rangle = 1$ if $\mathcal{M}$ can prove in j steps that for all $l \leq j$ the least upper bound of $\rho(\eta_p[u_m\mathbb{N}^{\mathbb{N}} \times w_l^{q_A}\mathbb{N}^{\mathbb{N}}])$ is smaller than $\nu_{\mathbb{Q}}(n)$. Otherwise $K_j\langle p, q_A \rangle \langle n, m \rangle = 0$. Then $K(\langle p, q_A \rangle) = \lim_{\mathbb{B}}(K_j(\langle p, q_A \rangle))_{j \in \mathbb{N}}$ provides the desired oracle. $\square$

Theorem 2. *There is a Σ^0_2-computable function t w.r.t. $(\delta^p_{\mathrm{MR}}, \delta_{\mathrm{MR}})$ mapping each partial continuous real function $f :\subseteq M \to [1,2]$ with closed domain and $\min(f) = 1, \max(f) = 2$ to some continuous total extension $g : M \to [1,2]$.*

Proof. Let $\mathrm{dom}(f) = A$. Consider the Dieudonné function $f \mapsto g$, where g is defined by:

$$g(x) = \begin{cases} f(x) & \text{if } x \in A \\ \dfrac{\inf_{y \in A}\{f(y)d(x,y)\}}{d_A(x)} & \text{otherwise.} \end{cases}$$

---

[3] Nevertheless, observe that s can be actually computed by considering some countable dense set in A which is known by q_A, see [3].

The function g is a total extension of f and $\min(f) = \min(g) = 1$, $\max(f) = \max(g) = 2$. Moreover, g is continuous, as proven by Dieudonné in [4]. Let $\langle p, q_A \rangle$ be a δ^p_{MR}-name of f. We show that there is a Σ^0_2-computable function mapping $\langle p, q_A \rangle$ to some δ_{MR}-name of g. Let H be a function such that $H\langle p, q_A \rangle$ is a δ_{MR}-name of the function $\lambda x. \inf_{y \in A}\{f(y)d(x,y)\}$, for $x \in M$. Lemma 2 gives us a suitable Σ^0_2-computable H (observe that a δ_{MMR}-name of the function $\lambda(x,y).f(y)d(x,y)$ can be computed using the utm-property).

By the proof of Lemma 1 we know that $d_A(x)$ is computable given q_A and some list (of all the ν_{M}-names) of the noted open balls not intersecting A. By the same proof we know that there is a suitable Σ^0_2-computable function L which provides such a list given q_A. The map $\lambda \langle p, q_A \rangle.(H\langle p, q_A \rangle, L(q_A))$ is therefore Σ^0_2-computable.

We then define a Turing machine $\mathcal{M}(p_1, p_2, p_3, p_4)$, which, on the input

$$(\langle p, q_A \rangle, H\langle p, q_A \rangle, L(q_A), r_x),$$

computes a ρ-name of $g(x)$, with r_x a δ_{M}-name of $x \in M$. For the sake of simplicity in the description of the algorithm, we show how to compute a list of open rational intervals in $\mathbb{R}$ with decreasing diameters and whose intersection is the singleton $\{g(x)\}$. From this, it is easy to compute a ρ-name of $g(x)$. To define $\mathcal{M}$ we partly modify the original proof of the continuity of g given by Dieudonné. Let

$$i(x) = \frac{\inf_{y \in A}\{f(y)d(x,y)\}}{d_A(x)}.$$

The intuitive idea is to apply always f to x, unless we realize at some stage that $x \notin A$. If so, the computation goes on applying i to x. The problem is to handle the process carefully, so that if we realize at a certain stage that the wrong function (i.e. f) has been applied to x, we are still in time to compute a name of $g(x) = i(x)$. This means that despite of having applied the wrong function, we have listed on the output tape only names of balls containing $i(x)$. If we succeed, we do not need to know at the beginning of the computation whether $x \in A$ or not, in order to make a choice between f and i. Such a knowledge may not be computably achievable with the information coded in the input and it is (partially) dependent on x. On the contrary, we want to find, by the smn-property, a possible name for a realization of the Dieudonné function, and this must be independent from x.

We define $\mathcal{M}$ by induction on the number of stages. At stage 0 the machine outputs nothing.

Stage $s > 0$) Suppose $\mathcal{M}$ has listed only $\nu_{\mathbb{R}}$-names of balls containing $g(x)$. Now $\mathcal{M}$ must write the name of a noted open ball $B^{\mathbb{R}} \subseteq \mathbb{R}$ such that $g(x) \in B^{\mathbb{R}}$ and $\mathrm{diam}\left(B^{\mathbb{R}}\right) \leq 2^{-s}$. Let u be the initial segment of r_x that has been considered until now by $\mathcal{M}$. Suppose any ball mentioned in u intersects A. Therefore we are sure that u is an initial segment of some δ_{M}-name of some point in A. Then $\mathcal{M}$ applies, by the utm-property, η_p to r_x until it finds some $n \in r_x, q_A$ such that $f\left[I^{\mathrm{M}}_n\right] \subseteq I^{\mathbb{R}}_m \subseteq \mathbb{R}$, where $\mathrm{diam}\left(I^{\mathbb{R}}_m\right) < 2^{-(s+2)}$.

If such a ball I_n^{M} exists, let $I_n^{\mathrm{M}} = B(c, \alpha)$ for $c \in Q, \alpha \in \mathbb{Q}^+$. Suppose $\mathcal{M}$ finds also another noted open ball $I_k^{\mathrm{M}} = B(e, \beta)$ such that $x \in I_k^{\mathrm{M}} \subseteq I_n^{\mathrm{M}}, I_k^{\mathrm{M}} \cap A \neq \emptyset$, and

$$\alpha - d(c, e) - \beta > 4\beta.$$

If $x \in A$ then both I_n^{M} and I_k^{M} exist. Now we see how to find a suitable ball B^{IR} containing $g(x)$ (independently on whether $x \in A$ or not). Let $C = A \cap I_n^{\mathrm{M}}$ and $D = A - C$. Since $x \in I_k^{\mathrm{M}} \subseteq I_n^{\mathrm{M}}$, $I_k^{\mathrm{M}} \cap A \neq \emptyset$ and $\operatorname{diam}\left(I_k^{\mathrm{M}}\right) = 2\beta$, there is a $y \in C$ such that $d(x, y) < 2\beta$. But for any $y \in D$:

$$d(x, y) \geq d(y, c) - d(c, e) - d(e, x) > \alpha - d(c, e) - \beta > 4\beta.$$

Therefore:

$$d_A(x) = d_C(x) = \inf_{y \in C} \{d(x, y)\}. \tag{2}$$

Moreover, for $y \in C \cap I_k^{\mathrm{M}}$ one has $f(y)d(x, y) < 4\beta$, whereas for $y \in D$ it holds $f(y)d(x, y) > 4\beta$. So

$$\inf_{y \in C} \{f(y)d(x, y)\} = \inf_{y \in A} \{f(y)d(x, y)\}. \tag{3}$$

Recall that for any $y, z \in C$: $|f(y) - f(z)| < 2^{-(s+2)}$, thus $f(z) - 2^{-(s+2)} < f(y) < f(z) + 2^{-(s+2)}$. Therefore, chosen a $z \in C$, for any $y \in C$:

$$\left(f(z) - 2^{-(s+2)}\right) d(x, y) \leq f(y)d(x, y) \leq \left(f(z) + 2^{-(s+2)}\right) d(x, y),$$

hence

$$\left(f(z) - 2^{-(s+2)}\right) \inf_{y \in C} \{d(x, y)\} \leq \inf_{y \in C} \{f(y)d(x, y)\}$$

and

$$\inf_{y \in C} \{f(y)d(x, y)\} \leq \left(f(z) + 2^{-(s+2)}\right) \inf_{y \in C} \{d(x, y)\}.$$

By (2), $d_A(x) = \inf_{y \in C}\{d(x, y)\}$, and so by (3) we conclude:

$$\left(f(z) - 2^{-(s+2)}\right) d_A(x) \leq \inf_{y \in A} \{f(y)d(x, y)\} \leq \left(f(z) + 2^{-(s+2)}\right) d_A(x), \tag{4}$$

which proves that $|g(x) - f(z)| \leq 2^{-(s+2)}$. Indeed if $g(x) = f(x)$ then $|g(x) - f(z)| < 2^{-(s+2)}$ by our hypothesis that $f\left[I_n^{\mathrm{M}}\right] \subseteq I_m^{\mathrm{IR}}$ and $\operatorname{diam}\left(I_m^{\mathrm{IR}}\right) < 2^{-(s+2)}$. Otherwise, by (4):

$$f(z) - 2^{-(s+2)} \leq \frac{\inf_{y \in A}\{f(y)d(x, y)\}}{d_A(x)} = g(x) \leq f(z) + 2^{-(s+2)}.$$

Let $\gamma \in \mathbb{Q}$ be the center of I_m^{IR}. Then $|\gamma - g(x)| \leq |\gamma - f(z)| + |f(z) - g(x)| < 2^{-(s+1)}$. The machine $\mathcal{M}$ puts then $B^{\mathrm{IR}} = B(\gamma, 2^{-(s+1)})$.
Suppose otherwise that either I_n^{M} or I_k^{M} is not defined. Then $x \notin A$ and $\mathcal{M}$

recognizes this, sooner or later, through $L(q_A)$. By induction hypothesis, any ball mentioned in the output tape at stage $s - 1$ contains $g(x)$. Using the computability of $\div$, the value $i(x)$ is computable via the utm-property applied to $(H\langle p, q_A\rangle, r_x)$ and $(q_A, L(q_A), r_x)$. So $\mathcal{M}$ computes $i(x)$ until it finds some ball with diameter smaller than or equal to 2^{-s} and writes its name on the output tape.

$\mathcal{M}$ proceeds to compute $i(x)$ similarly at any other stage $u > s$.

Let now $\xi(p_1, p_2, p_3, p_4)$ be the function computed by $\mathcal{M}$. By the smn-property there is a computable function S such that

$$\eta_{S(\langle p, q_A\rangle, H\langle p, q_A\rangle, L(q_A))}(r_x) = \xi(\langle p, q_A\rangle, H\langle p, q_A\rangle, L(q_A), r_x).$$

Then the function $\langle p, q_A\rangle \mapsto S(\langle p, q_A\rangle, H\langle p, q_A\rangle, L(q_A))$ is Σ_2^0-computable. $\qquad\Box$

Moreover it obtains:

Proposition 3. *The Dieudonné function t is not computable: in some cases it is Σ_2^0-complete (w.r.t . $(\delta_{\mathrm{MR}}^p, \delta_{\mathrm{MR}})$).*

Proof. Consider the computable function $f : \mathbb{R} \times \mathbb{R} \to \mathbb{R}$ such that $f(x, y) = |x - 1| + 1$ for all $x, y \in \mathbb{R}$. This function f has a computable $\delta_{\mathbb{R}\mathbb{R}\mathbb{R}}$-name, say $r \in \mathbb{N}^{\mathbb{N}}$. Let $x_n = (0; n)$, $y_n = (1; n)$, $z_n = (2; n)$ for all $n \in \mathbb{N}$ and take a computable function $H : \mathbb{N}^{\mathbb{N}} \to \mathbb{N}^{\mathbb{N}}$ such that for any $p \in \mathbb{N}^{\mathbb{N}}$:

$$\psi_+(H(p)) = \{x_n, y_n : n \in \mathbb{N}\} \cup \{z_n : \exists m(p\langle n, m\rangle \neq 0)\}.$$

Put $\psi_+(H(p)) = A$. Then $\langle r, H(p)\rangle$ is a $\delta_{\mathbb{R}\mathbb{R}\mathbb{R}}^p$-name of $f_{|A}$.
Consider the Dieudonné extension g of $f_{|A}$. For any $n \in \mathbb{N}$, if there is an m such that $p\langle n, m\rangle \neq 0$ then $g(z_n) = f_{|A}(z_n) = f(z_n) = 2$. If there is no such m then $g(z_n) = f_{|A}(y_n)d(y_n, z_n) = f(y_n)d(y_n, z_n) = 1$. $\qquad\Box$

References

1. Brattka, V.: Effective Borel measurability and reducibility of functions. Mathematical Logic Quarterly. **51** (2005) 19–44
2. Brattka, V.: On the Borel complexity of Hahn-Banach extensions. Electronic Notes in Theoretical Computer Science. **120** (2005) 3–16
3. Brattka, V., Presser, G.: Computability on subsets of metric spaces. Theoretical Computer Science. **305** (2003) 43–76
4. Dieudonné, J.: Foundations of Modern Analysis. Academic Press. New York. 1960
5. Weihrauch, K.: Computable Analysis. Springer. Berlin-Heidelberg-New York. 2000
6. Weihrauch, K.: On computable metric spaces Tietze-Urysohn extension is computable. Lecture Notes in Computer Science. **2064** (2001) 357–368

Enumeration Reducibility with Polynomial Time Bounds

Charles M. Harris

University of Leeds, Leeds LS2 9JT, England

Abstract. We introduce *polynomial time enumeration* reducibility ($\leq_{pe}$) and we retrace Selman's analysis of this reducibility and its relationship with *non deterministic polynomial time conjunctive* reducibility. We discuss the basic properties of the degree structure induced by $<_{pe}$ over the computable sets and we show how to construct meets and joins. We are thus able to prove that this degree structure is dense and to show the existence of two types of lattice embeddings therein.

1 Introduction

Polynomial time enumeration reducibility ($\leq_{pe}$) was defined by Selman in [Sel78] as a variant of enumeration reducibility ($\leq_e$) in terms of the non constructive formulation of the latter given in [Sel71]. Selman showed that $\leq_{pe}$ differs from the constructive polynomial time bounded variant of enumeration reducibility ($\leq_c^{NP}$) introduced by Ladner et al. in [LLS75]. However Selman also showed that $\leq_{pe}$ and $\leq_c^{NP}$ coincide over the class of sets computable in exponential time. Now, $\leq_c^{NP}$ is an *effective operator based* reducibility in the sense that there exists a computable enumeration of effective operators $\{\Phi_n \mid n \in \omega\}$ such that for any sets A and B, $A \leq_c^{NP} B$ iff $A = \Phi_n(B)$, for some $n \in \omega$. Accordingly the degree structure induced by this reducibility over the computable sets is amenable—see[1] [Cop97]—to many of the techniques used in the literature of the well known polynomial time bounded deterministic reducibilities. The fundamental definition of $\leq_{pe}$ is however not *effective operator based* and we possess no reformulation of this definition to suggest otherwise. Indeed, as the reader will observe, an *effective operator based* definition of $\leq_{pe}$ would appear highly implausible. Thus the study of its degree structure requires by definition a different approach. The primary purpose of the present paper is to elaborate on this point. In particular, we show that, with the use of results from [Sel78] joins and meets can be constructed in a uniform manner and that, accordingly, two results on lattice embeddings due to Ambos Spies [AS85b, AS87] apply in the context of the $\leq_{pe}$ degrees. However, from a more general viewpoint, this work also shows that there are two *distinct* and *viable* degree structures corresponding to the polynomial time bounded variants of enumeration reducibility. Moreover, the reader should note that the fact—as indicated above—that these degree structures coincide over the class of exponential time sets raises the

---

[1] Note that Copestake uses the pseudonym $\leq_e^P$ for $\leq_c^{NP}$.

A. Beckmann et al. (Eds.): CiE 2006, LNCS 3988, pp. 209–220, 2006.
© Springer-Verlag Berlin Heidelberg 2006

possibility of the existence of some structural property distinguishing the associated degrees within the $\leq_{pe}$ degree structure[2].

2   Background and Preliminaries

Basic Notation and Assumptions. Let $\Sigma = \{0, 1\}$. Our basic elements are
finite strings over Σ, the set of which we denote by Σ^*. $s, t, x, \ldots$ denote such
strings and $|s|$ denotes the length of string s. Sets of strings are denoted by
$A, B, C, \ldots$ and classes of sets by $\mathbf{A}, \mathbf{B}, \mathbf{C}, \ldots$ The complement of A in Σ^* is de-
noted $\overline{A}$ and the cardinality of any set S is written $\| S \|$. st is the concatenation
of strings s and t, sA is the set $\{ st \mid t \in A \}$ and $A \oplus B$ is the set $0A \cup 1B$.
The *semicharacteristic* function of A is defined to be the function s_A such that
$\text{Dom}(s_A) = A$ and $s_A(x) = 1$ for all $x \in A$. The *characteristic* function of A is
written c_A. We assume the standard length lexicographical ordering on strings
($\leq_L$) and we assume the reader to be conversant with the identification of Σ^*
with ω induced by $\leq_L$ (so that, for example $t = \log s$ makes sense). For any
(total) functions $f, g : \omega \to \omega$ we say that f is $\mathcal{O}(g)$ if there exists a constant c
such that $f(n) \leq c \cdot g(n)$ for all $n \in \omega$. We extend this notation in an obvious way
to time bounds. $\mathcal{P}$ denotes the class of polynomials in one variable. We assume
the fixed enumeration $\{p_i(n) \mid i \in \omega\}$ in $\mathcal{P}$ to be defined by $p_i(n) = n^i + i$ for
all $i \in \omega$. We assume the reader to be already familiar with the basic notions of
(time related) complexity theory and with the (oracle) Turing machine model
used in the time bounded context. Accordingly, we use $\mathbf{P}$ ($\mathbf{NP}$) to denote the
class of sets computable (*acceptable* non deterministically) in polynomial time
and $\mathbf{EXP}$ to denote the class of sets computable in exponential time. More
generally for any total function $t : \omega \to \omega$, $\mathbf{DTIME}(t(n))$ denotes the class of
sets computable in time $\mathcal{O}(t(n))$. Note that we assume an effective enumeration
$\{ V_i \mid i \in \omega \}$ of $\mathbf{NP}$ such that V_i is non deterministically computable in time
$p_i(n)$. We say that A is *polynomial time many one reducible* to B ($A \leq_m^P B$) if
$A = f^{-1}(B)$ for some total function f computable in polynomial time. We say
that (total) g is *polynomial time constructible* (p-constructible) if there exists
$p(n) \in \mathcal{P}$ such that $g(s)$ is computable in $p(|g(s)|)$ steps for all $s \in \Sigma^*$.

Coding Finite Sets and Pairs of Strings. We assume $\{ D_s \mid s \in \Sigma^* \}$ to be
a polynomial time computable and invertible/decodable enumeration of all finite
subsets of Σ^* (see for example the coding scheme in [Har06] Subsection 4.2.2).
Also, for any finite set D we define D^+ and D^- to be the sets $\{ s \mid 0s \in D \}$
and $\{ s \mid 1s \in D \}$ respectively.

To keep notation succinct we use $\langle \, , \, \rangle$ to denote polynomial time computable
and invertible bijections (I) from $\Sigma^* \times \Sigma^*$ to Σ^* and (II) from $\omega \times \Sigma^*$ to Σ^*
(where ω is identified with $\{0\}^*$). Note that the context always disambiguates
the meaning of this notation.

Enumeration Reducibility and Non Determinism. Assuming the identi-
fication of Σ^* with ω mentioned above, let A, B be any subsets of Σ^* and let

---

[2] If $\mathbf{NP} = \mathbf{EXP}$ this is trivial since $\mathbf{NP}$ is the zero degree in this structure.

f, g be any partial functions from Σ^* into Σ^*. We suppose the reader to be familiar with deterministic Turing reducibility $\leq_T$ between partial functions and its non deterministic counterpart $\leq_{NT}$ (for the latter see for example [Coo04] Section 11.1 or [Har06] Section 2.2.1). Accordingly, for any sets A, B we say that A is *Turing* reducible to B ($A \leq_T B$) if $c_A \leq_T c_B$. We say that A is *computably enumerable in B* (A c.e. in B) if there exists a partial function $f \leq_T c_B$ such that $A = \text{Ran}(f)$ or, equivalently, if $s_A \leq_T c_B$, and we say that A is *computably enumerable* (c.e.) if f is partial computable or, equivalently, if s_A is partial computable. We say that A is *enumeration reducible to B* ($A \leq_e B$) if there exists a c.e. set W such that, for all $s \in \Sigma^*$,

$$s \in A \quad \text{iff} \quad (\exists t \in \Sigma^*)[\langle s, t \rangle \in W \ \& \ D_t \subseteq B] \tag{2.1}$$

and we say that partial function f is *enumeration reducible* to partial function g ($f \leq_e g$) if $\text{Graph}(f) \leq_e \text{Graph}(g)$. McEvoy showed in [McE84] that, $f \leq_e g$ iff $f \leq_{NT} g$ for any such functions f and g. It follows that A is enumeration reducible to B iff $s_A \leq_{NT} s_B$. In contrast to this, we say that A is (setwise) *non deterministic Turing* reducible to B ($A \leq_T^N B$) if $s_A \leq_{NT} c_B$. Now, a straightforward argument shows that $f \leq_T g$ iff $f \leq_{NT} g$ provided that $\text{Dom}(g)$ is computable[3]. This means that A c.e. in B iff $A \leq_T^N B$ and, in particular, that the class of c.e. sets comprises precisely those sets *acceptable* by a non deterministic Turing machine. This also means that Selman's definition of enumeration reducibility [Sel71] is tantamount to saying that $A \leq_e B$ iff

$$(\forall X \subseteq \Sigma^*)[B \leq_T^N X \ \Rightarrow \ A \leq_T^N X] \tag{2.2}$$

Turing Reducibilities and Polynomial Time Bounds. Let A and B be any subsets of Σ^*. We suppose that the reader is conversant with the notion of a *polynomial time* (p-time) *bounded* Turing machine (in which the underlying program essentially contains a step counting *polynomial clock*). We say that a Turing reduction is *p-time bounded* or, is (effected) *in p-time* if there is a p-time bounded oracle Turing machine that witnesses the reduction. Accordingly we say that A is *p-time Turing* reducible to B ($A \leq_T^P B$) if $A \leq_T B$ (i.e. $c_A \leq_T c_B$) in p-time. We say that A is (setwise) *non deterministic p-time Turing* reducible to B ($A \leq_T^{NP} B$) if $A \leq_T^N B$ (i.e. $s_A \leq_{NT} c_B$) in p-time. Moreover, using a standard result due to Cook [Coo71] and Karp [Kar72] we can stratify the latter in terms of computation length and $\leq_T^P$. Thus, given any $q(n) \in \mathcal{P}$, we say that A is *size $q(n)$ non deterministic p-time* reducible to B ($A \leq_{T,q}^{NP} B$) if there exists $R \leq_T^P B$ such that for all $x \in \Sigma^*$, $x \in A$ iff $\exists w [|w| \leq q(|x|) \ \& \ R(x, w)]$. We say that A is *non deterministic p-time conjunctive* reducible to B ($A \leq_c^{NP} B$) if $s_A \leq_{NT} s_B$ in p-time. The reader will observe that the present definitions of $\leq_T^{NP}$ and $\leq_c^{NP}$ are straightforward reformulations of the definitions found in the literature, for example in [LLS75]. Moreover, Ladner et al. and Selman essentially showed that both reducibilities can be defined in terms of enumeration

---

[3] See Theorem 2.4.5 and Corollary 2.4.6. of [Har06] or, for the case when g is total, the intuitive argument in Section 11.1 (pages 174-5) of [Coo04].

type operators[4]. Accordingly, using the notation defined above (see page 210), we know that $A \leq_T^{NP} B$ iff there is a polynomial $p(n)$ and set $V \in \mathbf{NP}$ such that, for all $s \in \Sigma^*$,

$$s \in A \Leftrightarrow (\exists t \in \Sigma^*)\big(|t| \leq p(|s|) \ \& \ \langle s,t \rangle \in V \ \& \ D_t^+ \subseteq B \ \& \ D_t^- \subseteq \overline{B} \big) \quad (2.3)$$

Likewise $A \leq_c^{NP} B$ iff there is a polynomial $p(n)$ and set $V \in \mathbf{NP}$ such that, for all $s \in \Sigma^*$,

$$s \in A \Leftrightarrow (\exists t \in \Sigma^*)\big(|t| \leq p(|s|) \ \& \ \langle s,t \rangle \in V \ \& \ D_t \subseteq B \big) \quad (2.4)$$

Moreover, we can define $\leq_m^{NP}$, the non deterministic version of $\leq_m^P$ via (2.4) by simply replacing the conjunct "$D_t \subseteq B$" by "$t \in B$". We use $\Psi_{p,V}$ to denote the set of putative *axioms* induced by the polynomial $p(n)$ and the set $V \in \mathbf{NP}$ in the sense that

$$\Psi_{p,V} = \{ \langle s,t \rangle \mid |t| \leq p(|s|) \ \& \ \langle s,t \rangle \in V \} \quad (2.5)$$

and we refer to such sets as *np-operators*. It is important to note that that $\Psi_{p,V} \in \mathbf{NP}$ for *any* polynomial $p(n)$ and $V \in \mathbf{NP}$. We assume a fixed effective enumeration of np-operators $\{ \Phi_n \mid n \in \omega \}$ defined such that $\Phi_n = \Psi_{p_i, V_j}$ for $n = \langle i, j \rangle$. Notice that for any $n \in \omega$ and $s, t \in \Sigma^*$, if $\langle s, t \rangle \in \Phi_n$ then $|t| \leq p_n(|s|)$. Of course, according to the above definitions, these operators can be *used* in three different ways. We thus specify $\Psi_{p,V}$ to be an *np-T-operator* if $\Psi_{p,V}$ witnesses (2.3) and we write $A = \Psi_{p,V}^T(B)$ for this reduction. We specify *np-c-operators* and *np-m-operators* in a similar manner and we use the notation $A = \Psi_{p,V}^c(B)$ and $A = \Psi_{p,V}^m(B)$ respectively in this case. When no ambiguity arises we drop the superscripts. Accordingly, for $r \in \{T, c, m\}$ we can view $\{ \Phi_n \mid n \in \omega \}$ as an enumeration of np-r-operators such that $A \leq_r^{NP} B$ iff $A = \Phi_n(B)$ for some $n \in \omega$.

Notational Conventions. Our notation is based on that found in [AS99]. Thus for example we use $\mathbf{REC}$ to denote the class of computable sets and, for $(R, s) \in \{ (P, T), (NP, c) \}$, we use $\langle \mathbf{REC}_s^R, \leq \rangle$ to denote the degree structure induced by $\leq_s^R$. We also refer to the latter as *the computable r-s-degrees*. Likewise we use $\mathbf{a}_s^R, \mathbf{b}_s^R, \ldots$ to denote individual degrees and we drop super/subscripts if the context is unambiguous.

3   Basic Properties of $\leq_{pe}$

We noted in Section 2 that, for any $A, B \subseteq \Sigma^*$, $A \leq_e B$ iff $s_A \leq_{NT} s_B$ whereas $A \leq_c^{NP} B$ iff $s_A \leq_{NT} s_B$ in p-time. Moreover, this analogy is borne out by the operator orientated formulations of these reducibilities given by (2.1) and (2.4). Therefore $\leq_c^{NP}$ can be seen as a polynomial time bounded variant of $\leq_e$. However, as Selman showed in [Sel78], there exists another distinct polynomial time

---

[4] See [LLS75] page 120, [Sel78] page 454 or [Har06] Lemmas 4.2.8-4.2.9.

bounded variant of $\leq_e$ whose definition reflects the non constructive formulation of the latter given by (2.2). Selman gave this variant the name *polynomial time enumeration* reducibility ($\leq_{pe}$). We proceed below by introducing this reducibility. We then follow Selman's argument by defining a constructive version ($\leq_{pe'}$) and by explaining the equivalence of the two. We present Selman's results on the comparison of $\leq_c^{NP}$ with $\leq_{pe}$ and we discuss some of the basic properties of the degree structure induced by $\leq_{pe}$.

Definition 3.1 ([Sel78]). *For any $A, B \subseteq \Sigma^*$, A is said to be* polynomial time enumeration *reducible to B ($A \leq_{pe} B$) if*

$$(\forall X \subseteq \Sigma^*)[\, B \leq_T^{NP} X \;\Rightarrow\; A \leq_T^{NP} X \,] \tag{3.1}$$

The reader will observe the analogy that we mentioned above between this definition and the non constructive definition of $\leq_e$ given by (2.2). We proceed by defining approximations to $\leq_{pe}$.

Definition 3.2 ([Sel78]). *For any $A, B \subseteq \Sigma^*$, and polynomial $q(n)$, A is said to be* q(n) time enumeration *reducible to B ($A \leq_{pe}^q B$) if*

$$(\forall X \subseteq \Sigma^*)[\, B \leq_{T,q}^{NP} X \;\Rightarrow\; A \leq_T^{NP} X \,] \tag{3.2}$$

It is obvious that $\leq_{pe}$ can be derived from the above approximations.

Lemma 3.1 ([Sel78]). $\leq_{pe} \;=\; \bigcap_{q \in \mathcal{P}} \leq_{pe}^q$.

In the constructive approach, in contrast, we begin by defining the appropriate approximations. For the sake of succinctness we define the latter directly in terms of np-operators.

Definition 3.3. *For any $A, B \subseteq \Sigma^*$, and polynomial $q(n)$, A is said to be* constructive q(n) time enumeration *reducible to B ($A \leq_{pe'}^q B$) if there exists an np-operator $\Psi_{p,V}$ (see (2.5)) and $k \geq 0$ such that, for all $s \in \Sigma^*$,*

$$s \in A \;\Leftrightarrow\; (\exists t \in \Sigma^*)[\, \langle s, t \rangle \in \Psi_{p,V} \;\&\; D_t^+ \subseteq B \;\&\; D_t^- \subseteq \overline{B}$$
$$\&\; \forall z(\, z \in D_t^- \Rightarrow q(|z|) \leq k \cdot \log |s| \,) \,]$$

Note that we also refer to this as a pe'-reduction for $q(n)$ of A to B.

Note 3.1 ([Sel78]). Viewed as a non deterministic p-time Turing reduction, $A \leq_{pe'}^q B$ can be described as follows. Suppose that machine N witnesses $A \leq_{pe'}^q B$. Then for any input s, if $s \in A$ there exists an accepting computation of $N^B(s)$ such that all negative queries z in this computation satisfy $q(|z|) \leq k \cdot \log |s|$. Note that, for simplicity, we refer to such z as *relevant* negative queries.

Definition 3.4 ([Sel78]). *For any sets A and B, A is said to be* constructive polynomial time enumeration *reducible to B ($A \leq_{pe'} B$) if $A \leq_{pe'}^q B$ for every polynomial $q(n)$. In other words,*

$$\leq_{pe'} \;=\; \bigcap_{q \in \mathcal{P}} \leq_{pe'}^q$$

Note 3.2. For any polynomials $p(n)$, $q(n)$, set $V \in \mathbf{NP}$ and number $k \geq 0$, we define the set

$$\Psi^{\mathrm{pe}'}_{p,V,q,k} = \{\, \langle s,t\rangle \mid \langle s,t\rangle \in \Psi_{p,V} \ \& \ \forall z(\, z \in D_t^- \Rightarrow q(|z|) \leq k \cdot \log|s|\,)\,\}$$

to be a *pe'-operator for* $q(n)$ with *index* $(p(n), V, k)$ and we note that $\Psi^{\mathrm{pe}'}_{p,V,q,k}$ is obviously in $\mathbf{NP}$. We assume a fixed computable enumeration $\{\Phi_n \mid n \in \omega\}$ of pe'-operators such that $\Phi_n = \Psi^{\mathrm{pe}'}_{p_i,V_j,p_l,k}$ for $n = \langle i,j,l,k\rangle$. Accordingly it is now easily seen that, for any fixed polynomial $q(n)$ there exists an enumeration $\{\widehat{\Phi}_n \mid n \in \omega\}$ of pe'-operators for $q(n)$ such that, for any $A,B \subseteq \Sigma^*$, $A \leq^{\mathrm{q}}_{\mathrm{pe}'} B$ iff $A = \widehat{\Phi}_n(B)$ for some $n \in \omega$—i.e. that $\leq^{\mathrm{q}}_{\mathrm{pe}'}$ is *effective operator based*.

The next step in Selman's argument is to show that $\leq^{\mathrm{q}}_{\mathrm{pe}} \equiv \leq^{\mathrm{q}}_{\mathrm{pe}'}$ for all $q(n) \in \mathcal{P}$. Note firstly that $\leq^{\mathrm{q}}_{\mathrm{pe}} \supseteq \leq^{\mathrm{q}}_{\mathrm{pe}'}$ is intuitively obvious. Indeed suppose that $A \leq^{\mathrm{q}}_{\mathrm{pe}'} B$ and $B \leq^{\mathrm{NP}}_{\mathrm{T},q} X$. Then by definition there exists binary $R \leq_{\mathrm{T}} X$ such that, for all $z \in \Sigma^*$, $z \in B$ iff $\exists y\,[\,|y| \leq q(|z|) \ \& \ R(z,y)\,]$. Thus the query "$z \in B$?" can be *deterministically* computed with oracle R in $\mathcal{O}(2^{q(|z|)})$ steps. However, for any input $x \in \Sigma^*$ any *relevant* negative query z in the reduction $A \leq^{\mathrm{q}}_{\mathrm{pe}'} B$ satisfies $q(|z|) \leq k \cdot \log|x|$ for some fixed $k \geq 0$. Thus $A \leq^{\mathrm{NP}}_{\mathrm{T}} R$ via an appropriate simulation derived from the original reduction $A \leq^{\mathrm{q}}_{\mathrm{pe}'} B$ since all *relevant* negative queries can be deterministically computed relative to R in time $\mathcal{O}(2^{k \cdot \log n}) = \mathcal{O}(n^k)$. Since $R \leq^{\mathrm{P}}_{\mathrm{T}} X$ it follows that $A \leq^{\mathrm{NP}}_{\mathrm{T}} X$. On the other hand, in order to prove $\leq^{\mathrm{q}}_{\mathrm{pe}} \subseteq \leq^{\mathrm{q}}_{\mathrm{pe}'}$ it suffices to prove the contrapositive $\not\leq^{\mathrm{q}}_{\mathrm{pe}} \supseteq \not\leq^{\mathrm{q}}_{\mathrm{pe}'}$. This is our next result.

Lemma 3.2 ([Sel78]). *For any* $A,B \subseteq \Sigma^*$ *and* $q(n) \in \mathcal{P}$, *if* $A \not\leq^{\mathrm{q}}_{\mathrm{pe}'} B$ *then there exists* $C \leq_{\mathrm{T}} A \oplus B$ *such that* $B \leq^{\mathrm{NP}}_{\mathrm{T},q} C$ *and* $A \not\leq^{\mathrm{NP}}_{\mathrm{T}} C$. *(And so* C *witnesses the fact that* $A \not\leq^{\mathrm{q}}_{\mathrm{pe}} B$.*)*

Proof. See the proof of [Sel78] Theorem 10 or [Har06] Lemma 4.3.12. □

The equivalence of $\leq_{\mathrm{pe}}$ and $\leq_{\mathrm{pe}'}$ is now evident from the above results.

Theorem 3.1 ([Sel78]). $\leq_{\mathrm{pe}} \equiv \leq_{\mathrm{pe}'}$

Corollary 3.1. *For any sets* A *and* B *if* $A \not\leq_{\mathrm{pe}} B$ *then there exists* $C \leq_{\mathrm{T}} A \oplus B$ *such that* $B \leq^{\mathrm{NP}}_{\mathrm{T}} C$ *whereas* $A \not\leq^{\mathrm{NP}}_{\mathrm{T}} C$. *In particular* C *is computable if* A *and* B *are both computable.*

Proof. If $A \not\leq_{\mathrm{pe}} B$ then, by Theorem 3.1 $A \not\leq_{\mathrm{pe}'} B$. Thus, by definition, $A \not\leq^{\mathrm{q}}_{\mathrm{pe}'} B$ for some $q(n) \in \mathcal{P}$. Therefore, by Lemma 3.2, there exists a set $C \leq_{\mathrm{T}} A \oplus B$ such that $B \leq^{\mathrm{NP}}_{\mathrm{T}} C$ but $A \not\leq^{\mathrm{NP}}_{\mathrm{T}} C$. □

Note 3.3 (The pe-operator problem). We saw in Note 3.2 that the relation $\leq^{\mathrm{q}}_{\mathrm{pe}'}$ is *effective operator based* for all $q(n) \in \mathcal{P}$. However, despite Theorem 3.1, if $X \leq_{\mathrm{pe}} Y$ all that we know is that, for all $q(n) \in \mathcal{P}$ there exists a pe'-operator

Φ for $q(n)$ such that $A = \Phi(B)$. This might mean, in the worst case, that we need an infinite list of pe'-operators as witness to the single reduction $X \leq_{pe} Y$. (Of course in practice there are many cases in which only a single operator is required, for example if $x \in X$ iff $\log\log x \in \overline{Y}$ for all $x \in \Sigma^*$ or whenever $X \leq_c^{NP} Y$.)

Selman showed in Theorem 6 of [Sel78] that $\leq_{pe}$ is a maximal transitive subrelation[5] of $\leq_T^{NP}$ over Σ^* and we can deduce from Selman's argument and Corollary 3.1 that this property also holds over **REC**. In contrast Ladner et al. constructed in Lemma 4.3 of [LLS75] sets $A, B, C \in \mathbf{DTIME}(2^{2^n})$ such that $A \leq_T^{NP} B \leq_T^{NP} C$ whereas $A \nleq_T^{NP} C$ thus proving that $\leq_T^{NP}$ is not transitive[6]. In particular for us this means that $\leq_{pe}$ is properly contained in $\leq_T^{NP}$ over classes of relatively low time complexity. On the other hand it is easily seen that $\leq_m^P \subseteq \leq_m^{NP} \subseteq \leq_c^{NP}$ and also that $\leq_c^{NP} \subseteq \leq_{pe}$ (the latter by Theorem 3.1 since obviously $\leq_c^{NP} \subseteq \bigcap_{q \sqsubseteq p} \leq_{pe}^q$). On the other hand, in Theorem 11 of [Sel78] Selman constructed sets A and B (in elementary time[7]) such that $A \leq_{pe} B$ (via the pe-reduction $x \in A$ iff $\log\log x \in \overline{B}$) whereas $A \nleq_c^{NP} B$. Thus $\leq_c^{NP}$ is properly contained in $\leq_{pe}$. However Selman also showed that this is not the case over **EXP**. Indeed, suppose that $A \leq_{pe} B$ and $B \in \mathbf{EXP}$. Then $B \in \mathbf{DTIME}(2^{q(n)})$ for some $q(n) \in \mathcal{P}$. Now, it follows from the assumption that $A \leq_{pe} B$ and Theorem 3.1 that $A \leq_{pe'}^q B$. Suppose that Φ is a pe'-operator for $q(n)$ witnessing this reduction. Then, by definition, for any $\langle s, t \rangle \in \Phi$ we know that $(\forall z \in D_t^-)[\, q(|z|) \leq k \cdot \log|s|\,]$ for some fixed $k \geq 0$. Thus "$z \in \overline{B}$?" can be computed in $\mathcal{O}(2^{k \cdot \log|s|}) = \mathcal{O}(|s|^k)$ steps for all such z. It is therefore straightforward to construct an np-c-operator $\widehat{\Phi}$ witnessing $A \leq_c^{NP} B$ (see the proof of [Sel78] Theorem 12 or [Har06] Proposition 4.3.25). Furthermore, taking into account that **EXP** is closed under $\leq_{pe}$ we obtain our next result.

Proposition 3.1 ([Sel78]). $\leq_c^{NP}$ *and* $\leq_{pe}$ *coincide over* **EXP**.

Let $\langle \mathbf{REC}_{pe}, \leq \rangle$ —which we also refer to as *the computable pe-degrees*—denote the degree structure induced by $\leq_{pe}$ over **REC**. Then notice that Proposition 3.1 tells us that $\langle \mathbf{REC}_c^{NP}, \leq \rangle$ and $\langle \mathbf{REC}_{pe}, \leq \rangle$ are identical over **EXP**. Now, as mentioned earlier a number of results concerning $\langle \mathbf{REC}_c^{NP}, \leq \rangle$ were proved by Copestake in [Cop97]. But what can we say about $\langle \mathbf{REC}_{pe}, \leq \rangle$? Well, firstly it is easily seen that the latter is an upper semilattice with **NP** as zero degree (just as for $\langle \mathbf{REC}_c^{NP}, \leq \rangle$). Moreover, $\langle \mathbf{REC}_{pe}, \leq \rangle$ is not a lattice and is not distributive. (These properties are proved in Theorem 4.4.3 and Theorem 4.4.5 of [Har06] using straightforward adaptations of similar arguments used in [AS85a, AS99].) Also the computable pe-degrees display branching properties

---

[5] Note that in [Sel71] Theorem 2.7 Selman had shown that $\leq_e$ is a maximal transitive relation of the relation "*c.e. in*" or, in other words (as we saw in Section 2) $\leq_T^N$.

[6] In fact the construction is such that $A \leq_T^P B \leq_{pe} C$ and so it also follows—by transitivity of $\leq_T^P$ and $\leq_{pe}$—that $A \nleq_{pe} B$ and $B \nleq_T^P C$ (see [Har06] Corollary 4.3.20). Thus this proof also shows the separation of $\leq_T^P$ and $\leq_{pe}$ in both possible ways.

[7] This is clear from Selman's construction. See [Har06] Lemma 4.3.23 for an approximate time analysis.

(see [Har06] Proposition 4.4.7) similar to those displayed by the computable p-T-degrees. This brings us to the question of whether joins and meets can be constructed in the pe-degrees in a manner similar to that developed for the p-T-degrees, in particular by Ambos-Spies. Bearing in mind the "pe-operator problem" stated above, this question requires a slight change of methodology. It is also the principal subject of the work presented below.

4   Join and Meet Lemmas

In this Section we present two results taken from Ambos Spies' work [AS85a, AS85b, AS87] which we will adapt to the context of the pe-degrees. The reader will notice that we do this directly in the case of the *meet* lemma below. The *join* lemma on the other hand has more general scope and is adapted to the present context in Section 5. We begin by a reminder of the notion of *recursively presentable* class and related issues.

Notation. We use $\mathbf{P}_\omega$ to denote the class $\{ X \mid X \subseteq \omega\ \&\ X \in \mathbf{P} \}$ under the indentification of ω with the unary language $\{0\}^*$.

Definition 4.1. *A class* $\mathbf{C}$ *of computable sets is* recursively presentable *(r.p.) if* $\mathbf{C}$ *is empty or there exists a computable set* $U \subseteq \omega \times \Sigma^*$ *such that* $\mathbf{C} = \{ U_n \mid n \in \omega \}$, *where* $U_n =_{\text{def}} \{ s \mid \langle n, s \rangle \in U \}$. *Note that* $C \leq_{\text{m}}^{\text{P}} U$ *for all* $C \in \mathbf{C}$ *and observe that we call* U *a* universal set *for* $\mathbf{C}$. *A class* $\mathbf{D}$ *is* closed under finite variants *(c.f.v.) if, for all sets* A *and* B, *if* $A \in \mathbf{D}$ *and* $B \stackrel{*}{=} A$ *then* $B \in \mathbf{D}$ *also.*

Our next result is proved by modifying the proof of Lemma 2.1(c) in [AS85b].

Lemma 4.1. *Let* $\mathbf{C}$ *and* $\mathbf{D}$ *be r.p. classes of computable sets. Define*

$$[\mathbf{C}, \mathbf{D}]_{\text{NP}} = \{ A \mid (\exists C \in \mathbf{C})\,(\exists D \in \mathbf{D})\,(C \leq_{\text{T}}^{\text{NP}} A \leq_{\text{T}}^{\text{NP}} D) \}$$

Then $[\mathbf{C}, \mathbf{D}]_{\text{NP}}$ *is recursively presentable and closed under finite variants.*

Note 4.1. Any finite class of computable sets is recursively presentable. In particular, the class $[\{A\}, \{B\}]_{\text{NP}}$ is r.p. and c.f.v. for any $A, B \subseteq \Sigma^*$.

Notation. Let $f : \omega \to \omega$ be a strictly increasing function. The nth iteration f^n of f is defined inductively by: $f^0(m) = m$ and $f^{n+1}(m) = f(f^n(m))$. We use the denotation $I_n^f =_{\text{def}} \{ x \in \Sigma^* \mid f^n(0) \leq |x| < f^{n+1}(0) \}$ and we call this the $(n+1)^{st}$ f-*interval*. Since f is strictly increasing $\{ I_n^f \mid n \in \omega \}$ is a partition of Σ^* (i.e. $\Sigma^* = \bigcup \{ I_n^f \mid n \in \omega \}$ and $I_m^f \cap I_l^f = \emptyset$ for all $m \neq l$). For any set $\alpha \subseteq \omega$ the notation I_α^f is used as shorthand for the set $\bigcup \{ I_n^f \mid n \in \alpha \}$.

Note 4.2 ([AS85b]). If $f : \omega \to \omega$ is p-constructible and strictly increasing and $\alpha \in \mathbf{P}_\omega$ then $I_\alpha^f \in \mathbf{P}$, and therefore for any $X \subseteq \Sigma^*$, $I_\alpha^f \cap X \leq_{\text{m}}^{\text{P}} X$.

Note 4.3 ([AS85b]). Any computable function $g : \omega \to \omega$ is *dominated* by a strictly increasing p-constructible function f in the sense that $(\forall n)[\,g(n) < f(n)\,]$.

Lemma 4.2 (Join lemma [AS85b]). *Let C_0, C_1 be computable sets and let $\mathbf{C}_0, \mathbf{C}_1$ be r.p and c.f.v. classes such that $C_0 \cup C_1 \notin \mathbf{C}_0$ and $C_1 \notin \mathbf{C}_1$. Then there is a computable function $g_0 : \omega \to \omega$ such that the following holds. If g is a strictly increasing computable function that dominates g_0 and α is an infinite and co-infinite set of natural numbers then $(C_0 \cap I_\alpha^g) \cup C_1 \notin \mathbf{C}_0 \cup \mathbf{C}_1$.*

Lemma 4.3 (Meet lemma [AS85b]). *For any computable set B there is a computable function g_1 such that $g_1(n) > n$ and the following holds. If g is a p-constructible and strictly increasing function which dominates g_1, and if $\alpha, \beta \in \mathbf{P}_\omega$ and $C \subseteq \Sigma^*$ is computable, then*

$$deg_{pe}((B \cap I_{2\alpha \cap 2\beta}^g) \oplus C)= deg_{pe}((B \cap I_{2\alpha}^g) \oplus C) \cap deg_{pe}((B \cap I_{2\beta}^g) \oplus C) \quad (4.1)$$

Proof (Sketch). Given B, let g_1 be the stepcounting function of some deterministic Turing machine computing B such that $g_1(n) > n$. Fix g, α, β and C as in the premise of the Lemma. Now, it is easily seen that the pe-degree on the L.H.S. of (4.1) is *below* both of the pe-degrees mentioned on the R.H.S. of the latter. Thus we only need to show that for any $X \subseteq \Sigma^*$,

$$[\, X \leq_{pe} (B \cap I_{2\alpha}^g) \oplus C \ \& \ X \leq_{pe} (B \cap I_{2\beta}^g) \oplus C \Rightarrow X \leq_{pe} (B \cap I_{2\alpha \cap 2\beta}^g) \oplus C \,]$$

Accordingly, the proof now proceeds in a similar manner to that of Lemma 3.4 of [AS87] except that we replace p-T-reductions by pe′-reductions. Indeed, by the same argument—and bearing in mind that pe′-reductions are just *specialised* np-T-reductions—we find that for any $q(n) \in \mathcal{P}$, if (I) $X \leq_{pe'}^q (B \cap I_{2\alpha}^g) \oplus C$ and (II) $X \leq_{pe'}^q (B \cap I_{2\beta}^g) \oplus C$, then (III) $X \leq_T^{NP} (B \cap I_{2\alpha \cap 2\beta}^g) \oplus C$. Moreover, on any input s, all queries made in reduction (III) are either queries made in reduction (I) or queries made in reduction (II) on input s. So suppose that $k', k'' \geq 0$ witness respectively the fact that the np-T-reductions (I) and (II) are pe′-reductions for $q(n)$ in the sense of Definition 3.3. Then we know that $q(|z|) \leq k' \cdot \log |s|$ for all[8] *relevant* (see Note 3.1) negative queries of reduction (I) and that $q(|z|) \leq k'' \cdot \log |s|$ for all *relevant* negative queries of reduction (II). But this means that any *relevant* query in reduction (III) satisfies

$$q(|z|) \leq q(|z|) + q(|z|) \leq k' \cdot \log |s| + k'' \cdot \log |s| = (k' + k'') \cdot \log |s|$$

Now, since s was chosen arbitrarily we know that $k = k' + k''$ witnesses the fact that (III) is in fact a pe′-reduction for $q(n)$, i.e. that $X \leq_{pe'}^q (B \cap I_{2\alpha \cap 2\beta}^g) \oplus C$.

Now suppose that $X \leq_{pe} (B \cap I_{2\alpha}^g) \oplus C$ and $X \leq_{pe} (B \cap I_{2\beta}^g) \oplus C$. Then, by Theorem 3.1 we know that (I) and (II) apply for all $q(n) \in \mathcal{P}$. Thus we know, by the above argument, that (III) also applies for all $q(n) \in \mathcal{P}$. Therefore, by applying Theorem 3.1 once more we obtain that $X \leq_{pe} (B \cap I_{2\alpha \cap 2\beta}^g) \oplus C$. This proves the Lemma. (See Lemma 4.5.7 of [Har06] for a more formal proof.) $\quad \square$

---

[8] In a more formal argument using operators (as defined in Note 3.2) all *possible* negative queries satisfy this condition.

5 Lattice Embeddings

Ambos-Spies' *join* and *meet* lemmas provide us with the background tools for the construction of lattice embeddings in the pe-degrees. As the reader will observe, the *join* lemma is used indirectly in that it relies on the fact that np-T-reductions are *effective operator based* (something we cannot guarantee for pe-reductions—see Note 3.3), and Corollary 3.1, to perform the necessary diagonalisation. In other words it uses what is essentially a corollary of the (non constructive) definition of pe-reducibility. On the other hand, the construction of meets relies heavily on the constructive formulation of pe-reducibility (see the proof of Lemma 4.3) in combination with the methods used to construct joins. We begin by presenting the basic construction of joins and meets in the pe-degrees and, in so doing, we prove that the computable pe-degrees are dense. We then go on to state two of Ambos-Spies' lattice embedding theorems which are applicable in the present context due to our ability to construct joins and meets in the manner described below.

Theorem 5.1. *If A,B are computable sets such that $A <_{\mathrm{pe}} B$ then there exist computable sets B_0, B_1 such that $A <_{\mathrm{pe}} B_0, B_1 <_{\mathrm{pe}} B$ and $B_0 \oplus B_1 \equiv_{\mathrm{pe}} B$.*

Proof. Fix $A, B \subseteq \Sigma^*$ such that $A <_{\mathrm{pe}} B$. Then, since $B \not\leq_{\mathrm{pe}} A$, by Corollary 3.1 there exists computable $C \subseteq \Sigma^*$ such that $A \leq_{\mathrm{T}}^{\mathrm{NP}} C$ but $B \not\leq_{\mathrm{T}}^{\mathrm{NP}} C$.

Note that, by Lemma 4.1, the following classes of sets $\mathbf{C}_0$ and $\mathbf{C}_1$ are recursively presentable and closed under finite variants:

$$\mathbf{C}_0 =_{\mathrm{def}} \{ E \mid E \leq_{\mathrm{T}}^{\mathrm{NP}} C \}$$
$$\mathbf{C}_1 =_{\mathrm{def}} \{ E \mid B \leq_{\mathrm{T}}^{\mathrm{NP}} E \leq_{\mathrm{T}}^{\mathrm{NP}} B \oplus C \}$$

Now without loss of generality suppose that $B \subseteq 0\Sigma^*$ and $C \subseteq 1\Sigma^*$. Therefore $(B \cap E) \cup C \equiv_{\mathrm{m}}^{\mathrm{P}} (B \cap E) \oplus C$ for any set E. (For example, $B \oplus C \equiv_{\mathrm{m}}^{\mathrm{P}} B \cup C$.)

Then $B \cup C \notin \mathbf{C}_0$ because $B \cup C \equiv_{\mathrm{m}}^{\mathrm{P}} B \oplus C$ and $B \not\leq_{\mathrm{T}}^{\mathrm{NP}} C$, whereas $C \notin \mathbf{C}_1$ (again because $B \not\leq_{\mathrm{T}}^{\mathrm{NP}} C$). Now, let g_0 be the computable function stipulated by Lemma 4.2 and let g be a p-constructible strictly increasing function dominating g_0 (such functions always exist—see Note 4.3). Therefore, by Theorem 4.2 we have

$$E_0 =_{\mathrm{def}} (B \cap I_{2\omega}^g) \oplus C \notin \mathbf{C}_0 \cup \mathbf{C}_1$$
$$E_1 =_{\mathrm{def}} (B \cap I_{2\omega+1}^g) \oplus C \notin \mathbf{C}_0 \cup \mathbf{C}_1$$

Now also define $B_0 =_{\mathrm{def}} (B \cap I_{2\omega}^g) \oplus A$ and $B_1 =_{\mathrm{def}} (B \cap I_{2\omega+1}^g) \oplus A$. Then since $I_{2\omega}^g, I_{2\omega+1}^g \in \mathbf{P}$ (see Note 4.2) we know that $(B \cap I_{2\omega}^g) \leq_{\mathrm{m}}^{\mathrm{P}} B$ and $(B \cap I_{2\omega+1}^g) \leq_{\mathrm{m}}^{\mathrm{P}} B$ so it follows that $A \leq_{\mathrm{pe}} B_0, B_1 \leq_{\mathrm{pe}} B$ (since $\leq_{\mathrm{m}}^{\mathrm{P}} \subseteq \leq_{\mathrm{pe}}$ and $A \leq_{\mathrm{pe}} B$ by hypothesis).

We now show that B_0 and B_1 lie strictly (pe-) in between A and B. Accordingly fix $i \in \{0, 1\}$.

- Suppose that $B \leq_{\mathrm{pe}} B_i$. Clearly $B_i \leq_{\mathrm{T}}^{\mathrm{NP}} E_i$ (as $A \leq_{\mathrm{T}}^{\mathrm{NP}} C$) and so $B \leq_{\mathrm{T}}^{\mathrm{NP}} E_i$ by definition of $\leq_{\mathrm{pe}}$. However this contradicts the fact that $E_i \notin \mathbf{C}_1$ (since obviously $E_i \leq_{\mathrm{T}}^{\mathrm{NP}} B \oplus C$).

- Suppose that $B_i \leq_{pe} A$. Observe that $B \cap I^g_{2\omega+i} \leq^P_m B_i$ and it thus follows that $B \cap I^g_{2\omega+i} \leq_{pe} A$. However this implies, by definition of $\leq_{pe}$ that $B \cap I^g_{2\omega+i} \leq^{NP}_T C$ (since $A \leq^{NP}_T C$). Therefore $E_i = (B \cap I^g_{2\omega+i}) \oplus C \leq^{NP}_T C$ in contradiction with the fact that $E_i \notin \mathbf{C}_0$.

We conclude that $A <_{pe} B_0, B_1 <_{pe} B$. Clearly also $B_0 \oplus B_1 \equiv_{pe} B$.   □

Corollary 5.1 (Density and splitting). *For any (computable) pe-degrees a and b such that $a < b$ there exist pe-degrees b_0 and b_1 such that $a < b_0, b_1 < b$ and $b = b_0 \cup b_1$. In other words the (computable) pe-degrees are dense and every pe-degree splits.*

Theorem 5.2 (Meet reducibility). *For any (computable) pe-degrees a and b such that $a < b$ there exist pe-degrees a_0 and a_1 such that $a < a_0, a_1 < b$ and $a = a_0 \cap a_1$. Thus the (computable) pe-degrees are meet reducible.*

Proof. Let a,b be (computable) pe-degrees such that $a < b$ and let $A \subseteq 1\Sigma^*$ and $B \subseteq 0\Sigma^*$ be sets such that $A \in a$ and $B \in b$. Also let $C \subseteq 1\Sigma^*$ be a computable set such that $A \leq^{NP}_T C$ whereas $B \nleq^{NP}_T C$ (using Corollary 3.1). Apply Lemma 4.2 (the *join* lemma) to $C_0 = B$, $C_1 = C$, $\mathbf{C}_0 = \{ X \mid X \leq^{NP}_T C \}$ and $\mathbf{C}_1 = \{ X \mid B \leq^{NP}_T X \leq^{NP}_T B \oplus C \}$. Also apply Lemma 4.3 (the *meet* lemma) to B (i.e. as B in the wording of the Lemma). Let g_0, g_1 be the respective functions guaranteed by Lemma 4.2 and Lemma 4.3 and let g be a p-constructible function dominating both g_0 and g_1. Then we follow the same reasoning as in the proof of Theorem 5.1 where now we have

$$
\begin{aligned}
E_0 &=_{def} (B \cap I^g_{4\omega}) \oplus C \notin \mathbf{C}_0 \cup \mathbf{C}_1 \\
E_1 &=_{def} (B \cap I^g_{4\omega+2}) \oplus C \notin \mathbf{C}_0 \cup \mathbf{C}_1
\end{aligned}
$$

and we define $B_0 =_{def} (B \cap I^g_{4\omega}) \oplus A$ and $B_1 =_{def} (B \cap I^g_{4\omega+2}) \oplus A$ which ensures that $A <_{pe} B_0, B_1 <_{pe} B$. We combine this with a straightforward application of Lemma 4.3. Whence we are able to conclude the present Theorem (using the fact that $(B \cap I^g_{4\omega \cap 4\omega+2}) \oplus A = \emptyset \oplus A \equiv_{pe} A$) by taking

$$
a_0 = deg_{pe}((B \cap I^g_{4\omega}) \oplus A) \quad \text{and} \quad a_1 = deg_{pe}((B \cap I^g_{4\omega+2}) \oplus A)   \square
$$

Corollary 5.2. *Any non-zero (computable) pe-degree b bounds a minimal pair.*

With the above results in mind we can now see that two different types of lattice embeddings proved by Ambos-Spies to exist in the p-m and p-T-degrees [AS85a, AS85b, AS87] also exist in the pe-degrees.

Theorem 5.3. *Let $\mathcal{L} = \langle L, \leq \rangle$ be any countable distributive lattice. Let a and b be computable pe-degrees such that $a < b$. Then there exist lattice embeddings $f_0, f_1 : \mathcal{L} \to [a, b]$ of $\mathcal{L}$ into the interval $[a, b]$ such that f_0 maps the least element 0 of $\mathcal{L}$ (if any) to a and f_1 maps the greatest element 1 of $\mathcal{L}$ (if any) to b.*

Indeed, to prove Theorem 5.3, we proceed in a similar manner to the proof of Corollary 4.3 in [AS85b] except that we apply the *join* lemma (Lemma 4.2) in the indirect manner exemplified in the proof of Theorem 5.1. (See the proof of Theorem 4.6.7. in [Har06] for details.) Similar observations apply to our final result below, with regard to the proof of Theorem 7.1 in [AS87].

Theorem 5.4. *Let $\mathcal{L} = \langle L, \leq \rangle$ be any finite distributive lattice which is nowhere complemented (i.e. no $a \in L - \{0,1\}$ has a complement). Let a and b be computable pe-degrees such that $a < b$. Then there exists a lattice embedding $f : \mathcal{L} \rightarrow [a,b]$ of $\mathcal{L}$ into the interval $[a,b]$ such that f maps the least element 0 of $\mathcal{L}$ (if any) to a and the greatest element 1 of $\mathcal{L}$ (if any) to b.*

References

[AS85a] K. Ambos-Spies. On the structure of the polynomial time degrees of recursive sets. Habilitationsschrift 206, Lehrstuhl für Informatik II, Universität Dortmund, 1985.

[AS85b] K. Ambos-Spies. Sublattices of the polynomial time degrees. *Information and Control*, 65(1):63–84, 1985.

[AS87] K. Ambos-Spies. Polynomial time degrees of NP-sets. In E. Börger, editor, *Current Trends in Theoretical Computer Science*, pages 95–142. Computer Science Press, Maryland, 1987.

[AS99] K. Ambos-Spies. Polynomial time reducibilities and degrees. In E. R. Griffor, editor, *Handbook of Computability Theory*, pages 683–705. Elsevier Science B.V., 1999.

[Coo71] S. A. Cook. The complexity of theorem-proving procedures. In *Proceedings of the Third ACM Symposium on the Theory of Computing*, pages 151–158, Shaker Heights, Ohio, 1971.

[Coo04] S.B. Cooper. *Computability Theory*. Chapman and Hall, 2004.

[Cop97] K. Copestake. On nondeterminism, enumeration reducibility and polynomial bounds. *Mathematical Logic Quarterly*, 43:287–310, 1997.

[Har06] C.M. Harris. *Enumeration Reducibility and Polynomial Time Bounds*. PhD thesis, The University of Leeds, UK, January 2006. (Available online at http://www.maths.leeds.ac.uk/~charlie/).

[Kar72] R.M. Karp. *Reducibility among Combinatorial Problems*, pages 85–104. Plenum Press, New York, 1972.

[LLS75] R. Ladner, N. A. Lynch, and A. L. Selman. A comparison of polynomial time reducibilities. *Theoretical Computer Science*, 1:103–123, 1975.

[McE84] K. McEvoy. *The Structure of the Enumeration Degrees*. PhD thesis, The University of Leeds, UK, October 1984.

[Sel71] A. L. Selman. Arithmetical reducibilities I. *Zeitshrift Math. Logik Grundlagen Math.*, 17:335–360, 1971.

[Sel78] A. L. Selman. Polynomial time enumeration reducibility. *SIAM Journal on Computing*, 7:440–457, 1978.

Coinductive Proofs for Basic Real Computation

Tie Hou

University of Wales Swansea, Swansea, SA2 8PP, Wales UK
tyehou@googlemail.com

Abstract. We describe two representations for real numbers, signed digit streams and Cauchy sequences. We give coinductive proofs for the correctness of functions converting between these two representations to show the adequacy of signed digit stream representation. We also show a coinductive proof for the correctness of a corecursive program for the average function with regard to the signed digit stream representation. We implemented this proof in the interactive proof system Minlog. Thus, reliable, corecursive functions for real computation can be guaranteed, which is very helpful in formal software development for real numbers.

Keywords: Real computation, Coinductive proof, Signed digit streams, Computability, Minlog.

1 Introduction

Computers are widely used for scientific applications in different fields, such as mathematics, physics, engineering and so on. The modeling of problems in above areas with a desirable accuracy requires considerable amount of computational effort. As the computational complexity increases, the risk of round off errors also increases. No matter how much precision is offered, these computations are not guaranteed to produce reliable results. Such unreliable computational results obtained may be useless to real-life problems, even may cause serious consequences.

Therefore, a mathematical model of exact computation is highly desirable. This applies in particular to computations concerned with real numbers. In the current computer model of real numbers through floating point numbers, the computer memory stores the approximations of (possibly irrational) real numbers, which truncate at a fixed rate precision. The probability that this yields inaccurate results is high, especially if these numbers are used as intermediate results. Hence, it is necessary to have more accurate representation of real numbers and algorithms to implement the computations using these representations.

Aiming at the above purpose, a wealth of alternative approaches are proposed, including interval arithmetic, stochastic arithmetic, multiple-precision arithmetic and exact arithmetic. Exact real arithmetic is a method of performing arithmetic operations whose results are guaranteed to be completely accurate, based on potentially infinite data structures such as streams. There are a number of alternative representations used for exact real arithmetic, such as any integral

A. Beckmann et al. (Eds.): CiE 2006, LNCS 3988, pp. 221–230, 2006.

base with negative digits, base $2/3$ with binary digits, nested sequences of rational intervals, Cauchy sequences, continued fractions ([8]), base golden-ratio with binary digits, and linear fractional transformations ([6]). Meanwhile, many algorithms have been proposed for real computations using these representations. However, few give formal proofs for the algorithms. More recently, Chirimar and Howe ([4]) represented real numbers by Cauchy sequences and implemented real analysis in Nuprl based on the type theory. Plume ([13]) gave algorithms for the basic arithmetic operations, transcendental functions, integration, and function minimum and maximum. Only informal proofs of correctness for some algorithms were shown. Formalisation of real numbers using corecursive streams as a coinductive type was discussed in [5], [3], [1] in the logical framework Coq. Lenisa ([11]) introduced set-theoretic generalizations of the coinduction proof principle in the view of bisimulation. However, the usual coinduction, based on bisimulation, is not expressive enough for the equality on real numbers, due to the redundancy of representation. In contrast, our approach is based on classical set theory and conventional mathematical reasoning.

Coinduction is a method of growing importance in reasoning about functional languages, due to the increasing prominence of lazy data structure. What is more, the proof of coinductive assertions is easy to implement in proof assistants like Minlog, Coq and so on. The average function for signed digit streams in this paper has been implemented in the Minlog system. See also [14] for other proof developments in Minlog based on the Cauchy sequence representation of real numbers.

1.1   Contributions

The main contributions of this paper are:

- (a) We define (in Section 2) a general lemma on closure properties of coinductively defined relations.
- (b) We define (in Section 3) coinductive representations of real numbers by signed digit streams and Cauchy sequences.
- (c) We give (in Section 4) coinductive proofs for the correctness of functions converting between the representations in (b).
- (d) We give (in Section 5) coinductive proofs for the correctness of the average function for signed digit stream representation.

2   Coinduction

In this section we introduce the concept of coinduction from a classical set-theoretic point of view. We prove a general lemma on closure properties of coinductive sets, which will be useful later.

2.1   Coinductive Relations as Largest Fixed Points

Let A be a set and $\wp(A) := \{X | X \subseteq A\}$ its power set. An operation $\Phi : \wp(A) \to \wp(A)$, is *monotone* iff $X \subseteq Y$ implies $\Phi(X) \subseteq \Phi(Y)$.

For any $X, Y \subseteq A$, it is well-known that any monotone operation Φ has a least and a largest fixed point, X_Φ and X^Φ respectively, that is, $\Phi(X_\Phi) = X_\Phi, \Phi(X^\Phi) = X^\Phi$, and for any other fixed point $Y \subseteq A$ of Φ (i.e. $\Phi(Y) = Y$) we have $X_\Phi \subseteq Y \subseteq X^\Phi$. The sets X_Φ and X^Φ can be defined by

$$X_\Phi := \bigcap \{Y | Y \subseteq A, \Phi(Y) \subseteq Y\}$$
$$X^\Phi := \bigcup \{Y | Y \subseteq A, Y \subseteq \Phi(Y)\}$$

It is easy to see that the monotonicity of Φ implies the required properties of X_Φ and X^Φ.

In the following, we will concentrate on the largest fixed point, X^Φ. By definition of X^Φ, we have for any set $Y \subseteq A$ that $Y \subseteq \Phi(Y)$ implies $Y \subset X^\Phi$. This principle is called *coinduction*.

In applications, the operation Φ is usually described by a formula $F[X, a]$ as $\Phi(X) :- \{a \in A | F[X, a]\}$. In this case the monotonicity of Φ is guaranteed by the condition that X does only occur positively in $F[X, a]$. For our purposes it will suffice to consider formulae of the form $F[X, a] :\equiv X(f(a)) \land a \in B$, where $f : A \to A$ is a fixed function and B is a fixed subset of A, hence, $\Phi(X) := f^{-1}(X) \cap B$.

In this particular case the coinductive principle reads (setting $X^{f,B} := X^\Phi$),

$$coind_{f,B}(Y) \qquad \frac{\forall a \in A \quad (a \in Y \Rightarrow f(a) \in Y \land a \in B)}{\forall a \in A \quad (a \in Y \Rightarrow a \in X^{f,B})}$$

The closure condition, $X^\Phi \subseteq \Phi(X^\Phi)$, then reads

$$cl_{f,B}(X^{f,B}) \qquad \forall a \in A(a \in X^{f,B} \Rightarrow f(a) \in X^{f,B} \land a \in B).$$

Because in fact $X^\Phi = \Phi(X^\Phi)$, the reverse of this implication holds as well. We will simply say that $X^{f,B}$ is coinductively defined by $cl_{f,B}$.

2.2   Closure Properties of Coinductive Relations

Consider $r \subseteq A$, which is coinductively defined by $B \subseteq A$ and $f : A \to A$ (that is, $r = X^{f,B}$ in the notation above; we write interchangeably $r(a)$ for $a \in r$).

$$(r) \qquad \forall a \in A(r(a) \Rightarrow B(a) \land r(f(a))) \tag{1}$$

We are interested in the question under which conditions r is closed under a given function. The following lemma takes care of a slightly more general situation.

Lemma 1. *Given $g : X \to X, h : X \to A, s \subseteq X$, s.t. for all $x \in X$, if $s(x)$ then*

1. $s(g(x))$
2. $f(h(x)) = h(g(x))$
3. $B(h(x))$

Then $\forall x \in X \Big(s(x) \Rightarrow r(h(x)) \Big)$.

Proof. Set $\tilde{r}(a) :\equiv \exists x \in X\big(s(x) \wedge a = h(x) \wedge r(a)\big)$. We need to show that (1) holds when r is replaced by $\tilde{r}$. It is given that $s(x) \Rightarrow B\big(h(x)\big)$, that is, $B(a)$ holds. By (1) we know that $r(a) \Rightarrow r(f(a))$. It is also given that $f(a) = f\big(h(x)\big) = h\big(g(x)\big)$ and $s(x) \Rightarrow s\big(g(x)\big)$. Therefore, $\exists g(x) \in X\big(s\big(g(x)\big) \wedge f(a) = h\big(g(x)\big) \wedge r(f(a)))\big)$, that is, $\tilde{r}(f(a))$ holds. Hence, by coinduction

$$
\frac{(\tilde{r}) \qquad \forall a \quad \big(\tilde{r}(a) \Rightarrow B(a) \wedge \tilde{r}(f(a))\big)}{\tilde{r} \subseteq r \qquad \forall a \quad \big(\tilde{r}(a) \Rightarrow r(a)\big)}
$$

follows $\tilde{r} \subseteq r$.

Now assume $s(x)$. Set $a := h(x)$. Then $\tilde{r}(a)$ implies $r(a)$ since $\tilde{r} \subseteq r$, that is, $r(h(x))$. Hence we have shown $\forall x \in X\big(s(x) \Rightarrow r(h(x))\big)$.   □

Corollary 1. *Let $r \subseteq A$ be coinductively defined by f and B and assume that for all $a, b \in A$*

1. $f(h(a,b)) = h(f(a), f(b))$
2. $r(a) \wedge r(b) \Rightarrow B(h(a,b))$

Then $\forall a, b\big(r(a) \wedge r(b) \Rightarrow r(h(a,b))\big)$.

Proof. By Lemma 1 where $X = A \times A, s = r \times r, g = f \times f$ (that is, $g(a,b) = \big(f(a), f(b)\big)$).   □

3   Coinductive Representations of Real Numbers by Signed Digit Streams and Cauchy Sequences

In this section we show how to represent real numbers by streams of signed digits (-1,0,1) and Cauchy sequences of rational numbers using coinductively defined representation relations. We will prove that these representations are equivalent to the usual ones involving the notion of infinite sum and limits from analysis.

If X is a set, then $[X]$ denotes the set of infinite streams of elements in X (i.e. $[X] = X^{\mathbb{N}}$). If $xs = (x_0 : x_1 : x_2 : \ldots) \in [X]$, then we set $head(xs) = x_0, tail(xs) = (x_1 : x_2 : x_3 : \ldots)$.

3.1   Coinductive Representation by Signed Digit Streams

Let $SD := \{-1, 0, 1\}$ be the set of *signed digits* and $[SD]$ the set of signed digit streams. Let $ds = (d_0 : d_1 : d_2 : \ldots)$ be a signed digit stream. Then the real number r in the interval $[-1, 1]$ that is represented by ds will be

$$
r = \sum_{n=0}^{\infty} d_i \cdot 2^{-(n+1)} \tag{2}
$$

In order to represent *all* real numbers r, we use an exponential factor 2^k where $k \in \mathbb{Z}$ as follows.

$$
r = 2^k \cdot \sum_{n=0}^{\infty} d_i \cdot 2^{-(n+1)} \tag{3}
$$

Hence, we define the set of signed digit stream representations of real numbers as $SDR := [SD] \times \mathbb{Z}$.

In order to be able to convince ourselves that the above representation is correct, a function $SDToReal$, based on (3), converting from signed digit streams to real numbers is needed.

Definition 1. *For every $k \in \mathbb{Z}$, we define*

$$SDToReal : SDR \rightarrow \mathbb{R}, \quad SDToReal(ds, k) = 2^k \cdot \sum_{n=0}^{\infty} d_i \cdot 2^{-(n+1)}$$

According to Definition 1, we have the following lemma.

Lemma 2. *For every $k \in \mathbb{Z}$,*

$$SDToReal(ds, k) = 2^{k-1} \cdot head(ds) + SDToReal(tail(ds), k - 1).$$

Especially, $SDToReal(ds, 0) = \big(head(ds) + SDToReal(tail(ds), 0)\big)/2$, when $k = 0$, that is, $r \in [-1, 1]$.

Lemma 2 suggests the following coinductive definition of a relation $\sim \subseteq SDR \times \mathbb{R}$ with the intended meaning $(ds, k) \sim r \Leftrightarrow SDToReal(ds, k) = r$.

Definition 2 (Coinductive definition of $(ds, k) \sim r$). *We coinductively define a relation $\sim \subseteq SDR \times \mathbb{R}$ by*

$$(\sim) \quad (ds, k) \sim r \Rightarrow |r| \leq 2^k \wedge (tail(ds), k - 1) \sim r - 2^{k-1} \cdot head(ds) \quad (4)$$

For the case $k = 0$, for short, we use the following coinductive definition.

Definition 3 (Coinductive definition of $ds \sim' x$). *We coinductively define a relation $\sim' \subseteq [SD] \times [-1, 1]$ by*

$$(\sim') \quad ds \sim' x \Leftrightarrow |x| \leq 1 \wedge tail(ds) \sim' 2 \cdot x - head(ds) \quad (5)$$

We can prove the correctness of above coinductively defined representation relations by the following lemmas.

Lemma 3. *For every $x \in [-1, 1]$, $ds \sim' x \Leftrightarrow SDToReal(ds, 0) = x$.*

Proof. $\Longrightarrow$: That is to show

$$\forall n \in \mathbb{N} \quad ds \sim' x \Rightarrow |SDToReal(ds, 0) - x| \leq 2^{1-n} \quad (6)$$

By induction on n.

$\underline{n = 0}$: by Definition 3, we can get $ds \sim' x \Rightarrow |x| \leq 1$. By Definition 1, it is easy to see that $SDToReal(ds, k) \in [-2^k, 2^k] \Rightarrow SDToReal(ds, 0) \in [-1, 1]$. Hence, $|SDToReal(ds, 0) - x| \leq 2 = 2^{1-0}$, that is (6) holds.

$\underline{n = n + 1}$: Now assume $ds \sim' x \Rightarrow |SDToReal(ds, 0) - x| \leq 2^{1-n}$ holds, we need to show that $ds \sim' x \Rightarrow |SDToReal(ds, 0) - x| \leq 2^{1-(n+1)}$ holds. By Definition 3, we can get $ds \sim' x \Rightarrow tail(ds) \sim' 2 \cdot x - head(ds)$.

By I.H., we know that $|SDToReal(tail(ds), 0) - (2 \cdot x - head(ds))| \leq 2^{1-n}$. By Lemma 2, we know $2 \cdot SDToReal(ds, 0) = head(ds) + SDToReal(tail(ds), 0)$. Hence, $|SDToReal(ds, 0) - x| = |2 \cdot SDToReal(ds, 0) - 2 \cdot x|/2 =$

$|SDToReal(tail(ds), 0) + head(ds) - 2 \cdot x|/2 \leq 2^{1-n}/2 = 2^{1-(n+1)}$. Therefore, (6) is proved.

$\Longleftarrow$: By coinduction. Set $Y := \{(ds, x)|SDToReal(ds, 0) = x\}$. We need to show $Y \subseteq \sim'$. By the principle of coinduction it suffices to show

1. $f(ds, x) \in Y$, that is, $(tail(ds), 2 \cdot x - head(ds)) \in Y$
2. $(ds, x) \in B$, that is, $|x| \leq 1$

Condition 1 holds by Lemma 2. We have $|x| \leq 1$, so condition 2 holds. $\square$

Lemma 4. *For every* $k \in \mathbb{Z}, r \in \mathbb{R}$, $(ds, k) \sim r \Rightarrow ds \sim' 2^{-k} \cdot r$.

Proof. We apply Lemma 1. We define g, f, h by $g((ds, k), r) = ((tail(ds), k - 1), r - 2^{k-1} \cdot head(ds))$, $f(ds, x) = (tail(ds), 2 \cdot x - head(ds))$, $h((ds, k), r) = (ds, 2^{-k} \cdot r)$. Now we need to show that the three conditions in Lemma 1 hold.

Condition 1 holds by $cl_{g,C}(\sim)$, where $C((ds, k), r) :\equiv |r| \leq 2^k$. It is easy to see $f(h((ds, k), r)) = (tail(ds), 2^{-(k-1)} \cdot r - head(ds)) = h(g((ds, k), r))$. Hence, condition 2 holds. We know $B(h((ds, k), r)) = |2^{-k} \cdot r| \leq 1$, that is, condition 3 holds. Therefore, by Lemma 1, we get $(ds, k) \sim r \Rightarrow ds \sim' 2^{-k} \cdot r$. $\square$

Lemma 5. *For every* $k \in \mathbb{Z}, x \in [-1, 1]$, $ds \sim' x \Rightarrow (ds, k) \sim 2^k \cdot x$.

Proof. For lack of space, the proof which is similar to Lemma 4 is omitted. $\square$

3.2   Coinductive Representation by Cauchy Sequences

We call a sequence $xs = (xs_0 : xs_1 : ...)$ of rational numbers $(xs_i \in \mathbb{Q})$ an *l-Cauchy sequence* if $\forall n \forall m \geq n.|xs_n - xs_m| \leq 2^{l-n}$, where xs_i represents the i-th element of the Cauchy sequence. We set $CR = [\mathbb{Q}] \times \mathbb{Z}$. We coinductively define a relation $\sim^c \subseteq CR \times \mathbb{R}$ with the intended meaning $(xs, l) \sim^c r \Leftrightarrow xs$ *is an l-Cauchy sequence converging to* r.

Definition 4 (Coinductive definition of $(xs, l) \sim^c r$). *For every* $(xs, l) \in CR, r \in \mathbb{R}$*, we define a relation* $\sim^c \subseteq CR \times \mathbb{R}$ *by*

$$(\sim^c) \quad (xs, l) \sim^c r \Rightarrow |head(xs) - r| \leq 2^l \wedge (tail(xs), l - 1) \sim^c r \quad (7)$$

We can prove the correctness of this definition by the following lemma.

Lemma 6. *For every* $(xs, l) \in CR, r \in \mathbb{R}$, $(xs, l) \sim^c r \Leftrightarrow \forall n.|xs_n - r| \leq 2^{l-n}$.

Proof. Similar to the proof of Lemma 3. $\square$

4   Adequacy of the Signed Digit Stream Representation

We consider the Cauchy sequence representation of real numbers as the standard one. We call any other representation adequate if there are computable back-and-forth translations between these two representations. The concept of computability on infinite streams can be explained by means of 'Oracle Turing machine' (Alan Turing ([15])). More recent accounts of the complexity of stream functions are studied by e.g. Ko ([10]) and Weihrauch ([16]). Hence, in order to show that the signed digit stream representation is adequate, we need to provide computable functions $SDTC : SDR \to CR$ and $CTSD : CR \to SDR$, such that for all $r \in \mathbb{R}$,

1. $\forall (ds, k) \in SDR \quad ((ds, k) \sim r \Rightarrow SDTC(ds, k) \sim^c r)$
2. $\forall (xs, l) \in CR \quad ((xs, l) \sim^c r \Rightarrow CTSD(xs, l) \sim r)$

Definition 5. *For every $k \in \mathbb{N}$, we define*

$$SDTC' : \mathbb{N} \times \mathbb{Q} \times [SD] \to [\mathbb{Q}]$$
$$SDTC'(k, q, ds) = (q + 2^{k-1} \cdot head(ds))$$
$$: SDTC'(k - 1, q + 2^{k-1} \cdot head(ds), tail(ds)).$$

Then we set $SDTC(ds, k) := (SDTC'(k, 0, ds), k + 1)$.

The definition of $SDTC'$ is an instance of a well-known corecursion scheme for defining infinite streams. More general schemes of corecursion are discussed, for example, in recent work of Buchholz ([2]).

Lemma 7 (Convert from SD to Cauchy)

$$\forall ds, k, r, q[(ds, k) \sim r \Rightarrow (SDTC'(k, q, ds), k + 1) \sim^c q + r]$$

Proof. We use Lemma 1.

We define g, f, h by $g((ds, k), r) = ((tail(ds), k - 1), r - 2^{k-1} \cdot head(ds))$, $f((xs, l), r) = ((tail(xs), l-1), r), h((ds, k), r) = ((SDTC'(k, q, ds), k+1), q+r)$. Now we need to show that three conditions in Lemma 1 hold.

Condition 1 holds by $cl_{g,C}(\sim^c)$, where $C((xs, l), r) := |head(xs) - r| \leq 2^l$. We can get $f(h((ds, k), r)) = ((SDTC'(k - 1, q + 2^{k-1} \cdot head(ds), tail(ds)), k)$, $q + r) = h(g((ds, k), r))$. Hence, condition 2 holds. We know $B(h((ds, k), r)) = |head(SDTC'(k, q, ds)) - r| < 2^{k+1}$, that is, condition 3 holds. Therefore, by Lemma 1, we get $\forall ds, k, r, q[(ds, k) \sim r \Rightarrow (SDTC_k(q, ds), k + 1) \sim^c q + r]$. □

Since from the third element of an l-Cauchy sequence, it is easy to decide in which part of the interval r is, according to Lemma 6, function $CTSD$ can be defined as follows.

Definition 6. *For every $n \in \mathbb{N}$, we define*

$$CTSD' : \mathbb{N} \times CR \to SDR$$
$$CTSD'(n, (xs, l)) = (d_0 : ds, k + 1)$$

where

$$k = \max(l, 1 + log_2|y|)$$
$$y = head(xs) - n$$

$$d_0 = \begin{cases} 0 & \text{if } |y| \le 2^{k-1} \\ -1 & \text{if } y < -2^{k-1} \\ 1 & \text{if } y > 2^{k-1} \end{cases}$$

$$ds = fst(CTSD'((n + 2^{k-1} \cdot head(ds)), (tail(xs), l - 1)))$$

Then we set $CTSD(xs, l) := CTSD'(0, (xs, l))$.

Lemma 8 (Convert from Cauchy to SD)

$$\forall xs, l, r, n[(xs, l) \sim^c r \Rightarrow CTSD'(n, (xs, l)) \sim r - n]$$

Proof. By an application of Lemma 1, similar to the proof of Lemma 7.   □

Lemma 7 and Lemma 8 show the coherence between models of representations and their implementations. Hence, the adequacy of signed digit stream representation is proved.

5   Average of Signed Digit Streams

The average function plays an important role as a tool to get other computable functions, e.g. [7]. In the following we define the average function on real numbers in the interval [-1, 1]. Then we give the coinduction proof of its correctness.

In order to calculate the average of two signed digit streams, a *carry* function that takes two digits as the input should be defined as follows($a_0 = head(a)$, $b_0 = head(b), a_1 = head(head(a)), b_1 = head(head(b))$).

$$carry(a, b) = \begin{cases} 1 & \text{if } a_0 + b_0 = 2 \\ 0 & \text{if } a_0 + b_0 = 0 \\ -1 & \text{if } a_0 + b_0 = -2 \\ 1 & \text{if } a_0 + b_0 = 1 \wedge a_1 + b_1 > 0 \\ 0 & \text{if } a_0 + b_0 = 1 \wedge a_1 + b_1 \le 0 \\ -1 & \text{if } a_0 + b_0 = -1 \wedge a_1 + b_1 < 0 \\ 0 & \text{if } a_0 + b_0 = -1 \wedge a_1 + b_1 \ge 0 \end{cases}$$

The average of signed digit streams is defined via an auxiliary function:

Definition 7 (Corecursive definition of function avA). *For every $a, b \in [SD]$, we define the auxiliary function avA as follows.*

$$avA : [SD] \to [SD] \to [SD]$$
$$avA(a, b) = (head(a) + head(b) - 2 \cdot carry(a, b) + carry(tail(a), tail(b)))$$
$$: \quad avA(tail(a), tail(b))$$

Using avA, the average function can be easily defined by:

Definition 8 (Average function av). *For every $a, b \in [SD]$, we define average fuction $av : [SD] \to [SD] \to [SD]$, $av(a, b) = carry(a, b) : avA(a, b)$.*

We can prove the correctness of the auxiliary function of average by the following lemma.

Lemma 9 (Auxiliary function of average). *For every $a, b \in [SD], x, y \in [-1, 1]$, $a \sim' x \wedge b \sim' y \Rightarrow avA(a, b) \sim' x + y - carry(a, b)$.*

Proof. According to Corollary 1, define f, B, h by $f(a, x) = (tail(a), 2 \cdot x - head(a))$, $B(a, x) = |x| \le 1$, $h((a, x), (b, y)) = (avA(a, b), x + y - carry(a, b))$. We need to show

1. $tail(avA(a, b)) = avA(tail(a), tail(b))$
2. $2\,(x + y - carry(a, b)) - head(avA(a, b)) = (2 \cdot x - head(a)) + (2 \cdot y - head(b)) - carry(tail(a), tail(b))$
3. $|x + y - carry(a, b)| \le 1$

Obviously, condition 3 holds. By Definition 7, it is easy to find that

$$tail(avA(a, b)) = avA(tail(a), tail(b)).$$

Condition 1 is proved. Using Definition 7 to calculate $head(avA(a, b))$, condition 2 is also proved. Therefore, by Corollary 1, we get $a \sim' x \wedge b \sim' y \Rightarrow avA(a, b) \sim' x + y - carry(a, b)$.   □

The correctness of the average function is proved as follows.

Lemma 10 (Average function). *For every $a, b \in [SD]$, $x, y \in [-1, 1]$, $a \sim' x \wedge b \sim' y \Rightarrow av(a, b) \sim' (x + y)/2$.*

Proof. By Lemma 9, we can get $a \sim' x \wedge b \sim' y \Rightarrow avA(a, b) \sim' x + y - carry(a, b)$. By Definition 8, we can get $avA(a, b) \sim' x + y - carry(a, b) \Rightarrow tail(av(a, b)) \sim' x + y - head(av(a, b))$. Obviously, $|(x + y)/2| \le 1$. By Definition 3, we can get $tail(av(a, b)) \sim' x + y - head(av(a, b)) \Rightarrow av(a, b) \sim' (x + y)/2$.   □

6   Conclusion and Future Work

Using a general lemma on closure properties of coinductively defined relations, we have coinductively proved the correctness of the basic arithmetic operation average and operations that convert between signed digit streams and l-Cauchy sequences. Parts of these proofs have been implemented in the Minlog system. This shows that coinductive proofs are very helpful in developing correct functions for real computations. We hope the coinductive methods will further narrow the gap between theory and practice in the formal development of reliable software systems.

As future work within this topic, we intend to perform coinductive proofs of multiplication and division functions for the signed digit stream representation. Also left to future research is to compare the efficiency of different proof methods in finding logical errors which normal testing can not discover.

Acknowlegdements

I would like to thank Ulrich Berger for many discussions about this paper. I have benefited a lot from him. And also thanks to the anonymous referees for their comments.

References

1. Bertot, Y.: Coinduction in Coq. In Lecture Notes of TYPES Summer School 2005, August 15-26 2005, Sweden, vol. II (2005). http://www.cs.chalmers.se/Cs/Research/Logic/TypesSS05/Extra/bertot.pdf
2. Buchholz, W.: A term calculus for (co-)recursive definitions on streamlike data-structures. Annals of Pure and Applied Logic, Volume **136** (2005) 75–90.
3. Ciaffaglione, A., Gianantonio, Di: A certified, corecursive implementation of exact real numbers. Theoretical Computer Science, Volume **351** (2006) 39–51.
4. Chirimar, J., Howe, D.J.: Implementing constructive real analysis: preliminary report. LNCS **613** (1992) 165–178.
5. Ciaffaglione, A.: Certified reasoning on real numbers and objects in co-inductive type theory. PHD Thesis. Department of Mathematics and Computer Science, University of Udine, and INPL-ENSMNS, Nancy, France (2003).
6. Edalat, A., Heckmann, R.: Computing with real numbers - I. The LFT approach to real number computation - II. A domain framework for computational geometry. In: Barthe G, Dybjer P, Pinto L, Saraiva J, editors, International summer school on applied semantics, Caminha, Portugal, Berlin, Springer-Verlag (2002) 193–267.
7. Escardo, M.H., Simpson, A.: A universal characterization of the closed Euclidean interval (extended abstract). Proceedings of the 16th Annual IEEE Symposium on Logic in Computer Science, Boston, Massachusetts (2001) 115–125.
8. Gibbons, J.: Streaming Representation-Changers. LNCS **3125** (2004) 142–168.
9. Jones, C.: Completing the rationals and metric spaces in LEGO. In Huet, G. and Plotkin, G., editors, Proceedings of the Second Annual Workshop on Logical Frameworks (1992).
10. Ko, Ker-I: Complexity theory of real functions. Birkhauser, Boston (1991).
11. Lenisa, M.: From Set-theoretic Coinduction to Coalgebraic Coinduction: some results, some problems. Coalgebraic Methods in Computer Science CMCS'99 Conference Proceedings, B. Jacobs, J. Rutten eds., ENTCS vol. **19** (1999).
12. Niqui, M.: Formalising exact arithmetic in type theory. In S. B. Cooper, B. Lowe, and L. Torenvliet, editors, New Computational Paradigms: First Conference on Computability in Europe, CiE 2005, Amsterdam, The Netherlands, June 8 12, 2005. Proceedings, LNCS **3526** (2005) 368–377.
13. Plume, D.: A Calculator for Exact Real Number Computation. 4th year project. Departments of Computer Science and Artificial Intelligence, University of Edinburgh (1998).
14. Schwichtenberg, Helmut: Inverting monotone continuous functions in constructive analysis. To appear in Proc. CiE 2006, Swansea (2006).
15. Turing, A.M.: Systems of logic based on ordinals. Proc. London Math. Soc. **45** (1939) 161–228.
16. Weihrauch, K.: Computable analysis, an introduction. Springer-Verlag (2000).

A Measure of Space for Computing over the Reals

Paulin Jacobé de Naurois*

LIPN - Université Paris XIII
99, avenue Jean-Baptiste Clément
93430 Villetaneuse - France
denaurois@lipn.univ-paris13.fr

Abstract. We propose a new complexity measure of space for the BSS model of computation. We define LOGSPACE$_W$ and PSPACE$_W$ complexity classes over the reals. We prove that LOGSPACE$_W$ is included in NC$_{\mathbb{R}}^2 \cap$ P$_W$, i.e. is small enough for being relevant. We prove that the Real Circuit Decision Problem is P$_{\mathbb{R}}$-complete under LOGSPACE$_W$ reductions, i.e. that LOGSPACE$_W$ is large enough for containing natural algorithms. We also prove that PSPACE$_W$ is included in PAR$_{\mathbb{R}}$.

Keywords: BSS model of computation, weak model, algebraic complexity, space.

1 Introduction

The real number model of computation, introduced in 1989 by Blum, Shub and Smale in their seminal paper [BSS89], has proved very successful in providing a sound framework for studying the complexity of decision problems dealing with real numbers. A large number of complexity classes have been introduced, and many natural problems have been proved to be complete for these classes. A nice feature of this model is that it extends many concepts of the classical complexity theory to the broader setting of real computation; in particular a question P$_{\mathbb{R}} \neq$ NP$_{\mathbb{R}}$ has arisen, which seems at least as difficult to prove as the classical one, and several NP$_{\mathbb{R}}$-complete natural problems have been exhibited.

It has been soon pretty obvious, however, that all features of the classical complexity theory could not be brought to this setting. In particular, the only complexity measures considered so far were dealing with *time*, and not, say, *space*: in 1989, Michaux proved in [Mic89] that, under a straightforward notion of space, everything is computable in constant space. Therefore, no notion of logarithmic or polynomial space complexity exists so far over the reals. A way to deal with this situation has been to define parallel complexity classes in terms of algebraic circuits, such that the NC$_{\mathbb{R}}^i$ and the PAR$_{\mathbb{R}}$ classes.

This model of computation has also long been criticized for being unrealistic: the assumption that one could multiply two arbitrary real numbers in constant

---

* Partially supported by the ANR project NO CoST: New tools for complexity - semantics and types.

time was the usual target, that Koiran faced in [Koi97] by defining a notion of *weak* cost that increases the cost for repeatedly multiplying or adding numbers.

Inspired by his approach, we propose here a new measure of space for the real number model, denoted as *weak* space, such that a repeated sequence of multiplications or additions on a number increases its size. Our notion allows us to define a logarithmic space complexity class, that falls within $\mathsf{NC}^2_{\mathbb{R}}$. We also prove that this class is large enough for containing natural algorithms: in particular, we prove a $\mathsf{P}_{\mathbb{R}}$-completeness result under $\mathsf{LOGSPACE}_W$ reductions.

The paper is organized as follows: in Section 2, we recall concepts and notations from the BSS model of computation. We define machines, circuits, and some major complexity classes. In Section 3, we briefly recall Michaux's result, and sketch a proof. In Section 4, we briefly introduce Koiran's notion of weak cost, and state some of the major results related to this notion. Then, we introduce our notion of *weak size* in Section 5, and state our results.

2   A Short Introduction on the BSS Model

In this section, we list the notations used in the paper, and recall some basic notions and results on the BSS model. A comprehensive reference for these notions is [BCSS98].

2.1   Notations

For an integer $c \in \mathbb{Z}$ we define its height as $\lceil \log(|c|+1) \rceil$. The height of an integer is the number of digits of its binary encoding. We also define $\mathbb{R}^* = \bigcup_{n \in \mathbb{N}} \mathbb{R}^n$.

2.2   Real Machines

We consider BSS machines over $\mathbb{R}$ as they are defined in [BSS89, BCSS98]. Roughly speaking, such a machine takes an input from $\mathbb{R}^*$, performs a number of arithmetic operations and comparisons following a finite list of instructions, and halts returning an element in $\mathbb{R}^*$ (or loops forever). Such a machine can be seen as a Turing machine over $\mathbb{R}$. It essentially consists in a finite directed graph, whose nodes are *instructions*, together with an input tape, an output tape, and a bi-infinite work tape, equipped with scanning heads. The instructions can be of the following types: *Start*, *Input* (reads an input value), *Output* (writes an output value), *Computation* (performs one arithmetical operation on two elements on the work tape), *Constant* (writes a constant parameter $A_i \in \mathbb{R}$), *Branch* (compares two elements, and branches accordingly), *Shift*, *Copy* and *Halt*.

For a given machine M, the function φ_M associating its output to a given input $x \in \mathbb{R}^*$ is called the *input-output function*. We say that a function $f : \mathbb{R}^* \to \mathbb{R}^*$ is *computable* when there is a machine M such that $f = \varphi_M$.

Also, a set $L \subseteq \mathbb{R}^*$, or a *language* is *decided* by a machine M if its characteristic function $\chi_L : \mathbb{R}^* \to \{0,1\}$ coincides with φ_M.

This model of computation allows one to define complexity classes. In particular, $\mathsf{P}_{\mathbb{R}}$ is the set of subsets of $\mathbb{R}^*$ that are decided by a real machine that works in deterministic polynomial time. Similarly, $NP_\mathbb{R}$ is the set of subsets of $\mathbb{R}^*$ that are decided by a real machine that works in nondeterministic polynomial time i.e, $x \in \mathbb{R}^*$ is accepted if and only if there exists $y \in \mathbb{R}^*$ of polynomial size such that the machine accepts (x, y).

2.3   Configurations

Definition 1. Configurations
- *A configuration of a machine M is given by an instruction q of M along with the position of the heads of the machine and three words $w_{input} \in \mathbb{R}^*$, $w_{work} \in \mathbb{R}_*$, $w_{output} \in \mathbb{R}^*$ that give the contents of the input tape, of the work tape and of the output tape.*
- *A transition of a machine M is a couple (c_i, c_j) of configurations such that, whenever M is in configuration c_i, M reaches c_j in one computation step.*

Definition 2. Configuration Graph
For a given machine M and a given set $\mathcal{C}$ of configurations of M, we define the configuration graph of M on $\mathcal{C}$ to be the directed graph with vertexes all elements in $\mathcal{C}$, and edges all transitions of M between elements in $\mathcal{C}$.

2.4   Algebraic Circuits

We introduce the notion of algebraic circuits, that allows to denote parallel computations and to define complexity classes below $P_\mathbb{R}$.

Definition 3. Algebraic Circuit
An algebraic circuit $\mathcal{C}$ is a sequence of gates $(G_1, \ldots, G_m)$ of one of the following types:

1. *Input gates: $G_i = x_i$, takes the input x_i from $\mathbb{R}$,*
2. *Arithmetic gates: perform the operation $*$ to the outputs of gates G_j and G_l, $j, l < i$ and $* \in \{+, -, ., /\}$,*
3. *Constant gates: $G_i = A_i$, $A_i \in \mathbb{R}$,*
4. *Sign gates: If $G_j \geq 0$ then $G_i = 1$ else $G_i = 0$, $j < i$.*

If a circuit has n input gates, we can suppose that they are the first ones, $G_1, \ldots, G_n$. If moreover the last node G_m is a sign node, we shall say that $\mathcal{C}$ is a decision circuit.

An algebraic circuit is a finite directed graph with no loops: its *size* is the number of gates, and its *depth* is the length of its longest path, starting from an input gate.

Algebraic circuits extend the classical notion of boolean circuit to the BSS setting, and allows one to define the $NC_\mathbb{R}$ hierarchy of complexity classes:

Definition 4. $NC_\mathbb{R}$
For all $i \in \mathbb{N}$, $NC_\mathbb{R}^i$ is the class of real decision problems decided by a P-uniform family of circuits of polynomial size and of depth bounded by $\mathcal{O}(\log^i(n))$, and

$$NC_\mathbb{R} = \bigcup_{i \in \mathbb{N}} NC_\mathbb{R}^i.$$

As remarked by Poizat in [Poi95], in the definition above the uniformity of the family can be considered relative to the classical Turing model. Hence, a P-uniform family of algebraic decision circuit is such that there exists an finite enumeration of all the constant gates in the family. There exists then a P time Turing machine which, on input n, k, outputs a discrete description of the kth gate of the nth circuit of the family.

Proposition 1. *[Cuc92]*

$$\text{NC}_{\mathbb{R}} \subsetneq \text{P}_{\mathbb{R}}.$$

3   Michaux's Result

This section is devoted to a brief exposition of Michaux's Result [Mic89], which states that a straightforward measure of space fails in differentiating one algorithm from another. In this section, we will use the following notion of space as a complexity measure:

Definition 5. Unit Space
Let M be a machine over $\mathbb{R}$, and let c be a configuration of M. We define $USize(c)$, the unit size *of c to be number of non-empty cells on the work tape at configuration c. Assume that on an input $(x_1, \ldots, x_n)$, the computation of M ends within t computation steps. The computation follows a path $c_0, \ldots, c_t$. We define the* unit space *used by M on input $(x_1, \ldots, x_n)$ to be*

$$USpace(M, (x_1, \ldots, x_n)) = \max_{0 \leq k \leq t} USize(c_k).$$

Assume that the running time of M is bounded by a function t. We define the unit space *used by M on input size n to be*

$$USpace(M, n) = \max_{(x_1, \ldots, x_n) \in \mathbb{R}^n} USpace(M, (x_1, \ldots, x_n)).$$

This notion of *unit space* is essentially the same as the classical notion of space for Turing machines. While in the classical Turing model this notion gives rise to a whole hierarchy of complexity classes like LOGSPACE, PSPACE, interlaced with the time hierarchy, this is not the case in the real setting. In order to precise a bit how unit space behaves on the reals, let us begin with the following well known technical result.

Lemma 1. *Let M be a real machine with parameters $A_1, \ldots, A_m$, whose running time is bounded by a function t. Let $n \in \mathbb{N}$. On any input $x_1, \ldots, x_n \in \mathbb{R}^n$, at any computation step $k \leq t(n)$, any non-empty cell on the work tape, say e_l, contains the evaluation of a rational fraction $f_{l,k} \in \mathbb{Z}(X_1, \ldots, X_{n+m})$ on $(x_1, \ldots, x_n, A_1, \ldots, A_m)$.*

Proof. Details arguments can be found in [Mic89, Poi95, Koi97]. We only sketch a proof here. The key argument is that, at any computation step k, the content of any cell e_l is obtained from the input values and the parameter values by a finite sequence of arithmetical operations. Therefore, the value in e_l is the evaluation of a rational fraction $f_{l,k} \in \mathbb{Z}(X_1, \ldots, X_{n+m})$ on $(x_1, \ldots, x_n, A_1, \ldots, A_m)$.

Proposition 2. *[Mic89] Let $L \subseteq \mathbb{R}^*$ by a real language decided by a machine M in time bounded by a function t. There exists a constant $k \in \mathbb{N}$ and a machine M' deciding L in unit space k.*

Proof. We only sketch the proof here. The interested reader can find more explanations in [Mic89, Poi95].

Rational fractions with integer coefficients can be easily encoded in binary, therefore, by Lemma 1, any configuration of M can also be encoded in binary. It suffices to realize that this binary encoding can be embedded into the digits of only two real numbers. Then, there exists a machine M' simulating M with only a constant number of real registers, among which two are needed for encoding the configurations of M.

4   The Weak BSS Model by Koiran

4.1   Definitions

Definition 6. Weak Cost
Let M be a machine whose running time is bounded by a function t, and let $A_1, \ldots, A_m$ be its real parameters. On any input $x_1, \ldots, x_n$, the computation of M consists in a sequence $c_0, \ldots, c_t, t \leq t(n)$ of configurations. To a transition c_k, c_{k+1} in this sequence we associate its weak cost as follows:

- *If the current instruction of c_k is a computation node, let e_l be the current cell on the work tape: the transition c_k, c_{k+1} consists in the computation of a rational fraction $f_{l+1,k+1} = g_{l+1,k+1}/h_{l+1,k+1} \in \mathbb{Z}(x_1, \ldots, x_n, A_1, \ldots, A_m)$, which is placed on the cell e_{l+1} in c_{k+1}. The weak cost of the transition c_k, c_{k+1} is defined to be the maximum of $\deg(g_{l+1,k+1})$, $\deg(h_{l+1,k+1})$, and the maximum height of the coefficients of $g_{l+1,k+1}$ and $h_{l+1,k+1}$.*
- *Otherwise, the weak cost of the transition c_k, c_{k+1} is defined to be 1.*

The weak running time of M on input $x_1, \ldots, x_n$ is the sum of the weak costs of the transitions in the sequence $c_0, \ldots, c_t$.

The weak running time of M is the function that associates with every n the maximum over all $x \in \mathbb{R}^n$ of the running time of M on x.

4.2   Some Results

Lemma 2. *[Koi97] A function is polynomial-time in the weak BSS model if and only if it is polynomial-time computable in the standard BSS model and the rational fractions $f_{l,k}$ have polynomial degree and coefficients of polynomial bit-size.*

Let P_W (respectively NP_W) be the set of real languages decided in deterministic (resp. nondeterministic) weak polynomial time, and EXP_W be the set of real languages decided in weak exponential time by a real machine.

Proposition 3

$$\mathsf{NC}^2_{\mathbb{R}} \not\subseteq \mathsf{P}_W \subsetneq \mathsf{P}_{\mathbb{R}} \subseteq \mathsf{NP}_W = \mathsf{NP}_{\mathbb{R}} \subseteq \mathsf{PAR}_{\mathbb{R}} \subseteq \mathsf{EXP}_W.$$

$\mathsf{P}_W \subsetneq \mathsf{NP}_W = \mathsf{NP}_{\mathbb{R}}$ is from [CSS94], $\mathsf{NP}_{\mathbb{R}} \subseteq \mathsf{EXP}_W$ from [Koi97] (where it is shown that the inclusion is strict). The missing items can be found in [BCSS98].

5   Weak Size and Space

5.1   Definitions

Instead of considering a unit size for all values on the work tape, which allows one to decide every decidable language in constant space, we would like to have a notion of size for the values computed on the work tape. The weak size of a computed value is a reasonable upper bound for the size of a boolean description of the corresponding rational fraction with integer coefficients. The weak size of a configuration is then the sum of the weak sizes of all computed values on the work tape in this configuration.

Yet, we need to precise a bit more the idea. Our purpose is to have a "nice" measure of space, allowing one to define a reasonable logarithmic space class. A trivial rational fraction like $f_1(X_1, \ldots, X_n, A_1, \ldots, A_m) = X_1$ has clearly a boolean description of size 1, while, for describing $f_n(X_1, \ldots, X_n, A_1, \ldots, A_m) = X_n$, one would need $\lceil \log(n+1) \rceil$ digits (for encoding the variable index). It seems rather unsatisfactory that a logarithmic space configuration may have a logarithmic number of occurrences of f_1, but only a constant ones of f_n. This feature can be corrected by allowing a permutation of the input variables, provided the permutation is simple enough, i.e can be described in logarithmic boolean space. In this paper, we have restricted ourselves to circular permutations, that can be described by an offset in $\{0, \ldots, n-1\}$.

Definition 7. Weak Size
Let $A_1, \ldots, A_m \in \mathbb{R}^m$ be given real numbers, and $g \in \mathbb{Z}[X_1, \ldots, X_{n+m}]$ a real polynomial with integer coefficients. Define a real polynomial $g_{A_1,\ldots,A_m} = g[X_1, \ldots, X_n, A_1, \ldots, A_m]$, with free variables $X_1, \ldots, X_n$. Let $0 \leq \mathsf{O} < n$, $\mathsf{O} \in \mathbb{N}$ be a number, the offset. To g, $A_1, \ldots, A_m \in \mathbb{R}^m$ and O, we associate the following:

- $deg(g)$ is the degree of g. We will write $\mathsf{D}(g)$ for $\lceil \log(\deg(g)+1) \rceil$.
- $Var_{A_1,\ldots,A_m}(g) \subseteq \{X_1, \ldots, X_n\}$ is the set of input variables on which $g_{A_1,\ldots,A_m}$ effectively depends.
- $R_{A_1,\ldots,A_m,\mathsf{O}}(g) = \max\{i + \mathsf{O} \mod n\}$ for $X_i \in Var_{A_1,\ldots,A_m}(g)$, is the range of g. We will write $\mathsf{R}(g)$ for $\lceil \log(R_{A_1,\ldots,A_m,\mathsf{O}}(g)+1) \rceil$.
- $\mathsf{N}(g) \in \mathbb{N}$ is the number of non-zero monomials of g.
- $S(g) \in \mathbb{Z}$ is the maximal absolute value of the integer coefficients of g. We will write $\mathsf{S}(g)$ for $\lceil \log(2S(g)+1) \rceil$.

– $Vc_{A_1,\ldots,A_m}(g)$ is the maximum, for every monomial of g, of the number of input variables on which it effectively depends. We will write $V(g)$ for $Vc_{A_1,\ldots,A_m}(g)$.

The weak size $S_{A_1,\ldots,A_m,\mathsf{o}}(g)$ of g is defined as follows:

$$S_{A_1,\ldots,A_m,\mathsf{o}}(g) = \mathsf{N}(g)\,(\mathsf{S}(g) + \mathsf{V}(g).\mathsf{R}(g) + \mathsf{V}(g).\mathsf{D}(g)) \tag{1}$$

For a rational fraction $f = g/h$ we take the weak size of f to be the maximum of the weak sizes of g and h.

It is clear that the weak size of g bounds the size of a boolean encoding of g, where g is presented as a sum of monomials modulo a circular permutation of the variable indexes. We do not take into account succinct boolean descriptions of factorized polynomials to ensure the tractability of our measure.

This measure of size for an element on the work tape naturally yields a notion of weak space for the given work tape, as follows:

Definition 8. Weak Space
Let M be a machine with real parameters $A_1,\ldots,A_m$. Let c_k be a configuration of M, with the corresponding input $x_1,\ldots,x_n \in \mathbb{R}^n$. We define:

– $e_i,\ldots,e_j$ *to be the non-empty part of the work tape in the configuration c_k.*
– *For any non-empty cell e_l in c_k, $f_{l,k} \in \mathbb{Z}(x_1,\ldots,x_n,A_1,\ldots,A_m)$ denotes the rational fraction it contains.*

The weak size of the work tape at the configuration c_k is then:

$$Size_w(c_k) = \min_{0 \le \mathsf{o} < n} \sum_{l=i}^{j} S_{A_1,\ldots,A_m,\mathsf{o}}(f_{l,k})$$

Assume that the running time of M is bounded by a function t. For a given input $x_1,\ldots,x_n$, the computation of M consists in a sequence $c_0,\ldots,c_t, t \le t(n)$ of configurations.

The weak running space of M on input $x_1,\ldots,x_n$ is the maximum for all configurations $c_0,\ldots,c_t$ of their weak size.

The weak running space of M is the function that associates with every n the maximum over all $x \in \mathbb{R}^n$ of the running space of M on x.

Definition 9. Complexity Classes

– *A language $L \subseteq \mathbb{R}^*$ is in LOGSPACE$_W$ if and only if there exist a machine M and a constant $k \in \mathbb{N}$ such that, for all $n \in \mathbb{N}$, on input $x \in \mathbb{R}^n$, M decides whether $x \in L$ in weak space less than $k\log(n)$.*
– *A language $L \subseteq \mathbb{R}^*$ is in PSPACE$_W$ if and only if there exist a machine M and two constants $k,d \in \mathbb{N}$ such that, for all $n \in \mathbb{N}$, on input $x \in \mathbb{R}^n$, M decides whether $x \in L$ in weak space less than kn^d.*
– *A function $f : \mathbb{R}^* \to \mathbb{R}^*$ is in FLOGSPACE$_W$ if and only if there exist a machine M and two constant $k,m \in \mathbb{N}$ such that, for all $n \in \mathbb{N}$, on input $x \in \mathbb{R}^n$, and computation $c_0,\ldots,c_t$ of M on x:*

1. *M computes $f(x)$ in weak space less than $k \log(n)$.*
2. *for every configuration c_i with current node an output node and current cell e_l, the weak size of the content of e_l is less than m.*

In the definition of FLOGSPACE$_W$, the output consists in a sequence of real values of constant weak size in the input. This ensures that one can compose FLOGSPACE$_W$ algorithms, and that the result of the composition remains an FLOGSPACE$_W$ algorithm. This is necessary for defining notions like logarithmic space reductions and for obtaining completeness results.

5.2   What Michaux's Result Becomes

Lemma 3.   *There exists $L \subseteq \mathbb{R}^*$ such that:*

– $L \in P_W$
– *for all $k \in \mathbb{N}$, L is not decidable in weak space less than k.*

Proof. Let $p(X_1, \ldots, X_n) = X_1 + \ldots + X_n$, and consider the set L of points $(x_1, \ldots, x_n) \in \mathbb{R}^n$, such that $p(x_1, \ldots, x_n)$ equals 0. Assume L is decided by a machine M. It is well known that the set of inputs accepted by a BSS machine is semi-algebraic, therefore, L can be described as a finite union of sets given by systems of polynomials inequalities of the form

$$\bigwedge_{i=0}^{s} F_i(X_1, \ldots, X_n) = 0 \wedge \bigwedge_{j=0}^{t} G_j(X_1, \ldots, X_n) > 0,$$

where the values $F_i(x_1, \ldots, x_n)$ and $G_j(x_1, \ldots, x_n)$ are effectively computed by M. Since L has dimension $n - 1$, at least one of these sets must have dimension $n - 1$. Since the set described by the $G'_j s$ is open, it must be nonempty, and then it defines an open subset of $\mathbb{R}^n$. All the polynomials F_i vanish on that nonempty open subset of L. Since this open subset of L is clearly infinite, and p is an irreducible polynomial, all the polynomials F_i must vanish on the whole set L. It is then a well known result ([BCSS98], Proposition 2 p.362) that the polynomials F_i are multiples of p. Also, at least one of these F_i is a non-trivial multiple of p.

It is clear that $p(x_1, \ldots, x_n)$ has weak size at least $n \log(n)$, and so does this non-trivial multiple of p. Therefore, M decides L in weak space at least $n \log(n)$.

5.3   Structural Complexity Results

Theorem 1

$$\text{LOGSPACE}_W \subseteq P_W \cap \text{NC}_{\mathbb{R}}^2,$$
$$\text{PSPACE}_W \subseteq \text{PAR}_{\mathbb{R}}.$$

Proof. In a first step, we prove LOGSPACE$_W \subseteq P_{\mathbb{R}}$. The key argument is an upper bound for the number of configurations of weak size at most $k \log(n)$.

Consider a machine M, with t nodes. For a fixed input size n, and an offset O, simple counting arguments show that the number of rational fractions of weak size at most B for some $B \in \mathbb{N}$ is bounded by α^B, for some $\alpha \in \mathbb{N}$. It follows that the number of possible work tape contents of weak size B, for the same fixed offset, is bounded by $(2\alpha^2)^B$. Taking into account all possible values for the offset, the scanning head positions and the current node of the machine, the number of configurations of weak size at most B is then bounded by $tn^2B(2\alpha^2)^B$. When $B = k\log(n)$, this bound is polynomial.

$\mathsf{LOGSPACE}_W \subseteq \mathsf{P}_W$ follows then by Lemma 2, since all rational fractions of logarithmic weak size have clearly polynomial degrees and coefficient heights.

$\mathsf{LOGSPACE}_W \subseteq \mathsf{NC}^2_{\mathbb{R}}$ is then proven along the lines of [Bor77]: given a $\mathsf{LOGSPACE}_W$ machine M, we exhibit a $\mathsf{NC}^1_{\mathbb{R}}$ construction of its configuration graph. This construction involves some numeric computation, in order to check whether two given configurations are connected, and produces a boolean description of the configuration graph of M. Next, it suffices to decide whether the input and accepting configurations are connected in this graph: this is the classical reachability problem, which is decidable in the boolean class NC^2.

$\mathsf{PSPACE}_W \subseteq \mathsf{PAR}_{\mathbb{R}}$ is a corollary.

5.4   Completeness Results

Definition 10. *[CT92]* Real Circuit Decision Problem ($\mathsf{CDP}_{\mathbb{R}}$)
Input: $(\mathcal{C}, \overline{x})$, where C is an arithmetic circuit with k input gates and $\overline{x} \in \mathbb{R}^k$. Question: Does C output 1 on input $\overline{x}$?

It has been shown in [CT92] that $\mathsf{CDP}_{\mathbb{R}}$ is $\mathsf{P}_{\mathbb{R}}$-complete under $\mathsf{NC}^2_{\mathbb{R}}$-reductions.

Theorem 2. $\mathsf{CDP}_{\mathbb{R}}$ *is $\mathsf{P}_{\mathbb{R}}$-complete under* $\mathsf{FLOGSPACE}_W$-*reductions.*

Proof. The proof follows [CT92]. The reduction happens to be in $\mathsf{FLOGSPACE}_W$.

We have stated this completeness results under $\mathsf{FLOGSPACE}_W$ reductions. By Theorem 1, it is clear that $\mathsf{FLOGSPACE}_W$ reductions are in $\mathsf{P}_W \cap \mathsf{NC}^2_{\mathbb{R}}$. The problem considered has already been proven complete under first-order reductions [GM96], which also happen to be in $\mathsf{P}_W \cap \mathsf{NC}_{\mathbb{R}}$. Yet, it remains unclear how the two types of reductions compare.

6   Concluding Remarks and Open Questions

In the discrete model, space has proven to be a very relevant complexity measure. Many natural problems have been found in $\mathsf{LOGSPACE}$, and many others in NC^2 whose membership in $\mathsf{LOGSPACE}$ is unclear. We believe that weak space may play the same role in the real setting. An argument in this direction is the following remark: consider a real algorithm that reads an input, normalizes it to $\{0,1\}$ with some step function, and applies a boolean $\mathsf{LOGSPACE}$ procedure. Real complexity analysis until now only allowed one to say that such a real algorithm belongs to $\mathsf{P}_W \cap \mathsf{NC}^2_{\mathbb{R}}$: the algorithmic flavor behind it was lost. However, it is now clear that such an algorithm belongs to $\mathsf{LOGSPACE}_W$. An important task now is to exhibit some natural problems in $\mathsf{LOGSPACE}_W$. Others in $\mathsf{NC}^2_{\mathbb{R}}$ or P_W, not easily in $\mathsf{LOGSPACE}_W$, may also be of interest.

Structural results remain also to be found. In particular, it needs to be checked whether the following conjecture holds:

Conjecture 1

$$\mathsf{NC}^1_{\mathbb{R}} \not\subseteq \mathsf{LOGSPACE}_W,$$
$$\mathsf{LOGSPACE}_W \subseteq \mathsf{NC}^1_{\mathbb{R}} \Rightarrow \mathsf{LOGSPACE} \subseteq \mathsf{NC}^1.$$

Similar questions arise also for PSPACE_W.

Acknowledgements

We thank the anonymous referees for their helpful comments, and for pointing out references to the notion of first-order reductions.

References

[BCSS98] L. Blum, F. Cucker, M. Shub, and S. Smale. *Complexity and Real Computation*. Springer-Verlag, 1998.

[Bor77] A. Borodin. On relating time and space to size and depth. *SIAM J. Comp.*, 6:733–744, 1977.

[BSS89] L. Blum, M. Shub, and S. Smale. On a theory of computation and complexity over the real numbers: NP-completeness, recursive functions and universal machines. *Bulletin of the Amer. Math. Soc.*, 21:1–46, 1989.

[CSS94] F. Cucker, M. Shub, and S. Smale. Separation of complexity classes in Koiran's weak model. *Theoretical Computer Science*, 133(1):3–14, 11 October 1994.

[CT92] F. Cucker and A. Torrecillas. Two p-complete problems in the theory of the reals. *Journal of Complexity*, 8(4):454–466, 1992.

[Cuc92] F. Cucker. $\mathsf{P}_{\mathbb{R}} \neq \mathsf{NC}_{\mathbb{R}}$. *Journal of Complexity*, 8:230–238, 1992.

[GM96] E. Grädel and K. Meer. Descriptive complexity theory over the real numbers. *Lecture Notes in Applied Mathematics*, 32:381–404, 1996.

[Koi97] P. Koiran. A weak version of the blum, shub & smale model. *Journal of Computer and System Sciences*, 54:177–189, 1997.

[Mic89] C. Michaux. Une remarque à propos des machines sur $\mathbb{R}$ introduites par Blum, Shub et Smale. *C. R. Acad. Sci. Paris*, 309, Série I:435–437, 1989.

[Poi95] B. Poizat. *Les Petits Cailloux*. Aléas, 1995.

On Graph Isomorphism for Restricted Graph Classes[*]

Johannes Köbler

Institut für Informatik, Humboldt-Universität zu Berlin, Germany
koebler@informatik.hu-berlin.de

Abstract. Graph isomorphism (GI) is one of the few remaining problems in NP whose complexity status couldn't be solved by classifying it as being either NP-complete or solvable in P. Nevertheless, efficient (polynomial-time or even NC) algorithms for restricted versions of GI have been found over the last four decades. Depending on the graph class, the design and analysis of algorithms for GI use tools from various fields, such as combinatorics, algebra and logic.

In this paper, we collect several complexity results on graph isomorphism testing and related algorithmic problems for restricted graph classes from the literature. Further, we provide some new complexity bounds (as well as easier proofs of some known results) and highlight some open questions.

1 Introduction

In this section we briefly review some important complexity results for graph isomorphism as well as for related problems as, e.g., computing the automorphism group $\mathrm{Aut}(X)$ of a given graph X in terms of a generating set of automorphisms (we refer to this problem as AUT) or the canonization problem (i.e., renaming the vertices of a given graph in such a way that all isomorphic graphs become equal). It is easy to see that GI reduces to both problems (in fact, in the unrestricted case, GI and AUT are polynomial-time equivalent, whereas it is open whether canonization reduces to GI). Formal definitions of these and other concepts used in the paper are deferred to the next section. In some sense, graph isomorphism represents a whole class of algorithmic problems; for example, GI is polynomial-time equivalent to the isomorphism problem for semigroups as well as for finite automata [15]. For the interesting relationships between GI and isomorphism testing for other algebraic structures like groups and rings we refer the reader to the excellent surveys [1, 5].

Two graphs X and Y are *isomorphic* (denoted by $X \cong Y$) if there is a bijective mapping g between the vertices of X and the vertices of Y that preserves the adjacency relation, i.e., g relates edges to edges and non-edges to non-edges. *Graph Isomorphism* is the problem of deciding whether two given graphs are isomorphic. The problem has received considerable attention since it is one of

---

[*] Work supported by a DST-DAAD project grant for exchange visits.

A. Beckmann et al. (Eds.): CiE 2006, LNCS 3988, pp. 241–256, 2006.
© Springer-Verlag Berlin Heidelberg 2006

the few natural problems in NP that are neither known to be NP-complete nor known to be solvable in polynomial time.

There is some evidence that GI is not NP-complete. First of all, GI is polynomial-time equivalent to its counting version #GI which consists in computing the number of isomorphisms between two given graphs [40]. In contrast, the counting versions of NP-complete problems (like #SAT) are typically much harder; in fact they are #P-complete and hence at least as hard as any problem in the polynomial-time hierarchy [46]. More strikingly, the complement of GI belongs to the class AM of decision problems whose positive instances have short membership proofs checkable by a probabilistic verifier [7]. As a consequence, GI is not NP-complete unless the polynomial hierarchy collapses to its second level [16, 45].

A promising approach in tackling the graph isomorphism problem for general graphs is to design efficient algorithms for restricted graph classes. In fact, Luks' efficient GI algorithm for graphs of bounded degree [38] yields the fastest known general graph isomorphism algorithm due to Babai, Luks, and Zemlyachenko [6, 11, 49]. The strongest known hardness result due to Torán [47] says that GI is hard for the class DET of problems that are NC^1 reducible to the computation of the determinant of a given integer matrix (cf. [20]). DET is a subclass of NC^2 (even of TC^1) and contains NL as well as all logspace counting classes like Mod_kL, $C_=L$, PL and $L(\#L)$ [2, 17].

The first significant complexity result for restricted graph classes is the linear time canonization algorithm for trees, designed by Hopcroft and Tarjan [29], and independently by Zemlyachenko [48]. Miller and Reif [42] later gave an NC algorithm for tree canonization, based on tree contraction methods. Then Lindell came up with a logspace algorithm for tree canonization [37]. As shown in [34], this upper bound is optimal, since tree isomorphism is also hard for L under AC^0 reductions. If we consider complexity bounds below L, then the representation that we use to encode the input trees becomes important. For trees encoded in the string representation, Buss [18] located the canonization problem even in NC^1 (which is also optimal [34]).

Shortly after the linear time canonization algorithm for trees was found, Hopcroft, Tarjan and Wong designed a linear time canonization algorithm for planar graphs [28, 30]. This line of research has been pursued by Lichtenstein, Miller, Filotti, and Mayer, culminating in a polynomial-time GI algorithm for graphs of bounded genus [36, 41, 33]. In 1991, Miller and Reif [42] designed an AC^1 algorithm for planar graph isomorphism.

Using a group theoretic approach, Babai showed in 1979 that GI is decidable in random polynomial time for the class $\mathcal{CG}_b$ of colored graphs with constant color multiplicity b. More precisely, the vertices of a graph in $\mathcal{CG}_b$ are colored in such a way that at most b vertices have the same color and we are only interested in isomorphisms that preserve the colors. Inspired by Babai's work, Furst, Hopcroft and Luks [22] developed efficient solutions for various permutation group problems and as a byproduct they could eliminate the need for randomness in Babai's algorithm. Both algorithms exploit, in a significant manner, the fact that the automorphism group $\text{Aut}(X)$ of a graph X with constant color class size, is contained in the product of constant size symmetric groups. For such groups the pointwise stabilizer series can be used to successively compute generators for the groups in the series.

In a breakthrough result, Luks in 1982 was able to design an algorithm for computing $\text{Aut}(X)$ in polynomial time for graphs of bounded degree [38]. To achieve this result, Luks considerably refined the group-theoretic techniques used in earlier algorithms. By combining Luks' algorithm with a preprocessing procedure due to Zemlyachenko [49] (see also [6]) for reducing the color valence of the input graphs, Babai and Luks obtained an $2^{O(\sqrt{n \log n})}$ time-bounded GI algorithm, where n denotes the number of vertices in the input graphs (see [11]). This is the fastest algorithm known for the unrestricted graph isomorphism problem. In [11] it is also shown that for general graphs there is a $2^{O(n^{1/2+o(1)})}$ canonizing algorithm which closely matches the running time of the best known GI algorithm.

Later, Luks in [39] gave a remarkable NC algorithm for the bounded color class case. Building on [39], Arvind, Kurur and Vijayaraghavan further improved Luks' NC upper bound by showing that GI for graphs in $\mathcal{CG}_b$ (we denote this version of GI by GI_b) is in the Mod_kL hierarchy (and hence in TC^1), where the constant k and the level of the hierarchy depend on b [4]. Prior to this result, Torán showed that GI_{b^2} is hard for the logspace counting class Mod_bL [47]. Torán's lower bound has been extended in [4] where it is shown that for each level in the Mod_kL hierarchy there is a constant b such that GI_b is hard for this level.

The pointwise stabilizer series approach has also been applied by Babai, Grigoryev and Mount to compute the automorphism group for graphs with bounded eigenvalue multiplicity [10]. By applying group theory to a greater extent, Babai, Luks, and Séress were able to show that isomorphism testing for these graph classes is in NC [12, 8, 39]. However, it is still open whether also Luks' efficient GI algorithm for graphs with bounded degree is parallelizable.

Question 1. *Is GI for graphs with bounded degree in NC?*

Ponomarenko proved that GI for graphs with excluded minors is decidable in P [43]. In 1990, Bodlaender gave a polynomial-time GI algorithm for graphs of bounded treewidth [14]. This class contains all series-parallel graphs, all outerplanar graphs, all graphs with constant bandwidth (or cutwidth) and all chordal graphs with constant clique-size. Very recently, Grohe and Verbitsky [26] improved Bodlaender's upper bound by showing that GI for graphs of bounded treewidth is in TC^1. This follows by combining the following two results which are interesting on their own.

First, they show that a parallel version of the r-round k-dimensional Weisfeiler-Lehman algorithm (r-round WL^k for short) can be implemented as a logspace uniform family of TC circuits of depth $O(r)$ and polynomial size. As a consequence, for any class $\mathcal{C}$ for which the multidimensional WL algorithm correctly decides GI on $\mathcal{C}$ in $O(\log n)$ rounds, GI on $\mathcal{C}$ is decidable by a TC^1 algorithm.

As a second ingredient of the proof, Grohe and Verbitsky show that for $r = O(k \log n)$, the r-round WL^{4k+3} correctly decides GI on all graphs of treewidth at most k. This latter result is obtained by designing a winning strategy for a suitable Ehrenfeucht-Fraïssé game with $4k + 4$ pebbles and r moves. An interesting question in this context is whether this approach can be extended to the canonization version of WL.

Question 2. *Do graphs of bounded treewidth admit an* NC *(or even* TC^1*) canonization?*

As Grohe and Verbitsky use the WL algorithm to solve GI for graphs with bounded treewidth, it follows that these graphs have a TC^1 computable complete normalform (also called invariant). Although, as shown by Gurevich, canonization is polynomial-time reducible to computing a complete normalform [27], it is not clear whether such a reduction is computable in NC for graphs with bounded treewidth.

Another possibility to answer Question 2 affirmatively may be to use a variation of the WL algorithm to canonize the input graph. For example, in [32, Theorem 1.9.4] Immerman and Lander propose the following procedure: as soon as the refinement process stabilizes choose any vertex (or tuple) from the lexicographically smallest color class of size at least two and individualize it (i.e., give it a new color). Then restart WL and repeat the process until all color classes are singletons. The resulting (total) refinement induces unique names for all the vertices. An interesting question is whether this variant of WL indeed computes a canon for all graphs of bounded treewidth, and if the answer is yes, whether this task can be performed in a logarithmic number of rounds.

2   Preliminaries

In this section we fix the notation and give formal definitions for the concepts used in this paper. For other basic definitions we refer the reader to [35] or to any textbook on complexity like [13].

We denote the *symmetric group* of all permutations on a set A by $\text{Sym}(A)$ and by S_n in case $A = \{1, \ldots, n\}$. Let G be a subgroup of $\text{Sym}(A)$ and let $a \in A$. Then the set $\{b \in A \mid \exists g \in G : g(a) = b\}$ of all elements $b \in A$ reachable from a via a permutation $g \in G$ is called the *orbit* of a in G.

2.1   Colored Graphs

Let $X = (V, E)$ denote a (finite) hypergraph, i.e., E is a subset of the power set $\mathcal{P}(V)$ of V. We always assume that the vertex set is of the form $V = [n]$, where $[n]$ denotes the set $\{1, \ldots, n\}$. For a subset $U \subseteq V$, we use $X[U]$ to denote the *induced subgraph* $(U, E(U))$ of X, where $E(U) = \{e \in E \mid e \subseteq U\}$. For usual graphs, i.e., $E \subseteq \binom{V}{2} = \{e \subseteq v \mid \|e\| = 2\}$, we use $\Gamma_X(u)$ to denote the neighborhood $\{v \in V \mid \{u, v\} \in E\}$ of vertex u in the graph X (if X is clear from the context we omit the subscript). Further, for disjoint subsets $U, U' \subseteq V$, we use $X[U, U']$ to denote the *induced bipartite subgraph* $(U \cup U', E(U, U'))$, where $E(U, U')$ contains all edges $e \in E$ with $e \cap U \neq \emptyset$ and $e \cap U' \neq \emptyset$.

A *coloring* of X is given by a function $c : V \to [m]$. We represent colored hypergraphs as triples $X = (V, E, \mathcal{C})$, where $\mathcal{C} = (C_1, \ldots, C_m)$ is the color partition induced by c, i.e., $C_i = \{u \in V \mid c(u) = i\}$. We denote the class of all colored hypergraphs by $\mathcal{CHG}$ and the class of all colored graphs by $\mathcal{CG}$. Note that the class of uncolored (hyper)graphs can also be seen as a subclass of $\mathcal{CHG}$ where all nodes have color 1. In case $\|C_i\| \leq b$ for all $i \in [m]$, we refer to X as a *b-bounded (hyper)graph*. The class of all b-bounded graphs (hypergraphs) is denoted by $\mathcal{CG}_b$ (respectively, $\mathcal{CHG}_b$).

2.2   Isomorphisms and Automorphisms

Let $X = (V, E, \mathcal{C})$ and $Y = (V, E', \mathcal{C})$ be hypergraphs and let g be a permutation on V. We can extend g to a mapping on subsets $U = \{u_1, \ldots, u_k\}$ of V by

$$g(U) = \{g(u_1), \ldots, g(u_k)\}.$$

g is an *isomorphism* between hypergraphs X and Y, if g preserves the edge relation, i.e.,

$$\forall e \subseteq V : e \in E \Leftrightarrow g(e) \in E'$$

as well as the color relation,

$$\forall i \in [m] : g(C_i) = C_i.$$

We also say that g maps X to Y and write $g(X) = Y$. If $g(X) = X$, then g is called an *automorphism* of X. We use $\mathrm{Aut}(X)$ to denote the automorphism group of X. Note that the identity mapping on V is always an automorphism. Any other automorphism is called *nontrivial*.

The decision problem HGI_b consists of deciding whether two given b-bounded hypergraphs X and Y are isomorphic (GI_b denotes the restriction of this problem to graphs). A related problem is the automorphism problem HGA_b (GA_b) of deciding if a given b-bounded hypergraph (respectively, graph) has a nontrivial automorphism. For uncolored (hyper)graphs $X = (V, E)$ we denote these problems by HGI, GI, HGA and GA, respectively.

2.3   Normal Forms and Canonization

In the following we assume an appropriate binary encoding of colored (hyper)graphs and we identify each graph X with its encoding. Let $\mathcal{D} \subseteq \mathcal{CHG}$ be a graph class and let $f : \{0, 1\}^* \to \{0, 1\}^*$ be a function. We say that f computes a *normal form* for $\mathcal{D}$, if

$$\forall X, Y \in \mathcal{D} : X \cong Y \Rightarrow f(X) = f(Y).$$

If f also fulfils the backward implication, i.e.

$$\forall X, Y \in \mathcal{D} : X \cong Y \Leftrightarrow f(X) = f(Y),$$

f is called a *complete normal form* for $\mathcal{D}$. A normal form f for $\mathcal{D}$ that computes for any graph $X \in \mathcal{D}$ a graph $f(X)$ that is isomorphic to X, i.e.

$$\forall X, Y \in \mathcal{D} : X \cong f(X) \wedge [X \cong Y \Rightarrow f(X) = f(Y)],$$

is called a *canonization* for $\mathcal{D}$. Note that a canonization for $\mathcal{D}$ is also a complete normal form for $\mathcal{D}$. We call $f(X)$ the *canon* of X (w.r.t. f). Of course, $f(X)$ is uniquely determined by any isomorphism g between X and $f(X)$. We call any such g a *canonical relabeling* of X (w.r.t. f).

2.4   The Weisfeiler-Lehman Algorithm

For the history of this approach to GI we refer the reader to [9, 19, 21]. We will abbreviate *k-dimensional Weisfeiler-Lehman algorithm* by WL^k. WL^1 is commonly known as the *canonical labeling* or *color refinement algorithm*. On input a colored graph $X = (V, E, \mathcal{C})$, where $\mathcal{C} = (C_1, \ldots, C_m)$, the algorithm proceeds in rounds starting with the *initial coloring* $\mathcal{C}^0 = \mathcal{C}$, i.e., c^0 assigns to each node $v \in V$ its color $c(v)$. In each round, each node $v \in V$ receives a new color that depends on the previous colors of v and all its neighbors. More precisely, in the $(i + 1)$st round, WL^1 assigns to node v the color

$$c^{i+1}(v) = (c^i(v), \{\!\!\{\, c^i(u) \mid u \in \Gamma(v)\}\!\!\})$$

consisting of the preceding color $c^i(v)$ and the multiset $\{\!\!\{\, c^i(u) \mid u \in \Gamma(v)\}\!\!\}$ of colors $c^i(u)$ for all $u \in \Gamma(v)$. For example, $c^1(v) = c^1(w)$ if and only if for each color $i \in [m]$, v and w have the same number of neighbors with that color. To keep the color encoding short, after each round the colors are lexicographically sorted and renamed (hence the renamed colors are in the range $[m_i]$, where $m_i = \|\{c^i(v) \mid v \in V\}\| \leq n$). However, the algorithm retains a table that can be used to derive the old color names from the new ones. After r rounds, the r-round WL^1 stops and outputs the multiset $\{\!\!\{\, c^r(v) \mid v \in V\}\!\!\}$ of colors in the coloring C^r (together with the tables retained at each round). Note that as long as C^{i+1} is a proper refinement of C^i, the number of colors increases. Hence, the coloring stabilizes after at most n rounds, i.e. $C^{s+1} = C^s$ for some $s < n$. We call C^s the WL^1-*stable coloring* of X.

   Following the same idea, the k-dimensional version iteratively refines a coloring of V^k. The initial coloring of a k-tuple $\bar{v}$ is the isomorphism type of the subgraph induced by the vertices in $\bar{v}$ (viewed as a labeled graph where each vertex is labeled by its color and by the positions in the tuple where it occurs). The refinement step takes into account the colors of all neighbors of $\bar{v}$ in the Hamming metric (see [19, 26] for details).

   Since the coloring is stable after at most n^k rounds, WL^k can be implemented in polynomial time for each constant dimension k. Further, since the colorings computed by the WL algorithm in each round only depend on the isomorphism class of X, it is clear that WL computes a normal form on the class of all graphs. We say that the r-round WL^k *works correctly* for a graph X, if the output for X is distinct from all outputs produced for any nonisomorphic graph $Y \not\cong X$.

It is clear that the r-round WL^k computes a complete normalform on a graph class $\mathcal{D}$, provided that it works correctly for each graph $X \in \mathcal{D}$ (note that for some graph classes the latter condition might be stronger than the former).

Of course, WL^n needs at most one round to work correctly on all graphs with n vertices. In fact, already WL^1 works correctly on all trees and almost all graphs (in the $\mathcal{G}_{n,1/2}$ model), and WL^2 succeeds on all graphs of color class size 3 [32]. Thus there was some hope that a low dimensional WL algorithm may work correctly on all graphs. However, in 1990 Cai, Fürer and Immerman [19] proved a striking negative result: For any sublinear dimension $k = o(n)$, WL^k does not work correctly even on graphs of vertex degree 3 and color class size 4. Nevertheless, it was realized later that a constant-dimensional WL is still applicable to particular classes of graphs, including planar graphs [23], graphs of bounded genus [24], and graphs of bounded treewidth [25].

3   Hardness of HGA

To show that there are n-vertex graphs of vertex degree 3 and color class size 4 that are hard instances for $\mathrm{WL}^{o(n)}$, Cai, Fürer and Immerman used a graph gadget that originally appeared in [31]. This gadget has also been used by Torán in a significant manner to show that GI and GA are hard for various subclasses of TC^1 [47]. Here we use a hypergraph variant of this gadget to show that for any prime p, HGA_p is hard for $\mathrm{Mod}_p\mathrm{L}$. The proof given here simplifies a proof of a similar result in [3].

It is well-known that the following problem is $\mathrm{Mod}_p\mathrm{L}$ complete (cf. [17]). Given a homogenous system

$$\sum_{j \in [n]} a_{ij} x_j = 0, \, i \in [k] \tag{1}$$

of linear equations over the field $\mathbb{Z}_p = \mathbb{Z}/p\mathbb{Z}$, decide whether (1) has a nontrivial solution $\bar{x} \in \mathbb{Z}_p^n$. This problem remains $\mathrm{Mod}_p\mathrm{L}$ complete, if we require that the support $S_i = \{j \in [n] \mid a_{ij} \neq 0\}$ of each equation contains at most three elements and $S_j \neq S_k$ for $j \neq k$ (these restrictions are not really necessary but they simplify the reduction and keep the orbit size of the hyperedges in the reduced hypergraph small). Now consider the following hypergraph $X = (V, E, \mathcal{C})$ with

$$V = \bigcup_{j=1}^{n} V_j, \quad E = \bigcup_{j=1}^{n} Z_j \cup \bigcup_{i=0}^{k} E_i \text{ and } \mathcal{C} = (C_1, C_1', C_1'', \ldots, C_n, C_n', C_n''),$$

where

$$V_j = C_j \cup C_j' \cup C_j'',$$
$$C_j = \{u_x^j \mid x \in \mathbb{Z}_p\}, C_j' = \{v_x^j \mid x \in \mathbb{Z}_p\}, C_j'' = \{w_x^j \mid x \in \mathbb{Z}_p\},$$
$$Z_j = \{\{u_x^j, v_x^j\}, \{v_x^j, w_x^j\}, \{w_x^j, u_{x+1}^j\} \mid x \in \mathbb{Z}_p\}, \text{ and }$$
$$E_i = \{\{u_{x_j}^j \mid j \in S_i\} \mid \sum_{j \in [n]} a_{ij} x_j = 0\}.$$

In the hypergraph X we have for each variable x_j a cycle $X_j = X[V_j]$ such that $\mathrm{Aut}(X_j)$ is isomorphic to the additive group $(\mathbb{Z}_p, +)$. Fix any isomorphism φ between $\mathrm{Aut}(X_j)$ and $\mathbb{Z}_p$ and denote the automorphism $g \in \mathrm{Aut}(X_j)$ with $\varphi(g) = x$ by g_x^j. Then $\mathrm{Aut}(X_j)$ is represented as $\{g_x^j \mid x \in \mathbb{Z}_p\}$ and we have $g_x^j \circ g_{x'}^j = g_{x+x'}^j$.

For any vector $\bar{x} = (x_1, \ldots, x_n)$ we use $\bar{x}|_{S_i}$ to denote the s_i-dimensional projection $(x_j)_{j \in S_i}$ of $\bar{x}$ to S_i. Since E_i contains for each solution $\bar{x} = (x_1, \ldots, x_n)$ of the i-th equation in (1) the hyperedge $e(\bar{x}|_{S_i}) = \{u_{x_j}^j \mid j \in S_i\}$, E_i consists of exactly $p^{s_i - 1}$ hyperedges. We use L_i to denote the set of vectors $\bar{x} \in \mathbb{Z}_p^{s_i}$ with $e(\bar{x}) \in E_i$.

Of course, for $p = 2, 3$ we can simplify X to the graph $\hat{X} = X[C_1 \cup \cdots \cup C_n]$ since in these cases the groups $\mathrm{Aut}(\hat{X}_j)$ are cyclic anyway. Figure 1 shows the graph $\hat{X}$ corresponding to the equation $x_1 + x_2 - x_3 = 0$ over $\mathbb{Z}_3$.

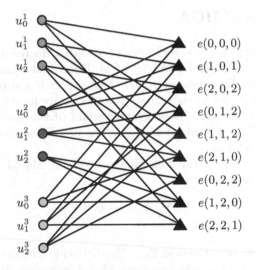

Fig. 1. The hypergraph gadget for the equation $x_1 + x_2 - x_3 = 0$ over $\mathbb{Z}_3$

Now it is easy to see that for each $i \in [k]$ the automorphism group $\mathrm{Aut}(Y_i)$ of the hypergraph $Y_i = (W_i, F_i, \mathcal{C}_i)$ where $W_i = \bigcup_{j \in S_i} V_j$, $F_i = E_i \cup \bigcup_{j \in S_i} Z_j$ and $\mathcal{C}_i$ is the restriction of the coloring $\mathcal{C}$ to W_i, is isomorphic to the solution space L_i of the equation $\sum_{j \in S_i} a_{ij} x_j = 0$. For example, if $S_i = \{1, 2, 3\}$ and the i-th equation of (1) is $x_1 + x_2 - x_3 = 0$, then

$$\mathrm{Aut}(Y_i) = \{(g_{x_1}^1, g_{x_2}^2, g_{x_3}^3) \mid x_1 + x_2 - x_3 = 0\}.$$

Hence, a permutation $g = (g_{x_1}^1, \ldots, g_{x_n}^n) \in \mathrm{Aut}(X_1) \times \cdots \times \mathrm{Aut}(X_n)$ is an automorphism of X if and only if for all $i \in [k]$, the restriction of g to W_i is an automorphism of Y_i, implying that

$$\mathrm{Aut}(X) = \{(g_{x_1}^1, \ldots, g_{x_n}^n) \mid (x_1, \ldots, x_n) \text{ is a solution of (1)}\}.$$

This shows that $X \in \mathrm{HGA}_p$ if and only if the system (1) has a nontrivial solution. Since the reduction from the given homogenous system (1) to the hypergraph X can be performed in AC^0, it follows that for any prime $q \leq p$, HGA_p is hard for the class $\mathrm{Mod}_q\mathrm{L}$ under AC^0 many-one reductions.

Moreover, since HGA_p has an easily computable or-function (just take the union of the graphs where we assume w.l.o.g. that the input graphs have no colors in common) and since any set in the class $\mathrm{Mod}_m\mathrm{L}$ can be represented as the union $A_1, \ldots, A_k$ of sets A_i in $\mathrm{Mod}_{p_i}\mathrm{L}$, where $p_1, \ldots, p_k$ are the prime factors of m [17], it immediately follows that HGA_p is even hard for $\mathrm{Mod}_{p!}\mathrm{L}$. Since the orbit size of the hyperedges in the reduced hypergraph is bounded by p^2, we also get that GA_{p^2} is $\mathrm{Mod}_{p!}\mathrm{L}$ hard.

Theorem 3. HGA_p and GA_{p^2} are hard for $\mathrm{Mod}_{p!}\mathrm{L}$.

In [3] it is shown that GA_4 (as well as GA_5 and HGA_2) in fact is complete for the class $\mathrm{Mod}_2\mathrm{L} = \oplus\mathrm{L}$. The best known upper bound for HGA_b, $b > 2$, is P [3]. We remark that if the hyperedges are all of constant size, i.e., $\|e\| \leq k$ for all $e \in E$, then HGA_b is reducible to $\mathrm{GA}_{b'}$ for $b' = b^k$ which is known to be in TC^1 [39,4]. However, when hyperedges are of unbounded size, it is not clear whether HGA_b is reducible to $\mathrm{GA}_{b'}$ for any constant b'.

Question 4. Is HGA_b in NC for some constant $b > 2$?

Torán's proof that GI and GA are hard for NL crucially hinges on the fact that the produced graphs have unbounded color classes. Since already in the 2-bounded case the orbits of the edges of a hypergraph can have exponential size it might be possible to reduce NL to HGA_b (or HGI_b) for a constant b. Note that the orbit size of the edges of a b-bounded graph is at most b^2.

Question 5. Is there any constant b for which HGA_b (or HGI_b) is NL hard?

4   Logspace Canonization of 3-Bounded Graphs

In this section we improve the result from [34] that GI for 2-bounded as well as for 3-bounded graphs is equivalent to undirected graph reachability (and therefore complete for L [44]). We first describe a logspace canonization algorithm for 2-bounded graphs. This algorithm performs a 1-round WL^1 and uses individualization to refine the remaining size two color classes. We also sketch how this algorithm can be improved to handle the 3-bounded case. For the complete proof we refer the reader to the journal version of [3] (in preparation).

Let $X = (V, E, \mathcal{C})$ be a b-bounded graph and let $\mathcal{C} = (C_1, \ldots, C_m)$. We use X_i to denote the graph $X[C_i]$ induced by C_i and X_{ij} to denote the bipartite graph $X[C_i, C_j]$ induced by the pair of color classes C_i and C_j. Since it suffices to compute a canonical relabeling for X we can assume that all vertices in the same color class C_i have the same degree and each graph X_i is regular of degree at most $(\|C_i\| - 1)/2$. Otherwise we can either canonically split C_i into smaller color classes or we can replace X_i by the complement graph. Further, we assume that the edge set E_{ij} of X_{ij} is of size at most $\|C_i\| \cdot \|C_j\|/2$, since otherwise, we can replace X_{ij} by the complement bipartite graph.

We say that two color classes C_i, C_j with $\|C_i\| = \|C_j\|$ are *directly linked*, if E_{ij} is a perfect matching in X_{ij}. C_i and C_j are *linked*, if C_i is reachable from C_j by a chain of directly linked color classes. We make use of some basic facts from [3].

Lemma 6. [3] *For any directly linked pair C_i, C_j of color classes there is a bijection $\pi_{ij} : \mathrm{Sym}(C_i) \to \mathrm{Sym}(C_j)$ such that for any automorphism $g = (g_i, g_j) \in \mathrm{Aut}(X_{ij})$ it holds that $g_j = \pi_{ij}(g_i)$.*

Let G_i be the intersection of $\mathrm{Aut}(X_i)$ with the projections of $\mathrm{Aut}(X_{ij})$ on C_i for all $j \neq i$. Any subgroup of the symmetric group $\mathrm{Sym}(C_i)$ of all permutations on C_i is called a *constraint* for C_i. We call G_i the *direct constraint* for C_i.

Lemma 7. [3] *For a given b-bounded graph, the direct constraints of each color class can be determined in deterministic logspace.*

We use Lemma 6 to define a symmetric relation on constraints. Let G_i and G_j be constraints of two directly linked classes C_i and C_j, respectively, and let g_{ij} be the bijection provided by Lemma 6. We say that G_i is *directly induced* by G_j, if g_{ij} is an isomorphism between G_i and G_j. Further, a constraint G *is induced by* a constraint H, if G is reachable from H via a chain of directly induced constraints. Note that the latter relation is an equivalence on the set of all constraints. We call the intersection of all constraints of C_i that are induced by some direct constraint the *induced constraint* of C_i and denote it by G_i'.

Lemma 8. [3] *For a given b-bounded graph, the induced constraints of each color class can be determined in deterministic logspace.*

Proof. Consider the undirected graph $X' = (V', E')$ where V' consists of all constraints G in X and $E' = \{(G, H) \mid G \text{ is directly induced by } H\}$. In this graph we mark all direct constraints computed by Lemma 7 as special nodes. Now, the algorithm outputs for each color class C_i the intersection of all constraints for C_i that are reachable from some special node, and since $\mathrm{SL} = \mathrm{L}$ [44], this can be done in deterministic logspace. $\square$

We define two special types of constraints. We say that C_i is *split*, if its induced constraint G_i' has at least two orbits, and we call the partition of C_i in the orbits of G_i' the *splitting partition* of C_i. Further, a class C_i of size b is called *whole*, if its induced constraint G_i' is the whole group $\mathrm{Sym}(C_i)$. The following lemma summarizes some properties of whole color classes.

Lemma 9. *Let C_i be a whole color class in a b-bounded graph X and let C_j be a color class such that $E_{ij} \neq \emptyset$. Then the following holds.*

- *$X[C_i, \Gamma_{X_{ij}}(C_i)]$ is semiregular,i.e., the degree of any node u in the bipartite graph only depends on its (non)membership to C_i.*

- If also C_j is whole, then $\|C_i\| = \|C_j\|$ and C_i, C_j are directly linked.
- If C_j is split or $\|C_j\| < b$, then all vertices in C_i have the same neighborhood in X_{ij}.

Lemma 9 tells us that the action of an automorphism on a whole color class C is not influenced by its action on color classes that are either smaller or split, i.e., only other whole color classes can influence C. Similarly, it follows that WL^1 will never refine any of the whole color classes in X. Let W be the union of all whole color classes. Then $\mathrm{Aut}(X[W])$ is computable in logspace.

Lemma 10. [34, 3] *A generating set for $\mathrm{Aut}(X[W])$ is computable in FL.*

Proof. The algorithm works by reducing the problem to reachability in undirected graphs. For each whole color class C_i we create a set P_i of $b!$ nodes (one for each permutation of C_i). Recall that if C_i and C_j are directly linked, then each $g \in P_i$ induces a unique permutation $h = \pi_{ij}(g)$ on C_j and hence, we put an undirected edge between g and h. This gives an undirected graph G with $(b-1)!\|W\|$ nodes.

A connected component P in G that picks out at most one element g_i from each set P_i defines a valid automorphism g for the graph $X[W]$, if P contains only elements $g_i \in \mathrm{Aut}(X_i)$. On the color classes C_i, for which P contains an element $g_i \in P_i$, g acts as g_i, and it fixes all nodes of the other color classes. By collecting these automorphisms we get a generating set for $\mathrm{Aut}(X[W])$ and since $\mathrm{SL} = \mathrm{L}$ [44], this can be done in deterministic logspace.   □

Now we prove that WL^1 on 2-bounded graphs can be implemented in logspace.

Theorem 11. *For graphs in $\mathcal{CG}_2$ the WL^1-stable coloring is computable in FL.*

Proof (sketch). Let $X = (V, E, \mathcal{C})$ be a 2-bounded graph with coloring $\mathcal{C} = (C_1, \ldots, C_m)$. The only way that a color class C_i gets directly split (i.e. by its direct constraint G_i) is that one node $a \in C_i$ is incident to some color class C_j whereas the other node $b \in C_i$ is not. Let C_j be the lexicographically smallest color class with this property. Then WL^1 refines C_i into $(\{b\}, \{a\})$. These are exactly the refinements that WL^1 performs by the initial coloring and they are clearly computable in logspace.

If C_i gets refined in a later round, then this refinement is caused by a direct link to a color class that has been refined earlier. Let C_j be the lexicographically smallest directly split color class which is linked to C_i by a chain $(C_j, \ldots, C_i)$ of directly linked color classes of minimal length. Then WL^1 transposes the refinement of C_j to C_i via the chain $(C_j, \ldots, C_i)$. Clearly, also these refinements are computable in logspace. Finally, observe that the whole color classes never get refined by WL^1. In fact, they form orbits in $\mathrm{Aut}(X)$.   □

As an easy consequence we get a logspace canonization for all 2-bounded graphs.

Theorem 12. *$\mathcal{CG}_2$ admits a logspace canonization.*

Proof (sketch). Let $X = (V, E, C)$ be a 2-bounded graph with coloring $C = (C_1, \ldots, C_m)$. By Theorem 11 the WL^1-stable coloring X' of X is computable in logspace. If X' assigns unique colors to all vertices, then a canonical labeling is determined.

Otherwise, for each connected component of linked color classes (of size 2), the algorithm determines the lexicographically smallest color class C in that component. Since the nodes of different C's can be flipped independently, the algorithm can select in each such color class an arbitrary node and give it a new color. Now it suffices to run WL^1 once more to compute the stable coloring for the modified graph which will provide unique colors for all vertices.   □

We notice that the above proof also shows that the canonization version of WL^1 (as proposed in [32, Theorem 1.9.4]) succeeds on the class CG_2 (despite the fact that WL^1 does not work correctly on CG_2 [32, Corollary 1.6.2]).

Question 13. *For which values of k and b does the canonization version of WL^k succeed in canonizing the graphs in CG_b?*

Similar to the proof of Theorem 11 it can be shown that also for graphs in CG_3 the WL^1-stable coloring is computable in logspace but it is not clear whether this generalizes.

Question 14. *What is the complexity of computing the WL^k-stable coloring for graphs in CG_b?*

Immerman and Lander have shown that WL^2 works correctly on all 3-bounded graphs, implying that the canonizing version of WL^2 succeeds on the class CG_3 [32]. Here we give a logspace canonization algorithm for this class.

Theorem 15. *CG_3 admits a logspace canonization.*

Proof (sketch). Let $X = (V, E, C)$ be a 3-bounded graph with coloring $C = (C_1, \ldots, C_m)$. Let C_w denote the subclass of C containing all whole color classes of size 3 that are linked to the lexicographically smallest whole class C_i and let W be the set of vertices in these color classes. W.l.o.g. let $i = 1$ and $C_w = (C_1, \ldots, C_l)$ for some $l \leq m$.

We first show how to refine the color classes in C_w in a canonical way. We define a (canonical) reflexive, transitive and connex relation $\preceq$ on C_1 such that u and v are in the same orbit of $\text{Aut}(X[W])$ if and only if $u \preceq v$ as well as $v \preceq u$. To define $\preceq$, for $u \in C_1$ consider the set $Z(u)$ of all cycles of color classes starting (and ending) at C_1 such that starting from vertex $u \in C_1$ it is possible to follow this cycle along the edges in E and come back to u. Now define $u \preceq v$ if $Z(u) = Z(v)$ or the lexicographically smallest cycle in $Z(u) \Delta Z(v)$ is in $Z(v)$. Then we can proof the following claim.

Claim. If the three nodes u_1, u_2, u_3 in C_1 are cycle-equivalent (i.e. $Z(u_1) = Z(u_2) = Z(u_3)$), then the permutation $g_1 : u_1 \mapsto u_2 \mapsto u_3$ is extendible to an automorphism of $X[W]$.

Proof of Claim. The permutation g_1 uniquely extends to a permutation $g = (g_1, \ldots, g_l) \in \text{Aut}(X_1) \times \cdots \times \text{Aut}(X_l)$ on $X[W]$, where we extend g successively by the lexicographically smallest color class that is linked to a color class on which g is already defined. If $g \notin \text{Aut}(X[W])$, then there must exist two vertices u, v in two color classes C_i, C_j, respectively, such that

$$\{u, v\} \in E_{ij} \Leftrightarrow (g_i(u), g_j(v)) \notin E_{ij}.$$

Now let u' be the vertex in C_j that is linked to u in the spanning tree T along which g has been extended.

In case $u' = v$ and $\{u, v\} \in E_{ij}$ it follows that there is a cycle starting at u following some path in T to $u' = v$ and then back to u. Starting at $g_i(u)$ we reach $g_j(v)$ following the same path through T but proceeding further to C_i we don't come back to $g_i(u)$, implying that u and $g_i(u)$ (and hence also the corresponding vertices in C_1) are not cycle-equivalent.

The other cases are similar. This completes the proof of the claim.   ◁

A similar argument shows that if exactly two of the three vertices u_1, u_2, u_3 are cycle-equivalent, then there is an automorphism flipping them. Now, we select any vertex $u \in C_1$ with $u \preceq v$ for all $v \in C_1$ and give it a new color. As in the proof of Theorem 12, this can be done in parallel for all connected components of linked color classes. Running WL^1 again on the graph with the individualized vertices will now refine all whole color classes. Thus we have transformed X into a canonical 2-bounded refinement and hence we can invoke Theorem 12.   □

It follows that for the graph classes GA_2 and GA_3 all problems related to GI are complete for L: GA, #GA, #GI, AUT, computing a complete normalform and canonization. Is this also true for the class of 2-bounded hypergraphs (or for GA_4), i.e., is the canonization problem for these graphs solvable in $FL(\oplus L)$?

Question 16. *What is the complexity of computing a canonizing function for the graph classes $\mathcal{CG}_b$ and $\mathcal{CHG}_b$?*

We remark that the TC^1 upper bound for GI_b given in [4] uses the group theoretic approach to compute a generating set for $\text{Aut}(X)$. Can this approach be adapted to give also an NC upper bound on the canonization problem for graphs with bounded color classes?

Acknowledgements

For helpful conversations and suggestions on this work I'm very grateful to V. Arvind, O. Beyersdorff and O. Verbitsky.

References

1. M. Agrawal and N. Saxena. Automorphisms of finite rings and applications to complexity of problems. In *Proc. 22nd Symposium on Theoretical Aspects of Computer Science*, volume 3404 of *Lecture Notes in Computer Science*, pages 1–17. Springer-Verlag, Berlin Heidelberg, 2005.

2. C. Àlvarez and B. Jenner. A very hard log-space counting class. *Theoretical Computer Science*, 107(1):3–30, 1993.

3. V. Arvind and J. Köbler. Hypergraph isomorphism testing for bounded color classes. In *Proc. 23rd Symposium on Theoretical Aspects of Computer Science*, volume 3884 of *Lecture Notes in Computer Science*, pages 384–395. Springer-Verlag, Berlin Heidelberg, 2006.

4. V. Arvind, P. Kurur, and T. Vijayaraghavan. Bounded color multiplicity graph isomorphism is in the #L hierarchy. In *Proc. 20th Annual IEEE Conference on Computational Complexity*, pages 13–27. IEEE Computer Society Press, 2005.

5. V. Arvind and J. Torán. Isomorphism testing: Pespective and open problems. *Bulletin of the European Association of Theoretical Computer Science (BEATCS)*, 86, 2005.

6. L. Babai. Moderately exponential bounds for graph isomorphism. In *Proc. International Symposium on Fundamentals of Computing Theory 81*, volume 117 of *Lecture Notes in Computer Science*, pages 34–50. Springer-Verlag, Berlin Heidelberg, 1981.

7. L. Babai. Trading group theory for randomness. In *Proc. 17th ACM Symposium on Theory of Computing*, pages 421–429. ACM Press, 1985.

8. L. Babai. A Las Vegas-NC algorithm for isomorphism of graphs with bounded multiplicity of eigenvalues. In *Proc. 27th IEEE Symposium on the Foundations of Computer Science*, pages 303–312. IEEE Computer Society Press, 1986.

9. L. Babai. Automorphism groups, isomorphism, reconstruction. In R. L. Graham, M. Grötschel, and L. Lovász, editors, *Handbook of Combinatorics*, pages 1447–1540. Elsevier Science Publishers, 1995.

10. L. Babai, D. Grigoryev, and D. Mount. Isomorphism of graphs with bounded eigenvalue multiplicity. In *Proc. 14th ACM Symposium on Theory of Computing*, pages 310–324. ACM Press, 1982.

11. L. Babai and E. Luks. Canonical labeling of graphs. In *Proc. 15th ACM Symposium on Theory of Computing*, pages 171–183, 1983.

12. L. Babai, E. Luks, and Á. Seress. Permutation groups in NC. In *Proc. 19th ACM Symposium on Theory of Computing*, pages 409–20. ACM Press, 1987.

13. J. L. Balcázar, J. Díaz, and J. Gabarró. *Structural Complexity I*. EATCS Monographs on Theoretical Computer Science. Springer-Verlag, Berlin Heidelberg, second edition, 1995.

14. H. Bodlaender. Polynomial algorithm for graph isomorphism and chromatic index on partial k-trees. *Journal of Algorithms*, 11:631–643, 1990.

15. K. Booth. Isomorphism testing for graphs, semigroups, and finite automata are polynomially equivalent problems. *SIAM Journal on Computing*, 7(3):273–279, 1978.

16. R. Boppana, J. Hastad, and S. Zachos. Does co-NP have short interactive proofs? *Information Processing Letters*, 25(2):27–32, 1987.

17. G. Buntrock, C. Damm, U. Hertrampf, and C. Meinel. Structure and importance of logspace-MOD classes. *Mathematical Systems Theory*, 25:223–237, 1992.

18. S. Buss. Alogtime algorithms for tree isomorphism, comparison, and canonization. In *Computational Logic and Proof Theory, 5th Kurt Gödel Colloquium'97*, volume 1289 of *Lecture Notes in Computer Science*, pages 18–33. Springer-Verlag, Berlin Heidelberg, 1997.

19. J. Cai, M. Fürer, and N. Immerman. An optimal lower bound for the number of variables for graph identification. *Combinatorica*, 12:389–410, 1992.

20. S. A. Cook. A taxonomy of problems with fast parallel algorithms. *Information and Control*, 64:2–22, 1985.

21. S. Evdokimov, M. Karpinski, and I. Ponomarenko. On a new high dimensional Weisfeiler-Lehman algorithm. *Journal of Algebraic Combinatorics*, 10:29–45, 1999.
22. M. Furst, J. Hopcroft, and E. Luks. Polynomial time algorithms for permutation groups. In *Proc. 21st IEEE Symposium on the Foundations of Computer Science*, pages 36–41. IEEE Computer Society Press, 1980.
23. M. Grohe. Fixed-points logics on planar graphs. In *Proceedings of the 13th Symposium on Logic in Computer Science*, pages 6–15, 1998.
24. M. Grohe. Isomorphism testing for embeddable graphs through definability. In *Proc. 32th ACM Symposium on Theory of Computing*, pages 63–172, 2000.
25. M. Grohe and J. Mariño. Definability and descriptive complexity on databases of bounded tree-width. In C. Beeri and P. Bunemann, editors, *Proceedings of the 7th Conference on Database Theory*, volume 1540, pages 70–82. Springer-Verlag, Berlin Heidelberg, 1999.
26. M. Grohe and O. Verbitsky. Testing graph isomorphism in parallel by playing a game. Manuscript, 2006.
27. Y. Gurevich. From invariants to canonization. *Bulletin of the European Association of Theoretical Computer Science (BEATCS)*, 63, 1997.
28. J. E. Hopcroft and R. E. Tarjan. Isomorphism of planar graphs (working paper). In R. Miller and J. Thatcher, editors, *Complexity of computer computations*, pages 131–152. Plenum Press, New York-London, 1972.
29. J. E. Hopcroft and R. E. Tarjan. Efficient planarity testing. *Journal of the ACM*, 21:549–568620, 1974.
30. J. E. Hopcroft and J. Wong. Linear time algorithm for isomorphisms of planar graphs. *Proc. 6th ACM Symposium on Theory of Computing*, pages 172–184, 1974.
31. N. Immerman. Number of quantifiers is better than number of tape cells. *Journal of Computer and System Sciences*, 22(3):384–406, 1981.
32. R. Impagliazzo and L. A. Levin. No better ways to generate hard NP-instances than picking uniformly at random. In *Proc. 31st IEEE Symposium on the Foundations of Computer Science*, pages 812–821. IEEE Computer Society Press, 1990.
33. J. M. I.S. Filotti. A polynomial-time algorithm for determining the isomorphism of graphs of fixed genus. In *Proc. 12th ACM Symposium on Theory of Computing*, pages 236–243. ACM Press, 1980.
34. B. Jenner, J. Köbler, P. McKenzie, and J. Torán. Completeness results for graph isomorphism. *Journal of Computer and System Sciences*, 66:549–566, 2003.
35. J. Köbler, U. Schöning, and J. Torán. *The Graph Isomorphism Problem: Its Structural Complexity*. Progress in Theoretical Computer Science. Birkhäuser, Boston, 1993.
36. D. Lichtenstein. Isomorphism for graphs embaddable on the projective plane. In *Proc. 12th ACM Symposium on Theory of Computing*, pages 218–224. ACM Press, 1980.
37. S. Lindell. A logspace algorithm for tree canonization. In *Proc. 24th ACM Symposium on Theory of Computing*, pages 400–404. ACM Press, 1992.
38. E. Luks. Isomorphism of bounded valence can be tested in polynomial time. *Journal of Computer and System Sciences*, 25:42–65, 1982.
39. E. Luks. Parallel algorithms for permutation groups and graph isomorphism. In *Proc. 27th IEEE Symposium on the Foundations of Computer Science*, pages 292–302. IEEE Computer Society Press, 1986.
40. R. Mathon. A note on the graph isomorphism counting problem. *Information Processing Letters*, 8:131–132, 1979.
41. G. Miller. Isomorphism testing for graphs of bounded genus. In *Proc. 12th ACM Symposium on Theory of Computing*, pages 225–235. ACM Press, 1980.

42. G. Miller and J. Reif. Parallel tree contraction, part 2: Further applications. *SIAM Journal on Computing*, 20:1128–1147, 1991.
43. I. Ponomarenko. The isomorphism problem for classes of graphs that are invariant with respect to contraction (Russian). *Zap. Nauchn. Sem. Leningrad. Otdel. Mat. Inst. Steklov. (LOMI)*, 174:147–177, 1988.
44. O. Reingold. Undirected st-connectivity in log-space. In *Proc. 37th ACM Symposium on Theory of Computing*, pages 376–385. ACM Press, 2005.
45. U. Schöning. Graph isomorphism is in the low hierarchy. *Journal of Computer and System Sciences*, 37:312–323, 1988.
46. S. Toda. PP is as hard as the polynomial-time hierarchy. *SIAM Journal on Computing*, 20:865–877, 1991.
47. J. Torán. On the hardness of graph isomorphism. *SIAM Journal on Computing*, 33(5):1093–1108, 2004.
48. V. N. Zemlyachenko. Canonical numbering of trees (Russian). *Proc. Seminar on Comb. Anal. at Moscow State University*, 1970.
49. V. N. Zemlyachenko, N. Konienko, and R. I. Tyshkevich. Graph isomorphism problem (Russian). *The Theory of Computation I, Notes Sci. Sem. LOMI 118*, 1982.

Infinite Time Register Machines

Peter Koepke

University of Bonn
Mathematisches Institut
Beringstraße 1
D 53115 Bonn
Germany
Koepke@Math.Uni-Bonn.de

Abstract. *Infinite time register machines* (ITRMs) are register machines which act on natural numbers and which may run for arbitrarily many ordinal steps. Successor steps are determined by standard register machine commands, at limits the register contents are defined as lim inf's of the previous register contents. We prove that a real number is computable by an ITRM iff it is hyperarithmetic.

1 Introduction

In [2], JOEL D. HAMKINS and ANDY LEWIS define *infinite time* TURING *machines* (ITTMs) by letting an ordinary TURING machine run for arbitrarily many ordinal steps, taking appropriate limits at limit times. An ITTM can compute considerably more functions than a standard TURING machine. In analogy, we let a standard *register* machine run for arbitrarily many ordinal steps and call it an *infinite time register machines* (ITRM). An ITRM can carry out infinitely many steps of an ordinary register machine and can thus compute the halting problem. Indeed we show in Lemma 1 that it can compute any Δ_1^1 real number. Conversely it will be shown in Lemma 4 that if a computation by an ITRM halts then it halts before the CHURCH-KLEENE ordinal ω_1^{CK}. Hence all computable reals are in the admissible set $L_{\omega_1^{\mathrm{CK}}}$ (Lemma 5). Since the Δ_1^1-reals coincide with the reals in $L_{\omega_1^{\mathrm{CK}}}$ and with the *hyperarithmetic* reals (see [8]) this yields a new characterisation of the hyperarithmetic reals:

Theorem 1. *A real $x \subseteq \omega$ is computable by an infinite time register machine iff it is hyperarithmetic.*

This result was inspired by discussions with JOEL HAMKINS and PHILIP WELCH at Oberwolfach in December 2005. Infinite time register machines belong to the following schema of machines which may all run for arbitrarily many ordinal steps. Let Ord be the class of ordinal numbers.

1.1: Infinite time Turing machines, ITTMs, with finitely many standard Turing tapes; every Σ_1^1 real and every Π_1^1 real is ITTM computable [2].

A. Beckmann et al. (Eds.): CiE 2006, LNCS 3988, pp. 257–266, 2006.

1.2: Ordinal Turing machines, OTMs, with finitely many Turing tapes of length ORD; a set of ordinals is OTM computable iff it is a constructible set of ordinals [3], [4], [6].

2.1: Infinite time register machines, ITRMs, as defined in this article; a real is ITRM computable iff it is hyperarithmetic (Δ_1^1).

2.2: Ordinal register machines, ORMs, with finitely many registers containing arbitrary ordinals; a set of ordinals is ORM computable iff it is a constructible set of ordinals [7].

2   Infinite Time Register Machines

We base our presentation of infinite time machines on the *unlimited register machines* as presented in [1].

Definition 1. *An* unlimited register machine *URM has registers* $R_0, R_1, \ldots$ *which can hold* natural numbers. *A register program consists of commands to increase or to reset a register. The program may jump on condition of equality between two registers.*

An URM program is a finite list $P = I_0, I_1, \ldots, I_{s-1}$ *of instructions each of which may be of one of four kinds:*

a) *the* zero instruction $Z(n)$ *changes the contents of* R_n *to 0, leaving all other registers unaltered;*

b) *the* successor instruction $S(n)$ *increases the natural number contained in* R_n *by 1, leaving all other registers unaltered;*

c) *the* transfer instruction $T(m, n)$ *replaces the contents of* R_n *by the natural number contained in* R_m, *leaving all other registers unaltered;*

d) *the* jump instruction $J(m, n, q)$ *is carried out within the program* P *as follows: the contents* r_m *and* r_n *of the registers* R_m *and* R_n *are compared, but all the registers are left unaltered; then, if* $R_m = R_n$, *the URM proceeds to the qth instruction of* P; *if* $R_m \neq R_n$, *the URM proceeds to the next instruction in* P.

The instructions of a register program can be addressed by their indices which are called program states. *At each ordinal time t the machine will be in a configuration consisting of a program state* $I(t) \in \omega$ *and the register contents which can be viewed as a function* $R(t) : \omega \to \omega$. *$R(t)(n)$ is the content of the register R_n at time t. We also write $R_n(t)$ instead of $R(t)(n)$.*

Definition 2. *Let* $P = I_0, I_1, \ldots, I_{s-1}$ *be an URM program. A pair*

$$I : \theta \to \omega, R : \theta \to ({}^\omega \omega)$$

is an (infinite time register) computation by P *if the following hold:*

a) θ *is an ordinal or* $\theta = \mathrm{Ord}$; θ *is the length of the computation;*

b) $I(0) = 0$; *the machine starts in state 0;*

c) *If* $t < \theta$ *and* $I(t) \notin s = \{0, 1, \ldots, s-1\}$ *then* $\theta = t+1$; *the machine stops if the machine state is not a program state of* P;

d) *If* $t < \theta$ *and* $I(t) \in$ state(P) *then* $t+1 < \theta$; *the next configuration is determined by the instruction* $I_{I(t)}$:

i. *if* $I_{I(t)}$ *is the zero instruction* $Z(n)$ *then let* $I(t+1) = I(t)+1$ *and define* $R(t+1) : \omega \to$ Ord *by*

$$R_k(t+1) = \begin{cases} 0, & \textit{if } k = n \\ R_k(t), & \textit{if } k \neq n \end{cases}$$

ii. *if* $I_{I(t)}$ *is the successor instruction* $S(n)$ *then let* $I(t+1) = I(t)+1$ *and define* $R(t+1) : \omega \to$ Ord *by*

$$R_k(t+1) = \begin{cases} R_k(t) + 1, & \textit{if } k = n \\ R_k(t), & \textit{if } k \neq n \end{cases}$$

iii. *if* $I_{I(t)}$ *is the transfer instruction* $T(m,n)$ *then let* $I(t+1) = I(t)+1$ *and define* $R(t+1) : \omega \to$ Ord *by*

$$R_k(t+1) = \begin{cases} R_m(t), & \textit{if } k = n \\ R_k(t), & \textit{if } k \neq n \end{cases}$$

iv. *if* $I_{I(t)}$ *is the jump instruction* $J(m,n,q)$ *then let* $R(t+1) = R(t)$ *and*

$$I(t+1) = \begin{cases} q, & \textit{if } R_m(t) = R_n(t) \\ I(t) + 1, & \textit{if } R_m(t) \neq R_n(t) \end{cases}$$

e) *If* $t < \theta$ *is a limit ordinal, the machine constellation at* t *is determined by taking inferior limits. If* $\liminf_{r \to t} R_k(r) = \omega$ *for some* $k \in \omega$ *then let* $\theta = t$; *the machine stops if one of the registers overruns; otherwise let*

$$\forall k \in \omega \; R_k(t) = \liminf_{r \to t} R_k(r);$$
$$I(t) = \liminf_{r \to t} I(r).$$

The computation is obviously determined recursively by the initial register contents $R(0)$ *and the program* P. *We call it the* (infinite time register) *computation by* P *with input* $R(0)$. *If the computation stops at a successor ordinal* $\theta = \beta + 1$ *then* $R(\beta)$ *is the final register content. In this case we say that* P *computes* $R(\beta)(0)$ *from* $R(0)$ *and write* $P : R(0) \mapsto R(\beta)(0)$.

The definition of $I(t)$ for limit t can be motivated as follows. Since a program is finite its execution will lead to some (complex) looping structure involving loops, subloops and so forth. This can be presented by pseudo code like:

```
    . . .
17:begin mainloop
    . . .
```

```
    21:    begin subloop
            . . .
    29:    end subloop
            . . .
32:end mainloop
       . . .
```

Assume that for times $r \to t$ the main loop $(17-32)$ with its subloop $(21-29)$ is traversed cofinally often. Then at time t it is natural to put the machine at the start of the "main loop". Assuming that the lines of the program are enumerated in increasing order this corresponds to the lim inf rule

$$I(t) = \liminf_{r \to t} S(r).$$

The interpretation of programs yields associated notions of computability.

Definition 3. *An n-ary partial function $F : \omega^n \rightharpoonup \omega$ is* (ordinal register) *computable if there is a register program P such that for every n-tuple $(a_0, \ldots, a_{n-1})$ $\in \mathrm{dom}(F)$ holds*

$$P : (a_0, \ldots, a_{n-1}, 0, 0, \ldots) \mapsto F(a_0, \ldots, a_{n-1}).$$

Definition 4. *A subset $x \subseteq \omega$, i.e., a real number, is* (ordinal register) *computable if there is a register program P such that for every $m \in \omega$ holds*

$$P : (m, 0, 0, \ldots) \mapsto \chi_x(m),$$

where χ_x is the characteristic function of x.

Obviously any standard recursive function is ordinal register computable.

3   Computing Δ_1^1-Reals

For $e \in \omega$ let R_e denote the e-th recursively enumerable, binary relation on ω. If R_e is wellfounded, let $|R_e|$ denote the ordinal rank of R_e. Consider a *hyperarithmetic* real number x, i.e., $\{x\}$ is a parameter-free Δ_1^1-singleton. By standard representation theorems for Π_1^1-reals there exists a recursive function $f : \omega \to \omega$ such that for all $n \in \omega$:

$$n \in x \text{ iff } R_{f(n)} \text{ is a wellfounded relation.} \tag{1}$$

Since x is also Σ_1^1 the boundedness property for parameter-free Σ_1^1-sets implies the existence of an ordinal α less than the CHURCH-KLEENE ordinal ω_1^{CK} such that for all $n \in \omega$:

$$n \in x \text{ iff } R_{f(n)} \text{ is a wellfounded relation of rank } |R_{f(n)}| < \alpha. \tag{2}$$

The ordinal α is the ordertype of some recursive wellorder (ω, S). The right-hand side of (2) holds iff there is an orderpreserving embedding from $(\omega, R_{f(n)})$ into (ω, S).

More generally, consider any infinite time register computable relations (ω, R) and (ω, S) where (ω, S) is a wellorder. We shall define a register program P uniformly in programs for R and S which computes whether (ω, R) can be embedded orderpreservingly into (ω, S). This shows that the right-hand side of the equivalence (2) is infinite time register computable and proves

Lemma 1. *If $x \subseteq \omega$ is a hyperarithmetic real then x is computable by an infinite time register machine.*

For $r \in \omega$ let $\mathrm{TC}_R(r)$ be the transitive closure of r in R, i.e. the $\subseteq$-smallest set which contains r and is closed under R-predecessors. Define $\mathrm{TC}_S(s)$ similarly. Define a relation $r \sim s$ iff there is an orderpreserving map

$$\pi : (\mathrm{TC}_R(r), R) \to (\mathrm{TC}_S(s), S) \text{ with } \pi(r) = s.$$

If the relations R and S both have 0 as their maximum element, i.e.,

$$\forall r \in \mathrm{dom}(R) \setminus \{0\}\ rR0 \text{ and } \forall s \in \mathrm{dom}(S) \setminus \{0\}\ sS0 ,$$

then (ω, R) can be embedded orderpreservingly into (ω, S) iff $0 \sim 0$. Since we may simply assume that R and S have maximum elements, this reduces the embeddablility property to the problem of computing $\sim$ with an ITRM. Since S is a wellorder the following lemma yields a recursive definition of $\sim$.

Lemma 2. *For every r and s, $r \sim s$ iff $\forall r'Rr\exists s'Ss\ r' \sim s'$.*

Proof. Assume $r \sim s$. Take an orderpreserving map

$$\pi : (\mathrm{TC}_R(r), R) \to (\mathrm{TC}_S(s), S) \text{ with } \pi(r) = s.$$

Let $r'Rr$. Let $s' = \pi(r')\ S\ s = \pi(r)$. Then $\mathrm{TC}_R(r') \subseteq \mathrm{TC}_R(r)$ and

$$\pi \restriction \mathrm{TC}_R(r') : \mathrm{TC}_R(r') \to \mathrm{TC}_S(s')$$

orderpreservingly with $\pi(r') = s'$. Thus $\forall r'Rr\exists s'Ss\ r' \sim s'$.

Conversely assume that $\forall r'Rr\exists s'Ss\ r' \sim s'$. For every $r'Rr$ choose a map $\pi_{r'} : \mathrm{TC}_R(r') \to \mathrm{TC}_S(s')$ witnessing $r' \sim s'$. Note that

$$\mathrm{TC}_R(r) = \{r\} \cup \bigcup_{r'Rr} \mathrm{TC}_R(r').$$

Thus we may define a map $\pi : \mathrm{TC}_R(r) \to \mathrm{TC}_S(s)$ by $\pi(r) = s$ and for $r'' \neq r$:

$$\pi(r'') = \min\{\pi_{r'}(r'')|r'Rr\}$$

where the minimum is formed with respect to the the wellorder S. Then π witnesses that $r \sim s$.

We shall compute $\sim$ on an ITRM using finite *stacks* of natural numbers. Code a stack $(r_0, \ldots, r_{m-1})$ by $r = 2^{r_0} \cdot 3^{r_1} \cdots p_{m-1}^{r_{m-1}+1}$. Standard stack operations like *pushing* and *popping* natural numbers or finding the *length* $m-1$ of the stack r are recursive and thus computable by an ITRM. Since the relations R and S are infinite time register computable the question whether the stack $(r_0, \ldots, r_{m-1})$ is strictly descending in R or S can also be computed by an ITRM. For the subsequent program we shall use two registers A and B as stacks with associated operations pushA, popA, lenghthA, A-is-decreasing-in-R and pushB, popB, lenghthB, B-is-decreasing-in-S. The specific coding of stack contents leeds to a controlled limit behaviour:

Proposition 1. *Let* $\alpha < t$ *where* t *is a limit ordinal. Assume that the stack* A *(or* B*) contains the contents* $r = (r_0, \ldots, r_{m-1})$ *for cofinally many times below* t *and that all contents in the time interval* (α, t) *are endextensions of* $r = (r_0, \ldots, r_{m-1})$. *Then at time* t *the stack contents are* $r = (r_0, \ldots, r_{m-1})$.

So let us assume that R and S both have 0 as their maximum element. Running the following program P on an ITRM outputs yes/no depending on whether R can be embedded order-preservingly into S. We present the program in simple pseudo-code and assume that it is translated into a register program according to Definition 1 so that the order of commands is kept. Also the stack commands like pushA are understood as *macros* which are inserted into the code with appropriate renaming of variables and statement numbers.

```
      pushA 0;
      pushB 0;
      FLAG := 1; %% ask whether 0 ~ 0
Loop: Case1: if FLAG=0 and lengthA=lengthB=1 %% 0 ~ 0
            then begin; output 'yes'; stop; end;
      Case2: if FLAG=0 and lengthA>lengthB=1 %% 0 !~ 0
            then begin; output 'no'; stop; end;
      Case3: if FLAG=0 and lengthA = lengthB > 1
      %% last element of A ~ last of B
            then begin; %% check next
      popA N;
      pushA N+1;
      popB N;
      pushB 0;
      FLAG:=1; %% ask whether last of A ~ last of B
      goto Loop;
      end;
      Case4: if FLAG=0 and lengthA>lengthB
      %% 2nd-but-last of A !~ last of B
            then begin;
      popA N;
      popB N;
      pushB N+1;
```

```
        FLAG:=1; %% ask whether last element of A ~ last of B
        goto Loop;
        end;
    Case5: if FLAG=1 and A-is-decreasing-in-R
        and B-is-decreasing-in-S
        then begin;
        pushA 0;
        pushB 0;
        FLAG:=0; FLAG:=1; %% flash the flag
        goto Loop;
        end;
    Case6: if FLAG=1 and A-is-decreasing-in-R
        and not B-is-decreasing-in-S
        then begin;
        popB N;
        pushB N+1;
        FLAG:=0; FLAG:=1; %% flash the flag
        goto Loop;
        end;
    Case7: if FLAG=1 and not A-is-decreasing-in-R
        then begin;
        popA N;
        pushA N+1;
        popB N;
        pushB 0;
        FLAG:=0; FLAG:=1; %% flash the flag
        goto Loop;
        end;
```

The next Lemma proves the correctness of the program. Note that the program will always loop back to Loop until the program stops.

Lemma 3. *Let*

$$I : \theta \to \omega, R : \theta \to ({}^{\omega}\omega)$$

be the computation by P with trivial input $(0, 0, \ldots)$. Then the computation satisfies:

a) *Suppose the machine is in state Loop and the stack contents of A and B are $(r_0, \ldots, r_{m-1})$ and $(s_0, \ldots, s_{m-1})$, $m \geqslant 1$ which descend strictly in R and S resp. Moreover suppose that Flag=1 and $r_{m-1} \sim s_{m-1}$. Then the machine will reach the state Loop with the same stack contents and Flag=0 after a certain interval of time; during that interval, $(r_0, \ldots, r_{m-1})$ and $(s_0, \ldots, s_{m-1})$ will always be initial segments of the stacks A and B resp.*

b) *Suppose the machine is in state Loop and the stack contents of A and B are $(r_0, \ldots, r_{m-1})$ and $(s_0, \ldots, s_{m-1})$, $m \geqslant 1$ which descend strictly in R and S resp. Moreover suppose that Flag=1 and $r_{m-1} \nsim s_{m-1}$. Let r_m be* the smallest integer such that $r_m R r_{m-1}$ for which there is no $s_m S s_{m-1}$ such that $r_m \sim s_m$. Then the machine will reach the state Loop with stack contents $(r_0, \ldots, r_{m-1}, r_m)$ and $(s_0, \ldots, s_{m-1})$ and Flag=0 after a certain interval of time; during that interval, $(r_0, \ldots, r_{m-1})$ and $(s_0, \ldots, s_{m-1})$ will always be initial segments of the stacks A and B resp.

c) If R can be embedded orderpreservingly into S then the computation stops with output 'yes'.

d) If R cannot be embedded orderpreservingly into S then the computation stops with output 'no'.

Proof. a) and b) are proved by simultaneous induction on s_{m-1} along the well-order S. So consider a situation $(r_0, \ldots, r_{m-1})$ and $(s_0, \ldots, s_{m-1})$ as in a) or b) and assume that a) and b) already hold for all appropriate stacks $(r'_0, \ldots, r'_{m'-1})$ and $(s'_0, \ldots, s'_{m'-1})$ with $s'_{m'-1} S s_{m-1}$.

We first prove a) for the given situation. So Flag=1 and $r_{m-1} \sim s_{m-1}$. Inspection of the program shows that the machine will successively enter the main loop with register A containing the stacks $(r_0, \ldots, r_{m-1}, i)$ for $i = 0, 1, \ldots$. Note that by Case7, only the strictly decreasing stacks with $i R r_{m-1}$ are relevant. For such a $(r_0, \ldots, r_{m-1}, i)$ in register A the machine will enter the main loop with register B containing stacks $(s_0, \ldots, s_{m-1}, j)$. Again, by Case6, only strictly decreasing stacks $(s_0, \ldots, s_{m-1}, j)$ with $j S s_{m-1}$ are relevant. In these cases, the main loop is entered with strictly descending stack contents $(r_0, \ldots, r_{m-1}, i)$ and $(s_0, \ldots, s_{m-1}, j)$ and Flag=1.

We can apply the inductive assumptions: If $i \sim j$ the machine will subsequently reach the state Loop with the same stack contents and Flag=0. If $i \nsim j$ the machine will reach the state Loop with stack contents $(r_0, \ldots, r_{m-1}, i, k)$, some $k < \omega$, and $(s_0, \ldots, s_{m-1}, j)$ and Flag=0; it will then set the stack contents to $(r_0, \ldots, r_{m-1}, i)$ and $(s_0, \ldots, s_{m-1}, j+1)$ with Flag=1. Since $r_{m-1} \sim s_{m-1}$ there is some j such that $i \sim j$ and so the machine will eventually reach the state Loop with stack contents $(r_0, \ldots, r_{m-1}, i)$ and $(s_0, \ldots, s_{m-1}, j)$, some $j < \omega$, and Flag=0. This will be the case in turn for all $i < \omega$. By the limit rules the limit of these configurations will be a machine configuration with stack contents $(r_0, \ldots, r_{m-1})$ and $(s_0, \ldots, s_{m-1})$, and Flag=0.

For b) assume that Flag=1 and $r_{m-1} \nsim s_{m-1}$. Let r_m be defined as above. Then the machine will proceed as in the proof of a), until it reaches the stack contents $(r_0, \ldots, r_{m-1}, r_m)$. We argue inductively that it will subsequently set the contents of B to $(s_0, \ldots, s_{m-1}, j)$ for $j = 0, 1, \ldots$ and enter the main loop with Flag=1.

For $j = 0$, an analysis of the program shows that when the contents of A are first set to $(r_0, \ldots, r_{m-1}, r_m)$, the contents of B are set to $(s_0, \ldots, s_{m-1}, 0)$ (Case3 or Case5). For the inductive step assume that the machine enters the main loop with stack contents $(r_0, \ldots, r_{m-1}, r_m)$ and $(s_0, \ldots, s_{m-1}, j)$ with Flag=1. If $(s_0, \ldots, s_{m-1}, j)$ is not strictly descending in S then Case6 will modify the contents of B to $(s_0, \ldots, s_{m-1})$ and $(s_0, \ldots, s_{m-1}, j+1)$ and enter the main loop with Flag=1. If $(s_0, \ldots, s_{m-1}, j)$ is strictly descending in S then we can apply the inductive assumptions. Since $r_m \nsim j$ the machine will reach the state

Loop with stack contents $(r_0, \ldots, r_{m-1}, r_m, k)$, some $k < \omega$, and $(s_0, \ldots, s_{m-1}, j)$ and **Flag=0** after a certain interval of time. Then **Case4** will modify the stack contents to $(r_0, \ldots, r_{m-1}, r_m)$ and $(s_0, \ldots, s_{m-1}, j+1)$, set **Flag:=0** and enter the main loop. This concludes the induction.

By the limit rules the limit of this inductive sequence of configurations will be a configuration with state **Loop**, **Flag=0**, and stack contents $(r_0, \ldots, r_{m-1}, r_m)$ and $(s_0, \ldots, s_{m-1})$, as required by b). Inspection of the algorithm shows that the desired configurations for a) and b) are first reached with the stack contents always endextending $(r_0, \ldots, r_{m-1})$ and $(s_0, \ldots, s_{m-1})$ resp.

c) Assume that R can be embedded orderpreservingly into S. Since 0 is the maximum element of both R and S, $0 \sim 0$. The computation will first reach state **Loop** with stack contents (0) and (0) and **Flag=1**. By a), it will later reach state **Loop** with stack contents (0) and (0) and **Flag=0**. By **Case1** of the main loop, the machine will output **'yes'** and stop.

d) is proved an analogy with c).

4   Admissible Sets and Infinite Register Computations

For the converse we show

Lemma 4. *Let* $I : \theta \to \omega, R : \theta \to (^\omega \omega)$ *be a computation by a program P which stops at some successor ordinal* $\theta = \beta + 1$. *Then* $\theta < \omega_1^{CK}$.

Proof. Assume that $\theta \geqslant \omega_1^{CK}$. Let $I(\omega_1^{CK}) = k$ and

$$R(\omega_1^{CK}) = (n_0, \ldots, n_{l-1}, 0, 0, \ldots)$$

where $R_0, \ldots, R_{l-1}$ includes all the registers mentioned in the program P. By the liminf rules for ITRMs there is some $\alpha < \omega_1^{CK}$ such that the sets

$$\{t \in (\alpha, \omega_1^{CK}) | I(t) = k\}$$

and

$$\{t \in (\alpha, \omega_1^{CK}) | R_j(t) = n_j\}$$

are closed unbounded in ω_1^{CK}. These sets are Σ_1-definable over the admissible set $L_{\omega_1^{CK}}$ in the parameter α. In $L_{\omega_1^{CK}}$ define a sequence $\alpha_0 = \alpha < \alpha_1 < \alpha_2 < \ldots$ such that

$$\exists t \in (\alpha_n, \alpha_{n+1}) \; I(t) = k \text{ and for } j = 0, \ldots, l-1 \exists t \in (\alpha_n, \alpha_{n+1}) \; R_j(t) = n_j.$$

Such a sequence may be defined by a Σ_1-definition over $L_{\omega_1^{CK}}$. By the Σ_1-bounding principle in $L_{\omega_1^{CK}}$, $\alpha^* = \bigcup_{n<\omega} \alpha_n < \omega_1^{CK}$. Also $I(\alpha^*) = k$ and $R(\alpha^*) = (n_0, \ldots, n_{l-1}, 0, 0, \ldots)$. So the constellation $I(t) = k$ and $R(t) = (n_0, \ldots, n_{l-1}, 0, 0, \ldots)$ occurs at times α^* and ω_1^{CK}. This means that the machine runs into a cycle and does not stop, contrary to our assumption.

Lemma 5. *Let $x \subseteq \omega$ be computable by an infinite time register machine. Then $x \in L_{\omega_1^{CK}}$.*

Proof. Let P be a register program such that such that for every $n < \omega$

$$P : (n, 0, 0, \ldots) \mapsto \chi_x(n).$$

For $n < \omega$ let the computation by P with input $(n, 0, 0, \ldots)$ stop at time θ_n. By the previous lemma, $\theta_n < \omega_1^{CK}$. Therefore the computation by P with input $(n, 0, 0, \ldots)$ is an element of $L_{\omega_1^{CK}}$. The characteristic function χ_x is Δ_1-definable over $L_{\omega_1^{CK}}$ by

$\chi_x(n) = 1$ iff there is a computation by P with input $(n, 0, 0, \ldots)$ and output 1

iff all computations by P with input $(n, 0, 0, \ldots)$ stop with output 1.

Since the admissible set $L_{\omega_1^{CK}}$ satisfies Δ_1-separation, $x \in L_{\omega_1^{CK}}$.

5   Further Considerations

One may consider variants of the ITRMs, where the registers can hold ordinals below a certain bound β. What is the collection of subsets of β computable by β-ITRMs? It is hoped that such interpolations between ITRMs and ORMs yield a stratification of the constructible sets which may lead to a fine structure theory of the class L of constructible sets (see [5]).

References

[1] Nigel J. Cutland. *Computability: An Introduction to Recursive Function Theory.* Perspectives in Mathematical Logic. Cambridge University Press, 1980.

[2] Joel D. Hamkins and Andy Lewis. Infinite Time Turing Machines. *J. Symbolic Logic*, 65(2):567–604, 2000.

[3] Peter Koepke. Turing computations on ordinals. *Bulletin of Symbolic Logic*, 11(3):377–397, 2005.

[4] Peter Koepke. Computing a model of set theory. In: S. Barry Cooper, Benedikt Löwe, Leen Torenvliet (editors). *New Computational Paradigms: First Conference on Computability in Europe, CiE 2005. Proceedings.* Lecture Notes in Computer Science 3526:223–232, 2005.

[5] Peter Koepke and Sy Friedman. An elementary approach to the fine structure of *L. Bulletin of Symbolic Logic*, 3(4):453–468, 1997.

[6] Peter Koepke and Martin Koerwien. Ordinal computations. To appear in: *Mathematics of Computation at CiE 2005.* Special issue of the journal *Mathematical Structures in Computer Science*: 17 pages.

[7] Peter Koepke and Ryan Siders. Computing the recursive truth predicate on ordinal register machines. Submission to CiE 2006, Swansea.

[8] Gerald E. Sacks. *Higher Recursion Theory.* Perspectives in Mathematical Logic. Springer-Verlag, Berlin, 1990.

Upper and Lower Bounds on Sizes of Finite Bisimulations of Pfaffian Hybrid Systems

Margarita Korovina[1] and Nicolai Vorobjov[2]

[1] Fachbereich Mathematik, Theoretische Informatik, Universität Siegen, D-57068, Germany, and IIS SB RAS, Novosibirsk, Russia
korovina@brics.dk

[2] Department of Computer Science, University of Bath, Bath BA2 7AY, England
nnv@cs.bath.ac.uk

Abstract. In this paper we study a class of hybrid systems defined by Pfaffian maps. It is a sub-class of o-minimal hybrid systems which capture rich continuous dynamics and yet can be studied using finite bisimulations. The existence of finite bisimulations for o-minimal dynamical and hybrid systems has been shown by several authors (see e.g. [3,4,13]). The next natural question to investigate is how the sizes of such bisimulations can be bounded. The first step in this direction was done in [10] where a double exponential upper bound was shown for Pfaffian dynamical and hybrid systems. In the present paper we improve this bound to a single exponential upper bound. Moreover we show that this bound is tight in general, by exhibiting a parameterized class of systems on which the exponential bound is attained. The bounds provide a basis for designing efficient algorithms for computing bisimulations, solving reachability and motion planning problems.

1 Introduction

One of the main complexities in the reasoning about hybrid systems arises from their uncountably infinite state spaces. To overcome this difficulty bisimulation by simpler systems was introduced. Informally, two hybrid systems are *bisimilar* if their behaviors are indistinguishable with respect to the properties we consider. It is desirable to have bisimulations on which we can verify basic properties (like reachability) effectively, in particular, finite bisimulations. A wide class of hybrid systems that admits finite bisimulations is formed by *o-minimal systems*, introduced and studied in [3,4,13]. This approach is based on the theory of o-minimal structures, intensively studied in model theory [15].

The existence of finite bisimulations for o-minimal hybrid systems has been shown by several authors (see e.g. [3,4,13]). The next natural question to investigate is how the sizes of such bisimulations can be bounded.

In order to give effective bounds on the sizes of the bisimulations we restrict ourselves to a particular case of o-minimal hybrid systems, namely to the class of Pfaffian hybrid systems introduced in [10], and represented by *Pfaffian functions*. Such functions naturally arise in applications as real analytic solutions of

A. Beckmann et al. (Eds.): CiE 2006, LNCS 3988, pp. 267–276, 2006.

triangular first order partial differential equations with polynomial coefficients, and include polynomials, algebraic functions, exponentials, and trigonometric functions in appropriate domains [9]. In our previous work [10] we gave a double exponential upper bound on the sizes of bisimulations of Pfaffian hybrid systems. In the present paper we improve that bound to a single exponential upper bound. Moreover we show that the bound is tight in general, by exhibiting a parameterized class of polynomial dynamical systems on which the exponential bound is attained. Let us note that previous bounds were obtained using *cylindrical cell decomposition*, which is of intrinsically double exponential complexity. In this paper we avoid cylindrical decomposition by using some finer tools from real analytic geometry.

These tools also provide framework for further studies of the behavior of Pfaffian hybrid systems. In [10] an algorithm was proposed for computing finite bisimulations with the double exponential complexity. The bounds obtained in the present paper provide a basis for computing bisimulations, and via them, reachability, motion planning, etc. problems, with the single exponential complexity.

This paper is organized as follows. In Section 2 we recall the notions of bisimulation of transition systems and Pfaffian dynamical systems. In Section 3 we construct an upper bound on sizes of finite bisimulations of a Pfaffian dynamical system. In Section 4 we show that this bound is tight in general, by exhibiting a parameterized class of Pfaffian dynamical systems on which the exponential bound is attained. We then conclude with the future work.

2   Basic Notions and Definitions

2.1   Transition Systems and Dynamical Systems

One of the approaches to study of a dynamical system uses the partition of the state space into finitely many equivalence classes, so that equivalent states exhibit similar properties. This special quotient of the original state space, called bisimulation, is reachability preserving, i.e., checking the reachability on the quotient system is equivalent to checking it on the original system. In this section we recall (following [3]) the notion of bisimulations of transition systems, and basic results concerning finite bisimulations of o-minimal dynamical systems.

The first group of definitions describe transition systems and bisimulations between the transition systems.

Definition 1. *Let Q be an arbitrary set and $\to$ be a binary relation on Q. In the context of dynamical systems theory we call Q the set of states, $\to$ the transition, and $T := (Q, \to)$ the transition system.*

Definition 2. *Given two transition systems $T_1 := (Q_1, \to_1)$ and $T_2 := (Q_2, \to_2)$ we define a simulation of T_1 by T_2 as a binary relation $\sim \subset Q_1 \times Q_2$ such that:*

- $\forall q_1 \in Q_1 \exists q_2 \in Q_2 (q_1 \sim q_2)$;
- $\forall q_1, q_1' \in Q_1 \forall q_2 \in Q_2 ((q_1 \sim q_2 \wedge q_1 \to_1 q_1') \Rightarrow \exists q_2' \in Q_2 (q_1' \sim q_2' \wedge q_2 \to_2 q_2'))$.

Definition 3. *A* bisimulation *between two transition systems* $T_1 := (Q_1, \rightarrow_1)$ *and* $T_2 := (Q_2, \rightarrow_2)$ *is a simulation* $\sim \subset Q_1 \times Q_1$ *of* T_1 *by* T_2 *such that the converse relation* $\sim^{-1} := \{(q_2, q_1) \in Q_2 \times Q_1 | q_1 \sim q_2\}$ *is a simulation of* T_2 *by* T_1.

Definition 4. *A* bisimulation *between a transition system* T *and itself is called a* bisimulation on T.

Definition 5. *Let* $\sim$ *be a bisimulation on* $T = (Q, \rightarrow)$ *and also an equivalence relation on* Q. *Let* $\mathcal{P}$ *be a partition of* Q. *We say that* $\sim$ *is a* bisimulation with respect to $\mathcal{P}$ *if every* $P \in \mathcal{P}$ *is the union of some equivalence classes of* $\sim$.

Normally, the partition $\mathcal{P}$ reflects regions of interest such as invariants and initial conditions of the dynamical system.

In this paper we are concerned with estimating cardinality of bisimulations in the sense of Definition 5. We now give some definitions concerning dynamical systems.

Definition 6. *Let* $G_1 \subset \mathbb{R}^m$ *and* $G_2 \subset \mathbb{R}^n$ *be open domains. A* dynamical system *is a map* $\gamma : G_1 \times (-1, 1) \rightarrow G_2$. *For a given* $\mathbf{x} \in G_1$ *the set* $\Gamma_{\mathbf{x}} = \{\mathbf{y} | \exists t \in (-1, 1) (\gamma(\mathbf{x}, t) = \mathbf{y})\} \subset G_2$ *is called the* trajectory *determined by* $\mathbf{x}$, *and the graph* $\widehat{\Gamma}_{\mathbf{x}} = \{(t, \mathbf{y}) | \gamma(\mathbf{x}, t) = \mathbf{y}\} \subset (-1, 1) \times G_2$ *is called the* integral curve *determined by* $\mathbf{x}$.
A dynamical system is called o-minimal *if it is definable in an o-minimal structure over* $\mathbb{R}$.

Definition 7. *The* transition system $T_\gamma = (Q, \rightarrow)$ *associated to the dynamical system* γ *is defined as follows:*

- $Q := G_2$, *and*
- $\mathbf{y}_1 \rightarrow \mathbf{y}_2$ *for* $\mathbf{y}_1, \mathbf{y}_2 \in Q$ *if and only if*

$$\exists \mathbf{x} \in G_1 \exists t_1, t_2 \in (-1, 1)((t_1 \leq t_2) \wedge (\gamma(\mathbf{x}, t_1) = \mathbf{y}_1) \wedge (\gamma(\mathbf{x}, t_2) = \mathbf{y}_2))$$

We now introduce following [3], a technique of encoding trajectories of dynamical systems by words. Let $\mathcal{P} := \{P_1, \ldots, P_s\}$ be a finite partition of $\gamma(G_1 \times (-1, 1))$ definable in the o-minimal structure. Fix $\mathbf{x} \in G_1$. Define the set $\mathcal{F}_{\mathbf{x}}$ of points and open intervals I in $(-1, 1)$ which are maximal with respect to inclusion for the property $\exists i \in \{1, \ldots, s\} \forall t \in I (\gamma(\mathbf{x}, t) \in P_i)$.

Let the cardinality $|\mathcal{F}_{\mathbf{x}}| = r$ and $y_1 < \cdots < y_r$ be representatives of $\mathcal{F}_{\mathbf{x}}$ such that $\gamma(\mathbf{x}, y_j) \in P_{i_j}$. Then define the word $\omega := P_{i_1} \cdots P_{i_r}$ in the alphabet $\mathcal{P}$. Informally, ω is the list of names of elements of the partition in the order they are visited by the trajectory $\Gamma_{\mathbf{x}}$.

Let $\mathbf{y} \in \Gamma_{\mathbf{x}}$. Then $\mathbf{y} \in P_{i_j}$ for some $1 \leq j \leq r$, where P_{i_j} is a letter in ω. We represent the location of $\mathbf{y}$ on the trajectory $\Gamma_{\mathbf{x}}$ by the *dotted word*

$$\dot{\omega} := P_{i_1} \cdots \dot{P}_{i_j} \cdots P_{i_r}$$

It will be convenient to use the operation undot$(\dot{w}) = w := P_{i_1} \cdots P_{i_j} \cdots P_{i_r}$. In the sequel we will always assume that the dynamical system γ is injective. In this case there is a unique dotted word associated to a given $\mathbf{y} \in \gamma(G_1 \times (-1, 1))$. Introduce sets of words $\Omega := \{w | \mathbf{x} \in G_1\}$, $\dot{\Omega} := \{\dot{w} | \mathbf{x} \in G_1\}$. The following statement is an easy consequence of o-minimality.

Lemma 1. [3]   *The set Ω is finite.*

An obvious (purely combinatorial) corollary is that $\dot{\Omega}$ is also finite.

Definition 8. *The transition system $T_{\dot{\Omega}}$ is defined as follows:*

- $Q := \dot{\Omega}$, *and*
- $\dot{w}_1 \to \dot{w}_2$ *for* $\dot{w}_1, \dot{w}_2 \in Q$ *if and only if* $w_1 = w_2$ *and the dot on* $\dot{w}_2$ *is to the right of (or in the same) position as the dot on* $\dot{w}_1$.

Theorem 1. [3]   *Let the o-minimal dynamical system γ be bijective, and the partition $\mathcal{P}$ be definable in the o-minimal structure. Then there is a finite bisimulation on T_γ with respect to $\mathcal{P}$.*

Proof. To prove the theorem one first shows that $T_{\dot{\Omega}}$ is a bisimulation of T_γ, and then considers the following equivalence relation $\sim$ on G_2: $\mathbf{y}_1 \sim \mathbf{y}_2$ iff for respective pre-images $(\mathbf{x}_1, t_1), (\mathbf{x}_2, t_2)$, the locations of $\mathbf{y}_1, \mathbf{y}_2$ on trajectories $\Gamma_{\mathbf{x}_1}, \Gamma_{\mathbf{x}_2}$ are described by the same dotted word $\dot{w}$. Then $\sim$ is the required bisimulation (see details in [3]). $\square$

2.2   Pfaffian Functions and Related Sets

In what follows, in order to give a quantitative refinement of Theorem 1 we will restrict our considerations of o-minimal dynamical systems to a particular case, the class of *Pfaffian* dynamical systems. This section is a digest of the theory of Pfaffian functions and sets definable with Pfaffian functions. The detailed exposition can be found in the survey [6].

Definition 9. *A Pfaffian chain of order $r \geq 0$ and degree $\alpha \geq 1$ in an open domain $G \subset \mathbb{R}^n$ is a sequence of real analytic functions $f_1, \ldots, f_r$ in G satisfying differential equations*

$$\frac{\partial f_j}{\partial x_i} = g_{ij}(\mathbf{x}, f_1(\mathbf{x}), \ldots, f_j(\mathbf{x})) \tag{1}$$

for $1 \leq j \leq r$, $1 \leq i \leq n$. Here $g_{ij}(\mathbf{x}, y_1, \ldots, y_j)$ are polynomials in $\mathbf{x} = (x_1, \ldots, x_n), y_1, \ldots, y_j$ of degrees not exceeding α.
  A function

$$f(\mathbf{x}) = P(\mathbf{x}, f_1(\mathbf{x}), \ldots, f_r(\mathbf{x})),$$

where $P(\mathbf{x}, y_1, \ldots, y_r)$ is a polynomial of a degree not exceeding $\beta \geq 1$, the sequence $f_1, \ldots, f_r$ is a Pfaffian chain of order r and degree α, is called a Pfaffian *function of order r and degree (α, β).*

Apart from polynomials, the class of Pfaffian functions includes real algebraic functions, exponentials, logarithms, trigonometric functions, their compositions, and other major transcendental functions in appropriate domains (see [5]).

Now we introduce classes of sets definable with Pfaffian functions. In the case of polynomials they reduce to *semialgebraic* sets whose quantitative and algorithmic theory is treated in [2].

Definition 10. *A set $X \subset \mathbb{R}^n$ is called* semi-Pfaffian *in an open domain $G \subset \mathbb{R}^n$ if it consists of the points in G satisfying a Boolean combination of some atomic equations and inequalities $f = 0, g > 0$, where f, g are Pfaffian functions having a common Pfaffian chain defined in G. A semi-Pfaffian set X is* restricted *in G if its topological closure lies in G.*

Definition 11. *A set $X \subset \mathbb{R}^n$ is called* sub-Pfaffian *in an open domain $G \subset \mathbb{R}^n$ if it is the image of a semi-Pfaffian set under a projection into a subspace.*

In the sequel we will be dealing with the following subclass of sub-Pfaffian sets.

Definition 12. *Consider the closed cube $[-1, 1]^{m+n}$ in an open domain $G \subset \mathbb{R}^{m+n}$ and the projection map $\pi : \mathbb{R}^{m+n} \to \mathbb{R}^n$. A subset $Y \subset [-1, 1]^n$ is called* restricted sub-Pfaffian *if $Y = \pi(X)$ for a restricted semi-Pfaffian set $X \subset [-1, 1]^{m+n}$.*

Note that a restricted sub-Pfaffian set need not be semi-Pfaffian.

Definition 13. *Consider a semi-Pfaffian set*

$$X := \bigcup_{1 \le i \le M} \{\mathbf{x} \in \mathbb{R}^n | f_{i1} = 0, \ldots, f_{il_i} = 0, g_{i1} > 0, \ldots, g_{ij_i} > 0\} \subset G \quad (2)$$

where f_{is}, g_{is} are Pfaffian functions with a common Pfaffian chain of order r and degree (α, β), defined in an open domain G. Its format *is the tuple (r, N, α, β, n), where $N \ge \sum_{1 \le i \le M}(l_i + j_i)$. For $n = m + k$ and a sub-Pfaffian set $Y \subset \mathbb{R}^k$ such that $Y = \pi(\tilde{X})$, its* format *is the format of X.*

We will refer to the representation of a semi-Pfaffian set in the form (2) as to the *disjunctive normal form (DNF)*.

Remark 1. In this paper we are concerned with upper and lower bounds on sizes of bisimulations as functions of the format. In the case of Pfaffian dynamical systems these sizes and complexities also depend on the domain G. So far our definitions have imposed no restrictions on an open set G, thus allowing it to be arbitrarily complex and to induce this complexity on the corresponding semi- and sub-Pfaffian sets. To avoid this we will always assume in the context of Pfaffian dynamical systems that G is "simple", like $\mathbb{R}^n$, or $(-1, 1)^n$.

Theorem 2. [6,17] *Consider a semi-Pfaffian set $X \subset G \subset \mathbb{R}^n$, where G is an open domain, represented in DNF with a format (r, N, α, β, n). Then the sum of the Betti numbers (in particular, the number of connected components) of X does not exceed $N^n 2^{r(r-1)/2} O(n\beta + \min\{n, r\}\alpha)^{n+r}$.*

Theorem 3. ([7], Section 5.2) *Consider a sub-Pfaffian set $Y = \pi(X)$ as described in Definition 12. Let X be closed and represented in DNF with a format $(r, N, \alpha, \beta, n + m)$. Then the kth Betti number $\mathrm{b}_k(Y)$ does not exceed*

$$k((k+1)N)^{n+(k+1)m}2^{(k+1)r((k+1)r-1)/2}O((n+km)\beta+\min\{kr,n+km\}\alpha)^{n+(k+1)(m+r)}.$$

Let $d > \alpha + \beta$. Relaxing the bound from Theorem 3, we get

$$\mathrm{b}_k(Y) \leq (kN)^{O(n+km)}2^{O((kr)^2)}((n+km)d)^{O(n+km+kr)}.$$

3   The Upper Bound on Sizes of Finite Bisimulation of Pfaffian Dynamical Systems

It was shown in [13] that in an o-minimal hybrid system the continuous and discrete components can be separated, and therefore the problem of finite bisimulation reduces to the same problem for a transition system associated with a continuous dynamical system. Moreover the size of the bisimulation is linear in the number of discrete components (locations) of the hybrid system.

It follows from [16] that the Pfaffian hybrid systems are a subclass of o-minimal hybrid systems, therefore we can restrict ourselves to Pfaffian dynamical systems and partitions defined by semi-Pfaffian sets. Our main results concern upper and lower bounds for finite bisimulations of Pfaffian dynamical systems with respect to partitions defined by semi-Pfaffian sets.

Definition 14. *A dynamical system $\gamma : G_1 \times (-1, 1) \rightarrow G_2$, where $G_1 \subseteq \mathbb{R}^m$ and $G_2 \subseteq \mathbb{R}^n$ are open and γ is a map with a semi-Pfaffian graph, is called a Pfaffian dynamical system.*

Let $\gamma : G_1 \times (-1, 1) \rightarrow G_2$, where $G_1 = I^{n-1} := (-1, 1)^{n-1}$ and $G_2 = I^n$, be a homeomorphism, defined by its graph $\widehat{\Gamma} := \{(\mathbf{x}, t, \mathbf{y}) | \gamma(\mathbf{x}, t) = \mathbf{y}\}$ which is a semi-Pfaffian set, and $\mathcal{P}$ be a partition of G_2 into semi-Pfaffian sets. Suppose the number of functions involved in the definitions of the graph $\widehat{\Gamma}$ and the partition $\mathcal{P}$ does not exceeds N, and each of these functions has the order r and the degree (α, β).

Theorem 4. *Let $T_\gamma = (G_2, \rightarrow)$ be the transition system associated to the dynamical system γ. Then there is a bisimulation on T_γ with respect to $\mathcal{P}$ consisting of at most $N^{O(n^4)}(n(\alpha + \beta))^{O(n^6 r^3)}$ equivalence classes.*

Remark 2. The best upper bound known until now [10] was double exponential:

$$N^{(r+n)^{O(n)}}(\alpha + \beta)^{(r+n)^{O(n^3)}}$$

These results show that, w.r.t the size of coarsest bisimulations, Pfaffian hybrid systems behave like timed automata (see [1]). We consider an elementary example illustrating techniques which we use to show the single exponential upper bounds in the general case. For the full proof of Theorem 4 we refer to [11].

Let $G_1 := (-1, 1)$, $G_2 := (-1, 1)^2$, and $\gamma : (x, t) \mapsto (y_1 = x, y_2 = t)$. (Note that this dynamical system corresponds to the system of differential equations $\dot{y}_1 = 0$, $\dot{y}_2 = 1$.) Consider the graph $\widehat{\Gamma} := \{(x, t, y_1, y_2) | x - y_1 = 0, \ t - y_2 = 0\}$ of the map γ. Note that $\widehat{\Gamma}$ is an intersection of the 4-cube $(-1, 1)^4$ with a 2-plane, and therefore is a smooth manifold. In the general case the graph of a dynamical system may not be smooth and we will need to separate smooth and singular parts of it. For a fixed $x \in G_1$ the set $\widehat{\Gamma}_x := \{(t, y_1, y_2) | x - y_1 = 0, \ t - y_2 = 0\}$ is the integral curve, and the set $\Gamma_x := \{(y_1, y_2) | \exists t \, (x - y_1 = 0, \ t - y_2 = 0)\}$ is the trajectory of γ. Thus, in our example, the trajectories are open segments of straight lines parallel to y_2-axis.

Introduce the projection $\pi : G_1 \times (-1, 1) \times G_2 \to G_1$ as $(x, t, y_1, y_2) \mapsto x$. Let $\pi_{\widehat{\Gamma}}$ be the restriction of π on $\widehat{\Gamma}$. For a fixed $x \in G_1$ the fiber $\pi_{\widehat{\Gamma}}^{-1}(x)$ coincides with $\widehat{\Gamma}_x$. Let the partition $\mathcal{P}$ of G_2 consist of the disc $\{(y_1, y_2) | f := y_1^2 + y_2^2 - 1/4 \le 0\}$ labelled by letter A and its complement in G_2 labelled by B. The aim is to determine the number of different words in the alphabet $\{A, B\}$ encoding the trajectories. Clearly, it is sufficient to consider only intersections of the trajectories with the open sets $\{(y_1, y_2) | f < 0\}$ and $\{(y_1, y_2) | f > 0\}$ (in the general case, the transition to open sets is less trivial).

Let $\widehat{S} := \{(x, t, y_1, y_2) | f(y_1, y_2) = 0\}$. Observe that $\widehat{S} \cap \widehat{\Gamma}$ is a smooth curve. Consider the partition $\widehat{\mathcal{P}}$ of $\widehat{\Gamma}$ consisting of $\{(x, t, y_1, y_2) | f := y_1^2 + y_2^2 - 1/4 \le 0\}$ labelled by letter A and its complement in $\widehat{\Gamma}$ labelled by B. Clearly, it is sufficient to find the number of distinct words encoding the intersections of integral curves with open sets $\{(x, t, y_1, y_2) | f < 0\} \cap \widehat{\Gamma}$ and $\{(x, t, y_1, y_2) | f > 0\} \cap \widehat{\Gamma}$.

Consider the restriction $\pi_{\widehat{\Gamma}\widehat{S}} : \widehat{\Gamma} \to G_1$ of $\pi_{\widehat{\Gamma}}$ to $\widehat{S} \cap \widehat{\Gamma}$. Let C be the set of all critical values of $\pi_{\widehat{\Gamma}\widehat{S}}$. By setting to 0 the appropriate Jacobian we find that the critical points of $\pi_{\widehat{\Gamma}\widehat{S}}$ are $(1/2, 0, 1/2, 0)$ and $(-1/2, 0, -1/2, 0)$, thus $C = \{1/2, -1/2\}$. Let $R := G_1 \setminus C$. This set consists of three connected components:

$$\{x \in (-1, 1) | x < -1/2\}, \ \{x \in (-1, 1) | -1/2 < x < 1/2\}, \ \{x \in (-1, 1) | 1/2 < x\}.$$

Proposition 1. *If x, x' belong to the same connected component R' of R, then $\widehat{\Gamma}_x$ and $\widehat{\Gamma}_{x'}$ are labelled by the same word.*

In the general case the proposition requires a careful proof. As applied to our example, this proof has the following scheme.

(1) The restriction of $\pi_{\widehat{\Gamma}\widehat{S}}$ to $\pi_{\widehat{\Gamma}\widehat{S}}^{-1}(R')$ is a *trivial covering*, i.e., for any $x' \in R'$ the pre-image $\pi_{\widehat{\Gamma}\widehat{S}}^{-1}(R')$ is homeomorphic to $\pi_{\widehat{\Gamma}\widehat{S}}^{-1}(x') \times R'$. In our example, in the only non-trivial case of $R' = \{x \in (-1, 1) | -1/2 < x < 1/2\}$, we have:

$$\pi_{\widehat{\Gamma}\widehat{S}}^{-1}(R') = (\widehat{S} \cap \widehat{\Gamma}) \setminus \{(1/2, 0, 1/2, 0), (-1/2, 0, -1/2, 0)\}$$

is an oval minus two points, which is homeomorphic to the Cartesian product of the pair of points $\pi_{\widehat{\Gamma}\widehat{S}}^{-1}(x')$ by the interval R'. In other words, the connected components of $\pi_{\widehat{\Gamma}\widehat{S}}^{-1}(R')$ are two open arcs of simple curves.

(2) These arcs are naturally ordered separating the difference $\pi_{\widehat{F}}^{-1}(R') \setminus \pi_{\widehat{FS}}^{-1}(R')$ into ordered connected components. In the case of $R' = \{x \in (-1,1)| -1/2 < x < 1/2\}$ the components are (in order):

$$\{(x,t,y_1,y_2) \in \widehat{\Gamma}| (-1/2 < x < 1/2) \wedge (f > 0) \wedge (y_2 < 0)\},$$
$$\{(x,t,y_1,y_2) \in \widehat{\Gamma}| (-1/2 < x < 1/2) \wedge (f < 0)\},$$
$$\{(x,t,y_1,y_2) \in \widehat{\Gamma}| (-1/2 < x < 1/2) \wedge (f > 0) \wedge (y_2 > 0)\}.$$

For any $x \in R'$ the integral curve $\widehat{\Gamma}_x$ intersects these connected components according to their order.

(3) Each connected component of the difference $\pi_{\widehat{F}}^{-1}(R') \setminus \pi_{\widehat{FS}}^{-1}(R')$ lies either in the component $\{(x,t,y_1,y_2)| f < 0\}$, or in $\{(x,t,y_1,y_2)| f > 0\}$, and, therefore can be naturally labelled by A or B respectively. Since the connected components are ordered, the difference $\pi_{\widehat{F}}^{-1}(R') \setminus \pi_{\widehat{FS}}^{-1}(R')$ itself is labelled by a word (in the case of $R' = \{x \in (-1,1)| -1/2 < x < 1/2\}$ by BAB). It follows that for any $x \in R'$ the integral curve $\widehat{\Gamma}_x$ is labelled by this word, and the proposition is proved.

Proposition 1 implies that the number of distinct realizable words does not exceed the number of all connected components of R. In our example the latter is 3, which equals to the cardinality of the discrete set C plus 1. The general case uses the far-reaching extension of such method of counting, Alexander's duality, Theorems 2, 3 and Sard's Theorem (see [11]).

4 Lower Bound

We construct a parametric example of a semi-algebraic dynamical system $G_1 \times (-1,1) \to G_2$ together with a semi-algebraic partition of G_2 such that the format of both of them is (d,n) (degrees, number of variables) while the number of different words (size of a bisimulation) is $d^{\Omega(n)}$.

Let $g(y)$ be a polynomial of degree d such that $|g(y)| < 1$ for every $y \in (-1,1)$ and for every $c \in (-\frac{1}{2}, \frac{1}{2})$ the polynomial $g(y) - c$ has d simple roots in $(-1,1)$.

First we illustrate the idea of the example by describing the case $n = 2$. Let the dynamical system be given by $G_1 := (-1,1)$, $G_2 := (-1,1)^2$, $\gamma : (\mathbf{x},t) \mapsto (t,\mathbf{x})$. The partition $\mathcal{P}$ consists of two sets A and $B = G_2 \setminus A$ where $A := \{(y_1,y_2)| g(y_1) = 0, y_1 + y_2 > 0\}$. Notice that there are exactly $d + 1$ distinct words encoding all trajectories of the defined dynamical system. These words are formed by alternating letters starting and ending with B, i.e., B, BAB, ..., $BABABAB$, ... For arbitrary n, let $G_1 := (-1,1)$, $G_2 := (-1,1)^n$. Define a curve

$$\Delta := \{(y_1,\ldots,y_{n-1}) \in (-1,1)^{n-1}| y_2 = g(y_1),\ldots,y_{n-1} = g(y_{n-2})\}$$

Observe that Δ is connected in $(-1,1)^{n-1}$, being the graph of a map $\mathbf{f} : (-1,1) \to (-1,1)^{n-1}$, $y_1 \mapsto (g(y_1),\ldots,g(g(\cdots g(y_1)\cdots)))$, and smooth.

Consider the polynomial $h(y_{n-1}) := (y_{n-1} - b_1)(y_{n-1} - b_2) \cdots (y_{n-1} - b_d)$ where all $b_i \in (-\frac{1}{2}, \frac{1}{2})$ and $b_i \neq b_j$ for $i \neq j$. Then $\Delta \cap \{h = 0\}$ consists of d^{n-1} points. Define $A := \{(y_1, \ldots, y_n) | (y_1, \ldots, y_{n-1}) \in \Delta, h(y_{n-1}) = 0, L > 0\}$, where $L(y_1, \ldots, y_n)$ is a generic linear homogeneous polynomial such that $\{L = 0\}$ intersects all d^{n-1} parallel straight lines of $\{(y_1, \ldots, y_n) | (y_1, \ldots, y_{n-1}) \in \Delta, h(y_{n-1}) = 0\}$. Notice that the projection of this intersection on the y_n-coordinate consists of d^{n-1} distinct points.

Finally, define the dynamical system γ and the partition $\mathcal{P}$ as follows. The function γ maps $\mathbf{x} \subset G_1$ and $t \in (-1,1)$ to the point $(\mathbf{f}(t), \mathbf{x}) \in G_2$. The partition $\mathcal{P}$ consists of A and $B = G_2 \setminus A$. Clearly, there are exactly $d^{n-1} + 1$ pairwise distinct words encoding all trajectories.

In order to meet the requirement: G_1 has to be homeomorphic to I^{n-1}, G_2 has to be homeomorphic to I^n, we can do the following modifications.

Observe that there is a small enough $\varepsilon > 0$ such that for any sequence $0 < \varepsilon_1, \ldots, \varepsilon_{n-2} \leq \varepsilon$ and any sequence $*_1, \ldots, *_{n-2} \in \{+, -\}$, the algebraic set

$$\Delta' := \{(y_1, \ldots, y_{n-1}) \in (-1,1)^{n-1} | y_2 = g(y_1) *_1 \varepsilon_1, \ldots, y_{n-1} = g(y_{n-2}) *_{n-2} \varepsilon_{n-2}\}$$

is a smooth connected curve. These curves are disjoint and their union is

$$\Delta'' := \bigcap_{1 \leq i \leq n-2} \{(y_1, \ldots, y_{n-1}) \in (-1,1)^{n-1} | -\varepsilon < y_{i+1} - g(y_i) < \varepsilon\}.$$

Let $G_1 := (-\varepsilon, \varepsilon)^{n-2} \times (-1,1)$, $G_2 = \Delta'' \times (-1,1)$ and $\gamma : G_1 \times (-1,1) \rightarrow G_2$, such that

$$(*_1 \varepsilon_1, \ldots, *_{n-2} \varepsilon_{n-2}, x, t) \mapsto (g(t) *_1 \varepsilon_1, \ldots, g(g(\cdots g(t) \cdots)) *_{n-2} \varepsilon_{n-2}, x)$$

Note that γ is a diffeomorphism. It is obvious that the modified γ still has at least $d^{\Omega(n)}$ trajectories with pairwise distinct word codes with respect to the partition $\mathcal{P}$.

Let us summarize the obtained lower bound in the following theorem.

Theorem 5. *There exists a family of Pfaffian dynamical systems such that the sizes of bisimulations are bounded from below by an exponential function on the parameters of the system.*

5   Future Work

In [10] the authors proposed an algorithm (a Blum-Shub-Smale type machine with an oracle for deciding non-emptiness of semi-Pfaffian sets) for computing a finite bisimulation. That algorithm is based on the cylindrical cell decomposition technique and, accordingly, has a double exponential upper complexity bound. It seems feasible to construct a bisimulation algorithm with *single exponential* complexity using the approach employed in the present paper. Once a bisimulation is computed, it can be used in efficient algorithms for fundamental computational problems such as deciding reachability or motion planning in definable dynamical systems.

Acknowledgements

The authors would like to thank A. Gabrielov, V. Grandjean, and K. Korovin for useful discussions. The second author was supported in part by the European RTN Network RAAG (contract HPRN-CT-2001-00271), Grant Scientific School-4413.2006.1, Deutsche Forschungsgemeinschaft, Project WE 843/17-1, and Project GZ: 436 RUS 113/850/01.

References

1. R.Alur, D.L.Dill. A Theory of Timed Automata. *Theoretical Computer Science*, 126:183–235, 1994.
2. S. Basu, R. Pollack, and M.-F. Roy. *Algorithms in Real Algebraic Geometry*. Springer, Berlin-Heidelberg, 2003.
3. T. Brihaye, C. Michaux. On expressiveness and decidability of o-minimal hybrid systems. *J. Complexity*, 21; 447–478, 2005.
4. J. M. Davoren. Topologies, continuity and bisimulations. *ITA*, 33(4/5):357–382, 1999.
5. A. Gabrielov and N. Vorobjov. Complexity of stratifications of semi-Pfaffian sets. *Discrete Comput. Geom.*, 14:71–91, 1995.
6. A. Gabrielov and N. Vorobjov. Complexity of computations with Pfaffian and Noetherian functions. In Yu. Ilyashenko et al., editors, *Normal Forms, Bifurcations and Finiteness Problems in Differential Equations*, volume 137 of *NATO Science Series II*, pages 211–250. Kluwer, 2004.
7. A. Gabrielov, N. Vorobjov, and T. Zell. Betti numbers of semialgebraic and sub-Pfaffian sets. *J. London Math. Soc.*, 69(1):27–43, 2004.
8. M. Hirsch. *Differential Topology*. Springer-Verlag, New York, 1976.
9. A. Khovanskii. *Fewnomials*. Number 88 in Translations of Mathematical Monographs. American Mathematical Society, Providence, RI, 1991.
10. M. Korovina and N. Vorobjov. Pfaffian hybrid systems. In *Springer Lecture Notes in Comp. Sci.*, volume 3210 of *Computer Science Logic'04*, pages 430–441, 2004.
11. M. Korovina and N. Vorobjov. Upper and lower Bounds on Sizes of Finite Bisimulations of Pfaffian Hybrid Systems. In *Technical Report Nr 05-01, Universitat Siegen*, pages 1–16, 2005, www.brics.dk/ korovina/upperbounds.pdf.
12. K. Kurdyka, P. Orro, and S. Symon. Semialgebraic sard theorem for generalized critical values. *J. Differential Geometry*, 56:67–92, 2000.
13. G. Lafferriere, G. Pappas, and S. Sastry. O-minimal hybrid systems. *Math. Control Signals Systems*, 13:1–21, 2000.
14. W. S. Massey. *A Basic Course in Algebraic Topology*. Springer-Verlag, New York, 1991.
15. L. van den Dries. *Tame Topology and O-minimal Structures*. Number 248 in London Mathematical Society Lecture Notes Series. Cambridge University Press, Cambridge, 1998.
16. A.J. Wilkie. Model completeness results for expansions of the ordered field of real numbers by restricted Pfaffian functions and the exponential function. *J. Amer. Math. Soc.*, 9,(4):1051–1094, 1996.
17. T. Zell. Betti numbers of semi-Pfaffian sets. *J. Pure Appl. Algebra*, 139:323–338, 1999.

Forcing with Random Variables
and Proof Complexity

Jan Krajíček[1,2]

[1] Mathematical Institute, Academy of Sciences, Prague
[2] Faculty of Mathematics and Physics, Charles University, Prague

A fundamental problem about the strength of non-deterministic computations is the problem whether the complexity class $\mathcal{NP}$ is closed under complementation. The set $TAUT$ (w.l.o.g. a subset of $\{0,1\}^*$) of propositional tautologies (in some fixed, complete language, e.g. DeMorgan language) is co$\mathcal{NP}$-complete. The above problem is therefore equivalent to asking if there is a non-deterministic polynomial-time algorithm accepting exactly $TAUT$.

Cook and Reckhow (1979) realized that there is a suitably general definition of propositional proof systems that encompasses traditional propositional calculi but links naturally with computational complexity theory. Namely, a propositional proof system is defined to be a binary relation (on $\{0,1\}^*$) $P(x,y)$ decidable in polynomial time such that $x \in TAUT$ iff $\exists y, P(x,y)$. Any y such that $P(x,y)$ is called a P-proof of x.

It is easy to see (viz Cook and Reckhow (1979)) that the fundamental problem becomes a lengths-of-proofs question: Is there a propositional proof system in which every tautology admits a proof whose length is bounded above by a polynomial in the length of the tautology?

Proving lower bounds for particular propositional proof systems appears rather difficult. For example, no non-trivial lower bounds are known even for the ordinary text-book calculus based on a finite number of axiom schemes and inference rules (a Frege system in the terminology of Cook and Reckhow (1979)).

Proof complexity applies methods from logic, from finite combinatorics, from complexity theory (in particular, from circuit complexity, communication complexity, cryptography, or derandomization), from classical algebra (field theory or representation theory of groups), and even borrows abstract geometrical concepts like Euler characteristic or Grothendieck ring.

However, the most stimulating for proof complexity are its multiple connections to bounded arithmetic. In particular, the task of proving lower bounds (for any particular proof system) is equivalent to the task of constructing suitably non-elementary extensions of models of a bounded arithmetic theory (the theory in question depends on the proof system we want lower bounds for). Most lower bounds can be explained very naturally as constructions of such extensions (and some of the most treasured ones were discovered in this way).

In particular, models M to be extended are cuts in models of true arithmetic (they can be "explicitly" obtained as bounded ultrapowers of $\mathbf{N}$). Extensions N of M we are after should preserve polynomial-time properties but should not be elementary w.r.t. $\mathcal{NP}$-properties. There are two things going against each other:

A. Beckmann et al. (Eds.): CiE 2006, LNCS 3988, pp. 277–278, 2006.
© Springer-Verlag Berlin Heidelberg 2006

Under how fast functions is M closed and how strong theory model N satisfies. The former issue influences the rate of the lower bound deduced, the latter one the strength of the proof system for which it is proved.

I shall describe a new method for constructing these extensions. The models are Boolean-valued and are formed by random variables.

References

1. Cook, S. A., and Reckhow, R. A.: The relative efficiency of propositional proof systems. J. Symbolic Logic **44(1)** (1979) 36–50
2. Krajíček, J.: *Bounded arithmetic, propositional logic, and complexity theory.* Encyclopedia of Mathematics and Its Applications, Vol. **60**, Cambridge University Press, (1995)
3. Krajíček, J.: Forcing with random variables. A draft of lecture notes available at http://www.math.cas.cz/~krajicek

Complexity-Theoretic Hierarchies

Lars Kristiansen[1,2]

[1] Oslo University College, Faculty of Engineering
PO Box 4, St. Olavs plass, NO-0130 Oslo, Norway
larskri@iu.hio.no
http://www.iu.hio.no/~larskri
[2] Department of Mathematics, University of Oslo

Abstract. We introduce two hierarchies of unknown ordinal height. The hierarchies are induced by natural fragments of a calculus based on finite types and Gödel's T, and all the classes in the hierarchies are uniformly defined without referring to explicit bounds. Deterministic complexity classes like LOGSPACE, P, PSPACE, LINSPACE and EXP are captured by the hierarchies. Typical subrecursive classes are also captured, e.g. the small relational Grzegorczyk classes $\mathcal{E}_*^0$, $\mathcal{E}_*^1$ and $\mathcal{E}_*^2$.

Keywords: Complexity theory, subrecursive classes, types, λ-calculi, Gödel's T.

1 Introduction

In this paper we introduce two hierarchies. Many of the well-known deterministic complexity classes, e.g. LOGSPACE, P, PSPACE, LINSPACE and EXP, can be found in the hierarchies. These classes are defined by imposing explicit resource bounds on Turing machines, but note that the classes are not uniformly defined as some are defined by imposing *time* bounds, whereas other are defined by imposing *space* bounds. Small subrecursive classes can also be found in our hierarchies, e.g. the relational Grzegorczyk classes $\mathcal{E}_*^0$, $\mathcal{E}_*^1$ and $\mathcal{E}_*^2$. In contrast to a complexity class, a subrecursive class is defined as the least class containing some initial functions and closed under certain composition and recursion schemes. Some of the schemes might contain explicit bounds, but no machine models are involved.

The two hierarchies are induced by neat and natural fragments of a calculus based on finite types and Gödel's T, and all the classes in the hierarchies are uniformly defined without referring to explicit bounds. Thus, one should not expect the hierarchies to capture such a wide variety of classes, that is, both time classes, space classes and subrecursive classes. This indicates that a further investigation of the hierarchies might be rewarding, and perhaps shed light upon some of the notoriously hard open problems involving the classes captured by the hierarchies, e.g. maybe some of these problems turn out to be related in some unexpected way. (We comment on some of these open problems in Section 4 and Section 5) Moreover, the ingredients of the theoretic framework nourishing the hierarchies are well known and thoroughly studied in the literature,

A. Beckmann et al. (Eds.): CiE 2006, LNCS 3988, pp. 279–288, 2006.

e.g. the ordinal numbers, the typed λ-calculi, cut-elimination, rewriting systems and Gödel's T. Advanced and well proven techniques of mathematical logic and computability theory will thus be available facilitating the investigations.

2   Types

Definition. We define the *types* recursively: $\mathbf{q}$ is a type (primitive type); ι is a type (primitive type); $\sigma \oplus \tau$ is a type if σ and τ are types (sum types); $\sigma \otimes \tau$ is a type if σ and τ are types (product types); $\sigma \to \tau$ is a type if σ and τ are types (arrow types). We use TYP to denote the set of all types. We use $\sigma, \sigma' \to \sigma''$ as alternative notation for $\sigma \to (\sigma' \to \sigma'')$. We interpret $\sigma \to \sigma' \to \sigma''$ by associating parentheses to the right, i.e. as $\sigma \to (\sigma' \to \sigma'')$.

We define the *cardinality of type* σ *at base* b, written $|\sigma|_b$, by recursion on the structure of the type σ: $|\mathbf{q}|_b = 1$; $|\iota|_b = b$; $|\rho \oplus \tau|_b = |\rho|_b + |\tau|_b$; $|\rho \otimes \tau|_b = |\rho|_b \times |\tau|_b$; and $|\rho \to \tau|_b = |\tau|_b^{|\rho|_b}$.

A type σ is of *level* n when $\mathrm{lv}(\sigma) = n$ where $\mathrm{lv}(\mathbf{q}) = 0$; $\mathrm{lv}(\iota) = 0$; $\mathrm{lv}(\sigma \oplus \tau) = \max(\mathrm{lv}(\sigma), \mathrm{lv}(\tau))$; $\mathrm{lv}(\sigma \otimes \tau) = \max(\mathrm{lv}(\sigma), \mathrm{lv}(\tau))$; and

$$\mathrm{lv}(\sigma \to \tau) = \begin{cases} \mathrm{lv}(\tau) & \text{if } \exists k \, \forall x \, (\, |\sigma|_x = k \,) \\ \max(\mathrm{lv}(\sigma) + 1, \mathrm{lv}(\tau)) & \text{otherwise.} \end{cases}$$

We define the relation $\prec \subseteq \mathrm{TYP} \times \mathrm{TYP}$ by

$$\sigma \prec \tau \;\Leftrightarrow_{\mathrm{def}}\; \exists x_0 \, \forall x > x_0 \, (\, |\sigma|_x < |\tau|_x \,)$$

and the relation $\preceq \subseteq \mathrm{TYP} \times \mathrm{TYP}$ by $\sigma \preceq \tau \Leftrightarrow_{\mathrm{def}} \sigma \prec \tau \vee \forall x \, (\, |\sigma|_x = |\tau|_x \,)$. $\square$

Skolem [25] conjectures that $\prec$ is a well-ordering of the set TYP, and he asks what the ordinal number of this well-ordering will be. Ehrenfeucht [4] proves that $\prec$ indeed is a well-ordering, and Levitz [19] proves that the least critical epsilon number is an upper bound for the ordinal of the well-ordering. It follows from the results in [25] that if we restrict the arrow types to types of the form $\sigma \to \iota$, then the ordinal will be ϵ_0. If we omit product types, the ordinal will also be ϵ_0. We will develop our theory without any such restrictions, and thus, all we know is that the actual ordinal corresponding to the well-ordering $\prec$, lies somewhere between ϵ_0 and the the least critical epsilon number. (Levitz [19] conjectures that the actual ordinal indeed is ϵ_0.)

3   Calculi

Definition. We define the *terms* of *the typed λ-calculus.*

- We have an infinite supply of variables $x_0^\sigma, x_1^\sigma, x_2^\sigma, \ldots$ for each type σ. A variable of type σ is a term of type σ;
- $\lambda x M$ is a term of type $\sigma \to \tau$ if x is a variable of type σ and M is a term of type τ (*λ-abstraction*)
- (MN) is a term of type τ if M is a term of type $\sigma \to \tau$ and N is a term of type σ (*application*)

- $\langle M, N \rangle$ is a term of type $\sigma \otimes \tau$ if M is a term of type σ and N is a term of type τ *(product)*
- $\mathbf{fst}\,M$ is a term of type σ if M is a term of type $\sigma \otimes \tau$ *(projection)*
- $\mathbf{snd}\,M$ is a term of type τ if M is a term of type $\sigma \otimes \tau$ *(projection)*
- $\mathbf{inl}_\tau M$ is a term of type $\sigma \oplus \tau$ if M is a term of type σ
- $\mathbf{inr}_\tau M$ is a term of type $\tau \oplus \sigma$ if M is a term of type σ
- $\delta(xM, yN, P)$ is a term of type ξ if M and N are terms of type ξ; P is a term of type $\sigma \oplus \tau$; x and y are variables of type σ and τ respectively *(case)*.

We define the reduction rules of the typed λ-calculus. We have the following β-conversions:

- $(\lambda x M)N \rhd M[x := N]$ if $x \notin FV(N)$
- $\mathbf{fst}\langle M, N \rangle \rhd M$ and $\mathbf{snd}\langle M, N \rangle \rhd N$
- $\delta(xM, yN, \mathbf{inl}(P)) \rhd M[x := P]$ if $x \notin FV(P)$
- $\delta(xM, yN, \mathbf{inr}(P)) \rhd N[y := P]$ if $y \notin FV(P)$

Further, we have standard α-conversion and all the other standard reduction rules $(MN) \rhd (MN')$ if $N \rhd N'$; $(MN) \rhd (M'N)$ if $M \rhd M'$; ... etcetera. We will use the standard conventions in the literature and e.g. $F(X, Y)$ means $((FX)Y)$.

The calculus T^- is the typed λ-calculus extended with the constants $q : \mathbf{q}$ and $1 : \iota$, and for each type σ, the *recursor* R_σ of type $\sigma, \iota \to \sigma \to \sigma, \iota \to \sigma$.

The calculus T is the calculus T^- extended with the constants $0 : \iota$ (zero) and $s : \iota \to \iota$ (successor), the reduction rule $1 \rhd s0$, and for each type σ, the reduction rules $R_\sigma(P, Q, 0) \rhd P$ and $R_\sigma(P, Q, sN) \rhd Q(N, R_\sigma(P, Q, N))$.

We use $\overline{n}$ to denote the *numeral* $s^n 0$ where $s^0 0 = 0$ and $s^{n+1}0 = s(s^n 0)$. We will use $\overset{*}{\rhd}$ to denote the transitive-reflexive closure of $\rhd$. □

It is *crucial* that the successor s cannot occur in a T^--term, and the reader should note that the calculus T^- has no reduction rules in addition to those of the standard typed λ-calculus. E.g., the term $R_\sigma(M, N, 1)$ is irreducible in the calculus T^- if M and N are irreducible. Reductions take place in the system T, and $\rhd$ is the standard reduction relation for Gödel's system T.

It is well known that any closed T-term of type ι normalises to a unique numeral. Thus, a closed term M of type $\iota \to \iota$ defines a function $f : \mathbb{N} \to \mathbb{N}$, and the value $f(n)$ can be computed by normalising the term $M\overline{n}$. Any function provably total in Peano Arithmetic is definable in T. (See [1] for more on the T-calculus and Gödel's T.) If we disallow occurrences of the successor s in the defining terms, the class of functions definable is of course severely restricted. (Indeed, at a first glance it is hard to believe that any interesting functions at all can be defined without the successor function.) Roughly speaking, T^- is the calculus T where successors are not admissible in the defining terms.

The constant q should be interpreted as the sole element in the type $\mathbf{q}$.

Definition. A *problem* is a subset of $\mathbb{N}$. A term $M : \iota \to \iota$ *decides* a problem A when $M\overline{n} \overset{*}{\rhd} \overline{0}$ iff $n \in A$. Let T_σ^- denote the set of T^- terms such that $M \in T_\sigma^-$ iff we have $\tau \preceq \sigma$ for every recursor R_τ occurring in M.

We define the set of problems $\mathcal{G}_\sigma$ by $A \in \mathcal{G}_\sigma$ iff A is decided by a T_τ^--term where $\tau \prec \sigma$. We define the hierarchy $\mathcal{G}$ by $\mathcal{G} = \bigcup_{\sigma \in \mathrm{TYP}} \mathcal{G}_\sigma$.

The hierarchy $\mathcal{G}^{\mathbf{b}} = \bigcup_{\sigma \in \mathrm{TYP}} \mathcal{G}_\sigma^{\mathbf{b}}$ is defined as the hierarchy $\mathcal{G}$, with one exception: we use dyadic notation for the numerals, that is, the hierarchy is induced by a calculus with two successors constants $s_0 : \iota \to \iota$ and $s_1 : \iota \to \iota$. The recursor should of course be adjusted accordingly, that is, for any type σ we have the recursor $R_\sigma^{\mathbf{b}} : \sigma, \iota \to \sigma \to \sigma, \iota \to \sigma \to \sigma, \iota \to \sigma$ and the reduction rules $R_\sigma^{\mathbf{b}}(P, Q_1, Q_0, 0) \rhd P$ and $R_\sigma^{\mathbf{b}}(P, Q_1, Q_0, s_i N) \rhd Q_i(N, R_\sigma^{\mathbf{b}}(P, Q_1, Q_0, N))$ (for $i = 0, 1$). □

4   Complexity Classes

Complexity classes are defined by imposing explicit resource bounds on Turing machines. We will assume that the reader is familiar with Turing machines and basic complexity theory. For more on the subject see e.g. Odifreddi [20].

Definition. A Turing machine M *decides* a problem A when M on input $x \in \mathbb{N}$ halts in a distinguished accept state if $x \in A$, and in a distinguished reject state if $x \notin A$. The input $x \in \mathbb{N}$ should be represented in binary on the Turing machine's input tape. We will use $|x|$ to denote the length of the standard binary representation of the natural number x. For $i \in \mathbb{N}$, we define TIME 2_i^{LIN} (SPACE 2_i^{LIN}) to be the set of problem decidable by a deterministic Turing machine working in time (space) $2_i^{c|x|}$ for some fixed $c \in \mathbb{N}$ (where $2_0^x = x$ and $2_{i+1}^x = 2^{2_i^x}$). □

It is trivial that TIME $2_i^{\mathrm{LIN}} \subseteq$ SPACE 2_i^{LIN} and SPACE $2_i^{\mathrm{LIN}} \subseteq$ TIME 2_{i+1}^{LIN}, and thus, we have an *alternating space-time hierarchy*

$$\mathrm{SPACE}\ 2_0^{\mathrm{LIN}} \subseteq \mathrm{TIME}\ 2_1^{\mathrm{LIN}} \subseteq \mathrm{SPACE}\ 2_1^{\mathrm{LIN}} \subseteq \mathrm{TIME}\ 2_2^{\mathrm{LIN}} \subseteq \mathrm{SPACE}\ 2_2^{\mathrm{LIN}} \subseteq \mathrm{TIME}\ 2_3^{\mathrm{LIN}} \subseteq \cdots .$$

The three classes at the bottom of the hierarchy are often called respectively LINSPACE, EXP, and EXPSPACE in the literature. It is well known, and quite obvious, that we have SPACE $2_i^{\mathrm{LIN}} \subset$ SPACE 2_{i+1}^{LIN} and TIME $2_i^{\mathrm{LIN}} \subset$ TIME 2_{i+1}^{LIN} for any $i \in \mathbb{N}$. Thus, we know that at least one of the two inclusions

$$\mathrm{SPACE}\ 2_i^{\mathrm{LIN}} \subseteq \mathrm{TIME}\ 2_{i+1}^{\mathrm{LIN}} \subseteq \mathrm{SPACE}\ 2_{i+1}^{\mathrm{LIN}}$$

are strict, similarly, we know that at least one of the inclusions

$$\mathrm{TIME}\ 2_i^{\mathrm{LIN}} \subseteq \mathrm{SPACE}\ 2_i^{\mathrm{LIN}} \subseteq \mathrm{TIME}\ 2_{i+1}^{\mathrm{LIN}}$$

are strict, and the general opinion is that all they all are. Still, no one has ever been able to prove that any particular of the inclusions actually is strict.

Definition. Let LOGSPACE denote the set of problems decided by a Turing machine working in logarithmic space. Let TIME 2_i^{POL} (SPACE 2_i^{POL}) denote the set of problems decided by a Turing machine working in time (space) $2_i^{p(|x|)}$ for some polynomial p. □

This definition yields another alternating space-time hierarchy

$$\text{LOGSPACE} \subseteq \text{TIME } 2_0^{\text{POL}} \subseteq \text{SPACE } 2_0^{\text{POL}} \subseteq \text{TIME } 2_1^{\text{POL}} \subseteq \text{SPACE } 2_1^{\text{POL}} \subseteq \text{TIME } 2_2^{\text{POL}} \subseteq \ldots$$

analogous to the hierarchy above. The analogous open problems do also emerge. Let C_i, C_{i+1}, C_{i+2} be three arbitrary consecutive classes in the hierarchy. It is well-known that $C_i \subset C_{i+2}$, so at least one of the two inclusions $C_i \subseteq C_{i+1}$ and $C_{i+1} \subseteq C_{i+2}$ will be strict. Still, for any fixed $j \in \mathbb{N}$, it is an open problem if C_j is strictly included in C_{j+1}. Note that $\text{TIME } 2_0^{\text{POL}}$ and $\text{SPACE } 2_0^{\text{POL}}$ are the classes usually denoted respectively P and PSPACE in the literature, so the notorious open problem $\text{LOGSPACE} \overset{?}{\subset} \text{P} \overset{?}{\subset} \text{PSPACE}$ emerges at the bottom of the hierarchy.

The relationship between the two alternating space-time hierarchies is also a bit of a mystery. The only thing known about the relationship between $\text{SPACE } 2_i^{\text{LIN}}$ and $\text{TIME } 2_i^{\text{POL}}$ is that the two classes cannot be equal. So, it is known that e.g. $\text{LINSPACE} \neq \text{P}$, but it is an open problem if LINSPACE is strictly included in P, or if P is strictly included in LINSPACE, or if neither of the two classes is included in the other.

Definition. We will use bold faced natural numbers $\mathbf{0}, \mathbf{1}, \mathbf{2}, \ldots$ to denote the *pure types*, that is, $\mathbf{0} = \iota$ and $\mathbf{n+1} = \mathbf{n} \to \iota$, and we will say that $\mathcal{G}_\mathbf{n}$ (respectively $\mathcal{G}_\mathbf{n}^\mathbf{b}$) is a *pure class* in the hierarchy $\mathcal{G}$ (respectively $\mathcal{G}^\mathbf{b}$).   □

It turns out that the pure classes in the hierarchies $\mathcal{G}$ and $\mathcal{G}^\mathbf{b}$ match the classes in the alternating time-space hierarchies. We state the next lemma without proof.

Lemma 1. *We have* $\text{lv}(\sigma) = n$ *iff* $\sigma \prec \mathbf{n+1}$.

The alternating space-time hierarchies have enjoyed some attention from researchers in finite model theory. Goerdt & Seidel [7] (and Goerdt [8]) use finite models to characterise the latter of the hierarchies. Inspired by Goerdt & Seidel's work, Kristiansen & Voda [16] show that the two hierarchies match, level by level, hierarchies induced by a successor-free fragment of Gödel's T. The next theorem follows straightforwardly from Lemma 1 and the results proved in [16]. The theorem also follows from results proved in Kristiansen & Voda [17]. The proofs in [17] are based on an adaption of Schwichtenberg's Trade-off Theorem to a complexity-theoretic context and are essentially different from those in [16]. For more on Schwichtenberg's Theorem see [24].

Theorem 1. *The pure classes in the hierarchy $\mathcal{G}$ match the classes in the alternating space-time hierarchy starting with LINSPACE, that is, we have*

$$\text{SPACE } 2_i^{\text{LIN}} = \mathcal{G}_{\mathbf{2i}} \quad and \quad \text{TIME } 2_{i+1}^{\text{LIN}} = \mathcal{G}_{\mathbf{2i+1}}$$

for any $i \in \mathbb{N}$. The pure classes in the hierarchy $\mathcal{G}^\mathbf{b}$ match the classes in the alternating space-time hierarchy starting with LOGSPACE, that is, LOGSPACE = $\mathcal{G}_\mathbf{0}^\mathbf{b}$ and

$$\text{SPACE } 2_i^{\text{POL}} = \mathcal{G}_{\mathbf{2i+2}}^\mathbf{b} \quad and \quad \text{TIME } 2_i^{\text{POL}} = \mathcal{G}_{\mathbf{2i+1}}^\mathbf{b}$$

for any $i \in \mathbb{N}$.

5 Subrecursive Classes

Definition. We will use some notation and terminology from Clote [3]. An *operator*, here also called *(definition) scheme*, is a mapping from functions to functions. Let $\mathcal{X}$ be a set of functions (possibly given in a slightly informal notation), and let OP be a collection of operators. The *function algebra* $[\mathcal{X}; \text{OP}]$ is the smallest set of functions containing $\mathcal{X}$ and closed under the operations of OP. COMP denotes the definition scheme called *composition*, i.e. the scheme $f(\vec{x}) = h(g_1(\vec{x}), \ldots, g_m(\vec{x}))$ where $m \geq 0$. BR denotes the scheme *bounded (primitive) recursion*, i.e. the scheme

$$f(\vec{x}, 0) = g(\vec{x}) \qquad f(\vec{x}, y+1) = h(\vec{x}, y, f(\vec{x}, y)) \qquad f(\vec{x}, y) \leq j(\vec{x}, y)$$

Let S denote the successor function, and let I_i^n denote the projection function, i.e. $I_i^n(\vec{x}) = x_i$ where $\vec{x} = x_1, \ldots, x_n$ and $1 \leq i \leq n$. Let I denote the set of all such projection functions. The small Grzegorczyk classes $\mathcal{E}^0, \mathcal{E}^1$ and $\mathcal{E}^2$ are defined by $\mathcal{E}^0 = [I, 0, S; \text{COMP}, \text{BR}]$, $\mathcal{E}^1 = [I, 0, S, +; \text{COMP}, \text{BR}]$ and $\mathcal{E}^2 = [I, 0, S, +, x^2 + 2; \text{COMP}, \text{BR}]$.

A unary number-theoretic function f decides the problem A when $f(x) = 0$ iff $x \in A$. For any set $\mathcal{F}$ of number-theoretic functions $\mathcal{F}_*$ denotes the set of problem decided by the functions in $\mathcal{F}$.

A problem A is *rudimentary* when there exist a Δ_0^0 statement $\phi(x)$ in Peano Arithmetic such that $x \in A$ iff $\mathbb{N} \models \phi(x)$, and $\Delta_0^{\mathbb{N}}$ denotes the class of rudimentary problems. □

Our use of the $*$ subscript differs slightly from the literature standard. Normally, $\mathcal{F}_*$ denotes the 0-1 valued functions in $\mathcal{F}$ whereas we use $\mathcal{F}_*$ to denote the set of problem decided by the functions in $\mathcal{F}$. This is a matter of convenience, and the deviation has no essential mathematical implications. Our definitions of the Grzegorczyk classes are the ones given in Rose [23]. Grzegorczyk [9] original definitions are slightly different, but yield the same classes of functions. The next lemma states some important and well known properties of the small Grzegorczyk classes. The proofs can be found in Rose [23].

Lemma 2. *(i) For any $f \in \mathcal{E}^0$ there exist fixed numbers i, k where $1 \leq i \leq n$ such that $f(x_1, \ldots, x_n) \leq x_i + k$. (ii) For any $f \in \mathcal{E}^1$ there exists a fixed number k such that $f(\vec{x}) \leq k \max(\vec{x}, 1)$. (iii) For any $f \in \mathcal{E}^2$ there exists a polynomial p such that $f(\vec{x}) \leq p(\vec{x})$.*

The next lemma is a consequence of Lemma 2 and will be used to prove the main result of this section.

Lemma 3. *(i) For any $f \in \mathcal{E}^1$ there exist $f' \in \mathcal{E}^0$ and fixed $k \in \mathbb{N}$ such that $f(\vec{x}) = f'(y, \vec{x})$ for any $y \geq k \max(\vec{x}, 1)$ (ii) For any $f \in \mathcal{E}^2$ there exist $f' \in \mathcal{E}^0$ and a fixed polynomial p such that $f(\vec{x}) = f'(y, \vec{x})$ for any $y \geq p(\vec{x})$. Moreover, in both (i) and (ii) we have $f'(y, \vec{x}) \leq y$.*

Proof. Assume $f \in \mathcal{E}^1$. It follows from Lemma 2 (ii) that there exists a fixed $k \in \mathbb{N}$ such that $f \in [I, 0, S, +_z, k \max(\vec{x}, 1); \text{COMP}, \text{BR}]$ where the ternary function

$+_z$ is addition modulo $z+1$, that is, $x +_z y = x + y \pmod{z+1}$. Now, $+_z \in \mathcal{E}^0$, and hence, it is easy to prove by induction over the build-up of f form the functions in $[I, 0, S, +_z, k \max(\vec{x}, 1); \text{COMP}, \text{BR}]$ that there exists a function f' with the required properties. Thus, (i) holds. The proof of (ii) is similar; use Lemma 2 (iii) and that the function $x \times y \pmod{z+1}$ belongs to $\mathcal{E}^0$.   □

It is well know, and rather obvious, that $\Delta_0^{\text{N}} \subseteq \mathcal{E}_*^0 \subseteq \mathcal{E}_*^1 \subseteq \mathcal{E}_*^2$, but it is not known whether any of the inclusions are strict, indeed it is open if the inclusion $\Delta_0^{\text{N}} \subseteq \mathcal{E}_*^2$ is strict. It is proved in Bel'tyukov [2] that $\mathcal{E}_*^1 = \mathcal{E}_*^2$ implies $\mathcal{E}_*^0 = \mathcal{E}_*^2$. Furthermore, we know that $\Delta_0^{\text{N}} = \mathcal{E}_*^0$ implies $\Delta_0^{\text{N}} = \mathcal{E}_*^2$ (see Kristiansen & Barra [13]). The open problems can be traced back to Grzegorczyk's initial paper [9] from 1953. For more on the Grzegorczyk classes and the rudimentary relations see Clote [3], Rose [23], Kutylowski [18], Esbelin & More [5], Gandy [6], Paris & Willie [21], Kristiansen & Barra [13].

Definition. Let BCOMP denote the definition scheme called *bounded composition*, i.e. the scheme

$$f(\vec{x}) = h(g_1(\vec{x}), \dots, g_m(\vec{x})) \qquad f(\vec{x}) \le j(\vec{z})$$

where every variable in the list $\vec{z}$ occurs in the list $\vec{x}$. (For technical reasons we cannot just state the bound as $f(\vec{x}) \le j(\vec{x})$. All the variables in a bound should be considered universally quantified, and e.g. the bound $f(x, y) \le j(x)$ should hold for all values of y.) We define the *bounded successor* function $\hat{S}$ by $\hat{S}(x, y) = x + 1$ if $x < y$; otherwise $\hat{S}(x, y) = y$. For each type σ, we define the function $\hat{\sigma}$ by $\hat{\sigma}(x) = |\sigma|_{\max(x,1)+1} - 1$.   □

Theorem 2. *We have* $\mathcal{G}_{\sigma \oplus \mathbf{q}} = [I, 0, \hat{S}, \hat{\sigma}; \text{BCOMP}, \text{BR}]_*$ *for any type* σ *such that* $0 \prec \sigma \prec 1$.

We are now ready to state and prove one of the main theorems of this paper. Clause (ii), (iii) and (iv) of the theorem are corollaries of Theorem 2. The long, and occasionally very technical, proof of Theorem 2 can be found in Kristiansen [11].

Theorem 3. *(i)* $\Delta_0^{\text{N}} \subseteq \mathcal{G}_{\iota \oplus \mathbf{q}}$. *(ii)* $\mathcal{G}_{\iota \oplus \iota} = \mathcal{E}_*^0$. *(iii)* $\mathcal{G}_{\iota \otimes \iota} = \mathcal{E}_*^1$. *(iv)* $\mathcal{G}_1 = \mathcal{E}_*^2$.

Proof. First we prove that we have the equivalence

$$f \in [I, 0, \hat{S}, x + k; \text{BCOMP}, \text{BR}] \text{ for some } k \in \mathbb{N} \quad \Leftrightarrow \quad f \in \mathcal{E}^0 . \qquad (*)$$

The left-right implication is trivial. To prove the right-left implication, let $f \in \mathcal{E}^0$, and chose any definition of f which witness membership in the function algebra $[I, 0, S; \text{COMP}, \text{BR}] = \mathcal{E}^0$ Let $f_1, \dots, f_m$ be the functions involved in the definition. Let us say that $f = f_m$. By Lemma 2 we have fixed $i_0, k_i \in \mathbb{N}$ such that $f_i(x_1, \dots, x_n) \le x_{i_0} + k_i$ where $1 \le i_0 \le n$ (for $i = 1, \dots, m$). Let $k = \max(k_1, \dots, k_m)$. It is easy to see that f can be defined in the function algebra $[I, 0, \hat{S}, x + k; \text{BCOMP}, \text{BR}]$, and thus $(*)$ holds.

To see that (ii) holds, note that for every $k \in \mathbb{N}$ there exists a type σ such that $\sigma \prec \iota \oplus \iota$ and $x + k < \hat{\sigma}(x)$; and for every σ such that $\sigma \prec \iota \oplus \iota$ there exists $k \in \mathbb{N}$ such that $\hat{\sigma}(x) < x + k$. Hence, (ii) follows from (*) and Theorem 2.

We prove (iii). Let A be a problem in $\mathcal{E}_*^1$. Thus, there exists $f_A \in \mathcal{E}^1$ deciding A. By Lemma 3 (i) there exists $f' \in \mathcal{E}^0$ and fixed $k \in \mathbb{N}$ such that $f_A(x) = f'(y, x)$ for any $y \geq k \max(x, 1)$. Furthermore we have $f'(y, x) \leq y$. There exists a type σ such that $k \max(x, 1) \leq \hat{\sigma}(x)$ and $\sigma \prec \iota \otimes \iota$, and since $f' \in \mathcal{E}^0$, it follows easily from (*) that $f' \in [I, 0, \hat{S}, \hat{\sigma}; \text{BCOMP}, \text{BR}]$. Now, the function algebra is closed under bounded composition, and hence we also have $f_A \in [I, 0, \hat{S}, \hat{\sigma}; \text{BCOMP}, \text{BR}]$ since $f_A(x) = f'(\hat{\sigma}(x), x) \leq \hat{\sigma}(x)$. By Theorem 2 we have $A \in \mathcal{G}_{\iota \otimes \iota}$. This proves $\mathcal{E}_*^1 \subseteq \mathcal{G}_{\iota \otimes \iota}$.

Let A be a problem in $\mathcal{G}_{\iota \otimes \iota}$. Thus, there exist a type σ and a function f_A deciding A such that $\sigma \prec \iota \otimes \iota$ and $f_A \in [I, 0, \hat{S}, \hat{\sigma}; \text{BCOMP}, \text{BR}]$. Since $\sigma \prec \iota \otimes \iota$ there exists a fixed $k \in \mathbb{N}$ such that $\hat{\sigma}(x) \leq k \max(x, 1)$. Now, it is easy to see that $f_A \in [I, 0, \hat{S}, k \max(x, 1); \text{BCOMP}, \text{BR}]$. Furthermore, it is easy to prove that $[I, 0, \hat{S}, k \max(x, 1); \text{BCOMP}, \text{BR}] \subseteq \mathcal{E}^1$. Hence, $f_A \in \mathcal{E}^1$, and thus $A \in \mathcal{E}_*^1$. This proves $\mathcal{G}_{\iota \otimes \iota} \subseteq \mathcal{E}_*^1$.

The proof of (iv) is similar to the proof of (iii). Use Lemma 3 (ii) in place of Lemma 3 (i); use the fact that for any polynomial $p(x)$ the exists a type σ such that $p(x) \leq \hat{\sigma}(x)$ and $\sigma \prec \iota \to \iota$; and use the fact that for any type σ such that $\sigma \prec \iota \to \iota$ there exists a polynomial $p(x)$ such that $\hat{\sigma}(x) \leq p(x)$. (i) follows straightforwardly from results proved in Kristiansen & Barra [13].   □

It is well known that LINSPACE $= \mathcal{E}_*^2$ (Ritchie [22]), and hence, Clause (iv) of Theorem 3 also follows from Theorem 1. The reader should note that the proof of Theorem 1 is based on Turing machines whereas the proof of Theorem 3 makes no detours via computations by machine models.

We expect a wide variety of more or less natural subrecursive classes to be captured by our hierarchies, and hence, the hierarchies might turn out as apt tools for analysing the relationship between subrecursive classes and complexity-theoretic classes.

6   Nondeterminism

In this section we discuss a notion of nondeterminism that might be worth further study. The basic idea is very simple: Let $\tilde{T}^-$ be the calculus T^- extended by

- $(M|N)$ is a term of type σ if M and N are terms of type σ
- $(M|N) \triangleright M$ and $(M|N) \triangleright N$.

All the definitions in the preceding sections will still make sense when the calculus $\tilde{T}^-$ replaces the calculus T^-.

If $\mathcal{C}$ denotes a class of problems induced by T^--terms, let $\tilde{\mathcal{C}}$ denotes the corresponding class induced by $\tilde{T}^-$-terms. Recall Theorem 1 states e.g. that $\mathcal{G}_2^b = \text{PSPACE}$, that $\mathcal{G}_1^b = \text{P}$, and that $\mathcal{G}_0^b = \text{LOGSPACE}$. Will it be the case that $\tilde{\mathcal{G}}_1^b = \text{NP}$ and that $\tilde{\mathcal{G}}_0^b = \text{NLOGSPACE}$? Presumably it is, but this immediate conjecture does definitely require a meticulous proof. Also, since $\mathcal{G}_2^{\mathsf{b}} = \mathrm{PSPACE} = \mathrm{NPSPACE}$, and since we presumably can prove $\tilde{\mathcal{G}}_2^{\mathsf{b}} = \mathrm{NPSPACE}$, we presumably have $\mathcal{G}_2^{\mathsf{b}} = \tilde{\mathcal{G}}_2^{\mathsf{b}}$. Still, it seems like e.g. $\mathcal{G}_{2\oplus\iota}^{\mathsf{b}} \neq \tilde{\mathcal{G}}_{2\oplus\iota}^{\mathsf{b}}$. For which types σ can we prove $\mathcal{G}_\sigma^{\mathsf{b}} = \tilde{\mathcal{G}}_\sigma^{\mathsf{b}}$, and for which types σ can we prove $\mathcal{G}_\sigma^{\mathsf{b}} \neq \tilde{\mathcal{G}}_\sigma^{\mathsf{b}}$? We do indeed have a nondeterministic version of any class in our hierarchies, even those classes we cannot characterise by imposing natural resource bounds on Turing machines. In particular, we have nondeterministic versions of the small Grzegorczyk classes $\mathcal{E}_*^0$ and $\mathcal{E}_*^1$. How do these nondeterministic Grzegorczyk classes fit into the picture?

7   References to Related Research

Some years ago the author (and others, e.g. Jones [10]) discovered that interesting things tend to happen when successor-like functions are removed from a standard computability theoretic framework. The present paper is the last in a series of papers investigating successor-free models of computation. In [14] we characterise well-known complexity classes, like e.g. LOGSPACE and P, by fragments of a first order imperative programming language, and in [12] we give function algebraic characterisations of LOGSPACE and LINSPACE.

Imperative and functional programming languages embodying higher types are investigated in [15] and [16]. The system T^- is introduced in [15], but no T^--hierarchies are introduced. Two T^--hierarchies of ordinal height ω are introduced in [16], and it is proved that these hierarchies capture the alternating space-time hierarchies discussed in Section 4. In [17] we relate the functionals of T^- to the Kleene-Kreisel functionals and undertake a further study of the hierarchies introduced in [16].

In [13] we study a T^--hierarchy of ordinal height ω where the classes in the hierarchy adds up to $\mathcal{E}_*^2$. Neither $\mathcal{E}_*^0$ nor $\mathcal{E}_*^1$ are captured by the hierarchy.

The full hierarchies induced by admitting both arrow types, product types and sum types, are introduced for the first time in the present paper. The main original technical result of this paper is the theorem stating that $\mathcal{E}_*^0$, $\mathcal{E}_*^1$ and $\mathcal{E}_*^2$ are captured by the hierarchy $\mathcal{G}$. A fairly complete proof of the theorem is available in [11].

References

1. Avigad, J., Feferman, S.: Gödel's functional interpretation. In Buss, S., ed.: Handbook of Proof Theory. Elsevier (1998)
2. Bel'tyukov, A.: A machine description and the hierarchy of initial Grzegorczyk classes. J. Soviet Math. (1982) Zap. Naucn. Sem. Leninigrad. Otdel. May. Inst. Steklov. (LOMI) 88 (1979) 30-46.
3. Clote, P.: Computation models and function algebra. In Griffor, E., ed.: Handbook of Computability Theory. Elsevier (1996)
4. Ehrenfeucht, A.: Polynomial functions with exponentiation are well ordered. Algebra Universalis 3 (1973) 261–262
5. Esbelin, M.A., More, M.: Rudimentary relations and primitive recursion: A toolbox. Theoretical Computer Science 193 (1998) 129–148

6. Gandy, R.: Some relations between classes of low computational complexity. Bulletin of London Mathematical Society (1984) 127–134

7. Goerdt, A., Seidl, H.: Characterizing complexity classes by higher type primitive recursive definitions, part II. In: Aspects and prospects of theoretical computer science. Volume 464 of LNCS., Springer (1990) 148–158 Smolenice, 1990.

8. Goerdt, A.: Characterizing complexity classes by higher type primitive recursive definitions. Theoretical Computer Science 100(1) (1992) 45–66

9. Grzegorczyk, A.: Some classes of recursive functions. Rozprawy Matematyczne, No. IV (1953) Warszawa.

10. Jones, N.: The expressive power of higher-order types or, life without CONS. Journal of Functional Programming 11 (2001) 55–94

11. Kristiansen, L.: Appendix to the present paper. Available from the author's home page http://www.iu.hio.no/~larskri (2006)

12. Kristiansen, L.: Neat function algebraic characterizations of LOGSPACE and LINSPACE. Computational Complexity 14(1) (2005) 72–88

13. Kristiansen, L., Barra, G.: The small Grzegorczyk classes and the typed λ-calculus. In: CiE 2005: New Computational Paradigms. Volume 3526 of LNCS., Springer-Verlag (2005) 252–262

14. Kristiansen, L., Voda, P.: Complexity classes and fragments of C. Information Processing Letters 88 (2003) 213–218

15. Kristiansen, L., Voda, P.: The surprising power of restricted programs and gödel's functionals. In Baaz, M., Makowsky, J., eds.: CSL 2003: Computer Science Logic. Volume 2803 of LNCS., Springer (2003)

16. Kristiansen, L., Voda, P.: Programming languages capturing complexity classes. Nordic Journal of Computing 12 (2005) 1–27 Special issue for NWPT'04.

17. Kristiansen, L., Voda, P.: The trade-off theorem and fragments of gödel's T. In: TAMC'06: Theory and Applications of Models of Computation. Volume 3959 of LNCS., Springer-Verlag (2006)

18. Kutylowski, M.: Small Grzegorczyk classes. Journal of the London Mathematical Society 36(2) (1987) 193–210

19. Levitz, H.: An ordinal bound for the set of polynomial functions with exponentiation. Algebra Universalis 8 (1978) 233–244

20. Odifreddi, P.: Classical recursion theory. Vol. II. North-Holland Publishing Co., Amsterdam (1999)

21. Paris, J., Wilkie, A.: Counting problems in bounded arithmetic. In: Methods in mathematical logic. Volume 1130 of Lecture Notes in Mathematics., Springer (1985) 317–340 Proceedings, Caracas 1983.

22. Ritchie, R.W.: Classes of predictably computable functions. Transactions of the American Mathematical Society 106 (1963) 139–173

23. Rose, H.: Subrecursion. Functions and hierarchies. Clarendon Press (1984)

24. Schwichtenberg, H.: Classifying recursive functions. In Griffor, E., ed.: Handbook of computability theory. Elsevier (1996) 533–586

25. Skolem, T.: An ordered set of arithmetic functions representing the least ε-number. Det Kongelige Norske Videnskabers Selskabs Forhandlinger 29(12) (1956) 54–59

Undecidability in the Homomorphic Quasiorder of Finite Labeled Forests

Oleg V. Kudinov[1,*] and Victor L. Selivanov[2,**]

[1] S.L. Sobolev Institute of Mathematics
Siberian Division of the Russian Academy of Sciences
kud@math.nsc.ru
[2] A.P. Ershov Institute of Informatics Systems
Siberian Division of the Russian Academy of Sciences
vseliv@nspu.ru

Abstract. We prove that the homomorphic quasiorder of finite k-labe-led forests has undecidable elementary theory for $k \geq 3$, in contrast to the known decidability result for $k = 2$. We establish also undecidablity (again for every $k \geq 3$) of elementary theories of two other relevant structures: the homomorphic quasiorder of finite k-labeled trees, and of finite k-labeled trees with a fixed label of the root element.

Keywords: Tree, labeled tree, forest, homomorphic quasiorder, undecidability, elementary theory.

1 Introduction

In [Se04] (see also [H96]) the stucture $(\mathcal{F}_k; \leq)$, $k < \omega$, of finite k-labeled forests with the homomorphic quasiorder was studied. The structure is interesting in its own right since the homomorphic quasiorder is one in a series relations on words, trees and forests relevant to computer science (see [Ku06] and references therein). The original interest to this structure [Se04] was motivated by its close relationship to the Boolean hierarchy of k-partitions [Ko00, KW00]. Throughout this paper, k denotes an arbitrary integer, $k \geq 2$, which is identified with the set $\{0, \ldots, k-1\}$.

As was observed in [Ku06], elementary theory $Th(\mathcal{F}_k; \leq)$ (and even the monadic second order theory) of this structure is decidable for $k = 2$. For $k > 2$ the question on decidability of this structure was left open in [Ku06]. In this paper we solve this question (and a couple of relevant questions) in the negative. Next we recall some necessary definitions and formulate our main result.

We use some standard notation and terminology on posets which may be found in any book on the subject, see e.g. [DP94]. We will not be very cautious when applying notions about posets also to quasiorders (known also as preorders); in such cases we mean the corresponding quotient-poset of the quasiorder.

A poset $(P; \leq)$ will be often shorter denoted just by P (this applies also to structures of other signatures in place of $\{\leq\}$). Any subset of P may be

---

* Partially supported by RFBR grant 05-01-00819a.
** Partially supported by a DAAD project within the program "Ostpartnerschaften".

A. Beckmann et al. (Eds.): CiE 2006, LNCS 3988, pp. 289–296, 2006.

considered as a poset with the induced partial ordering. In particular, this applies to the "cones" $\check{x} = \{y \in P | x \leq y\}$ and $\hat{x} = \{y \in P | y \leq x\}$ defined by any $x \in P$.

By *a forest* we mean a finite poset in which every lower cone $\hat{x}$ is a chain. A *tree* is a forest having the least element (called *the root* of the tree). Note that any forest is uniquely representable as a disjoint union of trees, the roots of the trees being the minimal elements of the forest. A *proper forest* is a forest which is not a tree. Notice that our trees and forests "grow bottom up", like the natural ones while trees in [Se04, Se06] grow in the opposite direction.

A *k-labeled poset* (or just a *k-poset*) is an object $(P; \leq, c)$ consisting of a poset $(P; \leq)$ and a labeling $c : P \to k$. Sometimes we simplify notation of a *k*-poset to (P, c) or even to P. A *morphism* $f : (P; \leq, c) \to (P'; \leq', c')$ between *k*-posets is a monotone function $f : (P; \leq) \to (P'; \leq')$ respecting the labelings, i.e. satisfying $c = c' \circ f$.

Let $\mathcal{F}_k$ and $\mathcal{T}_k$ be the classes of all finite *k*-forests and finite *k*-trees, respectively. Define [Ko00, KW00] a quasiorder $\leq$ on $\mathcal{F}_k$ as follows: $(P, c) \leq (P', c')$, if there is a morphism from (P, c) to (P', c'). By $\equiv$ we denote the equivalence relation on $\mathcal{F}_k$ induced by $\leq$. For technical reasons we consider also the empty *k*-forest $\emptyset$ (which is not assumed to be a tree) assuming that $\emptyset \leq P$ for each $P \in \mathcal{F}_k$. Note that in this paper (contrary to notation in [Se04]) we assume that $\emptyset \in \mathcal{F}_k$.

For arbitrary finite *k*-trees $T_0, \ldots, T_n$, let $F = T_0 \sqcup \cdots \sqcup T_n$ be their join, i.e. the disjoint union. Then F is a *k*-forest whose trees are exactly $T_0, \ldots, T_n$. Of course, every *k*-forest is (equivalent to) the join of its trees. Note that the join operation applies also to *k*-forests, and the join of any two *k*-forests is clearly their supremum under $\leq$. Hence, $(\mathcal{F}_k; \leq)$ is an upper semilattice.

For every finite *k*-forest F and every $i < k$, let $p_i(F)$ be the *k*-tree obtained from F by joining a new smallest element and assigning the label i to this element. In particular, $p_i(\emptyset)$ will be the singleton tree carrying the label i. In [Se06] some interesting properties of the operations $p_0, \ldots, p_{k-1}$ were established.

For each $i < k$, let $\mathcal{T}_k^i$ be the set of finite *k*-trees the roots of which carry the label i. Our interest to the sets $\mathcal{F}_k$, $\mathcal{T}_k$ and $\mathcal{T}_k^i$ is explained by the above-mentioned relation to the Boolean hierarchy of *k*-partitions. Namely, the sets $\mathcal{T}_k^i$ and $\mathcal{F}_k \setminus \mathcal{T}_k$ generalize respectively Σ- (and Π-) levels and Δ-levels of the Boolean hierarchy of *k*-partitions.

The main result of this paper is now formulated as follows.

Theorem 1. *For all $k > 2$ and $i < k$, the elementary theories of the quotient structures of $(\mathcal{F}_k; \leq)$, $(\mathcal{T}_k^i; \leq)$ and $(\mathcal{T}_k; \leq)$ are undecidable.*

In Section 2 we describe a general scheme of proving the three undecidability results. In Sections 3, 4 and 5 we prove the three results one by one. We conclude in Section 6 with mentioning some of remaining open questions.

2   Interpretation Scheme

In this section we isolate a general part in the proof of all the three undecidability results.

As is well known [E+65] (see also [E65]), for establishing undecidability of elementary theory of a structure, say of a partial order $(P; \leq)$, it suffices to show that the class of finite models of the theory of two equivalence relations is respectively elementary definable (or interpretable) in $(P; \leq)$ with parameters. It turns out that the following very particular interpretation scheme will work in our case.

It suffices to find first order formulas $\phi_0(x, \bar{p})$, $\phi_1(x, y, \bar{p})$ and $\phi_2(x, y, p)$ of signature $\{\leq\}$ (where x, y are variables and $\bar{p}$ is a string of variables called parameters) with the following property:

(*) for every $n < \omega$ and for all equivalence relations ξ, η on $\{0, \ldots, n\}$ there are values of parameters $\bar{p} \in P$ such that the structure $(\{0, \ldots, n\}; \xi, \eta)$ is isomorphic to the structure $(\phi_0(P, \bar{p}); \phi_1(P, \bar{p}), \phi_2(P, \bar{p}))$.

Here

$$\phi_0(P, \bar{p}) = \{a \in P | (P; \leq) \models \phi_0(a, \bar{p})\},$$

$$\phi_1(P, \bar{p}) = \{(a, b) \in P | (P; \leq) \models \phi_1(a, b, \bar{p})\}$$

and similarly for ϕ_2. So for each of the quotient structures $(\mathcal{F}_k; \leq)$, $(\mathcal{T}_k^i; \leq)$ and $(\mathcal{T}_k; \leq)$ it remains only to find suitable formulas ϕ_0, ϕ_1, ϕ_2 and to specify parameter values as described in (*).

3   Undecidability of $\mathcal{F}_k$

Before going to the proof of undecidability of $Th(\mathcal{F}_k; \leq)$ we recall some known facts about this structure established in [Se04].

Proposition 1. *For every $k \geq 2$, the quotient structure of $(\mathcal{F}_k; \leq)$ is a distributive lattice in which the non-zero join-irreducible elements are exactly the elements defined by the finite k-trees.*

In [Ko00] it was observed (this is actually an easy exercise) that for all $k > 2$ and $n < \omega$ there are repetition-free k-chains $C_0, \ldots, C_n$ (i.e. repetition-free words over the alphabet k) of the same length which are pairwise incomparable under $\leq$ and have the same label 0 on the roots.

Proof of Undecidability of $Th(\mathcal{F}_k; \leq)$. Let $ir(x)$ be a formula of signature $\{\leq\}$ which defines in every lattice exactly the non-zero join-irreducible elements. Such a formula is written easily in the signature $\{0, \leq, \cup\}$, namely

$$x \neq 0 \wedge \forall y \forall z (x \leq y \cup z \rightarrow (x \leq y \vee x \leq z)).$$

Since 0 and $\cup$ are first order definable in signature $\{\leq\}$ the last formula may be rewritten as an equivalent formula of $\{\leq\}$.

Let $\phi(x, u)$ be the formula

$$x \leq u \wedge ir(x) \wedge \neg \exists y > x (y \leq u \wedge ir(y))$$

which means that x is a maximal non-zero join-irreducible element below u. From Proposition 1 it follows that if u is a nonempty k-forest with pairwise incomparable trees then the set $\phi(\mathcal{F}_k, u)$ (see Section 2) coincides with the set of trees of u. In particular, if $u = C_0 \sqcup \cdots \sqcup C_n$ then

$$\phi(\mathcal{F}_k, u) = \{C_0, \ldots, C_n\}. \tag{1}$$

Let $\psi(x, y, u, v)$ be the formula

$$\phi(x, u) \wedge \phi(y, u) \wedge \exists t (\phi(t, v) \wedge x \leq t \wedge y \leq t).$$

To illustrate the meaning of ψ for an important particular case, fix the following values of parameters u, v in $\mathcal{F}_k$:

$$u = C_0 \sqcup \cdots \sqcup C_n, \quad v = p_1(\bigsqcup_{i \in \xi_0} C_i) \sqcup \cdots \sqcup p_1(\bigsqcup_{i \in \xi_m} C_i), \tag{2}$$

where $\xi \subseteq (n+1)^2$ is an equivalence relation on $n + 1 = \{0, \ldots, n\}$ and $(\xi_0, \ldots, \xi_m)$ is the partition of $n + 1$ to ξ-equivalence classes. By observation from the last paragraph, $\mathcal{F}_k \models \psi(x, y, u, v)$ iff $x = C_i$ and $y = C_j$ for some unique $i, j \leq n$ such that $C_i, C_j \leq t$ for some $t = p_1(\sqcup_{i \in \xi_l} C_i)$ and $l \leq m$. In other words, $\mathcal{F}_k \models \psi(x, y, u, v)$ iff $x = C_i$ and $y = C_j$ for some ξ-equivalent $i, j \leq n$. Therefore, for the values (2) we have

$$\psi(\mathcal{F}_k, u, v) = \{(C_i, C_j) | (i, j) \in \xi\}. \tag{3}$$

Now let $\bar{p}$ be the string of variables u, v, w, $\phi_0(x, \bar{p})$ be $\phi(x, u)$, $\phi_1(x, y, \bar{p})$ be $\psi(x, y, u, v)$ and $\phi_2(x, y, \bar{p})$ be $\psi(x, y, u, w)$ (the last formula is obtained from $\psi(x, y, u, v)$ by substituting w in place of v). We claim that formulas ϕ_0, ϕ_1, ϕ_2 satisfy the condition (*) from Section 2 for $P = \mathcal{F}_k$. Let equivalence relations ξ, η on $n + 1$ be given. Specify values of the parameters u, v as in (2), and the value of parameter w as

$$w = p_1(\bigsqcup_{i \in \eta_0} C_i) \sqcup \cdots \sqcup p_1(\bigsqcup_{i \in \eta_l} C_i),$$

where $(\eta_0, \ldots, \eta_l)$ is the partition of $n+1$ to η-equivalence classes. From (1) and (3) we obtain

$$\phi_0(\mathcal{F}_k, \bar{p}) = \{C_0, \ldots, C_n\}, \quad \phi_1(\mathcal{F}_k, \bar{p}) = \{(C_i, C_j) | (i, j) \in \xi\}$$

and

$$\phi_2(\mathcal{F}_k, \bar{p}) = \{(C_i, C_j) | (i, j) \in \eta\}.$$

This means that $i \mapsto C_i$ defines an isomorphism of $(n+1; \xi, \eta)$ onto the structure $(\phi_0(\mathcal{F}_k, \bar{p}); \phi_1(\mathcal{F}_k, \bar{p}), \phi_2(\mathcal{F}_k, \bar{p}))$. This completes the proof.

4   Undecidability of $\mathcal{T}_k^i$

Before proving the undecidability result, we recall some necessary facts established in [Se06].

Proposition 2. *(i) For all $i, j < k$, $(T_k^i; \leq)$ is isomorphic to $(T_k^j; \leq)$.*

(ii) For every $i < k$, the quotient structure of $(T_k^i; \leq)$ is a distributive lattice the non-zero join-irreducible elements of which are exactly the elements $p_i p_j(x)$, where $x \in \mathcal{F}_k$ and $j < k$, $j \neq i$. The supremum operation in this lattice is given by $x \cup y = p_i(x \sqcup y)$.

(iii) The operations $\sqcup, p_0, \ldots, p_{k-1}$ have the following properties in $(\mathcal{F}_k; \leq)$:
$x \leq p_i(x)$, $x \leq y \to p_i(x) \leq p_i(y)$ and $p_i(p_i(x)) \leq p_i(x)$;
for all distinct $i, j < k$, $p_i(x) \leq p_j(y) \to p_i(x) \leq y$;
$p_i(x) \leq y \sqcup z \to (p_i(x) \leq y \vee p_i(x) \leq z)$.

As in Section 3, we need for each $n < \omega$ the sequence of k-chains $C_0, \ldots, C_n$. W.l.o.g. we may assume additionally that the second element in every $(C_i; \leq)$ carries the label 1, i.e. $C_i = p_0 p_1(D_i)$ for suitable D_i, $i \leq n$.

Proof of Undecidability of $Th(T_k^i; \leq)$**.** By Proposition 2(i), it suffices to prove undecidability of T_k^0. We claim that the formulas ϕ_0, ϕ_1, ϕ_2 from the proof for $\mathcal{F}_k$ do the job, i.e. satisfy the condition (*) from Section 2 for $P = T_k^0$. Let equivalence relations ξ, η on $n + 1$ be given. We take now the following values for the parameters $(u, v, w) = \bar{p} \in T_k^0$:

$$u = p_0(C_0 \sqcup \cdots \sqcup C_n) = p_0(p_1(D_0) \sqcup \cdots \sqcup p_1(D_n)),$$

$$v = p_0(p_2(\bigsqcup_{i \in \xi_0} D_i) \sqcup \cdots \sqcup p_2(\bigsqcup_{i \in \xi_m} D_i))$$

and

$$w = p_0(p_2(\bigsqcup_{i \in \eta_0} D_i) \sqcup \cdots \sqcup p_2(\bigsqcup_{i \in \eta_l} D_i)).$$

As in the proof for $\mathcal{F}_k$, it remains to check that

$$\phi(T_k^0, u) = \{C_0, \ldots, C_n\}$$

and

$$\psi(T_k^0, u, v) = \{(C_i, C_j) | (i, j) \in \xi\}.$$

The first equation follows from Proposition 2(ii). In checking the second equation we use Proposition 2(iii). Let $T_k^0 \models \psi(x, y, u, v)$. Then $\phi(x, u)$, $\phi(y, u)$ and $x, y \leq t$ for some t with $\phi(t, v)$. Then $x = C_i$, $y = C_j$ for unique $i, j \leq n$ and

$$C_i, C_j \leq t = p_0 p_2(\bigsqcup_{i \in \xi_l} D_i)$$

for a unique $l \leq m$. We have

$$p_0 p_1(D_j) = C_j \leq t = p_0 p_2(\bigsqcup_{i \in \xi_l} D_i),$$

hence $p_1(D_j) \leq p_0 p_2(\sqcup_{i \in \xi_l} D_i)$, hence $p_1(D_j) \leq \sqcup_{i \in \xi_l} D_i$ and therefore $j \in \xi_l$. A similar computation shows $i \in \xi_l$, so $(i, j) \in \xi$. A converse argument settles the inclusion $\{(C_i, C_j) | (i, j) \in \xi\} \subseteq \psi(T_k^0, u, v)$. This completes the proof.

5 Undecidability of $\mathcal{T}_k$

Again before proving the undecidability we cite a fact from [Se06].

Proposition 3. *For every finite sequence $x_0, \ldots, x_n \in \mathcal{T}_k$ there exist elements $u_0, \ldots, u_{k-1} \in \mathcal{T}_k$ with the following properties:*
(i) $\forall i \le n \forall j < k(x_i \le u_j)$;
(ii) for every $x \in \mathcal{T}_k$, $\forall i \le n(x_i \le x) \to \exists j < k(u_j \le x)$;
(iii) for every $x \in \mathcal{T}_k$, $\forall j < k(x \le u_j) \to \exists i \le n(x \le x_i)$.

To see this it suffices to set $u_j = p_j(x_0 \sqcup \cdots \sqcup x_n)$, $j < k$, and apply Proposition 2(iii). Note that if there is no greatest element in $(\{x_0, \ldots, x_n\}; \le)$ then the elements $u_0, \ldots, u_{k-1}$ are pairwise incomparable. Note also that any of the sets $\{x_0, \ldots, x_n\}$, $\{u_0, \ldots, u_{k-1}\}$ is definable through the other; we use this fact below.

Proof of Undecidability of $Th(\mathcal{T}_k; \le)$. We again use the chains $C_0, \ldots, C_n$ from Section 3. We use also the strings $\bar{u} = (u_0, \ldots, u_{k-1})$, $\bar{v} = (v_0, \ldots, v_{k-1})$ and $\bar{w} = (w_0, \ldots, w_{k-1})$ of different variables.

Let $\phi(x, \bar{u})$ be the formula

$$(\bigwedge_{i<k}(x \le u_i)) \wedge \neg \exists y > x(\bigwedge_{i<k}(y \le u_i))$$

which means that x is a maximal lower bound for $\{u_0, \ldots, u_{k-1}\}$. If we fix the values

$$u_j = p_j(C_0 \sqcup \cdots \sqcup C_n) \in \mathcal{T}_k, \quad j < k, \tag{4}$$

of parameters $\bar{u}$ then, by Proposition 3,

$$\phi(\mathcal{T}_k, \bar{u}) = \{C_0, \ldots, C_n\}. \tag{5}$$

Let $\psi(x, y, \bar{u}, \bar{v})$ be the formula

$$\phi(x, \bar{u}) \wedge \phi(y, \bar{u}) \wedge \exists t(\phi(t, \bar{v}) \wedge x \le t \wedge y \le t).$$

Let us fix the values $\bar{u}$ as in (4), and values of $\bar{v} \in \mathcal{T}_k$ as follows:

$$v_j = p_j(p_0(\bigsqcup_{i \in \xi_0} C_i) \sqcup \cdots \sqcup p_0(\bigsqcup_{i \in \xi_m} C_i)), \quad j < k, \tag{6}$$

where ξ is an equivalence relation on $n + 1$ and $(\xi_0, \ldots, \xi_m)$ is the partition of $n + 1$ to ξ-equivalence classes. From Propositions 2(iii) and 3 it follows that for these values we have

$$\psi(\mathcal{T}_k, \bar{u}, \bar{v}) = \{(C_i, C_j) | (i, j) \in \xi\}. \tag{7}$$

Now let $\bar{p}$ be the string of $3k$ variables $(\bar{u}, \bar{v}, \bar{w})$, $\phi_0(x, \bar{p})$ be $\phi(x, \bar{u})$, $\phi_1(x, y, \bar{p})$ be $\psi(x, y, \bar{u}, \bar{v})$ and $\phi_2(x, y, \bar{p})$ be $\psi(x, y, \bar{u}, \bar{w})$ (the last formula is obtained from

$\psi(x, y, \bar{u}, \bar{v})$ by substituting $\bar{w}$ in place of $\bar{v}$). We claim that formulas ϕ_0, ϕ_1, ϕ_2 satisfy the condition (*) from Section 2 for $P = \mathcal{T}_k$. Let equivalence relations ξ, η on $n + 1$ be given. Specify values of the parameters $\bar{u}, \bar{v}$ as in (4), (6), and values of parameter $\bar{w}$ as

$$w_j = p_j(p_0(\bigsqcup_{i \in \eta_0} C_i) \sqcup \cdots \sqcup p_0(\bigsqcup_{i \in \eta_l} C_i)), \; j < k,$$

where $(\eta_0, \ldots, \eta_l)$ is the partition of $n + 1$ to η-equivalence classes. From (5) and (7) we obtain

$$\phi_0(\mathcal{T}_k, \bar{p}) = \{C_0, \ldots, C_n\}, \; \phi_1(\mathcal{T}_k, \bar{p}) = \{(C_i, C_j) | (i, j) \in \xi\}$$

and

$$\phi_2(\mathcal{T}_k, \bar{p}) = \{(C_i, C_j) | (i, j) \in \eta\}.$$

This means that $i \mapsto C_i$ defines an isomorphism of $(n+1; \xi, \eta)$ onto the structure $(\phi_0(\mathcal{T}_k, \bar{p}); \phi_1(\mathcal{T}_k, \bar{p}), \phi_2(\mathcal{T}_k, \bar{p}))$. This completes the proof.

Remark. From Proposition 1 it follows that $\mathcal{T}_k$ is first-order definable in $(\mathcal{F}_k; \leq)$ without parameters, hence the undecidability of $Th(\mathcal{T}_k; \leq)$ implies the undecidability of $Th(\mathcal{F}_k; \leq)$. Nevertheless, we included the proof of the second fact for methodical reasons because it is a bit simpler and its ideas are used in the two other proofs. In contrast, the undecidability of $Th(\mathcal{T}_k^i; \leq)$ does not imply the other two undecidability results because, by Proposition 2(i), $\mathcal{T}_k^i$ is not definable without parameters in the other two structures.

6   Open Questions

There are many natural open questions related to this paper. From our proofs it follows that for each of the three structures there exists an $n < \omega$ such that the n-quantifier theory of this structure is undecidable. It would be nice to find for each structure the least such n.

All the three structures are clearly recursively presentable, hence their elementary theories are interpretable in arithmetics, and therefore they are m-reducible to $\emptyset^\omega$. Our proofs show that $\emptyset'$ is m-reducible to any of the theories. The question remains to characterize the m-degree of each of the three theories. We expect that all three theories are recursively isomorphic to $\emptyset^\omega$.

There are also several interesting questions on definability (or non-definability) in any of the structures. E.g., we do not know whether the set $\mathcal{C}_k$ of finite k-chains is definable in $(\mathcal{T}_k; \leq)$ or in $(\mathcal{F}_k; \leq)$. This question is related to questions from the last paragraph because in [Ku06] it was shown that $Th(\mathcal{C}_k; \leq)$ is recursively isomorphic to $\emptyset^\omega$. So if $\mathcal{C}_k$ were e.g. first order definable without parameters in $(\mathcal{F}_k; \leq)$ then the theory of the last structure would also be recursively isomorphic to $\emptyset^\omega$.

References

[DP94]   B.A. Davey and H.A. Pristley. *Introduction to Lattices and Order*. Cambridge, 1994.

[E+65]   Yu.L. Ershov, I.A. Lavrov, A.D. Taimanov and M.A. Taitslin. Elementary theories. *Uspechi Mat. Nauk*, 20, N 4 (1965), 37–108, in Russian.

[E65]   Yu.L. Ershov. Undecidability of some fields. *DAN SSSR (Reports of AS USSR)*, 161, N 1 (1965), 27–29, in Russian.

[H96]   P. Hertling. *Unstetigkeitsgrade von Funktionen in der effectiven Analysis.* PhD thesis, FernUniversität Hagen, Informatik-Berichte 208–11, 1996.

[Ko00]   S. Kosub. On NP-partitions over posets with an application of reducing the set of solutions of NP problems. *Lecture Notes of Computer Science*, 1893 (2000), 467–476, Berlin, Springer.

[KW00]   S. Kosub and K. Wagner. The boolean hierarchy of NP-partitions. In: *Proc. 17th Symp. on Theor. Aspects of Comp. Sci., Lecture Notes of Computer Science*, 1770 (2000), 157—168, Berlin, Springer.

[Ku06]   D. Kuske. Theories of orders on the set of words. *Theoretical Informatics and Applications*, 40 (2006), 53–74.

[Se04]   V. L. Selivanov. Boolean hierarchy of partitions over reducible bases. *Algebra and Logic*, 43, N 1 (2004), 77–109 (see also http://www.informatik.uni-wuerzburg.de, *Technical Re-port 276*, Institut für Informatik, Universität Würzburg, 2001.

[Se06]   V. L. Selivanov. The Algebra of Labeled Forests Modulo Homomorphic Equivalence. *Schriften zur Theoretischen Informatik*, Technical Report 06-01, The University of Siegen, 2006.

Lower Bounds Using Kolmogorov Complexity

Sophie Laplante

LRI, Université Paris-Sud XI, 91405 Orsay CEDEX, France
laplante@lri.fr

Abstract. In this paper, we survey a few recent applications of Kolmogorov complexity to lower bounds in several models of computation. We consider KI complexity of Boolean functions, which gives the complexity of finding a bit where inputs differ, for pairs of inputs that map to different function values. This measure and variants thereof were shown to imply lower bounds for quantum and randomized decision tree complexity (or query complexity) [LM04]. We give a similar result for deterministic decision trees as well. It was later shown in [LLS05] that KI complexity gives lower bounds for circuit depth. We review those results here, emphasizing simple proofs using Kolmogorov complexity, instead of strongest possible lower bounds.

We also present a Kolmogorov complexity alternative to Yao's min-max principle [LL04]. As an example, this is applied to randomized one-way communication complexity.

Keywords: Lower bounds, Kolmogorov complexity, circuit complexity, query complexity, communication complexity.

1 Introduction

Kolmogorov complexity has been used in a variety of settings to prove lower bounds and other complexity results. However, until recently, the methods have been *ad hoc*, tailored to a particular problem and a particular computational model. In the past few years, techniques have been developed that apply to any Boolean function, and to a wide variety of computational models, so that a single analysis yields lower bounds in multiple models. In this paper, we review these results and present them in a unified setting, called KI complexity. We also present a Kolmogorov-based alternative to Yao's min-max principle, and apply it to one-way randomized communication complexity.

2 Preliminaries

Kolmogorov complexity is the main tool that is used to prove lower bounds in this paper, and we recall the main notions here. We also present the models of computation used in the paper.

A. Beckmann et al. (Eds.): CiE 2006, LNCS 3988, pp. 297–306, 2006.

2.1   Kolmogorov Complexity

Kolmogorov complexity captures well the information theoretic component of many lower bound arguments. We review a few of its main properties in this section.

Definition 1. *Let M be a Turing machine. Let x and y be finite strings.*

1. *The* Kolmogorov complexity *of x given y with respect to M is denoted $C_M(x|y)$, and defined as follows:*

$$C_M(x|y) = \min(|P| \text{ such that } M(P,y) = x).$$

2. *A set of strings is* prefix-free *if no string is a prefix of another in the set.*
3. *A Turing machine M' is* prefix-free *if the set of programs is prefix-free, that is, the set $\{P : \exists x\, M'(P,x) \neq \epsilon\}$, where ϵ is the empty string, is prefix-free.*
4. *The* prefix-free Kolmogorov complexity *of x given y with respect to a prefix-free Turing Machine M' is denoted $K_{M'}(x|y)$, and defined as follows:*

$$K_{M'}(x|y) = \min(|P| \text{ such that } M'(P,y) = x),$$

In the rest of the paper M is a fixed prefix-free universal Turing machine, and we will write K instead of $K_{M'}$. When y is the empty string, we write $K(x)$ instead of $K(x|y)$. To simplify notation we omit additive terms in the upper bounds.

Incompressibility. Perhaps the most important property of Kolmogorov complexity that we use for lower bounds is the existence of incompressible strings, that is, strings whose shortest description is maximal.

Proposition 1. *[Incompressibility] For any finite set $A \subseteq \{0,1\}^*$, and any string σ, there exists $x \in A$ such that $K(x|\sigma) \geq \log(\#A)$.*

The proposition is proved by comparing the number of succinct programs ($2^l - 1$ have length strictly less than l), with the number of strings ($\#A$) that these programs are purported to describe, and conclude by applying the pigeonhole principle.

This should be compared with the corresponding upper bound.

Proposition 2. *For any finite set $A \subseteq \{0,1\}^*$, $\exists \sigma, \forall x \in A$, $K(x|\sigma) \leq \log(\#A)$.*

To describe x, it suffices to give an index into some pre-determined enumeration of the set A, which can be encoded in σ.

We will also need Kraft's inequality.

Proposition 3 (Kraft's inequality). *Let S be any prefix-free set of finite strings. Then $\sum_{x \in S} 2^{-|x|} \leq 1$.*

We shall also use the following bound on conditional Kolmogorov complexity.

Proposition 4. *There is a constant $c \geq 0$ such that for any three strings x, y, z,*

$$K(z|x) \geq K(x,y) - K(x) - K(y|z,x) + K(z|x,y,K(x,y)) - c.$$

The proof uses symmetry of information in an essential way.

2.2   Decision Trees and Query Complexity

A decision tree is a rooted binary tree, where each internal node is labeled with an integer i referencing an input variable, one of the outgoing edges of an internal node is labeled 0 and the other is labeled 1, and each leaf is labeled with an output value. The tree is evaluated on an input $x = x_1 \cdots x_n$, starting at the root, by evaluating x_i if the node is labeled i and following the corresponding edge, and so on, until a leaf is reached, and outputing the value at the leaf. A decision tree T computes f if the output on x equals $f(x)$, for all x. The decision tree complexity of f, written $\mathsf{DT}(f)$, is the depth of the shallowest decision tree that computes f.

We also consider quantum and randomized analogues of decision trees. In these models, the complexity measure is the number of queries to the input, but unlike the classical case, queries can be made in superposition, in the quantum case, or according to some distribution, in the randomized case. Access to the input is achieved by way of a query operator O_x, which behaves like a classical query on classical states, but in the quantum case, it is defined as a unitary matrix O_x that satisfies $O_x|i, z, w\rangle = |i, z \oplus x_i, w\rangle$, for every i, z, w, where i represents a query, z is a register to hold the answer to the query, and w is the remainder of the workspace of the algorithm. Randomized queries can be defined similarly, except the matrix is stochastic. The *query complexity* of an algorithm is the number of calls to O_x. Details of the model can be found for example in [LM04], but they are not necessary for this paper.

We say that the algorithm A ε-computes a function $f : \{0,1\}^n \to \{0,1\}$, if the observation of the last bits of the work register equals $f(x)$ with probability at least $1 - \varepsilon$, for every $x \in S$. Then $\mathsf{QQC}(f)$ (resp., $\mathsf{RQC}(f)$) is the minimum query complexity of quantum (resp., randomized) query algorithms that ε_0-compute f, where ε_0 is a fixed positive constant no greater than $\frac{1}{3}$.

2.3   Communication Complexity

Communication complexity is a model of computation widely used to prove lower bounds in various models of computation. Here we will appeal to this model for lower bounds for circuit depth. We also consider one-way communication complexity in Section 4.

Let X, Y, Z be finite sets, and $R \subseteq X \times Y \times Z$. In the communication game for R, Alice is given some $x \in X$, Bob is given some $y \in Y$ and their goal is to find some $z \in Z$ such that $(x, y, z) \in R$, if such a z exists. A communication protocol determines what message each player sends in each round, and by convention, Bob produces an output at the end of the protocol. The cost of a protocol is the total number of bits exchanged in the worst case, and the communication complexity of R, written $\mathsf{D}(R)$, is the minumum cost of a protocol computing R.

There are many variants of communication complexity, and we will also consider one-way communication complexity of boolean functions. In a one-way communication protocol, two players, A and B wish to compute the value of a two-argument function $f : X \times Y \to Z$. Player A receives an input $x \in X$, and sends a message m to Player B. Player B receives an input $y \in Y$, as well as A's message m and should output the value of the function $f(x,y)$. The protocol is successful if B's output equals $f(x,y)$, for all x,y.

In the randomized model, a protocol is δ-correct if for all inputs x,y, the error probability on x,y is at most δ. The probability is taken over the random choices made by the players. $R_\delta(R)$ is the minumum cost of a protocol computing R in this way.

In the distributional model, we consider deterministic protocols, together with a distribution of the inputs μ, and an error threshold δ. A distributional protocol is δ-correct if the probability taken over μ that the output differs from the function is at most δ. The *distributional communication complexity for μ*, $D_{\delta,\mu}(f)$, is the maximum number of bits exchanged for the best δ-correct protocol for f when the input is chosen according to μ. The distributional complexity $D_\delta(f)$ of f is the maximum, over all probability distributions μ on the inputs, of $D_{\delta,\mu}(f)$.

2.4 Circuits and Formulae

A Boolean formula over the standard basis $\{\vee, \wedge, \neg\}$ is a binary tree where each internal node is labeled with $\vee$ or $\wedge$, and each leaf is labeled with a literal, that is, a Boolean variable or its negation. The size of a formula is its number of leaves.

Definition 2. *Let $f : \{0,1\}^n \to \{0,1\}$ be a Boolean function. The formula size of f, denoted $L(f)$, is the size of the smallest formula which computes f. The formula depth of f, denoted $d(f)$ is the minimum depth of a formula computing f.*

It is clear that $L(f) \le 2^{d(f)}$. Spira has also shown that $d(f) \le O(\log L(f))$ [Spi71].

Karchmer and Wigderson [KW88] give an elegant characterization of formula size and depth in terms of communication complexity.

Definition 3. *For any Boolean function f, the relation $R_f = \{(x,y,i) : f(x) = 0, f(y) = 1, x_i \ne y_i\}$.*

Theorem 1 (Karchmer-Wigderson). *For any Boolean function f, $d(f) = D(R_f)$.*

3   KI Complexity, Its Variants, and Applications

In order to prove a lower bound for a Boolean function f, consider two inputs that are mapped by f to different values. Then these two inputs must differ in some position and if the computation is correct, it must implicitly or explicitly have found one of these positions where the inputs differ. This is the principle which we will show how to exploit in this section, to obtain lower bounds in various models of computation.

3.1   Decision Trees and KI Complexity

Proposition 5. *Let f be a Boolean function, x,y be inputs such that $f(x) \ne f(y)$. Then*

$$DT(f) \ge \min_{\alpha \in \{0,1\}^*} \max_{\substack{x,y \\ f(x) \ne f(y)}} \min_{i:x_i \ne y_i} \{\max\{2^{K(i|x,\alpha)}, 2^{K(i|y,\alpha)}\}\}$$

Proof. Let T be a decision tree for f. If $f(x) \neq f(y)$, then the computation paths on x and y must diverge at some level of the decision tree. Let i be the variable queried at this level. Since the computation paths diverge at this point, $x_i \neq y_i$. So $\mathsf{K}(i|x,T) \leq log(depth(T))$ since it suffices to give an index into the depth of the tree, and similarly, $\mathsf{K}(i|y,T) \leq log(depth(T))$. Therefore, $\exists \alpha = T, \forall x, y : f(x) \neq f(y), \exists i, \mathsf{DT}(f) \geq \max\{2^{\mathsf{K}(i|x,T)}, 2^{\mathsf{K}(i|y,T)}\}$, which concludes the proof. □

Similar results hold for various models of computation, but with somewhat different combinations of the terms $K(i|x)$ and $K(i|y)$, for $f(x) \neq f(y)$ and $x_i \neq y_i$. We introduce a general definition that captures the known lower bounds in a common framework.

Definition 4. *Let $f : \{0,1\}^n \to [0,1]$. Let $\Lambda : \mathbb{R}^? \to \mathbb{R}$ (Λ takes an arbitrary number of real inputs, such as $\max$ or Σ, which we will take over all terms parameterized by i where $x_i \neq y_i$) and $\star : \mathbb{R} \times \mathbb{R} \to \mathbb{R}$ (where we sometimes use infix notation, e.g. $A \star B$). Define*

$$\mathsf{KI}^{\Lambda,\star}(f) = \min_{\alpha \in \{0,1\}^*} \max_{\substack{x,y \\ f(x) \neq f(y)}} \frac{1}{\Lambda_{i:x_i \neq y_i} 2^{-K(i|x,\alpha)} \star 2^{-K(i|y,\alpha)}}.$$

Reformulating Proposition 5 in terms of KI, we have

Proposition 6. $\mathsf{DT}(f) \geq \mathsf{KI}^{\max,\min}(f)$.

3.2 Randomized and Quantum Query Complexity Lower Bounds

Proposition 6 can be extended to randomized and quantum query complexity. The intuition is the same, but one has to analyze the the contribution of making a "useful" query much more carefully, since in these models, a query can be made with some probability or some amplitude.

Theorem 2. *[LM04] Let $f : \{0,1\}^n \to \{0,1\}$.*

1. *$\mathsf{QQC}(f) \geq \Omega(KI^{\Sigma,geom}(f))$ where Σ denotes sum over i such that $x_i \neq y_i$ and geom is the geometric average: $geom(A,B) = \sqrt{A \cdot B}$.*
2. *$\mathsf{RQC}(f) \geq \Omega(KI^{\Sigma,\min}(f))$.*

The theorem is proved by analyzing the overall contribution of each query towards disinguishing pairs of inputs with different values. Roughly speaking, the sum appears as a result of considering progress over all input pairs x, y such that $f(x) \neq f(y)$. The $\star$ operation is not so easily explained but the difference can be attributed to the fact that in the quantum case we operate under the ℓ_2 norm whereas in the randomized case, the ℓ_1 norm is used.

It turns out that this lower bound on query complexity implies all so-called adversary techniques for proving lower bounds in quantum query complexity, including the quantum and randomized weighted methods [Amb03, Aar04] and the spectral method [BSS03].

To give an idea of why this is the case we give an proof of Ambainis' un-weighted adversary method, which is given in terms of the combinatorial structure of the graph that represents pairs (edges) x, y such that $f(x) \neq f(y)$. This graph is thought of as containing the pairs of instances that are hard to distinguish. Furthermore, the pairs x, y that differ on some index i are those that a query to i can be helpful to distinguish x from y. Comparing the graph R with the subgraph R_i where the ith query is useful allows us to establish lower bounds on the number of queries required to distinguish all the pairs in R.

Theorem 3. *[Amb02, Aar04, LM04] Let $R \subseteq X \times Y$, be a relation on pairs of instances, where $X = f^{-1}(0)$ and $Y = f^{-1}(1)$, and let R_i be the restriction of R to pairs x, y for which $x_i \neq y_i$. Viewing the relation R as a bipartite graph, then if*

- *m is a lower bound on the degree of all $x \in X$,*
- *m' is a lower bound on the degree of all $y \in Y$,*
- *for any fixed $i, 1 \leq i \leq n$, the degree of any $x \in X$ in R_i is at most l,*
- *for any fixed $i, 1 \leq i \leq n$, the degree of any $y \in Y$ in R_i is at most l',*

then $\mathsf{QQC}(f) = \Omega\left(\sqrt{\frac{mm'}{ll'}}\right)$ and $\mathsf{RQC}(f) = \Omega\left(\max\{\frac{m}{l}, \frac{m'}{l'}\}\right)$.

Proof. We make the following observations.

1. $|R| \geq \max\{m|X|, m'|Y|\}$, so $\exists x, y \; \mathsf{K}(x, y) \geq \max\left(\log(m|X|), \log(m'|Y|)\right)$.
2. $\forall x \in X, \mathsf{K}(x) \leq \log(|X|)$ and $\mathsf{K}(y) \leq \log(|Y|)$, for all $y \in Y$.
3. $\forall x, y, i$ with $(x, y) \in R_i, \mathsf{K}(y|i, x) \leq \log(l)$ and similarly, $\mathsf{K}(x|i, y) \leq \log(l')$.

For any i with $x_i \neq y_i$, by Proposition 4,

$$\mathsf{K}(i|x) \geq \mathsf{K}(x, y) - \mathsf{K}(x) - \mathsf{K}(y|i, x) + \mathsf{K}(i|x, y, \mathsf{K}(x, y))$$
$$\geq \log(m|X|) - \log(|X|) - \log(l) + \mathsf{K}(i|x, y, \mathsf{K}(x, y))$$
$$= \log(\tfrac{m}{l}) + \mathsf{K}(i|x, y, \mathsf{K}(x, y))$$

The same proof works to show that $\mathsf{K}(i|y) \geq \log(\frac{m'}{l'}) + \mathsf{K}(i|x, y, \mathsf{K}(x, y))$. We can conclude by Theorem 2 and Kraft's inequality. □

3.3   Circuit Depth and Formula Size

Another model where KI can be used to obtain lower bounds is boolean formulas. We give a simple proof that KI gives a lower bound on circuit depth.

Theorem 4. *For any Boolean function f, $\mathsf{d}(f) \geq \mathsf{KI}^{\max, \cdot}(f)$.*

Proof. Let P be a protocol for R_f. Fix x, y with different values under f, and let T_A be a transcript of the messages sent from A to B, on input x, y. Similarly, let T_B be a transcript of the messages sent from B to A. Let i be the output of the protocol, therefore $x_i \neq y_i$. To print i given x, simulate P using x and T_B. To print i given y, simulate P using y and T_A. This shows that $\forall x, y :$ $f(x) \neq f(y), \exists i : x_i \neq y_i, K(i|x, \alpha) + K(i|y, \alpha) \leq |T_A| + |T_B| \leq \mathsf{D}(R_f)$, where α is a description of A's and B's algorithms. The theorem then follows from Theorem 1. □

3.4   A Few Examples

We give a few elementary examples to demonstrate how the technique can be applied to specific functions. To apply the adversary method, we have to give a relation R of hard instances; however, when applying KI, it suffices to exhibit a single hard pair of inputs.

Example 1: OR. The OR function is 0 on the all-0 input and 1 everywhere else. Consider inputs x, y of length n, where x is the all-0 string, and y is 0 everywhere except in bit i, where i is chosen so that $K(i) \geq \log(n)$. (More exactly, for any α we choose i such that $K(i|\alpha) \geq \log(n)$.) Such an i exists by incompressibility (Proposition 1). Therefore, by Theorems 2 and 4, and Proposition 6,

1. $DT(OR) \geq \Omega(n)$,
2. $RQC(OR) \geq \Omega(n)$,
3. $QQC(OR) \geq \Omega(\sqrt{n})$,
4. $d(OR) \geq \Omega(\log n)$.

Example 2: PARITY. The parity function is defined as $\oplus(x) = \Sigma_i x_i \ (mod\ 2)$. Consider inputs x, y chosen as follows. Take x, i so that $K(x, i) \geq n + \log(n)$, and let $y = x^i$ (x with the ith bit flipped). It is easy to show that $K(i|x) \geq \log(n)$ and $K(i|y) \geq \log(n)$.

1. $DT(\oplus) \geq \Omega(n)$,
2. $RQC(\oplus) \geq \Omega(n)$,
3. $QQC(\oplus) \geq \Omega(n)$,
4. $d(\oplus) \geq \Omega(\log n)$.

Several examples relating to graph properties are also given in [LM04].

4   Kolmogorov Alternative to the Min-Max Principle

Usually, lower bounds for randomized complexity are proven by first applying Yao's min-max principle, and proving a lower bounds in the distributional model where the algorithms are deterministic and the inputs are chosen at random according to some distribution. We propose an alternative to (or perhaps only a reformlation of) Yao's min-max principle, which makes use of Kolmogorov complexity. (To be precise, we only give an analogue of the "easy direction" that is generally used for lower bounds.) We illustrate how it can be applied, by proving a very general statement about one-way communication complexity. In this case, the proof is somewhat simpler than the previous proof of Bar-Yossef, Jayram, Kumar and Sivakumar [BYJKS02] that used information theory.

4.1   Yao in the Style of Kolmogorov

Yao's min-max principle consists in replacing randomness in the algorithm, with randomness in the inputs. Our approach is to replace randomness in the algorithm by a Kolmogorov random string, resulting in a deterministic algorithm.

It remains to see that the errors made on this random string are not too many. This is what is proven in the following lemma. The lemma is stated for private coin communication complexity but a similar statement can be made for other models of computation.

We assume, without loss of generality, that the players use a random string r_A, r_B taken uniformly at random from finite sets R_A, R_B, and that this is the same distribution regardless of the players' inputs x, y.

Lemma 1. *Let* $f : X \times Y \to Z$. *Fix any* δ-*correct randomized communication complexity protocol* P *for* f, *and consider any subset of inputs* $S \subseteq X \times Y$. *Fix* $(r_A^*, r_B^*) \in R_A \times R_B$ *such that* $C(r_A*, r_B^*|P, S) \geq \log(|R_A|) + \log(|R_B|)$. *Then when the protocol is run using* r_A^*, r_B^* *as random choices, the output is incorrect on at most* $2\delta|S|$ *inputs in* $|S|$.

Proof. For any r_A, r_B, let $\tilde{S}$ represent the inputs on which the outcome of the protocol is incorrect, that is, $\tilde{S}_{r_A,r_B} = \{\tilde{x}, \tilde{y} \in S : P(\tilde{x}, \tilde{y}, r_A, r_B) \neq f(\tilde{x}, \tilde{y})\}$. Also define the set of "much-worse-than-average" random choices for inputs in S to be $\tilde{R} = \{r_A, r_B : |\tilde{S}_{r_A,r_B}| > 2\delta|S|\}$.

Because at most half the inputs can have more than double the average number of errors, $|\tilde{R}| \leq \frac{|R_A||R_B|}{2}$, therefore by incompressibility, $r_A^*, r_B^* \notin \tilde{R}$. (Otherwise, describe r_A^*, r_B^* by giving an index into the set $\tilde{R}$. using $\log(|R|) < \log(|R_A|) + \log(|R_B|)$ bits, a contradiction.) Therefore $|\tilde{S}_{r_A^*,r_B^*}| \leq 2\delta|S|$. □

4.2   Shatter Coefficients Lower Bound

To give an example of how this method is applied, we give a proof of a general theorem on one-way communication complexity.

First we define VC dimension and its generalization, shatter coefficients. Let F be a set of strings of length n, and I be a set of indices, $I \subseteq [n], I = i_1, \cdots, i_{|I|}$. For any string $x = x_0, \cdots x_{n-1}$ of length n, $x|_I$ denotes the string $x_{i_1} \cdots x_{i_{|I|}}$. Likewise, $F|_I = \{x|_I : x \in F\}$ A set of strings F is *shattered* by a set of indices I if $F|_I$ is the set of all possible strings of length $|I|$. The VC *dimension* of F, denoted $VC(F)$, is the size of the largest I that shatters F.

The lth shatter coefficient of F (for any $l > VC(F)$), denoted $SC(F, l)$ is the maximum, over all I of size l, of $|F|_I|$. Let $F' \subseteq F$ be a subset of F for which $F'|_I$ takes on this maximal number of distinct values. We say that $F' \times I$ is a witness for $SC(F, l)$.

We give a new proof of a well-known result about one-way communication complexity. Recall that in this model, Alice sends one message to Bob and Bob produces the output. We use the superscript $A \to B$ to specify this model.

Theorem 5 ([KNR99, BYJKS02]). *For every function* $f : X \times Y \to \{0, 1\}$, *every* $l \geq VC(f)$, *and every* $\delta > 0$, $R_\delta^{A \to B}(f) \geq \log(SC(f|_X, l)) - lH_2(2\delta)$, *where* $H_2(p) = -p\log(p) - (1-p)\log(1-p)$.

Proof. Let $row_f(x, Y') = f(x, y_1) \cdots f(x, y_{|Y'|})$ be the string of consecutive values of f when x is fixed, where $Y' = \{y_1, \ldots y_{|Y'|}\}$. We denote by $f|_{X,Y}$ the set of strings $\{row_f(x, Y) : x \in X\}$. Let $S' = X' \times Y'$ be a witness for $SC(F, l)$ where $F = f|_{X,Y}$. Fix $x^* \in X', r_A^* \in R_A, r_B^* \in R_B$ with $C(x^*, r_A^*, r_B^*|P, S') \geq \log(|X'|) + \log(|R_A|) + \log(|R_B|)$ and let $S = \{x^*\} \times Y'$. Notice that $|S| = l$. By Lemma 1, when the protocol is run using r_A^*, r_B^* as random choices, the output is incorrect on at most $2\delta|S|$ inputs in $|S|$. To correct these errors we can just describe their location. This requires $\log(\binom{l}{2\delta l}) \approx l \cdot H_2(2\delta)$ additional bits.

All $\{row_f(x, Y') : x \in X'\}$ are unique, so x^* is uniquely determined within X' by $row_f(x, Y')$. This allows us to conclude that

$$
\begin{aligned}
\log(SC(f|_X, l)) &\leq C(x^*|P, r_A^*, r_B^*) \\
&\leq C(row_f(x^*, Y')|P, r_A^*, r_B^*) \\
&\leq R_\delta^{A \to B}(f) + lH_2(2\delta).
\end{aligned}
$$
□

5 Concluding Remarks

We have presented two different frameworks based on Kolmogorov complexity in which many lower bound techniques can be expressed. One might naturally ask what other models of computation these techniques can be applied to. One consequence of studying the KI lower bounds is that it brings to light the shared limitations of these techniques (see for example [LLS05]. Hopefully, understanding these limitations better will be a first step towards breaking the current lower bound barriers.

In the case of the min-max proofs using Kolmogorov complexity, it turns out in many cases that after rewriting the proofs in terms of Kolmogorov complexity, one can the remove Kolmogorov complexity entirely. An important role of Kolmogorov complexity is that the intuition it provides to help highlight the essential parts of the argument.

Acknowledgments

Many thanks are due to Marc Kaplan, Troy Lee, Frédéric Magniez, and Mario Szegedy, for many enlightening discussions on these results. Special thanks to Troy Lee for permission to include the unpublished results of Section 4. This work is funded in part by ANR AlgoQP, EU QAP, and ACI SI Réseaux quantiques.

References

[Aar04]   S. Aaronson. Lower bounds for local search by quantum arguments lower bounds for local search by quantum arguments. In *Proceedings of the thirty-sixth annual ACM symposium on Theory of computing*, pages 465–474, 2004.

[Amb02]   A. Ambainis. Quantum lower bounds by quantum arguments. *Journal of Computer and System Sciences*, 64:750–767, 2002.

[Amb03]   A. Ambainis. Polynomial degree vs. quantum query complexity. In *Proceedings of 44th IEEE Symposium on Foundations of Computer Science*, pages 230–239, 2003.

[BSS03]   H. Barnum, M. Saks, and M. Szegedy. Quantum decision trees and semi-definite programming. In *Proceedings of the 18th IEEE Conference on Computational Complexity*, pages 179–193, 2003.

[BYJKS02]   Z. Bar-Yossef, T. S. Jayram, R. Kumar, and D. Sivakumar. Information theory methods in communication complexity. In *Proceedings of the 17th Annual IEEE Conference on Computational Complexity*, pages 93–102, 2002.

[KNR99]   I. Kremer, N. Nisan, and D. Ron. On randomized one-round communication complexity. *Computational Complexity*, 8(1):21–49, 1999.

[KW88]   M. Karchmer and A. Wigderson. Monotone connectivity circuits require super-logarithmic depth. In *Proceedings of the 20th STOC*, pages 539–550, 1988.

[LL04]   S. Laplante and T. Lee. A few short lower bounds in one-way communication complexity. Unpublished manuscript, July 2004.

[LLS05]   S. Laplante, T. Lee, and M. Szegedy. The quantum adversary method and classical formula size lower bounds. In *Proceedings of the Twentieth Annual IEEE Conference on Computational Complexity*, pages 76–90, 2005.

[LM04]   S. Laplante and F. Magniez. Lower bounds for randomized and quantum query complexity using kolmogorov arguments. In *Proceedings of the Nineteenth Annual IEEE Conference on Computational Complexity*, pages 294–304, 2004.

[Spi71]   P. Spira. On time-hardware complexity tradeoffs for Boolean functions. In *Proceedings of the 4th Hawaii Symposium on System Sciences*, pages 525–527. Western Periodicals Company, North Hollywood, 1971.

The Jump Classes of Minimal Covers

Andrew E.M. Lewis*

Dipartimento di Scienze Matematiche ed Informatiche, Siena

Abstract. We work in $\mathcal{D}[< 0']$. Given the jump class of any (Turing) degree a, the jump classes of the minimal covers of a is a matter which is entirely settled unless a is $high_2$. We show that there exists a c.e. degree which is $high_2$ with no $high_1$ minimal cover.

1 Introduction

Simply by relativizing the construction of a minimal degree below any degree which is c.e. [CY2] it can be seen that in $\mathcal{D}[< 0']$ every degree has a minimal cover. If we are given the jump class of any degree in $\mathcal{D}[< 0']$, a say, then the jump classes of the minimal covers of a is a matter which has been almost entirely settled since 1978. For any $n \geq 1$ we regard a degree as being properly low_n if it is in $low_n - low_{n-1}$ and we regard a degree as being properly $high_n$ if it is in $high_n - high_{n-1}$. First let us consider the least degree 0. Since there are c.e. degrees which are low_1, it follows by Yates' construction of a minimal degree below any given c.e. degree that there are minimal degrees which are low_1. Sasso [SA], Cooper and Epstein have shown that there minimal degrees which are properly low_2 and it follows by the result of Jockusch and Posner, that every degree not in GL_2 bounds a 1-generic, that all minimal degrees below $0'$ are low_2. By relativizing we can conclude that any degree which is low_1 has minimal covers which are low_1 and minimal covers which are properly low_2. Given a degree in any proper jump class other than low_0, low_1 and $high_2$ we may conclude that all minimal covers are of the same proper jump class as that degree (in considering the minimal covers of degrees in $high_1$ recall that $0'$ is not a minimal cover). Given a degree which is properly $high_2$, however, it is not known whether this degree will have minimal covers which are $high_1$. In [AL] we show that there exists a c.e. degree which is $high_2$ and which has no $high_1$ minimal cover. What appears here is an abbreviated version of that paper, stopping short of the technical details of the construction. In order to prove the result we introduce a new technique involving what are called 'modifiers', which enable us to make use of the recursion theorem where it would otherwise be impossible to do so.

2 The Intuition

Let κ be a computable bijection $\omega \to 2^{<\omega}$. For any $e \in \omega$ we say that W_e specifies a convergent approximation if $\not\exists m[\exists^\infty n(n \in W_e \wedge \kappa(n)(m) \downarrow= 0) \wedge \exists^\infty n(n \in$

---

* The author was supported by EPSRC grant No. GR /S28730/01.

A. Beckmann et al. (Eds.): CiE 2006, LNCS 3988, pp. 307–318, 2006.

$W_e \wedge \kappa(n)(m) \downarrow = 1)]$. Define Conv to be the set of all those $e \in \omega$ such that W_e specifies a convergent approximation (so that Fin $\subset$ Conv). We show that Conv is Π_3 complete as follows. Fix $A \in \Sigma_3$. Now for some computable function g,

$$n \in A \leftrightarrow (\exists m)[W_{g(n,m)} \text{ is infinite}].$$

Given any $n \in \omega$ we define a set $W_{f(n)}$ according to the construction below.

Stage 0. Define $\sigma_0(0) = 0$ and for all $m > 0$ let $\sigma_0(m)$ be undefined. Let n_0 be such that $\kappa(n_0) = \sigma_0$ and enumerate n_0 into $W_{f(n)}$.
Stage $s > 0$. We define a string σ_s of length $s + 1$. For all $m < s$ let $\sigma_s(m) = \sigma_{s-1}(m)$ if $W_{g(n,m),s} - W_{g(n,m),s-1} = \emptyset$ and let $\sigma_s(m) = |1 - \sigma_{s-1}(m)|$ otherwise. Let $\sigma_s(s) = 0$. Let n_s be such that $\kappa(n_s) = \sigma_s$ and enumerate n_s into $W_{f(n)}$.
   Clearly f is computable and $n \in A \leftrightarrow f(n) \notin$ Conv.
   In order to construct a c.e. set A which is $high_2$ we shall enumerate a set S_j for all $j \in \omega$. We shall show that there exists $f \leq_T A''$ such that, for all $e \in \omega$, $e \in$ Conv iff there exist an infinite number of n such that $n \in S_{f(e)}$ and $A(n) = 0$. At stage s of the construction it is convenient to be able to consider strings defined on all arguments $\leq s$. Thus for any $e \in \omega$ let the computable sequence of finite binary strings $\{\tau_{e,s}\}_{s \in \omega}$ be defined inductively as follows. We define $\tau_{e,0} = 0$. Suppose we are given $\tau_{e,s}$. If there does not exist $n \in W_{e,s+1} - W_{e,s}$ then for all $m \leq s$ define $\tau_{e,s+1}(m) = \tau_{e,s}(m)$ and define $\tau_{e,s+1}(s + 1) = 0$. Otherwise we assume that there can only be one such n. For all $m \leq min\{s + 1, |\kappa(n)| - 1\}$ define $\tau_{e,s+1}(m) = \kappa(n)(m)$. For all m such that $|\kappa(n)| \leq m \leq s$ define $\tau_{e,s+1}(m) = \tau_{e,s}(m)$ and if $|\kappa(n)| \leq s+1$ then define $\tau_{e,s+1}(s+1) = 0$. It is not difficult to see that, for all $e \in \omega$, W_e specifies a convergent approximation iff $\forall n \exists \tau \exists s[[|\tau| = n] \wedge (\forall s' \geq s)[\tau \subset \tau_{e,s'}]]$. For all $e \in \omega$, if W_e specifies a convergent approximation then let B_e be the set which the sequence $\{\tau_{e,s}\}_{s \in \omega}$ approximates, and otherwise let B_e be undefined.

Definition 2.1 *Given $\sigma \in 2^{<\omega}$ we let $\sigma^\star$ be defined as follows; for all $n \in \omega$ if $\sigma(2n) \downarrow$ then $\sigma^\star(n) \downarrow = \sigma(2n)$ and $\sigma^\star(n) \uparrow$ otherwise. Given $C \subseteq \omega$ we let $C^\star$ be defined as follows; for all $n \in \omega$, $C^\star(n) = C(2n)$. Let $\sigma^\dagger$ and $C^\dagger$ be defined similarly with $2n + 1$ in place of $2n$.*

Now suppose that W_e specifies a convergent approximation and that B_e is of high degree. Then there will exist $z \in \omega$ such that we enumerate axioms for a Turing functional Γ_g, where $g = \langle e, z \rangle$, so that if $C = \Gamma_g^{A \oplus B_e}$ then $C^\star = A$ and,

$$(\forall k)(\exists \sigma \subset C)[[\Psi_k^\sigma(k) \downarrow] \vee (\forall \sigma' \supseteq \sigma)[\sigma'^\star \subset A \rightarrow \Psi_k^{\sigma'}(k) \uparrow]].$$

To the parameter z we will be able to apply the recursion theorem relative to an oracle for B_e. We must show that this suffices to prove that if $b = deg(B_e)$ and $a = deg(A)$ then $a \vee b$ is not a minimal cover for a. Let C_0 and C_1 be defined as follows.

C_0: for all $n \in \omega$, $C_0(2n) = C(2n)$ and $C_0(2n + 1) = C(4n + 1)$.
C_1: for all $n \in \omega$, $C_1(2n) = C(2n)$ and $C_1(2n + 1) = C(4n + 3)$.

Then C_0 and C_1 are Turing incomparable and of degree above a and below $a \vee b$. In order to see that $\Psi_e^{C_0} \neq C_1$, for example, consider the Turing functional Ψ_k such that $\Psi_k^\sigma(x) = 1$ if $\exists m(\Psi_e^{\sigma_0}(m) \downarrow \neq \sigma_1(m))$ and is otherwise undefined, where σ_0 and σ_1 are defined in terms of σ in exactly the same way that we defined C_0 and C_1 in terms of C.

2.1 Having outlined the basic framework we shall now explain the intuition behind the construction by considering a series of simplified situations. In each case techniques will be described and an approach given which is not yet sufficient as it stands. We shall attempt to describe the principal ideas required in order to overcome the basic obstacles - any remaining technical problems are dealt with in [AL]. So let us begin by supposing that for some $e, j \in \omega$ we wish to enumerate a set S_j and that we wish to ensure there are an infinite number of $n \in S_j$ such that $A(n) = 0$ iff W_e specifies a convergent approximation. Then we might proceed simply as follows:

Stage $s = 0$. Enumerate j into S_j and define $\Delta_0(j) = \lambda$ (where λ is the empty string).

Stage $s > 0$.

Step 0. For each n that we have enumerated into S_j check to see whether $\Delta_0(n) \subset \tau_{e,s}$ and if not then enumerate n into A.

Step 1. Let n be the greatest number which we have enumerated into S_j such that $A(n) = 0$ and suppose that $\Delta_0(n) = \tau$. Choose n' larger than any number yet mentioned during the course of the construction, enumerate n' into S_j and define $\Delta_0(n') = \tau'$, where τ' is the initial segment of $\tau_{e,s}$ of length $|\tau| + 1$.

2.2 Let us suppose, for now, that for each $j \in \omega$ we will enumerate S_j and A so as to ensure that there are an infinite number of $n \in S_j$ such that $A(n) = 0$ iff W_j specifies a convergent approximation. Thus S_0 and A are enumerated so as to ensure that there are an infinite number of $n \in S_0$ such that $A(n) = 0$ iff W_0 specifies a convergent approximation. That this should be the case will be our requirement of highest priority, $\mathcal{H}_0$. The requirement of next highest priority will be $\mathcal{G}_0$, the first of the 'genericity' requirements $\mathcal{G}_h$ where $h = \langle e, z, k \rangle$. The demands of this requirement are as follows; if $B_e \downarrow$, is of high degree and if z is some suitable fixed point when we apply the recursion theorem relative to an oracle for B_e then $(\exists \sigma \subset \Gamma_g^{A \oplus B_e})[[\Psi_k^\sigma(k) \downarrow] \vee (\forall \sigma' \supseteq \sigma)[\sigma'^* \subset A \rightarrow \Psi_k^{\sigma'}(k) \uparrow]]$, where $g = \langle e, z \rangle$.

Convention 2.1 *We adopt the convention that for any j, σ and any m, $\Psi_j^\sigma(m) \downarrow$ only if the computation converges in $\leq |\sigma|$ steps.*

In order that the demands of the requirement $\mathcal{G}_h$ should be made precise we must describe what use we intend to make of the recursion theorem. Let us assume for now that the axioms we enumerate for any Γ_g will be consistent. If it is the case that $B_e \downarrow$ and is of high degree then we might immediately try to make use of this fact in the following kind of way. Once we have defined the construction and having fixed such an $e \geq 0$ there will exist a computable function f_e such that, for all z and all k, $\lambda n.[\Psi_{f_e(z)}^{B_e}(k, n)]$ approximates whether there exists $\psi \subset \Gamma_g^{A \oplus B_e}$ (where $g = \langle e, z \rangle$) and a stage s such that at no stage $s' \geq s$ is there a string $\psi' \supseteq \psi$ of length $\leq s'$ such that $\psi'^* \subset A_{s'}$ and such that $\Psi_k^{\psi'}(k) \downarrow$. In order to see this we argue as follows. For fixed e such that $B_e \downarrow$ and is of high degree we can assume given z_0 such that $\Psi_{z_0}^{B_e}$ approximates $\emptyset''$. In order to compute $f_e(z)$ first produce z_1 such that Ψ_{z_1} operates as follows when provided with an oracle for $\emptyset'$ and on input k. It enumerates all $\psi \subset \Gamma_g^{A \oplus B_e}$ and as it does so it dovetails through the asking of all questions of the the following form for any $s \in \omega$, 'does there exist $s' \geq s$ such that there is a string $\psi' \supseteq \psi$ of length $\leq s'$ such that $\psi'^* \subset A_{s'}$ and such that $\Psi_k^{\psi'}(k) \downarrow$'. If it finds the answer to one of these questions is 'no' then it terminates. Then it is clear how to use z_0 in order to produce $f_e(z)$. By the recursion theorem there exists z such that $\Psi_z = \Psi_{f_e(z)}$, so that Ψ_z will satisfy the (potentially) useful properties satisfied by $\Psi_{f_e(z)}$.

Before expanding upon these ideas let us introduce some terminology and methodology that will be used in the construction. When, at any stage s, we enumerate an axiom $\Gamma_g^\phi = \psi$ for some $\phi \subset A_s \oplus \tau_{e,s}$ and $g = \langle e, z \rangle$ we shall declare $\beta = (\phi, n)$ to be a node for g, for some $n \in \omega$. The n parameter here is just a counter and indicates that there are precisely n nodes (ϕ', n') that we have defined for g such that $\phi' \subset \phi$ (and $n' < n$). So suppose that at some point of the construction we are working above the node $\beta = (\phi, n)$ that we have declared for 0 $(0 = \langle 0, 0 \rangle)$, since of all the strings ϕ' such that we have declared a node (ϕ', n') for 0 and $\phi' \subset A_s \oplus \tau_{0,s}$ it is the case that ϕ is the longest, and that we are looking to satisfy the requirement $\mathcal{G}_0$, $(0 = \langle 0, 0, 0 \rangle)$. Suppose that we have already enumerated the axiom $\Gamma_0^\phi = \psi$. If it is the case that $\Psi_0^\psi(0) \downarrow$ then we need do nothing for the sake of requirement $\mathcal{G}_0$. So suppose that this is not the case. At every subsequent stage s at which we work above the node β we might look to see whether there exists $\psi' \supset \psi$ of length $\leq s$ such that $\psi'^* \subset A_s$ and $\Psi_0^{\psi'}(0) \downarrow$. If so then we enumerate some axiom $\Gamma_0^{\phi'} = \psi'$, where $\phi' \supset \phi$ is of at least the same length as ψ' and declare another node for 0, $\beta' = (\phi', n + 1)$. If not then we look to see whether there exists $n' > n$ such that $\Psi_0^{\tau_{0,s}}(0, n') \downarrow = 1$ (here we use Ψ_0 since we are considering the case $z = 0$). If so then we let τ be some suitable initial segment of $\tau_{0,s}$ on which this computation converges. If $n = 0$ then we define a node for 0, $\beta' = (\phi', 1)$ where ϕ' is the initial segment of $A_s \oplus \tau_{0,s}$ of length $2|\tau|$. If $n > 0$ we acknowledge that at any stage $s' \geq s$ when we work above the node β and $\tau \subset \tau_{0,s'}$ we shall now be prepared to try and satisfy one of the lower priority requirements $\mathcal{G}_h$ such that $h = \langle 0, 0, k \rangle$ for $1 \leq k \leq n$ (starting with that of highest priority) before defining another node for 0 and paying attention to the needs of the requirement $\mathcal{G}_0$ again.

2.3 The first and most basic obstacle now presents itself. Let us suppose just for now that $B_0 \downarrow$ and is of high degree, that $z = 0$ is a fixed point of f_0 and that, for $\beta = (\phi, n)$ as above, ϕ is an initial segment of the final value $A \oplus B_0$. It may be the case that at some stage s, when working above the node β we do find $\psi' \supset \psi$ such that $\psi'^* \subset A_s$ and $\Psi_0^{\psi'}(0) \downarrow$ and that we declare $(\phi', n + 1)$ to be a node for 0. At some subsequent stage s' we may then find that there is m, enumerated into S_0 before stage s, such that $\phi'^*(m) \downarrow = \psi'^*(m) \downarrow = 0$ but which we now must enumerate into A for the sake of the requirement $\mathcal{H}_0$. It may be the case, in fact, that there does not exist $\psi' \subset \Gamma_0^{A \oplus B_0}$ and a stage s such that at no stage $s' \geq s$ is there a string $\psi'' \supseteq \psi'$ of length $\leq s'$ such that $\psi''^* \subset A_{s'}$ and such that $\Psi_0^{\psi''}(0) \downarrow$, and that there does not exist $n' > n$, $\Psi_0^{B_0}(0, n') \downarrow = 1$, but that every time we define a node $\beta' = (\phi', n+1)$ for 0 we subsequently find that this action is spoiled by action we have to take on behalf of $\mathcal{H}_0$. The following eventuality is also possible. It may be the case that there does not exist $n' > n$, $\Psi_0^{B_0}(0, n') \downarrow = 1$, so that there are an infinite number of stages s such that there exists $\psi' \supset \psi$ of length $\leq s$, $\psi'^* \subset A_s$ and $\Psi_0^{\psi'}(0) \downarrow$, but that for all but finitely many of such stages either we do not work above β but work above $\beta' = (\phi', n')$ such that $\phi' \supset \phi$ and $n' > n$ and ϕ' is not an initial segment of the final value $A \oplus B_0$, or we do work above β but look to satisfy a lower priority requirement. This latter problem is easily dealt with once we have developed a good approach to the former.

Definition 2.2 *For $j \subset \omega$ we define T_j to be the set of all those n enumerated into S_j such that the final value $A(n) = 0$.*

The following slightly more sophisticated approach takes us closer to a solution. Firstly we shall demand that a little more should be satisfied by the function f_e. During the course of the construction we make use of various varieties of tuple. Alpha tuples will be of the form $\alpha = (q, i_0, .., i_y)$ where $q \in 2^{<\omega}$, $i_0, .., i_y \in \omega$ and $|\{j : q(j) \downarrow = 0\}| = y + 1$. Here q should be thought of as a guess as to which $j < |q|$ will satisfy the condition that T_j is infinite. Thus $q(j) \downarrow = 1$ corresponds to the guess that T_j will be infinite while $q(j) \downarrow = 0$ corresponds to the guess that it will not be. Let $\{j : q(j) \downarrow = 0\} = j_0 < .. < j_y$. For each $y' \leq y$, $i_{y'}$ reflects the guess that $T_{j_{y'}} = D_{i_{y'}}$ (and where $\{D_i\}_{i \in \omega}$ is some effective listing of all finite sets of natural numbers).

Definition 2.3 *Given an alpha tuple $\alpha = (q, i_0, .., i_y)$, let $\{j : q(j) \downarrow = 0\} = j_0 < .. < j_y$. We say that σ complies with α at stage s if $\sigma^* \subseteq A_s$ and for all $y' \leq y$ and $n \in \omega$, if $\sigma^*(n) \downarrow$ and n has been enumerated into $S_{j_{y'}}$ then $\sigma^*(n) = 0$ iff $n \in D_{i_{y'}}$.*

The function f_e, then, must satisfy the following for all $z, k, j \in \omega$:

a) If T_j is finite then for all $n \in \omega$, $\Psi_{f_e(z)}^{B_e}(0, j, n) \downarrow$ and $lim_{n \to \infty} \Psi_{f_e(z)}^{B_e}(0, j, n) \downarrow = i$ such that $T_j = D_i$.

b) Let $g = \langle e, z \rangle$ and $h = \langle e, z, k \rangle$. For any q of length $\Sigma_{i=0}^h 2^i$ and any alpha tuple $\alpha = (q, i_0, .., i_y)$ we have that $\lambda n.[\Psi_{f_e(z)}^{B_e}(1, k, \alpha, n)]$ approximates whether there exists $\psi \subseteq \Gamma_g^{A \oplus B_e}$ and a stage s such that at no stage $s' \geq s$ is there a string $\psi' \supseteq \psi$ of length $\leq s'$ which complies with α at stage s' and such that $\Psi_k^{\psi'}(k) \downarrow$. The length of the string q here requires some comment. We have stated previously that we shall enumerate S_0 so as to ensure that T_0 is infinite iff W_0 specifies a convergent approximation and that the requirement of next highest priority will be $\mathcal{G}_0$. In enumerating S_1 and S_2, however, it will be the case (for reasons that will be explained subsequently) that we have to proceed according to a guess as regards whether the action of $\mathcal{G}_0$ satisfies a certain property. Thus we enumerate S_1 – with corresponding requirement $\mathcal{H}_1$ – according to the guess that the action of $\mathcal{G}_0$ does not satisfy this property and S_2 according to the guess that it does. Whether or not the action of $\mathcal{G}_0$ does satisfy this (presently mysterious) property will be a fact deducible from an oracle for A''. After $\mathcal{H}_2$ the requirement of next highest priority is then $\mathcal{G}_1$. In enumerating S_3, S_4, S_5 and S_6 we have to proceed according to a guess as regards whether the actions of $\mathcal{G}_0$ and $\mathcal{G}_1$ each satisfy the property alluded to. Whether or not the action of $\mathcal{G}_1$ satisfies this property will be a fact deducible from an oracle for A''. After $\mathcal{H}_6$ the requirement of next highest priority will be $\mathcal{G}_2$, and so on.

In working above the node β that we have defined for 0 (as above) we shall proceed for various pairs (h, q) in turn. First we proceed for the pair $(0, q)$ such that q is the finite binary string which is a single zero (indicating the guess that T_0 will be finite). If it is the case that $\Psi_0^\psi(0) \downarrow$ then we need do nothing for the sake of requirement $\mathcal{G}_0$, so suppose that this is not the case. At every stage s subsequent to that at which we declare β to be a node for 0 and such that $\phi \subset A_s \oplus \tau_{0,s}$ (and irrespective of whether we work above β at stage s) we look to see whether the value $\hat{\psi}(\beta, 0)$ is defined – the use of the function $\hat{\psi}$ is in the process of being described. The second argument which takes the value 0 here will, in general, take the value of the parameter $h = \langle e, z, k \rangle$. If so but $(\hat{\psi}(\beta, 0))^* \not\subset A_s$ then make it undefined. If, subsequent to this action, $\hat{\psi}(\beta, 0)$ is now undefined we look to see whether there exists $\psi' \supset \psi$ of length $\leq s$ such that $\psi'^* \subset A_s$ and such that $\Psi_0^{\psi'}(0) \downarrow$. If so then we define $\hat{\psi}(\beta, 0)$ to be the shortest such string and we try to preserve $(\hat{\psi}(\beta, 0))^*$ as an initial segment of A for the sake of $(0, q)$ (and with the priority afforded $\mathcal{G}_0$). Now let us describe how to proceed for the pair $(0, q)$ when actually working above the node β. The first such stage we use Ψ_0, which we hope of course is a fixed point of f_0, in order to produce a guess as regards i such that $T_0 = D_i$. More precisely we define i, then, as follows. If there exists n' such that $\Psi_0^{\phi^\dagger}(0, 0, n') \downarrow$ then let n' be the greatest such and define $i = \Psi_0^{\phi^\dagger}(0, 0, n')$. Otherwise define $i = 0$. At every subsequent stage s at which we work above the node β we look to see whether $\hat{\psi}(\beta, 0) \downarrow$ and complies with $\alpha = (q, i)$. If so then we enumerate some axiom $\Gamma_0^{\phi'} = \hat{\psi}(\beta, 0)$, where ϕ' is of at least the same length as $\hat{\psi}(\beta, 0)$ and declare another node for $0, \beta' = (\phi', n + 1)$. If not then we look to see whether there exists $n'' > n$ such that $\Psi_0^{\tau_{0,s}}(1, 0, \alpha, n'') \downarrow = 1$. If so then we let τ be some suitable initial segment of $\tau_{0,s}$ on which this computation converges and we acknowledge that at any stage $s' \geq s$ when we work above the node β and $\tau \subset \tau_{0,s'}$ we shall now proceed for the pair $(0, q')$ such that q' is the finite binary string which is a single one.

Let us briefly discuss the possible outcomes of this activity. Just for now and only for the sake of simplicity, let us assume that $B_0 \downarrow$ and is of high degree, that $z = 0$ is a fixed point of f_0 and that, for $\beta = (\phi, n)$ as above, ϕ is an initial segment of the final value $A \oplus B_0$. Suppose first that there exist an infinite number of stages at which $\hat{\psi}(\beta, 0) \downarrow$ and complies with α. Then there exists a stage at which $\hat{\psi}(\beta, 0)$ is defined and is never subsequently made undefined, so that $(\hat{\psi}(\beta, 0))^* \subset A$. We shall therefore be able to define a node $\beta' = (\phi', n + 1)$

for g such that $\phi' \subset A \oplus B_0$ and ensure that the requirement $\mathcal{G}_0$ is satisfied, unless possibly there exists $n'' > n$ such that $\Psi_0^{B_0}(1,0,\alpha,n'') \downarrow = 1$. Now suppose that there exist a finite number of stages at which $\hat{\psi}(\beta,0) \downarrow$ and complies with α. Then there exists $n'' > n$ such that $\Psi_0^{B_0}(1,0,\alpha,n'') \downarrow = 1$ and for some $\tau \subset B_0$ and at some stage s we shall acknowledge that at any stage $s' \geq s$ when we work above the node β and $\tau \subset \tau_{0,s'}$ we shall now proceed for the pair $(0,q')$ such that q' is the finite binary string which is a single one.

Definition 2.4 *For all $j \in \omega$ we define $e^\star(j) = \mu e.(\Sigma_{i=0}^e 2^i > j)$. Let j' be the least such that $e^\star(j') = e^\star(j)$. We define $p^\star(j)$ to be the $(j - j' + 1)^{th}$ string of length $e^\star(j)$ (and where $2^{<\omega}$ is considered to be ordered lexicographically).*

Definition 2.5 *Let $\{\rho_j\}_{j\in\omega}$ be a uniformly computable sequence of functions such that for all $j,n \in \omega$, $\rho_j(n)$ is of the form (h,q) where $h \geq e^\star(j)$, q is of length $\Sigma_{i=0}^h 2^i$, $q(j) = 1$ and such that for fixed j and any pair (h,q) of this form there exist an infinite number of n, $\rho_j(n) = (h,q)$.*

Given $j \in \omega$, the value $p^\star(j)$ should be thought of as the guess that j makes as regards which of the genericity requirements of higher priority will satisfy the (still mysterious) property mentioned previously and expanded upon in 2.5. The value $e^\star(j)$ tells us that, if this guess is correct, then T_j is infinite iff $W_{e^\star(j)}$ specifies a convergent approximation. Although we do not directly need these concepts at this point in the description of the intuition behind the construction, their definition needs to be made in order that we can properly define those terms which are in the process of being explained. This approach seems preferable to that in which definitions are made and then constantly revised.

In order to describe how to proceed for the pair $(0,q')$, where q' is the finite binary string which is a single one, we must explain what shall be meant by the term 'modifier'. The basic idea behind the use of modifiers is as follows. Suppose that n_0 is enumerated into S_0. When we make this enumeration, suppose that we define $\Delta_0(n_0) = \tau$ for some $\tau \in 2^{<\omega}$ – then we shall declare that n_0 is a modifier for $\rho_0(|\tau|)$. If at a subsequent stage n_0 is enumerated into A then we shall declare that n_0 is no longer a modifier (for $\rho_0(|\tau|)$). Now at every stage s such that $\phi \subset A_s \oplus \tau_{0,s}$ we decide how and whether $\hat{\psi}(\beta,0)$ should be defined, in the manner described above. If this value is defined then we have stated already that we try to preserve $(\hat{\psi}(\beta,0))^\star$ as an initial segment of A with the priority afforded $\mathcal{G}_0$, for the sake of the pair $(0,q)$. We also try to preserve this string as an initial segment of A for the sake of the pair $(0,q')$ but with a certain advantage. We shall find n_0 which is the least number which has been enumerated into S_0, which is a modifier for $(0,q')$ and which was declared to be such after stage n such that $\beta = (\phi,n)$ (if there exists such). For any $n_1 > n_0$ which have been enumerated into S_0, such that $A(n_1) = (\hat{\psi}(\beta,0))^\star(n_1) = 0$, we then agree that n_1 will not be enumerated into A so long as n_0 is not enumerated into A or, what amounts to the same at this point, so long as $(\hat{\psi}(\beta,0))^\star$ looks to be an initial segment of A. Of course we must ensure that each n_0 which is declared to be a modifier is responsible for the failure to enumerate only a finite number of

$n_1 > n_0$ into A. This problem will be addressed more thoroughly in subsection 2.4, but the following problem is worth considering now. If we proceed in the manner described thus far then it may be the case that n_0 which is declared to be a modifier prevents the enumeration of $n_1 > n_0$ into A – that, were it not for the use of n_0 as a modifier, n_1 would have been enumerated into A. Since n_1 will have been declared to be a modifier it may then be the case that n_1 prevents the enumeration of $n_2 > n_1$ into A, and so on. This problem is easily dealt with, but in order to do so it is convenient to enumerate an auxiliary set A^*. A^*, then, will be enumerated in exactly the same manner as A, except that a modifier n_0 cannot prevent the enumeration of $n_1 > n_0$ into A^*. Thus we shall only allow n_0, n_1, n_2 to be modifiers just so long as they have not been enumerated into A^*.

Now let us describe how to proceed for the pair $(0, q')$ when actually working above the node β. The first such stage we shall find n_0 which is the least number which has been enumerated into S_0, which is a modifier for $(0, q')$ and which was declared to be such after stage n such that $\beta = (\phi, n)$. The terminology used in the construction is that we define $\Omega_1(\beta, 0, q') = n_0$ if there exists such and leave this value undefined otherwise. Now suppose that we have already proceeded for the pair $(0, q)$ and that for some $\tau \subset B_0$ we have acknowledged that at stages s such that $\tau \subset \tau_{0,s}$ we shall proceed with the pair $(0, q')$. In what follows, then, we consider only the stages latterly described. At every subsequent stage s of the relevant variety we first check to see whether $\Omega_1(\beta, 0, q')$ is still a modifier and if not, or if this value is undefined, then we proceed as follows. If $n = 0$ then we define a node for 0, $\beta' = (\phi', 1)$ where ϕ' is some initial segment of $A_s \oplus \tau_{0,s}$ of length at least $2|\tau|$. If $n > 0$ we acknowledge that at any stage $s' \geq s$ when we work above the node β and $\tau \subset \tau_{0,s'}$ we shall now be prepared to try and satisfy one of the lower priority requirements $\mathcal{G}_h$ such that $h = \langle 0, 0, k \rangle$ for $1 \leq k \leq n$. Otherwise we proceed in almost exactly the same way as we did for the pair $(0, q)$. Let $\alpha' = (q')$. We look to see whether $\hat{\psi}(\beta, 0) \downarrow$ (and complies with α'). If so then we enumerate some axiom $\Gamma_0^{\phi'} = \hat{\psi}(\beta, 0)$, where ϕ' is of at least the same length as $\hat{\psi}(\beta, 0)$ and declare another node for 0, $\beta' = (\phi', n + 1)$. If not then we look to see whether there exists $n'' > n$ such that $\Psi_0^{\tau_{0,s}}(1, 0, \alpha', n'') \downarrow= 1$. If so then we let $\tau' \supset \tau$ be some suitable initial segment of $\tau_{0,s}$ on which this computation converges. If $n = 0$ then we define a node for 0, $\beta' = (\phi', 1)$ where ϕ' is the initial segment of $A_s \oplus \tau_{0,s}$ of length $2|\tau'|$. If $n > 0$ we acknowledge that at any stage $s' \geq s$ when we work above the node β and $\tau' \subset \tau_{0,s'}$ we shall now be prepared to try and satisfy one of the lower priority requirements $\mathcal{G}_h$ such that $h = \langle 0, 0, k \rangle$ for $1 \leq k \leq n$.

So once again let us assume, just for now, that $B_0 \downarrow$ and is of high degree, that $z = 0$ is a fixed point of f_0, that ϕ is an initial segment of the final value $A \oplus B_0$ and that τ is an initial segment of B_0. Hopefully it is clear that it will be easy to engineer a situation in which we declare an infinite number of nodes for 0 of the form $\beta' = (\phi', n')$ such that ϕ' is an initial segment of the final value $A \oplus B_0$. It is convenient to ensure that if $\beta' = (\phi', n')$ and $\beta'' = (\phi'', n'')$ are nodes of this kind and $n'' > n'$ then $\phi'' \supset \phi'$. Now if T_0 is infinite then it will be the case for an infinite number of such nodes β' that $\Omega_1(\beta', 0, q')$ is defined and is never declared not to be a modifier (except that actually we shant bother defining $\Omega_1(\beta', 0, q')$ at nodes β' where the requirement $\mathcal{G}_0$ already looks to be satisfied). So suppose that β is a node of this kind and suppose that there does not exist $n'' > n$ such that $\Psi_0^{B_0}(1, 0, \alpha', n'') \downarrow = 1$. Then after the first stage at which $\hat{\psi}(\beta, 0)$ becomes defined it will never become undefined. Thus, we shall be able to declare a node $\beta' = (\phi', n+1)$ for 0 such that ϕ' is an initial segment of the final value $A \oplus B_0$ and enumerate an axiom $\Gamma_0^{\phi'} = \psi'$ such that $\Psi_0^{\psi'}(0) \downarrow$. Of course this is a particularly simplified situation and in the general context there will be a little more complexity to deal with.

2.4 If n is a modifier then we must ensure that it is responsible for the failure to enumerate only a finite number of n' into A. In the general context we shall actually have to consider 'modifier groups' – in order to proceed for the pair (h, q) such that $|q| = \Sigma_{i=0}^h 2^i$ we shall need a group of modifiers consisiting of one modifier from each S_j such that $q(j) \downarrow = 1$. We must ensure that each group of modifiers is responsible for the failure to enumerate only a finite number of n' into A. This will suffice in order to ensure that each modifier is only responsible for the failure to enumerate a finite number of n' into A since we shall be able to ensure that each modifier only belongs to one modifier group at any given stage of the construction and that, if we change the modifier group to which n belongs at an infinite number of stages then, for all but a finite number of the modifier groups to which n is declared to belong at some stage, there is some modifier in the group which is subsequently declared not to be a modifier. From what follows it will be clear that this last condition is sufficient to ensure that the modifier group is not responsible for the failure to enumerate any n' into A.

We are yet to specify the exact nature of the mechanism by which a modifier group may prevent the enumeration of n' into A, but in choosing such a mechanism we must be careful in order to prevent the possibility that, for a given n', while it is the case that no single modifier group seems responsible for the failure to enumerate n' into A (since, for example, any such group only prevents enumeration for a finite number of stages), the combined effect of the action of all modifier groups is to prevent the enumeration of n' into A for the entire duration of the construction. In order to avoid such complications we take the following approach. Suppose that at stage s we declare some modifier group consisiting of $n_0, .., n_m$ and we declare that this is a modifier group for the pair (h, q). Now suppose that at some subsequent stage we are looking to preserve $(\hat{\psi}(\beta, h))^\star$ as an initial segment of A for the sake of (h, q), where $\beta = (\phi, n)$ is a node that we have declared for $g = \langle e, z \rangle$ (and where $h = \langle e, z, k \rangle$). If we are to be able to use this particular modifier group consisiting of $n_0, .., n_m$ we insist that it must be the case $n \leq s$. We are only interested that this preservation should be successful in the case that ϕ is an initial segment of the final value $A \oplus B_e$ (should it be the case that $B_e \downarrow$). Therefore the first thing that we do is to find n_{m+1} which has been enumerated into S_j such that $e^\star(j) = e$ and which will be enumerated into $A^\star$ at any subsequent stage s' such that $\phi^\dagger \not\subset \tau_{e,s'}$. Let n_{m+2} be the largest number which has been enumerated into $S_{j'}$ such that $q(j') \downarrow = 0$ and such that

$(\hat{\psi}(\beta,h))^\star(n_{m+2}) \downarrow = 0$. Now let $n_{m+3} = max\{n_0, ..., n_{m+2}\}$ and let σ_0, σ_1 be the initial segments of $A^\star, A$ of length $n_{m+3} + 1$ respectively. For each n' such that $(\hat{\psi}(\beta,h))^\star(n') \downarrow = 0$ and either a) n' has been enumerated into $S_{j'}$ such that $q(j') \downarrow = 1$ and n' is greater than the corresponding modifier n_i (also enumerated into $S_{j'}$), or b) n' has been enumerated into $S_{j'}$ such that $q(j') \uparrow$, we enumerate $\sigma_0 \oplus \sigma_1$ into the set $\Delta_1(n')$ – this means that n' cannot be enumerated into A so long as it seems that $\sigma_0 \oplus \sigma_1$ is an initial segment of $A^\star \oplus A$. Now it is worth noting that at the end of section 2.3. we were able to remark that, according to the use of modifiers described up to that point (and in the situation there described), the first stage at which $\hat{\psi}(\beta,0)$ becomes defined it will never become undefined. This does not reflect the more general situation and approach that we have now described. All that we actually require, however, is that the following should be true:

(†) Suppose that $B_e \downarrow$, $\beta = (\phi,n)$ is a node for $g = \langle e, z \rangle$ such that $\phi \subset A \oplus B_e$, that $q \in 2^{<\omega}$ is of length $\Sigma_{i=0}^h 2^i$, where $h = \langle e, z, k \rangle$ and that the alpha tuple $\alpha = (q, i_0, .., i_y)$ is 'correct' as a guess about A i.e. letting $\{j : q(j) \downarrow = 0\} = \{j_0, .., j_y\}$ we have that if $q(j) \downarrow = 1$ then T_j is infinite and if $q(j) \downarrow = 0$ then $T_j = D_{i_{y'}}$ where $j = j_{y'}$. If there are an infinite number of stages at which $\hat{\psi}(\beta,h)$ is defined and complies with α then there is a stage after which this value never becomes undefined.

In order to see that (†) will hold we can argue as follows. We are yet to describe precisely the manner in which modifier groups will be declared, but hopefully it is clear that if α is correct as a guess about A (according to the precise definition of this terminology given in the above) then there will be an infinite number of stages at which we are able to declare a modifier group for (h, q) which will never subsequently be declared not to be a modifier group. Thus there will be a stage, s_0 say, after which, whenever $\hat{\psi}(\beta,h)$ is defined and complies with α, the values $n_0, .., n_{m+2}$ (as defined above) will always take the same value as at the last such stage. At such stages then, there is a point after which any strings $\sigma_0 \oplus \sigma_1$ which we enumerate into some set $\Delta_1(n')$ will be of a fixed length, l say. So let $s_1 > s_0$ be large enough such that the initial segment of $A^\star_{s_1} \oplus A_{s_1}$ of length l is an initial segment of $A^\star \oplus A$. Let $s_2 > s_1$ be a stage at which $\hat{\psi}(\beta,h)$ is defined and complies with α. Then $\hat{\psi}(\beta,h)$ is never made undefined subsequent to stage s_2.

If $n' \in A^\star - A$ is enumerated into S_j and a modifier group to which $n \in S_j$ belongs is responsible for the enumeration of $\sigma_0 \oplus \sigma_1$ which is an initial segment of $A^\star \oplus A$ into $\Delta_1(n')$ then we say that this modifier group is responsible for the failure to enumerate n' into A. If we are looking to preserve $(\hat{\psi}(\beta,h))^\star$ when this enumeration is made we also say that (β,h) is responsible for the failure to enumerate n' into A. Now the use of the parameter n_{m+1} described above in determining the length of the string $\sigma_0 \oplus \sigma_1$ means that (β,h), with $\beta = (\phi,n)$ say, cannot be responsible for the failure to enumerate n' into A unless there is a stage after which it is always the case $\phi \subset A_s \oplus \tau_{e,s}$ (and where $h = \langle e, z, k \rangle$). If the modifier group is declared at stage s then there can only be a finite number of nodes $\beta' = (\phi', n')$ declared for $g = \langle e, z \rangle$ of this kind such that $n' \leq s$. The modifier group is only availble for use at nodes $\beta' = (\phi', n')$ such that $n' \leq s$ and, hopefully it is clear that, any individual node can only be responsible for the failure to enumerate a finite number of n' into A.

Definition 2.6 *Given strings σ_0, σ_1 such that $|\sigma_0| \leq |\sigma_1|$, let σ_2 be the initial segment of σ_1 of length $|\sigma_0|$. We call $\sigma_0 \oplus \sigma_2$ the 'reduction' of $\sigma_0 \oplus \sigma_1$.*

In actual fact, various technicalities mean that it is convenient to be able to enumerate strings into sets $\Delta_1(n')$ of the form $\sigma_0 \oplus \sigma_1$ such that $|\sigma_0| \leq |\sigma_1|$. We agree, however, that the modifier group responsible for such an enumeration must take responsibility for the failure to enumerate n' if the reduction of $\sigma_0 \oplus \sigma_1$ is an initial segment of $A^* \oplus A$. It is then easy to argue that if $n' \in A^* - A$ then there is $\sigma_0 \oplus \sigma_1$ enumerated into $\Delta_1(n')$ such that the reduction of this string is an initial segment of $A^* \oplus A$.

2.5 There are two significant problems which remain to be addressed. The first of these problems we shall consider in this subsection. The second is dealt with in [AL].

Definition 2.7 *Suppose that $\beta = (\phi, n)$ is a node which is declared for $g = \langle e, z \rangle$, that $h = \langle e, z, k \rangle$, that at some stage of the construction we define $\hat{\psi}(\beta, h) = \psi$ and that ψ^* is an initial segment of the final value of A. Then we say that (β, h) fixes an initial segment of A at stage s. We also say that h fixes an initial segment of A at stage s.*

Now suppose we were to proceed simply by enumerating A and each S_j in order to try and ensure that T_j is infinite iff W_j specifies a convergent approximation. The immediate problem arising is that there may be an infinite number of $(\beta, 0)$, for example, which fix an initial segment of A and that this may cause infinite injury to $\mathcal{H}_1$. In order to overcome this problem we enumerate A and S_1 so as to ensure that if there are a finite number of stages at which 0 fixes an initial segment of A then T_1 is infinite iff W_1 specifies a convergent approximation. We enumerate A and S_2 so as to ensure that if there are an infinite number of stages at which 0 fixes an initial segment of A then T_2 is infinite iff W_1 specifies a convergent approximation.

The enumeration of S_1. In order to enumerate S_1 we can proceed basically as decribed in 2.1 except that we now allow elements of S_1 which are modifiers to prevent the enumeration of numbers into A, as described in 2.3. Modifications which we are about to introduce mean that we have to be a little more careful in arguing that, if 0 fixes only a finite number of initial segments of A, then there exist only a finite number of n such that there exists $(\beta, 0)$ which is responsible for the failure to enumerate n. These considerations are dealt with in the verification of [AL].

The enumeration of S_2. Let $\beta = (\phi, n)$ be a node which is declared for 0. When this node is declared and at every subsequent stage s at which $\phi \not\subset A_s \oplus \tau_{0,s}$ this node will be 'initialized'; for all j, l such that $e^*(j) > 0$ and either $\langle j, l \rangle > s$ or $p^*(j)(0) = 0$ we declare that the pair (j, l) lacks freedom at $(\beta, 0)$ (the second argument here taking the value $h = \langle e, z, k \rangle)$ and for all other j, l we declare that the pair (j, l) does not lack freedom at $(\beta, 0)$. Now suppose that at some stage s, $\phi \subset A_s \oplus \tau_{0,s}$, we are trying to preserve $(\hat{\psi}(\beta, 0))^{\star}$ as an initial segment of A for the sake of $(0, q)$ or $(0, q')$ (as defined previously in 2.3), $(\hat{\psi}(\beta, 0))^{\star}(n') \downarrow = 0$, n' has been enumerated into S_j such that $j > 0$ and that $\Delta_0(n') = \tau$ for some $\tau \in 2^{<\omega}$. Only if the pair $(j, |\tau|)$ lacks freedom at $(\beta, 0)$ shall we enumerate strings into $\Delta_1(n')$. If the enumeration of n' into A subsequently causes $\hat{\psi}(\beta, 0)$ to become undefined then we shall declare that the pair $(j, |\tau|)$ lacks freedom at $(\beta, 0)$. Clearly such modifications are unproblematic where the satisfaction of (†) is concerned – providing only finite injury. In enumerating S_2, then, we can proceed as follows. Suppose that we wish to enumerate some new n' into S_2 and define $\Delta_0(n')$ to be the initial segment of $\tau_{1,s}$ of length l. Before doing so we shall wait for a stage at which it is the case, for all nodes β that we have declared for 0, $\hat{\psi}(\beta, 0) \downarrow$ or $(2, l)$ does not lack freedom at $(\beta, 0)$. If it really is the case that there are an infinite number of stages at which 0 fixes an initial segment of A then there must exist such a stage s. In fact we must revise the form of the function Δ_0. When we enumerate n' into S_2 at stage s we let σ be the longest string such that, for some node β that we have declared for 0 and at which $(2, l)$ lacks freedom, $\hat{\psi}(\beta, 0) = \sigma$ and we define $\Delta_0(n') = (\tau, \sigma)$ where $\tau \subset \tau_{1,s}$ is of length l. Then at the first stage $s' \geq s$ at which *either* $\tau \not\subset \tau_{1,s'}$ or $\sigma \not\subset A$ we shall enumerate n' into $A^{\star}$ – and into A unless prevented from doing so by the appropriate modifiers.

References

[BC]   Cooper,S.B. 1973 *Minimal degrees and the jump operator*, J. Symbolic Logic 38, 249-271.

[RE]   Epstein,R.L. 1975 *Minimal degrees of unsolvability and the full approximation construction*, Memoirs of the American Mathematical Society, n. 163.

[JP]   Jockusch,C.G. and Posner,D. 1978 *Double-jumps of minimal degrees*, J. Symbolic Logic 43, 715-724.

[AL]   Lewis,A.E.M. 2005 *The jump classes of minimal covers*, (unabbreviated).

[SA]   Sasso, 1974 *A minimal degree not realising the least possible jump*, J,Symbolic Logic 39, 571-4.

[CY1]   Yates,C.E.M. 1967 *Recursively enumerable degrees and the degrees less than 0′*, Models and Recursion Theory, Crossley et al. eds., North Holland, 264-271.

[CY2]   Yates,C.E.M. 1970 *Initial segments of the degrees of unsolvability, part 2: minimal degrees*, J. Symbolic Logic, 35, 243-266.

Space Bounds for Infinitary Computation

Benedikt Löwe*

Institute for Logic, Language and Computation, Universiteit van Amsterdam,
Plantage Muidergracht 24, 1018 TV Amsterdam, The Netherlands
Mathematisches Institut, Rheinische Friedrich-Wilhelms-Universität Bonn,
Beringstraße 1, 53115 Bonn, Germany
Fachbereich Mathematik, Universität Hamburg, Bundesstrasse 55,
20146 Hamburg, Germany
bloewe@science.uva.nl

Infinite Time Turing Machines (or Hamkins-Kidder machines) have been introduced in [HaLe00] and their computability theory has been investigated in comparison to the usual computability theory in a sequence of papers by Hamkins, Lewis, Welch and Seabold: [HaLe00], [We00a], [We00b], [HaSe01], [HaLe02], [We04], [We05] (*cf.* also the survey papers [Ha02], [Ha04] and [Ha05]). Infinite Time Turing Machines have the same hardware as ordinary Turing Machines, and almost the same software. However, an Infinite Time Turing Machine can continue its computation if it still hasn't reached the HALT state after infinitely many steps (for details, see § 1).

In [Sc03], Schindler started the investigation of the corresponding complexity theory by defining natural time complexity classes for Infinite Time Turing Machines. Schindler, Welch, Hamkins and Deolalikar have proved with methods of descriptive set theory that the big open questions of standard complexity theory $\mathbf{P} \stackrel{?}{=} \mathbf{NP}$ and $\mathbf{P} \stackrel{?}{=} \mathbf{NP} \cap \mathbf{coNP}$ have negative answers for Infinite Time Turing Machines [Sc03, DeHaSc05, HaWe03].

For an ordinary Turing machine that stops in a finite number t of steps, it is easy to define its space usage: during its computation, it has used at most t cells of the tape, possibly less. This finite number of used cells can serve as a measure of space usage. A halting computation will have used a finite amount of time and space; if, however, time or space usage are infinite, then this corresponds to usage of order type ω and automatically implies that the computation was non-halting.

In this paper, we shall consider both Hamkins-Kidder machines and Koepke's *Ordinal Machines* as described in [Ko₀05a] and [Ko₀05b]. Koepke machines can not only extend their computation into transfinite ordinal time, but they also have ordinal-indexed cells on their tapes. Therefore, there is a natural notion of space usage for computations on Koepke machines that corresponds to the classical idea of space constraints on Turing Machines: just count the number (order type) of cells being used.

---

* The author thanks Joel Hamkins (New York NY), Peter Koepke (Bonn), Philip Welch (Bristol) and Joost Winter (Amsterdam) for discussions about infinitary computation, and the anonymous referees for important comments.

A. Beckmann et al. (Eds.): CiE 2006, LNCS 3988, pp. 319–329, 2006.
© Springer-Verlag Berlin Heidelberg 2006

This is very different for Hamkins-Kidder machines whose space is constrained to a tape of order type ω whereas time can have arbitrary ordinals as order type. This asymmetry makes is hard to give a definition of space usage that can be compared to time usage.

In this paper, we discuss the basics of possible definitions for space constraints for the mentioned two types of infinitary Turing machine computations. In §1, we give all definitions needed in the paper and then briefly discuss space constraints for Koepke's machines in §2 and space constraints for Hamkins-Kidder machines in §3. Finally, in §4, we discuss nondeterministic computation.

This paper raises very general questions about infinitary algorithms. We list them here and will explain the questions in more detail in the respective sections:

1. Are there any algorithms for Koepke's ordinal machines that use the additional transfinite space in order to compute more than Hamkins-Kidder machines within time restrictions? (§2; Theorem 3 gives an example of a use of the additional space, but it is not time efficient.)
2. Are there any algorithms for Hamkins-Kidder machines that compute complicated sets with unlimited time but very simple snapshots on the scratch tape? (§3; Question 10 provides a very basic test question.)
3. Are there any nondeterministic algorithms that are space efficient? (§4; Proposition 11 gives a general description of nondeterministic algorithms that mimic *guess nondeterminism*, but they are not space efficient.)

1   Definitions

In the following, we shall give a description of both Hamkins-Kidder machines and Koepke's ordinal machines.

Like ordinary Turing machines, both types of infinitary Turing machines consist of an **input tape**, a **scratch tape** and an **output tape**, a **reading/writing head**, a finite set of **states** and a program δ that assigns to a state s and the content of a bit on the scratch tape and the input tape an **action**. The action consists of moving the head right, moving the head left, writing on the output tape, writing on the scratch tape, erasing on the scratch tape, or a combination of these actions. Note that we may not erase on the output tape; this is to make sure that the program doesn't abuse the output tape as a scratch tape (in our definition of the space complexity, we shall only count the complexity of snapshots on the scratch tape).[1]

In the case of the Hamkins-Kidder machines, all of the three tapes have order type ω (as for ordinary Turing machines), in the case of the Koepke machines, the tapes are class-sized with a cell for every ordinal. If we have a class-sized tape, then we have to say what the machine will do if it in a cell indexed with a limit ordinal and receives the comment "move left". In that case, we'll move the reading/writing head to the 0th cell.

---

[1] Since we're only discussing decision problems here, *i.e.*, the output is either 0 or 1, this is equivalent to saying that the output tape has only one bit.

If we fix a machine T, an input x and an appropriate time α, then we write $s_\alpha^T(x)$ for the state that the machine is in and $h_\alpha^T(x)$ for the position of the head at time α with initial input x. If β is an index for a cell on the scratch tape, then we write $c_\alpha^T(x, \beta)$ for its content at time α. We also use $c_\alpha^T(x)$ for the function $\beta \mapsto c_\alpha^T(x, \beta)$ which we call the **snapshot at time** α. Note that this is a function with domain ω for Hamkins-Kidder machines, and a class function with domain Ord for Koepke machines.

For finitary computation, the times α mentioned in the last definitions are always finite. Infinitary Turing machines differ from a normal Turing machines in that they are allowed to continue their computation beyond ω many stages of computation. At a limit step of the computation, all the cells on the tape are adjusted according to the limit behaviour of the entries along the infinite computation: If 0 occurred cofinally often, the cell will get value 0 in the limit step. If on the other hand, from a point on, 1 was written in the cell, the cell will get value 1 (this corresponds to taking the *liminf* of the cell values). The state of the machine at a limit stage λ will also be the *liminf* of the states below λ. Note that while this is the definition from [Ko₀05a] for Koepke machines, it is not the standard definition for Hamkins-Kidder machines: in [HaLe00], these have a designated LIMIT state that is assumed in all limit stages. In terms of computational power, the two definitions for Hamkins-Kidder machines don't make a difference (as long as we have more than one tape).

The position of the reading/writing head at a limit stage λ is where our two infinitary models differ: For the Hamkins-Kidder machines, the head will always be moved to cell 0 at a limit stage: consequently, the head will never move to a cell indexed with an infinite ordinal. For the Koepke machines, the head will be moved to

$$h_\lambda^T(x) := \liminf_{s_\alpha^T(x) = s_\lambda^T(x)} h_\alpha^T(x).$$

Let us summarize the behaviour of the three types of machines in the following table:

	time	tape(s)	cells at limit	head at limit
Turing machines	ω	ω	N.A.	N.A.
Hamkins-Kidder machines	Ord	ω	lim inf	first cell
Koepke machines	Ord	Ord	lim inf	lim inf

We say a machine **accepts** an input x if it reaches the HALT state and has 1 on the output tape at that time. If it yields 0, we say that it **rejects** the input x. A set A is **(Turing, Hamkins-Kidder, Koepke) decidable** if there is a (Turing, Hamkins-Kidder, Koepke) machine that accepts exactly the elements of A and rejects exactly the reals not in A. Let us denote the set of Turing (Hamkins-Kidder, Koepke) decidable sets by $\mathbf{Dec}^T$ ($\mathbf{Dec}^{HK}$, $\mathbf{Dec}^K$).

For all of the three types of machines, there is an obvious definition of time usage for a halting computation: if T is a machine of the appropriate type that reaches the HALT state at input x, then we write $\mathbf{time}(x, T)$ for the first α such that $s_\alpha^T(x) = $ HALT. If $f : \mathbb{R} \to$ Ord is a function assigning ordinals to inputs, we say that T is an **time f machine** (or more simply, an f-machine) if for all x, we have $\mathbf{time}(x, T) < f(x)$.

Following Schindler [Sc03], we write $\mathbf{P}_f$ for the class of all sets of reals that are decidable by f-machines. If f is the constant function $f(x) = \xi$, we also call f-machines ξ-machines. We write $\mathbf{P}_\xi$ for the class of all sets decidable by an η-machine for some $\eta < \xi$. In order to distinguish the time classes for our types of machines, we write $\mathbf{P}_f^{HK}$ and $\mathbf{P}_\xi^{HK}$ for the time classes for Hamkins-Kidder machines and $\mathbf{P}_f^K$ and $\mathbf{P}_\xi^K$ for the time classes for Koepke machines.

Note that for Turing machines and Hamkins-Kidder machines, there is only a set of snapshots, whereas for Koepke machines, there is a proper class of snapshots. This simple observation has a portentous consequence for Hamkins-Kidder machines: they have far more time at their disposal than there are possible computation situations. If there are two limit ordinals $\lambda < \lambda^*$ such that the computation at λ and λ^* has the same state and snapshot and none of the cells with the value 1 at λ changes its value between λ and λ^*, we call this a **looping situation**.

Observation 1 (Hamkins-Lewis). *A Hamkins-Kidder machine does not halt if and only if its computation has a looping situation.*

Proof. [HaLe00, Corollary 1.2].[2]   q.e.d.

The analogue of Observation 1 is not true for Turing and Koepke machines, as they have exactly as much time as there are snapshots. In both cases consider the empty input and the machine that writes 1 and moves right if it hits a 0. This machine will never halt nor loop.

Another observation that will be important is that infinitary computations can be done in initial segments of the constructible hierarchy. For a Hamkins-Kidder machine T and any time α and input x, let $c_\alpha^T(x)$ be the content of the full tape (of order type ω) at time α (with input x and machine T).

Observation 2. *For any Hamkins-Kidder machine T, any ordinals $\alpha < \xi$ such that ξ is admissible, and any $x \in 2^\omega$, we have that*

$$c_\alpha^T(x) \in \mathbf{L}_\xi[x].$$

2   Koepke's Ordinal Machines

Koepke's analysis of ordinal machines from [Koo05a, Koo05b, KooKo1∞] does not pay attention to either computing resources or real numbers. Whereas we are interested in decision problems, he is interested in creating (characteristic functions of) sets on the output tape and allows as input finite sets of ordinals as parameters.

---

[2] The diligent reader checking this again [HaLe00] might notice that they write "the cells which are 0 at the limit never subsequently turn into 1 (we allow the 1s to turn to 0 and back again)". This is due to the fact that [HaLe00] uses a *limsup* rule instead of our *liminf* rule.

However, if you restrict your attention to decision problems, Koepke machines are still more expressive than Hamkins-Kidder machines as the following result shows:

Theorem 3. *The halting problem for Hamkins-Kidder machines is Koepke decidable. Hence, the set of Koepke decidable sets of reals is strictly bigger than that of Hamkins-Kidder decidable sets.*

Proof. By [HaLe00, Theorem 4.1], the Hamkins-Kidder halting problem is not Hamkins-Kidder decidable but semi-decidable. We only have to give an algorithm to decide the complement of the Hamkins-Kidder halting problem with a Koepke machine.

We shall be using Observation 1. In order to find out whether a computation doesn't halt, we can just check whether a looping situation occurred in the computation.

Since a Koepke machine has an unlimited amount of space, and every Hamkins-Kidder situation can be coded as a real, we can simulate the run of a Hamkins-Kidder machine while keeping track of the entire computation on the class-sized tape. It is now easy to check whether a looping situation occurred.   q.e.d.

If you look at the algorithm used to decide the Hamkins-Kidder halting problem in this proof, you'll notice that it is neither time nor space efficient. This is a general problem with complexity theory for Koepke machines: while Theorem 3 proves that you can use the size of the tape to compute more, there is no known technique to use it in order to compute faster.

As a consequence, we do not know any non-trivial separation results of time complexity for Hamkins-Kidder machines and Koepke machines.

Proposition 4. *If $\omega^2 \leq \alpha \leq \omega_1^{\mathrm{CK}}$, then $\mathbf{P}_\alpha^{\mathrm{K}} = \mathbf{P}_\alpha^{\mathrm{HK}}$.*

Proof. Fix $\eta < \alpha$. Given a set A that is Koepke decidable by an η-machine, we will describe how we decide it with a Hamkins-Kidder η-machine. Note that the original η-machine can never use more than the first η many cells of the class-size tape.

Since $\eta < \omega_1^{\mathrm{CK}}$, we can produce a code of η on the scratch tape within ω steps. After that, we use that code in order to think of the ω-tape as an η-tape and run the Koepke algorithm on it. This combined algorithm takes $\omega + \eta = \eta$ steps.   q.e.d.

What if we allow a Koepke machine more than a constant amount of time? Let $f_0(x) := \omega_1^x$. Then what is $\mathbf{P}_{f_0}^{\mathrm{K}}$? By [DeHaSc05, Theorem 4.2 (ii)], $\mathbf{P}_{f_0}^{\mathrm{HK}} = \mathbf{P}_{\omega_1^{\mathrm{CK}}}^{\mathrm{HK}}$; this was strengthened by Welch [We06, Proposition 2] to the following result:

Proposition 5 (Welch). *Every f_0-machine is an ω_1^{CK}-machine.*

An analogue of Proposition 5 for Koepke machines together with Proposition 4 would show that $\mathbf{P}_{f_0}^{\mathrm{K}} = \mathbf{P}_{f_0}^{\mathrm{HK}}$.

For a Koepke machine T, an input x and a time α, we write $u_\alpha^T(x) :=$ $\sup\{\beta\,;\,c_\alpha^T(x,\beta) \neq 0\}$ for the space used at time α. If a Koepke machine with input x reaches the HALT state at time $\mathbf{time}(x,T)$, we write

$$\mathbf{space}(x,T) := \sup\{u_\xi^T(x)\,;\,\xi < \mathbf{time}(x,T)\}.$$

If $f : \mathbb{R} \to \mathrm{Ord}$ is a function assigning ordinals to inputs, we say that T is an **space** f **machine** if for all x, we have $\mathbf{space}(x,T) < f(x)$. In analogy to Schindler's **P**-notation for time complexity, we write $\mathbf{PSPACE}_f^K$ for the class of all sets decidable by space f machines and $\mathbf{PSPACE}_\xi^K$ for the class of all sets decidable by space η machines for some $\eta < \xi$.

As for ordinary Turing machines, we immediately get $\mathbf{P}_f^K \subseteq \mathbf{PSPACE}_f^K$ (for all functions f), as each new used cell on the scratch tape requires one unit of time to be used or skipped.

Since Hamkins-Kidder machines are essentially just space ω Koepke machines, we can mimic arbitrary Hamkins-Kidder computations with space-bounded Koepke machines:

Proposition 6. $\mathbf{Dec}^{HK} \subseteq \mathbf{PSPACE}_{\omega+2}^K$.

Proof. Let T be a Hamkins-Kidder machine deciding A. Based on T, we construct a Koepke machine that can recognize when it is in a limit stage and that, whenever it is in a limit stage, moves to cell 0. (If it happens to be in cell ω, moving one step left will move the head to cell 0.) This machine mimics the limit behaviour of T and uses the cells up to the ωth cell, *i.e.*, is a space $\omega + 1$ machine.   q.e.d.

Corollary 7. $\mathbf{P} \neq \mathbf{PSPACE}$ *for Koepke machines, i.e.,* $\mathbf{P}_{\omega^\omega}^K \neq \mathbf{PSPACE}_{\omega^\omega}^K$.

Proof. Propositions 4 and 6 yield this simple separation result as follows:

$$\mathbf{P}_{\omega^\omega}^K = \mathbf{P}_{\omega^\omega}^{HK} \subsetneq \mathbf{Dec}^{HK} \subseteq \mathbf{PSPACE}_{\omega+2}^K \subseteq \mathbf{PSPACE}_{\omega^\omega}^K.$$

q.e.d.savit

3   Hamkins-Kidder Machines

For Hamkins-Kidder machines, the function $u_\alpha^T(x) := \sup\{\beta\,;\,c_\alpha^T(x,\beta) \neq 0\}$ as defined above will be equal to ω as soon as the entire tape is being used, and consequently for almost all non-trivial T and x, we'll have $\mathbf{space}(x,T) = \omega$. Thus, we have to use a different approach in order to get an informative measure for space usage.

In this section, we shall give two different definitions of space usage for Hamkins-Kidder machines. As above, let $c_\alpha^T(x)$ be the content of the full tape (of order type ω) at time α (with input x and machine T). We can define

$$\ell_\alpha^T(x) := \min\{\eta\,;\,c_\alpha^T(x) \in \mathbf{L}_\eta[x]\}, \text{ and}$$

$$\mathbf{space}^0(x,T) := \sup\{\ell_\xi^T(x)\,;\,\xi < \mathbf{time}(x,T)\}.$$

As before, we define a notion of **space^0 f machine** for Hamkins-Kidder machines and derive a definition of the class **PSPACE$_f^{\text{HK},0}$** from it. Let us call a function $f : \mathbb{R} \to \text{Ord}$ **admissible** if for all x, the ordinal $f(x)$ is x-admissible.

Proposition 8. *For any admissible function f, we have $\mathbf{P}_f^{\text{HK}} \subseteq \mathbf{PSPACE}_f^{\text{HK},0}$.*

Proof. Let T be a machine deciding A in time f. That means that for all inputs x, the computation has length shorter than $f(x)$. Fix some $\xi < f(x)$. By Observation 2, we have that $c_\xi^T(x) \in \mathbf{L}_{f(x)}[x]$.   q.e.d.

Alternatively, we define

$$\mathbf{space}^1(x, T) := \sup \left\{ \omega_1^{c_\xi^T(x)} \, ; \, \xi < \mathbf{time}(x, T) \right\} + 1,$$

and a notion of **space^1 f machine** and the class **PSPACE$_f^{\text{HK},1}$**.

Proposition 9. *For any admissible function f, we have $\mathbf{P}_f^{\text{HK}} \subseteq \mathbf{PSPACE}_f^{\text{HK},1}$.*

Proof. Let T be a machine deciding A in time f. Fix x, then the computation with input x has length shorter than $f(x)$. As in the proof of Proposition 8, we know that $\{ c_\xi^T(x) \, ; \, \xi < \mathbf{time}(x, T) \} \subseteq \mathbf{L}_{f(x)}[x]$.

Towards a contradiction, assume that $\mathbf{space}^1(x, T) \geq f(x)$, so there must be some ξ such that $\omega_1^{c_\xi^T(x)} = f(x)$. But then there is a code z for $f(x)$ that is (Turing)-recursive in $c_\xi^T(x) \in \mathbf{L}_{f(x)}[x]$, and hence $z \in \mathbf{L}_{f(x)}[x]$, contradicting the x-admissibility of $f(x)$.   q.e.d.

Propositions 8 and 9 are instances of the slogan "Using space costs time". This is equally true for classical, finitary complexity theory. Even for finite time Turing machines, it is not known whether $\mathbf{P} \subsetneq \mathbf{PSPACE}$. This question can be rephrased as

"Are there space-efficient algorithms for problems that cannot be solved quickly?"

Of course, this question can be applied to the three relations

$$\mathbf{P}_f^{\text{K}} \subseteq \mathbf{PSPACE}_f^{\text{K}},$$
$$\mathbf{P}_f^{\text{HK}} \subseteq \mathbf{PSPACE}_f^{\text{HK},0}, \text{ and}$$
$$\mathbf{P}_f^{\text{HK}} \subseteq \mathbf{PSPACE}_f^{\text{HK},1}$$

as well. The first algorithm that comes to mind is the Hamkins-Lewis algorithm for deciding Π_1^1 sets [HaLe00, Count-Through Theorem 2.2]: it is not in $\mathbf{P}_{f_0}$ for $f_0 : x \mapsto \omega_1^x$; however, it is easily seen that this algorithm produces the ill-founded part of the relation coded by x on the scratch tape which is not in general in $\mathbf{L}_{f(x)}[x]$. As a consequence, the algorithm uses both a lot of time and a lot of space, and is no answer to the above question.

This example is illustrative in the following sense: looking at the different infinitary algorithms that are at our disposal, the only way that they use their infinite time is to produce more complicated reals on the scratch tape. This observation might lead to the conjecture that time and space complexity for infinitary computation are the same.

Let us highlight this bold conjecture with a precise test question: Call a Hamkins-Kidder machine **recursive** if it halts on all inputs and for all x and α, the real $c_\alpha^T(x)$ is (Turing-)recursive. We define **PSPACE*** to be the class of sets of reals that are decidable by recursive Hamkins-Kidder machines. Clearly,

$$\mathbf{PSPACE}^\star \subseteq \mathbf{PSPACE}_{\omega^\omega}^{\mathrm{HK},0} =: \mathbf{PSPACE}.$$

Question 10. *Can we prove that* **PSPACE**$^\star \subseteq \mathbf{P} := \mathbf{P}_{\omega^\omega}^{\mathrm{HK}}$?

4   Nondeterministic Computation

Savitch's Theorem [Pa94, Theorem 7.5] tells us that for Turing computations, nondeterminism does not increase space efficiency (in other words, **PSPACE** = **NPSPACE**).

In this section, we briefly look at the interaction between nondeterminism and our space complexity classes for Hamkins-Kidder machines and Koepke machines.

In [Sc03], Schindler defined the class **NP**$_f$ without introducing a notion of nondeterministic Hamkins-Kidder computation. For reals x and y, we define as usual

$$y * x(n) := \begin{cases} y(k) \text{ if } n = 2k, \text{ and} \\ x(k) \text{ if } n = 2k+1. \end{cases}$$

We call a machine T a **$*$-time f machine** if it halts for all inputs and for all x and y, we have that $\mathbf{time}(y * x, T) < f(x)$. A set A is in **NP**$_f$ if there is a $*$-time f machine T such that

$$x \in A \iff \exists y (T(y * x)\!\downarrow = 1).$$

If f is a constant function, we can replace the "$*$-time f machine" with a "time f machine".

Schindler's notion naturally connects to a notion of nondeterministic computation: a nondeterministic Hamkins-Kidder machine is a machine with the Hamkins-Kidder architecture but instead of a program δ that is a function it has a relation that gives a set of allowed actions. A nondeterministic Hamkins-Kidder machine T is called a **nondeterministic time f machine** if all possible T-computations with input x halt before time $f(x)$. A set A is nondeterministically decidable by a machine T if there is at least one possible T-computation that accepts x.

Proposition 11. *Let $f : \mathbb{R} \to \mathrm{Ord}$ be a function such that for all x, we have $\omega \cdot f(x) = f(x)$. Then the following are equivalent for a set of reals A:*

1. *$A \in \mathbf{NP}_f$, and*
2. *A is nondeterministically decidable by a nondeterministic time f machine.*

Proof. "⇒": Let T be a ∗-time f machine for deciding A. At input x, we use the first ω stages of the computation to generate arbitrary witnesses by the simple program described by "write either 0 or 1 and move on". Thus, a nondeterministic Hamkins-Kidder machine can produce at stage ω of the computation all possible values of y on the scratch tape. Now run T on the arrangement of x on the input tape and y on the scratch tape as if it were $y * x$ on the input tape. We know that T will reach the HALT state in less than $f(x)$ steps. So the entire computation uses less than $\omega + f(x) = f(x)$ steps and one of the branches of the computation accepts x.

"⇐": Let T be a nondeterministic time f machine deciding A. Then we can see the computation of T at input x as a finitely branching tree of height at most $f(x) < \omega_1$. The branching pattern in each branch b of the tree can be coded into a real y_b (the code is an element of WO coding the length of the computation in the branch b, thus identifying each step of the computation with a natural number, and a function assigning the behaviour of T at the computation step coded by n in the branch b).

We can now define a ∗-time f machine T^* as follows: on input $y * x$, the machine checks whether y is a code (in the sense of the previous paragraph) for a T-computation with input x, and –as long as it is–, follows this computation. Note that each step of the T-computation may take ω steps in the T^*-computation, as T^* has to search for the next command to execute in the code y. If at any point it turns out that y is not a code for a T-computation, the machine HALTs and returns 0.

If b is an accepting branch of T, then $y_b * x$ will be accepted by T^*, and for each y, the computation with input $y * x$ will take at most $\omega \cdot f(x)$ steps.   q.e.d.

From the point of view of space constraints, it is easy to see that the proof of Proposition 11 is highly inefficient: the nondeterministic computation contains every single real as a potential snapshot of the scratch tape. This raises the third general question: can we come up with a nondeterministic algorithm that is space efficient?

More precisely, if T is a nondeterministic Hamkins-Kidder machine and b is a branch through its computation tree at input x with the sequence $\langle b_\gamma ; \gamma < \xi \rangle$ of snapshots occurring on the scratch tape during the computation along b, we write

$$\ell_\gamma(b) := \min\{\eta; b_\gamma \in \mathbf{L}_\eta[x]\},$$
$$\mathbf{space}^0(b) := \sup\{\ell_\gamma(b); \gamma < \xi\}, \text{ and}$$
$$\mathbf{space}^1(b) := \sup\{\omega_1^{b_\gamma}; \gamma < \xi\} + 1.$$

For $i \in \{0, 1\}$, we say that a Hamkins-Kidder machine T is a **nondeterministic space^i f machine** if all possible T-computations with input x halt for all branches b, we have $\mathbf{space}^i(b) < f(x)$. We say that $A \in \mathbf{NPSPACE}_f^{HK,i}$ if it is decidable by a nondeterministic space^i f machine.

Question 12. *For what functions f do we have*

$$\mathbf{PSPACE}_f^{HK} = \mathbf{NPSPACE}_f^{HK,i}?$$

Analogously, we can define nondeterministic space classes $\mathbf{NPSPACE}_f^K$ for Koepke machines. Hamkins and Welch have noticed [HaWe03, Theorem 1.7] that in general, nondeterministic Hamkins-Kidder computation can be more powerful than deterministic Hamkins-Kidder computation. Their proof shows that the Hamkins-Kidder halting problem is in $\mathbf{NP}_{\omega_1}^{HK}$. Combining this result with Propositions 6 and 11, we get that

$$\mathbf{PSPACE}_{\omega+1}^K \subsetneq \mathbf{NPSPACE}_{\omega+1}^K.$$

References

[CoLöTo05] S. Barry **Cooper**, Benedikt **Löwe**, Leen **Torenvliet** (*eds.*), CiE 2005: New Computational Paradigms, Papers presented at the conference in Amsterdam, June 8-12, 2005, Heidelberg 2005 [Lecture Notes in Computer Science 3526]

[DeHaSc05] Vinay **Deolalikar**, Joel D. **Hamkins**, Ralf-Dieter **Schindler**, $\mathbf{P} \neq \mathbf{NP} \cap$ **coNP** for Infinite Time Turing Machines, **Journal of Logic and Computation** 15 (2005), p. 577–592

[Ha02] Joel D. **Hamkins**, Infinite time Turing machines, **Minds and Machines** 12 (2002), p. 521–539

[Ha04] Joel D. **Hamkins**, Supertask Computation, *in:* [LöPiRä04, p. 141–158]

[Ha05] Joel D. **Hamkins**, Infinitary Computability with Infinite Time Turing Machines *in:* [CoLöTo05, p. 180–187]

[HaLe00] Joel D. **Hamkins**, Andy **Lewis**, Infinite time Turing machines, **Journal of Symbolic Logic** 65 (2000), p. 567–604

[HaLe02] Joel D. **Hamkins**, Andy **Lewis**, Post's problem for supertasks has both positive and negative solutions, **Archive for Mathematical Logic** 41 (2002), p. 507–523

[HaSe01] Joel D. **Hamkins**, Daniel E. **Seabold**, Infinite time Turing machines with only one tape, **Mathematical Logic Quarterly** 47 (2001), p. 271–287

[HaWe03] Joel D. **Hamkins**, Philip D. **Welch**, $\mathbf{P}^f \neq \mathbf{NP}^f$ for almost all f, **Mathematical Logic Quarterly** 49 (2003), p. 536–540

[Koo05a] Peter **Koepke**, Turing Computations on Ordinals, **Bulletin of Symbolic Logic** 11 (2005), p. 377–397

[Koo05b] Peter **Koepke**, Computing a model of set theory, *in:* [CoLöTo05, p. 223–232]

[KooKo1∞] Peter **Koepke**, Martin **Koerwien**, Ordinal computations, *to appear in:* Barry Cooper, Benedikt Löwe, Dag Normann (*eds.*), Mathematics of Computation at CiE 2005, special issue of the journal **Mathematical Structures in Computer Science**

[LöPiRä04] Benedikt **Löwe**, Boris **Piwinger**, Thoralf **Räsch** (*eds.*), Classical and New Paradigms of Computation and their Complexity Hierarchies, Papers of the conference "Foundations of the Formal Sciences III", Dordrecht 2004 [Trends in Logic 23]

[Pa94] Christos H. **Papadimitriou**, Computational Complexity, Reading MA 1994

[Sc03] Ralf **Schindler**, $\mathbf{P} \neq \mathbf{NP}$ for infinite time Turing machines, **Monatshefte der Mathematik** 139 (2003), p. 335–340

[We00a]   Philip D. **Welch**, The length of infinite time Turing machine computations, **Bulletin of the London Mathematical Society** 32 (2000), p. 129–136

[We00b]   Philip D. **Welch**, Eventually infinite time Turing machine degrees: Infinite time decidable reals, **Journal of Symbolic Logic** 65 (2000), p. 1193–1203

[We04]   Philip D. **Welch**, Determinacy and Post's Problem for Infinite Time Turing Machines, *in:* [LöPiRä04, p. 223–237]

[We05]   Philip D. **Welch**, Arithmetical Quasi-inductive definitions and the transfinite action of 1-tape Turing Machines, *in:* [CoLöTo05, p. 532–539]

[We06]   Philip D. **Welch**, Non-deterministic halting times for Hamkins-Kidder Turing machines, THIS VOLUME

From a Zoo to a Zoology:
Descriptive Complexity for Graph Polynomials

J.A. Makowsky

Department of Computer Science
Technion–Israel Institute of Technology, Haifa, Israel
janos@cs.technion.ac.il

Abstract. We outline a general theory of graph polynomials which covers all the examples we found in the vast literature, in particular, the chromatic polynomial, various generalizations of the Tutte polynomial, matching polynomials, interlace polynomials, and the cover polynomial of digraphs. We introduce the class of (hyper)graph polynomials definable in second order logic, and outline a research program for their classification in terms of definability and complexity considerations, and various notions of reducibilities.

1 Introduction

During the last ten years I have studied questions of computability of graph polynomials, summarized in [36, 38, 34, 41, 39, 40, 37, 8]. I found uncharted territory with plenty of amazing theorems, surprising results, and the more I got into it, the more I was perplexed. I feel that we do not have a comprehensive understanding of graph polynomials, although about particular polynomials, such as the characteristic polynomial, the chromatic polynomial, the matching polynomials and the Tutte polynomial there is more known than what could be told in several books. It is noteworthy that many authors speak in their papers of *the* graph polynomial, suggesting that theirs is the one and only one worth studying. It is also noteworthy, that very few authors who study a particular graph polynomial P, have more than this particular polynomial and possibly some immediate relatives of P, in mind.

In this paper I try to sketch a research program of how a general theory of graph polynomials could be developed. The collection of graph polynomials I have gathered from the literature looks like a zoo[1]. There are prominent animals like the elephant, the giraffe, the gorilla, and there are exotic animals defying classification, like the lamprey (petromyzon marinus, not really a fish) or platypus (ornithorhynchus anatinus, not really a water bird, not really a mammal). Some animals look different, but are related, like the elephant and the rock hyrax (procavia capensis); some look alike, but are not related, like the hedgehog (erinaceus europus) and the echidna (tachyglossus aculeatus). *Zoology* is the science

---

[1] It was T. Zaslavsky who suggested the titel "From a zoo to a zoology" for this research program.

A. Beckmann et al. (Eds.): CiE 2006, LNCS 3988, pp. 330–341, 2006.

of comparing and classifying animals. *Graphpolynomology* would be the art of comparing and classifying graph polynomials.

2   Graph Polynomials

Let $\mathcal{G}$ be the class of graphs $G = (V, E)$ without loops and multiple edges. Let $\mathcal{R}$ be a ring and $\bar{X}$ be a (not necessarily finite) set of indeterminates. A *graph polynomial* is a function

$$p : \mathcal{G} \to \mathcal{R}[\bar{X}]$$

such that for isomorphic graphs $G_1 \simeq G_2$ we have $p(G_1) = p(G_2)$. If we consider labeled graphs, the notion of isomorphism has to be correspondingly modified. If $p(G)$ takes only values 0 or 1 in $\mathcal{R}$ we speak of *graph properties*.

There are plenty of graph polynomials which have been discussed in the literature, although no systematic treatment on graph polynomials in general is available[2]. To put our results into perspective we discuss briefly four classical graph polynomials, the *chromatic polynomial* $\chi(G, \lambda)$, the *characteristic polynomial* $P(G, \lambda)$, the *acyclic generating matching polynomials* $m(G, \lambda)$ *and* $g(G, \lambda)$ and the *Tutte polynomial* $T(G, X, Y)$. For historic reasons we also discuss briefly the very first polynomial introduced into graph theory, the *edge difference polynomial*. We also add to our discussion two more recent examples, the two *interlace polynomials*, and the *cover polynomial* defined on digraphs.

The Edge-Difference Polynomial. The historically first polynomial in graph theory was introduced by J.J. Sylvester in 1878, [51] and further studied by J. Peterson in 1891. It is the multivariate polynomial depending on the ordering of the vertices $V = \{v_1, v_2, \ldots, v_n\}$ and defined as

$$P_G(X_1, X_2, \ldots, X_n) = \sum_{\substack{i < j \\ (v_i, v_j) \in E}} (X_i - X_j)$$

This polynomial is not a graph invariant, but it was used as a tool in studying regularity and colorability questions of graphs. In particular, N. Alon and M. Tarsi [3] observed that it can be used to study list colorings. For a survey, cf. Z. Tuza [55]. In our context, however, the edge-difference polynomial does not play a prominent rôle.

The Chromatic Polynomial. Let $\chi(G, \lambda)$ denote the number of proper vertex colorings of G with at most λ colors. G. Birkhoff, [7], observed in 1912 that $\chi(G, \lambda)$ is, for a fixed graph G, a polynomial in λ, which is now called the *chromatic polynomial of G*. The chromatic polynomial is the oldest graph polynomial to appear in the literature, which is a graph invariant. Since then a substantial body of knowledge about the chromatic polynomial of graphs and its applications has been accumulated. The recent book by F.M. Dong, K.M. Koh and

---

[2] I have found over 250 entries in MathSciNet querying "graph polynomial" or "polynomial of a graph" in the review text.

K.L. Teo [18] gives an excellent and extensive survey. One of the surprising facts is a theorem of R.P. Stanley, [49], which states that $\chi(G, -1)$ is the number of acyclic orientations of G.

The Tutte Polynomial. Interesting generalizations of the chromatic polynomial were introduced by H. Whitney in 1932 and W.T. Tutte in 1947. The most prominent among them is now called the *Tutte polynomial* $T(G, X, Y)$ which is a two variable polynomial from which the chromatic polynomial can be obtained via a simple substitution and multiplication with a prefactor. The exact relationship is given by

$$\chi(G, X) = (-1)^{r(G)} X^{k(G)} T(G; 1 - X, 0)$$

Here $k(K)$ is the number of connected components of G and $r(G) = \mid V \mid -k(G)$.

For a modern exposition the reader is referred to [10, chapter X], [24] or [58]. Tutte's own account on how he got involved with his polynomial is very enjoyable, [54]. For this extended abstract we do not need a full definition of the Tutte polynomial. We only note that it is a polynomial in two variables related to the rank generating function of matroids. But we should note that in the years after 1980 the Tutte polynomial found important interpretations in statistical mechanics and quantum field theory, knot theory, and biology, cf. [58] and [48]. F. Jaeger in [29] showed that the Jones polynomial of knot theory on alternating knots is just an instance of the Tutte polynomial of the knot diagram viewed as a graph. L. Kauffman in [31] introduced first a generalization of the Tutte polynomial which gives the Jones polynomial for arbitrary knots. Different approaches to multivariate versions of the Tutte polynomial are discussed in [11, 48].

Other univariate graph polynomials were introduced after 1955, often first motivated by problems from chemistry and physics. The two most prominent are the characteristic and the matching polynomial (which comes in two versions).

The Characteristic Polynomial of a Graph G, denoted by $P(G, \lambda)$ is the characteristic polynomial of the adjacency matrix M_G of the graph G, $P(G, \lambda) = \det(\lambda \cdot 1 - M_G)$ and is completely determined by the eigenvalues of M_G, which are all real, as the matrix is symmetric.

The Matching Polynomials. The *acyclic polynomial of* G is the polynomial $m(G, \lambda) = \sum_k (-1)^k \cdot m_k(G) \cdot \lambda^{n-2k}$, where the coefficients $m_k(G)$ count k-matchings. A chemical point of view of these polynomials is given in [17] and [53], where algorithmic aspects are also touched. A close relative of the acyclic polynomial is the *matching generating polynomial of a graph* G defined as $g(G, \lambda) = \sum_k m_k(G)\lambda^k$, where $m(G, \lambda) = \lambda^n g(G, (-\lambda^{-2}))$. An excellent survey on these two matching polynomials may be found in [25, Chapter 1] and [35, Chapter 8.5]. We shall refer to both as *matching polynomials*. Somewhat surprisingly we have $m(G, \lambda) = P(G, \lambda)$ if and only if G is a forest.

The Interlace Polynomials. Two of the more interesting recent graph polynomials were introduced by R. Arratia, B. Bollobás and G. Sorkin in [5, 6]. They are called *interlace polynomials* and there is a univariate and a two-variable version. M. Aigner and H. van der Holst [2] studied these polynomials from a matrix point of view and derived various combinatorial interpretations.

The Cover Polynomial of Directed Graphs. An interesting recent graph polynomial on directed graphs is the *cover polynomial* introduced by F.R.K. Chung and R.L. Graham [14], and independently in the context of rook polynomials, by Gessel, [23]. In [14] it is presented as an attempt to create a Tutte-like polynomial for directed graphs, and is closely related to the chromatic polynomial. There is also related work by R.P. Stanley [50] and T. Chow [13], and very recently, by P. Pitteloud [46].

A Zoo of Graph Polynomials. Without giving all the necessary references, we list a few of the many graph polynomials we found in the literature. There are variations of matching polynomials, like the rook polynomials, cf. [47]. There are polynomials counting the number of (induced) subgraphs of a certain kind. Let $\mathcal{H}$ be a graph property and put $ind_{\mathcal{H}}(G, k)$ be number of induced subgraphs of size k having property $\mathcal{H}$ in a given graph G. Then we can look at the polynomial

$$gen_{\mathcal{H}}(G, \lambda) = \sum_k ind_{\mathcal{H}}(G, k)\lambda^k$$

For $\mathcal{H}$ consisting of all the K_n's (cliques), E_n's (isolated points), C_n's (cycles), P_n's (paths) the corresponding polynomials have been studied. Instead of graph properties one can also use subsets of graphs with desirable properties such as vertex covers, coverings with subgraphs of special type etc. Some of these have been studied in a very general context as Farrell polynomials, cf. [22, 39]. There are interlace polynomials [5], Go-polynomials [21], Penrose polynomials [1], and many more. It is worth searching for all these at `scholar.google.com`.

3 Recursive Definitions

One of the outstanding features of the more prominent graph polynomials are recursive definitions with respect to some order independent way of deconstructing the input graph. The main paradigm stems from the chromatic polynomial and the Tutte polynomial. We first note that for the chromatic polynomial we have

$$\chi(G) = \chi(G - e) - \chi(G/e)$$

where e is an edge and $G - e$ and G/e denotes the deletion respectively contraction of the edge e. Furthermore, for the disjoint union we have $\chi(G_1 \sqcup G_2) = \chi(G_1) \cdot \chi(G_2)$, for the graph consisting of n isolated verices E_n $\chi(E_n) = \lambda^n$. One easily verifies that $\chi(G - e - f) = \chi(G - f - e)$, $\chi(G/e/f) = \chi(G/f/e)$, $\chi(G - e/f) = \chi(G/f - e)$ and $\chi(G/e - f) = \chi(G - f/e)$, which is a kind of *Church-Rosser property* or *confluence property*. From this we get a recursive definition of $\chi(G)$ by choosing any order of the edges. Similarly, for the Tutte polynomial we have

$$T(G, X, Y) = \begin{cases} X \cdot T(G/e, X, Y) & \text{if } e \text{ is a bridge} \\ Y \cdot T(G - e, X, Y) & \text{if } e \text{ is a loop} \\ T(G/e, X, Y) + T(G - e, X, Y) & else \end{cases}$$

together with multiplicativity for disjoint unions and $T(E_n, X, Y) = 1$. Again one can verify the Church-Rosser property, and gets a recursive defintion for the Tutte polynomial. In [11] this recursive defintion was used as the starting point for the defintion of the *colored Tutte polynomial*.

In [5, 6] similar but more complicated recursive definitions are given for the various interlace polynomials. Here the recursion also involves a *pivot operation* G^{ab} on a graph G and two vertices a, b. In [14] such recursive definitions are given for the cover polynomial of directed graphs. Even for the matching polynomial one can give such rules: for the acyclic polynomial we have $m(E_n) = \lambda^n$, multiplicativity for disjoint unions and, for an edge $e = (u, v)$

$$m(G, \lambda) = m(G - e) - m(G - u - v, \lambda)$$

It is a curious fact that the literature does not explore this aspect of the matching polynomial further, and does not even note the Church-Rosser property, although it is easily verified.

The recursive definition of a graph polynomial gives an easy but slow way of computing the graph polynomial. As the recursion unwinds a number of subtasks exponential in the size of the graph has to be computed. But the nature of the recursion usually gives deeper insights. Furthermore, the various Tutte and interlace polynomials can be proven to be, in a certain sense, the most general graph polynomials satisfying their specific recursion scheme. Similar characterizations very recently shown for generalizations of the cover and the matching polynomials in [15].

Although some particular recursion schemes based on the behaviour of the graph polynomial under deletion of vertices or edges, contractions of edges, pivoting, etc. are well studied, no general theory has emerged so far, and it remains an interesting challenge to develop a satisfactory framework for recursion schemes for graph invariants.

4   Complexity

It is natural to ask how difficult it is to compute the various graph polynomials. The characteristic polynomial is computable in polynomial time using classical algorithms for the determinant of a matrix. Computing the chromatic polynomial is $\sharp$**P**-hard due to its connection to counting colorings. This also makes computing the Tutte polynomial $\sharp$**P**-hard. The same is true for the acyclic polynomial due to its connection to counting matchings, cf. [57]. Furthermore, F. Jaeger, D. Vertigan and D. Welsh, [30], have characterized completely the points (a, b)

in the complex plane $\mathbb{C}^2$, where evaluating the Tutte polynomial $T(G, a, b)$ is difficult for arbitrary graphs. J. Oxley and D. Welsh [45] also noted that the Tutte polynomial for series–parallel graphs, which are graphs of tree-width at most 2, can be computed in polynomial time. This was extended to arbitrary fixed tree-width k independently by A. Andrejak [4] and S. Noble [43], and therefore also holds for the chromatic polynomial. Actually, they showed that computing the Tutte polynomial is fixed parameter tractable **FPT** on graph classes of tree-width at most k. In other words, it is computable in time $f(k)n^d$, where d is independent of k and n is the size of the input. For an extensive discussion of the complexity class **FPT**, cf. [19].

5 Enter Logic

Already D. Courcelle in [16] observed that graph properties definable in Monadic Second Order Logic (MSOL) are in **FPT** on graph classes of tree-width at most k, cf. also [19]. This approach was extended to graph polynomials by the author in [37]. The fact that the Tutte polynomial is in **FPT** also follows from [37], which also covers the acyclic and the matching polynomial and a wide range of other graph polynomials where summations are restricted to families of subsets of edges which are definable in MSOL. Without going into the more delicate details, such polynomials are in a polynomial ring $\mathcal{R}[\bar{X}]$ and are of the form

$$g(G, \bar{X}) = \sum_{A:\phi(A)} \prod_{v:v \in A} t(v)$$

where A is a unary relation variable, $\phi(A)$ is an MSOL-definable property of the graph, and $t(v)$ is a term in $\mathcal{R}[\bar{X}]$ which may depend uniformly on v. We speak here of MSOL-polynomials.

In the same paper [37], it is shown that, in combination with the work of P. Seymour and S. Oum [44], graph polynomials, where summations are restricted to families of subsets of vertices which are MSOL-definable, are in **FPT** for graph classes of clique-width at most k. However, this method does not apply to the chromatic polynomial, the Tutte polynomial and the matching polynomials.

6 The Need for a General Framework

Although there is a large *zoo of graph polynomials*, there is no *general zoology*. We offer here an outline of what such a zoology could look like. Our general framework is somewhat inspired by [26], but both the scope and the emphasis are quite different. Initial work in this direction may be found in [41].

The literature on Turing complexity or algebraic complexity does not provide a natural framework to develop a complexity theory of graph polynomials. In particular there is no agreed upon notion of *efficient reducibility between graph polynomials*. The existing framworks do allow the formulation of hardness results by reductions to ♯**P**-hard instances which are easily recognizable as polynomial time computable in an intuitive sense. But in the existings frameworks no *hardest graph polynomial* could be identified.

6.1 SOL-Polynomials as a Complexity Class

We propose a class of graph polynomials which covers all the examples, so far, from the literature, which has sufficient closure properties, and which guarantees that all its members are computable in exponential time in the unit cost computational model over the underlying ring $\mathcal{R}$, in the sense of the Blum-Shub-Smale model of computation, [9]. For this purpose we allow (full) Second Order Logic in the definition of the polynomials:

$$g(G, \bar{X}) = \sum_{R:\psi(R)} \prod_{\bar{v}:\bar{v}\in R} t(\bar{v})$$

R now can be a relation variable of any fixed arity, and ψ any formula of Second Order Logic (SOL). We speak then of SOL-polynomials. If $\mathcal{L}$ is a sublogic of SOL and ψ is an $\mathcal{L}$ formula, we speak of $\mathcal{L}$-polynomials.

7 Towards a General Framework

The purpose of the general framework is to initiate a comparative study of the many graph, digraph and hypergraph polynomials which have appeared in the literature. For an extensive list of references cf. [37] and [39]. In particular, we address the following:

Universality. All the polynomials we have encountered in the literature can be put into the framework of SOL-polynomials. Sometimes this is obvious. The matching polynomial can be written as

$$g(G, \lambda) = \sum_{\substack{M:M\subseteq E \\ M \text{ is a matching}}} \prod_{e:e\in M} \lambda$$

Sometimes this needs a non-trivial proof, which is the case for the interlace polynomial.

Definability. Not all SOL-definable properties of graphs are MSOL-definable, though, and sometimes it is useful to look at variations of MSOL which allow quantification over edge sets or subsets of fixed relations, which we call MSOL-2. Guarded Second Order Logic is the fragment of Second Order Logic in which the relation variables have to range over subsets of the relations specified in the vocabulary. If the edge relation of the graph is the only relation specified by the vocabulary Guarded SOL is just MSOL-2.

As graph properties can be viewed as polynomials with constant value **true** or **false** (or **1** or **0**) this gives (too) easy examples of SOL-polynomials which are not MSOL-polynomials. The definition of the matching polynomial given above shows it is an MSOL-2-polynomial. It is hard to believe that it could be an MSOL-polynomial, but we do not know how to prove this. The interlace polynomial is an SOL-polynomial, of which we do not know whether it is an MSOL-polynomial. In [15] it shown to be definable in MSOL with a parity quantifier, but we do not know whether this can be avoided.

Recursion schemes and definability. The existence of a recursion scheme and SOL-definability both garantee that a graph polynomial is computable in exponential time. In the examples we know, every graph polynomial defined by a recursion scheme is also SOL-definable. Is this always true?

On the other hand it is unlikely that every SOL-polynomial has such a recursion scheme. How can we prove that for a given SOL-polynomial no recursion scheme exists? Can we give sufficient conditions which assure that an SOL-polynomial has a certain recursion scheme?

Comparability. Given two graph polynomials $f(G, \bar{x})$ and $g(G, \bar{x})$, we say that g is weaker than f, and write $g \preceq f$, if for any two graphs G_1, G_2 with $f(G_1, \bar{x}) = f(G_2, \bar{x})$ we also have $g(G_1, \bar{x}) = g(G_2, \bar{x})$. If $g \preceq f$ and $f \preceq g$, we say the polynomials are graph-equivalent. Comparability of graph polynomials is undecidable. This follows from the undecidablity of the conse-quence problem of First Order Logic if restricted to finite graphs, which was proven by M. Taitslin [52] and sharpened by I. Lavrov [32], cf. [27, Theorem 5.5.1]. We note that the two matching polynomials $m(G, \lambda)$ and $g(G, \lambda)$ are graph equivalent, but incomparable with respect to the characteristic polynomial $P(G, \lambda)$, and also with respect to the chromatic polynomial, and the Tutte polynomial. The study of this partial order among SOL-polynomials, MSOL-polynomials, or other restricted classes of graph polynomials is a natural topic of investigation. In particular, one can ask: is there a strongest SOL-polynomial, what are its additional structural properties, is it a lattice, etc.

Complexity. For a logic $\mathcal{L}$ which captures a complexity class $\mathbf{C}$ on ordered structures[3], we speak also of **C**-polynomials. Interesting cases for **C** are deterministic and non-deterministic Log-Space, denoted by **L** and **NL** respectively, and determinsitic polynomial time **P**. All examples in the literature actually are **P**-polynomials, most actually are **NL**-polynomials. For example, to see that the matching polynomial

$$g(G, \lambda) = \sum_{\substack{M : M \subseteq E \\ M \text{ is a matching}}} \prod_{e : e \in M} \lambda$$

is a **P**-polynomial, it suffices to note that "$M \subseteq E$ is a matching" is a property recognizable in polynomial time.

We have seen before that the chromatic polynomial is $\sharp\mathbf{P}$-hard to compute, hence **P**-polynomials are usually not computable in polynomial time. Neither are they in $\sharp\mathbf{P}$, as they can have arbitrary values in the polynomial ring. We propose two classes of graph polynomials as *natural complexity classes* for graph polynomials: the class of **P**-polynomials which we call $\mathbf{P} - POL$, and the class of SOL-polynomials, which we call $SOL - POL$. Clearly we have, $\mathbf{P} - SOL \subseteq SOL - POL$. Is the inclusion proper? Do these classes have complete problems with respect to some notion of reduction (see below)?

---

[3] In the sense of descriptive complexity, [28, 20, 33].

On the other hand we have seen that the characteristic polynomial $P(G, \lambda)$ is computable in polynomial time over the ring $\mathcal{R}$. Therefore it is computable in polynomial time in the unit cost model BSS over the ring $\mathcal{R}$ in the sense of Blum, Shub and Smale, in short, it is in $\mathbf{P}_\mathcal{R}$, [9]. Can we characterize the graph polynomials in **P-POL** which are in $\mathbf{P}_\mathcal{R}$?

Reducibilities. Reducibilities have now two components:

(i) Computations in the ring, performed on the polynomial, in the uniform computational model BSS, or in the non-uniform model of L. Valiant [56, 12]. Algebraic circuits (straight-line programs) or $\mathbf{P}_\mathcal{R}$-programs are natural choices, where $\mathcal{R}$ is the underlying ring.

(ii) Transductions of the graphs (relational structures), expressible in the logic $\mathcal{L}$ for suitably chosen $\mathcal{L}$, or computable by Turing machine transducers in the corresponding complexity class $\mathbf{C}$.

For **P**-polynomials over $\mathcal{R}$, $\mathbf{P}_\mathcal{R}$ and **P**-transductions, respectively transductions definable in Fixed Point Logic, are natural choices. For details see [28, 20, 33]. In this case we speak of **P**-reducibility between two graph polynomials f, g and write $g \preceq_P f$. The comparability and reducibility relations between graph polynomials do not coincide. The chromatic polynomial $\chi(G, \lambda)$ is **P**-reducible to the Tutte polynomials, but it is not comparable to the Tutte polynomial. This can be easily seen from the formula

$$\chi(G, X) = (-1)^{r(G)} X^{k(G)} T(G; 1 - X, 0)$$

Recall that $k(K)$ is the number of connected components of G and $r(G) = | V | - k(G)$. The formula shows that the chromatic polynomial is computable in polynomial time from the Tutte polynomial, but the Tutte polynomial remains invariant under the addition of isolated vertices to the graph G, whereas the chromatic polynomial does not.

It is open whether the matching polynomials $m(G, \lambda)$ and $g(G, \lambda)$ are **P**-reducible to the Tutte polynomial.

Easy loci. The Tutte polynomial is $\sharp\mathbf{P}$-hard to evaluate on all the points of the complex plane with the exception of a quasi-algebraic set of lower dimension, cf. [30]. M. Bläser and J.A. Makowsky, [8], have generalised this for the colored Tutte polynomial studied in [11]. Is a similar phenomenon also observable for arbitrary SOL-polynomials for which evaluation is $\sharp\mathbf{P}$-hard at least at some point?

Graph invariants. We say a graph polynomial g is a graph invariant, if, whenever G_1 and G_2 are isomoprrphic, then $g(G_1) = g(G_2)$. We say g is a *complete* graph invariant, if additionally, whenever $g(G_1) = g(G_2)$, then G_1 and G_2 are isomorphic. There are artificial graph polynomials even in one variable, which are complete graph invariants. They are artificial, because they use coding tricks, are expensive to compute, and computing other graph invariants from such a complete polynomial may be very hard. It remains open whether there are "natural" complete graph invariants, in particular, it is not obvious what we could mean by "natural".

Reduction-complete polynomials. A graph polynomial is *reduction complete* in a complexity class $\mathbf{C}$ equipped with a notion of reducibility, if every other **C**-polynomial is reducible to it. To speak about reduction-complete polynomials it may be reasonable to fix the number of variables of the polynomials under consideration. Are there any reduction-complete **P**-polynomials? Is the Tutte polynomial reduction-complete? We note that the Tutte polynomial has been shown to be the most general graph polynomial with respect to certain reduction rules (contraction and deletion of edges), cf. [10, Chapter X]. But this excludes the matching polynomial from the discussion.

Acknowledgements

I am indebted to my colleagues, coauthors and students of my seminar on graph polynomials, with whom I could discuss, during the last ten years, various aspects of graph polynomials. Among them are Y. Altschuler, I. Averbouch, M. Bläser, P. Bürgisser, B. Courcelle, B. Dubrov, E. Fischer, B. Godlin, M. Lotz, J. Mariño, A. Matsliach, K. Meer, E. Ravve, and U. Rotics. In particular, I would like to thank B. Courcelle for inviting me in February 2006 to spend time in Bordeaux working with him on graph polynomials.

Certain passages in section 2 of the paper are, with the explicit consent of my co-author, literally taken from [42]. I thank T. Zaslavsky for suggesting the title of the paper. I am also indebted to two anonymous referees for many valuable suggestions for improving the clarity of the presented ideas, and to the editors of the proceedings, who allowed me to exceed the space limit originally alloted for my contribution.

References

1. M. Aigner. The Penrose polynomial of graphs and matroids. In *Surveys in Combinatorics, 2001 (Sussex)*, volume 288 of *London Mathematical Society Lecture Note Series*, pages 11–46. Cambridge University Press, 2001.
2. M. Aigner and H. van der Holst. Interlace polynomials. *Linear Algebra and Applications*, 377:11–30, 2004.
3. N. Alon and M. Tarsi. Colorings and orientations of graphs. *Combinatorica*, 12:125–134, 1992.
4. A. Andrzejak. An algorithm for the Tutte polynomials of graphs of bounded treewidth. *Discrete Mathematics*, 190:39–54, 1998.
5. R. Arratia, B. Bollobás, and G.B. Sorkin. The interlace polynomial: a new graph polynomial. *Journal of Combinatorial Theory, Series B*, 92:199–233, 2004.
6. R. Arratia, B. Bollobás, and G.B. Sorkin. A two-variable interlace polynomial. *Combinatorica*, 24.4:567–584, 2004.
7. G.D. Birkhoff. A determinant formula for the number of ways of coloring a map. *Annals of Mathematics*, 14:42–46, 1912.
8. M. Bläser and J.A. Makowsky. Where is computing the colored Tutte polynomial hard? in preparation, 2006.
9. L. Blum, F. Cucker, M. Shub, and S. Smale. *Complexity and Real Computation*. Springer Verlag, 1998.
10. B. Bollobás. *Modern Graph Theory*. Springer, 1999.

11. B. Bollobás and O. Riordan. A Tutte polynomial for coloured graphs. *Combinatorics, Probability and Computing*, 8:45–94, 1999.
12. P. Bürgisser. *Completeness and Reduction in Algebraic Complexity*, volume 7 of *Algorithms and Computation in Mathematics*. Springer, 2000.
13. T.Y. Chow. The path-cycle symmetric function of a digraph. *Advances in Mathematics*, 118:71–98, 1996.
14. F.R.K. Chung and R.L. Graham. On the cover polynomial of a digraph. *Journal of Combinatorial Theory, Ser. B*, 65(2):273–290, 1995.
15. B. Courcelle and J.A. Makowsky. Recursive definitions of graph polynomials. in preparation, 2006.
16. Bruno Courcelle. The monadic second-order logic of graphs III: Treewidth, forbidden minors and complexity issues. *Informatique Théorique*, 26:257–286, 1992.
17. D.M. Cvetković, M. Doob, and H. Sachs. *Spectra of Graphs*. Johann Ambrosius Barth, 3rd edition, 1995.
18. F.M. Dong, K.M. Koh, and K.L. Teo. *Chromatic Polynomials and Chromaticity of Graphs*. World Scientific, 2005.
19. R.G. Downey and M.F Fellows. *Parametrized Complexity*. Springer, 1999.
20. H.D. Ebbinghaus and J. Flum. *Finite Model Theory*. Perspectives in Mathematical Logic. Springer, 1995.
21. G.E. Farr. The Go polynomials of a graph. *Theoretical Computer Science*, 306:1–18, 2003.
22. E.J. Farrell. On a general class of graph polynomials. *Journal of Combinatorial Theory, Series B*, 26:111–122, 1979.
23. I. Gessel. Generalized rook polynomials and orthogonal polynomials. In *IMA Volumes in Mathematics and Its Applications*, volume 18, pages 159–176. Springer, 1989.
24. C. Godsil and G. Royle. *Algebraic Graph Theory*. Graduate Texts in Mathematics. Springer, 2001.
25. C.D. Godsil. *Algebraic Combinatorics*. Chapman and Hall, 1993.
26. E. Grädel and Y. Gurevich. Metafinite model theory. *Information and Computation*, 140:26–81, 1998.
27. W. Hodges. *Model Theory*, volume 42 of *Encyclopedia of Mathematics and its Applications*. Cambridge University Press, 1993.
28. N. Immerman. *Descriptive Complexity*. Graduate Texts in Computer Science. Springer, 1999.
29. F. Jaeger. Tutte polynomials and link polynomials. *Proceedings of the American Mathematical Society*, 103:647–654, 1988.
30. F. Jaeger, D.L. Vertigan, and D.J.A. Welsh. On the computational complexity of the Jones and Tutte polynomials. *Math. Proc. Camb. Phil. Soc.*, 108:35–53, 1990.
31. L.H. Kauffman. A Tutte polynomial for signed graphs. *Discrete Applied Mathematics*, 25:105–127, 1989.
32. I.A. Lavrov. Effective inseparability of the set of identically true formulas and the set of formulas with finite counterexamples for certain elementary theories (in russian). *Algebra i Logika*, 2:5–18, 1962.
33. L. Libkin. *Elements of Finite Model Theory*. Springer, 2004.
34. M. Lotz and J.A. Makowsky. On the algebraic complexity of some families of coloured Tutte polynomials. *Advances in Applied Mathematics*, 32(1-2):327–349, 2003.
35. L. Lovasz and M. Plummer. *Matching Theory*. North Holland, 1986.

36. J.A. Makowsky. Colored Tutte polynomials and Kauffman brackets on graphs of bounded tree width. In *Proceedings of the 12th Symposium on Discrete Algorithms*, pages 487–495. SIAM, 2001.
37. J.A. Makowsky. Algorithmic uses of the Feferman-Vaught theorem. *Annals of Pure and Applied Logic*, 126:1–3, 2004.
38. J.A. Makowsky. Colored Tutte polynomials and Kauffman brackets on graphs of bounded tree width. *Disc. Appl. Math.*, 145(2):276–290, 2005.
39. J.A. Makowsky and J.P. Mariño. Farrell polynomials on graphs of bounded treewidth. *Advances in Applied Mathematics*, 30:160–176, 2003.
40. J.A. Makowsky and J.P. Mariño. The parametrized complexity of knot polynomials. *Journal of Computer and System Sciences*, 64(4):742–756, 2003.
41. J.A. Makowsky and K. Meer. On the complexity of combinatorial and metafinite generating functions of graph properties in the computational model of Blum, Shub and Smale. In *CSL'00*, volume 1862 of *Lecture Notes in Computer Science*, pages 399–410. Springer, 2000.
42. J.A. Makowsky and U. Rotics. Computing the chromatic polynomial on graphs of bounded clique-width. Preprint, 2006.
43. S.D. Noble. Evaluating the Tutte polynomial for graphs of bounded tree-width. *Combinatorics, Probability and Computing*, 7:307–321, 1998.
44. S. Oum. Approximating rank-width and clique-width quickly. In *Graph Theoretic Concepts in Computer Science, WG 2005*, volume 3787 of *Lecture Notes in Computer Science*, pages 49–58, 2005.
45. J.G. Oxley and D.J.A. Welsh. Tutte polynomials computable in polynomial time. *Discrete Mathematics*, 109:185–192, 1992.
46. P. Pitteloud. Chromatic polynomials and the symmetric group. *Graphs and Combinatorics*, 20:131–144, 2004.
47. J. Riordan. *An Introduction to Combinatorial Analysis*. Wiley, 1958.
48. A. Sokal. The multivariate Tutte polynomial (alias Potts model) for graphs and matroids. In *Survey in Combinatorics, 2005*, volume 327 of *London Mathematical Society Lecture Notes*, pages 173–226, 2005.
49. R. P. Stanley. Acyclic orientations of graphs. *Discrete Mathematics*, 5:171–178, 1973.
50. R.P. Stanley. A symmetric function generalization of the chromatic polynomial of a graph. *Advances in Mathematics*, 111:166–194, 1995.
51. J.J. Sylvester. On an application of the new atomic theory to the graphical presentation of the invariants and covariants of binary quantics, with three appendices. *American Journal of Mathematics*, 1:161–228, 1878.
52. M.A. Taitslin. Effective inseparability of the sets of identically true and finitely refutable formulae of elementary lattice theory (in Russian). *Algebra i Logika*, 1:24–38, 1961.
53. N. Trinajstić. *Chemical Graph Theory*. CRC Press, 2nd edition, 1992.
54. W. T. Tutte. Graph-Polynomials. *Advances in Applied Mathematics*, 32:5–9, 2004.
55. Z. Tuza. Graph colorings with local constraints – a survey. *Discussiones Mathematicae - Graph Theory*, 17.2:161–228, 1997.
56. L.G. Valiant. Completeness classes in algebra. In *Proceedings of 11th STOC*, pages 249–261, 1979.
57. L.G. Valiant. The complexity of enumeration and reliability problems. *SIAM Journal on Computing*, 8(3):410–421, 1979.
58. D.J.A. Welsh. *Complexity: Knots, Colourings and Counting*, volume 186 of *London Mathematical Society Lecture Notes Series*. Cambridge University Press, 1993.

Towards a Trichotomy for Quantified H-Coloring

Barnaby Martin and Florent Madelaine

Department of Computer Science, University of Durham, DH1 3LE, U.K.
b.d.martin@durham.ac.uk

Abstract. Hell and Nešetřil proved that the H-colouring problem is NP-complete if, and only if, H is bipartite. In this paper, we investigate the complexity of the quantified H-colouring problem (a restriction of the quantified constraint satisfaction problem to undirected graphs). We introduce this problem using a new two player colouring game. We prove that the quantified H-colouring problem is:

1. tractable, if H is bipartite;
2. NP-complete, if H is not bipartite and not connected; and,
3. Pspace-complete, if H is connected and has a unique cycle, which is of odd length.

We conjecture that the last case extends to all non-bipartite connected graphs.

1 Introduction

A very natural generalisation of graph colouring problems is defined in terms of graph homomorphism; the problem takes as input a graph G and accepts it if, and only if, there exists a homomorphism into a fixed graph H. This problem is known as the *H-colouring problem*. In [1], Hell and Nešetřil proved that the class of H-colouring problems exhibits dichotomy: the problem is tractable if H is bipartite and NP-complete otherwise. Constraint satisfaction problems (CSPs) are closely related to H-colouring problems. In the case of CSPs with boolean domains, known as *generalised satisfiability*, Schaefer used an exhaustive analysis of the types of relations expressible to prove a dichotomy between those cases that are tractable and those that are NP-complete [2]. An algebraic approach has been successful in identifying certain tractable and NP-complete cases (see for example [3,4]), and has enabled Schaefer's dichotomy to be extended to domains of size three [5]. However, the *dichotomy conjecture* [6], that states that every (non-uniform) CSP is either tractable or NP-complete, is still open. Building on the result from [2], dichotomy results were proved independently in [7] and [8] for *quantified generalised satisfiability* problems without constants (in [2], the result was proved only in the case where the boolean constants were included): they are either tractable or Pspace-complete. The algebraic approach to the CSP has successfully been applied to *quantified constraint satisfaction problems* (QCSPs), to determine sufficient conditions for tractability and Pspace-completeness (see [9,10]). A partial trichotomy result was proved for

A. Beckmann et al. (Eds.): CiE 2006, LNCS 3988, pp. 342–352, 2006.

QCSPs in [9]: for a restricted class the authors prove that every problem is either tractable, NP-complete or Pspace-complete. A full classification for QCSP appears to be even harder than the classification for CSP. Hell and Nešetřil's dichotomy motivates us to concentrate on the restriction of QCSPs to graphs, which we call *quantified H-colouring*. We introduce quantified H-colouring by means of a natural game. We show that the quantified H-colouring problem is tractable if H is bipartite; NP-complete if H is not bipartite and not connected; and, Pspace-complete if H is connected and contains a unique cycle which is of odd length. Thus, as a corollary, we obtain the main contribution of this paper: a trichotomy when H is a graph with at most one cycle. In the NP-complete and tractable cases, we prove that if certain paths are present in the input then it must necessarily be rejected, otherwise the problem collapses to NP. Moreover, when H is a bipartite graph, we provide a reduction to 2-colourability, which demonstrates our problem's tractability. Note that bipartite graphs are not closed under majority functions [11], and are not known to be closed under any of the functions of [9, 10] that would guarantee QCSP tractability. When H is not connected and not bipartite, we derive from Hell and Nešetřil's theorem that the quantified H-colouring problem is NP-complete. In the Pspace-complete case, we first study *odd Catherine wheels*, which are graphs that consists of an odd cycle together with disjoint paths, where each path is attached to a different vertex of the cycle. Using some structural properties of these graphs, we prove that a variant of quantified satisfiability, known also to be Pspace-complete, reduces to the quantified H-colouring problem, whenever H is an odd Catherine wheel. As a corollary, we prove that the quantified H-colouring problem is Pspace-complete, whenever H has only one cycle, which is of odd length. The paper is organised as follows. In Section 2, we introduce a new colouring game and use this game to define the quantified H-colouring problem. In Section 3, we study cases where the quantified H-colouring problem is in NP. In Section 4, we prove that the quantified H-colouring problem is Pspace-complete, whenever H is a connected graph with a unique cycle which is of odd length. Finally, in Section 5, we derive the main result and conclude with a conjecture.

2   A New Colouring Game

In this paper, we only ever consider finite undirected graphs without self-loops. Given graphs G and H, a *homomorphism* f from G to H, denoted $G \xrightarrow{f} H$, is a function $f : V(G) \to V(H)$ such that $\{x, y\} \in E(G)$ implies $\{f(x), f(y)\} \in E(H)$. We write $G \longrightarrow H$ if there exists a homomorphism from G to H. The H-*colouring problem* takes as input a graph G, which is a yes-instance if, and only if, there exists a homomorphism from G to H. For $n \geq 0$, an n-*partitioned graph* $\mathscr{G}$ consists of a graph G together with a partition $\{U_1, X_2, \ldots, U_{2n+1}, X_{2n+2}\}$ of $V(G)$. In the following, G will always designate the underlying graph of $\mathscr{G}$. Let $\mathscr{G}$ be an n-partitioned graph and H a (non-partitioned) graph. The $(\mathscr{G}, H)$-*game* is a two-player game, that pitches Adversary (male) against Prover (female). Adversary plays on the universal partitions (the sets U_i) and Prover plays on the existential partitions (the sets X_i). They play alternate partitions, in ascending order, until all the partitions have been played. The game goes as follows. For $0 \leq i \leq n$:

- for every vertex in partition U_{2i+1}, Adversary chooses a vertex in H: i.e. he gives a function $f_{U_{2i+1}} : U_{2i+1} \rightarrow V(H)$; and,
- for every vertex in partition X_{2i+2}, Prover chooses a vertex in H: i.e. she gives a function $f_{X_{2i+2}} : X_{2i+2} \rightarrow V(H)$.

Prover wins if, and only if, the function $f := f_{U_1} \cup f_{X_2} \cup \ldots \cup f_{X_{2n+2}}$ is a homomorphism from G to H. A *strategy* for Prover (resp., Adversary) tells her (resp., him) how to play a partition given what has been played before. A strategy for Prover is *winning* if it beats all possible strategies of Adversary.

We say there is an *alternating-homomorphism* from the n-partitioned graph $\mathcal{G}$ to the (non-partitioned) graph H, and we write $\mathcal{G} \xrightarrow{alt} H$ if, and only if, for all functions $f_{U_1} : U_1 \rightarrow V(H)$, there exists a function $f_{X_2} : X_2 \rightarrow V(H)$, such that, $\ldots$, for all functions $f_{U_{2n+1}} : U_{2n+1} \rightarrow V(H)$, there exists a function $f_{X_{2n+2}} : X_{2n+2} \rightarrow V(H)$, such that, $f_{U_1} \cup f_{X_2} \cup \ldots f_{U_{2n+1}} \cup f_{X_{2n+2}}$ is a homomorphism from G to H.

If the n-partitioned graph $\mathcal{G}$ is viewed as a quantified sentence, then our game is exactly a model-checking, or Hintikka, game [12] over the model H. In this guise our game is closely related to that used in the analysis of QCSP by Chen [10]. In any case, the following is a direct consequence of the above definitions.

Proposition 1. *Let $\mathcal{G}$ be an n-partitioned graph and H be a graph. $\mathcal{G} \xrightarrow{alt} H$ if, and only if, Prover has a winning strategy in the $(\mathcal{G}, H)$-game.*

We define the *quantified H-colouring* problem as the decision problem which takes as input a partitioned graph $\mathcal{G}$ (n-partitioned, for some n) and whose yes-instances are those $\mathcal{G}$ for which $\mathcal{G} \xrightarrow{alt} H$. We refer to H as the problem's *template*.

2.1   Restricting Partitions

We will be particularly interested in templates H whose quantified H-colouring yes-instances are partitioned inputs $\mathcal{G}$ that in some way collapse to NP. We will now formalise that notion of collapse.

Let $\mathcal{G}$ be a partitioned graph. We say that $\mathcal{G}$ is in Π_2-*form* (resp., Σ_1-*form*), if the only non-empty partitions are among $\{U_1, X_2\}$ (resp., $\{X_2\}$). If $\mathcal{G}$ is in Π_2-form and there is at most one vertex in U_1, then we say that $\mathcal{G}$ is in Π_2-*fan form*. Finally, we say that $\mathcal{G}$ is in Π_2-*multifan form*, if $\mathcal{G}$ is the finite disjoint union of graphs in Π_2-fan form.

Theorem 1. *The restriction of the quantified H-colouring problem to partitioned graphs in Π_2-multifan form is in NP.*

Proof. Let $\mathcal{G}$ be the disjoint union of $\mathcal{G}_1, \mathcal{G}_2, \ldots, \mathcal{G}_m$, all in Π_2-fan form. Note that $\mathcal{G} \xrightarrow{alt} H$ if, and only if, $\mathcal{G}_i \xrightarrow{alt} H$, for every $1 \leq i \leq m$. To test whether

$\mathscr{G}_i \xrightarrow{alt} H$ we may consider all possible maps for the vertex in U_1 (if there is one) and then guess the remainder of the homomorphism and verify in polynomial time.   □

We describe two partitioned graphs $\mathscr{G}$ and $\mathscr{G}'$ as *problem-equivalent* if, and only if, for all templates H, $\mathscr{G} \xrightarrow{alt} H$ iff $\mathscr{G}' \xrightarrow{alt} H$. From a partitioned graph $\mathscr{G}$, we derive the *reduced* graph $\overline{\mathscr{G}}$ by collapsing all universal partitions to U_1 and all existential partitions to X_2. Note that $\mathscr{G}$ and $\overline{\mathscr{G}}$ share the same underlying graph G.

3   Cases in NP

In this section, we will find that, for certain H, every input which is not essentially in Π_2-multifan form can be discarded.

Let H be *either* a bipartite graph *or* a non-connected graph, and let $\mathscr{G}$ be a partitioned graph. We first prove that certain paths are forbidden in yes-instances $\mathscr{G}$ of the quantified H-colouring problem. Later, we show that if $\mathscr{G}$ does not have such paths then it must essentially be in Π_2-multifan form. Finally, we use this important structural property to prove that, for such instances, the quantified H-colouring problem essentially collapses to the H-colouring problem. The following lemma begins with a special non-connected case.

Lemma 1. *Suppose H has an isolated vertex, then: if there is an edge in $\mathscr{G}$ between x in U_i and y in X_j (for any i, j), or between x in U_i and y in U_j (for any i, j) then $\mathscr{G} \xrightarrow{alt}\!\!\!\!/\;\; H$.*

Let H be any bipartite or non-connected graph. If there is a path in $\mathscr{G}$ between any x in X_i and y in U_j (for $i < j$) then $\mathscr{G} \xrightarrow{alt}\!\!\!\!/\;\; H$. If there is a path in $\mathscr{G}$ between any x in U_i and y in U_j ($x \neq y$; any i, j), then $\mathscr{G} \xrightarrow{alt}\!\!\!\!/\;\; H$.

Proof. (Isolated vertex.) Let s be an isolated vertex of H. Regardless of what is played before, when Adversary plays x on s he wins.

(Bipartite.) We prove the first claim; the second may be done similarly. Let a be any vertex in H on which Prover plays x. If a is isolated then Adversary plays y on a and wins. Assume that a is not isolated. If the path in $\mathscr{G}$ between x and y is of even length then Adversary plays y on b, where b is adjacent to a. A winning strategy for Prover would imply the existence of an odd cycle in H, which would contradict the fact that H is bipartite. Thus, it follows that Adversary wins. If the path in $\mathscr{G}$ is of odd length then Adversary plays y on a and wins by the same argument.

(Non-connected.) We prove the first claim; the second may be done similarly. Wherever Prover plays x in H, Adversary answers y in a different connected component of H. Since there is no path between these vertices in H, Adversary wins.   □

Theorem 2. *If H has an isolated vertex then the quantified H-colouring problem is equivalent to the H-colouring problem under logspace reduction.*

Proof. The H-colouring problem reduces trivially to the quantified H-colouring problem. We define the converse reduction as follows. Let N be a fixed no-instance of the H-colouring problem (say, H augmented with one vertex adjacent to every vertex of H). If $\mathscr{G}$ has an edge as in the previous lemma then we know that it is a no-instance and we reduce $\mathscr{G}$ to N (with all vertices in X_2). If $\mathscr{G}$ has no such edge then any vertex in a universal partition is isolated, and $\mathscr{G}$ is essentially in Σ_1-form. We may reduce $\mathscr{G}$ to its underlying graph G.   □

The following result shows that every partitioned graph that does not have the paths mentioned in Lemma 1 is essentially in Π_2-multifan form.

Lemma 2. *If there is no path in a partitioned graph $\mathscr{G}$ between any x in X_i and y in U_j (for $i < j$), or between any x in U_i and y in U_j ($x \neq y$; any i, j), then $\mathscr{G}$ is in Π_2-multifan form and is problem-equivalent to $\mathscr{G}$.*

Proof. It suffices to prove, for every connected component $\mathscr{G}'$ of $\mathscr{G}$, that $\overline{\mathscr{G}'}$ is in Π_2-fan form and is problem-equivalent to $\mathscr{G}'$. Let $\mathscr{G}'$ be such a component and let $0 < i \leq n$ be the largest integer such that U_{2i+1} is non-empty. Take x in U_{2i+1}, and let y be any element of $\mathscr{G}'$ connected to x via a path. It follows from the second assumption that y can not be in a universal partition (U_{2i+1} included). Thus, it follows from the first assumption that y belongs to an existential partition of index at least $2i + 2$. It is not hard to see that we can move x to U_1 and all other [existential] vertices of $\mathscr{G}'$ to X_2, preserving problem-equivalence, and generating $\overline{\mathscr{G}'}$ in Π_2-fan form.   □

The following result is essential. We show that, provided the input is in Π_2-fan form and H has no isolated vertex, the quantified H-colouring problem essentially coincides with the K_2-colouring problem.

Lemma 3. *Let H be a bipartite graph that has no isolated vertex and let $\mathscr{G}$ be in Π_2-multifan form. The following are equivalent:*

   (i) $\mathscr{G} \xrightarrow{alt} H$

   (ii) $\mathscr{G} \xrightarrow{alt} K_2$

   (iii) $G \longrightarrow K_2$

Proof. In the following, for any a in H, let a' be some adjacent vertex. Note that H is homomorphically equivalent to K_2, and thus *(iii)* is equivalent to $G \longrightarrow H$. Denote by 0 and 1 the two vertices of K_2. Let $H \xrightarrow{h} K_2$ be some homomorphism such that $h(a) = 0$ and $h(a') = 1$. Let $K_2 \xrightarrow{h'} H$ be the homomorphism defined by $h'(0) = a$ and $h'(1) = a'$.

    It suffices to prove the result for each connected component of $\mathscr{G}$. Thus, assume w.l.o.g. that $\mathscr{G}$ is in Π_2-fan form. If $\mathscr{G}$ has no vertex in U_1 then the result holds trivially. Otherwise, let x be the unique vertex in U_1.

   – *(i)* $\Rightarrow$ *(ii)*: Consider a winning strategy for Prover in the $(\mathscr{G}, H)$-game, where Adversary plays x on a (resp., on a'). Since $\mathscr{G} \xrightarrow{alt} H$, there exists a homomorphism $G \xrightarrow{s_a} H$ such that $s_a(x) = a$ (resp., $G \xrightarrow{s_{a'}} H$ such that $s_{a'}(x) = a'$).

We now construct a winning strategy for Prover, in the $(\mathscr{G}, K_2)$-game, as follows. If Adversary plays x on 0 (resp., on 1) in K_2 then all remaining moves are Prover's: she plays every vertex y in X_2 on $h \circ s_a(y)$ (resp., $h \circ s_{a'}(y)$) and wins.

- $(ii) \Rightarrow (i)$: Consider a winning strategy for Prover in the $(\mathscr{G}, K_2)$-game, where Adversary plays x on 0. Since $\mathscr{G} \xrightarrow{alt} K_2$, there exists a homomorphism $G \xrightarrow{s_0} K_2$ satisfying $s_0(x) = 0$.

  We now construct a winning strategy for Prover in the $(\mathscr{G}, H)$-game. When Adversary plays x on some a in H, then Prover plays each vertex y of X_2 according to $h' \circ s_0(y)$, and wins.

- $(ii) \Rightarrow (iii)$: clear.
- $(iii) \Rightarrow (ii)$: follows from the symmetry of K_2.

□

Theorem 3. *Let H be a bipartite graph. The quantified H-colouring problem is tractable.*

Proof. We propose the following algorithm to solve the quantified H-colouring problem. If H has an isolated vertex then by Theorem 2 the problem reduces in logspace to the H-colouring problem, known to be tractable by Hell and Nešetřil's theorem. If H has no isolated vertex then we use the following algorithm. The input $\mathscr{G}$ is first scanned to check whether it has any of the forbidden paths of Lemma 1. If there are any then the input is rejected. Otherwise, the algorithm accepts the input if, and only if, it is bipartite.

This algorithm is clearly polynomial. We now prove its correctness. We know that if $\mathscr{G}$ has none of the forbidden paths, then it is problem-equivalent to the reduced $\overline{\mathscr{G}}$ in Π_2-multifan form, by Lemma 2. Thus, $\mathscr{G} \xrightarrow{alt} H$ iff $\overline{\mathscr{G}} \xrightarrow{alt} H$ iff $G \longrightarrow K_2$ (by Lemma 3), and we are done. □

Remark 1. It is well-known that two graphs H and H' give rise to identical H-colouring and H'-colouring problems iff they are homomorphically equivalent (alternatively, that they have the same *core*). Let H be a bipartite graph. There are precisely two possible H-colouring problems, corresponding to the cores K_1 and K_2. It follows immediately from Theorem 2 and Lemma 3 that there are precisely three possible quantified H-colouring problems, corresponding to K_1, K_2 and, say, $K_2 \uplus K_1$.

Theorem 4. *Let H be a non-connected graph. The quantified H-colouring problem is in NP. Moreover, if H is not bipartite then the quantified H-colouring problem is NP-complete.*

Proof. (Membership of NP.) Let $\mathscr{G}$ be the input partitioned graph. If $\mathscr{G}$ has any of the paths of Lemma 1 then we know it is a no-instance. Otherwise, by Lemma 2, $\overline{\mathscr{G}}$ is problem-equivalent and in Π_2-multifan form, and we can use the algorithm of Theorem 1.

(NP-hardness.) If H is not bipartite then, by Hell and Nešetřil's theorem [1] the H-colouring problem is NP-complete. Since it reduces trivially to the quantified H-colouring problem, the NP-hardness of that problem follows. □

4   Pspace-Complete Cases

In this section, we prove that the quantified H-colouring problem is Pspace-complete whenever H is a connected graph with a unique cycle, which is of odd length. Firstly, we study the special case of what we call *odd Catherine wheels*. Secondly, we reduce a variant of the quantified satisfiability problem that is also Pspace-complete, *quantified not-all equal $2k + 1$-sat* (qnae $2k + 1$-sat) to the quantified H-colouring problem, where H is an odd Catherine wheel. Finally, we extend this reduction to any connected graph with a unique cycle, which is of odd length.

4.1   Odd Catherine Wheels

Let W be a graph that consists of an odd cycle C_{2k+1} (where $k > 0$) together with $2k + 1$ disjoint paths $P^0, P^1, \ldots, P^{2k}$, each of any finite length (but potentially trivial, i.e. the path of length 0 consists of a single vertex) such that the end of each path is identified with a different vertex of the odd cycle. We say that W is an *odd Catherine wheel*. Ordering the cycle and choosing an initial vertex, we represent W by the corresponding sequence of path lengths, which we call a *listing*. For any distinct vertices u and v, we denote by $d(u, v)$ the length of the shortest path from u to v. Set

$$\omega := \min\{D(\{u, v\}) : \{u, v\} \text{ is an edge on the cycle of } W\}$$

where $D(\{u, v\}) := \max\{\min\{d(u, w), d(v, w)\} : w \in V(W)\}$, for any $\{u, v\}$ in $E(W)$. We say that a listing of W is *minimal* if $D(\{k, k + 1\}) = \omega$. We need the following two lemmas to prove the completeness of the quantified W-colouring problem. The proofs, though technical, present no difficulties and due to restrictions on space are omitted.

Lemma 4. *Let W be an odd Catherine wheel with a cycle of size $2k + 1$. Let $P^0, P^1, \ldots, P^{2k}$ be a minimal listing of W. There exists t in $V(W)$ such that there is an ω-walk from t to k, but no ω-walk from t to $k + 1$, and there exists s in $V(W)$ such that there is an ω-walk from s to $k + 1$, but no ω-walk from s to k.*

Lemma 5. *Let W be an odd Catherine wheel with a cycle of size $2k + 1$. Let x and y be any vertices on this cycle. There is a walk from x to y of length $2k - 1$ if, and only if x is distinct from y.*

4.2   Reduction from Quantified Not-all-Equal $2k + 1$-sat

Let W be an odd Catherine wheel with a cycle of size $2k + 1$. In this section we prove that the quantified W-colouring problem is Pspace-complete. The proof involves a reduction from the problem qnae $2k + 1$-sat, inspired by that given in [9]. This problem takes as input a formula φ of the form

$$Q_1 v_1, Q_2 v_2, \ldots, Q_n v_n \bigwedge_{i=1}^{c} N(v_{i_1}, v_{i_2}, \ldots, v_{i_{2k+1}}),$$

where $n, c \geq 1$ and for every $1 \leq j \leq n$, Q_j is either the quantifier $\exists$ or $\forall$ and v_j is a variable symbol. Such a formula is accepted if, and only if, it is true, where the semantics of N is "not-all-equal" (that is, under a Boolean valuation a clause is satisfied if, and only if, at least two variables are given different values).

Let ω be defined as in the previous section. Throughout this section, let φ be an input of qnae $2k + 1$-sat. We build a partitioned graph $\mathscr{G}$ as follows.

The Underlying Graph

- For each variable v_i of φ, we add a copy of C_{2k+1}; we fix a vertex w_i and two adjacent vertices x_i and y_i, both at distance exactly k from w_i; and, we add a path of length ω which we attach at the vertex y_i, and call z_i the other extremity of this path. We call this graph, together with the specified vertices x_i, y_i and w_i, *the variable gadget* associated with v_i. Next, we identify every vertex w_i, of every variable gadget, as a single vertex w, common to every cycle of every variable gadget.
- For each clause $N(v_{i_1}, v_{i_2}, \ldots, v_{i_{2k+1}})$ of φ, we add a copy of C_{2k+1}, whose vertices are labelled by $1, 2 \ldots, 2k + 1$; each vertex of the cycle represents a position in the clause; and, for $1 \leq j \leq 2k + 1$, we add a path of length $2k - 1$ from the vertex j of the cycle to the vertex x_{i_j}. We call this graph the *clause gadget* associated with $N(v_{i_1}, v_{i_2}, \ldots, v_{i_{2k+1}})$.

The Partition. We set U_1 to be empty. We add w to X_2. Next, we read the quantifiers in φ from left to right, and proceed inductively. Let X_{l_i} be the current last (existential) partition used. Let v_i be the next quantified variable. There are two cases:

1. v_i is universally quantified. We add a universal partition U_{l_i+1} and set $U_{l_i+1} := \{z_i\}$; and, we add an existential partition X_{l_i+2} and add to X_{l_i+2} the rest of the variable gadget associated with v_i: that is, the rest of z_i's path, and every vertex of the cycle C_{2k+1} (apart from $w_i = w$, already in X_2).
2. v_i is existentially quantified. We add the rest of the variable gadget associated with v_i to X_{l_i}: that is, z_i, its path, and the rest of the corresponding C_{2k+1} (just like above $w_i = w$ is already in X_2).

Once we have read all the quantifiers of φ, we add all of the remaining vertices, i.e. those in the cycles and paths that encode clauses, to the last existential partition X_l.

Preliminary Observations. The construction specified is well-defined and the partitioned graph can be built in time polynomial in the size of the formula φ. It remains for us to prove that it is indeed a reduction. We may assume that φ has at least one universal variable, since qnae $2k + 1$-sat reduces trivially to this restriction, under the reduction which adds a dummy universal variable.

Note that almost all vertices of $\mathscr{G}$ are in existential partitions: the only vertices in universal partitions are the z_i that belong to the variable gadgets associated with universal variables v_i of φ. Recall that Prover plays these existential vertices and wants to build a homomorphism to W, for each choice made by Adversary. In particular, every cycle must be played to the cycle of W, whether they are cycles from a clause gadget or a variable gadget, otherwise we would not have a homomorphism. Moreover, since C_{2k+1} is a core (see [13]) every endomorphism of C_{2k+1} is an automorphism. Thus, distinct elements of a cycle in $\mathscr{G}$ must be played on distinct elements of the cycle in W.

We will need only to consider the case where w is played on the vertex labelled by 0 in some minimal listing of W. Indeed, if Prover has a winning strategy then w must be played on such a vertex. Otherwise, by minimality of ω, there would exists a vertex u in W that does not have an ω-walk to either k or $k+1$ and if Adversary plays a vertex z_i (from a variable gadget associated with a universal variable v_i) on this vertex u then Prover would not be able to map y_i to k or $k+1$ (which she has to, see previous remark on the play of cycles), and she would lose. Conversely, as we shall prove shortly, if φ is a yes-instance then we can build a winning strategy for Prover, where w is played on the vertex labelled by 0 in some minimal listing of W.

Assignment vs. Homomorphism. In the following we assume that w is played on the vertex labelled 0, according to some minimal listing of W. Consider the variable gadget associated with a variable v_i of φ and ignore, for now, the variable z_i and its path of length ω. A homomorphism from this partial gadget to W corresponds to a truth assignment of v_i according to the following encoding: if the vertex x_i is played on k then the variable is assigned to true and if the vertex x_i is played on $k+1$ then the variable is assigned to false. Note that this encoding establishes a one-to-one correspondence between assignments to the variables of φ and the homomorphism h from the partial variable gadgets that satisfy $h(w) = 0$. Thus, under the assumption that w is played on the vertex labelled 0, for notational simplicity, given such a homomorphism h, we feel free to also write h for the corresponding boolean assignment. The following may be derived from Lemma 5.

Lemma 6. *Let $N(v_{i_1}, \ldots, v_{i_{2k+1}})$ be a clause of φ. $N(v_{i_1}, \ldots, v_{i_{2k+1}})$ is satisfied by h if, and only if, there exists a homomorphism from the clause gadget associated with $N(v_{i_1}, \ldots, v_{i_{2k+1}})$ together with the partial gadgets associated with the variables $v_{i_1}, \ldots, v_{i_{2k+1}}$ such that not all the vertices $x_{i_1}, \ldots, x_{i_{2k+1}}$ are mapped on the same vertex in W.*

Correctness of the Reduction. Assume that φ is a yes-instance. We build a winning strategy for Prover accordingly, such that she plays w on the vertex 0 in a minimal listing of W. For every vertex z_i in a variable gadget that corresponds to a universal variable v_i of φ, no matter where z_i is played by Adversary in W, by definition of ω, there exists an ω-walk from this vertex to one of the vertices k or $k+1$. Thus, Prover may play x_i on k or $k+1$ and ensure that the correspondence with boolean assignment still holds. However, by Lemma 4, by choosing carefully the vertex on which he plays, Adversary can force $h(v_i)$ to be false or true, at his will. Similarly, if a variable is quantified existentially then it follows from Lemma 4 that Prover can play z_i on a vertex such that x_i is played on k or $k+1$, at her will. This extends the correspondence between boolean assignments and homomorphisms to take the quantifiers into account. Assume a homomorphism h has been played by Adversary and Prover, where Prover mimics the assignment induced by the fact that φ is a yes-instance. Since h satisfies φ, for every clause, not all variables have the same value and by Lemma 6, we can extend h to the whole graph.

Conversely, assume that $\mathscr{G} \xrightarrow{alt} W$. We have already observed that Prover must play w on [some] 0. We build a winning assignment h for φ accordingly. This time we use the converse implication in Lemma 6.

Theorem 5. *Let H be a connected graph that has a unique cycle, which is odd. Then the quantified H-colouring problem is* Pspace*-complete.*

Proof. The problem qnae 3-sat is Pspace-complete [9] and reduces trivially to the problem qnae $2k+1$-sat. Together with the reduction presented above, this proves that if W is an odd Catherine wheel then the quantified W-colouring problem is Pspace-complete.

Let H be a connected graph with a unique cycle that is odd. H induces an odd Catherine wheel W as follows. For every tree T^i attached to the odd cycle, replace T^i by a path P^i of the same length as the height of T^i. If φ is an instance of qnae $2k+1$-sat, we build $\mathscr{G}$ as above, and we claim that $\mathscr{G} \xrightarrow{alt} H$ if, and only if $\mathscr{G} \xrightarrow{alt} W$. This is due to the fact that in $\mathscr{G}$ any vertex z_i in a universal partition lies on the tip of a path that lies in the next existential partition. Thus, if Adversary plays z_i in some path P' within a tree T^i, at distance d from the cycle of H, then Prover can forget every other path and act as if Adversary had played on the vertex at distance d from the cycle in the path P^i of W. The converse holds trivially and the result follows.   □

5   Conclusion

Combining the results of this paper in a single statement, we get a trichotomy for the quantified H-colouring problem, when H has at most one cycle.

Theorem 6 (main result). *Let H be a graph with at most one cycle. The quantified H-colouring problem exhibits a trichotomy.*

- *If H is bipartite then the quantified H-colouring problem is tractable.*
- *If H is not bipartite and not connected then the quantified H-colouring problem is* NP*-complete.*
- *If H is not bipartite and connected then the quantified H-colouring problem is* Pspace*-complete.*

Proof. The first case follows from Theorem 3 and the third case follows from the previous theorem. The second case remains. Membership of NP follows from Theorem 4. For completeness, note that the core of H is some odd cycle C_{2k+1}: there is a trivial reduction from the C_{2k+1}-colouring problem, known to be NP-complete (e.g. [1]).   □

We note that the method of forbidden paths, which allowed us to derive results for templates that are either bipartite or non-connected, can not be extended further. This is because, for a non-bipartite connected H, there exists a number M such that, for all vertices $x, y \in V(H)$ and for all $m \geq M$, there is an m-walk from x to y. This leads us to the following conjecture that would imply a full trichotomy for quantified H-colouring.

Conjecture 1. If H is not bipartite and connected then the quantified H-colouring problem is Pspace-complete.

References

1. Hell, P., Nešetřil, J.: On the complexity of H-coloring. J. Combin. Theory Ser. B **48**(1) (1990) 92–110
2. Schaefer, T.J.: The complexity of satisfiability problems. In: 10th Annual ACM Symp. on Theory of Computing. (1978)
3. Jeavons, P.: On the algebraic structure of combinatorial problems. Theoret. Comput. Sci. **200**(1-2) (1998) 185–204
4. Bulatov, A., Jeavons, P., Krokhin, A.: Classifying the complexity of constraints using finite algebras. SIAM J. Comput. **34**(3) (2005)
5. Bulatov, A.: A dichotomy theorem for constraints on a three-element set. In: Proceedings 43rd IEEE Symposium on Foundations of Computer Science, FOCS'02. (2002) To appear
6. Feder, T., Vardi, M.Y.: The computational structure of monotone monadic SNP and constraint satisfaction: a study through Datalog and group theory. SIAM J. Comput. **28**(1) (1999) 57–104 (electronic)
7. Creignou, N., Khanna, S., Sudan, M.: Complexity classifications of Boolean constraint satisfaction problems. SIAM Monographs on Discrete Mathematics and Applications. (2001)
8. Dalmau, V.: A dichotomy theorem for learning quantified boolean formulas. Machine Learning **35**(3) (1999)
9. Börner, F., Bulatov, A., Jeavons, P., Krokhin, A.: Quantified constraints: algorithms and complexity. In: Computer science logic. Volume 2803 of LNCS. Springer, Berlin (2003) 58–70
10. Chen, H.: Collapsibility and consistency in quantified constraint satisfaction. In: Procs. 19th National Conference on AI. (2004)
11. Bjäreland, M., Jonsson, P.: Exploiting bipartiteness to identify yet another tractable subclass of CSP. In: Principles and Practice of Constraint Programming-CP'99. Volume 1713 of LNCS. (1999)
12. Gradel, E.: Model checking games. Electronic Notes in Theoretical Computer Science **67** (2002)
13. Hell, P., Nešetřil, J.: Graphs and homomorphisms. Volume 28 of Oxford Lecture Series in Maths and its Applications. OUP (2004)

Two Open Problems on Effective Dimension

Elvira Mayordomo*

Dept. de Informática e Ing. de Sistemas, Universidad de Zaragoza,
María de Luna 1, 50018 Zaragoza, Spain
elvira@unizar.es

1 Introduction

Effective fractal dimension was defined by Lutz [13] in order to quantitatively analyze the structure of complexity classes. The dimension of a class X inside a base class $\mathcal{C}$ is a real number in [0,1] corresponding to the relative size of $X \cap \mathcal{C}$ inside $\mathcal{C}$. Basic properties include monotonicity, so dimension 1 classes are maximal and dimension 0 ones are minimal, and the fact that dimension is defined for *every class* X, making effective dimension a precise quantitative tool.

The first goal of such quantitative methods is to extend existence results of the form "there is a problem in $\mathcal{C}$ that is in X" to abundance results of the form "a non-negligible part of the problems in $\mathcal{C}$ are in X" formally expressed as "the class X has positive dimension in $\mathcal{C}$". Another application is in relation with the probabilistic method, proving that X has positive dimension can be simpler than proving non-emptiness, the easiness here comes from proving abundance as opposed to constructing a particular object. A third aspect of effective dimension is as a formal tool in Computational Complexity, allowing us to consider new working hypothesis such as "NP has positive dimension in exponential time", that can imply plausible consequences that haven't been derived from P$\neq$NP.

The concept of effective dimension is a generalization of classical fractal or Hausdorff dimension, one of the most powerful tools of fractal geometry, an extensively developed subfield of geometric measure theory with applications throughout the sciences [5, 7, 6]. Tricot [19] and Sullivan [18] independently developed a dual of Hausdorff dimension called packing dimension, that now rivals Hausdorff dimension's importance in such investigations.

In 2000 Lutz proved a new characterization of Hausdorff dimension in terms of gales [13], that are betting strategies that generalize martingales. The most important benefit of this characterization is that it enables one to define effective versions of fractal dimension by imposing various computability and complexity constraints on the gales. Four years later Athreya, Lutz, Hitchcock and Mayordomo [3] proved that packing dimension also admits a gale characterization, with a different notion of gale success. We can now define versions of Hausdorff and packing dimensions that are meaningful inside complexity classes such as exponential time and exponential space.

* This research was supported in part by Spanish Government MEC project TIN 2005-08832-C03-02.

A. Beckmann et al. (Eds.): CiE 2006, LNCS 3988, pp. 353–359, 2006.

Effective dimension has indeed proven to be very fruitful in Computational Complexity for obtaining useful results in the three aspects mentioned above. A very recent summary of the main achievements can be found in [10].

In this note we propose two interesting open problems on Computational Complexity, both related to polynomial-time reductions. Complete or partial solutions of these problems imply a big advance in what we know on the classes NP and BPP. In both cases quantitative methods such as resource-bounded measure have given initial answers in the past, and the fact that dimension is defined for every class can overcome non-measurability obstacles.

2   Effective Dimension

For the sake of completeness we include the basic definitions of resource-bounded or effective dimension, based on the notion of s-gale. A more detailed treatment, motivation, references and historical introduction can be found in [10], [15], and [14].

We work in the Cantor space $\mathbf{C}$ that is the set of all infinite binary sequences. $\{0,1\}^*$ is the set of finite binary strings.

Formally, if $s \in [0, \infty)$, then an s-gale is a function $d : \{0,1\}^* \to [0, \infty)$ satisfying the condition

$$d(w) = 2^{-s}[d(w0) + d(w1)] \tag{1}$$

for all $w \in \{0,1\}^*$ [13]. A martingale is a 1-gale.

A gale d succeeds on a sequence S if

$$\limsup_{w \to S} d(w) = \infty$$

and succeeds strongly on S if

$$\liminf_{w \to S} d(w) = \infty.$$

The success set $S^\infty[d]$ of a gale d is the set of all sequences on which d succeeds. The strong success set $S^\infty_{\mathrm{str}}[d]$ is the set of all sequences on which d succeeds strongly.

Intuitively, we think of a gale d as a strategy for betting on the successive bits of a sequence S. The quantity $d(w)$ is interpreted as the capital (amount of money) that a gambler using the strategy d has after betting on the successive bits of the prefix w of S. The parameter s regulates the fairness of the payoffs via identity (1). If $s = 1$, the payoffs are fair in the usual sense that the conditional expectation of the gambler's capital $d(wb)$, given that w has occurred, is precisely $d(w)$, the gambler's actual capital before betting on the last bit of wb. If $s < 1$, then the payoffs are unfair, and the smaller s is, the more unfair the payoffs are.

Theorem 1. *(Gale characterization of fractal dimension) Let X be a set of sequences.*

1. (Lutz [13]) $\dim_H(X) = \inf\{s \mid$ *there is an s-gale d such that* $X \subseteq S^\infty[d]\}$.
2. (Athreya et al. [3]) $\dim_P(X) = \inf\{s \mid$ *there is an s-gale d such that* $X \subseteq S^\infty_{\text{str}}[d]\}$.

Intuitively, Theorem 1 says that the fractal dimension of a set X of sequences is the *most hostile environment* (i.e., most unfair payoff parameter s) in which a gambler can win on every sequence in X. Of course, the word "win" here means "succeed" in the case of Hausdorff dimension and "succeed strongly" in the case of packing dimension.

It is easy to see that $0 \leq \dim_H(X) \leq \dim_P(X) \leq 1$ holds in any case. Both of these fractal dimensions are *monotone* (i.e., $X \subseteq Y$ implies $\dim(X) \leq \dim(Y)$), *countably stable* (i.e., $\dim(\bigcup_{i=0}^\infty X_i) = \sup_i \dim(X_i)$), and *nonatomic* (i.e., $\dim(\{S\}) = 0$ for each sequence S) [6]. In particular, every countable set of sequences has Hausdorff and packing dimension 0.

We say that a gale $d : \{0,1\}^* \to [0, \infty)$ is p-*computable* if there is a function $\hat{d} : \{0,1\}^* \times \mathbb{N} \to \mathbb{Q}$ such that $\hat{d}(w,r)$ is computable in time polynomial in $|w|+r$ and $|\hat{d}(w,r) - d(w)| \leq 2^{-r}$ holds for all w and r. Gales that are p_2-computable are defined analogously, with $\hat{d}(w,r)$ required to be computable in $2^{(\log(|w|+r))^{O(1)}}$ time.

We are finally ready to bring this all home to complexity classes. We identify each language (i.e., decision problem) $A \subseteq \{0,1\}^*$ with its characteristic sequence, whose n^{th} bit is 1 if the nth string in $\{0,1\}^*$ (in the standard enumeration $\lambda, 0, 1, 00, 01, \ldots$) is an element of A, and 0 otherwise. We say that a gale succeeds on A if it succeeds on the characteristic sequence of A and similarly for strong success. We now show how to define fractal dimension in the complexity classes $\mathrm{E} = \mathrm{TIME}(2^{\text{linear}})$ and $\mathrm{EXP} = \mathrm{TIME}(2^{\text{polynomial}})$.

Definition 1. *[13, 3] Let X be a set of languages.*

1. *If Δ is any of the resource bounds* p, p_2, *then the Δ-dimension of X is*

$$\dim_\Delta(X) = \inf\{s \mid \text{there is a } \Delta\text{-computable } s\text{-gale } d \text{ such that} X \subseteq S^\infty[d] \},$$

and the strong Δ-dimension *of X is*

$$\mathrm{Dim}_\Delta(X) = \inf\{s \mid \text{there is a } \Delta\text{-computable } s\text{-gale } d \text{ such that} X \subseteq S^\infty_{\text{str}}[d]\}.$$

2. *The dimension of X in E is* $\dim(X \mid \mathrm{E}) = \dim_p(X \cap \mathrm{E})$.
3. *The dimension of X in EXP is* $\dim(X \mid \mathrm{EXP}) = \dim_{p_2}(X \cap \mathrm{EXP})$.
4. *The strong dimensions* $\mathrm{Dim}(X \mid \mathrm{E})$, $\mathrm{Dim}(X \mid \mathrm{EXP})$ *are defined analogously.*

By Theorem 1, $\dim(X \mid \mathcal{C})$ and $\mathrm{Dim}(X \mid \mathcal{C})$ are analogs of Hausdorff and packing dimension, respectively. It was shown in [13, 3] that these analogs are in fact well-behaved, *internal* dimensions in the classes $\mathcal{C}$ that we have mentioned. In all these classes, $0 \leq \dim(X \mid \mathcal{C}) \leq \mathrm{Dim}(X \mid \mathcal{C}) \leq 1$ hold, with $\dim(\mathcal{C} \mid \mathcal{C}) = 1$.

2.1   Scaled Dimension

Scaled dimension [9] are versions of resource-bounded dimension that have been "rescaled" to better fit the phenomena that they are measuring. They correspond to the concept of generalized dimension already suggested by Hausdorff.

Scaled dimension arises using more general factors than 2^{-s} in the definition of gale in equation 1. In the general theory, there is a natural hierarchy of scales $g_i(s, n)$, one for each integer $i \in \mathbb{Z}$, built around the standard scale

$$g_0(m, s) = ms.$$

The ith scale gives us i^{th}-order scaled dimension.

The first scales are the following, for $0 \leq s \leq 1$,

$$g_3(m, s) = 2^{2^{(\log\log m)^s}}$$
$$g_2(m, s) = 2^{(\log m)^s}$$
$$g_1(m, s) = m^s$$
$$g_0(m, s) = ms$$
$$g_{-1}(m, s) = m + 1 - m^{1-s}$$
$$g_{-2}(m, s) = m + 2 - 2^{(logm)^{1-s}}$$
$$g_{-3}(m, s) = m + 2^2 - 2^{2^{(\log\log m)^{1-s}}}$$

We refer to [9] for a justification of the choice of these scales, related for instance to complexity classes such as $\text{SIZE}(2^{n^\alpha})$ and $\text{SIZE}(2^{n^\alpha})$.

An $s^{(i)}$-gale is a function $d : \{0, 1\}^* \to [0, \infty)$ satisfying

$$d(w) = 2^{-g_i(|w|+1, s) + g_i(|w|, s)}[d(w0) + d(w1)]$$

for all $w \in \{0, 1\}^*$.

The concept of success and strong success of an $s^{(k)}$-gale on a sequence is defined exactly as in the previous section, the corresponding limsup (liminf) must be infinity.

Definition 2. *[9] Let X be a set of languages.*

1. *If Δ is any of the resource bounds* p, p$_2$, *then the ith-order scaled Δ-dimension of X is*

$$\dim_\Delta^{(i)}(X) = \inf\{s \mid \text{there is a } \Delta\text{-computable } s^{(i)}\text{-gale } d \text{ such that}$$
$$X \subseteq S^\infty[d]\}.$$

2. *The ith-order scaled dimension of X in E is* $\dim^{(i)}(X \mid E) = \dim_\text{p}^{(i)}(X \cap E)$.
3. *The ith-order scaled dimension of X in EXP is* $\dim^{(i)}(X \mid \text{EXP}) = \dim_{\text{p}_2}^{(i)}(X \cap \text{EXP})$.

The ith-*order scaled strong Δ-dimension of X, written* $\text{Dim}_\Delta^{(i)}(X)$, *is defined in the same way, instead requiring strong success of the scaled-gale. We also define*

$\mathrm{Dim}^{(i)}(X \mid \mathrm{E}) = \mathrm{Dim}_{\mathrm{p}}^{(i)}(X \cap \mathrm{E})$, $\mathrm{Dim}^{(i)}(X \mid \mathrm{EXP}) = \mathrm{Dim}_{\mathrm{p}_2}^{(i)}(X \cap \mathrm{EXP})$, etc. (analogously to the definitions in Section 2).

The 0th-order scaled dimension is the standard (unscaled) dimension. The other scaled dimensions have similar properties. For example, $0 \le \dim_{\Delta}^{(i)}(X) \le \mathrm{Dim}_{\Delta}^{(i)}(X) \le 1$ and if $\dim_{\Delta}^{(i)}(X) < 1$, then X has Δ-measure 0. The following theorem states two important facts about the scaled dimensions.

Theorem 2. *[9] The scaled dimension $\dim_{\Delta}^{(i)}(X)$ is nondecreasing in the order i. There is at most one order i for which $\dim_{\Delta}^{(i)}(X)$ is not 0 or 1.*

In particular, the sequence of scaled dimensions must have one of the following four forms.

(i) For all i, $\dim_{\Delta}^{(i)}(X) = 0$.	(ii) For all i, $\dim_{\Delta}^{(i)}(X) = 1$.
(iii) There is an order i^* such that – $\dim_{\Delta}^{(i)}(X) = 0$ for all $i \le i^*$ and – $\dim_{\Delta}^{(i)}(X) = 1$ for all $i > i^*$.	(iv) There is an order i^* such that – $\dim_{\Delta}^{(i)}(X) = 0$ for all $i < i^*$, – $0 < \dim_{\Delta}^{(i^*)}(X) < 1$, and – $\dim_{\Delta}^{(i)}(X) = 1$ for all $i > i^*$.

We find (iv) to be the most interesting case. Then i^* is the "best" order at which to measure the Δ-dimension of X because $\dim_{\Delta}^{(i^*)}(X)$ provides much more quantitative information about X than is provided by $\dim_{\Delta}^{(i)}(X)$ for $i \ne i^*$.

3   Small Span Theorems and BPP

We classify an apparently intractable problem A by identifying and studying the class of all problems that are efficiently reducible to A. Efficiently reducible can be taken as polynomial-time many-one reducible ($\le_{\mathrm{m}}^{\mathrm{P}}$-reductions), polynomial-time Turing reducible ($\le_{\mathrm{T}}^{\mathrm{P}}$-reductions) or any of the intermediate reductions obtained by restricting the query mechanism in polynomial-time Turing reducibilities.

The lower $\le_{\mathrm{m}}^{\mathrm{P}}$-span of A is the set of problems that are $\le_{\mathrm{m}}^{\mathrm{P}}$-reducible to A

$$\mathrm{P}_{\mathrm{m}}(A) = \left\{ B \mid B \le_{\mathrm{m}}^{\mathrm{P}} A \right\}$$

and similarly for other reductions, and the upper $\le_{\mathrm{m}}^{\mathrm{P}}$-span is the set $\mathrm{P}_{\mathrm{m}}^{-1}(A)$ consisting of those decision problems B to which A is $\le_{\mathrm{m}}^{\mathrm{P}}$-reducible.

A Small Span Theorem for a reduction $\le_{r}^{\mathrm{P}}$ in a class $\mathcal{C}$ is the assertion that for every $A \in \mathcal{C}$ it must be the case that either $\mathrm{P}_r(A)$ or $\mathrm{P}_r^{-1}(A)$ have minimal size. This kind of result implies that the degree structure of $\mathcal{C}$ is very fine, so whenever a subclass is not minimal it must contain problems in several degrees.

Juedes and Lutz [12] obtained the first Small Span Theorem for the reduction $\leq_m^P$ and both classes E and EXP, using resource-bounded measure. Other authors pushed this result to $\leq_{k-tt}^P$ in E and $\leq_{n^{o(1)}}^P$ in EXP [2, 4].

In dimension the situation is far more complicated, sice Ambos-Spies et al. [1] and later Hitchcock [8] proved that for scales $i \geq -2$ there are degrees of maximal p-dimension 1. For scale -3, Hitchcock [8] proved a Small Span Theorem for $\leq_m^P$ and E.

An important application of Small Span Theorems is related to BPP, the class corresponding to probabilistic polynomial time, since the class of hard problems for BPP has maximal size in all quantitative settings (Martin-Löf tests, resource-bounded measure, effective dimensions) also when restricted to the subclass of hard sets for BPP in EXP. The existence of a Small Span Theorem for $\leq_T^P$ or $\leq_{tt}^P$ would imply the separation of BPP and EXP, since the degree of Turing-complete sets for EXP would then be minimal size.

Open question. Prove that for every $A \in$ EXP

$$\dim^{(-3)}(P_T(A) \mid \text{EXP}) = 0 \quad \text{or} \quad \dim^{(-3)}(P_T^{-1}(A) \mid \text{EXP})$$

Alternatively prove that $\dim^{(-3)}(\deg_T^P(A) \mid \text{EXP}) = 0$. Similar questions for E in the place of EXP and for $\leq_{tt}^P$ in the place of $\leq_T^P$ are relevant.

Notice that a solution to this question would be the least exigent form of a Small Span Theorem for dimension and polynomial-time reductions. Any statement about other scaled or strong dimensions is either false or would imply an affirmative answer to this. It is also weaker, thus an affirmative answer is more plausible, than a resource-bounded measure version.

4   Completeness Separations

Even if we assume that P≠NP, many open questions in Computational Complexity remain open. Lutz [17] proposed investigation of various strong, measure-theoretic hypothesis, the most notable of which is the hypothesis that NP does not have resource-bounded measure 0 in EXP, that is known now to have many interesting consequences (not known to follow from P≠NP).

One of these consequences is the separation of the notions of $\leq_m^P$ and $\leq_T^P$-completeness for the class NP [16], a statement that seems far stronger than P versus NP.

The answer to each of the following questions would be an improvement over the result in [16].

Open questions. Prove that $\dim_p(NP) > 0$ implies the separation of $\leq_T^P$ and $\leq_{tt}^P$ completeness notions for NP.

Hitchcok et al. have proven in [11] that the separation of $\leq_m^P$ and $\leq_T^P$-completeness for NP can be obtained from the hypothesis $\dim_p^{(-3)}(NP) > 0$, which is weaker than the original measure hypothesis. Can this be improved to a bigger scale (-2, -1, 0, ...)?

References

1. K. Ambos-Spies, W. Merkle, J. Reimann, and F. Stephan. Hausdorff dimension in exponential time. In *Proceedings of the 16th IEEE Conference on Computational Complexity*, pages 210–217, 2001.
2. K. Ambos-Spies, H.-C. Neis, and S. A. Terwijn. Genericity and measure for exponential time. *Theoretical Computer Science*, 168(1):3–19, 1996.
3. K. B. Athreya, J. M. Hitchcock, J. H. Lutz, and E. Mayordomo. Effective strong dimension in algorithmic information and computational complexity. *SIAM Journal on Computing*. To appear.
4. H. Buhrman and D. van Melkebeek. Hard sets are hard to find. *Journal of Computer and System Sciences*, 59(2):327–345, 1999.
5. K. Falconer. *The Geometry of Fractal Sets*. Cambridge University Press, 1985.
6. K. Falconer. *Fractal Geometry: Mathematical Foundations and Applications*, John Wiley & sons, 2003.
7. K Falconer. *Techniques in Fractal Geometry*. John Wiley & sons, 2003.
8. J. M. Hitchcock. Small spans in scaled dimension. *SIAM Journal on Computing*, 34:170–194, 2004.
9. J. M. Hitchcock, J. H. Lutz, and E. Mayordomo. Scaled dimension and nonuniform complexity. *Journal of Computer and System Sciences*, 69:97–122, 2004.
10. J. M. Hitchcock, J. H. Lutz, and E. Mayordomo. The fractal geometry of complexity classes. *SIGACT News Complexity Theory Column*, 36:24–38, 2005.
11. J. M. Hitchcock, A. Pavan, and N. V. Vinodchandran. Partial bi-immunity, scaled dimension, and NP-completeness. *Theory of Computing Systems*. To appear.
12. D. W. Juedes and J. H. Lutz. The complexity and distribution of hard problems. *SIAM Journal on Computing*, 24(2):279–295, 1995.
13. J. H. Lutz. Dimension in complexity classes. *SIAM Journal on Computing*, 32:1236–1259, 2003.
14. J. H. Lutz. The dimensions of individual strings and sequences. *Information and Computation*, 187:49–79, 2003.
15. J. H. Lutz. Effective fractal dimensions. *Mathematical Logic Quarterly*, 51:62–72, 2005.
16. J. H. Lutz and E. Mayordomo. Cook versus Karp-Levin: Separating completeness notions if NP is not small. *Theoretical Computer Science*, 164(1–2):141–163, 1996.
17. J. H. Lutz and E. Mayordomo. Twelve problems in resource-bounded measure. In G. Paun, G. Rozenberg, and A. Salomaa, editors, *Current Trends in Theoretical Computer Science: Entering the 21st Century*, pages 83–101. World Scientific, 2001.
18. D. Sullivan. Entropy, Hausdorff measures old and new, and limit sets of geometrically finite Kleinian groups. *Acta Mathematica*, 153:259–277, 1984.
19. C. Tricot. Two definitions of fractional dimension. *Mathematical Proceedings of the Cambridge Philosophical Society*, 91:57–74, 1982.

Optimization and Approximation Problems Related to Polynomial System Solving

Klaus Meer

Department of Mathematics and Computer Science
Syddansk Universitet, Campusvej 55, 5230 Odense M, Denmark
meer@imada.sdu.dk

Abstract. We outline some current work in real number complexity theory with a focus on own results. The topics discussed are all located in the area of polynomial system solving. First, we concentrate on a combinatorial optimization problem related to homotopy methods for solving numerically generic polynomial systems. Then, approximation problems are discussed in relation with Probabilistically Checkable Proofs over the real numbers.

1 Introduction

Whereas the classical theory of computability focuses on discrete problems (over bits, integers, or fractions that is), Turing's original goal of introducing 'his' machine was the issue of real number computability [24] — the birth of Recursive Analysis [25]. Here, computing some $x \in \mathbb{R}$ amounts to the generation of an infinite sequence of rational numbers q_n converging to x with error at most 2^{-n}. We point out that the approach is thus based both on discrete computability on the set of fractions $\mathbb{Q}$, regarded as an ordered field, and on its approximations to $\mathbb{R}$ as a topological space. This bi-category raises the question for an intrinsic approach to real and complex computability formulated in terms of $\mathbb{R}$ or $\mathbb{C}$ only.

Indeed, a natural yet very different type of real/complex number algorithm is illustrated by the Gaussian elimination method for linear equations and matrices: a finite sequence of operations $+, -, \times, \div$ and branches based on comparisons (pivoting). The latter approach reflects the *modus operandi* of both mathematicians and computer algebra systems. Its formalization [5, 4], now referred to as Blum-Shub-Smale or BSS model, has exhibited a rich variety of structural properties analogous to (though often by entirely different arguments than) well-known results for discrete (i.e., Turing-) computation. Many of them pertain the area of Algebraic Complexity Theory [6], that is, they focus on complexity aspects such as real/complex counter-parts to the classical NP $\overset{?}{=}$ P and similar class separation questions [21, 7].

The present work will review some of the author's (and his coworkers) recent contributions to this field. We discuss different aspects related to certain optimization and approximation problems that are important in the BSS setting. In Section 2 the focus is put on the question whether a polynomial system has

A. Beckmann et al. (Eds.): CiE 2006, LNCS 3988, pp. 360–367, 2006.

a real solution and how to compute solutions numerically. Currently used numerical methods naturally lead to the combinatorial problem of computing the minimal multi-homogeneous Bézout number for a given system. We discuss some negative results concerning that problem obtained in [17]. In Section 3 we turn to structural complexity results involving probabilistically checkable proofs over the reals. Again, starting point will be the question whether a given polynomial system is solvable.

We suppose the reader to be familiar with the basics of real number computability as given in [5, 4]. For the interested reader the two volumes [22] and [12] give a good insight into further research topics.

2   Computation of Solutions of Polynomial Systems

The main result in the seminal paper of Blum, Shub, and Smale [5] is the introduction of real and complex analogues of the classical P versus NP question together with the proof that in this framework both $NP_\mathbb{R}$- and $NP_\mathbb{C}$-complete problems exist and are decidable. More precisely, the following is shown

Theorem 1. *([5]) a) Given $n, m \in \mathbb{N}$ and a quadratic polynomial system*

$$p_1(x) = 0 , \ \ldots , \ p_m(x) = 0,$$

where each $p_i \in \mathbb{R}[x_1, \ldots, x_n]$, all p_i of degree at most 2, the question whether a common real solution $x^* \in \mathbb{R}^n$ exists is $NP_\mathbb{R}$-complete.

b) All decision problems in class $NP_\mathbb{R}$ are decidable in the real number model by an algorithm that runs in simple exponential time.

c) Analogous statements hold for computations over the complex numbers.

In particular statement b) is much more involved than its classical counterpart. Since the search space becomes uncountable deep results about quantifier elimination algorithms in real and algebraically closed fields are needed. For more on this see the recent textbook [3] and the literature cited in there.

The above theorem indicates how real and complex number complexity theory is closely related to a lot of classical problems that have been studied at least during decades, if not centuries. This includes both theoretical results about quantifier elimination as well as the huge area of practical algorithms for polynomial systems solving. Let us first concentrate on the latter.

In many practical applications one is not only interested in the existence of solutions of a polynomial system but also in their (numerical) computation. Numerical methods of choice that have turned out to be successfull in such applications are homotopy methods, see [14]. For target systems $f : \mathbb{C}^n \to \mathbb{C}^n$ of a certain generic structure the idea is to first choose a simple start system g for which all the zeros (solutions) are known. Then the zeros of g are (hopefully) followed by a Newton method along the linear (or another) homotopy $H(x, t) = (1 - t) \cdot g(x) + t \cdot f(x)$ into the zeros of f. Though this principle idea sounds easy there are a lot of interesting and difficult problems related to it, both from the theoretical and from the practical side. For more on this see [14, 11, 23].

We shall now focus on one such problem related to the choice of a good start system g above. One general aim is to fit the zero structure of the start system as good as possible to that of the target system f. An easy way would be to apply Bézout's theorem and construct g such that it has as many (isolated) zeros in $\mathbb{C}^n$ as Bézout's theorem would give as upper bound. Whereas such a g is easy to compute because Bézout's bound is given as the product of the degrees of the involved polynomials, the number of zeros can be far away from the correct one, thus resulting in many superfluous paths that are followed by the homotopy approach. On the "opposite end" of the scale we find start systems that are computed according to the requirement to have the same *mixed volume* as the target system. For generic polynomial systems the mixed volume gives exactly the number of solution in $(\mathbb{C} \setminus \{0\})^n$; however, it is extremely hard to compute.

A third way to design a suitable start system are *multi-homogeneous* Bézout numbers. They are based on a variant of Bézout's theorem where the set of variables is divided into several groups and Bézout's theorem then is applied to each such group. It is used quite successfully in practice. Multi-homogeneous Bézout numbers as well have very interesting relations to analyzing interior point methods as recently shown in [13]. However, until recently [17] the complexity of computing the best multi-homogeneous partition of a generic polynomial system was not known. Let us now describe the main results obtained in [17].

Definition 1. *Let $n \in \mathbb{N}$ and a finite $A \subset \mathbb{N}^n$ be given as input. Find the minimal multi-homogeneous Bézout number, among all choices of a multi-homogeneous structure for a polynomial system with support A.*

$$\begin{cases} f_1(z) = \sum_{\alpha \in A} f_{1\alpha} z_1^{\alpha_1} z_2^{\alpha_2} \cdots z_n^{\alpha_n} \\ \quad \vdots \\ f_n(z) = \sum_{\alpha \in A} f_{n\alpha} z_1^{\alpha_1} z_2^{\alpha_2} \cdots z_n^{\alpha_n} \end{cases} \tag{1}$$

where the $f_{i\alpha}$ are non-zero complex coefficients.

Here, the multi-homogeneous Bézout numbers *of the above system are defined as follows.*

A multi-homogeneous structure is given by a partition of $\{1, \ldots, n\}$ into (say) k sets $I_1, \ldots, I_k$. Then for each set $I_j, 1 \le j \le k$, we consider the group of variables $Z_j = \{z_i : i \in I_j\}$.

The degree of each f_i in the group of variables Z_j is

$$d_j \stackrel{def}{=} \max_{\alpha \in A} \sum_{l \in I_j} \alpha_l$$

When for some j the maximum d_j is attained for all $\alpha \in A$, we say that (1) is homogeneous in the variables Z_j. The dimension of the projective space associated to Z_j is

$$a_j \stackrel{def}{=} \begin{cases} \#I_j - 1 & \text{if (1) is homogeneous in } Z_j, \text{ and} \\ \#I_j & \text{otherwise.} \end{cases}$$

We assume that $n = \sum_{j=1}^{k} a_j$. Now the multi-homogeneous Bézout number $\text{Béz}(A; I_1, \ldots, I_k)$ related to partition $(I_1, \ldots, I_k)$ is defined as[1]

$$\text{Béz}(A; I_1, \ldots, I_k) = \begin{pmatrix} n \\ a_1 \ a_2 \ \cdots \ a_k \end{pmatrix} \prod_{j=1}^{k} d_j^{a_j} \tag{2}$$

where the multinomial coefficient

$$\begin{pmatrix} n \\ a_1 \ a_2 \ \cdots \ a_k \end{pmatrix} \overset{def}{=} \frac{n!}{a_1! \ a_2! \ \cdots \ a_k!}$$

is the coefficient of $\prod_{j=1}^{k} \zeta_j^{a_k}$ in $(\zeta_1 + \cdots + \zeta_k)^n$ (recall that $n = \sum_{j=1}^{k} a_j$).

Though our starting question arose from BSS complexity theory over $\mathbb{R}$ and $\mathbb{C}$ the above question is a purely combinatorial optimization problem. The main result now states that this problem is not only hard to solve exactly unless P equals NP classically, but also hard to approximate.

Theorem 2. *([17]) Computing the minimal multi-homogeneous Bézout number of a generic polynomial system f, when given $n \in \mathbb{N}$ and a support set $A \subset \mathbb{N}^n$ as above, is classically NP-hard. Moreover, even approximating this minimal number within an arbitrary fixed constant C is NP-hard, i.e. the problem does not belong to class APX unless P=NP (for a definition of class APX see [2]).*

The proof of the theorem is based on a reduction from the three coloring problem of graph theory. Given a graph a polynomial system is constructed such that the support of the latter reflects nodes, edges, and triangles in the given graph in a particular way. For the non-approximability part a special multiplicativity structure of multi-homogeneous Bézout numbers has to be established.

For practical purposes it has to be decided whether one prefers to design a start system according to the mixed volume approach or wants to use a heuristic for approximating the best multi-homogeneous Bézout number. For some such heuristics see [15, 16]

3   Probabilistically Checkable Proofs

The optimization problem discussed in the previous section of course is only one particular problem related to polynomial system solving. Theorem 1 substantiates that *deciding* existence of zeros of polynomial systems is computationally hard, Theorem 2 showed that already in a generic situation *computing* the exact number of solutions is hard – even when restricting the computational goal to approximate that number only. A bunch of similar other questions are of importance and lead to different problems and methods. For example, instead of

---

[1] We remark that a more general definition is possible for systems where each polynomial has its own support. However, since our complexity results are negative the restricted version of polynomial systems considered here is sufficient.

approximating the number of zeros for generic systems one can ask for a concise structural complexity theory of counting geometric quantities. A real number version $\#P_\mathbb{R}$ of Valiant's class $\#P$ of counting problems was introduced in [18] and studied under a logical point of view. It is not surprising (nor hard to see) that counting the number of zeros for arbitrary polynomials is a hard problem in class $\#P_\mathbb{R}$. In a series of papers Bürgisser, Cucker, and Lotz recently started a concise development of analyzing the computational complexity of the above and further real and complex counting problems [8, 9, 10]. They obtained a lot of interesting completeness results both in the full BSS model and its linearly restricted version. Most of their results are centered around important problems from algebraic and semi-algebraic geometry such as computing the (modified) Euler characteristic of a semi-algebraic set or the degree of an algebraic variety. In view of developing a theory of approximation algorithms for the BSS model it would be interesting to ask for the complexity of approximating these quantities.

Yet another problem related to approximation algorithms is the following. Once again, it is closely related to another area, namely *Probabilistically Checkable Proofs* over the reals. As before, consider a system of polynomial equations $p_1(x) = 0, \ldots, p_m(x) = 0$ over n real variables $x \in \mathbb{R}^n$ to be given. We know already that deciding solvability or computing the number of solutions are hard problems, and thus also the following is $NP_\mathbb{R}$-hard: Compute the maximal number of polynomials among $p_1, \ldots, p_m$ that have a zero $x^* \in \mathbb{R}^n$ in common. But what about approximating that number? For general polynomial systems we cannot expect efficient algorithms that approximate this maximal number up to a constant factor due to the result below.

Theorem 3. *[20] Consider the following approximation problem over the reals: Given a system of polynomials $p_1, \ldots, p_m \in \mathbb{R}[x_1, \ldots, x_n]$ in n variables such that*

- *each p_i has degree 2 and*
- *the total number of non-vanishing terms in all p_i's is bounded by $O(m^2)$.*

Then there is no polynomial time BSS algorithm approximating the value $\max_{x \in \mathbb{R}^n} |\{i | p_i(x) = 0\}|$ within a constant factor $C > 1$ unless $P_\mathbb{R} = NP_\mathbb{R}$.

However, the question is more interesting for restricted systems like the ones that actually are already sufficient for the completeness result of Theorem 1. An example would be to restrict each p_i to depend on at most 3 variables. Let us denote the related decision problem by *QPS* and its maximization version by *MAX-QPS*.

As it is the case in classical combinatorial optimization, questions like the above directly lead into the area of PCPs, i.e. probabilistically checkable proofs. In the BSS setting the first PCP result was given in [19]. Let us briefly recall the crucial definitions from [19] necessary to understand the further discussion.

Definition 2. *a) Let $r, q : \mathbb{N} \mapsto \mathbb{N}$ be two functions. A $(r(n), q(n))$-restricted verifier V in the BSS model is a particular randomized real number algorithm* *working as follows. For an input $x \in \mathbb{R}^*$ of algebraic size n and another vector $y \in \mathbb{R}^*$ representing a potential membership proof of x in a certain language, the verifier first produces a sequence of $O(r(n))$ many random bits (under the uniform distribution on $\{0,1\}^{O(r(n))}$). Given x and these $O(r(n))$ many random bits V computes in a deterministic fashion the indices of $O(q(n))$ many components of y. Finally, V uses the input x together with the values of the chosen components of y in order to perform a deterministic polynomial time algorithm (in the BSS model). At the end of this algorithm V either accepts or rejects x. For an input x, a guess y and a sequence of random bits ρ we denote by $V(x,y,\rho)$ the result of V supposed the random sequence generated for (x,y) was ρ.*

b) Let $\mathcal{R}, \mathcal{Q}$ be two classes of functions from $\mathbb{N} \mapsto \mathbb{N}$; a real number decision problem $L \subseteq \mathbb{R}^*$ is in class $PCP_\mathbb{R}(\mathcal{R}, \mathcal{Q})$ iff there exist $r \in \mathcal{R}, q \in \mathcal{Q}$ and a $(r(n), q(n))$-restricted verifier V such that conditions i) and ii) below hold:

i) For all $x \subset L$ exists a $y \in \mathbb{R}^*$ such that for all randomly generated strings $\rho \in \{0,1\}^{O(r(size_\mathbb{R}(x)))}$ the verifier accepts: $V(x,y,\rho) =$ 'accept'. In other words:

$$\Pr_\rho\{V(x,y,\rho) = 'accept'\} = 1$$

ii) For any $x \notin L$ and for each $y \in \mathbb{R}^*$

$$\Pr_\rho\{V(x,y,\rho) = 'reject'\} \geq \frac{1}{2}$$

The distribution is the uniform over all random strings $\rho \in \{0,1\}^{O(r(size_\mathbb{R}(x)))}$.

Then the existence of transparent long proofs for QPS can be established:

Theorem 4. *[19] The $NP_\mathbb{R}$-complete decision problem QPS belongs to class $PCP_\mathbb{R}(poly, 1)$. Here, poly denotes the class of polynomials and 1 denotes the class of constant functions.*

The proof is based on a new technique for self-testing linear functions on arbitrary finite subsets of some $\mathbb{R}^N$. The challenging open question is

Problem 1. Is $NP_\mathbb{R} = PCP_\mathbb{R}(\log, 1)$? (with log the class of functions $const \cdot \log$).

Coming back to the optimization version MAX-QPS a positive answer to the above problem would have impact on the (non)-existence of efficient algorithms approximating the maximal number of polynomials in a QPS instance that have a zero in common.

Theorem 5. *[20] If $NP_\mathbb{R} = PCP_\mathbb{R}(\log, 1)$ was true, then there exists no fully polynomial time approximation scheme for MAX-QPS in the BSS model unless $P_\mathbb{R} = NP_\mathbb{R}$.*

Problem 2. Can the above theorem be extended to conclude the non-existence of a *polynomial time approximation scheme* for MAX-QPS if

$NP_{\mathbb{R}} = PCP_{\mathbb{R}}(\log, 1)$ is assumed? More generally: What kind of efficient approximation algorithms at all can be designed for MAX-QPS?

Just as it is the case with the MAX-3-SAT maximization problem in the Turing model [1] there are maximization problems in the BSS setting for which the existence of polynomial time approximation schemes is equivalent to the full $PCP_{\mathbb{R}}$ conjecture of Problem 1 above. One such example involving algebraic circuits was given in [20]. However, a concise analysis of approximation algorithms in the real number model is still waiting to be started.

Acknowledgement

This work was done while the author spent a sabbatical at the Forschungsinstitut für Diskrete Mathematik at the university of Bonn, Germany. The hospitality during the stay is gratefully acknowledged. The author is partially supported by the IST Programme of the European Community, under the PASCAL Network of Excellence, IST-2002-506778 and by the Danish Natural Science Research Council SNF. This publication only reflects the authors' views.

References

1. ARORA S., LUND C.: "Hardness of Approximation", pp. 399– 446 in *Approximation Algorithms for NP-hard problems*, D. Hochbaum (ed.), PWS Publishing, Boston (1996).
2. AUSIELLO G., CRESCENZI P., GAMBOSI G.,KANN V., MARCHETTI-SPACCA-MELA A., PROTASI M.: *Complexity and Approximation: Combinatorial Optimization Problems and Their Approximability Properties*, Springer (1999).
3. BASU S., POLLACK R., ROY M.F.: *"Algorithms in Real Algebraic Geometry"*, Springer (2003).
4. BLUM L., CUCKER F., SHUB M., SMALE S.: *"Complexity and Real Computation"*, Springer (1998).
5. BLUM L., SHUB M., SMALE S.: "On a Theory of Computation and Complexity over the Real Numbers: NP–completeness, Recursive Functions and Universal Machines", pp.1–46 in *Bull. Amer. Math. Soc.* vol.21 (1989).
6. BÜRGISSER P., CLAUSEN M., SHOKROLLAHI A.: *"Algebraic Complexity Theory"*, Springer (1997).
7. BÜRGISSER P.: *"Completeness and Reduction in Algebraic Complexity Theory"*, Springer (2000).
8. BÜRGISSER P., CUCKER F.: "Counting Complexity Classes for Numeric Computations I: Semilinear Sets", pp. 227–260 in *SIAM Journal on Computing*, vol. 33, nr.1 (2003).
9. BÜRGISSER P., CUCKER F.: "Counting Complexity Classes for Numeric Computations II: Algebraic and Semialgebraic Sets", to appear in *Journal of Complexity*.
10. BÜRGISSER P., CUCKER F., LOTZ M.: "Counting Complexity Classes for Numeric Computations III: Complex Projective Sets", to appear in *Foundations of Computational Mathematics*.

11. Cox D., Sturmfels B. (eds.): "Applications of Computational Algebraic Geometry", *Proc. Sympos. Appl. Math.*, Vol. 53, American Mathematical Society, Providence, RI (1998).

12. Cucker F., Rojas J.M. (eds.): Proceedings of the Smalefest 2000, World Scientific (2002).

13. Dedieu J.P., Malajovich G., Shub M.: "On the curvature of the central path of linear programming theory", to appear in *Foundations of Computational Mathematics.*

14. Li T.Y.:"Numerical solution of multivariate polynomial systems by homotopy continuation methods", pp. 399–436 in *Acta Numerica*, 6 (1997).

15. Liang H., Bai F., Shi L.: "Computing the optimal partition of variables in multi-homogeneous homotopy methods", pp. 825–840 in *Appl. Math. Comput.*, vol. 163, nr. 2 (2005).

16. Li T., Lin Z., Bai F.: "Heuristic methods for computing the minimal multi-homogeneous Bézout number", pp. 237–256 in *Appl. Math. Comput.*, vol. 146, Nr. 1 (2003).

17. Malajovich G., Meer K.: " Computing Multi-Homogeneous Bézout Numbers is Hard", to appear in *Theory of Computing Systems*. Extended Abstract in: *22nd Symposium on Theoretical Aspects of Computer Science STACS 2005, Stuttgart,* Lecture Notes in Computer Science vol. 3404, pp. 244–255 (2005).

18. Meer K.: "Counting Problems over $\mathbb{R}$", pp. 41–58 in *Theor. Computer Science*, vol. 242, no. 1-2 (2000).

19. Meer K.:"Transparent long proofs: A first PCP theorem for $NP_{\mathbb{R}}$", to appear in *Foundations of Computational Mathematics*, Springer.

20. Meer K.: "On some relations between approximation and PCPs over the real numbers", to appear in *Theory of Computing Systems*, Springer.

21. Meer K., Michaux C.: "A Survey on Real Structural Complexity Theory", pp.113–148 in *Bull. Belgian Math. Soc.* vol.4 (1997).

22. Renegar J., Shub M., Smale S. (eds.): Proceedings of the AMS Summer Seminar on "Mathematics of Numerical Analysis: Real Number Algorithms", Park City 1995, Lectures in Applied Mathematics (1996).

23. Shub M., Smale S.: "Complexity of Bézout's Theorem I: Geometric aspects", pp. 459–501 in *Journal of the American Mathematical Society*, Vol. **6** (1993).

24. Turing A.M.: "On Computable Numbers, with an Application to the Entscheidungsproblem", pp.230–265 in *Proc. London Math. Soc.* vol.42:2 (1936).

25. Weihrauch K.: "*Computable Analysis*", Springer (2001).

Uncomputability Below the Real Halting Problem*

Klaus Meer[1] and Martin Ziegler[2]

[1] IMADA, Syddansk Universitet, Campusvej 55, 5230 Odense M, Denmark
meer@imada.sdu.dk
[2] University of Paderborn, 33095 Germany
ziegler@upb.de

Abstract. Most of the existing work in real number computation theory concentrates on complexity issues rather than computability aspects. Though some natural problems like deciding membership in the Mandelbrot set or in the set of rational numbers are known to be undecidable in the Blum-Shub-Smale (BSS) model of computation over the reals, there has not been much work on different degrees of undecidability. A typical question into this direction is the real version of Post's classical problem: Are there some explicit undecidable problems below the real Halting Problem?

In this paper we study three different topics related to such questions: First an extension of a positive answer to Post's problem to the linear setting. We then analyze how additional real constants increase the power of a BSS machine. And finally a real variant of the classical word problem for groups is presented which we establish reducible to and from (that is, complete for) the BSS Halting problem.

1 Introduction

We consider the model of real number computation introduced by Blum, Cucker, Shub, and Smale [BSS89, BCSS98]. As opposed to Type-2 machines [Wei01], these so-called BSS machines treat each real (or complex) number as an entity which can be processed (read, stored, compared, added, and so on) exactly and in a single step. They are thus sometimes referred to as algebraic model of real number computation.

It seems fair to state that most of the research in this model so far has been on complexity issues. We on the other hand are interested in associated computability questions. It is well known that the real version $\mathbb{H}$ of the Halting Problem, i.e.

---

* Part of this work was done while K. Meer visited the Forschungsinstitut für Diskrete Mathematik at the University of Bonn, Germany. The hospitality during the stay is gratefully acknowledged. He was partially supported by the IST Programme of the European Community, under the PASCAL Network of Excellence, IST-2002-506778 and by the Danish Natural Science Research Council SNF. This publication only reflects the authors' views.

M. Ziegler is supported by DFG (project Zi1009/1-1) and by JSPS (ID PE 05501).

A. Beckmann et al. (Eds.): CiE 2006, LNCS 3988, pp. 368–377, 2006.
© Springer-Verlag Berlin Heidelberg 2006

asking whether a given BSS machine terminates on a given input, is undecidable in this model. Other undecidable decision problems such as membership to the Mandelbrot set or to the rational numbers have been established, basically by taking into account structural properties of semi-algebraic sets and their close relation to decidable sets in the BSS model over the reals. A few more results of that type where given in [Cuc92], namely an investigation of the real counterpart to the classical arithmetical hierarchy, that is an infinite sequence of (classes of) problems of strictly increasing difficulty extending beyond $\mathbb{H}$. Our present focus is on undecidable real number problems below (and up to) $\mathbb{H}$:

- Regarding the question (raised by Emil Post in 1944) whether there actually *are* undecidable problems strictly easier than the Halting problem, Section 2 reviews and extends classical and recent affirmative results.
- The capability of a BSS machine to store a finite number of real constants in its code makes it more powerful than a Turing machine. Section 3 proves that the power of the BSS model indeed increases strictly with the number of constants it is allowed to store.
- As opposed to classical recursion theory, undecidability proofs in the BSS framework (of the Mandelbrot set, say) typically do not (and, at least for the rationals, cannot) proceed by reduction from $\mathbb{H}$. Section 4 presents a natural problem reducible both to and from (that is, equivalent to) $\mathbb{H}$.

2   Post's Problem in the Linear BSS Model

In 1944 Emil Post [Pos44] asked whether there exist problems in the Turing machine model which are undecidable yet strictly easier than the discrete Halting Problem H. Here a problem P is considered easier than H if H cannot be solved by a Turing machine having access to an oracle for P. Post's question was answered in the affirmative independently by Muchnik [Muc58] and Friedberg [Fri57]. However, so far there are no *explicit* problems with this behaviour known.

In [MZ05] the authors began to study Post's problem over the real numbers. The real Halting problem $\mathbb{H}$ is known to be BSS-undecidable [BSS89]. Thus Post's question makes perfect sense here as well, asking for the existence of BSS–undecidable problems which are semi-decidable but strictly easier than the real Halting Problem. It has turned out that the answer is not only positive, as in the discrete case, but even constructively witnessed by an explicitly statable problem. In fact, the following can be shown:

Fact 1 ([MZ05]). *No BSS algorithm can decide the real Halting Problem, even given access to an oracle for membership to the (undecidable) set $\mathbb{Q}$ of rationals.*

Since the rationals are easily shown to be semi- yet undecidable over $\mathbb{R}$ we thus have an explicit problem which is easier than $\mathbb{H}$. In [MZ05] it is also explicitly given an infinite number of incomparable problems below $\mathbb{H}$.

Remark[1] 2. *BSS- or, more generally, algebraic* [TZ00, SECTIONS 6.3+6.4]
*computability, reducibility, and particularly degree theory are of course sensitive
to the class of operations—including the (number of) constants, cf. Section 3—
permitted.*

In the last ten years, the linearly restricted version $(\mathbb{R}, +, -, 0, 1, <)$ of the full
BSS model over $\mathbb{R}$ has received increasing interest [Koi94, CK95] due to its
relation with the classical (i.e., discrete) "$\mathcal{P}=\mathcal{NP}$?" question [FK00]. Here only
additions, subtractions and comparisons as well as the constants 0 and 1 are
allowed but no multiplication $\times$ nor division $\div$. Thus, all computed intermediate
results on inputs $x \in \mathbb{R}$ have the form $ax + b$ for some $a, b \in \mathbb{Z}$. Analogously
to the full model, the Halting Problem $\mathbb{H}^\ell$ for linear machines is undecidable by
a linear machine; and Post's problem as well makes sense in the linear version.
The main result of the present section is an explicit solution to it.

Theorem 3. *Let* $\mathbb{SQ} := \{q^2 : q \in \mathbb{Q}\}$ *denote the set of quadratic rationals.
Then* $\mathbb{SQ} \preceq^\ell \mathbb{Q} \preceq^\ell \mathbb{H}^\ell$, *where* "$\preceq^\ell$" *and all similar notions refer to Turing
reducibility in the linear model.*

We have space to handle the easy claims: Both $\mathbb{Q}$ and $\mathbb{SQ}$ are undecidable in
the linear model since this already holds in the full model. Both sets are semi-
decidable: For input $x \in \mathbb{R}$ enumerate all pairs $(r, s) \in \mathbb{Z} \times \mathbb{N}$ and check for each
pair whether $x \cdot s = r$. Note that both the enumeration and the 'multiplication'
$x \cdot s$ can be performed in $(\mathbb{R}, +, -, 0, 1, <)$; similarly for semi-deciding $\mathbb{SQ}$ by
enumerating all pairs (r^2, s^2) based for instance on the recursion $(r + 1)^2 =
r^2 + r + r + 1$. Next, $\mathbb{SQ} \preceq^\ell \mathbb{Q}$: On input $x \in \mathbb{R}$, first check $x \geq 0$ and ask the
$\mathbb{Q}$-oracle whether $x \in \mathbb{Q}$. If this is the case use the above enumeration to find
$(r, s) \in \mathbb{N}^2$ with $xs = r$. Then test whether some of the (finitely many) pairs
$(\tilde{r}^2, \tilde{s}^2) \leq (r, s)$ satisfies $x \cdot \tilde{s}^2 = \tilde{r}^2$ or not.

Note that in the full BSS model the converse relation $\mathbb{Q} \preceq \mathbb{SQ}$ is also valid:
Having access to a $\mathbb{SQ}$–oracle one can decide $\mathbb{Q}$ by simply squaring the input
$x \in \mathbb{R}$. By the nontrivial claim of Theorem 3, this reduction does not hold in the
linear model.

Question 1. *In the linear setting, is* $\mathbb{Q}$ *as hard as the Halting Problem?*

In the full BSS model, Question 1 has a negative answer according to Fact 1.

3   The Benefit of Additional Real Constants

Already the paper [BSS89] revealed that the capability of BSS machines to
store a finite number of arbitrary real constants gives it super-recursive power.
Specifically, *any* $A \subseteq \mathbb{N}$ and in particular the discrete Halting Problem H can
be encoded into some $r \in \mathbb{R}$ and thus decided by a BSS machine.

This raises the question whether and to what extent real or complex constants
may be exploited in terms of complexity as well, that is, in order to accelerate

---

[1] We gladly follow an anonymous referee's suggestion to point this out explicitly.

solution of computational problems. We briefly mention some interesting results
in that respect. In the complex BSS model for rational decision problems one
can eliminate complex constants in potential decision algorithms with a polynomial slow down only, see [BCSS98, Koi96]. For the real number model it is an
important open question whether real constants can be eliminated without too
a high increase of the running time. Some aspects of this question are discussed
in [Cha99, BMM00]. In certain restrictions of the BSS model, however, it was
shown that the use of real constants introduces non-uniformity, see [Koi93]. Similar results where obtained for several complexity classes, for example in [CG97].

The present section deals with the *computational* power of BSS machines. We
want to gauge the *degree* of super-recursiveness yielded by one, two, or more real
constants. In the discrete realm, a pairing function like $\langle x, y \rangle := (2x + 1) \cdot 2^y$
admits a computable en- and decoding of several integers into a single one. This
significantly differs from the real case where, as a consequence to the *domain-
invariance theorem* in Algebraic Topology [Dei85, THEOREM 4.3], a bijection
$\mathbb{R} \times \mathbb{R} \to \mathbb{R}$ cannot be locally continuous, not to mention BSS-computable[2].
Thus the impossibility to effectively en- and decode two reals into a single one
should, at least intuitively, imply that two constants yield strictly more power
than a single one.

Our first result is based on a tool related to [TZ00, SECTIONS 6.3+6.4]:

Lemma 4. *For $A \subseteq \mathbb{R}^\infty$ and $c_1, \ldots, c_i \in \mathbb{R}$, consider the following claims:*

a) A is semi-decidable by a BSS Machine with constants $c_1, \ldots, c_i \in \mathbb{R}$.

b) There is an integer sequence $(d_n)_n$ such that A is a countable union $A = \bigcup_n A_n$ of sets $A_n \subseteq \mathbb{R}^{d_n}$ semi-algebraic over the field extension $\mathbb{Q}(c_1, \ldots, c_i)$.

c) There exists $c_{i+1} \in \mathbb{R}$ such that A is semi-decidable by a BSS Machine with constants $c_1, \ldots, c_i, c_{i+1}$.

Then a) implies b) from which in turn c) follows.

Proof. implicit in [Cuc92, THEOREM 2.4 and REMARK 2.5]; cf. also [Mic90]. $\square$

Theorem 5. *a) Let $(p_i)_i = (2, 3, 5, 7, 11, \ldots)$ denote the sequence of primes
and $c_i := \exp(\sqrt{p_i}) \in \mathbb{R}$. Then, $c_1, \ldots, c_i$ are algebraically independent.*

*b) Let $c_1, \ldots, c_i$ be algebraically independent. The finite set $A := \{c_1, \ldots, c_i\} \subseteq
\mathbb{R}$ is decidable with i real constants but not semi-decidable with $i - 1$ real
constants.*

In other words, the computational power of the BSS model strictly increases
with every further admitted constant.

Proof. a) Apply Lindemann-Weierstraß [Bak75, THEOREM 1.4] to the linearly independent numbers $\sqrt{p_i}$ (math.niu.edu/~rusin/known_math/00_incoming/sqrt_q).
b) Suppose A is semi-decidable by a machine with $i - 1$ real constants. By
Lemma 4 (a⇒b), it is semi-algebraic over some rational field extension $K =
\mathbb{Q}(\tilde{c}_1, \ldots, \tilde{c}_{i-1})$. A finite discrete set, *in*equalities can be eliminated revealing

---

[2] Although $(x, n) \mapsto \langle \lfloor x \rfloor, n \rangle + (x - \lfloor x \rfloor)$ is a bi-computable bijection $\mathbb{R} \times \mathbb{N} \to \mathbb{R}$.

that A is even algebraic over K—contradicting that A's transcendence degree exceeds that of K.   □

Next, let $\mathbb{H}_i$ denote the real Halting Problem for BSS-machines with i constants. Obviously $\mathbb{H}_i$ is no harder than $\mathbb{H}_{i+1}$—simply choose $c_{i+1} = c_i$. We want to show that $\mathbb{H}_i$ is in fact *strictly* easier than $\mathbb{H}_{i+1}$. This claim does not follow from Theorem 5 because the (anyway a bit artificial) sets A_i constructed there are neither reducible to nor from any $\mathbb{H}_j$. Formally:

Definition 6. *Let different versions of the Halting Problem be defined as*
$H := \{\langle M \rangle : M$ *is a Turing machine that terminates on input* $0\}$
$\mathbb{H}_0 := \{\langle M \rangle : M$ *is a constant-free BSS machine that terminates on input* $0\}$
$\mathbb{H}_1 := \{\langle M, c_1 \rangle : M$ *is a BSS machine with constant* c_1 *terminating on* $0\}$
$\mathbb{H}_2 := \{\langle M, c_1, c_2 \rangle : BSS$ *machine* M *with constants* c_1, c_2 *terminating on* $0\}$
and so on. Here, $\langle M \rangle \in \mathbb{N}$ *denotes a reasonable encoding of (the discrete, i.e. control part of) machine* M *by an integer number* [BSS89, SECTION 8].

Note that indeed the control part of a BSS machine (except for the machine constants, that is) can easily be coded by a single integer. In particular, the code of an instance for $\mathbb{H}_i$ varies piecewise continuously with the machine constants $c \in \mathbb{R}^i$ used. Since $\mathbb{H}_i \subseteq \mathbb{R}^i$ by virtue of footnote 2, Definition 6 describes the dimensional decomposition of the real BSS Halting Problem $\mathbb{H} = \bigcup_i \mathbb{H}_i \subseteq \mathbb{R}^\infty$. That $\mathbb{H}_i$ is strictly easier than $\mathbb{H}_{i+1}$ has the interesting consequence of yielding, in addition to the set $\mathbb{Q}$ of rationals according to [MZ05], a vast variety of further explicit solutions to Post's Problem over the Reals:

Corollary 7. *Fix* $i \in \mathbb{N}$ *and let* $A \subseteq \mathbb{R}^i$ *be undecidable yet recursively enumerable. Then* A *is strictly easier than* $\mathbb{H} \subseteq \mathbb{R}^\infty$. *In particular, unbounded dimension is unavoidable for any BSS-complete problem.*

For the underlying notion of reducibility to make sense here, the use of real constants must be limited in computing the corresponding reduction function.

Definition 8. *Let* $A, B \subseteq \mathbb{R}^\infty$ *be two real decision problems. A is called many-one reducible to B with i constants if there exists a BSS machine M having constants* $c_1, \ldots, c_i \in \mathbb{R}$ *that reduces A to B in the usual sense. We denote this by "$A \preceq_m^i B$"; similarly for equivalence "$\equiv_m^i$". Regarding Turing-reduction, write "$A \preceq_T^i B$" if a BSS machine with at most i constants can decide A given oracle access to B.*

It is well-known that the three basic classical characterizations of recursive enumerability of some $A \subseteq \mathbb{N}$—halting set of a Turing machine (semi-decidability), many-one reducibility to H, and range of a computable integer function—carry over to the real setting[3] [Mic91]. The same holds for $\mathbb{H}$'s single 'slices' $\mathbb{H}_i$:

Lemma 9. *For any decision problem* $A \subseteq \mathbb{R}^n$, *the following are equivalent:*

a) *A is the halting set of some BSS machine with i real constants;*
b) *$A \preceq_m^i \mathbb{H}_{i+n}$;*

---

[3] Notice that enumerability of a countable $A \subseteq \mathbb{R}$ does *not* mean $A = \text{range}(f)$ for a computable $f : \mathbb{N} \to \mathbb{R}$; compare Lemma 9c).

c) $A = \text{range}(f)$ *for some partial function* $f :\subseteq \mathbb{R}^n \to \mathbb{R}^n$ *computable by a BSS machine with i real constants.*

For the rest of this section, we show that the hierarchy $\mathbb{H}_i$ from Definition 6 is indeed strict. For the lowest two levels it is not hard to show

Proposition 10. $H \equiv_m^0 \mathbb{H}_0 \lneq_T^0 \mathbb{H}_1$.

The next result applies to all levels of the hierarchy but takes into account only many-one reductions (or, more generally, Turing-reductions permitted only one oracle query).

Theorem 11. *For all $i \in \mathbb{N}$ it is* $\mathbb{H}_{i+1} \not\leq_m^0 \mathbb{H}_i$ *and* $\mathbb{H}_{i+1} \not\leq_{T[1]}^0 \mathbb{H}_i$.

Proof Suppose that a constant-free machine M^* decides $\mathbb{H}_{i+1}$ making a single oracle call to $\mathbb{H}_i$. Consider an instance $(M, c_1^*, \ldots, c_{i+1}^*)$ for $\mathbb{H}_{i+1}$, where all c_i^* are algebraically independent. Then in a small ball $U(c^*)$ around $c^* := (c_1^*, \ldots, c_{i+1}^*)$ the reduction algorithm will use the same path for inputs of the form $(M, c), c \in U(c^*)$ and thus it computes instances of $\mathbb{H}_i$ having the form $(M', \mathcal{P}(c^*))$. Here, M' will be the same machine for all $c \in U(c^*)$ due to the remark preceding the theorem, and $\mathcal{P} : \mathbb{R}^{i+1} \to \mathbb{R}^i$ is a polynomial map, say $\mathcal{P} = (p_1, \ldots, p_i)$.

The main task of the proof now is to construct a situation where we can guarantee that both a yes- and a no-instance of the given problem $\mathbb{H}_{i+1}$ have to be reduced to the same instance of $\mathbb{H}_i$. This can be achieved by using the implicit function theorem together with the cylindrical algebraic decomposition of semi-algebraic sets.

The above arguments hold for any machine M having $i+1$ machine constants which are algebraically independent. We now specify M to be a machine that uses machine constants $(c_1, \ldots, c_{i+1})$ and halts on input 0 iff all its constants are algebraically independent. There are two cases to analyze:

Case 1. In $U(c^*)$ there exists a point $\tilde{c}$ such that $det D\mathcal{P}(\tilde{c}) \neq 0$. Without loss of generality we can assume that the components of such an $\tilde{c}$ are algebraically independent. Otherwise, continuity of the determinant and of $\mathcal{P}$ together with density of the tuples of algebraically independent numbers in $\mathbb{R}^{i+1}$ would yield a contradiction.

According to the implicit function theorem there exist a neighborhood V of $\tilde{c}_{i+1}$ and an implicit function $g : V \to \mathbb{R}^i$ such that for all $c_{i+1} \in V$ the vector $\mathcal{P}(g(c_{i+1}), c_{i+1})$ is constantly equal to $\mathcal{P}(g(\tilde{c}_{i+1}), \tilde{c}_{i+1})$. It is then clear that there is a rational point c_{i+1} in V such that the yes-instance $(M, \tilde{c})$ and the no-instance $(M, g(c_{i+1}), c_{i+1})$ both are mapped to the same instance of $\mathbb{H}_i$ by the reduction algorithm. Thus the reduction fails.

Case 2. Now suppose that the above matrix is singular in all points of $U(c^*)$. Since $\mathcal{P}$ is a polynomial and since $U(c^*)$ as ball is semi-algebraic the image $\mathcal{P}(U(c^*))$ is semi-algebraic as well. By Sard's theorem the image $\mathcal{P}(U(c^*))$ has measure 0 in $\mathbb{R}^i$ and thus is of dimension m for some $0 \leq m < i$. For notational simplicity below we set $m = i + 1 - j$ for a $j > 1$.

Using the well known properties of semi-algebraic sets, in particular the existence of a cylindrical algebraic decomposition we can find a set $W \subset \mathcal{P}(U(c^*))$ of dimension m such that the following holds:

- W is semi-algebraic in $\mathbb{R}^i$ and its projection onto the, say, m final components is an open ball in $\mathbb{R}^m$;
- thus W is diffeomorphic to a ball $K \subseteq \mathbb{R}^m$ via a map $\phi : W \to K$;
- there exists a $\hat{c} \in \mathcal{P}^{-1}(W)$ such that the matrix $\left\{ \dfrac{\partial (\phi \circ \mathcal{P})_k}{\partial c_\ell}(\hat{c}) \right\}_{\substack{1 \le k \le j \\ 1 \le \ell \le j}}$ has rank j in a neighborhood of $\hat{c}$. Using the same argument as for Case 1 we can without loss of generality assume all components of $\hat{c}$ to be algebraically independent.

Once again, the implicit function theorem yields existence of a neighborhood $V \subset \mathbb{R}^m$ of $(\hat{c}_{j+1}, \ldots, \hat{c}_{i+1})$ and a function $g : V \to \mathbb{R}^j$ such that

$$\phi \circ \mathcal{P}(g(c_{j+1}, \ldots, c_{i+1}), c_{j+1}, \ldots, c_{i+1}) = \phi \circ \mathcal{P}(g(\hat{c}_{j+1}, \ldots, \hat{c}_{i+1}), \hat{c}_{j+1}, \ldots, \hat{c}_{i+1})$$

for all $(c_{j+1}, \ldots, c_{i+1}) \in V$. Again, this neighborhood V contains a point with a rational component c_{i+1} and the reducing machine will fail on either $\hat{c}$ or that point with rational last component. $\qquad\square$

The above proof cannot be applied to yield the same result with respect to Turing reductions. It would only be possible to "fool" as many oracle calls of a reduction machine as the dimension of the image $\mathcal{P}(U)$ is. Thus we have the following open

Question 2. *Does Theorem 11 generalize to arbitrary Turing reductions?*

4   Completeness and the Real Word Problem: An Outline

Classical recursion theory knows a variety of natural problems equivalent (that is, reducible from *and to*) the discrete Halting Problem H: Post's Correspondence Problem, Hilbert's Tenth Problem, and the Word Problem for finitely presented groups [Nov59, Boo58] all are undecidable. The corresponding proofs proceed by reduction from H.

In the real setting of BSS machines on the other hand, most undecidability proofs involve algebraic and/or topological arguments. This is the case with the Mandelbrot set [BCSS98, THEOREM 2.4.2], convergence of Newton's iteration [BCSS98, THEOREM 2.4.4], and the sets $\mathbb{Q}$ and $\mathbb{A}$ of rationals and of algebraic reals, respectively [MZ05]. And indeed, these problems are (provably for the latter, the others probably as well) *strictly* easier than the real Halting Problem $\mathbb{H}$; cf. Section 2. This raises the question for BSS-complete problems other than $\mathbb{H}$, that is, for further systems capable of universal real number computation. Observe that the above discrete examples all are BSS-decidable by [BSS89, EXAMPLE 6]. In the present section we present a natural extension of the classical word problem to the reals which is provably many-one reducible to *and from*

ℍ, that is, a new BSS-complete problem in the sense of universal computation. Here we basically only present the formulation of the related problem. It is in full length discussed in [MZ06].

Definition 12. *a) Let X be a set. The* free group generated by X, *denoted by $F = (\langle X \rangle, \circ)$ or more briefly $\langle X \rangle$, is the set $(X \cup X^{-1})^*$ of all finite sequences $\bar{w} = x_1^{\alpha_1} \cdots x_n^{\alpha_n}$ with $n \in \mathbb{N}$, $x_i \in X$, $\alpha_i \in \{-1, +1\}$, equipped with concatenation $\circ$ as group operation subject to the rules*

$$x \circ x^{-1} \quad = \quad 1 \quad = \quad x^{-1} \circ x \qquad \forall x \in X \qquad (1)$$

where $x^1 := x$ and where 1 denotes the empty word, that is, the unit element.
b) For sets X and $R \subseteq \langle X \rangle$, consider the quotient $\langle X \rangle / \langle R \rangle_n =: \langle X | R \rangle$ of $\langle X \rangle$ with respect to the normal subgroup $\langle R \rangle_n$ *of $\langle X \rangle$ generated by R. If both X and R are finite, the tuple (X, R) will be called a* finite presentation *of G*
c) The word problem *for $\langle X | R \rangle$ is the task of deciding, given $\bar{w} \in \langle X \rangle$, whether $\bar{w} = 1$ holds in $\langle X | R \rangle$.*

The famous work of Novikov and, independently, Boone establishes

Fact 13. *a) The word problem for any fixed finitely presented group is semi-decidable by a Turing Machine.*
b) There is a finitely presented group whose associated word problem is many-one reducible by a Turing machine from the discrete Halting Problem H.

Proof. a) is immediate. For b) see e.g. the great textbook [LS77].   □

The word problem for discrete groups is decidable by a BSS-machine. Therefore we now consider *real* groups and their associated word problems. This approach differs significantly from other work dealing with groups G in the BSS setting which treat such G as underlying structure of the computational model, that is, not over the reals $\mathbb{R}$ and its arithmetic structure. [Tuc80] considers the question of computational realizing G and its operation, not of deciding properties of (elements of) G. [DJK05] does consider BSS-decidability (and -complexity) of properties of a real group, but *given* by some matrix generators and lacking completeness results. For instance, finiteness of the multiplicative subgroup of $\mathbb{C}$ generated by $\exp(2\pi i / x)$, $x \in \mathbb{R}$, is equivalent to $x \in \mathbb{Q}$ and thus undecidable; whereas any *fixed* such group is isomorphic either to $(\mathbb{Z}, +)$ or to some $(\mathbb{Z}_n, +)$, both with easy word problem.

Regarding that the BSS-machine is the natural extension of the Turing machine from the discrete to the reals, the following is equally natural a generalization of Definition 12b):

Definition 14. *Let $X \subseteq \mathbb{R}^N$ for some $N \in \mathbb{N}$ and $R \subseteq (X \cup X^{-1})^*$. We call the group $G = \langle X | R \rangle$* effectively presented *if both X and R are BSS-decidable.*

Remark 15. a) Though X inherits from $\mathbb{R}^N$ algebraic structure, Definition 12a) of the free group $G = (\langle X \rangle, \circ)$ considers X as a plain set only. Thus, (group-) inversion in G must not be confused with (multiplicative) inversion: $5 \circ \frac{1}{5} \neq 1 = 5 \circ 5^{-1}$ for $X = \mathbb{R}$. This difference may be stressed by writing 'abstract' generators $x_{\bar{a}}$ indexed with real vectors $\bar{a}$; here, 'obviously' $x_5^{-1} \neq x_{1/5}$.

b) BSS-computation of course refers to encoding input (and, if present, also output) as (finite sequences of) vectors of real numbers, that is, of $\bar{w} \in (X \cup X^{-1})^*$ as, e.g., $(w_1, \alpha_1, \ldots, w_n, \alpha_n) \in (\mathbb{R}^N \times \mathbb{Z})^n$.

c) While Definition 12c) requires the set X of generators to be finite, it must in the real setting be a finite-*dimensional* subset of $\mathbb{R}^\infty$. Considerable effort in the proof of Theorem 18b) is spent on asserting this condition. $\square$

Example 16. Let $\mathbb{S}$ denote the unit circle in $\mathbb{C}$ with complex multiplication. The following is an effective presentation $\langle X | R_1 \cup R_2 \rangle$ of $\mathbb{S}$ (with decidable word problem):

$$X := \left\{ x_{r,s} : (r,s) \in \mathbb{R}^2 \setminus \{0\} \right\} ,$$
$$R_1 := \left\{ x_{r,s} \circ x_{a,x}^{-1} : (r,s), (a,b) \neq 0, rb = sa \wedge ar > 0 \right\} ,$$
$$R_2 := \big\{ x_{r,s} \circ x_{a,b} \circ x_{u,v}^{-1} : (r,s), (a,b), (u,v) \neq 0,$$
$$r^2 + s^2 = 1 \wedge a^2 + b^2 = 1 \wedge u = ra - sb \wedge v = rb + sa \big\}   \square$$

Example 17. (Undecidable) real membership "$t \in \mathbb{Q}$" is reducible to the word problem of an effectively presented real group: Consider $X = \{x_r : r \in \mathbb{R}\}$, $R = \big\{ x_{nr} = x_r, x_{r+k} = x_k : r \in \mathbb{R}, n \in \mathbb{N}, k \in \mathbb{Z} \big\}$; then $x_r = x_0 \Leftrightarrow r \in \mathbb{Q}$. $\square$

The latter example does not establish BSS-*hardness* of the real word problem because $\mathbb{Q}$ is provably easier than the real Halting Problem $\mathbb{H}$ [MZ05]. It is the main result of the present section to provide a BSS counterpart to Fact 13.

Theorem 18.   *a) For any effectively presented real group $G = \langle X | R \rangle$, the associated word problem is BSS semi-decidable.*

b) There exists an effectively presented real group $\langle X | R \rangle$ whose associated word problem is many-one reducible from $\mathbb{H}$ by a BSS machine.

Notice that already Claim a) requires Tarski's quantifier elimination. We also point out that, in accordance with Definition 14 and as opposed to X, the set R of relations in b) lives in $\mathbb{R}^\infty$, that is, has unbounded dimension.

References

[Bak75]   A. Baker: "*Transcendental Number Theory*", Camb. Univ. Press (1975).

[BCSS98]   L. Blum, F. Cucker, M. Shub, S. Smale: "*Complexity and Real Computation*", Springer (1998).

[BMM00]   S. Ben-David, K. Meer, C. Michaux: "A note on non-complete problems in $NP_\mathbb{R}$", pp.324–332 in *Journal of Complexity* vol. **16**, no. 1 (2000).

[Boo58]   W.W. Boone: "*The word problem*", pp. 265–269 in *Proc. Nat. Acad. Sci. U.S.A*, vol.**44** (1958).

[BSS89]   L. Blum, M. Shub, S. Smale: "On a Theory of Computation and Complexity over the Real Numbers: $\mathcal{NP}$-Completeness, Recursive Functions, and Universal Machines", pp.1–46 in *Bulletin of the American Mathematical Society* (AMS Bulletin) vol.**21** (1989).

[Cha99]   O. Chapuis, P. Koiran: "Saturation and stability in the theory of computation over the reals", pp.1–49 in *Annals of Pure and Applied Logic*, vol.**99** (1999).

[CG97]   F. CUCKER, D.Y. GRIGORIEV: "On the power of real turing ma-
            chines over binary inputs", pp.243–254 in *SIAM Journal on Computing*,
            vol.**26**, no.1 (1997).

[CK95]   F. CUCKER, P. KOIRAN: "Computing over the Real with Addition and
            Order: Higher Complexity Classes", pp.358–376 in *Journal of Complexity*
            vol.**11** (1995).

[Cuc92]   F. CUCKER: "The arithmetical hierarchy over the reals", pp.375–395 in
            Journal of Logic and Computation vol.**2(3)** (1992).

[Dei85]   K. DEIMLING: "*Nonlinear Functional Analysis*", Springer (1985).

[DJK05]   H. DERKSEN, E. JEANDEL, P. KOIRAN: "Quantum automata and alge-
            braic groups", pp.357–371 in *J. Symbolic Computation* vol.**39** (2005).

[FK00]   H. FOURNIER, P. KOIRAN: "Lower Bounds Are Not Easier over the Re-
            als", pp.832–843 in *Proc. 27th International Colloqium on Automata, Lan-
            guages and Programming* (ICALP'2000), vol.**1853** in Springer LNCS.

[Fri57]   R.M. FRIEDBERG: "Two recursively enumerable sets of incompara-
            ble degrees of unsolvability", pp.236–238 in *Proc. Natl. Acad. Sci.*
            vol.**43** (1957).

[Koi93]   P. KOIRAN: "A weak version of the Blum-Shub-Smale model", pp. 486–495
            in *Proceedings FOCS'93*, (1993).

[Koi94]   P. KOIRAN: "Computing over the Reals with Addition and Order",
            pp.35–48 in *Theoretical Computer Science* vol.**133** (1994).

[Koi96]   P. KOIRAN: "Elimination of Constants from Machines over Algebraically
            Closed Fields", pp. 65–82 in *Journal of Complexity*, vol.**13** (1997).

[LS77]   R.C. LYNDON, P.E. SCHUPP: "Combinatorial Group Theory", Springer
            (1977).

[Mic90]   C. MICHAUX: "Machines sur les réels et problèmes $\mathcal{N}\mathcal{P}$–complets",
            Séminaire de Structures Algébriques Ordonnées, Prépublications de
            l'equipe de logique mathématique de Paris 7 (1990).

[Mic91]   C. MICHAUX: "Ordered rings over which output sets are recursively enu-
            merable", pp. 569–575 in *Proceedings of the AMS 111* (1991).

[Muc58]   A.A. MUCHNIK: "Solution of Post's reduction problem and of certain other
            problems in the theory of algorithms", pp. 391–405 in *Trudy Moskov Mat.
            Obsc.*, vol.**7** (1958).

[MZ05]   K. MEER, M. ZIEGLER: "An explicit solution to Post's problem over the
            reals", pp. 456–467 in *Proc. 15th International Symposium on Fundamen-
            tals of Computation Theory, Lübeck*, LNCS vol. 3623 (2005).

[MZ06]   K. MEER, M. ZIEGLER: "On the word problem for a class of groups with
            infinite presentations", Preprint (2006).

[Nov59]   P.S. NOVIKOV: "*On the algorithmic unsolvability of the word problem in
            group theory*", pp. 1–143 in *Trudy Mat. Inst. Steklov*, vol.**44** (1959).

[Pos44]   E.L. POST: "Recursively enumerable sets of positive integers and their
            decision problems", pp.284–316 in *Bull. Amer. Math. Soc.* vol.**50** (1944).

[Tuc80]   J.V. TUCKER: "Computability and the algebra of fields", pp.103–120 in
            J. Symbolic Logic vol.**45** (1980).

[TZ00]   J.V. TUCKER, J.I. ZUCKER: "Computable functions and semicomputable
            sets on many sorted algebras", pp317–523 in (S. Abramskz, D. Gabbay,
            T. Maibaum Eds.) *Handbook of Logic for Computer Science* vol.**V** (Logic
            and Algebraic Methods), Oxford University Press (2000).

[Wei01]   WEIHRAUCH K.: "*Computable Analysis*", Springer (2001).

Constraints on Hypercomputation

Greg Michaelson[1] and Paul Cockshott[2]

[1] Heriot Watt University
G.Michaelson@hw.ac.uk
[2] University of Glasgow
wpc@dcs.gla.ac.uk

Abstract. Wegner and Eberbach[16] have argued that there are fundamental limitations to Turing Machines as a foundation of computability and that these can be overcome by so-called superTuring models. In this paper we contest their claims for interaction machines and the π-calculus.

1   Introduction

The Turing machine (TM) [1] has been the dominant paradigm for Computer Science for 70 years: Ekdahl[12] likens an attack on it to a "challenge to the second law of thermodynamics".

The roots of Turing's work lie in debates about the notion of computability in the pre-computer age[10]. Just before the Second World War, in an outstanding period of serendipity, Turing, Church and Kleene all developed independent notions of computability which were quickly demonstrated to be formally equivalent. These seminal results form the basis for the Church-Turing Thesis that all notions of computability will be equivalent. Until now, the Church-Turing Thesis has remained unshaken.

A central concern of these pre-computer Mathematical Logicians was to formalise precisely the concept of effective computation. For Church, this is a matter of *definition*, explicitly identifying effective calculability with recursive or lambda-definable functions over the positive integers. Church[5] states that:

> If this interpretation or some similar one is not allowed it is difficult to see how the notion of an algorithm can be given any exact meaning at all.(p356)

Turing subsequently outlined a proof of the equivalence of his notion of "computability" with Church's "effective calculability".

A fundamental distinction of Turing's approach is that he identifies a human-independent mechanism to embody his procedure, by explicit analogy with a human being:

> We may compare a man in the process of computing a real number to a machine which is only capable of a finite number of conditions... ([1]Section 1).

A. Beckmann et al. (Eds.): CiE 2006, LNCS 3988, pp. 378–387, 2006.

In a 1939 paper discussing the unity of these different approaches, Turing is explicit about the mechanical nature of effective calculation:

> A function is said to be "effectively calculable" if its values can be found by some purely mechanical process ... We may take this statement literally, understanding by a purely mechanical process one which may be carried out by a machine. It is possible to give a mathematical description, in a certain normal form, of the structures of these machines. ([2]p166).

In a late paper he makes the same point with reference to digital computers:

> The idea behind digital computers may be explained by saying that these machines are intended to carry out any operations which could be done by a human computer. ([3]Section 4).

For Church and Kleene the presence of a human mathematician that applies the rules seems to be implicit, but the ability to give an explicit procedure for applying rules to symbols and to physically realise these procedures, was central to all three conceptions of effectiveness. The corollary is that a computation which is not physically realisable is not effective.

2   Wegner and Eberbach's SuperTuring Computers

There has been robust debate in Mathematics, Philosophy, Physics and, latterly, Computer Science about the possibility of *hypercomputation* which seeks to transcend the limits of classic computability. Copeland [6] and Cotogno[8] provide useful summaries.

Thus, Wegner and Eberbach[16] assert that the fundamental limitations to the paradigmatic conception of computation can be overcome by more recent "superTuring" approaches. They draw heavily on the idea of an algorithm as an essentially closed activity. That is, while the TM realising an algorithm may manipulate an unbounded memory, the initial memory configuration is pre-given and may only be changed by the action of the machine itself. Furthermore, an effective computation may only consume a finite amount of the unbounded memory and of time, the implication being that an algorithm must terminate to be effective.

They say that the TM model is too weak to describe the Internet, evolution or robotics. For the Internet, web clients initiate interactions with servers without any knowledge of the server history.

Wegner and Eberbach claim that there is a class of superTuring computations (sTC) which are a superset of TM computations. That is sTC includes computations which are not realisable by a TM. A superTuring computer is "any system or device which can carry out superTuring computation". They give discursive presentations of interaction machines (IM), the π-calculus and the $-calculus, and explore why they transcend the TM. here, we do not consider the $-calculus as its sTC properties appear to depend on those alleged for interaction machines and π-calculus.

3   How Might the TM Paradigm Be Displaced?

In general, a demonstration that a new system is more powerful than a C-T system involves showing that while all terms of some C-T system can be reduced to terms of the new system, there are terms of the new system which cannot be reduced to terms of that C-T systemMore concretely, we think that requirements for a new system to conclusively transcend C-T are, in increasing order of strength:

1. demonstration that some problem known to be semi-decidable in a C-T system is decidable in the new system;
2. demonstration that some problem known to be undecidable in a C-T system is semi-decidable in the new system;
3. demonstration that some problem known to be undecidable in a C-T system is decidable in the new system;
4. characterisations of classes of problems corresponding to 1-3;
5. canonical exemplars for classes of problems corresponding to 1-3.

Above all, we require that the new system actually encompasses effective computation; that is, that it can be physically realised in some concrete machine. While we are not unduly troubled by systems that require potentially unbounded resources such as an unlimited TM tape, we reject systems whose material realisations conflict with the laws of physics, or which require actualised infinities as steps in the calculation process.

4   Physical Realism and Computation

A key point about the Universal Computers proposed by Turing is that they are material apparatuses which operate by finite means. Turing assumes that the computable numbers are those that are computable by finite machines, and initially justifies this only by saying that the memory of a human computer is necessarily limited.

Turing is careful to construct his machine descriptions in such a way as to ensure that the machine operates entirely by finite means and uses no techniques that are physically implausible. His basic proposition remained that :"computable numbers may be described briefly as the real numbers whose expressions as a decimal are calculable by finite means."

Turing rules out computation by infinite means as a serious proposition. Most proposals for superTuring computation rest on the appeal of the infinite. Copeland[6] proposes the idea of accelerating Turing machines whose operation rate increases exponentially so that if the first operation were performed in a microsecond, the next would be done in $\frac{1}{2}\mu s$, the third in $\frac{1}{4}\mu s$, etc. The result would be that within a finite interval it would be able to perform an infinite number of steps. This evades all possibility of physical realisation. A computing machine must transfer information between its component parts in order to perform an operation. If the time for each operation is repeatedly halved. then one soon reaches the point at which signals traveling at the speed of light have insufficient time to propagate from one part to another within an operation step. Hamkins[14] discusses what could be computed on Turing machines if they were allowed to operate for an infinite time. He hypothesises a relativistic experiment in which a researcher moving near the speed of light experiences a finite duration whilst back on Earth his graduate student has an infinite duration to solve a problem. Hamkins fails to suggest which immortal graduate student he has in mind for this task. Another theme of those advocating Super-Turing computation is the use of analogue computation over real numbers. For a review see [7]. The idea of being able to physically represent real numbers is highly questionable in view of the quantum of action h. This poses fundamental and finite limits on the accuracy with which a physical system can approximate real numbers.

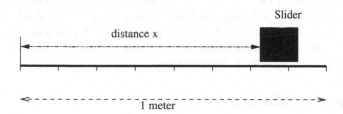

Fig. 1. An analogue representation of a real number using a physical version of the real number line

Suppose we want to use analogue encoding real numbers as spatial separation, as shown in Figure 1,to encode a real number x that could be used by some Super Turing analogue computational process. We first set up the the distance x and then measure it during the process of the computation.This raises two questions:

1. How precisely can we, in principle at least, measure the distance x?
2. How stable is such an analogue memory. For how long can it store the information?

Because of Heisenberg's equation

$$\Delta p \Delta x = \frac{h}{4\pi}$$

there is a tradeoff between the accuracy to which x can be represented as distance and the period for which x can be stored. If the mass of the slider is m the uncertainty in its velocity is given by $\Delta v = \frac{h}{4\pi m \Delta x}$. This implies that the persistence time of the analogue store T_p is constrained such that

$$T_p \approx \frac{\Delta x}{\Delta v} = \frac{4\pi m \Delta x^2}{h}$$

It is clear that for a practical device, T_p must be chosen exceed the time taken to measure the distance x. This in turn, must be greater than $\frac{2x}{c}$, since any distance measurement will involve at least one reflection off the slider. If $T_p < \frac{2x}{c}$ then the computer would be unable to read the real. We thus have the constraint

$$\frac{2x}{c} < \frac{4\pi m \Delta x^2}{h}$$

which implies

$$\Delta x^2 > \frac{2xh}{4\pi mc}$$

For a 1 kilo slider adjustable over a range of 1 meter which allows the representation of reals in the range 0..1 as $x \pm \Delta x$, we find that $\Delta x > 5.93 \times 10^{-22}$meters.

This corresponds to a real number stored with about 70 bits of precision.

To add an additional bit of precision to our real number we would have to quadruple the mass of the slider. The analogue encoding of the reals requires a mass which grows with Oe^{2b} where b is the number of bits of precision to which the reals are stored. For a Turing Machine the mass required grows linearly with the number of bits. Thus proposals to incorporate the full mathematical abstraction of real numbers into computing devices so as to allow them to outperform Turing machines are physically implausible.

5   Interaction Machines

Wegner and Eberbach claim that a Turing Machine is restricted by having to have all its inputs appear on the tape prior to the start of computation. Interaction machines on the contrary can perform input output operations to the environment in which they are situated. Interaction Machines, whose canonical model is the Persistent Turing Machine(PTM) of Goldin [9], are not limited to a pre-given finite input tape, but can handle potentially infinite input streams. These arguments have been thoroughly criticised by Ekdahl[12]. Rather than rehearse his arguments we shall focus on additional weaknesses of Wegner and Eberbach's claims.

5.1   Turing's Own Views

Turing's Test for machine intelligence is probably as well known as his original proposal for the Universal Computer. He proposed in a very readable paper[3], that a computer could be considered intelligent if it could fool a human observer into thinking they were interacting with another human being. It is clear that his putative intelligent machine would be an Interaction Machine in Wegner's sense. Rather than being cut off from the environment and working on a fixed tape, it receives typed input and sends printed output to a person.

Turing did not find it necessary to introduce a fundamental new class of computing machine for this Gedankenexperiment. He describes the machine using what is a paraphrase (Turing 1950, page 436) of his description of the computing machine in his 1936 paper. It is clear that Turing is talking about the same general category of machine in 1950 as he had in 1936. He says he is concerned with discrete state machines, and that a special property of such digital computers was their universality:

This special property of digital computers, that they can mimic any discrete state machine, is described by saying that they are universal machines. The existence of machines with this property has the important consequence that, considerations of speed apart, it is unnecessary to design various new machines to do various computing processes. They can all be done with one digital computer, suitably programmed for each case. It will be seen that as a consequence of this all digital computers are in a sense equivalent.(Turing 1950, page 442)

This is clearly a recapitulation of the argument in section 6 of his 1936 paper where he introduced the idea of the Universal Computer. Turing argued that such machines were capable of learning and that with a suitable small generalised learning program and enough teaching, then the computer would attain artificial intelligence.

5.2   Equivalence of Interaction Machines and Turing Machines

Consider first a digital computer interacting in the manner forseen by Turing in his 1950 paper, with teletype input/output. Suppose then we have a computer initialised with a simple learning program following which it is acquires more sophisticated behaviour as a result of being 'taught'. As the computer is taught we record every keystroke onto paper tape.

We initialise a second identical computer with the same program and at the end of the first computer's working life we give to the second machine as an input, the tape on which we have recorded all the data fed to the first machine. With the input channel of the second machine connected to the tape reader it then evolves through the same set of states and produce the same outputs as the original machine did. The difference between interactive input from a teletype and tape input is essentially trivial.

A small modification to the program of a conventional TM will transform it into a PTM. Like Goldin we will assume a 3 tape TM, M_1 with one tape T_1 purely for input, one tape T_2 purely for output and one tape T_3 used for working calculations. We assume that tapes T_1, T_2 are unidirectional, T_3 is bidirectional. M_1 has a distinguished start state S_0 and a halt state S_h. On being set to work it either goes into some non-terminating computation or eventually outputs a distinguished termination symbol τ to T_2, branches to state S_h and stops. We assume that all branches to S_h are from a state that outputs τ. Once τ has been output, the sequence of characters on T_2 up to to τ are the number computed by the machine.

We now construct a new machine M_2 from M_1 as follows: replace all branches to S_h with branches to S_0. From here it will start reading in further characters from T_1 and may again evolve to a state where it outputs a further τ on T_2. Machine M_2 now behaves as one of Goldin's PTMs. It has available to it the persisting results of previous computation on T_3 and these results will condition subsequent computations. It is still a classic TM, but a non-terminating one. It follows that PTM's, and thus Interaction Machines of which they are the canonical example, are a sub-class of TM programs and do not represent a new model of computation.

6 π-Calculus

The π-calculus is not a model of computation in the same sense as the TM: there is a difference in level. The TM is a specification of a material apparatus that is buildable. Calculi are rules for the manipulation of strings of symbols and these rules will not do any calculations unless there is some material apparatus to interpret them.

Is there any possible physical apparatus that can implement the π-calculus and, if so, is a conventional computer such an apparatus. Since it is possible to write a conventional computer program that will apply the formal term re-write rules of the π-calculus to strings of characters representing terms in the calculus [15], then it would appear that the π-calculus can have no greater computational power than the von Neumann computer. A possible source of confusion is the language used to describe the π-calculus: channels, processes, evolution which imply that one is talking about physically separate, but communicating entities evolving in space/time. There is a linguistic tension between what is strictly laid down as the rules of a calculus and the rather less specific physical system that is suggested by the language. One has to be very careful before accepting that the existence the π-calculus as a formal system implies a physically realisable distributed computing apparatus.

Consider two of the primitives: synchronisation and mobile channels.

Is π-calculus synchronisation in its general sense physically realistic?

Does it not imply the instantaneous transmission of information - faster than light communication if the processes are physically separated?

If the processors are in relative motion, there can be no unambiguous synchronisation shared by the different moving processes. It thus follows that the processors can not be physically mobile if they are to be synchronised with at least 3 way synchronisation (see [11] pp 25-26).

Suppose we have the following pi calculus terms

$$\alpha \equiv (\bar{a}v.Q) + (by.R[y]) \tag{1}$$

$$\beta \equiv (\bar{b}z.S) + (ax.T[x]) \tag{2}$$

In the above α and β are processes. The process α tries to either output the value v on channel a or to read from channel b into the variable y. The $+$ operator means non deterministic composition, so $A + B$ means that either A occurs or B occurs but not both. The notation $\bar{a}v$ means output v to a, whilst av would mean input from a into v. If α succeeds in doing an output on channel a it then evolves into the abstract process Q, if alternatively, it succeeds in doing an input from b into y, then it evolves into the process $R[y]$ which uses the value y in some further computation.

We can place the two processes in parallel:

$$(\bar{a}v.Q) + (by.R[y])|(\bar{b}z.S) + (ax.T[x]) \tag{3}$$

This should now evolve to

$$(Q|T[v]) \text{ or to } (S|R[z]) \tag{4}$$

where either Q runs in parallel with $T[v]$ after the communication on channel a or where S runs in parallel with $R[z]$ after the value z was transfered along channel b from process β to process α.

Since the two processes are identical mirror images of one another any deterministic local rule by which process β commits to communication on one of the channels, must cause α to commit to the other channel and hence synchronisation must fail.

Thus if α commits to communication on channel a then its mirror image β must commit to communicate on b leading to: $T[x]|R[y]$, but this is not permitted according to the π-calculus.

The argument is a variant of the Liar Paradox, but it is not a paradox within the π-calculus itself. It only emerges as a paradox once you introduce the constraints of relativity theory prohibiting the instantaneous propagation of information. Nor does abandoning determinism help. If the commitment process is non-deterministic, then on some occasions synchronisation will succeed, but on other occasions the evolution both processes will follow the same rule, in which case synchronisation will fail.

A global arbitration machine solves the problem at the loss of parallelism. A worse loss of parallelism, in terms of complexity order, is entailed by distributed broadcast protocols such as Asynchronous Byzantine Agreement[4].

In conclusion it is not possible to build a reliable mechanism that will implement in a parallel distributed fashion any arbitrary composition of π-calculus processes.

6.1   Wegner and Eberbach's Argument

Wegner's argument for the super-Turing capacity of the π-calculus rests on there being an implied infinity of channels and an implied infinity of processes. Taking into account the restrictions on physical communications channels the implied infinity could only be realised if one had an actual infinity of fixed link computers. At this point we are in the same situation as the Turing machine tape - a finite but unbounded resource. For any actual calculation a finite resource is used, but the size of this is not specified in advance. W&E then interprets 'as many times as is needed' in the definition of replication in the calculus as meaning an actual infinity of replication. From this he deduces that the calculus could implement infinite arrays of cellular automata for which he cites Garzon [13] to the effect that they are more powerful than TMs.

7   Conclusion

In Section 3, we gave criteria that must be met for the Church-Turing thesis to be displaced. In general, a demonstration that all terms in C-T systems should have equivalent terms in the new system, but there should be terms in the new system which do not have equivalents in C-T systems. In particular, the new system should be able to solve decision problems that are semi-decidable or undecidable in C-T systems. Finally, we require that a new system be physically realisable. We think that, under these criteria, Wegner and Eberbach's claims that Interaction Machines, the π-calculus and the $-calculus are super-Turing are not adequately substantiated.

First of all, Wegner and Eberbach do not present a concrete instance of terms in any of these three systems which do not have equivalents in C-T systems. Secondly, they do not identify decision problems which are decidable or semi decidable in any of these systems but semi-decidable or undecidable respectively in C-T systems. Finally, they do not explain how an arbitrary term of any of these three systems may be embodied in a physical realisation. We have shown that the synchronisation primitive of the calculus is not physically realistic. The modeling of cellular automata in the calculus rests on this primitive. Furthermore, the assumption of an infinite number of processes implies an infinity of mobile channels, which are also unimplementable. We therefore conclude that whilst the π-calculus can be practically implemented on a single computer, infinite distributed implementations of the sort that W&E rely upon for their argument can not be implemented.

Acknowledgments

Greg Michaelson wishes to thank the School of Information Technology at the University of Technology, Sydney for their hospitality during the writing of this paper.

References

1. A. M. Turing. On Computable Numbers with an Application to the Entschiedung-sproblem. *Proc. London Mathematical Soc.*, 42:230–265, 1936.
2. A. M. Turing. Systems of Logic Based on Ordinals. *Proc. London Mathematical Soc.*, 45, 1939.
3. A. M. Turing. Computing Machinery and Intelligence. *Mind*, 39:433–60, 1950.
4. G. Bracha and S. Toueg. Asynchronous consensus and byzantine protocol in a faulty environment. Technical Report TR-83-559, CS Dept., Cornell University, Ithaca,NY 14853, 1983.
5. A. Church. An Unsolvable Problem of Elementary Number Theory. *American Journal of Mathematics*, 58:345–363, 1936.
6. B. J. Copeland. Hypercomputation. *Minds and Machines*, 12:461–502, 2002.
7. B. J. Copeland and R. Sylvan. Beyond the universal turing machine. *Australasian Journal of Philosophy*, 77(1):46..66, 1999.

8. Paolo Cotogno. Hypercomputation and the Physical Church-Turing Thesis. *Brit. J. Phil. Sci.*, 54:181–223, 2003.
9. D. Goldin, S. Smolka and P.Wegner. Turing Machines, Transition Systems, and Interaction. *Information and Computation*, 194(2):101–128, November 2004.
10. Martin Davis. *Engines of Logic : Mathematicians and the Origins of the Computer*. Norton, 2001.
11. A. Einstein. *Relativity*. Methuen and Company, London, 1920.
12. B. Ekdahl. Interactive Computing does not supersede Church's thesis. In *Proc. Computer Science*, pages 261–265, 17th Int. Conf. San Diego, 1999. Association of Management and the International Association of Management,.
13. M Garzon. *Models of Massive Parallelism: Analysis of Cellular Automata and Neural Networks*. EATCS. Springer-Verlag, 1995.
14. J. Hamkins and A. Lewis. Infinite Time Turing Machines. *Journal of Symbolic Logic*, 65(2):567–604, 2000.
15. D. Turner and B. Pierce. Piçt; A programming language based on the picalculus. In G. Plotkin, C. Stirling, and M. Tofte, editors, *Proof, Language and Interaction: Essays in Honour of Robin Milner*, pages 455–494. MIT Press, 2000.
16. P. Wegner and E. Eberbach. New models of computation. *Computer Journal*, 47:4–9, 2004.

Martingale Families and Dimension in P

Philippe Moser*

Dept. de Informática e Ingeniería de Sistemas, María de Luna 1, 50018 España

Abstract. We introduce a new measure notion on small complexity classes (called F-measure), based on martingale families, that gets rid of some drawbacks of previous measure notions: it can be used to define dimension because martingale families can make money on all strings, and it yields random sequences with an equal frequency of 0's and 1's. As applications to F-measure, we answer a question raised in [1] by improving their result to: for almost every language A decidable in subexponential time, $\mathsf{P}^A = \mathsf{BPP}^A$. We show that almost all languages in PSPACE do not have small non-uniform complexity. We compare F-measure to previous notions and prove that martingale families are strictly stronger than Γ-measure [1], we also discuss the limitations of martingale families concerning finite unions. We observe that all classes closed under polynomial many-one reductions have measure zero in EXP iff they have measure zero in SUBEXP. We use martingale families to introduce a natural generalization of Lutz resource-bounded dimension [13] on P, which meets the intuition behind Lutz's notion. We show that P-dimension lies between finite-state dimension and dimension on E. We prove an analogue to the Theorem of Eggleston in P, i.e. the class of languages whose characteristic sequence contains 1's with frequency α, has dimension the Shannon entropy of α in P.

1 Introduction

Resource-bounded measure has been successfully used to understand the structure of the exponential time classes E and EXP, see [12] for a survey. Recently resource-bounded measure has been refined via effective dimension which is an effectivization of Hausdorff dimension, yielding applications in a variety of topics, including algorithmic information theory, computational complexity, prediction, and data compression [13, 17, 14, 4, 2, 6].

Unfortunately both Lutz's resource-bounded measure and dimension formulations [10, 13] only work on classes containing E (apart from finite-state dimension). One reason for this is that when a martingale is to bet on some string x depending on the history of the language for strings $y < x$, the history itself is exponentially larger than the string x. Thus even reading the history is far above the computational power of P.

One way to overcome this difficulty was proposed in [1], with a measure notion (called Γ-measure) defined via martingales betting only on a sparse subset of

---

* This work was partially supported by Programa Europa CAI-Gobierno de Aragón UZ-T27, and Spanish Government MEC Project TIC 2002-04019-C03-03.

A. Beckmann et al. (Eds.): CiE 2006, LNCS 3988, pp. 388–397, 2006.

strings of the history, with the drawback that the class of sparse languages does not have measure zero. Nevertheless it seems that sparse languages and more generally languages whose characteristic sequences satisfy some frequency property should be small for an appropriate measure notion on P, because there exists simple (exponential-time computable) martingales always making the same fixed bet that succeed on such languages. Such martingales are relatively "simple": exponential computational power is only required to keep track of the current capital. This also shows how important it is for a martingale to be able to bet on all strings, in order to succeed. This "betting on all strings" property becomes crucial in Lutz's recent formulation of effective Hausdorff dimension [13].

A stronger measure notion called dense martingale measure (denoted Γ_d) was then proposed in [22], with the surprising result that the polynomial time version of Lutz's hypothesis "NP does not have measure zero in E" does *not* hold [3]. Γ_d-measure does not satisfy the finite union property though; it was then shown that a restricted version (denoted $\Gamma/(P)$) of it does, unfortunately $\Gamma/(P)$-measure has some unnatural properties: a language with infinitely many easy instances can still be random.

Another limitation of previous martingale-based measure notions on P from [1, 22] and on PSPACE [18] is the inability of the corresponding martingales to bet on all strings. Γ-martingales can only bet on a polynomial number out of the exponentially many strings of length n, whereas Γ_d and $\Gamma/(P)$ martingales can only double their capital a polynomial number of times while betting on (the exponentially many) strings of size n, with the direct consequence that neither can be used to define a dimension notion, because the ability to bet on every string is essential for this purpose (notice that simply keeping track of the capital won by a martingale doubling its capital on every string is impossible in polynomial time). Moreover the random sequences yielded by either of those two measure notions do not necessarily have an equal frequency of 0's and 1's in the limit, whereas this property is captured by Lutz's resource-bounded measure notion on E, corresponding to the intuitive idea of a random sequence.

In this paper we introduce a measure notion on P based on martingale families (called F-measure), where martingale families can double their capital on all strings, thus enabling us to define dimension in P. F-measure gets rid of the unnatural random sequences of $\Gamma/(P)$-measure [22], and yields random sequences with an equal frequency of 0's and 1's, similarly to Lutz resource-bounded measure [10]. Moreover F-measure is strictly stronger than Γ-measure. *United, we stand; divided, we fall* is the key idea behind F-measure, i.e. whereas a single polynomial time computable martingale is not able to make money on all exponentially many strings of size n, a family of martingales working together and sharing their capital *can*. The idea is to separate the exponentially many strings of size n into groups of polynomial size, where each member of the family bets on one of these groups of strings. The family shares a common bank account: When such a martingale bets on a string x, the capital at its disposal amounts to the capital currently gathered by its family on predecessors of x, although it has no information about how much this (possibly) exponentially large capital is.

Constructing the perfect measure on P has turned out to be much more difficult as previously thought; it is now widely admitted that this perfect measure on P might be very difficult to achieve, and that for any measure notion on P some desirable properties must be abandoned; and F-measure is no exception. Similarly to Γ_d-measure [22], martingale families do not satisfy the finite union property, but only satisfy the union property in some non-general sense: we can only guarantee the union property for families with the same bank account structure; however this is usually enough to prove theorems where the union property is needed.

We show in Section 3.1 that except for general unions, martingale families satisfy the basic measure properties, i.e. every single language has measure zero, and the whole class P does not have measure zero, we then introduce uniform P-unions and show that the union property holds for those. We observe that it is easy to derive a F-measure notion on classes between P and E like QUASIPOLY, SUBEXP and PSPACE; for BPP see [20].

Next we show that the concept of randomness yielded by F-measure is optimal regarding frequency: every language L such that there are infinitely many n with $|L[1 \cdots n]| \leq \epsilon n$ (with $\epsilon < 1/2$), has measure zero in P (Section 3.2).

As applications to F-measure, we answer a question raised in [1], improving their result to: almost all (all except a measure zero class) languages computable in subexponential time, are hard enough to derandomize BPP, i.e. a polynomial time algorithm can use almost every language $L \in$ SUBEXP to derandomize every probabilistic polynomial time algorithm, even if the probabilistic algorithm has also oracle access to L.

We also investigate the nonuniform complexity of languages of PSPACE, and show that almost all languages in PSPACE do not have small nonuniform complexity, thus reducing the resource-bounds of a similar result in [11].

Next we compare F-measure to previous measure notions on P, and show that F-measure is strictly stronger than Γ-measure, i.e. every Γ-measure zero set has F-measure zero, and there are classes with Γ-measure non-zero that have F-measure zero. Due to their intrinsic differences, we cannot compare Γ_d-measure and $\Gamma/(\mathsf{P})$-measure [22] to F-measure. Nevertheless all sets proved to be small for $\Gamma/(\mathsf{P})$-measure in [22] are also small for F-measure. Regarding density arguments, F-measure performs better; indeed a (Lebesgue) random language has $(1/2 - o(1))2^n$ words of length n (with high probability), and this property is captured by F-measure, whereas for $\Gamma/(\mathsf{P})$-measure, the set of languages having $o(2^n)$ words of length n has $\Gamma/(\mathsf{P})$-measure zero. The advantage of $\Gamma/(\mathsf{P})$-measure over F-measure is that it satisfies the finite union property. Concerning Γ_d-measure and F-measure, both their respective strengths are different, whereas Γ_d-measure cannot be used to define dimension in P, F-measure fails to capture the Γ_d-measure zero sets in [3].

We also show that all classes closed under polynomial many-one reductions have measure zero in EXP iff they have F-measure zero in E_α, which reduces the time bounds of many results [8, 23, 8, 7] from measure on E to measure on SUBEXP.

The second part of the paper is devoted to dimension in P. Lutz resource-bounded dimension [13], has been introduced on a wide variety of complexity classes ranging from finite state automata, exponential time and space up to the class of recursively enumerable languages [17], with the exception of small classes like P.

Hausdorff dimension is a refinement of Lebesgue measure, where every measure zero class of languages is assigned a real number between 0 and 1, called its Hausdorff dimension. The key idea of Lutz is to receive a tax after each round (even if the martingale did not bet during that round): the largest tax rate which can be received without preventing the martingale from succeeding on a given class represents the dimension of the class.

Trying to bridge the gap between finite state automata and exponential time requires a measure notion which is able to bet and double the capital at every round. Whereas all previous measure notions on P [1, 22] are unable to do so, it is not a problem for martingale families. This leads to a natural generalization of Lutz resource-bounded dimension [13] on P, which meets the idea behind Lutz's notion.

We give some evidence that P-dimension is a natural extension to P of previously existing dimension notions, by showing that it lies exactly between finite-state dimension and dimension on E, i.e. we show that for any sequence S, $\dim_{\mathsf{FS}}(S) \geq \dim_{\mathsf{P}}(S) \geq \dim_{\mathsf{E}}(S)$.

Finally we prove an analogue to the Theorem of Eggleston [5] in P, i.e. the class of languages whose characteristic sequences contain 1's with frequency α, has strong dimension the Shannon entropy of α in P.

2   Preliminaries

$s_0, s_1, s_2 \ldots$ denotes the standard enumeration of the strings in $\{0,1\}^*$ in lexicographical order, where $s_0 = \lambda$ denotes the empty string. Note that $|w| = 2^{O(|s_{|w|}|)}$. A *sequence* is an element of $\{0,1\}^\infty$. If F is a string or a sequence and $1 \leq i \leq |w|$ then $w[i]$ and $w[s_i]$ denotes the ith bit of F. Similarly $w[i \ldots j]$ and $w[s_i \ldots s_j]$ denote the ith through jth bits.

For two string x, y, the concatenation of x and y is denoted xy. If x is a string and y is a string or a sequence extending x i.e. $y = xu$, where u is a string or a sequence, we write $x \sqsubseteq y$. We write $x \sqsubset y$ if $x \sqsubseteq y$ and $x \neq y$.

A *language* is a set of strings. A *class* is a set of languages. The cardinal of a language L is denoted $|L|$. Let n be any integer. The set of strings of size n of language L is denoted $L^{=n}$. Similarly $L^{\leq n}$ denotes the set of strings in L of size at most n.

We identify language L with its characteristic function χ_L, where χ_L is the sequence such that $\chi_L[i] = 1$ iff $s_i \in L$. Thus a language can be seen as a sequence in $\{0,1\}^\infty$. $L \upharpoonright s_n$ denotes the initial segment of L up to s_n given by $L[s_0 \cdots s_n]$.

We use standard notation for traditional complexity classes; see for instance [21]. For $\epsilon > 0$, denote by E_ϵ the class $\mathsf{E}_\epsilon = \bigcup_{\delta < \epsilon} \mathrm{DTIME}(2^{n^\delta})$. SUBEXP is the class $\cap_{\epsilon>0}\mathsf{E}_\epsilon$, and quasi polynomial time refers to the class $\mathsf{QUASIPOLY} = \cup_{k\geq 1}\mathsf{DTIME}(n^{\log^k n})$.

Lutz measure on E [11] is obtained by imposing appropriate resource-bounds on a game theoretical characterization of classical Lebesgue measure, via martingales. A martingale is a function $d : \{0,1\}^* \to \mathbb{R}_+$ such that, for every $w \in \{0,1\}^*$, $2d(w) = d(w0) + d(w1)$.

3   A New Measure on P Via Martingale Families

The following equivalent alternative to martingales will be useful.

Definition 1. *A rate-martingale is a function* $D : \{0,1\}^* \to [0,2]$ *such that for every* $w \in \{0,1\}^*$ $D(w0) + D(w1) = 2$.

A rate-martingale outputs the factor by which the capital is increased after the bet, whereas a martingale outputs the current capital.

The key idea to define our measure on small complexity classes is that instead of considering a single martingale as usual, we consider families of rate-martingales which share their wins. These rate-martingales are computed by Turing machines that have oracle access to their input and can query any bit of it. To enable such machines to compute the length of their input F without reading it, we also provide them with $s_{|w|}$; this convention is denoted by $M^w(s_{|w|})$. We assume their output to be two binary numbers (a, b) corresponding to the rational number $\frac{a}{b}$. With this convention, rational numbers such as $1/3$ can be said to be computed exactly. Here is a definition of such a family of rate-martingales.

Definition 2. *A* P-*family of rate-martingales* $(\{D_i\}_i, \{Q_i\}_i, \text{ind})$, *is a family of rate-martingales* $\{D_i\}_i$, *where* $Q_i : \mathbb{N} \to \mathcal{P}(\{0,1\}^*)$ *are disjoint polynomial-printable query sets (i.e. there is a Turing machine that on input* $(i, 1^n)$ *outputs all strings in* $Q_i(n)$ *in time polynomial in* n*), i.e.* $Q_i(n) \cap Q_j(n) = \emptyset$ *and* $Q_i(m) \subseteq Q_i(n)$ *for* $m < n$, $\text{ind} : \{0,1\}^* \to \mathbb{N}$ *is a polynomial time computable function, such that* $D_i(L \upharpoonright x)$ *is computable by a random access Turing machine* M *in time polynomial in* $|x|$ *i.e.* $M^{L\upharpoonright x}(x, i) = D_i(L \upharpoonright x)$ *where* M *queries its oracle only on strings in* $Q_i(|x|)$, *and* $\text{ind}(x)$ *is an index* i *such that* $x \notin Q_j(|x|)$ *for every* $j \neq i$.

For simplicity we omit the indexes and denote the family of rate-martingales by (D, Q, ind), unless needed. Each rate-martingale D_i of the family only bets on strings inside its query set Q_i. The function ind on input a string x, outputs which rate-martingale is to (possibly) bet on x. The idea is that the rate-martingales share their wins, and have the ability to divide the bets along all members of the family. We are interested in the total capital such a family wins.

Definition 3. *Let* (D, Q, ind) *be a* P-*family of rate-martingales such that* $D_i(\lambda) \leq 1$ *for every* i. *The wins of a* P-*family of rate-martingales is the function* $W_D : \{0,1\}^* \to \mathbb{Q}$, *where* $W_D(L \upharpoonright x) = \prod_{i \leq 2^{|x|}} \prod_{y \leq x} D_i(L \upharpoonright y)$.

For simplicity we simply write i for the index of the first product, unless needed. Remember that $D_i(L \upharpoonright x)$ is the factor by which the capital is multiplied after the bet on x. Thus the product in Definition 3 is exactly the total capital the whole family of rate-martingales would win, would they be able to share their wins after each bet. Note that the function W_D is not polynomial, but only exponential time computable. This is a major difference to previous measure notions on P: computing the global wins of the family of rate-martingales is above the computational power of P.

A class has measure zero if there is a family of rate-martingales whose wins on the languages of the class are unbounded. Here is a definition.

Definition 4. *A class C of languages has P-measure zero, denoted $\mu_P(C) = 0$, if there is a P-family of rate-martingales (D, Q, ind) such that for every $L \in C$, $\limsup_{n \to \infty} W_D(L \upharpoonright s_n) = \infty$.*

Whenever D's capital grows unbounded on L, we say that the family of rate-martingales succeeds on L, and write $L \in S^\infty[D]$. We call our measure notion F-measure.

It is easy to see that at higher complexity levels such as EXP, F-measure is equivalent to Lutz's measure notion [10], by taking a family containing a unique rate-martingale.

To prove a non-general union property we consider win-functions that succeed whatever small the starting capital of the corresponding rate-martingales is.

Definition 5. *The independent success set of a P-family of rate-martingales (D, Q, ind) denoted $S_I^\infty[D]$ is the set of languages L such that for every $\alpha > 0$, $\limsup_{n \to \infty} \prod_i \alpha \prod_{y \le s_n} D_i(L \upharpoonright y) = \infty$.*

It is sometimes more convenient to output the current capital of a rate-martingale, rather than the factor of increase. It is easy to check that Definition 2 can be reformulated by taking families of martingales instead of rate-martingales. We call such a family a P-family of martingales. Both definitions are equivalent, i.e. if (D, Q, ind) is a P-family of rate-martingales then (d, Q, ind) with $d_i(L \upharpoonright x) = \prod_{\{y|y \le x \text{ and } y \in Q_i(|x|)\}} D_i(L \upharpoonright y)$ is a P-family of martingales with the same win function. For the other direction take $D_i(L \upharpoonright x) = \frac{d_i(L \upharpoonright x)}{d_i(L \upharpoonright x - 1)}$. Since both definitions are equivalent we shall switch from one to the other depending on which is the most appropriate in a given context.

Sometimes we need approximable martingales instead of exactly computable ones. Here is a definition.

Definition 6. *A P-approximable family of martingales $(\{d_i\}_i, \{Q_i\}_i, \mathrm{ind})$, is a family of martingales $\{d_i\}_i$, where Q_i and ind are as in Definition 2 and such that $d_i(L \upharpoonright x)$ is k-approximable by a random access Turing machine M in time polynomial in $|x| + k$, i.e. $|M^{L \upharpoonright x}(x, i, k) - d_i(L \upharpoonright x)| \le 2^{-k}$ where M queries its oracle only on strings in $Q_i(|x|)$.*

3.1 The Basic Measure Properties

We can show the union property for the following non-general case, where the query sets Q_i are the same for each family of rate-martingales to be considered for the union.

Definition 7. *A* P-*union of measure zero sets is a family of classes* $\{C_j\}_j$ *such that there exists a* P-*family of rate-martingales* $(\{D_{i,j}\}_{i,j}, \{Q_i\}_i, \text{ind})$ *such that for every* $j \geq 1$, $C_j \subseteq S_I^\infty[\{D_{i,j}\}_i]$.

As the following result shows, the basic measure properties hold for F-measure, as long as we restrict ourselves to P-unions.

Theorem 1. *1. Let L be any language in* P, *then* $\{L\}$ *has* P-*measure zero.*
2. P *does not have* P-*measure zero.*
3. Let $\{C_j\}_j$ *be a* P-*union of measure zero sets, and let* $C = \bigcup_j C_j$, *then* C *has* P-*measure zero.*

It is easy to check that F-measure on P can be extended to a measure notion on QUASIPOLY, E_ϵ, and PSPACE, by taking the corresponding time and space bounds. For a measure on BPP we refer the reader to [20].

3.2 Applications: Some Classes of Measure Zero

3.3 Smallness of Languages with Low Density

As mentioned earlier, martingale families can bet on every string, thus yielding a randomness notion which is optimal in terms of density of random languages.

Theorem 2. *Let $0 \leq \epsilon < 1/2$. The set D_ϵ of languages L such that for infinitely many n $|L[s_1, s_2, \cdots, s_n]| \leq \epsilon n$, has* P-*measure zero.*

An immediate Corollary of Theorem 2 is that the class SPARSE of languages containing few information is small in P, as opposed to Γ-measure [1].

Corollary 1. SPARSE *has* P-*measure zero.*

3.4 Almost Every Language in SUBEXP Can Derandomize BPP

We improve a former result of [1] by showing that almost every language A in E_ϵ can derandomize BPP^A.

Theorem 3. *For every $\epsilon > 0$, the set of languages A such that* $P^A \neq BPP^A$ *has* E_ϵ *-measure zero.*

3.5 Almost Every Language in PSPACE Does Not Have Small Circuit Complexity

The following result shows that almost every language in PSPACE does not have small nonuniform complexity.

Theorem 4. *Let $c > 0$,* SIZE(n^c) *has* PSPACE-*measure zero.*

3.6   Comparison with Previous Measure Notions

The following result shows that F-measure is strictly stronger than Γ-measure [1].

Theorem 5. μ_P *is stronger than* μ_Γ, *i.e. for every class* C, $\mu_\Gamma(C) = 0$ *implies* $\mu_P(C) = 0$ *and there are classes* C *such that* $\mu_\Gamma(C) \neq 0$ *and* $\mu_P(C) = 0$.

We cannot compare F-measure to $\Gamma/(P)$-measure [22] directly, due to their intrinsic differences: a language L is said to have $\Gamma/(P)$-measure zero if there exists a "game strategy" which succeeds on *any* subsequence of L. This leads to the unnatural situation where for any random language L, $L \cup \{0\}^*$ does not have $\Gamma/(P)$-measure zero, although there are infinitely many easy instances. It is easy to check that such a set has P-measure zero. Nevertheless all sets proved to be small for $\Gamma/(P)$-measure in [22] are also small for F-measure. Regarding density arguments, F-measure performs better; indeed a (Lebesgue) random language has with high probability $(1/2 - o(1))2^n$ words of length n, and this property is captured by F-measure in Theorem 2, whereas for $\Gamma/(P)$-measure, the set of languages having $o(2^n)$ words of length n has $\Gamma/(P)$-measure zero. The advantage of $\Gamma/(P)$-measure over F-measure is that it satisfies the finite union property. Since $\Gamma/(P)$-measure is derived from Γ_d-measure [22], we cannot compare Γ_d-measure to F-measure, and both their respective strengths are different: whereas Γ_d-measure cannot be used to define dimension in P, F-measure fails to capture the Γ_d-measure zero sets in [3].

3.7   Equivalence Between Measure on EXP and SUBEXP

Many results have been obtained from the plausible hypothesis $\mu_E(NP) \neq 0$ see for instance [16, 8], and the E-measure of all classes ZPP, RP, BPP, SPP is now well understood, [23, 8, 7]. The following theorem shows that all these results follow from the a priori weaker assumption in terms of measure in E_ϵ.

Theorem 6. *Let* C *be a class downward closed under* $\leq_m^p$*-reducibilities, and let* $\alpha > 0$. *We have* $\mu_{E_\alpha}(C) \neq 0$ *iff* $\mu_{EXP}(C) \neq 0$.

4   Dimension on P

To define a dimension notion from F-measure, we need some minor modification for technical reasons. From now on we only consider P-families where the query sets of Definition 2 cover all strings of some size, and where the number of martingales allowed to bet on strings of size n is bounded, i.e. we require $\cup_{i \leq 2^n/n} Q_i(n) = \{0,1\}^{\leq n}$.

Lutz's key idea to define resource-bounded dimension is to tax the martingales' wins. The following definition formalizes this tax rate notion.

Definition 8. *Let* $s \in [0,1]$ *and* (D, Q, ind) *be a* P-*family of rate-martingales, and let* L *be a language. We say* D s-*succeeds on* L, *if* $\limsup_{n \to \infty} 2^{(s-1)n} W_D(L \upharpoonright n) = \infty$.

Similarly D s-succeeds on class C, if D s-succeeds on every language in C.

The dimension of a complexity class is the highest tax rate that can be received on the martingales' wins without preventing them from succeeding on the class.

Definition 9. *Let C be a class of languages. The P-dimension of C is defined as $\dim_P(C) = \inf\{s : \exists \text{ a } P\text{-family of rate-martingales } D \text{ that } s\text{-succeeds} \text{ on} C\}$.*

We say C has dimension s in P denoted $\dim(C|P)$ if $\dim_P(C \cap P) = s$. If lim sup is replaced with lim inf in Definition 8, we say strongly s-succeed, and denote by Dim_P the associated dimension notion. This is similar to the packing dimension notion from [2].

P-dimension satisfies a non-general union property, as shown in the following result.

Theorem 7. *Let $\{C_j\}_j$ be a family of classes, and let $\{s_j\}_j$ with $s_j \in [0,1]$ such that for every $\epsilon > 0$ there exists a P-family of martingales $\{d_{i,j}\}_{i,j}$ such that $\{d_{i,j}\}_i$ $(s_j + \epsilon)$-succeeds on C_j. Let $C = \bigcup_j C_j$, then $\dim_P(C) \le \sup_j\{s_j\}$.*

It is easy to check that P-dimension can be extended to classes above P like QUASIPOLY, subexponential time and PSPACE; for BPP see [20].

4.1 Finite-State Dimension Versus P-Dimension

The following result gives some evidence that P-dimension is a natural extension of previous dimension notions to the class P.

Theorem 8. *Let S be a language. Then $\dim_{FS}(S) \ge \dim_P(S) \ge \dim_E(S)$.*

4.2 Application: Connecting Frequency and Shannon Entropy

In this section we show a polynomial time version of the Theorem of Eggleston [5], i.e. we prove that the class of languages with asymptotic frequency α have strong dimension the Shannon entropy of α in P. Analogue version of the theorem of Eggleston have been proved for various resource bounds [4,13].

Let us introduce the following notations. First the Shannon entropy refers to the following continuous function $H : [0,1] \to [0,1]$, $H(\alpha) = \alpha \log \frac{1}{\alpha} + (1-\alpha) \log \frac{1}{1-\alpha}$.

For a language A and $n \in \mathbb{N}$, let $\text{freq}_A(n) = \frac{\#(1, A[0 \ldots n-1])}{n}$, where $\#(1, A[0 \ldots n-1])$ is the number of 1's in $A[0 \ldots n-1]$. For $\alpha \in [0,1]$, let $\text{FREQ}(\alpha) = \{A \in \{0,1\}^\infty | \lim_{n\to\infty} \text{freq}_A(n) = \alpha\}$.

The following is a polynomial time version of the Theorem of Eggleston [5].

Theorem 9. *For all E-computable $\alpha \in [0,1]$, we have $\text{Dim}(\text{FREQ}(\alpha)|P) = H(\alpha)$.*

References

1. E. Allender and M. Strauss. Measure on small complexity classes, with application for BPP. *Proc. of the 35th Ann. IEEE Symp. on Found. of Comp. Sci.*, pages 807–818, 1994.
2. K. B. Athreya, J. M. Hitchcock, J. H. Lutz, and E. Mayordomo. Effective strong dimension in algorithmic information and computational complexity. *Proc. of the 21st Symposium on Theo. Aspects of Comp. Sci.*, pages 632–643, 2004.
3. J.-Y. Cai, D. Sivakumar, and M. Strauss. Constant-depth circuits and the Lutz hypothesis. *Proc. 38'th Found. of Comp. Sci. Conf.*, pages 595–604, 1997.
4. J. Dai, J. Lathrop, J. Lutz, and E. Mayordomo. Finite-state dimension. *Theor. Comp. Sci.*, 310:1–33, 2004.
5. H. Eggleston. The fractional dimension of a set defined by decimal properties. *Quart. Jour. of Math.*, 20:31–36, 1949.
6. L. Fortnow and J. Lutz. Prediction and dimension. *Jour. of Comp. and Syst. Sci.*, 70:570–589, 2005.
7. J. M. Hitchcock. The size of SPP. *Th. Comp. Sci.*, 320:495–503, 2004.
8. R. Impagliazzo and P. Moser. A zero-one law for RP. *Proc. of the 18th Conf. on Comp. Comp.*, pages 48–52, 2003.
9. A. Klivans and D. van Melkebeek. Graph nonisomorphism has subexponential size proofs unless the polynomial hierarchy collapses. *Proc. of the 31st Ann. ACM Symp. on Theo. of Comp.*, pages 659–667, 1999.
10. J. Lutz. Category and measure in complexity classes. *SIAM Jour. on Comp.*, 19:1100–1131, 1990.
11. J. Lutz. Almost everywhere high nonuniform complexity. *Jour. of Comp. and Sys. Sci.*, 44:220–258, 1992.
12. J. Lutz. The quantitative structure of exponential time. In L. Hemaspaandra and A. Selman, editors, *Compl. Theo. Retro. II*, pages 225–260. Springer, 1997.
13. J. Lutz. Dimension in complexity classes. *Proc. of the 15th Ann. IEEE Conf. on Comp. Compl.*, pages 158–169, 2000.
14. J. Lutz. The dimensions of individual strings and sequences. *Inform. and Comp.*, 187:49–79, 2003.
15. J. Lutz. Introduction to normal sequences. *Lect. notes*, 2004.
16. J. Lutz and E. Mayordomo. Cook versus Karp-Levin: Separating completeness notions if NP is not small. *SIAM Jour. on Comp.*, 164:141–163, 1996.
17. J. H. Lutz. Effective fractal dimensions. *Math. Log. Quar.*, 51:62–72, 2005.
18. E. Mayordomo. Measuring in PSPACE. *Proc. of the 7th Int. Meet. of Young Comp. Sci. (IMYCS'92)*, 136:93–100, 1994.
19. P. Moser. Baire's categories on small complexity classes. *14th Int. Symp. Fund. of Comp. Theo.*, pages 333–342, 2003.
20. P. Moser. Resource-bounded measure on probabilistic classes. *subm.*, 2004.
21. C. Papadimitriou. *Computational complexity*. Add.-Wes., 1994.
22. M. Strauss. Measure on P- strength of the notion. *Inform. and Comp.*, 136:1:1–23, 1997.
23. D. van Melkebeek. The zero-one law holds for BPP. *Theo. Comp. Sci.*, 244(1-2):283–288, 2000.
24. J. Ziv and A. Lempel. Compression of individual sequences via variable-rate coding. *IEEE Trans. on Info. Theo.*, pages 530–536, 1978.

Can General Relativistic Computers Break the Turing Barrier?

István Németi and Hajnal Andréka

Rényi Institute of Mathematics, Budapest P.O. Box 127, H-1364 Hungary
nemeti@renyi.hu, andreka@renyi.hu
http://www.renyi.hu/pub/algebraic-logic/nhompage.htm

Abstract. - Can general relativistic computers break the Turing barrier? - Are there final limits to human knowledge? - Limitative results versus human creativity (paradigm shifts). - Gödel's logical results in comparison/combination with Gödel's relativistic results. - Can Hilbert's programme be carried through after all?

1 Aims, Perspective

The Physical Church-Turing Thesis, PhCT, is the conjecture that whatever physical computing device (in the broader sense) or physical thought experiment will be designed by any future civilization, it will always be simulatable by a Turing machine. The PhCT was formulated and generally accepted in the 1930's. At that time a general consensus was reached declaring PhCT valid, and indeed in the succeeding decades the PhCT was an extremely useful and valuable maxim in elaborating the foundations of theoretical computer science, logic, foundation of mathematics and related areas. But since PhCT is partly a physical conjecture, we emphasize that this consensus of the 1930's was based on the physical world-view of the 1930's. Moreover, many thinkers considered PhCT as being based on mathematics + common sense. But "common sense of today" means "physics of 100 years ago". Therefore we claim that the consensus accepting PhCT in the 1930's was based on the world-view deriving from Newtonian mechanics. Einstein's equations became known to a narrow circle of specialists around 1920, but around that time the consequences of these equations were not even guessed at. The world-view of modern black hole physics was very far from being generally known until much later, until after 1980.

Our main point is that in the last few decades (well after 1980) there has been a major paradigm shift in our physical world-view. This started in 1970 by Hawking's and Penrose's singularity theorem firmly establishing black hole physics and putting general relativity into a new perspective. After that, discoveries and new results have been accelerating. About 10 years ago astronomers obtained firmer and firmer evidence for the existence of larger and larger more exotic black holes [18],[17] not to mention evidence supporting the assumption that the universe is not finite after all [20]. Nowadays the whole field is in a state of constant revolution. If the background foundation on which PhCT was based

A. Beckmann et al. (Eds.): CiE 2006, LNCS 3988, pp. 398–412, 2006.

has changed so fundamentally, then it is desirable to re-examine the status and scope of applicability of PhCT in view of the change of our general world-picture. Cf. also [5] for a related perspective.

A special feature of the Newtonian world-view is the assumption of an absolute time scale. Indeed, this absolute time has its mark on the Turing machine as a model for computer. As a contrast, in general relativity there is no absolute time. Kurt Gödel was particularly interested in the exotic behavior of time in general relativity (GR). Gödel [8] was the first to prove that there are models of GR to which one cannot add a partial order satisfying some natural properties of a "global time". In particular, in GR various observers at various points of spacetime in different states of motion might experience time radically differently. Therefore we might be able to speed up the time of one observer, say C (for "computer"), relatively to the other observer, say P (for "programmer").

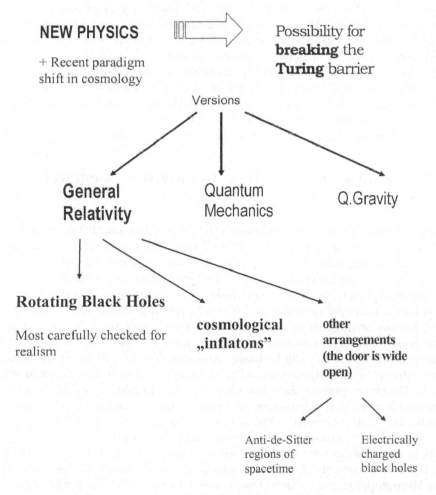

Fig. 1. Summary

Thus P may observe C computing very fast. The difference between general relativity and special relativity is (roughly) that in general relativity this speed-up effect can reach, in some sense, infinity assuming certain conditions are satisfied. Of course, it is not easy to ensure that this speed-up effect happens in such a way that we could utilize it for implementing some non-computable functions.

In [7], [15] we prove that it is consistent with Einstein's equations, i.e. with general relativity, that by certain kinds of relativistic experiments, future generations might find the answers to non-computable questions like the halting problem of Turing machines or the consistency of Zermelo Fraenkel set theory (the foundation of mathematics, abbreviated as ZFC set theory from now on). For brevity, we call such thought experiments *relativistic computers*. Moreover, the spacetime structure we assume to exist in these experiments is based in [7],[15] on huge slowly rotating black holes the existence of which is made more and more likely (almost certain) by recent astronomical observations [18],[17].

We are careful to avoid basing the beyond-Turing power of our computer on "side-effects" of the idealizations in our mathematical model/theory of the physical world. For example, we avoid relying on infinitely small objects (e.g. pointlike test particles, or pointlike bodies), infinitely elastic balls, infinitely (or arbitrarily) precise measurements, or anything like these. In other words, we make efforts to avoid taking advantage of the idealizations which were made when GR was set up. Discussing physical realizability and realism of our design for a computer is one of the main issues in [15].

The diagram in Figure 1 summarizes the ideas said so far.

2   An Intuitive Idea for How Relativistic Computers Work

In this section we would like to illuminate the ideas of how relativistic computers work, without going into the mathematical details. The mathematical details are elaborated, among others, in [7], [9], [15]. To make our narrative more tangible, here we use the example of huge slowly rotating black holes for our construction of relativistic computers. But we emphasize that there are many more kinds of spacetimes suitable for carrying out essentially the same construction (these are called Malament-Hogarth spacetimes in the physics literature). So, relativistic computers are not tied to rotating black holes, there are other general relativistic phenomena on which they can be based. An example is anti-de Sitter spacetime which attracts more and more attention in explaining recent discoveries in cosmology. We chose rotating black holes because they provide a tangible example for illustrating the kind of reasoning underlying general relativistic approaches to breaking the "Turing barrier". Astronomical evidence for their existence makes them an even more attractive choice for our didactic purposes.

Let us start out from the so-called Gravitational Time Dilation effect (GTD). The GTD is a theorem of relativity which says that gravity makes time run slow. More sloppily: gravity slows time down. Clocks that are deep within gravitational fields run slower than ones that are farther out. We will have to explain

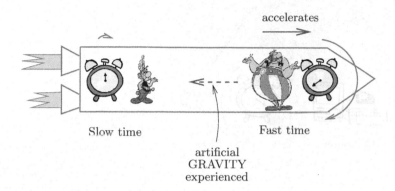

Fig. 2. GTD is a theorem of Special Relativity (SR) (easily proved in first-order logic version of SR)

what this means, but before explaining it we would like to mention that GTD is not only a theorem of general relativity. This theorem, GTD, can be already proved in (an easily understandable logic-based version of) special relativity in such a way that we simulate gravity by acceleration [11], [13]. So one advantage of GTD is that actually why it is true can be traced down by using only the simple methods of special relativity. Another advantage of GTD is that it has been tested several times, and these experiments are well known.

Roughly, GTD can be interpreted by the following thought experiment. Choose a high enough tower on the Earth, put precise enough (say, atomic) clocks at the bottom of the tower and the top of the tower, then wait enough time, and compare the readings of the two clocks. Then the clock on the top will run faster (show more elapsed time) than the one in the basement, at each time one carries out this experiment. Figure 2 represents how GTD can be proved in special relativity using an accelerated spaceship for creating artificial gravity and checking its effects on clocks at the two ends of the spaceship. Detailed purely logical formulation and proofidea is found in [12]. The next picture, Figure 3, represents the same GTD effect as before, but now using a tall tower on the Earth experiencing the same kind of gravity as in the spaceship. Gravity causes the clock on the top ticking faster. Therefore computers there also compute faster. Assume the programmer in the basement would like to use this GTD effect to speed up his computer. So he sends the computer to the top of the tower. Then he gets some speed-up effect, but this is too little. The next two pictures, Figure 4 and Figure 5, are about the theoretical possibility of increasing this speed-up effect.

How could we use GTD for designing computers that compute more than Turing Machines can? In the above outlined situation, by using the gravity of the Earth, it is difficult to make practical use of GTD. However, instead of the Earth, we could choose a huge black hole, cf. Figure 6. A black hole is a region of spacetime with so big "gravitational pull" that even light cannot escape from this region. There are several types of black holes, an excellent source is Taylor

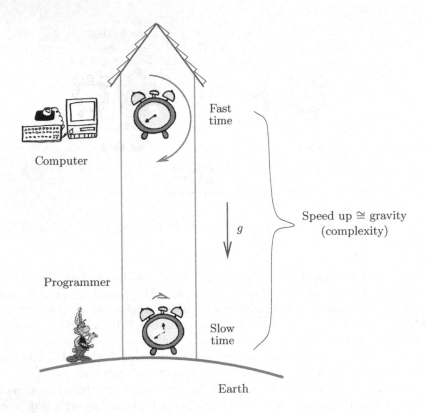

Fig. 3. TIME WARP (Tower Paradox, effects of gravity on time). Clocks higher in a gravitational well tick faster

and Wheeler [19]. For our demonstration of the main ideas here, we will use huge, slowly rotating black holes. (These are called slow-Kerr in the physics literature.) These black holes have two so-called *event horizons*, these are bubble-like surfaces one inside the other, from which even light cannot escape (because of the gravitational pull of the black hole). See Figures 7–9.

As we approach the outer event horizon from far away outside the black hole, the gravitational "pull" of the black hole approaches infinity as we get closer and closer to the event horizon. This is rather different from the Newtonian case, where the gravitational pull also increases but remains finite even on the event horizon.[1] For a while from now on "event horizon" means "outer event horizon".

Let us study observers suspended over the event horizon. Here, suspended means that the distance between the observer and the event horizon does not change. Equivalently, instead of suspended observers, we could speak about ob-

---

[1] The event horizon also exists in the Newtonian case, namely, in the Newtonian case, too, the event horizon is the "place" where the escape velocity is the speed of light (hence even light cannot escape to infinity from inside this event horizon "bubble").

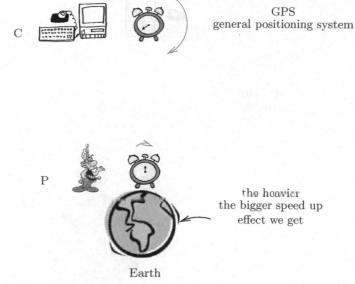

C

GPS
general positioning system

P

the heavier
the bigger speed up
effect we get

Earth

Fig. 4. Thought experiment for fast computation: The programmer "throws" his slave-computer to a high orbit. Communicates via radio.

servers whose spaceship is hovering over the event horizon, using their rockets for maintaining altitude. Assume one suspended observer H is higher up and another one, L, is suspended lower down. So, H sees L below him while L sees H above him. Now the gravitational time dilation (GTD) will cause the clocks of H run faster than the clocks of L. Moreover, they both agree on this if they are watching each other e.g. via photons. Let us keep the height of H fixed. Now, if we gently lower L towards the event horizon, this ratio between the speeds of their clocks increases. Moreover, as L approaches the event horizon, this ratio approaches infinity. This means that for any integer n, if we want H's clocks to run n times as fast as L's clocks, then this can be achieved by lowering L to the right position.

Let us see what this means for computational complexity. If the programmer wants to speed up his computer with an arbitrarily large ratio, say n, then he can achieve this by putting the programmer to the position of L and putting the computer to the position of H. Already at this point we could use this arrangement with the black hole for making computers faster. The programmer goes very close to the black hole, leaving his computer far away. Then the programmer has to wait a few days and the computer does a few million year's job of computing and then the programmer knows a lot about the consequences of, say, ZFC set theory or whatever mathematical problem he is investigating. So we could use GTD for just speeding up computation which means dealing with complexity issues. However, we do not want to stop at complexity issues. Instead, we would like to see whether we can attack somehow the "Turing barrier".

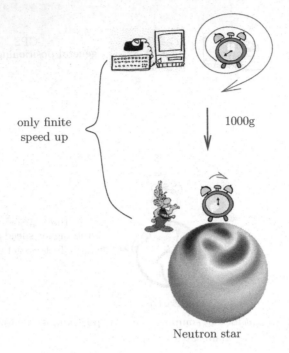

only finite
speed up

1000g

Neutron star

Fig. 5. By using a neutron star we still get only a finite speed-up

The above arrangement for speeding the computer up raises the question of how the programmer avoids consequences of the fact that the whole manoeuver will slow down the programmer's own time relative to the time on his home planet, e.g. on the Earth. We will deal with this problem later. Let us turn now to the question of how we can use this effect of finite (but unbounded) speed-up for achieving an infinite speed-up, i.e. for breaking the Turing barrier.

If we could suspend the lower observer L on the event horizon itself then from the point of view of H, L's clocks would freeze, therefore from the point of view of L, H's clocks (and computers!) would run infinitely fast, hence we would have the desired infinite speed-up upon which we could then start our plan for breaking the Turing barrier. The problem with this plan is that it is impossible to suspend an observer on the event horizon. As a consolation for this, we can suspend observers arbitrarily close to the event horizon. To achieve an "infinite speed-up" we could do the following. We could lower and lower again L towards the event horizon such that L's clocks slow down (more and more, beyond limit) in such a way that there is a certain finite time-bound, say b, such that, roughly, throughout the whole history of the universe L's clocks show a time smaller than b. More precisely, by this we mean that whenever H decides to send a photon to L, then L will receive this photon before time b according to L's clocks. This is possible. See Figure 9.

Are we done, then? Not yet, there is a remaining task to solve. As L gets closer and closer to the event horizon, the gravitational pull or gravitational

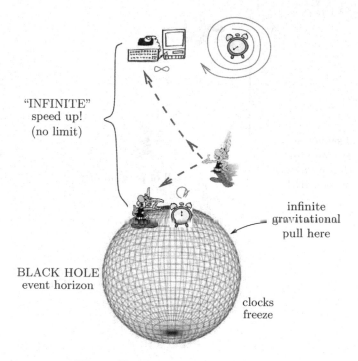

"INFINITE"
speed up!
(no limit)

infinite
gravitational
pull here

BLACK HOLE
event horizon

clocks
freeze

Fig. 6. Getting "infinite" speed-up

acceleration tends to infinity. If L falls into the black hole without using rockets to slow his fall, then he does not have to withstand the gravitational pull of the black hole. He would only feel the so-called tidal forces which can be made negligibly small by choosing a large enough black hole. However, his falling through the event horizon would be so fast that some photons sent after him by H would not reach him outside the event horizon. Thus L has to approach the event horizon relatively slowly in order that he be able to receive all possible photons sent to him by H. In theory he could use rockets for this purpose, i.e. to slow his fall (assuming he has unlimited access to fuel somehow). Because L approaches the event horizon slowly, he has to withstand this enormous gravity (or equivalently acceleration). The problem is that this increasing gravitational force (or acceleration) will kill L before his clock shows time b, i.e. before the planned task is completed.

At the outer event horizon of our black hole we cannot compromise between these two requirements by choosing a well-balanced route for L: no matter how he will choose his route, either L will be crashed by the gravitational pull, or some photons sent by H would not reach him. (This is the reason why we can not base our relativistic computer on the simplest kind of black holes, called Schwarzschild ones, which have only one event horizon and that behaves as we described as above.)

To solve this problem, we would like to achieve slowing down the "fall" of L not by brute force (e.g. rockets), but by an effect coming from the structure of

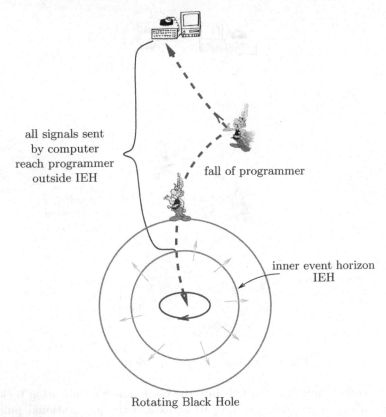

all signals sent
by computer
reach programmer
outside IEH

fall of programmer

inner event horizon
IEH

Rotating Black Hole

Fig. 7. Rotating Black Hole has two event horizons. Programmer can survive forever. (Ring singularity can be avoided.)

spacetime itself. In our slowly rotating black hole, besides the gravitational pull of the black hole (needed to achieve the time dilation effect) there is a counteractive repelling effect coming from the revolving of the black hole. This repelling effect is analogous to "centrifugal force" in Newtonian mechanics and will cause L to slow down in the required rate. So the idea is that instead of the rockets of L, we would like to use for slowing the fall of L this second effect coming from the rotation of the black hole. In some black holes with such a repelling force, and this is the case with our slowly rotating one, two event horizons form, see Figures 7–9. The outer one is the result of the gravitational pull and behaves basically like the event horizon of the simplest, so-called Schwarzschild hole, i.e. as described above. The *inner event horizon* marks the point where the repelling force overcomes the gravitational force. So inside the inner horizon, it is possible again to "suspend" an observer, say L, i.e. it becomes possible for L to stay at a constant distance from the center of the black hole (or equivalently from the event horizons).

Let us turn to describing how a slowly rotating black hole implements the above outlined ideas, and how it makes possible to realize our plan for "infinite

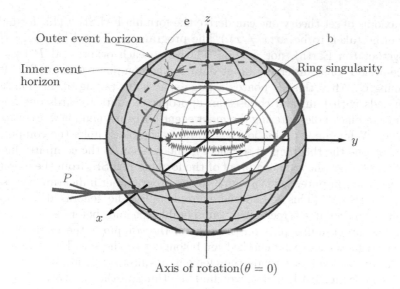

Fig. 8. A slowly rotating (Kerr) black hole has two event horizons and a ring-shape singularity (the latter can be approximated/visualized as a ring of extremely dense and thin "wire"). The ring singularity is inside the inner event horizon in the "equatorial" plane of axes x, y. Time coordinate is suppressed. Figure 9 is a spacetime diagram with x, y suppressed. Rotation of ring is indicated by an arrow. Orbit of in-falling programmer P is indicated, it enters outer event horizon at point e, and meets inner event horizon at point b.

speed-up". Figure 8 represents a slowly rotating huge Kerr black hole and Figure 9 represents its spacetime structure. As we said, there are two event horizons, the inner one surrounded by the outer one. The source of gravity of the black hole is a ring shaped singularity situated inside the inner horizon. The path of the in-falling observer L can be planned in such a way that the event when L reaches the inner horizon corresponds to the time-bound b (on the wristwatch of L) mentioned above before which L receives all the possible messages sent out by H. In Figures 8,9 the world-lines of L and H are denoted as P and C because we think of L as the programmer and we think of H as L's computer.

By this we achieved the infinite speed-up we were aiming for. This infinite speed-up is represented in Figure 9 where P measures a finite proper time between its separation from the computer C (which is not represented in the figure) and its touching the inner horizon at proper time b (which point also is not represented in Figure 9). It can be seen in the figure that whenever C decides to send a photon towards P, that photon will reach P before P meets the inner horizon. The above outlined intuitive plan for creating an infinite speed-up effect is elaborated in more concrete mathematical detail in [7], [15].

Let us see how we can use all this to create a computer that can compute tasks which are beyond the Turing limit. Let us choose the task, for an example, to decide whether ZFC set theory is consistent. I.e. we want to learn whether from the axioms of set theory one can derive the formula FALSE. (This formula FALSE can be taken to be $\exists x(x \neq x)$.) The programmer P and his computer C are together (on Earth), not moving relative to each other, and P uses a finite time-period for transferring input data to the computer C as well as for programming C. After this, P boards a huge spaceship, taking all his mathematical friends with him, and chooses an appropriate route towards our huge slowly rotating black hole, entering the inner event horizon when his wrist-watch shows time b. While he is on his journey towards the black hole, the computer checks one by one the theorems of set theory, and as soon as the computer finds a contradiction in set theory, i.e. a proof of the formula FALSE, from the axioms of set theory, the computer sends a signal to the programmer indicating that set theory is inconsistent. (This is a special example only. The general idea is that the computer enumerates a recursively enumerable set and, before starting the computer, the programmer puts on the tape of the computer the name of the element which he wants to be checked for belonging to the set. The computer will search and as soon as it finds the element in question inside the set, the computer sends a signal.) If it does not find the thing in the set, the computer does nothing.

What happens to the programmer P from the point of view of the computer C? This is represented in Figure 9. Let C's coordinate system be the one represented in Figure 9. By saying "from the point of view of C" we mean "in this particular coordinate system (adjusted to C) in Fig.9". In this coordinate system when the programmer goes closer and closer to the inner horizon of the black hole, the programmer's clock will run slower and slower and slower, and eventually on the inner event horizon of the black hole the time of the programmer stops. Subjectively, the programmer does not experience it this way, this is how the computer will coordinatize it in the distance, or more precisely, how the coordinate system shown in Figure 9 represents it. If the computer thinks of the programmer, it will see in its mind's eye that the programmer's clocks stop and the programmer is frozen motionless at the event horizon of the black hole. Since the programmer is frozen motionless at the event horizon of the black hole, the computer has enough time to do the computation, and as soon as the computer has found, say, the inconsistency in set theory, the computer can send a signal and the computer can trust that the programmer—still with his clock frozen—will receive this signal before it enters the inner event horizon.

What will the programmer experience? This is represented in Figure 8. The programmer will see that as he is approaching the inner event horizon, his computer in the distance is running faster and faster and faster. Then the programmer falls through the inner event horizon of the black hole. If the black hole is enormous, the programmer will feel nothing when he passes either event horizon of the black hole—one can check that in case of a huge black hole the so-called tidal forces on the event horizons of the black hole are negligibly small [16]. So the programmer falls into the inner event horizon of the black hole and either the programmer will experience that a light signal arrives from the direction of the computer, of an agreed color and agreed pattern, or the programmer will observe that he falls in through the inner event horizon and the light signal does not arrive. After the programmer has crossed the inner event horizon, the programmer can evaluate the situation. If a signal arrives from the computer, this means that the computer found an inconsistency in ZFC set theory, therefore the programmer will know that set theory is inconsistent. If the light signal does not arrive, and the programmer is already inside the inner event horizon, then he will know that the computer did not find an inconsistency in set theory, did not send the signal, therefore the programmer can conclude that set theory is consistent. So he can build the rest of his mathematics on the secure knowledge of the consistency of set theory.

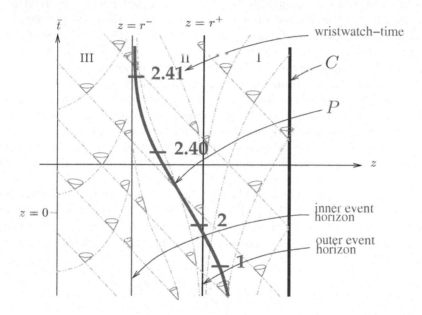

Fig. 9. The "tz-slice" of spacetime of slowly rotating black hole in coordinates where z is the axis of rotation of black hole. The pattern of light cones between the two event horizons r^- and r^+ illustrates that P can decelerate so much in this region that he will receive outside of r^- all messages sent by C. r^+ is the outer event horizon, r^- is the inner event horizon, $z = 0$ is the "center" of the black hole as in Figure 8. The tilting of the light cones indicates that not even light can escape through these horizons. That there is an outward push counteracting gravity can be seen by the shape of the light-cones in region III (central region of the black hole). The time measured by P is finite (measured between the beginning of the experiment and the event when P meets the inner event horizon at b) while the time measured by C is infinite.

The next question which comes up naturally is whether the programmer can use this new information, namely that set theory is consistent, or whatever he wanted to compute, for his purposes. A pessimist could say that OK they are inside a black hole, so—now we are using common sense, we are not using relativity theory—common sense says that the black hole is a small unfriendly area and the programmer will sooner or later fall into the middle of the black hole where there is a singularity and the singularity will kill the programmer. The reason why we chose our black hole to be a huge slowly rotating one, say of mass $10^{10} m_\odot$, is the following. If the programmer falls into a black hole which is as big as this and it rotates slowly, then the programmer will have quite a lot of time inside the black hole because the center of the black hole is relatively far from the event horizon. But this is not the key point. If it rotates, the "matter content", the so-called singularity, which is the source of the gravitational field of the black hole so-to-speak, is not a point but a ring. So if the programmer chooses his route in falling into the black hole in a clever way, say, relatively close to the north pole instead of the equatorial plane, then the programmer can comfortably pass through the middle of the ring, never get close to the singularity and happily live on forever. We mean, the rules of relativity will not prevent him from happily living forever. He may have descendants, he can found society, he can use the so obtained mathematical knowledge.

Technical details of realizability of this general plan are checked in [15], [7]. The above outlined train of thought can be pushed through to show that any recursively enumerable set can be decided by a relativistic computer [7]. Actually, more than that can be done by relativistic computers, but it is not the purpose of the present paper to check these limits. These limits are addressed in [9], [10], [21].

For the nonspecialist of general relativity, we include here the mathematical description of a double black hole with 2 event horizons suitable for the above outlined thought experiment. Instead of rotation, here we use an electric charge for "cushioning". The spacetime geometry of our black hole is described by the metric

$$ds^2 = A(r)dt^2 - \frac{1}{A(r)}dr^2 - r^2 d\varphi^2 \tag{1}$$

where φ is the space angle coordinate. Here $A(r) = (1 - \frac{1}{r} + \frac{e}{r^2})$ for some $0 \le e < 1/2$. (The event horizons form at $r = \frac{1}{2} \pm \sqrt{\frac{1}{4} - e}$. In our choice of $A(r)$, the "$-\frac{1}{r}$" part is responsible for gravitational attraction, while the "$\frac{e}{r^2}$" part for the cushioning caused by charge $\sqrt{e}$.) The tr-slice of the spacetime determined by the simple metric (1) above is basically the same as the one represented in Figure 9. (What was denoted as z coordinate should be denoted as r, now.) For completeness, in (1) above, r is the radial coordinate, $r = $ distance from the center of black hole.

3   Conclusion

A virtue of the present research direction is that it establishes connections between central questions of logic, foundation of mathematics, foundation of physics, relativity theory, cosmology, philosophy, particle physics, observational

astronomy, computer science and AI [21]. E.g. it gives new kinds of motivation
to investigating central questions of these fields like "is the universe finite or infinite (both in space and time) and in what sense", "exactly how do Kerr black
holes evaporate" (quantum gravity), "how much matter is needed for coding
one bit of information (is there such a lower bound at all)", questions concerning the statuses of the various cosmic censor hypotheses, questions concerning
the geometry of rotating black holes [4], to mention only a few. The interdisciplinary character of this direction was reflected already in the 1987 course given
by the present authors [14] during which the idea of relativistic hypercomputers
emerged and which was devoted to connections between the above mentioned areas. Tangible data underlying the above interconnections and also more history,
references are available in [15]. The book Earman [6, p.119, section 4.9] regards
the same interdisciplinary perspective as described above to be one of the main
virtues of the present research direction. It is the unifying power of logic which
makes it viable to do serious work on such a diverse collection of topics. One of
the main aims of the research direction represented by [3], [2], [1], [11]–[13] is to
make relativity theory accessible for anyone familiar with logic.

Acknowledgements

Special thanks are due to Gyula Dávid for intensive cooperation and ideas on this
subject. We are grateful for enjoyable discussions to Attila Andai, John Earman,
Gábor Etesi, Balázs Gyenis, Mark Hogarth, Judit Madarász, Chris Wüthrich.
We gratefully acknowledge support by the Hungarian National Foundation for
scientific research grant No. T43242, as well as by COST grant No. 274.

References

1. Andréka, H., Madarász, J. X. and Németi, I., Logic of Spacetime. In: Logic of Space,
   eds: M. Aiello, J. van Benthem, and I. Hartman-Pratt, Kluwer. In preparation.
2. Andréka, H., Madarász, J. X. and Németi, I., Logical axiomatizations of spacetime.
   In: Non-Euclidean Geometries, János Bolyai Memorial Volume. Ed. A. Prékopa and
   E. Molnár, Mathematics and its Applications Vol. 581, Springer 2006. pp.155-185.
3. Andréka, H., Madarász, J. X. and Németi, I. with contributions from Andai, A.,
   Sain, I., Sági, G., Tőke, Cs. and Vályi, S., On the logical structure of relativity
   theories. Internet book, Budapest, 2000. http://www.math-inst.hu/pub/algebraic-
   logic/olsort.html
4. Andréka, H., Németi, I. and Wüthrich, A twist in the geometry of rotating black
   holes: seeking the cause of acausality. Manuscript, 2005. 15pp.
5. Cooper, S. B., How can Nature Help Us Compute? In SOFSEM 2006: Theory and
   Practice of Computer Science - 32nd Conference on Current Trends in Theory and
   Practice of Computer Science, Merin, Czech Republic, January 2006 (Jiri Wiedermann, Julius Stuller, Gerard Tel, Jaroslav Pokorny, Maria Bielikova, editors),
   Springer Lecture Notes in Computer Science No. 3831, 2006, pp. 1-13.
6. Earman, J., Bangs, crunches, whimpers, and shrieks. Singularities and acausalities
   in relativistic spacetimes. Oxford university Press, Oxford, 1995.

7. Etesi, G. and Németi, I., Turing computability and Malament-Hogarth spacetimes. International Journal of Theoretical Physics 41,2 (2002), 342-370.
8. Gödel, K., Lecture on rotating universes. In: Kurt Gödel Collected Works, Vol. III. Eds.: Feferman, S., Dawson, J. S., Goldfarb, W., Parson, C. and Solovay, R. N., Oxford University Press, New York Oxford 1995. pp. 261-289.
9. Hogarth, M. L., Predictability, computability, and spacetime. PhD Dissertation, University of Cambridge, UK, 2000. http://ftp.math-inst.hu/pub/algebraiclogic/Hogarththesis.ps.gz
10. Hogarth. M. L., Deciding arithmetic using SAD computers. Brit. J. Phil. Sci. 55 (2004), 681-691.
11. Madarász, J. X., Németi, I. and Székely, G., Twin Paradox and the logical foundation of space-time. Foundation of Physics, to appear. arXiv:gr-qc/0504118.
12. Madarász, J. X., Németi, I. and Székely, G., First-order logic foundation of relativity theories. In: Mathematical problems from applied logic II, International Mathematical Series, Springer-Verlag, to appear.
13. Madarász, J. X. and Székely, G., The effects of gravitation on clocks, proved in axiomatic relativity. Abstracts for the conference "Logic in Hungary 2005", http://atlas-conferences.com/cgi.bin/abstract.caqb-41.
14. Németi, I., On logic, relativity, and the limitations of human knowledge. Iowa State University, Department of Mathematics, Ph. D. course during the academic year 1987/88.
15. Németi, I. and Dávid, Gy., Relativistic computers and the Turing barrier. Journal of Applied Mathematics and Computation, to appear.
16. Ori, A., On the traversability of the Cauchy horizon: Herman and Hiscock's argument revisited, in: Internal Structures of Black Holes and Spacetime Singularitites (ed. A. Ori, L.M. Ori), Ann. Isra. Phys. Soc. 13, IOP (1997).
17. Reynolds, C. C., Brenneman, L. W. and Garofalo, D., Black hole spin in AGN and GBHCs. Oct. 5, 2004. arXiv:astro-ph/0410116. (Evidence for rotating black holes.)
18. Strohmayer, T. E., Discovery of a 450 HZ quasi-periodic oscillation from the microquasar GRO J1655-40 wtih the Rossi X-Ray timing explorer. The Astrophysical Journal, 553,1 (2001), pp.L49-L53. (Evidence for rotating black holes.) arXiv:astro-ph/0104487.
19. Taylor, E. F. and Wheeler, J. A., Black Holes. Addison, Wesley, Longman, San Francisco, 2000.
20. Tegmark, M., Parallel Universes. Scientific American May 2003, pp.41-51.
21. Wiedermann, J. and van Leeuwen, J., Relativistic computers and non-uniform complexity theory. In: Calude et al (eds.) UMC 2002. Lecture Notes in Computer Science Vol. 2509, Springer-Verlag, Berlin, 2002. pp.287-299.

Degrees of Weakly Computable Reals*

Keng Meng Ng[1], Frank Stephan[2], and Guohua Wu[3]

[1] School of Mathematics, Statistics and Computer Science
Victoria University of Wellington, New Zealand
[2] School of Computing and Department of Mathematics
National University of Singapore, Singapore 117543
[3] School of Physical and Mathematical Sciences
Nanyang Technological University, Singapore 639798

Abstract. This paper studies the degrees of weakly computable reals. It is shown that certain types of limit-recursive reals are Turing incomparable to all weakly computable reals except the recursive and complete ones. Furthermore, it is shown that an r.e. Turing degree is array-recursive iff every real in it is weakly computable.

1 Introduction

A real α is *left-computable* if we can effectively generate α from below. That is, the left Dedkind cut of α, $L(\alpha) = \{q \in \mathbb{Q} : q \leq \alpha\}$, forms a r.e. set. Equivalently, a real α is left-computable if it is the limit of a converging recursive increasing sequence of rational numbers. *If we can also compute the radius of convergence effectively, then α is recursive.*

Left-computable reals are the *measures* of the domains of prefix-free Turing machines, or halting probabilities. These reals occupy a central place in the study of algorithmic randomness in the same way as recursively enumerable *sets* occupy a central place in classical recursion theory. However, the collection of left-computable reals does not behave well algebraically since it is not closed under subtraction. Because of this, in [1], Ambos-Spies, Weihrauch and Zheng introduced the collection of weakly computable reals, where a real α is *weakly computable* if there are left-computable reals β and γ such that $\beta - \gamma$ equals to α. Ambos-Spies, Weihrauch and Zheng [1] proved that the collection of weakly computable reals is closed under the arithmetic operations, and hence forms a field. The following proposition gives an analytical characterization of weakly computable reals:

Theorem 1.1 [1, Ambos-Spies, Weihrauch and Zheng]. *A real number x is weakly computable iff there is a recursive sequence $\{x_s\}_{s \in \mathbb{N}}$ of rational numbers converging to x such that $\sum_{s \in \mathbb{N}} |x_s - x_{s+1}| \leq c$ for a constant c.*

---

* F. Stephan is supported in part by NUS grant number R252–000–212–112. G. Wu is partially supported by the Start-up grant number M48110008 from Nanyang Technological University and International Collaboration grant number 60310213 of NSFC from China.

A. Beckmann et al. (Eds.): CiE 2006, LNCS 3988, pp. 413–422, 2006.

In this paper, we will study the Turing degrees of weakly computable reals. The following is known:

Theorem 1.2 [3, Downey, Wu and Zheng]. (1) *Any ω-r.e. degree contains a weakly computable real.* (2) *There are Turing degrees below $\mathbf{0}'$ containing no weakly computable reals.*

In this paper, we first generalize the notion of degrees considered in Theorem 1.2 (2), by introducing a generalized notion of those degrees constructed in [3]. Say that a nonzero degree $\mathbf{a}$ is *nonbounding* if every nonzero degree $\leq \mathbf{a}$ contains no weakly computable reals. The existence of such nonbounding degrees can be proved by an oracle construction.

Theorem 1.3. *There is a degree below $\mathbf{0}'$ such that every nonzero degree below it contains no weakly computable reals.*

Our construction can be easily modified to make the nonbounding degrees 1-generic. However, if we let $\mathbf{c}$ be any r.e. and strongly contiguous degree, then every degree below $\mathbf{c}$ is ω-r.e., and hence contains a weakly computable real (by Theorem 1.2 (1)). Thus, not every 1-generic degree below $\mathbf{0}'$ is nonbounding.

The notion of f-limit-genericity will be introduced, and an alternative proof of Theorem 1.3 by using f-limit-genericity will be given. This proof can be modified to prove:

Theorem 1.4. *There are degrees $\mathbf{a}$ below $\mathbf{0}'$ such that the degrees containing weakly computable reals comparable with $\mathbf{a}$ are only $\mathbf{0}$ and $\mathbf{0}'$.*

Theorem 1.4 improves Yates' result in [14].

We will also consider those Turing degrees on the other extreme, those degrees containing only weakly computable reals. A Turing degree is called *completely weakly computable* if every set in this degree is weakly computable. We will provide in this paper another characterization of array recursive degrees. For more information on array (non)recursive degrees, see [6]. We need some background of Chaitin's Ω numbers.

Chaitin [2] introduced Ω as the halting probability of a universal prefix-free machine and Kučera and Slaman [10] showed these Ω-numbers cover all the left computable Martin-Löf random sets. Indeed, it is sufficient for the further investigations and definitions to fix Ω as one of these possible numbers as the notions defined below turn out to be the same, independently of the choice of Ω. Ω has the following properties:

- Ω has a recursive approximation $\Omega_0, \Omega_1, \ldots$ from the left as it is left-computable.
- The convergence module c_Ω defined as

$$c_\Omega(n) = \min\{s : \forall m \leq n\,(\Omega_s(m) = \Omega(m))\}$$

dominates all total-recursive functions and furthermore $c_\Omega(n)$ is larger than the time that any terminating computation of the underlying universal machine takes to halt on any input of length n or less.

– There are nonrecursive sets A such that Ω is random relative to A. These sets are called low for Ω.

In particular the subclass of those sets low for Ω which are reducible to K has several natural characterizations [5,11]. Downey, Hirschfeldt, Miller and Nies [4, Corollary 8.6] showed that every Δ^0_2 degree low for Ω is completely weakly computable, and that such degrees can be nonrecursive.

Theorem 1.5 [4, Downey, Hirschfeldt, Miller and Nies]. *If a set $A \leq_T K$ is low for Ω, then it is weakly computable.*

One could generalize the notion "low for Ω" to the notion that c_Ω dominates every A-recursive function. This class of degrees is indeed an old friend and there are several characterizations for it [4,5,8], one of which adapts "Ω is Martin-Löf random relative to A" to "Ω is Schnorr random relative to A". So for every r.e. set A the following statements are equivalent:

– c_Ω dominates every A-recursive function;
– Ω is Schnorr random relative to A;
– the Turing degree of A is array recursive;
– the Turing degree of A has a strong minimal cover;
– A is r.e. traceable, that is, for every $f \leq_T A$ and almost all n, the Kolmogorov complexity of $f(n)$ is at most n.

Theorem 1.6. *For any r.e. set A, the following are equivalent:*

1. *The Turing degree of A is array recursive;*
2. *Every $B \leq_T A$ is weakly computable;*
3. *The Turing degree of A is completely weakly computable.*

Our notation and terminology are standard and generally follow Soare [13].

2   Nonbounding Degrees

In this section, we prove Theorem 1.3. We will construct a real A such that weakly computable reals Turing reducible to it are all recursive. A is constructed satisfying the following requirements:

$\mathcal{P}_e$: $A \neq \{e\}$;
$\mathcal{R}_{e,i,j}$: if $\{e\}^A$ is total, then either $\{e\}^A$ is recursive or $\{e\}^A \neq \alpha_i - \alpha_j$,

where $\{\alpha_i\}_{i \in \mathbb{N}}$ is an effective list of all left-computable reals.

$\mathcal{P}_e$ requirement can be satisfied by the Kleene-Post's diagonalization. That is, at stage s, given a finite approximation σ_s, we can ask whether there is some number $m > |\sigma_s|$ such that $\{e\}(m)$ converges. This is a Σ_1 question, and we can get the answer from oracle K. If the answer is "yes", then we can define σ_{s+1} as an extension of σ_s such that $|\sigma_{s+1}| = m + 1$ and $\sigma_{s+1}(m) \neq \{e\}(m)$. Otherwise, we extend σ_s to σ_{s+1} by just letting $\sigma_{s+1} = \sigma_s \hat{\ } 0$. Obviously, $\mathcal{P}_e$ is satisfied in both cases.

Now we describe the strategy satisfying the requirement $\mathcal{R}_{e,i,j}$. For convenience, we omit the subscript, and it will not cause any confusion. Suppose that at stage $s+1$, σ_s is given, and we want to satisfy $\mathcal{R}$. We ask K whether there is a number n and strings τ_1, τ_2 extending σ_s such that for all $m \leq n$, $\{e\}^{\tau_1}(m), \{e\}^{\tau_2}(m)$ converge and that $|\{e\}^{\tau_1} \restriction (n+1) - \{e\}^{\tau_2} \restriction (n+1)| \geq 2^{-(n-1)}$.

If the answer is "no", then we claim that if $\{e\}^A$ is total, then $\{e\}^A$ is recursive. To see this, for any n, to calculate $\{e\}^A(n)$, we find a string τ extending σ_s such that for all $m \leq n+2$, $\{e\}^{\tau}(m)$ converges (by the assumption that $\{e\}^A$ is total, such a τ exists). Then $\{e\}^A(n) = \{e\}^{\tau}(n)$, because by our assumption, $|\{e\}^A \restriction (n+3) - \{e\}^{\tau} \restriction (n+3)|$ is less than $2^{-(n+1)}$. In this case, we let $\sigma_{s+1} = \sigma_s \widehat{\ } 0$.

On the other hand, if the answer is "yes", then for any real number x,

$$|\{e\}^{\tau_1} \restriction (n+1) - x \restriction (n+1)| + |\{e\}^{\tau_2} \restriction (n+1) - x \restriction (n+1)|$$

is bigger than $|\{e\}^{\tau_1} \restriction (n+1) - \{e\}^{\tau_2} \restriction (n+1)|$ and hence is bigger than $2^{-(n-1)}$. As a consequence, $|\{e\}^{\tau_1} \restriction (n+1) - x \restriction (n+1)|$ or $|\{e\}^{\tau_2} \restriction (n+1) - x \restriction (n+1)|$ must be bigger than 2^{-n}. If we know that $|\{e\}^{\tau_1} \restriction (n+1) - x \restriction (n+1)| > 2^{-n}$, then we can define σ_{s+1} as τ_1, and we will have $\{e\}^A \restriction (n+1) = \{e\}^{\tau_1} \restriction (n+1)$. As a consequence, $\{e\}^A$ differs from x in the first $n+1$ digits.

Then, how can we decide which one of τ_1 and τ_2 is the one we want to satisfy $\mathcal{R}$? Since α_i and α_j are left-computable reals, there are effective approximations of α_i, α_j from the left, $\{\alpha_{i,s}\}_{s \in \mathbb{N}}, \{\alpha_{j,s}\}_{s \in \mathbb{N}}$ say, and hence, we can use K as oracle to find a stage s such that $\alpha_{i,s} \restriction (n+3) = \alpha_i \restriction (n+3)$, $\alpha_{j,s} \restriction (n+3) = \alpha_j \restriction (n+3)$. Thus $\alpha_i \restriction (n+3) - \alpha_{i,s} \restriction (n+3) \leq 2^{-(n+2)}$, $\alpha_j \restriction (n+3) - \alpha_{j,s} \restriction (n+3) \leq 2^{-(n+2)}$ and hence $|(\alpha_i - \alpha_j) \restriction (n+3) - (\alpha_{i,s} - \alpha_{j,s}) \restriction (n+3)| \leq 2^{-(n+1)}$. Now if we let x above be $\alpha_{i,s} - \alpha_{j,s}$, then we can know which one of $|\{e\}^{\tau_1} \restriction (n+1) - (\alpha_{i,s} - \alpha_{j,s}) \restriction (n+1)|$ and $|\{e\}^{\tau_2} \restriction (n+1) - (\alpha_{i,s} - \alpha_{j,s}) \restriction (n+1)|$ is bigger than 2^{-n}. Suppose that $|\{e\}^{\tau_1} \restriction (n+1) - (\alpha_{i,s} - \alpha_{j,s}) \restriction (n+1)| \geq 2^{-n}$. Then

$$
\begin{aligned}
& |\{e\}^{\tau_1} \restriction (n+1) - (\alpha_i - \alpha_j) \restriction (n+1)| \\
\geq\ & ||\{e\}^{\tau_1} \restriction (n+1) - (\alpha_{i,s} - \alpha_{j,s}) \restriction (n+1)| - \\
& |(\alpha_i - \alpha_j) \restriction (n+1) - (\alpha_{i,s} - \alpha_{j,s}) \restriction (n+1)|| \\
\geq\ & 2^{-n} - 2^{-(n+1)} = 2^{-(n+1)}.
\end{aligned}
$$

Therefore, we can satisfy $\mathcal{R}$ by extending σ_s to τ_1 (that is, define $\sigma_{s+1} = \tau_1$).

The whole construction of A is a finite extension argument, with K as oracle, where at each stage, one requirement is satisfied. ∎

3 f-Limit-Generic Degrees

In [14], Yates proved that there are degrees $\mathbf{d}$ below $\mathbf{0}'$ such that the r.e. degrees comparable with $\mathbf{d}$ are exactly $\mathbf{0}$ and $\mathbf{0}'$. Actually, as noticed later, Yates' degree $\mathbf{d}$ can be 1-generic, and can be minimal. In [15], Wu proved that Yates' degree $\mathbf{d}$ can appear in every jump class. In this section, we construct a degree $\mathbf{a}$ below $\mathbf{0}'$ such that the degrees containing weakly computable reals which are comparable with **a** are exactly **0** and **0'**. To do this, we first need the following notion of f-limit-genericity.

Definition 3.1. (1) A set A is called f-limit-generic iff for each e, if there are infinitely many m such that $W^K_{e,f(m)}$ contains an extension of $A(0)A(1)\dots A(m)$, then there is an n such that $A(0)A(1)\dots A(n) \in W^K_e$.

(2) A set A is called f-limit-semigeneric iff for each e, if for almost all m, $W^K_{e,f(m)}$ contains an extension of $A(0)A(1)\dots A(m)$, then there is an n such that $A(0)A(1)\dots A(n) \in W^K_e$.

Here W^K_e is the set of all strings enumerated by the e-th algorithm using the oracle K and $W^K_{e,f(m)}$ is the set of those strings in W^K_e which are enumerated in time $f(m)$.

An f-limit-generic set A forces membership in W^K_e only if for infinitely many prefixes $A(0)A(1)\dots A(m)$ of A an extension in W^K_e can be found within time $f(m)$, so this notion differs from the 1-genericity (see [9]) by having an oracle and bounding the search. Nevertheless, if f is growing fast sufficiently, then f-limit-genericity implies 1-genericity.

We note that for a real α, the notions of computable (by approximation) and recursive (by computing all digits) coincides. But this does no longer hold for sequences of reals, $\{\alpha_i\}_{i \in \mathbb{N}}$ as one can have that they are uniformly computable in the sense that there is a computable function $g : \mathbb{N} \times \mathbb{N} \to \mathbb{Q}$ such that $|\alpha_i - g(i,j)| < 2^{-j}$ for all i,j while they are not uniformly recursive in the sense that the function $i,j \to \alpha_i(j)$ which computes the digit $j+1$ of α_i after the dot is not computable.

This fact relativizes to the oracle K. While there is a uniformly K-recursive sequence of all left-computable reals, there is no uniformly K-recursive sequence containing all weakly computable reals. But there is still an enumeration $\alpha_0, \alpha_1, \alpha_2, \dots$ of all weakly computable reals and a K-recursive function $g : \mathbb{N} \times \mathbb{N} \to \mathbb{Q}$ such that the approximation condition $|\alpha_i - g(i,j)| < 2^{-j}$ holds for all i,j.

Theorem 3.2. Assume that $\alpha_0, \alpha_1, \alpha_2, \dots$ is a list of weakly computable reals such that there is a K-recursive function $g : \mathbb{N} \times \mathbb{N} \to \mathbb{Q}$ with $\forall i,j\, (|\alpha_i - g(i,j)| < 2^{-j})$. Then there is a function $f \leq_T K$ such that: (1) every f-limit-generic set A is 1-generic, (2) for all i, if $\alpha_i \leq_T A$, then α_i is recursive, and (3) for all i, if $\alpha_i \geq_T A$, then α_i is complete. Furthermore, one can choose A such that $A \leq_T K$ and hence A can be low.

Proof. Below in Propositions 3.3, 3.4 and 3.5, we will construct functions $f_1, f_2, f_3 \leq_T K$ respectively such that: every f_1-limit-semigeneric set satisfies (1), every f-limit-semigeneric set satisfies (2) and every f_3-limit-generic set satisfies (3).

Let f be defined as $f(n) = f_1(n) + f_2(n) + f_3(n)$ for all n. Then every f-limit-generic set A is f_1-limit-semigeneric, f_2-limit-semigeneric and f_3-limit-generic, and hence A satisfies all the requirements. ∎

Proposition 3.3. There is a function $f_1 \leq_T K$ such that every f_1-limit-semigeneric set is also 1-generic.

Proof. In an acceptable numbering, there are indices for algorithms and not only for sets. Thus every r.e. set has an index in the enumeration $W_0^K, W_1^K, \ldots$ such that the oracle is not accessed during the enumeration of this r.e. set. So it is reasonable to make the following definition.

Let $f_1(n)$ be the time needed to find for every $e \leq n$ and every string $\sigma \in \{0,1\}^*$ with $|\sigma| \leq n+1$ a string $\tau \succeq \sigma$ which is enumerated into W_e^K without having accessed the oracle K whenever such a τ exists.

As the search in an W_e^K is aborted for this e whenever the oracle K is accessed, the oracle K does not play any role in the definition of f_1 and so $f_1 \leq_T K$.

Now let A be any f_1-limit-generic set and consider any r.e. set V of strings. There is an index e such that $W_e^K = V$ and the enumeration procedure does not access the oracle K at all. Suppose that there are infinitely many n for which $A(0)A(1)\ldots A(n)$ has an extension in V. Then by the definition of f_1, it is easy to see that for all n, there is an extension of $A(0)A(1)\ldots A(n)$ in $W_{e,f_1(n)}^K$. Since A is f_1-limit-semigeneric, there is an m with $A(0)A(1)\ldots A(m) \in W_e^K$. Thus A meets V and hence, A is 1-generic. ∎

Proposition 3.4. *Let $\alpha_0, \alpha_1, \alpha_2, \ldots$ and g be the same as in Theorem 3.2. Then there is a function $f_2 \leq_T K$ such that for every f_2-limit-semigeneric set A and every i, if $\alpha_i \leq_T A$, then α_i is recursive.*

Proof. Given any binary string τ, for any e, let $\sigma_{\tau,e,0}$, $\sigma_{\tau,e,1}$, $k_{\tau,e}$ and t be the first data found such that (1) $\sigma_{\tau,e,0}$, $\sigma_{\tau,e,1}$ both extend τ and have length t; (2) $\{e\}_t^{\sigma_{\tau,e,0}}(m)$ and $\{e\}_t^{\sigma_{\tau,e,1}}(m)$ are defined for all $m < k_{\tau,e}$; (3) there are at least two binary strings of length $k_{\tau,e}$ lexicographically between $\{e\}_t^{\sigma_{\tau,e,0}}(m) \restriction k_{\tau,e}$ and $\{e\}_t^{\sigma_{\tau,e,1}}(m) \restriction k_{\tau,e}$.

Let h be a recursive function such that $W_{h(e,i)}^K$ contains one of $\sigma_{\tau,e,0}$ and $\sigma_{\tau,e,1}$, $\sigma_{\tau,e,j}$ say, such that α_i does not extend $\{e\}_t^{\sigma_{\tau,e,j}}(m)$, if the above search terminates for e, i, τ (*such a string can be found since the third condition guarantees that the restriction of $g(i, k_{\tau,e}+2)$ to its $k_{\tau,e}$ first bits cannot be identical with or be a neighbour of both computed strings*). In other words, the function h searches for the "real number variant" of an e-splitting, as described in Theorem 1.3.

Let $f_2(n)$ be the time needed to find with oracle K for each $e, i \leq n$ and each $\tau \in \{0,1\}^{n+1}$ an extension of τ in $W_{h(e,i)}^K$ whenever $\sigma_{\tau,e,0}, \sigma_{\tau,e,1}$ exist.

Assume that A is f_2-limit-semigeneric and $\alpha_i = \{e\}^A$. If there is an n with $A(0)A(1)\ldots A(n) \in W_{h(e,i)}^K$, then, as a real, α_i does not extend $\{e\}^{A(0)A(1)\ldots A(n)}$, contradicting the assumption. Thus there is an $n \geq e+i$ such that no extension of $A(0)A(1)\ldots A(n)$ is in $W_{h(e,i),f_2(n)}^K$. Then no extension of $A(0)A(1)\ldots A(n)$ is in $W_{h(e,i)}^K$ and α_i is the unique real such that for every $\eta \succeq A(0)A(1)\ldots A(n)$ there is a binary representation of α_i extending $\{e\}_{|\eta|}^{\eta}$. This means that α_i is recursive. ∎

Proposition 3.5. *Let $\alpha_0, \alpha_1, \alpha_2, \ldots$ and g be the same as in Theorem 3.2. Then there is a function $f_3 \leq_T K$ such that for every f_3-limit-generic set A and every i, if $\alpha_i \geq_T A$, then $K \leq_T \alpha_i$.*

Proof. Let $c(n)$ be the convergence module of K, that is, the time to enumerate all elements of $\{0, 1, \ldots, n\} \cap K$ into K. Now let $\tilde{h}(e, i)$ be a recursive function such that $W^K_{\tilde{h}(e,i)}$ contain all strings σ of length $n + 1$ for which there are m, j, η such that (1) $m < n$, $j < c(n)$, (2) η is the binary representation of the first $j + 3$ bits of $g(i, j + 4)$, (3) $\eta(j) = 0$ and $\eta(j + 1) = 1$, and (4) $\{e\}^\eta_j(m)$ converges to a value different from $\sigma(m)$ without querying the oracle at j or beyond.

Since the sets $W^K_{\tilde{h}(e,i)}$ are uniformly K-recursive, there is a K-recursive function f_3 such that $f_3(n)$ is the time needed to enumerate relative to K all members of length $\leq n + 2$ of sets $W^K_{\tilde{h}(e,i)}$ with $e, i \leq n$.

Let A be a f_3-limit-generic set. Suppose that α_i is irrational and that $A = \{e\}^{\alpha_i}$. By the construction of $W^K_{\tilde{h}(e,i)}$, for any n, $A(0)A(1) \ldots A(n)$ is not in $W^K_{\tilde{h}(e,i)}$. Consequently, there are only finitely many n such that $A(0)A(1) \ldots A(n)$ has extensions in in $W^K_{\tilde{h}(e,i), f_3(n)}$. By the choice of f_3, for almost all n, no extension of $A(0)A(1) \ldots A(n)$ is in $W^K_{\tilde{h}(e,i)}$.

Now let $u(n)$ be the first j such that for all $m \leq n$, $\alpha_i(j) = 0$, $\alpha_i(j + 1) = 1$, and $\{e\}^{\alpha_i}(m)$ converges within j steps without querying the oracle at j or above.

Since α_i is is assumed to be irrational, the function u is total. Obviously, $u \leq_T \alpha_i$. Since for almost all n, no initial segment of A of length n is in $W^K_{\tilde{h}(e,i)}$, it follows that $u(n) \geq c(n)$ for these n. Thus $K \leq_T \alpha_i$.

The case when α_i is rational is trivial. ∎

Proposition 3.6. *If $f \leq_T K$, then there is a set $A \leq_T K$ such that A is f-limit-generic.*

Proof. We assume that f is strictly monotonically increasing; if it is not, then one replaces f by $\hat{f}$ with $\forall n \, (\hat{f}(n) = f(0) + f(1) + \ldots + f(n) + n)$ and uses that every f-limit-generic set is also f-limit-generic.

First define a partial K-recursive function h with K-recursive domain such that, for all σ, e, if $W^K_{e,f(e+|\sigma|)}$ contains a proper extension of σ, then $h(e, \sigma)$ is that proper extension of σ which is enumerated into W^K_e first, and if $W^K_{e,f(e+|\sigma|)}$ contains no proper extension of σ, then $h(e, \sigma)$ is undefined.

Obviously, $\tau \prec \sigma \prec h(e, \tau)$ implies that $h(e, \sigma) = h(e, \tau)$.

Now we construct set A relative to oracle K. Assume that $A(m)$ for all $m < n$ is already defined and let σ be the string $A(0)A(1) \ldots A(n - 1)$ (we let σ be the empty string if $n = 0$). $A(n)$ is defined as follows: find the least e such that $h(e, \sigma)$ is defined and there is no $m < n$ with $h(e, A(0)A(1) \ldots A(m)) \preceq \sigma$ and define $A(n) = h(e, \sigma)(n)$.

The first step in the definition of $A(n)$ can be satisfied since there are a t and infinitely many programs e with $W^K_e = W^K_{e,t} = \{0, 1\}^{n+2}$. The second step is again satisfied since $n = |\sigma|$ and $h(e, \sigma) \succ \sigma$. Thus $h(e, \sigma)$ is so long that the bit $h(e, \sigma)(n)$ exists and can be copied into $A(n)$. Therefore, the definition above never runs into an undefined place. Obviously, $A \leq_T K$.

Now we verify that A is f-limit-generic. For the sake of contradiction, assume that there is an index e such that there are infinitely many n such that $W^K_{e,f(n)}$

contains an extension of $A(0)A(1)\ldots A(n)$ but no element of W_e^K is of the form $A(0)A(1)\ldots A(n)$. Let e be the least such index. Then there is a length n for which $h(e, A(0)A(1)\ldots, A(n))$ is defined and for all $e' < e$, either there is an $m < n$ with $A(0)A(1)\ldots A(m) \in W_{e',f(n+e')}^K$ or there is no $m \geq n$ such that $W_{e',f(n+e')}^K$ contains a proper extension of $A(0)A(1)\ldots A(m)$.

Now for any m with $n < m < |h(e, A(0)A(1)\ldots A(n))|$, the first step of the algorithm takes e as the parameter of the same name and assigns to $A(m)$ the value $h(e, A(0)A(1)\ldots A(n))(m)$. This is done since $h(e, A(0)A(1)\ldots A(m-1)) = h(e, A(0)A(1)\ldots A(n))$. Thus if $k = |h(e, A(0)A(1)\ldots A(n))|$, then

$$A(0)A(1)\ldots A(k-1) = h(e, A(0)A(1)\ldots A(n))$$

and this string is a member of W_e^K, contradicting our assumption on e. Therefore, A is f-limit-generic. This completes the proof.   ∎

Following Theorem 3.2 and Proposition 3.6, the following is obvious:

Theorem 3.7. *There is degree* **a** *below* $\mathbf{0}'$ *such that* $\mathbf{0}$ *and* $\mathbf{0}'$ *are the only degrees containing weakly computable reals and comparable with* **a**.

Actually, the degree **a** in Theorem 3.7 occurs in every jump class (work in progress). This genearlize results in Wu [15] and Yates [14]. We end this section by stating the following result without giving a proof:

Theorem 3.8. *If* $A, f \leq_T K$ *and* A *is nonlow₂, then there are* f-*limit-semi-generic sets* A_0, A_1 *such that* $A \equiv_T A_1 \oplus A_2$. *Particularly, every nonlow₂ degree below* $\mathbf{0}'$ *is the join of nonbounding degrees.*

4   Completely Weakly Computable Degrees

In this section, we prove Theorem 1.6.

Proof. $(1 \Rightarrow 2)$: Assume that A is r.e. and domination low for Ω. Let $B \leq_T A$. B has a recursive approximation of rationals $\beta_0, \beta_1, \ldots$ such that the convergence module

$$c_B(n) = \min\{t : \forall s \geq t \forall m \leq n\,(\beta_s(m) = \beta_t(m))\}$$

is A-recursive. Furthermore define the recursive function g inductively by $g(0) = 0$ and $g(s+1)$ being the minimal element of $\Omega_{s+1} - \Omega_s$ which, by choice, is indeed the first element where these two numbers differ; without loss of generality such an element always exists. Since c_Ω dominates the function $n \rightarrow c_B(2n)$, with a change of finitely many β_n, one can achieve that c_Ω actually majorizes this function. Now define a subsequence $\gamma_0, \gamma_1, \ldots$ of $\beta_0, \beta_1, \ldots$ by the following recursive algorithm:

1. Let $t = 0$. Let $s = 0$.
2. Let $\gamma_t = \beta_s$.
3. While $(\exists m \leq 2g(s+1) - 2)[\beta_{s+1}(m) \neq \gamma_t(m)]$ Do $s = s + 1$.
4. Let $t = t + 1$. Let $s = s + 1$. Goto 2.

First one shows by induction that for all n and all $s = c_\Omega(n)$ there is a t with $\gamma_t = \beta_s$.

If $c_{\Omega(0)} = 0$, then $\gamma_0 = \beta_0$ and the assumption holds; if $c_{\Omega(0)} > 0$, then $\Omega(0) = 1$, $g(c_\Omega(n)) = 0$ and the existential quantifier in step 3 becomes false when $s = c_\Omega(0) - 1$, thus the while loop stops and $\beta_{c_\Omega(0)}$ is added into the γ_e.

Assume now that the assumption is true for n, so there is a t' with $\gamma_{t'} = \beta_{c_\Omega(n)}$. If $c_\Omega(n+1) = c_\Omega(n)$, then there is nothing to show. If $c_\Omega(n+1) > c_\Omega(n)$, then let s, t be the values of the variables of the same name in the algorithm when $s = c_\Omega(n+1) - 1$. In this case $g(s+1) = n+1$, $t \geq t'$ and $\gamma_t(m) = B(m)$ for all $m \leq 2n = g(n+1) - 2$ since γ_t equals to some $\beta_{s''}$ with $s'' \geq c_B(2n)$. So $\beta_{s+1}(m) = \gamma_t(m)$ for all $m \leq g(s+1)$ and the loop in Step 3 terminates. Thus β_{s+1} will become γ_{t+1} and the assumption is verified again.

So it follows that the sequence of all γ_t is infinite and converges to B. Furthermore, for every number t there is a unique number s with $\gamma_{t+1} = \beta_{s+1}$ and $\forall m \leq g(s+1) - 2 \, [\gamma_{t+1}(m) = \gamma_t(m)]$. Thus $|\gamma_{t+1} - \gamma_t| \leq 2^{3-g(s+1)}$ and $\sum_{t \in \mathbb{N}} |\gamma_{t+1} - \gamma_t| \leq \sum_{s \in \mathbb{N}} 2^{g(s+1)-2} \leq \sum_{n \in \mathbb{N}} 2^n \cdot 2^{3-2n} \leq 16$ and the sequence $\gamma_0, \gamma_1, \ldots$ witnesses that B is weakly computable.

$(2 \Rightarrow 3)$: Obvious.

$(3 \Rightarrow 1)$: Let $f \leq_T A$ be a strictly increasing function. Define a set $B \equiv_T A$ as follows: $B(0) = 0$; $B(2^n) = A(n)$; $B(2^n + m) = \Omega_{f(2^{n+3})}(2^{n+1} + m)$ for $m = 1, 2, \ldots, 2^n - 1$.

Since $f \leq_T A$, it is obvious that $B \equiv_T A$. Thus B is weakly computable. Suppose that $B = \beta_0 - \beta_1$ for two left-computable reals β_0 and β_1. Now one has the following four facts:

(1) $\exists c_0 \forall n \, C(\beta_0(0)\beta_0(1) \ldots \beta_0(n) | \Omega(0)\Omega(1) \ldots \Omega(n)) \leq c_0$;
(2) $\exists c_1 \forall n \, C(\beta_1(0)\beta_1(1) \ldots \beta_1(n) | \Omega(0)\Omega(1) \ldots \Omega(n)) \leq c_1$;
(3) $\exists c_2 \forall n \, C(B(0)B(1) \ldots B(n) | \Omega(0)\Omega(1) \ldots \Omega(n)) \leq c_2$;
(4) $\forall c_3 \forall^\infty n \, (\Omega(2^{n+1}+1)\Omega(2^{n+1}+2) \ldots \Omega(2^{n+1}+2^n-1) | \Omega(0)\Omega(1) \ldots \Omega(2^n)) > c_3$.

Here the first two statements follow from the fact that Ω is complete among the left-computable sets with respect the so called rK-reducibility, the third statement follows from the fact that $B = \beta_0 - \beta_1$ and the fourth statement from the fact that Ω is Martin-Löf random and that the digits between 2^{n+1} and $2^{n+1} + 2^n$ cannot be predicted from those up to 2^{n+1}. Therefore, for almost all n, there is an $m \in \{1, 2, \ldots, 2^n - 1\}$ with $B(2^n + m) \neq \Omega(2^{n+1} + m)$. Thus, for almost all n, $f(2^{n+3}) < c_\Omega(2^{n+1} + 2^n - 1)$. The fact that both functions are monotone gives the following inequality:

$$\forall^\infty k \exists n \, (2^{n+2} \leq k < 2^{n+3} \text{ and } f(k) < f(2^{n+3}) < c_\Omega(2^{n+1}+2^n-1) < c_\Omega(k)).$$

So c_Ω dominates f.   ∎

In this proof, the direction $(1 \Rightarrow 2)$ needs that A has r.e. Turing degree, while the directions $(2 \Rightarrow 3)$ and $(3 \Rightarrow 1)$ work for all sets $A \leq_T \emptyset'$. Note that there are r.e. sets such that their Turing degree is array-recursive and low_2 but not low; thus these sets are not low for Ω. So the above characterization shows that there are more completely weakly computable degrees than those found in [4].

Corollary 4.1. *There is a completely weakly computable and r.e. Turing degree which is not low for Ω.*

References

1. K. Ambos-Spies, K. Weihrauch and X. Zheng, *Weakly computable real numbers,* Journal of Complexity **16** (2000), 676-690.
2. G. Chaitin. *A theory of program size formally identical to information theory,* Journal of the Association for Computing Machinery **22** (1975), 329-340.
3. R. Downey, G. Wu and X. Zheng. *Degrees of d.c.e. reals,* Mathematical Logic Quarterly **50** (2004),345-350.
4. R. Downey, D. Hirschfeldt, J. Miller and A. Nies. *Relativizing Chaitin's halting probability,* Journal of Mathematical Logic **5** (2005), 167-192.
5. R. Downey, D. Hirschfeldt, A. Nies and F. Stephan. *Trivial reals,* Proceedings of the 7th and 8th Asian Logic Conferences (2003), 103-131. World Scientific.
6. R. Downey, C. Jockusch Jr. and M. Stob. *Array nonrecursive sets and multiple permitting arguments,* Recursive Theory Week (Proceedings, Oberwolfach 1989) (Ambos-Spies, et al., eds.), Lecture Notes in Math. **1432** (1990), 141-173, Springer.
7. R. Downey, C. Jockusch Jr. and M. Stob. *Array nonrecursive sets and genericity,* Computability, Enumerability, Unsolvability: Directions in Recursion Theory. London Math. Soc. Lecture Notes Series (Cambridge University Press) **224** (1996), 93-104.
8. S. Ishmukhametov. *Weak recursive degrees and a problem of Spector,* Recursion Theory and Complexity (Proceedings of the Kazan 1997 workshop) (1999), 81-88. Walter de Gruyter.
9. C. Jockusch Jr. *Simple proofs of some theorems on high degrees of unsolvability,* Canadian Journal of Mathematics **29** (1977), 1072-1080.
10. A. Kučera and T. A. Slaman. *Randomness and recursive enumerability,* SIAM Journal on Computing **31** (2001), 199–211.
11. A. Nies. *Lowness properties of reals and randomness,* Advances in Mathematics **197** (2005), 274-305.
12. P. Odifreddi. *Classical Recursion Theory,* North-Holland / Elsevier, Amsterdam, Vol. I in 1989 and Vol. II in 1999.
13. R. I. Soare. *Recursively Enumerable Sets and Degrees,* Springer, Heidelberg, 1987.
14. C. E. Mike Yates. *Recursively enumerable degrees and the degrees less than $0'$,* Models and Recursion Theory, North Holland, 264-271, 1967.
15. G. Wu. *Jump operators and Yates degrees.* The Journal of Symbolic Logic **71** (2006), 252-264.

Understanding and Using Spector's Bar Recursive Interpretation of Classical Analysis

Paulo Oliva[*]

Department of Computer Science
Queen Mary, University of London
Mile End Road
London E1 4NS, UK
pbo@dcs.qmul.ac.uk

Abstract. This note reexamines Spector's remarkable computational interpretation of full classical analysis. Spector's interpretation makes use of a rather abstruse recursion schema, so-called bar recursion, used to interpret the double negation shift DNS. In this note bar recursion is presented as a generalisation of a simpler primitive recursive functional needed for the interpretation of a finite (intuitionistic) version of DNS. I will also present two concrete applications of bar recursion in the extraction of programs from proofs of $\forall\exists$-theorems in classical analysis.

1 Introduction

In [3], Gödel presents an interpretation of first-order intuitionistic arithmetic **HA** into a quantifier-free calculus of higher-order primitive recursive functionals. Gödel's interpretation, nowadays called *Dialectica interpretation*, can be naturally extended to an interpretation of $\mathbf{HA}^\omega$, intuitionistic arithmetic in the language of finite types (see [7]). Moreover, via the negative translation (e.g. Kuroda's [4]) of classical into intuitionistic logic, Dialectica interpretation is also applicable to $\mathbf{PA}^\omega$, classical arithmetic in the language of finite types.

One of the nicest features of Gödel's interpretation is that (in the intuitionistic context) it trivialises both Markov principle

$$\mathsf{MP} \quad : \quad \neg\forall n^{\mathbb{N}} A_{\mathsf{qf}}(n) \to \exists n \neg A_{\mathsf{qf}}(n),$$

$A_{\mathsf{qf}}(n)$ a decidable formula, and the axiom of choice

$$\mathsf{AC} \quad : \quad \forall x^\rho \exists y^\tau A(x, y) \to \exists f^{\rho \to \tau} \forall x A(x, fx).$$

Given that the interpretation strengthens $\forall\exists$-formulas[1], this gives a simple proof that $\mathbf{HA}^\omega + \mathsf{MP} + \mathsf{AC}$ is $\forall\exists$-conservative over $\mathbf{HA}^\omega$. The same cannot be said about $\mathbf{PA}^\omega + \mathsf{AC}$, however, since the Dialectica interpretation of $\mathbf{PA}^\omega + \mathsf{AC}$ will have to interpret the negative translation of AC, which is not necessarily weaker than AC itself. In fact, $\mathbf{PA}^\omega$ extended with a weaker form of choice

$$\mathsf{cAC} \quad : \quad \forall n^{\mathbb{N}} \exists y^\sigma A(n, y) \to \exists f^{\mathbb{N} \to \sigma} \forall n A(n, fn)$$

---

[*] This research has been supported by the UK EPSRC grant GR/S31242/01.
[1] If A is a $\forall\exists$-formula then the interpretation of A implies A, provably in $\mathbf{HA}^\omega$.

A. Beckmann et al. (Eds.): CiE 2006, LNCS 3988, pp. 423–434, 2006.
© Springer-Verlag Berlin Heidelberg 2006

so-called *countable choice*, is already strong enough to prove the *comprehension schema*

CA   : $\exists f^{\mathbb{N}\to\sigma}\forall n^{\mathbb{N}}(fn = 0 \leftrightarrow A(n))$.

Simply apply cAC to the classically valid statement $\forall n \exists k (k = 0 \leftrightarrow A(n))$. This shows that AC as stated above is innocuous in an intuitionistic context, but not in a classical one.

Therefore, we choose $\mathbf{PA}^\omega + \mathsf{cAC}$ as the formal system of *classical analysis*. Note that $\mathbf{PA}^\omega + \mathsf{cAC}$ (and even $\mathbf{PA}^\omega + \mathsf{CA}$) clearly contains second-order arithmetic $\mathbf{PA}^2$, if we represent sets via their characteristic functions. In [6], Spector observes that classical analysis has a negative translation into $\mathbf{HA}^\omega + \mathsf{cAC} + \mathsf{DNS}$, where

DNS   : $\forall n^{\mathbb{N}} \neg\neg A(n) \to \neg\neg \forall n A(n)$

is called the *double negation shift*. This is the case since, after classical logic is eliminated via the negative translation, one is left with the system $\mathbf{HA}^\omega + \mathsf{cAC}^N$ to be interpreted, where

cAC^N : $\forall n \neg\neg \exists y A(n, y) \to \neg\neg \exists f \forall n A(n, fn)$

is the schema sufficient for proving the negative translation of each cAC instance. On the other hand, cAC^N follows (in $\mathbf{HA}^\omega$) from cAC + DNS.

It is worth noting that the double negation shift is intuitionistically equivalent to the double negation of the *generalised Markov principle*

GMP   : $\neg\forall n^{\mathbb{N}} A(n) \to \exists n \neg A(n)$

for arbitrary formulas $A(n)$, i.e. $\mathbf{HA}^\omega \vdash \mathsf{DNS} \leftrightarrow \neg\neg\mathsf{GMP}$.

This note concerns Spector's [6] Dialectica interpretation of DNS (and hence, full classical analysis). Spector's interpretation makes use of a new recursion schema, called *bar recursion*. Our aim is to present bar recursion as a generalisation of a primitive recursive functional needed for the interpretation of $\wedge_{i=0}^{n} \neg\neg A_i \to \neg\neg \wedge_{i=0}^{n} A_i$, an intuitionistically valid special case of DNS. We will also present two concrete applications of bar recursion in the analysis of two $\forall\exists$-theorems in classical analysis.

1.1   Dialectica Interpretation

In this section we will shortly recall Gödel's Dialectica interpretation, using the unifying notation of [5]. Sequences of variables $x_0, \ldots, x_n$ will be abbreviated as $\boldsymbol{x}$, and a functional application of two sequences of variables $\boldsymbol{xy}$ is a shorthand for $x_0 y, \ldots, x_n y$. In the Dialectica interpretation, each formula A of $\mathbf{HA}^\omega$ is associate with a new (quantifier-free) formula $|A|_{\boldsymbol{y}}^{\boldsymbol{x}}$, with two distinguished tuples of free-variables $\boldsymbol{x}$ (so-called *witness variables*) and $\boldsymbol{y}$ (so-called *counter-example variables*), other than the variables already free in A. Intuitively, the formula A is interpreted by the formula $\exists \boldsymbol{x} \forall \boldsymbol{y} |A|_{\boldsymbol{y}}^{\boldsymbol{x}}$.

For an atomic formula P we set $|P| :\equiv P$, with empty tuples of both witness and counter-example variables. Assume we have already defined $|A|_{\boldsymbol{y}}^{\boldsymbol{x}}$ and $|B|_{\boldsymbol{w}}^{\boldsymbol{v}}$, we define

$$|A \wedge B|_{y,w}^{x,v} \; :\equiv \; |A|_y^x \wedge |B|_w^v$$

$$|A \vee B|_{y,w}^{x,v,n} \; :\equiv \; (n = 0 \wedge |A|_y^x) \vee (n \neq 0 \wedge |B|_w^v)$$

$$|A \rightarrow B|_{x,w}^{f,g} \; :\equiv \; |A|_{gxw}^x \rightarrow |B|_w^{fx}$$

$$|\forall z A(z)|_{y,z}^f \; :\equiv \; |A(z)|_y^{fz}$$

$$|\exists z A(z)|_y^{x,z} \; :\equiv \; |A(z)|_y^x$$

For instance, the interpretation of the formula $A \equiv \forall n \exists m (m \geq n \wedge P(m))$ has Dialectica interpretation $fn \geq n \wedge P(fn)$, where f is the witness variable and n is the counter-example variable. In our notation $|A|_n^f \equiv fn \geq n \wedge P(fn)$.

The soundness theorem for the Dialectica interpretation guarantees that if **HA**$^\omega$ proves a formula A, then there exists a sequence of terms t, over the free-variables of A, such that (the quantifier-free fragment of) **HA**$^\omega$ proves $|A|_y^t$. In order to extend Gödel's Dialectica interpretation of Heyting arithmetic with a new principle B, we must be able to produce a witnessing term s for the interpretation of B, i.e. we must be able to show $\forall w |B|_w^s$.

In this paper I will be using Kuroda's [4] negative translation of classical into intuitionistic logic, which simply places double negation after each universal quantifiers and in front of the whole formula.

Notation. We use $\mathbb{N}$ for the basic finite type, and $\rho \rightarrow \tau$ for functional types. For convenience we will also use sequence types, i.e. ρ^* denotes the type of sequences of elements of type ρ. The length of a finite sequence is represented as $|s|$, while $\langle x_0, \ldots, x_n \rangle$ denotes the finite sequence of length $n + 1$ with elements $x_0, \ldots, x_n$. The concatenation of two finite sequences, a finite sequence and an element, or a finite sequence with an infinite sequence will all be denote via the $*$ construction, i.e. $s * t$, $s * x$ and $s * f$, respectively. Whenever possible we omit parenthesis in functional application, i.e. fxy stands for $f(x, y)$. The following primitive recursive constructions will be used. If $s : \rho^*$ then

- $\hat{s}(i)$ is s_i if $i < |s|$ and 0^ρ otherwise

- if $i \leq |s|$ then $\overline{(s,i)}(n)$ is s_n when $n < i$ and 0^ρ otherwise

- if $i \leq |s|$ then $\bar{s}i$ is $\langle s_0, \ldots, s_{i-1} \rangle$.

2   Double Negation Shift

The principle DNS can be viewed as the infinite counterpart of the following intuitionistic schema

$$\text{fDNS} \quad : \quad \bigwedge_{i=0}^n \neg\neg A_i \rightarrow \neg\neg \bigwedge_{i=0}^n A_i$$

which we will refer to as the *finite double negation shift*. In order to understand Spector's interpretation of DNS, we will first look at the Dialectica interpretation of fDNS. A possible proof of fDNS in minimal logic is shown in Table 1.

$$[A_0]_{\alpha_0} \quad \cdots \quad [A_n]_{\alpha_n}$$

$$\dfrac{\displaystyle\bigwedge_{i=0}^{n} A_i \qquad [\neg \displaystyle\bigwedge_{i=0}^{n} A_i]_\beta}{\dfrac{\bot}{\neg A_0}\,(\alpha_0)} \qquad \dfrac{[\displaystyle\bigwedge_{i=0}^{n} \neg\neg A_i]_\gamma}{\neg\neg A_0}$$

$$\dfrac{\dfrac{\bot}{\neg A_1}\,(\alpha_1) \qquad \dfrac{[\displaystyle\bigwedge_{i=0}^{n} \neg\neg A_i]_\gamma}{\neg\neg A_1}}{}$$

$$\dfrac{\dfrac{\bot}{\neg A_n}\,(\alpha_n) \qquad \dfrac{[\displaystyle\bigwedge_{i=0}^{n} \neg\neg A_i]_\gamma}{\neg\neg A_n}}{}$$

$$\dfrac{\dfrac{\bot}{\neg\neg \displaystyle\bigwedge_{i=0}^{n} A_i}\,(\beta)}{\displaystyle\bigwedge_{i=0}^{n} \neg\neg A_i \to \neg\neg \displaystyle\bigwedge_{i=0}^{n} A_i}\,(\gamma)$$

Table 1. Proof of fDNS

Assume that each A_i has Dialectica interpretation $|A_i|_{y_i}^{x_i}$ (although x_i and y_i are potentially tuples of functionals, I will write them as single functionals for simplicity). Given that $\neg A$ is a shorthand for $A \to \bot$, the Dialectica interpretation of each $\neg\neg A_i$ is

$$|\neg\neg A_i|_{g_i}^{\Phi_i} \equiv \neg\neg |A_i|_{g_i(\Phi_i g_i)}^{\Phi_i g_i}$$

while $\neg\neg \bigwedge_{i=0}^{n} A_i$ has Dialectica interpretation

$$|\neg\neg \textstyle\bigwedge_{i=0}^{n} A_i|_{\boldsymbol{\Delta}}^{\boldsymbol{x}} \equiv \neg\neg \textstyle\bigwedge_{i=0}^{n} |A_i|_{\Delta_i(\boldsymbol{x}\boldsymbol{\Delta})}^{x_i \boldsymbol{\Delta}}.$$

By the interpretation of implication, the interpretation of fDNS asks for sequences of functionals $\boldsymbol{g}$ and $\boldsymbol{x}$ satisfying

$$|\mathsf{fDNS}|_{\boldsymbol{v}}^{\boldsymbol{g},\boldsymbol{x}} \equiv \textstyle\bigwedge_{i=0}^{n} \neg\neg |A_i|_{g_i(\boldsymbol{v},\Phi_i(g_i\boldsymbol{v}))}^{\Phi_i(g_i\boldsymbol{v})} \to \neg\neg \textstyle\bigwedge_{i=0}^{n} |A_i|_{\Delta_i(\boldsymbol{x}\boldsymbol{v})}^{x_i \boldsymbol{v}}$$

where $\boldsymbol{v} \equiv \boldsymbol{\Delta}, \boldsymbol{\Phi}$. Such functionals can be produced if we can solve (on the parameters $\boldsymbol{\Delta}, \boldsymbol{\Phi}$) the following set of equations

$$x_i \overset{\rho_i}{=} \Phi_i g_i$$
$$g_i(x_i) \overset{\tau_i}{=} \Delta_i(\boldsymbol{x}) \tag{1}$$

for $i \in \{0, \ldots, n\}$, where the parameters $\boldsymbol{\Delta}, \boldsymbol{\Phi}$ have been omitted for clarity. There is an apparent circularity since each x_i needs the definition of g_i, and each g_i seems to need the previous definition of all $x_0, \ldots, x_n$. Consider the case where $n = 2$, i.e.

$$x_0 = \Phi_0 g_0$$

$$x_1 = \Phi_1 g_1$$

$$x_2 = \Phi_2 g_2$$

$$g_0(x_0) = \Delta_0(x_0, x_1, x_2)$$

$$g_1(x_1) = \Delta_1(x_0, x_1, x_2)$$

$$g_2(x_2) = \Delta_2(x_0, x_1, x_2)$$

(2)

The analysis of the proof gives us the following solution. First define x_2 and g_2 assuming that x_0, x_1 have already been defined:

$$G_2[x_0, x_1] := \lambda x.\Delta_2(x_0, x_1, x)$$

$$X_2[x_0, x_1] := \Phi_2(G_2[x_0, x_1])$$

Next, we give a parametrised definition of x_1 and g_1, assuming only that x_0 has already been defined, as

$$G_1[x_0] := \lambda x.\Delta_1(x_0, x, X_2[x_0, x])$$

$$X_1[x_0] := \Phi_1(G_1[x_0])$$

We then define x_0 and g_0, using the parametrised definitions of x_1 and x_2, as

$$g_0 := \lambda x.\Delta_0(x, X_1[x], X_2[x, X_1[x]])$$

$$x_0 := \Phi_0 g_0$$

Finally, x_1, x_2 and g_1, g_2 can be defined, using the definition of x_0, as

$$g_1 := G_1[x_0]$$

$$x_1 := \Phi_1 g_1$$

$$g_2 := G_2[x_0, x_1]$$

$$x_2 := \Phi_2 g_2$$

The cunning solution above breaks the apparent circularity of (2) by making use of the fact that the definition of g_i does not require x_i to be already defined, although g_i does require all other x_j, for $j \neq i$ to be defined.

The general algorithm we get from the proof shown in Table 1 (following Dialectica interpretation) can be succinctly presented as follows.

Let us define a master functional fB (*finite bar recursion*), which is supposed to return the suffix $\langle x_i, \ldots, x_n \rangle$ assuming we already have an approximate solution $\langle x_0, \ldots, x_{i-1} \rangle$, as follows:

$$\mathsf{fB}(\langle x_0, \ldots, x_{i-1} \rangle) := \begin{cases} \langle \, \rangle & n < i \\ X_{\langle x_0, \ldots, x_{i-1} \rangle} * \mathsf{fB}(\langle x_0, \ldots, x_{i-1}, X_{\langle x_0, \ldots, x_{i-1} \rangle} \rangle) & n \geq i \end{cases}$$

where

$$G_{\langle x_0,\ldots,x_{i-1}\rangle} := \lambda x.\Delta_i(\langle x_0,\ldots,x_{i-1},x\rangle * \mathsf{fB}(\langle x_0,\ldots,x_{i-1},x\rangle))$$

$$X_{\langle x_0,\ldots,x_{i-1}\rangle} := \Phi_i G_{\langle x_0,\ldots,x_{i-1}\rangle}.$$

We can then take $\langle x_0,\ldots,x_n\rangle := \mathsf{fB}(\langle\ \rangle)$ and $g_i := G_{\langle x_0,\ldots,x_{i-1}\rangle}$. One can show by induction on i that if $\mathsf{fB}(\langle\ \rangle) = \langle x_0,\ldots x_n\rangle$ then $\mathsf{fB}(\langle x_0,\ldots,x_{i-1}\rangle) = \langle x_i,\ldots,x_n\rangle$. This implies that $x_i = X_{\langle x_0,\ldots,x_{i-1}\rangle}$, i.e. $x_i = \Phi_i g_{\langle x_0,\ldots,x_{i-1}\rangle} = \Phi_i g_i$. Moreover, $g_i(x_i) = \Delta_i(x)$ since

$$\langle x_0,\ldots,x_i\rangle * \mathsf{fB}(\langle x_0,\ldots,x_i\rangle) = \langle x_0,\ldots,x_n\rangle.$$

2.1   Example

We present here a concrete example of a proof whose Dialectica interpretation can be solved using the finite bar recursion fB. Let $m \in S$ be a shorthand for $Sm = 0$, $\mathsf{INF}(S)$ be the predicate stating that the set S is infinite, i.e. $\forall n \exists m \geq n(m \in S)$, and $\bar{S}$ denote the set complement of S. We will also use $\mathsf{INF}_0(S,g,n)$ as an abbreviation for the quantifier-free part of the Skolemised $\mathsf{INF}(S)$, namely $(gn \geq n) \wedge (gn \in S)$. Consider the following simple $\mathbf{PA}^\omega$-theorem.

Theorem 1. *For all $r \in \mathbb{N}$ and r-partition P of $\mathbb{N}$ (namely, $P : \mathbb{N} \to \{0,\ldots,r\}$) there exist a P-homogeneous infinite subset $H \subseteq \mathbb{N}$, i.e. $\exists k \forall n \in H(Pn = k)$.*

Let the P-homogeneous sets H_k be defined as $n \in H_k$ whenever $Pn = k$, for $k \leq r$. Therefore, Theorem 1 can be written formally as:

$$\forall r, P^{\mathbb{N}\to\{0,\ldots,r\}} \exists k \leq r\, \mathsf{INF}(H_k).$$

After the negative translation, the Dialectica interpretation asks for witnessing functionals for the following:

$$\forall r, P^{\mathbb{N}\to\{0,\ldots,r\}}, \Phi \exists g, k \leq r\, \mathsf{INF}_0(H_k, g, \Phi kg).$$

We will produce for each $k \leq r$ a function g_k (on parameters r, P and Φ) such that for some $0 \leq k \leq r$, we have $\mathsf{INF}_0(H_k, g_k, \Phi kg_k)$. It is easy to see that if those functions satisfy the following set of equations then we are done:

$$x_k = \Phi k g_k$$
$$g_k(x_k) = \max\{x_0,\ldots,x_r\} \tag{3}$$

since, for all $0 \leq k \leq r$, all $g_k(\Phi kg_k)$ would have the same value, namely $\max\{x_0,\ldots,x_r\}$. Moreover, $g_k(x_k) \geq x_k$. Therefore, once those have been computed, we can produce k and g as

$$k, g := \begin{cases} 0, g_0 & \text{if } \max\{x_0,\ldots,x_r\} \in H_0 \\ \cdots & \\ r, g_r & \text{if } \max\{x_0,\ldots,x_r\} \in H_r \end{cases}$$

But (3) is exactly the kind of system of equations that can be solved via finite bar recursion, as shown above.

2.2   Spector's Interpretation

In this section we present Spector's [6] bar recursive Dialectica interpretation of DNS. Let $|A(n)|_y^x$ be the interpretation of $A(n)$. The Dialectica interpretation of the double negation shift DNS is as follows

$$|DNS|_{\Phi,\Psi}^{n,g,f} \equiv \neg\neg|A(n)|_{g(\Phi ng)}^{\Phi ng} \to \neg\neg|A(\Psi f)|_{\Delta f}^{f(\Psi f)}$$

where, for clarity, we have omitted the parameters Φ, Ψ of the witnessing functionals n, g, f. Such functionals can be produced if we can solve the following generalisation of (1)

$$n \overset{N}{=} \Psi f$$

$$fn \overset{\rho}{=} \Phi ng \tag{4}$$

$$y(fn) \overset{\tau}{=} \Delta f$$

Therefore, instead of a finite tuple $\langle x_1, \ldots, x_n \rangle$, we must produce an infinite sequence f. Intuitively, if we were to assume the continuity of Ψ we would only need to produce a finite initial segment of f. Based on this intuition, Spector then solves the following (more general) problem

$$|s| > \Psi\hat{s}$$

$$s_i \overset{\rho}{=} \Phi i g_i \tag{5}$$

$$g_i s_i \overset{\tau}{=} \Delta\hat{s}$$

which asks for a finite sequence s and a sequence of g_i, for all $i \leq \Psi\hat{s}$. Given a sequence s and a family $(g_i)_{i \leq \Psi\hat{s}}$ satisfying (5), we can take $f := \hat{s}$, $n := \Psi f$ and $g := g_n$, in order to solve (4).

Problem (5) is almost the same as (1) of Section 2, except that we do not know the required length of the sequence s. The only thing we have is a lower bound $\Psi\hat{s}$ on that length which depends on s itself! We must somehow be able to produce as long a sequence s as it takes to satisfy $|s| > \Psi\hat{s}$. Let $B(s)$ be defined recursively as follows

$$B(s) := \begin{cases} \langle\rangle & \Psi\hat{s} < |s| \\ X_s * B(s * X_s) & \text{otherwise} \end{cases}$$

where

$$g_s := \lambda x.\Delta(s * x * B(s * x)) * (\lambda n.0^\rho))$$

$$X_s := \Phi(|s|, g_s)$$

Lemma 1. *Let $s := B(\langle\rangle)$. The following holds for all $0 \leq i \leq |s|$*

(a) $s = (\bar{s}i) * B(\bar{s}i)$

(b) $\Psi(\overline{s,i}) \geq i$, *if* $i < |s|$

(c) $\Psi(\bar{s}, i) < i$, if $i = |s|$ (i.e. $\Psi\hat{s} < |s|$)

(d) $s_i = X_{\bar{s}i}$, if $i < |s|$

Proof. Points (b) and (c) follow easily from (a), given that $B(\bar{s}i) = \langle \ \rangle$ if and only if $\Psi(\bar{s}, i) < i$. Point (d) follows from (a) and (b), simply by the definition of B. We prove (a) by induction on i. If $i = 0$ the result is trivial since $\bar{s}0 = \langle \ \rangle$. Assume $s = (\bar{s}i) * B(\bar{s}i)$ and $i + 1 \leq |s|$. In this case $B(\bar{s}i)$ cannot be the empty sequence, i.e. $B(\bar{s}i) = X_{\bar{s}i} * B(\bar{s}i * X_{\bar{s}i})$. By induction hypothesis, $s_i = X_{\bar{s}i}$, which implies $s = (\bar{s}(i + 1)) * B(\bar{s}(i + 1))$. □

Given the definition of $X_{(.)}$ and $g_{(.)}$ above, points (c) and (d) of Lemma 1 imply that $s := B(\langle \ \rangle)$ and $g_i := g_{\bar{s}i}$ solve (5), i.e.

$$s_i = \Phi i g_{\bar{s}i} \wedge g_{\bar{s}i}(s_i) = \Delta\hat{s}$$

for all $i \leq \Psi\hat{s}$. Spector stated the following general recursion schema, so-called *bar recursion*

$$\mathsf{BR}(\Psi, \Delta, \Phi, s^{\rho^*}) \stackrel{\tau}{=} \begin{cases} \Phi s & \text{if } \Psi\hat{s} < |s| \\ \Delta(s, \lambda x^\rho.\mathsf{BR}(\Psi, \Delta, \Phi, s * x)) & \text{otherwise} \end{cases}$$

where ρ^* is a finite sequence of objects of type ρ and the types of Ψ, Δ and Φ can be inferred from the context.

3   Bar Recursion in Use

In this section we present two examples of how bar recursion can be used in the computational interpretation of $\forall\exists$-theorem in classical analysis. In each case we will not go into details of how the bar recursive programs have been extracted from the proof. Our focus is on the characteristics of the final programs.

3.1   Example 1: no injection from $\mathbb{N} \to \mathbb{N}$ to $\mathbb{N}$

Our first example is a simple theorem which states that there can not be an injection from $\mathbb{N} \to \mathbb{N}$ to $\mathbb{N}$. We analyse its straightforward classical proof.

Theorem 2. $\forall\Psi^2\exists\alpha^1, \beta^1(\alpha \neq \beta \wedge \Psi\alpha = \Psi\beta)$.

Proof. From the axiom $\forall\Psi, k(\exists\beta(k = \Psi\beta) \to \exists\alpha(k = \Psi\alpha))$ we get by classical logic

$$\forall\Psi, k\exists\alpha(\exists\beta(k = \Psi\beta) \to k = \Psi\alpha).$$

By countable choice cAC^1 we can prove the existence of a functional f satisfying

$$\forall\Psi\exists f\forall k(\exists\beta(k = \Psi\beta) \to k = \Psi(fk)),$$

i.e. f is an enumeration of functions such that $\Psi(fk) = k$, whenever k is in the image of Ψ. Let $\delta_f := \lambda k.(f(k)(k) + 1)$ and $k := \Psi\delta_f$. We have

$$\forall\Psi\exists f(\exists\beta(\Psi\delta_f = \Psi\beta) \to \Psi\delta_f = \Psi(f(\Psi\delta_f))).$$

The premise being provable we get

$$\forall\Psi\exists f(\delta_f \neq f(\Psi\delta_f) \wedge \Psi\delta_f = \Psi(f(\Psi\delta_f))),$$

since $\delta_f \neq f(\Psi\delta_f)$ (they differ at point $\Psi\delta_f$). So we get the desired result for $\alpha := \delta_f$ and $\beta := f(\Psi\delta_f)$. □

The reader is invited to reflect upon a *general* effective procedure for computing α and β given the functional Ψ. Obviously, if we assume that Ψ is continuous then we can simply consider the point of continuity n of Ψ on the constant zero function 0^1. Any two distinct functions which coincide up to n will have the same value for Ψ. On the other hand, if Ψ is not continuous, but is assumed to be majorizable, then we can again consider the point of weak continuity (see Lemma 5 of [2]) n of Ψ on the constant zero function. By the pigeon hole principle, any set of $n+1$ distinct functions which coincide with 0^1 up to n will have two with the same value. We present, however, a bar recursive solution which does not rely on the specific properties of the model for solving the problem.

The proof of Theorem 2 presented above can be formalised in the system $\mathbf{PA}^\omega + \mathrm{cAC}$. A bar recursive analysis of the negative translation of the proof (which we omit here for lack of space) will lead us to consider the finite sequence $t := \mathsf{B}(\langle\,\rangle)$, where

$$\mathsf{B}(s) \stackrel{1^*}{:=} \begin{cases} \langle\,\rangle & \Psi\delta_{\hat{s}} < |s| \\ \delta_{\hat{r}} * \mathsf{B}(s * \delta_{\hat{r}}) & \Psi\delta_{\hat{r}} = |s| \\ 0^1 * \mathsf{B}(s * 0^1) & \text{otherwise,} \end{cases}$$

and $r := s * 0^1 * \mathsf{B}(s * 0^1)$. Intuitively, we wish $t := \mathsf{B}(\langle\,\rangle)$ to be an enumeration of functions t_k such that $(+)$ $\Psi t_k = k$, whenever k is in the image of Ψ, i.e.

$$\forall k < |t|(\exists\beta(\Psi\beta = k) \to \Psi t_k = k).$$

Obviously, we cannot do that effectively for all indices of t. The trick is to test at each stage k whether $(+)$ holds for the diagonal function $\delta_{\hat{r}}$, i.e. $\Psi\delta_{\hat{r}} = k$? If that is the case then we add that function in current position k. If, however, that is not the case, we simply make sure that the whole sequence t equals r, so that $\Psi\delta_{\hat{t}} \neq k$. Therefore, the enumeration t will be such that for each $k \neq \Psi\delta_{\hat{t}}$ we have $\Psi t_k = k$. Let $n := \Psi\delta_{\hat{t}}$. By the stopping condition we have $n < |t|$. It follows that $\Psi t_n = n$, which gives us

$$\Psi\delta_f = \Psi(f(\Psi\delta_f))$$

taking $f := \hat{t}$.

3.2 Example 2: update procedures

In [1], Avigad shows that the 1-consistency of arithmetic is equivalent to the existence of solutions for a particular class of recursive equations, so-called *update equations.* In this second example we show how bar recursion can be used to compute the solution of one single update equation. The bar recursive program we present has been extracted from the proof of Lemma 2.1 in [1] (presented below).

In the following σ and τ will denote finite partial functions from $\mathbb{N}$ to $\mathbb{N}$, i.e. partial functions which are defined on a finite domain. A partial function which is everywhere undefined is denoted by $\langle\,\rangle$, whereas a partial function defined only at position k (with value n) is denoted by $\langle k, n\rangle$. The finite partial functions can be viewed as finite sequences of pairs of natural numbers. For a given partial function σ, we define $\hat{\sigma}$ as the total function which is obtained from σ by defining the output to be 0 (zero) wherever σ is undefined. We say that τ extends σ, written as $\sigma \sqsubseteq \tau$, if τ is defined wherever σ is defined, and on those points they coincide in value. We denote the domain of σ as $\mathsf{dom}(\sigma)$. For a finite partial function σ and $k, n \in \mathbb{N}$ we define the finite partial function $\sigma \oplus \langle k, n\rangle$ which maps k to n and agrees with σ otherwise, i.e.

$$(\sigma \oplus \langle k, n\rangle)(i) := \begin{cases} n & i = k \\ \sigma(i) & i \neq k \wedge i \in \mathsf{dom}(\sigma) \\ \uparrow & \text{otherwise.} \end{cases}$$

Let $\Psi : (\mathbb{N} \to \mathbb{N}) \to \mathbb{N}$ be a *continuous* functional with respect to the standard topology on the Baire space. Let Φ be also of type $(\mathbb{N} \to \mathbb{N}) \to \mathbb{N}$. We say that the pair $\langle\Psi, \Phi\rangle$ forms a unary *update procedure* if whenever τ extends $\sigma \oplus \langle\Psi\hat{\sigma}, \Phi\hat{\sigma}\rangle$ and $\Psi\hat{\sigma} = \Psi\hat{\tau}$ then $\Phi\hat{\sigma} = \Phi\hat{\tau}$.

Theorem 3 (Lemma 2.1, [1]). *Every unary update procedure has a finite fixed point, i.e.*

$$\forall\Psi, \Phi(\mathsf{Update}(\Psi, \Phi) \to \exists\sigma(\sigma = \sigma \oplus \langle\Psi\hat{\sigma}, \Phi\hat{\sigma}\rangle)).$$

Proof. Define the sequence of partial functions $\sigma_{(0)}, \sigma_{(1)}, \sigma_{(2)}, \dots$ as

$$\sigma_{(0)} := \langle\,\rangle$$

$$\sigma_{(i+1)} := \sigma_{(i)} \oplus \langle\Psi\hat{\sigma}_{(i)}, \Phi\hat{\sigma}_{(i)}\rangle.$$

The fact that $\langle\Psi, \Phi\rangle$ is an update procedure implies that $\sigma_{(0)} \sqsubseteq \sigma_{(1)} \sqsubseteq \sigma_{(2)} \dots$. Let g be the partial function extending all the $\sigma_{(i)}$, that is $g := \bigcup_{i \in \mathbb{N}} \sigma_{(i)}$. The continuity of Ψ implies that for some i we have

$$\Psi\hat{g} = \Psi\hat{\sigma}_{(i)} = \Psi\hat{\sigma}_{(i+1)} = \dots.$$

Since $\langle\Psi, \Phi\rangle$ forms an update pair we get

$$\sigma_{(i+1)} = \sigma_{(i)} \oplus \langle\Psi\hat{\sigma}_{(i)}, \Phi\hat{\sigma}_{(i)}\rangle = \sigma_{(i+1)} \oplus \langle\Psi\hat{\sigma}_{(i+1)}, \Phi\hat{\sigma}_{(i+1)}\rangle.$$

So, $\sigma_{(i+1)}$ is the desired fixed point. □

Comprehension is used in the proof above in order to obtain the function $\hat{g}$ as

$$\hat{g}(k) := \begin{cases} n & \exists i(\langle k, n \rangle \in \sigma_{(i)}) \\ 0 & \text{otherwise.} \end{cases}$$

If we could effectively build a functional f satisfying

$$\forall k (\exists i (k \in \text{dom}(\sigma_{(i)}))) \rightarrow (k \in \text{dom}(\sigma_{(fk)})))$$

we could produce $\hat{g}$ as

$$g_f(k) := \begin{cases} \sigma_{(fk)}(k) & k \in \text{dom}(\sigma_{(fk)}) \\ 0 & \text{otherwise} \end{cases}$$

since (the oracle) f tells us at which stages $i = fk$ each k is guaranteed to be defined, i.e. $k \in \text{dom}(\sigma_{(i)})$, if k does eventually become defined. The point of continuity n of Ψ on g guarantees that Ψ only looks at positions $k \leq n$. Therefore, the fixed point is surely obtained at point $\max\{fk : k \leq n\}$. Unfortunately, no such functional f can be built effectively, uniformly on all parameters. What we can build using bar recursion is an approximation for f, which as we will see, is sufficient for computing the position where the fixed point is attained. Define (uniformly on parameters ψ, ϕ) the following bar recursive functional:

$$\mathsf{B}_{\psi,\phi}(s) := \begin{cases} \langle \rangle & \psi \hat{s} < |s| \\ (\phi \hat{r}) * \mathsf{B}_{\psi,\phi}(s * (\phi \hat{r})) & |s| \subset \text{dom}(\sigma_{(\phi \hat{r})}) \\ 0^1 * \mathsf{B}_{\psi,\phi}(s * 0^1) & \text{otherwise} \end{cases}$$

where $r := s * 0^1 * \mathsf{B}_{\psi,\phi}(s * 0^1)$. In the definition above, $\sigma_{(i)}$ denotes the i-th element of the inductive sequence used in the proof of Theorem 3. Intuitively, taking $t := \mathsf{B}_{\psi,\phi}(\langle \rangle)$, the procedure above makes sure that $k \in \text{dom}(\sigma_{(t_k)})$, whenever $k \in \text{dom}(\sigma_{(\phi \hat{t})})$, i.e.

(+) $\forall k < |t| (k \in \text{dom}(\sigma_{(\phi \hat{t})}) \rightarrow k \in \text{dom}(\sigma_{(t_k)}))$

Finally, define

$$\phi f := \max\{fk : k \leq \omega_\Psi(g_f)\} + 1$$

$$\psi f := \omega_\Psi(g_f)$$

where ω_Ψ is the modulus of pointwise continuity of Ψ and g_f has been defined above. Notice that ω_Ψ is part of the witnessing information for the assumption that Ψ is continuity (see definition of $\mathsf{Update}(\Psi, \Phi)$), and therefore, it is one of the "inputs" for our effective procedure.

Since $\omega_\Psi(g_f)$ tells us what initial segment of g_f is necessary to compute Ψg_f, the functional ϕf computes (using f as an oracle) how far in the iteration $\{\sigma_{(i)}\}_{i \in \mathbb{N}}$ we need to go to get that much of initial segment. Let $t := \mathsf{B}_{\psi,\phi}(\langle \rangle)$ and $n := \phi \hat{t}$. Since $\psi \hat{t} < |t|$, by (+), we have

$$\forall k \leq \omega_\Psi(g_{\hat{t}})(k \in \text{dom}(\sigma_{(n)}) \rightarrow k \in \text{dom}(\sigma_{(t_k)})).$$

434   P. Oliva

Moreover, by the definition of ϕ we have that $n = \phi\hat{t} > t_k$, for all $k \leq \omega_\Psi(g_{\hat{t}})$. This implies that all positions in the domain of $\sigma_{(n)}$ have been defined before, since $\langle \Psi, \Phi \rangle$ is assumed to form an update procedure. Therefore, by the continuity of Ψ, it must be the case that $\Psi(\hat{\sigma}_{(n)}) \in \mathsf{dom}(\sigma_{(n)})$, and $\sigma_{(n-1)} = \sigma_{(n)}$.

Acknowledgements. I would like to thank the two anonymous referees for valuable comments and suggestions on the preliminary version of this paper. This research has been supported by the UK EPSRC grant GR/S31242/01.

References

1. J. Avigad. Update procedures and the 1-consistency of arithmetic. *Mathematical Logic Quarterly*, 48:3–13, 2002.
2. U. Berger and P. Oliva. Modified bar recursion and classical dependent choice. *Lecture Notes in Logic*, 20:89–107, 2005.
3. K. Gödel. Über eine bisher noch nicht benützte Erweiterung des finiten Standpunktes. *Dialectica*, 12:280–287, 1958.
4. S. Kuroda. Intuitionistische Untersuchungen der formalistischen Logik. *Nagoya Mathematical Journal*, 3:35–47, 1951.
5. P. Oliva. Unifying functional interpretations. To appear in: Notre Dame Journal of Formal Logic, 2006.
6. C. Spector. Provably recursive functionals of analysis: a consistency proof of analysis by an extension of principles in current intuitionistic mathematics. In F. D. E. Dekker, editor, *Recursive Function Theory: Proc. Symposia in Pure Mathematics*, volume 5, pages 1–27. American Mathematical Society, Providence, Rhode Island, 1962.
7. A. S. Troelstra (ed.). *Metamathematical Investigation of Intuitionistic Arithmetic and Analysis*, volume 344 of *Lecture Notes in Mathematics*. Springer, Berlin, 1973.

A Subrecursive Refinement
of the Fundamental Theorem of Algebra

Peter Peshev and Dimiter Skordev*

University of Sofia, Faculty of Mathematics and Computer Science,
5 blvd. J. Bourchier, 1164 Sofia, Bulgaria
ppeshev@gmail.com, skordev@fmi.uni-sofia.bg

Abstract. Let us call an *approximator* of a complex number α any sequence $\gamma_0, \gamma_1, \gamma_2, \ldots$ of rational complex numbers such that

$$|\gamma_t - \alpha| \le \frac{1}{t+1}, \quad t = 0, 1, 2, \ldots$$

Denoting by $\mathbb{N}$ the set of the natural numbers, we shall call a *representation* of α any 6-tuple of functions $f_1, f_2, f_3, f_4, f_5, f_6$ from $\mathbb{N}$ into $\mathbb{N}$ such that the sequence $\gamma_0, \gamma_1, \gamma_2, \ldots$ defined by

$$\gamma_t = \frac{f_1(t) - f_2(t)}{f_3(t) + 1} + \frac{f_4(t) - f_5(t)}{f_6(t) + 1} i, \quad t = 0, 1, 2, \ldots,$$

is an approximator of α. For any representations of the members of a finite sequence of complex numbers, the concatenation of these representations will be called a representation of the sequence in question (thus the representations of the sequence have a length equal to 6 times the length of the sequence itself). By adapting a proof given by P. C. Rosenbloom we prove the following refinement of the fundamental theorem of algebra: for any positive integer N there is a 6-tuple of computable operators belonging to the second Grzegorczyk class and transforming any representation of any sequence $\alpha_0, \alpha_1, \ldots, \alpha_{N-1}$ of N complex numbers into the components of some representation of some root of the corresponding polynomial $P(z) = z^N + \alpha_{N-1} z^{N-1} + \cdots + \alpha_1 z + \alpha_0$.

Keywords: Fundamental theorem of algebra, Rosenbloom's proof, computable analysis, computable operator, second Grzegorczyk class.

1 Introduction

In the paper [4] a proof is given of the fact that for any positive integer N and any complex numbers $\alpha_0, \alpha_1, \ldots, \alpha_{N-1}$ the polynomial

$$P(z) = z^N + \alpha_{N-1} z^{N-1} + \cdots + \alpha_1 z + \alpha_0 \tag{1}$$

has at least a root in the complex plane, and the proof is constructive in some sense. Making use of the notion of approximator considered in the abstract,

---

* Corresponding author.

A. Beckmann et al. (Eds.): CiE 2006, LNCS 3988, pp. 435–444, 2006.

we may describe the constructive character of the proof as follows: the proof shows implicitly (after some small changes) that for any positive integer N there is a computable procedure for transforming any approximators of any $\alpha_0, \alpha_1, \ldots, \alpha_{N-1}$ into some approximator of some root of the corresponding polynomial $P(z)$.[1] Clearly the following more rigorous formulation of this can be given, where F consists of all total mappings of $\mathbb{N}$ into $\mathbb{N}$: for any positive integer N there are recursive operators $\Gamma_1, \Gamma_2, \Gamma_3, \Gamma_4, \Gamma_5, \Gamma_6$ with domain F^{6N} such that whenever an element $\bar{f}$ of F^{6N} is a representation of some N-tuple $\alpha_0, \alpha_1, \ldots, \alpha_{N-1}$ of complex numbers, then $\Gamma_k(\bar{f})$, $k = 1, 2, 3, 4, 5, 6$, belong to F and form a representation of some root of the corresponding polynomial $P(z)$.[2]

The present paper is devoted to the fact that one can replace the words "recursive operators" in the above formulation by "computable operators belonging to the second Grzegorczyk class" (the fact was established in the first author's master thesis [3] written under the supervision of the second author).

2   The Notion of Computable Operator of the Second Grzegorczyk Class

For any natural number k let F_k be the set of all total k-argument functions in the set $\mathbb{N}$ (thus $F_1 = F$). For any natural numbers n and k we shall consider operators acting from F^n into F_k. The ones among them that are computable operators of the second Grzegorczyk class will be called $\mathcal{E}^2$-*computable operators* for short. The class of these operators can be defined by means of a natural extension of a definition of the class of functions $\mathcal{E}^2$ from the hierarchy introduced in [1] (such a step would be similar to the extension in [2] of the definition of $\mathcal{E}^3$ by introducing the notion of elementary recursive functional). Roughly speaking, we can use the same initial functions and the same ways of construction of new functions as in the definition of $\mathcal{E}^2$, except that we must add to the initial functions also the function arguments of the operator and to consider only constructions that are uniform with respect to these arguments. Skipping the details of the definition[3], we note the following properties of the $\mathcal{E}^2$-computable operators, where $\bar{f}$ is used as an abbreviation for the n-tuple $f_1, \ldots, f_n$ of functions from F.

1. For any k-argument function g belonging to the class $\mathcal{E}^2$ the mapping $\lambda\bar{f}.g$ of F^n into F_k is an $\mathcal{E}^2$-computable operator.

2. The mappings $\lambda\bar{f}.f_j$, $j = 1, \ldots, n$, of F^n into F are $\mathcal{E}^2$-computable operators.

---

[1] By certain continuity reasons, a dependence of this root not only on the coefficients $\alpha_0, \alpha_1, \ldots, \alpha_{N-1}$, but also on the choice of their approximators, cannot be excluded in the case of $N > 1$.

[2] This statement holds also for a more usual notion of approximator based on the inequality $|\gamma_t - \alpha| \leq 2^{-t}$ instead of the inequality $|\gamma_t - \alpha| \leq \frac{1}{t+1}$ (cf. for example the approach to computable analysis by Cauchy representations in [6]). However, the main result of the present paper would be not valid in that case, as it can be seen by means of an easy application of Liouville's approximation theorem.

[3] See, however, the remark on the next page.

3. Whenever Γ_0 is an $\mathcal{E}^2$-computable operator from F^n into F_m, $\Gamma_1, \ldots, \Gamma_m$ are $\mathcal{E}^2$-computable operators from F^n into F_k, and Γ is the mapping of F^n into F_k defined by

$$\Gamma(\bar{f})(x_1, \ldots, x_k) = \Gamma_0(\bar{f})(\Gamma_1(\bar{f})(x_1, \ldots, x_k), \ldots, \Gamma_m(\bar{f})(x_1, \ldots, x_k)),$$

then Γ is also an $\mathcal{E}^2$-computable operator.

4. Whenever Γ_0 is an $\mathcal{E}^2$-computable operator from F^n into F_m, Γ_1 is an $\mathcal{E}^2$-computable operator from F^n into F_{m+2}, Γ_2 is an $\mathcal{E}^2$-computable operator from F^n into F_{m+1}, the mapping Γ of F^n into F_{m+1} is defined by

$$\Gamma(\bar{f})(0, x_1, \ldots, x_m) = \Gamma_0(\bar{f})(x_1, \ldots, x_m),$$
$$\Gamma(\bar{f})(t+1, x_1, \ldots, x_m) = \Gamma_1(\bar{f})(\Gamma(\bar{f})(t, x_1, \ldots, x_m), t, x_1, \ldots, x_m),$$

and for all $\bar{f}, t, x_1, \ldots, x_m$ the inequality

$$\Gamma(\bar{f})(t, x_1, \ldots, x_m) \leq \Gamma_2(\bar{f})(t, x_1, \ldots, x_m)$$

holds, then Γ is also an $\mathcal{E}^2$-computable operator.

5. Whenever Γ_0 is an $\mathcal{E}^2$-computable operator from F^n into F_{m+1}, and the mapping Γ of F^n into F_{m+1} is defined by

$$\Gamma(\bar{f})(t, x_1, \ldots, x_m) = \min \left\{ s \mid s = t \vee \Gamma_0(\bar{f})(s, x_1, \ldots, x_m) = 0 \right\},$$

then Γ is also an $\mathcal{E}^2$-computable operator.

6. Whenever Γ_0 is an $\mathcal{E}^2$-computable operator from F^m into F_k, $\Gamma_1, \ldots, \Gamma_m$ are $\mathcal{E}^2$-computable operators from F^n into F_{l+1}, and Γ is the mapping of F^n into F_{k+l} defined by

$$\Gamma(\bar{f})(x_1, \ldots, x_k, y_1, \ldots, y_l) =$$
$$\Gamma_0(\lambda t. \Gamma_1(\bar{f})(y_1, \ldots, y_l, t), \ldots, \lambda t. \Gamma_m(\bar{f})(y_1, \ldots, y_l, t))(x_1, \ldots, x_k),$$

then Γ is also an $\mathcal{E}^2$-computable operator.

7. If Γ is an $\mathcal{E}^2$-computable operator from F^n into F_k, and the functions $f_1, \ldots, f_n$ belong to Grzegorczyk class $\mathcal{E}^m$, where $m \geq 2$, then the function $\Gamma(\bar{f})$ also belongs to $\mathcal{E}^m$.

Remark. The properties 1–4 can be used as the clauses of an inductive definition of the notion of $\mathcal{E}^2$-computable operator. Moreover, in such a case one can reduce the property 1 to its instances when g is the function $\lambda xy. (x+1) \cdot (y+1)$ or some of the functions $\lambda x_1 \ldots x_k. x_j$, $j = 1, \ldots, k$. In order to eliminate the not effectively verifiable domination requirement in the clause corresponding to property 4, one could omit this requirement and replace the right-hand side of the second equality by the expression

$$\min\{\Gamma_1(\bar{f})(\Gamma(\bar{f})(t, x_1, \ldots, x_m), t, x_1, \ldots, x_m), \Gamma_2(\bar{f})(t, x_1, \ldots, x_m)\}.$$

3 $\mathcal{E}^2$-Computable Functions in the Set of the Complex Numbers

Let $\mathbb{C}$ be the set of the complex numbers. A function φ from $\mathbb{C}^N$ into $\mathbb{C}$ will be called $\mathcal{E}^2$-*computable* if six $\mathcal{E}^2$-computable operators $\Gamma_1, \Gamma_2, \Gamma_3, \Gamma_4, \Gamma_5, \Gamma_6$ from F^{6N} into F exist such that, whenever an element $\bar{f}$ of F^{6N} is a representation of some N-tuple $\zeta_1, \ldots, \zeta_N$ of complex numbers, then the corresponding 6-tuple $\Gamma_1(\bar{f}), \Gamma_2(\bar{f}), \Gamma_3(\bar{f}), \Gamma_4(\bar{f}), \Gamma_5(\bar{f}), \Gamma_6(\bar{f})$ is a representation of the complex number $\varphi(\zeta_1, \ldots, \zeta_N)$. The following properties are evident or easily provable.[4]

1. All functions φ from $\mathbb{C}^N$ into $\mathbb{C}$ that have the form $\varphi(z_1, \ldots, z_N) = z_j$ with $j \in \{1, \ldots, N\}$ are $\mathcal{E}^2$-computable.
2. For any rational complex number γ the constant function $\varphi(z_1, \ldots, z_N) = \gamma$ is $\mathcal{E}^2$-computable.
3. The functions $\lambda z.\,\bar{z}$, $\lambda z_1 z_2.\, z_1 + z_2$ and $\lambda z_1 z_2.\, z_1 \cdot z_2$ are $\mathcal{E}^2$-computable.
4. If φ is an $\mathcal{E}^2$-computable function from $\mathbb{C}^m$ into $\mathbb{C}$, and $\psi_1, \ldots, \psi_m$ are $\mathcal{E}^2$-computable function from $\mathbb{C}^N$ into $\mathbb{C}$ then the function θ defined by

$$\theta(z_1, \ldots, z_N) = \varphi(\psi_1(z_1, \ldots, z_N), \ldots, \psi_m(z_1, \ldots, z_N))$$

   is also $\mathcal{E}^2$-computable.
5. The real-valued function $\lambda z.\, |z|^2$ is $\mathcal{E}^2$-computable.
6. For any given positive integer N, the value of a polynomial $P(z)$ of the form (1) as well as the corresponding value of $|P(z)|^2$ are $\mathcal{E}^2$-computable functions of the coefficients $\alpha_0, \alpha_1, \ldots, \alpha_{N-1}$ and the argument z.
7. For any given positive integer N, if $P(z)$ is an arbitrary polynomial of the form (1), and α is an arbitrary complex number, then the coefficients of the polynomial that is the quotient of $P(z) - P(\alpha)$ and $z - \alpha$ are $\mathcal{E}^2$-computable functions of $\alpha_0, \alpha_1, \ldots, \alpha_{N-1}$ and α.
8. If φ is an $\mathcal{E}^2$-computable function from $\mathbb{C}^{N+1}$ into $\mathbb{C}$, then there are $\mathcal{E}^2$-computable operators $\Gamma_1, \Gamma_2, \Gamma_3, \Gamma_4, \Gamma_5, \Gamma_6$ from F^{6N} into F_7 such that, whenever an element $\bar{f}$ of F^{6N} is a representation of some N-tuple $\zeta_1, \ldots, \zeta_N$ of complex numbers, then for any natural numbers $y_1, y_2, y_3, y_4, y_5, y_6$ the 6-tuple of the functions

$$\lambda u.\, \Gamma_j(\bar{f})(u, y_1, y_2, y_3, y_4, y_5, y_6), \quad j = 1, 2, 3, 4, 5, 6,$$

   is a representation of the complex number

$$\varphi\left(\zeta_1, \ldots, \zeta_N, \frac{y_1 - y_2}{y_3 + 1} + \frac{y_4 - y_5}{y_6 + 1} i\right).$$

---

[4] The property 5 can be derived from the properties 3, 4 and the equality $|z|^2 = z \cdot \bar{z}$. The properties 3, 4 and 5 imply the property 6, and it implies the property 7. The proof of the property 8 makes use of property 6 from section 2 (with $k = 1$, $l = 6$) and of the fact that for any natural numbers $y_1, y_2, y_3, y_4, y_5, y_6$ the 6-tuple of the constant functions $\lambda t.\, y_1, \lambda t.\, y_2, \lambda t.\, y_3, \lambda t.\, y_4, \lambda t.\, y_5, \lambda t.\, y_6$ is a representation of the rational complex number

$$\frac{y_1 - y_2}{y_3 + 1} + \frac{y_4 - y_5}{y_6 + 1} i.$$

Remark. Except for the case of $N = 1$, there is no $\mathcal{E}^2$-computable function φ from $\mathbb{C}^N$ into $\mathbb{C}$ such that for any complex numbers $\alpha_0, \alpha_1, \ldots, \alpha_{N-1}$ the number $\varphi(\alpha_0, \alpha_1, \ldots, \alpha_{N-1})$ is a root of the corresponding polynomial (1). This follows from the non-existence of a continuous function with such a property.

4   On Rosenbloom's Proof of the Fundamental Theorem of Algebra

P. C. Rosenbloom's proof in [4] of the fundamental theorem of algebra makes use of an analogue of Cauchy's theorem from the theory of analytic functions. In its complete form the result obtained in the proof of this analogue can be formulated as follows (see Lemma 2, Theorems 1, 2 and Corollary 1 in [4]).

THEOREM R. *Let N be a positive integer, $\alpha_0, \alpha_1, \ldots, \alpha_{N-1}$ be complex numbers, $P(z)$ be the corresponding polynomial (1), and ε be a positive real number. If*

$$A = \max\{|\alpha_0|, |\alpha_1|, \ldots, |\alpha_{N-1}|, 1\}, \quad \gamma = \binom{N+1}{[(N+1)/2]},$$

a is a real number not less than $5NA$, $K = 2^{(3N/2)+6}\gamma^3 A^3 a^{3N+3}$, and n is an integer greater than K/ε^3, then

$$\left| P\left(\frac{(u+vi)a}{n} \right) \right| < \varepsilon$$

for some integers u and v with $|u| \leq n$, $|v| \leq n$.

The further presentation in [4] goes through the following statement (its formulation here coincides with the original one up to inessential details).

LEMMA 3. *Let N be a positive integer, $\alpha_0, \alpha_1, \ldots, \alpha_{N-1}$ be complex numbers, $P(z)$ be the corresponding polynomial (1), and ε be a positive real number less than 1. Then we can find points $z_1, \ldots, z_N$ such that*

$$|P(z_j)| < \varepsilon, \quad j = 1, \ldots, N,$$

and such that if $|P(z)| < \delta$, where $\varepsilon \leq \delta < 1$, then

$$\min_{1 \leq j \leq N} |z_j - z| < 2\delta^{1/2^N}.$$

The proof of the lemma in the paper (after the elimination of a small problem[5]) can be adapted to the needs of the present paper. However, a strengthening of the lemma is possible that is more convenient for us, namely by adding

---

[5] The problem is in the induction used for actually proving a strengthening of the lemma with the factor $2^{1-1/2^N}$ in place of 2 in the last inequality. Namely the inequality $(2\delta)^{1/2} < 1$ is needed for being able to use the inductive hypothesis at the final step, but the assumption $\delta < 1$ is not sufficient for the truth of this inequality. Fortunately, as the first author observed, this problem can be eliminated by replacing the inequality $\delta < 1$ in the formulation of the lemma with the inequality $\delta < 2$.

the words "with rational coordinates" after the phrase "we can find points $z_1, \ldots, z_N$". We shall prove constructively even the slightly stronger statement with $2\delta^{1/2^{N-1}}$ instead of $2\delta^{1/2^N}$.

LEMMA 3'. *Let N be a positive integer, $\alpha_0, \alpha_1, \ldots, \alpha_{N-1}$ be complex numbers, $P(z)$ be the corresponding polynomial (1), and ε be a positive real number less than 1. Then we can find rational complex numbers $z_1, \ldots, z_N$ such that $|P(z_j)| < \varepsilon$, $j = 1, \ldots, N$, and such that if $|P(z)| < \delta$, where $\varepsilon \leq \delta < 1$, then*

$$\min_{1 \leq j \leq N} |z_j - z| < 2\delta^{1/2^{N-1}}.$$

Proof. Our reasoning will be very close to the proof of Lemma 3 in [4]. We see as there that all complex numbers z with $|P(z)| \leq 1$ satisfy the inequality

$$|z| < 1 + N \max_{0 \leq k < N} |\alpha_k|.$$

If $N = 1$ then we take a rational complex number z_1 such that $|z_1 + \alpha_0| < \varepsilon$. Clearly $|P(z_1)| < \varepsilon$, and if $|P(z)| < \delta$, where $\varepsilon \leq \delta < 1$, then

$$|z_1 - z| \leq |z_1 + \alpha_0| + |z + \alpha_0| < \varepsilon + \delta \leq 2\delta = 2\delta^{1/2^{N-1}}.$$

Suppose now $N > 1$, and the statement of Lemma 3' is true for $N - 1$. Let (as in the original proof) $\varepsilon_1 = \varepsilon/C$, where $C = 3 + NA + (N-1)NA(1 + NA)^{N-1}$, and $A = \max\{|\alpha_0|, |\alpha_1|, \ldots, |\alpha_{N-1}|, 1\}$. Clearly $\varepsilon_1 < \varepsilon$. By Theorem R (applied with some rational number a) we find a rational complex number z_1 such that $|P(z_1)| < \varepsilon_1$, hence $|P(z_1)| < \varepsilon < 1$, and therefore $|z_1| < 1 + NA$. Now

$$P(z) = P(z_1) + (z - z_1)P_1(z),$$

where

$$P_1(z) = \sum_{m=0}^{N-1} \beta_m z^m, \quad |\beta_m| = \left| \sum_{k=m+1}^{N} \alpha_k z_1^{k-m-1} \right| < NA(1 + NA)^{N-1}.$$

By the inductive assumption we can find rational complex numbers $z_2, \ldots, z_N$ such that $|P_1(z_j)| < \varepsilon_1$, $j = 2, \ldots, N$, and

$$\min_{2 \leq j \leq N} |z_j - z| < 2\delta^{1/2^{N-2}},$$

whenever $|P_1(z)| < \delta$ and $\varepsilon_1 \leq \delta < 1$. If j is any of the numbers $2, \ldots, N$, then $|P_1(z_j)| < 1$, hence $|z_j| < 1 + (N-1)NA(1 + NA)^{N-1}$, and therefore

$$|P(z_j)| \leq |P(z_1)| + (|z_j| + |z_1|)|P_1(z_j)| < \varepsilon_1 + (|z_j| + |z_1|)\varepsilon_1 < C\varepsilon_1 = \varepsilon.$$

Now let $|P(z)| < \delta$, where $\varepsilon \leq \delta < 1$. Then

$$|z_1 - z||P_1(z)| = |P(z_1) - P(z)| \leq |P(z_1)| + |P(z)| < \varepsilon + \delta \leq 2\delta \leq 2\delta^{1/2^{N-1}}\delta^{1/2},$$

hence $|z_1 - z| < 2\delta^{1/2^{N-1}}$ or $|P_1(z)| < \delta^{1/2}$. In the case of $|P_1(z)| < \delta^{1/2}$ we have the inequality

$$\min_{2 \leq j \leq N} |z_j - z| < 2(\delta^{1/2})^{1/2^{N-2}} = 2\delta^{1/2^{N-1}}$$

since $\varepsilon_1 < \delta < \delta^{1/2} < 1$. Therefore in both cases

$$\min_{1 \leq j \leq N} |z_j - z| < 2\delta^{1/2^{N-1}}. \qquad \square$$

The concluding part of Rosenbloom's proof is in his Theorem 3. Making use of Lemma 3′ instead of Lemma 3, we can strengthen Theorem 3 by constructing a sequence of rational complex numbers converging to a root of the polynomial. In the original proof Lemma 3 is applied with values of ε of the form 2^{-n2^N}, $n = 1, 2, \ldots$, and this leads to the inequality $|z_{n+1} - z_n| < 2^{1-n}$ for the members of the constructed sequence $z_1, z_2, \ldots$ Of course this can be done also through Lemma 3′, and the rate of the convergence is quite good. Unfortunately the exponential dependence of 2^{-n2^N} on n is an obstacle to realize such a construction of the sequence by means of $\mathcal{E}^2$-computable operators. Therefore it is appropriate to change the construction. Namely an inequality $|z_{n+1} - z_n| < (n+1)^{-2}$ would still give an admissible rate of convergence, and this inequality can be achieved by using values of ε of the form $2^{-2^{N-1}}(n+1)^{-2^N}$, $n = 1, 2, \ldots$ (since these values are used also as values of δ when $n + 1$ is considered instead of n, and we have $2\delta^{1/2^{N-1}} = (n+1)^{-2}$ for $\delta = 2^{-2^{N-1}}(n+1)^{-2^N}$).

5   Construction of the Needed $\mathcal{E}^2$-Computable Operators

We shall first formulate three theorems, and then we shall sketch their proofs. The first two of these theorems (corresponding to Theorem R and to Lemma 3′ from the previous section) describe the major preliminary steps in the construction of the $\mathcal{E}^2$-computable operators needed to get the promised refinement of the fundamental theorem of algebra. The third theorem is the refinement itself.

THEOREM 1. *For any positive integer N there are $\mathcal{E}^2$-computable operators $\Gamma_1, \Gamma_2, \Gamma_3, \Gamma_4, \Gamma_5, \Gamma_6$ from F^{6N} into F such that, whenever an element $\bar{f}$ of F^{6N} is a representation of an N-tuple $\alpha_0, \alpha_1, \ldots, \alpha_{N-1}$ of complex numbers, and $P(z)$ is the polynomial (1) corresponding to this N-tuple, then*

$$\left| P\left(\frac{\Gamma_1(\bar{f})(t) - \Gamma_2(\bar{f})(t)}{\Gamma_3(\bar{f})(t) + 1} + \frac{\Gamma_4(\bar{f})(t) - \Gamma_5(\bar{f})(t)}{\Gamma_6(\bar{f})(t) + 1} i \right) \right| < \frac{1}{t+1}, \quad t = 0, 1, 2, \ldots$$

THEOREM 2. *For any positive integer N there are $\mathcal{E}^2$-computable operators $\Gamma_{1j}, \Gamma_{2j}, \Gamma_{3j}, \Gamma_{4j}, \Gamma_{5j}, \Gamma_{6j}$, $j = 1, 2, \ldots, N$, from F^{6N} into F such that, whenever an element $\bar{f}$ of F^{6N} is a representation of an N-tuple $\alpha_0, \alpha_1, \ldots, \alpha_{N-1}$ of complex numbers, and $P(z)$ is the polynomial (1) corresponding to this N-tuple, then for any natural number t and*

$$z_j = \frac{\Gamma_{1j}(\bar{f})(t) - \Gamma_{2j}(\bar{f})(t)}{\Gamma_{3j}(\bar{f})(t) + 1} + \frac{\Gamma_{4j}(\bar{f})(t) - \Gamma_{5j}(\bar{f})(t)}{\Gamma_{6j}(\bar{f})(t) + 1} i, \quad j = 1, 2, \ldots, N,$$

the inequalities $|P(z_j)| < (t+1)^{-1}$, $j = 1, 2, \ldots, N$, *hold, and*

$$\min_{1 \leq j \leq N} |z_j - z| < 2\delta^{1/2^{N-1}}$$

for all δ *and* z *satisfying the inequalities* $(t+1)^{-1} \leq \delta < 1$, $|P(z)| < \delta$.

THEOREM 3. *For any positive integer* N *there are* $\mathcal{E}^2$-*computable operators* $\Gamma_1, \Gamma_2, \Gamma_3, \Gamma_4, \Gamma_5, \Gamma_6$ *from* F^{6N} *into* F *such that, whenever an element* $\bar{f}$ *of* F^{6N} *is a representation of an* N-*tuple* $\alpha_0, \alpha_1, \ldots, \alpha_{N-1}$ *of complex numbers, and* $P(z)$ *is the polynomial (1) corresponding to this* N-*tuple, then the 6-tuple of the functions* $\Gamma_1(\bar{f}), \Gamma_2(\bar{f}), \Gamma_3(\bar{f}), \Gamma_4(\bar{f}), \Gamma_5(\bar{f}), \Gamma_6(\bar{f})$ *is a representation of some root of* $P(z)$.

The proof of Theorem 1 is based on the statement of Theorem R and does not use any details from its proof. Let N be a positive integer. One easily constructs $\mathcal{E}^2$-computable operators Δ_1 and Δ_2 from F^{6N} into F_0 such that, whenever an element $\bar{f}$ of F^{6N} is a representation of an N-tuple $\alpha_0, \alpha_1, \ldots, \alpha_{N-1}$ of complex numbers, and $P(z)$ is the polynomial (1) corresponding to this N-tuple, the natural number $\Delta_1(\bar{f})$ is not less than the number $5NA$ from Theorem R for the given numbers $N, \alpha_0, \alpha_1, \ldots, \alpha_{N-1}$, and the natural number $\Delta_2(\bar{f})$ is not less than the number K for the given numbers $N, \alpha_0, \alpha_1, \ldots, \alpha_{N-1}$ and for $a = \Delta_1(\bar{f})$. In this situation, if t is an arbitrary natural number then an application of Theorem R with $\varepsilon = \frac{1}{2(t+1)}$, $a = \Delta_1(\bar{f})$, $n = 8(t+1)^3 \Delta_2(\bar{f}) + 1$ and with the substitution $u = r - n$, $v = s - n$ allows concluding that

$$\left| P\left(\frac{r\Delta_1(\bar{f}) - \Delta_1'(\bar{f})(t)}{\Delta_2'(\bar{f})(t) + 1} + \frac{s\Delta_1(\bar{f}) - \Delta_1'(\bar{f})(t)}{\Delta_2'(\bar{f})(t) + 1} i \right) \right|^2 < \frac{1}{4(t+1)^2} \qquad (2)$$

for some natural numbers r and s not greater than $\Delta_2'(\bar{f})(t)$, where Δ_1' and Δ_2' are the mappings of F^{6N} into F defined by

$$\Delta_2'(\bar{f})(t) = 8(t+1)^3 \Delta_2(\bar{f}), \quad \Delta_1'(\bar{f})(t) = (\Delta_2'(\bar{f})(t) + 1)\Delta_1(\bar{f})$$

(clearly Δ_1' and Δ_2' are also $\mathcal{E}^2$-computable operators). By the properties 6 and 8 from Section 3, there are $\mathcal{E}^2$-computable operators Γ and Δ from F^{6N} to F_4 such that, whenever an element $\bar{f}$ of F^{6N} is a representation of an N-tuple $\alpha_0, \alpha_1, \ldots, \alpha_{N-1}$ of complex numbers, and $P(z)$ is the polynomial (1) corresponding to this N-tuple, then the absolute value of the difference

$$\frac{\Gamma(\bar{f})(u, r, s, t)}{\Delta(\bar{f})(u, r, s, t) + 1} - \left| P\left(\frac{r\Delta_1(\bar{f}) - \Delta_1'(\bar{f})(t)}{\Delta_2'(\bar{f})(t) + 1} + \frac{s\Delta_1(\bar{f}) - \Delta_1'(\bar{f})(t)}{\Delta_2'(\bar{f})(t) + 1} i \right) \right|^2$$

is not greater than $(u+1)^{-1}$ for any r, s, t, u in $\mathbb{N}$. With $u = 4(t+1)^2 - 1$ we get that

$$\frac{\Gamma(\bar{f})(4(t+1)^2 - 1, r, s, t)}{\Delta(\bar{f})(4(t+1)^2 - 1, r, s, t) + 1} < \frac{1}{2(t+1)^2} \qquad (3)$$

for all natural numbers r, s, t satisfying the inequality (2), and

$$\left| P\left(\frac{r\Delta_1(\bar{f}) - \Delta_1'(\bar{f})(t)}{\Delta_2'(\bar{f})(t) + 1} + \frac{s\Delta_1(\bar{f}) - \Delta_1'(\bar{f})(t)}{\Delta_2'(\bar{f})(t) + 1} i \right) \right|^2 < \frac{3}{4(t+1)^2} < \frac{1}{(t+1)^2}$$

for any r, s, t in $\mathbb{N}$ that satisfy (3). It is a routine work (making use of the property 5 from Section 2) to construct two $\mathcal{E}^2$-computable operators Δ_3 and Δ_4 from F^{6N} to F such that $\Delta_3(\bar{f})$ and $\Delta_4(\bar{f})$ transform any natural number t into natural numbers r and s not greater than $\Delta_2'(\bar{f})(t)$ and satisfying (3), whenever such r and s exist. Then Theorem 1 will hold with $\Gamma_2 = \Gamma_5 = \Delta_1'$, $\Gamma_3 = \Gamma_6 = \Delta_2'$ and $\Gamma_1(\bar{f})(t) = \Delta_3(\bar{f})(t)\Delta_1(\bar{f})$, $\Gamma_4(\bar{f})(t) = \Delta_4(\bar{f})(t)\Delta_1(\bar{f})$.

The proof of Theorem 2 is actually an operator refinement of the one of Lemma 3' and follows closely it. In the case of $N = 1$ we, roughly speaking, use rational approximations of the number $-\alpha_0$ that can be constructed by means of the representation $\bar{f}$ of α_0. For the inductive step, we suppose the existence of the needed $6(N - 1)$-tuple of $\mathcal{E}^2$-computable operators for the case of polynomials of degree $N - 1$, and use them to construct the needed $6N$-tuple of ones for polynomials of degree N, making use also of the $\mathcal{E}^2$-computable operators from Theorem 1 for this case and of the properties 7 and 8 from Section 3.

Of course the operators $\Gamma_{1j}, \Gamma_{2j}, \Gamma_{3j}, \Gamma_{4j}, \Gamma_{5j}, \Gamma_{6j}$, $j = 1, 2, \ldots, N$, from Theorem 2 are used in the proof of Theorem 3. For any element $\bar{f}$ of F^{6N} and any natural number n we set

$$\gamma_{n,j}^{\bar{f}} = \frac{\Gamma_{1j}(\bar{f})(t_n) - \Gamma_{2j}(\bar{f})(t_n)}{\Gamma_{3j}(\bar{f})(t_n) + 1} + \frac{\Gamma_{4j}(\bar{f})(t_n) - \Gamma_{5j}(\bar{f})(t_n)}{\Gamma_{6j}(\bar{f})(t_n) + 1} i, \quad j = 1, 2, \ldots N, \quad (4)$$

where $t_n = 2^{2^{N-1}}(n + 1)^{2^N} - 1$. Then, for any element $\bar{f}$ of F^{6N}, we define a sequence $j_0^{\bar{f}}, j_1^{\bar{f}}, j_2^{\bar{f}}, \ldots$ of integers from the set $\{1, 2, \ldots, N\}$ in the following recursive way: we set $j_0^{\bar{f}} = 1$ and, whenever $j_n^{\bar{f}}$ is already defined, we set $j_{n+1}^{\bar{f}}$ to be the first $j \in \{1, 2, \ldots, N\}$ such that $j = N$ or $\left| \gamma_{n+1,j}^{\bar{f}} - \gamma_{n,j_n^{\bar{f}}}^{\bar{f}} \right| < (n + 1)^{-2}$. Finally, we set

$$\Gamma_k(\bar{f})(n) = \Gamma_{k j_n^{\bar{f}}}(\bar{f})(t_n), \quad k = 1, 2, 3, 4, 5, 6, \ n = 0, 1, 2, \ldots \quad (5)$$

The $\mathcal{E}^2$-computability of the constructed operators is easily verifiable, thus it remains only to prove the other property formulated in Theorem 3. Let an element $\bar{f}$ of F^{6N} be a representation of an N-tuple $\alpha_0, \alpha_1, \ldots, \alpha_{N-1}$ of complex numbers, and $P(z)$ be the polynomial (1) corresponding to this N-tuple. Then for any natural number n the inequalities $|P(\gamma_{n,j}^{\bar{f}})| < (t_n + 1)^{-1}$, $j = 1, 2, \ldots, N$, hold, and whenever $|P(z)| < \delta$, where $(t_{n+1} + 1)^{-1} \le \delta < 1$, then

$$\min_{1 \le j \le N} |\gamma_{n+1,j}^{\bar{f}} - z| < 2\delta^{1/2^{N-1}}.$$

By the first part of this statement $\lim_{n \to \infty} P(\gamma_{n,j_n^{\bar{f}}}^{\bar{f}}) = 0$. Applying the second one with $z = \gamma_{n,j_n^{\bar{f}}}^{\bar{f}}$, $\delta = (t_n + 1)^{-1}$, we see that $|\gamma_{n+1,j_{n+1}^{\bar{f}}}^{\bar{f}} - \gamma_{n,j_n^{\bar{f}}}^{\bar{f}}| < (n + 1)^{-2}$, since

$2\delta^{1/2^{N-1}} = (n+1)^{-2}$ for this value of δ. To complete the proof, it is sufficient to use the equality (4) with $j = j_n^{\bar{f}}$, as well as the equality (5) and the inequality $(n+1)^{-2} + (n+2)^{-2} + \cdots + (n+p)^{-2} \leq (n+1)^{-1}$.

6 Some Corollaries from Theorem 3

By induction on N one easily proves

COROLLARY 1. *For any positive integer N there are $\mathcal{E}^2$-computable operators $\Gamma_{1j}, \Gamma_{2j}, \Gamma_{3j}, \Gamma_{4j}, \Gamma_{5j}, \Gamma_{6j}$, $j = 1, 2, \ldots, N$, from F^{6N} into F such that, whenever an element $\bar{f}$ of F^{6N} is a representation of an N-tuple of complex numbers $\alpha_0, \alpha_1, \ldots, \alpha_{N-1}$, and $P(z)$ is the polynomial (1) corresponding to this N-tuple, then the N-tuples $\Gamma_{1j}(\bar{f}), \Gamma_{2j}(\bar{f}), \Gamma_{3j}(\bar{f}), \Gamma_{4j}(\bar{f}), \Gamma_{5j}(\bar{f}), \Gamma_{6j}(\bar{f})$, $j = 1, 2, \ldots, N$, are representations of some complex numbers $z_1, z_2, \ldots, z_N$ with the property that for all z the equality $P(z) = (z - z_1)(z - z_2) \cdots (z - z_N)$ holds.*

This implies the following statement (derivable also from Theorem 2.5 of [5]).

COROLLARY 2. *If an N-tuple of complex numbers $\alpha_0, \alpha_1, \ldots, \alpha_{N-1}$ has a representation consisting of functions from Grzegorczyk class $\mathcal{E}^m$, where $m \geq 2$, then any root of the corresponding polynomial (1) has a representation consisting of functions from the same class $\mathcal{E}^m$.*

Remark. Neither of the indicated two proofs of Corollary 2 yields an interpretation of the existential statement in the conclusion via recursive operators using as input an $\mathcal{E}^m$-representation of the sequence $\alpha_0, \alpha_1, \ldots, \alpha_{N-1}$ and an arbitrary representation of the considered root. The existence of such operators (even of $\mathcal{E}^2$-computable ones) was additionally shown by the second author.

7 Acknowledgments

Thanks are due to members of the Mathematical Logic Department at Sofia University and to anonymous referees for their useful suggestions.

References

1. A. Grzegorczyk. Some Classes of Recursive Functions. Dissertationes Math. (Rozprawy Mat.), **4**, Warsaw, 1953.
2. A. Grzegorczyk. Computable functionals. Fund. Math., **42**, 1955, 168–202.
3. P. Peshev. A subrecursive refinement of the fundamental theorem of algebra. Sofia University, Sofia, 2005 (master thesis, in Bulgarian).
4. P. C. Rosenbloom. An elementary constructive proof of the fundamental theorem of algebra. The Amer. Math. Monthly, **52**, 1945, no. 10, 562–570.
5. D. Skordev. Computability of real numbers by using a given class of functions in the set of the natural numbers. Math. Logic Quarterly, **48**, 2002, Suppl. 1, 91–106.
6. K. Weihrauch. Computable Analysis. An Introduction. Springer–Verlag, Berlin/Heidelberg, 2000.

An Introduction to
Program and Thread Algebra

Alban Ponse and Mark B. van der Zwaag

Programming Research Group, Informatics Institute, University of Amsterdam
Kruislaan 403, 1098 SJ Amsterdam, The Netherlands
alban@science.uva.nl, mbz@science.uva.nl

Abstract. We provide an introduction to Program Algebra (PGA, an algebraic approach to the modeling of sequential programming) and to Thread Algebra (TA). PGA is used as a basis for several low- and higher-level programming languages. As an example we consider a simple language with *goto*'s. Threads in TA model the execution of programs. Threads may be composed with services which model (part of) the execution environment, such as a stack. Finally, we discuss briefly the expressiveness of PGA and allude to current work on multithreading and security hazard risk assessment.

Keywords: PGA, Program Algebra, Thread Algebra.

1 Introduction

In this paper we report on a recent line of programming research conducted at the University of Amsterdam. This research comprises *program algebra* and *thread algebra*. A first major publication about this project is [7] (2002).

Program algebra (PGA, for ProGram Algebra) provides a rigid framework for the understanding of imperative sequential programming. Starting point is the perception of a *program object* as a possibly infinite sequence of primitive instructions. PGA programs are composed from primitive instructions and two operators: sequential composition and iteration. Based on this, a family of programming languages is built, containing well-known constructs such as labels and goto's, conditionals and while-loops, etc. These languages are defined with a projection to PGA which defines the program object described by a program expression.

Execution of a program object is single-pass: the instructions are visited in order and are dropped after having been executed. Execution of a basic instruction or test is interpreted as a request to the execution environment: the environment processes the request and replies with a Boolean value. This has lead to the modeling of the behavior of program objects as threads, i.e., as elements of Thread Algebra (TA). The primary operation of TA is postconditional composition:

$$P \trianglelefteq a \trianglerighteq Q$$

stands for the execution of action a which is followed by execution of P if **true** is returned and by execution of Q if **false** is returned. Threads can be composed with services which model (part of) the environment.

A. Beckmann et al. (Eds.): CiE 2006, LNCS 3988, pp. 445–458, 2006.

In Section 2 we present PGA, and in Section 3 we overview thread algebra and the interpretation of programs as threads. Then, in Section 4 we go into the expressiveness of PGA. Finally, in Section 5 we allude briefly to current work on multithreading and security hazard risk assessment. For further discussion on the *why*'s and *why not*'s of PGA, see [2], and of TA, see [3].

2   Program Algebra

Program Algebra (PGA) is based on a parameter set A. The *primitive instructions* of PGA are the following:

Basic instruction. All elements of A, written, typically, as $a, b, \ldots$ are *basic instructions*. These are regarded as indivisible units and execute in finite time. Furthermore, a basic instruction is viewed as a request to the environment, and it is assumed that upon its execution a boolean value (**true** or **false**) is returned that may be used for subsequent program control. The associated behavior may modify a state.

Termination instruction. The termination instruction ! yields termination of the program. It does not modify a state, and it does not return a boolean value.

Test instruction. For each element a of A there is a *positive test* instruction $+a$ and a *negative test* instruction $-a$. When a positive test is executed, the state is affected according to a, and in case **true** is returned, the remaining sequence of actions is performed. If there are no remaining instructions, inaction occurs. In the case that **false** is returned, the next instruction is skipped and execution proceeds with the instruction following the skipped one. If no such instruction exists, inaction occurs. Execution of a negative test is the same, except that the roles of **true** and **false** are interchanged.

Forward jump instruction. For any natural number k, the instruction $\#k$ denotes a jump of length k and k is called the counter of this instruction. If $k = 0$, this jump is to the instruction itself and inaction occurs (one can say that $\#0$ *defines* divergence, which is a particular form of inaction). If $k = 1$, the instruction skips itself, and execution proceeds with the subsequent instruction if available, otherwise inaction occurs. If $k > 1$, the instruction $\#k$ skips itself and the subsequent $k - 1$ instructions. If there are not that many instructions left in the remaining part of the program, inaction occurs.

PGA *program terms* are defined inductively as follows:

1. Primitive instructions are program terms.
2. If X and Y are program terms, then $X; Y$, called the *concatenation* of X and Y, is a program term.
3. If X is a program term, then X^ω (the *repetition* of X) is a program term.

2.1   Instruction Sequence Congruence and Canonical Forms

On PGA, different types of equality can be discerned, the most simple of which is *instruction sequence congruence*, identifying programs that execute identical sequences of instructions. Such a sequence is further called a *program object*. For programs not containing repetition, instruction sequence congruence boils down to the associativity of concatenation, and is axiomatized by

$$(X;Y);Z = X;(Y;Z). \tag{PGA1}$$

As a consequence, brackets are not meaningful in repeated concatenations and will be left out.

Now let $X^1 = X$ and $X^{n+1} = X; X^n$ for $n > 0$. Then instruction sequence congruence for infinite program objects is further axiomatized by the following axioms (schemes):

$$(X^n)^\omega = X^\omega, \tag{PGA2}$$
$$X^\omega; Y = X^\omega, \tag{PGA3}$$
$$(X;Y)^\omega = X;(Y;X)^\omega. \tag{PGA4}$$

It is straightforward to derive from PGA2–4 the unfolding identity of repetition:

$$X^\omega = (X;X)^\omega = X;(X;X)^\omega = X;X^\omega.$$

Instruction sequence congruence is decidable [7].

Every PGA program can be rewritten into one of the following forms:

1. X not containing repetition, or
2. $X;Y^\omega$, with X and Y not containing repetition.

Any program in one of the two above forms is said to be in *first canonical form*. For each PGA program there is a program in first canonical form that is instruction sequence congruent [7]. Canonical forms are useful as input for further transformations.

2.2   Structural Congruence and Second Canonical Forms

PGA programs in first canonical form can be converted into *second canonical form*: a first canonical form in which no chained jumps occur, i.e., jumps to jump instructions (apart from #0), and in which each non-chaining jump into the repeating part is minimized. The associated congruence is called *structural congruence* and is axiomatized by PGA1–4 presented above, plus the following axiom schemes, where the u_i and v_i range over primitive instructions:

$$\#n+1; u_1; \ldots; u_n; \#0 = \#0; u_1; \ldots; u_n; \#0, \tag{PGA5}$$
$$\#n+1; u_1; \ldots; u_n; \#m = \#n+m+1; u_1; \ldots; u_n; \#m, \tag{PGA6}$$
$$(\#k+n+1; u_1; \ldots; u_n)^\omega = (\#k; u_1; \ldots; u_n)^\omega, \tag{PGA7}$$

and

$$\#n+m+k+2; u_1; \ldots; u_n; (v_1; \ldots; v_{m+1})^\omega = $$
$$\#n+k+1; u_1; \ldots; u_n; (v_1; \ldots; v_{m+1})^\omega. \quad \text{(PGA8)}$$

Two examples, of which the right-hand sides are in second canonical form:

$$\#2; a; (\#5; b; +c)^\omega = \#4; a; (\#2; b; +c)^\omega,$$
$$+a; \#2; (b; \#2; -c; \#2)^\omega = +a; \#0; (+b; \#0; -c; \#0)^\omega.$$

Second canonical forms are not unique. However, if in $X; Y^\omega$ the number of instructions in X and Y is minimized, they are. In the first example above, the right-hand side is the unique minimal second canonical form; for the second example it is $+a; (\#0; +b; \#0; -c)^\omega$.

2.3   PGA-Based Languages

On the basis of PGA, a family of programming languages has been developed [7]. The programming constructs in these languages include backward jumps, absolute jumps, labels and goto's, conditionals and while loops, etc. All of these languages are given a projection semantics, that is, they come with a translation to PGA which determines their semantics (together with the semantics of PGA, see Section 3.1). Vice versa, PGA can be embedded in each of these languages, which shows that they share the same expressiveness. As an example we present the language PGLDg and its projection semantics.[1]

PGLDg is a program notation with label and goto instructions as primitives instead of jumps. Repetition is not available, so a PGLDg program is just a finite sequence of instructions. In PGLDg termination takes place when the last instruction has been executed, when a goto to a non-existing label is made, or when a termination instruction ! is executed.

A label in PGLDg is just a natural number. Label and goto-instructions are defined as follows:

Label instruction. The instruction $£k$, for k a natural number, represents a visible label. As an action it is a skip in the sense that it will not have any effect on a state space.

Goto instruction. For each natural number k the instruction $\#\#£k$ represents a jump to the (beginning of) the first (i.e. the left-most) instruction $£k$ in the program. If no such instruction can be found termination of the program execution will occur.

An example of a PGLDg program is: $£0; -a; \#\#£1; \#\#£0; £1$. In this program a is repeated until it yields reply value false. That is also the functionality of the simpler program $£0; +a; \#\#£0$.

---
[1] The languages presented in [7] are called PGLA, PGLB, PGLC, etc.

A projection from PGLDg to PGA works as follows:

$$\texttt{pgldg2pga}(u_1; \ldots; u_k) = (\psi_1(u_1); \ldots; \psi_k(u_k); !; !)^\omega,$$

where the u_i range over primitive instructions, the two added termination instructions serve the case that u_k is a test-instruction, and the auxiliary functions ψ_j are defined as follows:

$$\psi_j(\#\#\pounds n) = \begin{cases} ! & \text{if } \texttt{target}(n) = 0, \\ \#\texttt{target}(n){-}j & \text{if } \texttt{target}(n) \geq j, \\ \#k{+}2{-}j{+}\texttt{target}(n) & \text{otherwise,} \end{cases}$$

$$\psi_j(\pounds n) = \#1,$$

$$\psi_j(u) = u \quad \text{otherwise.}$$

The auxiliary function $\texttt{target}(k)$ produces for k the smallest number j such that the j-th instruction of the program is of the form $\pounds k$, if such a number exists and 0 otherwise. Projecting the two example programs above yields

$$\texttt{pgldg2pga}(\pounds 0; -a; \#\#\pounds 1; \#\#\pounds 0; \pounds 1) = (\#1; -a; \#2; \#4; \#1; !; !)^\omega,$$

$$\texttt{pgldg2pga}(\pounds 0; +a; \#\#\pounds 0) = (\#1; +a; \#3; !; !)^\omega.$$

The projection $\texttt{pgldg2pga}$ results from the composition of a number of projections defined in [7] and a tiny bit of smart reasoning.

3   Basic Thread Algebra

Basic Thread Algebra (BTA) is a form of process algebra which is tailored to the description of sequential program behavior. Based on a set A of *actions*, it has the following constants and operators:

- the *termination* constant S,
- the *deadlock* or *inaction* constant D,
- for each $a \in A$, a binary *postconditional composition* operator $\_ \trianglelefteq a \trianglerighteq \_$.

We use *action prefixing* $a \circ P$ as an abbreviation for $P \trianglelefteq a \trianglerighteq P$ and take $\circ$ to bind strongest. Furthermore, for $n \geq 1$ we define $a^n \circ P$ by $a^1 \circ P = a \circ P$ and $a^{n+1} \circ P = a \circ (a^n \circ P)$.

The operational intuition is that each action represents a command which is to be processed by the execution environment of the thread. The processing of a command may involve a change of state of this environment.[2] At completion of the processing of the command, the environment produces a reply value true or

---

[2] For the definition of threads we completely abstract from the environment. In Section 3.2 we define services which model (part of) the environment, and thread-service composition.

`false`. The thread $P \trianglelefteq a \trianglerighteq Q$ proceeds as P if the processing of a yields `true`, and it proceeds as Q if the processing of a yields `false`.

Every thread in BTA is finite in the sense that there is a finite upper bound to the number of consecutive actions it can perform. The *approximation operator* $\pi : \mathbb{N} \times \text{BTA} \rightarrow \text{BTA}$ gives the behavior up to a specified depth. It is defined by

1. $\pi(0, P) = \mathsf{D}$,
2. $\pi(n + 1, \mathsf{S}) = \mathsf{S}, \ \pi(n + 1, \mathsf{D}) = \mathsf{D}$,
3. $\pi(n + 1, P \trianglelefteq a \trianglerighteq Q) = \pi(n, P) \trianglelefteq a \trianglerighteq \pi(n, Q)$,

for $P, Q \in \text{BTA}$ and $n \in \mathbb{N}$. We further write $\pi_n(P)$ instead of $\pi(n, P)$. We find that for every $P \in \text{BTA}$, there exists an $n \in \mathbb{N}$ such that

$$\pi_n(P) = \pi_{n+1}(P) = \cdots = P.$$

Following the metric theory of [1] in the form developed as the basis of the introduction of processes in [6], BTA has a completion BTA^∞ which comprises also the infinite threads. Standard properties of the completion technique yield that we may take BTA^∞ as the cpo consisting of all so-called *projective* sequences:[3]

$$\text{BTA}^\infty = \{(P_n)_{n \in \mathbb{N}} \mid \forall n \in \mathbb{N} \ (P_n \in \text{BTA} \ \& \ \pi_n(P_{n+1}) = P_n)\}.$$

For a detailed account of this construction see [4] or [19].

Overloading notation, we now define the constants and operators of BTA on BTA^∞:

1. $\mathsf{D} = (\mathsf{D}, \mathsf{D}, \ldots)$ and $\mathsf{S} = (\mathsf{D}, \mathsf{S}, \mathsf{S}, \ldots)$;
2. $(P_n)_{n \in \mathbb{N}} \trianglelefteq a \trianglerighteq (Q_n)_{n \in \mathbb{N}} = (R_n)_{n \in \mathbb{N}}$ with $R_0 = \mathsf{D}$ and $R_{n+1} = P_n \trianglelefteq a \trianglerighteq Q_n$;
3. $\pi_n((P_m)_{m \in \mathbb{N}}) = (P_0, \ldots, P_{n-1}, P_n, P_n, P_n \ldots)$.

The elements of BTA are included in BTA^∞ by a mapping following this definition. It is not difficult to show that the projective sequence of $P \in \text{BTA}$ thus defined equals $(\pi_n(P))_{n \in \mathbb{N}}$. We further use this inclusion of finite threads in BTA^∞ implicitly and write $P, Q, \ldots$ to denote elements of BTA^∞.

We define the set $Res(P)$ of *residual threads* of P inductively as follows:

1. $P \in Res(P)$,
2. $Q \trianglelefteq a \trianglerighteq R \in Res(P)$ implies $Q \in Res(P)$ and $R \in Res(P)$.

A residual thread may be reached (depending on the execution environment) by performing zero or more actions. A thread P is *regular* if $Res(P)$ is finite.

A finite linear recursive specification over BTA^∞ is a set of equations

$$x_i = t_i$$

for $i \in I$ with I some finite index set, variables x_i, and all t_i terms of the form S, D, or $x_j \trianglelefteq a \trianglerighteq x_k$ with $j, k \in I$. Finite linear recursive specifications

---

[3] The cpo is based on the partial ordering $\sqsubseteq$ defined by $\mathsf{D} \sqsubseteq P$, and $P \sqsubseteq P', Q \sqsubseteq Q'$ implies $P \trianglelefteq a \trianglerighteq Q \sqsubseteq P' \trianglelefteq a \trianglerighteq Q'$.

represent continuous operators having unique fixed points [19]. In reasoning with finite linear specifications, we shall identify variables and their fixed points. For example, we say that P *is* the thread defined by $P = a \circ P$ instead of stating that P equals the fixed point for x in the finite linear specification $x = a \circ x$.

Theorem 1. *For all $P \in \mathrm{BTA}^\infty$, P is regular iff P is the solution of a finite linear recursive specification.*

The proof is easy:

Proof. $\Rightarrow$: Suppose P is regular. Then $Res(P)$ is finite, so P has residual threads $P_1, \ldots, P_n$ with $P = P_1$. We construct a linear specification with variables $x_1, \ldots, x_n$ as follows:

$$
x_i = \begin{cases}
\mathsf{D} & \text{if } P_i = \mathsf{D}, \\
\mathsf{S} & \text{if } P_i = \mathsf{S}, \\
x_j \trianglelefteq a \trianglerighteq x_k & \text{if } P_i = P_j \trianglelefteq a \trianglerighteq P_k.
\end{cases}
$$

$\Leftarrow$: Assume that P is the solution of a finite linear recursive specification. Because the variables in a finite linear specification have unique fixed points, we know that there are threads $P_1, \ldots, P_n \in \mathrm{BTA}^\infty$ with $P = P_1$, and for every $i \in \{1, \ldots, n\}$, either $P_i = \mathsf{D}$, $P_i = \mathsf{S}$, or $P_i = P_j \trianglelefteq a \trianglerighteq P_k$ for some $j, k \in \{1, \ldots, n\}$. We find that $Q \in Res(P)$ iff $Q = P_i$ for some $i \in \{1, \ldots, n\}$. So $Res(P)$ is finite, and P is regular. $\qquad\square$

Example 1. The regular threads $a^n \circ \mathsf{D}$, $a^n \circ \mathsf{S}$, and $a^\infty = a \circ a \circ \cdots$ are the respective fixed points for x_1 in the specifications

1. $x_1 = a \circ x_2, \ldots, x_n = a \circ x_{n+1}, x_{n+1} = \mathsf{D}$,
2. $x_1 = a \circ x_2, \ldots, x_n = a \circ x_{n+1}, x_{n+1} = \mathsf{S}$,
3. $x_1 = a \circ x_1$.

3.1   Extraction of Threads from Programs

The *thread extraction operator* $|\_|$ assigns a thread to a program. This thread models the behavior of the program. Note that the resulting behavioral equivalence is not a congruence: from $|X|$ equals $|Y|$, one cannot infer that, e.g., $|X; Z|$ equals $|Y; Z|$.

Thread extraction on PGA, notation $|X|$ with X a PGA program, is defined by the following thirteen equations (where a ranges over the basic instructions, and u over the primitive instructions):[4]

$$
\begin{aligned}
|a| &= a \circ \mathsf{D} & |!| &= \mathsf{S} \\
|+a| &= a \circ \mathsf{D} & |!; X| &= \mathsf{S} \\
|-a| &= a \circ \mathsf{D} & |\#k| &= \mathsf{D} \\
|a; X| &= a \circ |X| & |\#0; X| &= \mathsf{D}
\end{aligned}
$$

---

[4] We generally consider PGA programs modulo instruction sequence congruence, i.e., as program objects, so $|X^\omega| = |X; X^\omega|$.

$$|+a; X| = |X| \trianglelefteq a \trianglerighteq |\#2; X| \qquad\qquad |\#1; X| = |X|$$
$$|-a; X| = |\#2; X| \trianglelefteq a \trianglerighteq |X| \qquad\qquad |\#k + 2; u| = \mathsf{D}$$
$$|\#k + 2; u; X| = |\#k + 1; X|$$

Observe that we interpret basic instructions as actions.

For PGA programs in second canonical form, these equations yield either finite threads, or regular threads (in the case that a non-empty loop occurs, which can be captured by a system of recursive equations).

Example 2. Computation of $Q = |a; (+b; \#2; \#3; c; \#4; +d; !; a)^\omega|$ yields the following regular thread:[5]

$$Q = a \circ R, \quad R = c \circ R \trianglelefteq b \trianglerighteq (S \trianglelefteq d \trianglerighteq Q).$$

This thread can be depicted as follows:

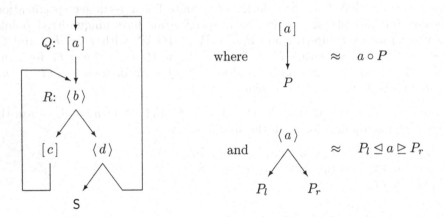

Example 3. Observe that thread extraction following the equations does not terminate for the program term

$$+a; \#2; (+b; \#2; -c; \#2)^\omega.$$

However, thread extraction on its second canonical form $+a; (\#0; +b; \#0; -c)^\omega$ yields the thread P defined by

$$P = \mathsf{D} \trianglelefteq a \trianglerighteq Q, \quad Q = \mathsf{D} \trianglelefteq b \trianglerighteq (Q \trianglelefteq c \trianglerighteq \mathsf{D}).$$

Any PGA program defines a regular thread, and conversely, every regular thread can be defined in PGA, see Section 4. Behavioral equivalence is decidable for PGA programs [7].

---

[5] Note that a *linear* recursive specification of Q requires (at least) five equations.

3.2   Services

A *service*, or a *state machine*, is a pair $\langle \Sigma, F \rangle$ consisting of a set Σ of so-called *co-actions* and a reply function F. The reply function is a mapping that gives for each non-empty finite sequence of co-actions from Σ a reply true or false.

Example 4. A *stack* can be defined as a service with co-actions *push:i*, *topeq:i*, and *pop*, for $i = 1, \ldots, n$ for some n, where *push:i* pushes i onto the stack and yields true, the action *topeq:i* tests whether i is on top of the stack, and *pop* pops the stack with reply true if it is non-empty, and it yields false otherwise.

Services model (part of) the execution environment of threads. In order to define the interaction between a thread and a service, we let actions be of the form $c.m$ where c is the so-called *channel* or *focus*, and m is the co-action or *method*. For example, we write $s.pop$ to denote the action which pops a stack via channel s. For service $\mathcal{H} = \langle \Sigma, F \rangle$ and thread P, $P /_c \mathcal{H}$ represents P *using the service* $\mathcal{H}$ via channel c. The defining rules are:

$$S /_c \mathcal{H} = S,$$
$$D /_c \mathcal{H} = D,$$
$$(P \unlhd c'.m \unrhd Q) /_c \mathcal{H} = (P /_c \mathcal{H}) \unlhd c'.m \unrhd (Q /_c \mathcal{H}) \quad \text{if } c' \neq c,$$
$$(P \unlhd c.m \unrhd Q) /_c \mathcal{H} = P /_c \mathcal{H}' \quad \text{if } m \in \Sigma \text{ and } F(m) = \text{true},$$
$$(P \unlhd c.m \unrhd Q) /_c \mathcal{H} = Q /_c \mathcal{H}' \quad \text{if } m \in \Sigma \text{ and } F(m) = \text{false},$$
$$(P \unlhd c.m \unrhd Q) /_c \mathcal{H} = D \quad \text{if } m \notin \Sigma,$$

where $\mathcal{H}' = \langle \Sigma, F' \rangle$ with $F'(\sigma) = F(m\sigma)$ for all co-action sequences $\sigma \in \Sigma^+$.

In the next example we show that the use of services may turn regular threads into non-regular ones.

Example 5. We define a thread using a stack as defined in Example 4. We only push the value 1 (so the stack behaves as a counter), and write $S(n)$ for a stack holding n times the value 1. By the defining equations for the use operator it follows that for any thread P,

$$(s.push:1 \circ P) /_s S(n) = P /_s S(n{+}1),$$
$$(P \unlhd s.pop \unrhd S) /_s S(0) = S,$$
$$(P \unlhd s.pop \unrhd S) /_s S(n{+}1) = P /_s S(n).$$

Now consider the regular thread Q defined by

$$Q = s.push:1 \circ Q \unlhd a \unrhd R, \quad R = b \circ R \unlhd s.pop \unrhd S,$$

where actions a and b do not use focus s. Then, for all $n \in \mathbb{N}$,

$$Q /_s S(n) = (s.push:1 \circ Q \unlhd a \unrhd R) /_s S(n)$$
$$= (Q /_s S(n{+}1)) \unlhd a \unrhd (R /_s S(n)).$$

It is not hard to see that $Q/_s S(0)$ is an infinite thread with the property that for all n, a trace of $n+1$ a-actions produced by n positive and one negative reply on a is followed by $b^n \circ S$. This yields an *non-regular* thread: if $Q/_s S(0)$ were regular, it would be a fixed point of some finite linear recursive specification, say with k equations. But specifying a trace $b^k \circ S$ already requires $k+1$ linear equations $x_1 = b \circ x_2, \ldots, x_k = b \circ x_{k+1}, x_{k+1} = S$, which contradicts the assumption. So $Q/_s S(0)$ is not regular.

3.3   Classes of Threads

We shall see in Section 4 that *finite* threads (the elements of BTA) correspond exactly to the threads that can be expressed in PGA without iteration, and that *regular* threads (threads definable by finite linear specifications) are exactly those that can be expressed in PGA. Equality is decidable for regular threads [7]. We mention two classes of non-regular threads: *pushdown* threads and *computable* threads. In both cases the non-regularity can be obtained by composing regular threads with certain services.

We call a regular thread that uses a stack as described in Example 4 a *pushdown* thread. In Example 5 we have seen that a pushdown thread may be non-regular. Equality is decidable for pushdown threads, but inclusion (the ordering $\sqsubseteq$ defined in Section 3) is not [5].[6]

Finally, a thread is *computable* if it can be represented by an identifier P_0 and two computable functions f and g as follows ($k \in \mathbb{N}$):

$$P_k = \begin{cases} D & \text{if } g(k) = 0, \\ S & \text{if } g(k) = 1, \\ P_{\langle k+f(k),1 \rangle} \unlhd a_{g(k)} \unrhd P_{\langle k+f(k),2 \rangle} & \text{if } g(k) > 1, \end{cases}$$

where $\langle \_, \_ \rangle$ is a bijective, computable pairing function.

Obviously computable threads can, in general, not be expressed by PGA programs. However, infinite sequences of primitive PGA instructions are universal: for every computable thread P there is such an infinite sequence with P as its behavior [12]. Computable threads can be obtained by composition of regular threads with a Turing machine tape as a service [12].

---

[6] In [5], the undecidability of inclusion for pushdown threads is proved using a reduction of the halting problem for Minsky machines. In this construction one of the counters is "weakly simulated". This method was found by Jančar and recorded first in 1994 [15], where it was used to prove various undecidability results for Petri nets. In 1999, Jančar et al. [16] used the same idea to prove the undecidability of simulation preorder for processes generated by one-counter machines, and this is most comparable to the approach in [5]. However, in the case of pushdown threads the inclusion relation itself is a little more complex than in process simulation or language theory because $D \sqsubseteq P$ for any thread P. Moreover, threads have restricted branching, and therefore transforming a regular (control) thread into one that simulates one of the counters of a Minsky machine is more complex than in the related approaches referred to above. See [5] for a further discussion.

4   On the Expressiveness of PGA

We present some expressiveness results for PGA.

Proposition 1. PGA *without repetition characterizes BTA, that is, each program without repetition defines a finite thread, and all finite threads can be expressed.*

Proof. It follows immediately from the equations for thread extraction that PGA programs without repetition define finite threads. Vice versa, we give a mapping [_] from BTA to PGA:

$$[\mathsf{D}] = \#0,$$
$$[\mathsf{S}] = !,$$
$$[P \unlhd a \unrhd Q] = +a; \#2; \#(n[P] + 1); [P]; [Q],$$

where $n[P]$ is the number of instructions in $[P]$.   □

Proposition 2. PGA *characterizes the regular threads.*

Proof. It follows immediately from the equations for thread extraction that PGA programs define regular threads. Vice versa, any regular thread can be given by a finite linear recursive specification by Theorem 1. Assume a specification with variables $x_1, \ldots, x_n$. We obtain the PGLDg program $[x_1]; [x_2]; \ldots; [x_n]$ for x_1 by the mapping [_] which is defined as follows:

$$[x_i] = \begin{cases} \pounds i; +a_i; \#\#\pounds j; \#\#\pounds k & \text{if } x_i = x_j \unlhd a \unrhd x_k, \\ \pounds i; ! & \text{if } x_i = \mathsf{S}, \\ \pounds i; \#\#\pounds i & \text{if } x_i = \mathsf{D}. \end{cases}$$

The resulting PGLDg expression for the thread is mapped to a PGA program by `pgldg2pga` (see Section 2.3).   □

Corollary 1. *Basic instructions and negative tests instructions do not enhance the expressive power of* PGA.

Proof. Take any PGA program. By thread extraction and the method sketched in the proof of Proposition 2 we find an equivalent PGA program without occurrences of basic instructions or negative tests.   □

This corollary establishes that PGA's set of primitive instructions is not minimal with respect to its expressiveness. The next proposition shows that that unbounded jump counters are necessary for the expressiveness of PGA.

Proposition 3. *For $n \in \mathbb{N}$, let* PGA_n *denote the set of* PGA *expressions not containing jump counters strictly greater than n. For every $n \geq 2$, there is a* PGA *behavior that cannot be expressed in* PGA_n.

Proof. Take $n \geq 2$ and a basic instruction a. Consider the PGA program X defined by

$$X = Y_1; \ldots; Y_{n+1};!; (Z_1; \ldots; Z_{n+1})^\omega,$$
$$Y_i = +a; \#k_i,$$
$$Z_i = a^i; +a;!; \#l_i,$$

where $k_i = 2n + 1 + i(i+1)/2$, and $l_i = (n+4)(n+5)/2 - (i+8)$.

Note that $\#k_i$ jumps from Y_i to the first instruction of subexpression Z_i, and that $\#l_i$ jumps from Z_i also to the first instruction of Z_i. For example, if $n = 2$, then X equals

$$+a; \#6; +a; \#8; +a; \#11;!; (a; +a;!; \#12; a^2; +a;!; \#11; a^3; +a;!; \#10)^\omega.$$

We oberve that X has these properties:

1. After the execution of $Y_1; \ldots; Y_{n+1};!$, any of the Z_i can be the first part of the iteration that is executed.
2. Execution of the iterative part is completely determined by one of the Z_i and distguished from the execution determined by another Z_j.

We show that $|X|$ cannot be expressed in PGA_n. First, for $i \leq n+1$, define threads

$$Q_i = a^i \circ (\mathsf{S} \trianglelefteq a \trianglerighteq Q_i),$$
$$P_i = Q_i \trianglelefteq a \trianglerighteq P_{i+1},$$
$$P_{n+2} = \mathsf{S}.$$

We find that

$$P_i = |Y_i; \ldots; Y_{n+1};!; (Z_1; \ldots; Z_{n+1})^\omega|,$$
$$Q_i = |Z_i; \ldots; Z_{n+1}; (Z_1; \ldots; Z_{n+1})^\omega|,$$

and in particular that $|X| = P_1$.

Now suppose that $|X|$ can be expressed in PGA_n (we shall derive a contradiction). Then there must be a first canonical form

$$u_1; \ldots; u_m; (v_1; \ldots; v_k)^\omega$$

in PGA_n with this behavior. We can picture the iteration of $v_1; \ldots; v_k$ as a circle of k instructions:

Note that each of the v_j serves at most one Q_i.

By the restriction on the values of jump counters, we know that between any two subsequent Q_i-instructions on the circle, there are at most $n - 1$ other instructions. Hence, for any i there are at least $\lceil k/n \rceil$ Q_i-instructions on the circle, so in total the circle contains at least $(n + 1) \cdot \lceil k/n \rceil$ instructions. Since

$$(n + 1) \cdot \lceil k/n \rceil \geq (n + 1) \cdot (k/n) > k,$$

this contradicts the fact that the circle contains k instructions. □

5   Current Work

Current work includes research on multithreading. In thread algebra, a multithread consists of a number of basic threads together with an interleaving operator which executes the threads in parallel based on a certain interleaving strategy [8, 9]. This theory is applied in the setting of processor architectures, in particular of so-called *micro-grids* executing micro-threads [17]. For the mathematical modeling of processor architectures, so-called Maurer computers are used [18, 10, 11].

Another branch of research is about the forecasting of certain actions, given the program to be executed. The main purpose of this research is a formal modeling of security hazard risk assessment (or virus detection) [14, 13]. For pushdown threads this type of forecasting is decidable: rename the action(s) to be forecasted and decide whether the thread thus obtained equals the original one. Forecasting becomes much more complicated if a program may contain test instructions that yield a reply according to the result of this type of forecasting. For example, assume that the action to be forecasted is named *risk* and that there is a test action *test* that yields true if its true-branch does not execute *risk*, and false otherwise. Then a current *test* action in the code to be inspected may yield true because a future one will yield false. The reply to these *test* actions can be modeled with a use-application. For regular threads, the associated service has a decidable reply function [13], while for pushdown threads this is still an open question.

References

1. J.W. de Bakker and J.I. Zucker. Processes and the denotational semantics of concurrency. *Information and Control*, 54(1-2):70–120,1982.
2. J.A. Bergstra. www.science.uva.nl/~janb/ta/, Februari 2006.
3. J.A. Bergstra. www.science.uva.nl/~janb/pga/, Februari 2006.
4. J.A. Bergstra and I. Bethke. Polarized process algebra and program equivalence. In J.C.M. Baeten, J.K. Lenstra, J. Parrow, G.J. Woeginger, eds., *Proceedings of ICALP 2003*, Springer-Verlag, LNCS 2719:1–21, 2003.
5. J.A. Bergstra, I. Bethke, and A. Ponse. Decision problems for pushdown threads. Report PRG0502, Programming Research Group, University of Amsterdam, June 2005. Available at www.science.uva.nl/research/prog/publications.html.

6. J.A. Bergstra and J.W. Klop. Process algebra for synchronous communication. *Information and Control*, 60(1-3):109–137, 1984

7. J.A. Bergstra and M.E. Loots. Program algebra for sequential code. *Journal of Logic and Algebraic Programming*, 51(2):125–156, 2002.

8. J.A. Bergstra and C.A. Middelburg. A Thread Algebra with Multi-Level Strategic Interleaving. Report 200441, Computing Science Department, Eindhoven University of Technology, 2004.

9. J.A. Bergstra and C.A. Middelburg. Thread Algebra for Strategic Interleaving. Report PRG0404, Programming Research Group, University of Amsterdam, 2004. Available at www.science.uva.nl/research/prog/publications.html.

10. J.A. Bergstra and C.A. Middelburg. Simulating Turing Machines on Maurer Machines. Report 200528, Computing Science Department, Eindhoven University of Technology, 2005.

11. J.A. Bergstra and C.A. Middelburg. Maurer Computers with Single-Thread Control. Report 200517, Computing Science Department, Eindhoven University of Technology, 2005.

12. J.A. Bergstra and A. Ponse. Execution architectures for program algebra. To appear in *Journal of Applied Logic*, 2006. An earlier version appeared as Logic Group Preprint Series 230, Department of Philosophy, Utrecht University, 2004.

13. J.A. Bergstra and A. Ponse. A bypass of Cohen's impossibility result. In P.M.A. Sloot, A.G. Hoekstra, T. Priol, A. Reinefeld, and M. Bubak, eds., *Advances in grid computing – EGC 2005*, Springer-Verlag, LNCS 3470:1097–1106, 2005.

14. F. Cohen. Computer viruses — theory and experiments. *Computers & Security*, 6(1):22–35, 1984. Available as http://vx.netlux.org/lib/afc01.html.

15. P. Jančar. Decidability questions for bisimilarity of Petri nets and some related problems. *Proceedings of STACS94*, Springer-Verlag, LNCS 775:581-592, 1994.

16. P. Jančar, F. Moller, and Z. Sawa. Simulation problems for one-counter machines. *Proceedings of SOFSEM99: The 26th Seminar on Current Trends in Theory and Practice of Informatics*, Springer-Verlag, LNCS 1725:398-407, 1999.

17. C. Jesshope. Implementing an efficient vector instruction set in a chip multiprocessor using micro-threaded pipelines. *Australian Computer Science Communications*, 23(4):80–88, 2001.

18. W.D. Maurer. A theory of computer instructions. *Journal of the ACM*, 13(2):226–235, 1966.

19. T.D. Vu. Metric denotational semantics for BPPA. Report PRG0503, Programming Research Group, University of Amsterdam, July 2005. Available at www.science.uva.nl/research/prog/publications.html.

Fast Quantifier Elimination Means P = NP

Mihai Prunescu[1,2]

[1] "Dr. Achim Hornecker — Software-Entwicklung und I.T.-Dienstleistungen",
Freiburg im Breisgau, Germany
[2] Institute of Mathematics "Simion Stoilow" of the Romanian Academy,
Bucharest, Romania

Abstract. The first part is a survey of Poizat's theory about fast elimination of quantifiers and the P = NP question according to the unit-cost model of computation, as developed along the book [7]. The second part is a survey of the structure with fast elimination constructed by the author in [9].

1 Introduction

In [9] a structure with fast elimination is constructed. Here I intend to recall the whole context of this construction and to explain why it is said that the structure constructed there satisfies P = NP for the unit-cost model of computation over algebraic structures. The construction itself will be also shortly presented here, but I will emphasize exactly the steps which hasn't been presented with too much details in the cited paper. With this occasion I shall try to answer to some frequently asked questions. This extended abstract should be seen as a complement to [9].

The notation used is the standard notation for mathematical logic. Bold-faced letters like u denote tuples $(u_1, u_2, \ldots, u_n)$.

2 Machines, Circuits and Existential Formulas with Parameters

The unit-cost complexity over algebraic structures was born with the paper of L. Blum, M. Shub and S. Smale dedicated to unit-cost computations over the ordered field of the reals. The approach presented here belongs to B. Poizat and was developed by him along the lines of the book "Les petits cailloux", [6].

Let L be an abstract finite signature for algebraic structures. L consists in a set of constant-symbols (c_i), a set of relation-symbols (R_j) with arities (n_j) and a set of operation-symbols (f_k) with arities (m_k). We fix an L-structure S. The interpretation of L in S shall be done using the same symbols. (We do not make a notational difference between symbol and interpretation here.)

Definition 1. A Turing machine working over the L-structure S is a multi-tape Turing machine with finitely many states and the following ideal abilities:

A. Beckmann et al. (Eds.): CiE 2006, LNCS 3988, pp. 459–470, 2006.

- The machine works with a possibly infinite alphabet consisting in the elements of S. In other words, every element of S has a unique name. This name can be written in one cell of the machine as component of the input or as result of a computation. During a computation, the name of a given element may arrise several times in different cells. Equivalently, you can think about the underlying set of S like about a (possibly infinite) alphabet used by a Turing machine.
- Let $x_i \in S$ be the content of a cell of the tape number i which is read at this moment by the head H_i. Following program lines can occur: stop; H_i+, H_i- to move a head on a tape; $x_i := f(x_k, x_l, \ldots, x_s)$, where f is a function symbol in L or the identity (this means $x_i := x_j$); if $R(x_k, x_l, \ldots x_s)$ then continue with state q, else continue with state q', where R is a relation symbol in L or the equality; if H_i is reading an empty cell then continue with state q, else continue with state q'.
- Any such step is performed in a unit of time.

This will be simply called a machine over S.

To say that "a cell is empty" is the same as saying "a cell contains a blank symbol" — I will not insist here on this. The multi-tape formulation given here is directly used used by Poizat in order to prove his Theorem 1. One can define the notion of computability over algebraic functions using only one-tape Turing machines: if L is finite and all relations (functions) have a finite arity, then there is a translation of multi-tape Turing machines in one-tape Turing machines, that does not change the defined class P. Indeed, if k is the number of tapes, consider the $mk + i$-th cell of the one-tape machine to be the m-th cell of the i-th tape $(m \in \mathbb{Z}, 0 \leq i < k)$, and multiply the number of states with k.

Definition 2. A problem is a subset of $S^* := \coprod_{n \in \mathbb{N}} S^n$ seen as set of finite inputs which are accepted by a machine over S. For an input $x \in S^*$ let $|x|$ be its length. Note: the symbol $\coprod$ used here is meant as disjoint union. There are no identifications between S^n and factors of S^m when $n \leq m$.

Definition 3. For an L-structure S we define the complexity class $P(S)$ as the set of all problems over S accepted by deterministic machines over S in polynomial time. This means the following: there is a polynomial $p(n)$ with natural coefficients such that for all inputs $x \in S^*$ the decision is taken in less that $p(|x|)$ units of time.

Definition 4. A problem $B \subset S^*$ is said to belong to the class $NP(S)$ if and only if there is a problem A in $P(S)$ and a polynomial $q(n)$ such that for all $x \in S^*$:
$$x \in B \leftrightarrow \exists\, y \in S^* \;\; |y| = q(|x|) \land xy \in A.$$
By xy we mean the concatenation of the strings x and y.

At this point I must make some commentaries about parameters. In the literature unit-cost Turing machines are allowed to contain a finite tuple of elements of the structure, and to use them in the computations. The classical notation for the complexity classes with allowed parameters is P and NP. For the situation described here, where a finite signature is fixed and the machines are not allowed to contain other parameters than the fixed interpretation of the given constants, the classical notation is P^0 and NP^0. I like to work with the definitions and the notations as given here because I consider them more clear and more resonant with the model-theoretic point of view. It is worth to remark that by writing down the things in this way, I didn't really introduce a restriction. As an anonymous referee pointed out *using the classical notation*: if a structure has $P^0 = NP^0$ then it has P = NP, because all tuples in a problem can be completed with the tuple of suplementary parameters. On the other hand, if the structure has P = NP, one can expand the structure with a finite number of constants such that the new one has $P^0 = NP^0$: the new constants are the parameters used to solve some NP-complete problem over the structure. This notion is explained in the sequel, together with the equivalent of the work of Cook for the classes P and NP over a an L-structure S.

Definition 5. Suppose from now on that the language L contains at least two constants, which will be called 0 and 1. An L-circuit is a finite directed graph without cycles. The vertices of the graph are called gates, and the directed edges are called arrows. There are input-gates, constant-gates, operation-gates, relation-gates, selection-gates and output-gates; at least one input-gate and one output-gate must be present. We call fan-in of a gate the number of arrows going into the gate. The fan-out is the number of arrows going from the gate outside. All gates have an unbounded fan-out. The gates input and output elements of S.

— An input-gate has fan-in 1. It just copies the input element and sends it along the outgoing arrows.
— A constant-gate has fan-in 0. It sends copies of the corresponding (in S interpreted) constant along the outgoing arrows.
— Operation-gates and relation-gates have a fan-in equal to the arity of the corresponding operation (relation). The operation-gate for f computes the value $f(x_1, \ldots, x_s) \in S$. The relation-gate checks if the relation $R(x_1, \ldots, x_s)$ is true and outputs the constant 1 then, else it outputs the constant 0.
— The selection-gate has fan-in 3 and computes the function $s(x, y, z)$, where $s(0, y, z) = y$, $s(1, y, z) = z$ and $s(x, y, z) = x$ if $x \neq 0$ and $x \neq 1$.
— In the case of the so-called decision circuits there is only one output gate that outputs 0 or 1.

The complexity-measure of a circuit C is its number of gates $|C|$.

Considering the circuits to be compressed first-order formulas, they are a good instrument for making concepts like problem or complexity class independent of a special type of computing device.

Theorem 1. *Let M be a machine with k tapes over S, working in a bounded time $\leq t(n)$, where $t(n) \geq n$ is a function of the length $|x|$. Then there is a* *recursive sequence $(C_n(x_1, \ldots, x_n))$ with $|C_n| \leq t(n)^{k+1}$, such that $C_n(\boldsymbol{x})$ gives for input $\boldsymbol{x}$ of length n the same result as the machine M, and C_n are uniformly constructed by a classical Turing machine in polynomial time $p(n)$.*

Definition 6. The satisfiability problem for L-circuits with parameters in S:
Given a string $\boldsymbol{wa} \in S^*$, such that the subword $\boldsymbol{w}$ is a binary word made up by the special boolean constants $0, 1 \in S$ and encoding a decision L-circuit $C(\boldsymbol{x}, \boldsymbol{y})$;
It is asked if there is $\boldsymbol{b} \in S^*$ of appropriate length, such that

$$(S, \boldsymbol{a}, \boldsymbol{b}) \models C(\boldsymbol{a}, \boldsymbol{b}) = 1.$$

Don't wonder about our use of the symbol "models" ($\models$) in this context. As already said, circuits are first-order formulas written down compactly.

It follows directly from the theorem that the satisfiability problem for L-circuits with parameters in S is NP(S)-complete. This problem belongs to P(S) if and only if S has the property P = NP.

We now come to the most delicate point of the reduction: from the satisfaction of circuits to the satisfaction of first-order formulae.

Definition 7. The satisfiability problem for quantifier-free L-formulae with parameters in S:
Given a string $\boldsymbol{wa} \in S^*$, such that the subword $\boldsymbol{w}$ is a binary word made up by the special boolean constants $0, 1 \in S$ and encoding a quantifier-free L-formula $\varphi(\boldsymbol{x}, \boldsymbol{y})$;
It is asked whether there is a $\boldsymbol{b} \in S^*$ of appropriate length, such that

$$(S, \boldsymbol{a}, \boldsymbol{b}) \models \varphi(\boldsymbol{a}, \boldsymbol{b}).$$

At first sight there is no big difference between this satisfiability problem and the satisfiability problem concerning circuits. In fact, there is an important difficulty here. It is true that every circuit is logically equivalent with a quantifier-free formula, *but the translation might not be possible in polynomial time!*

Example. (Poizat) Let S be a structure possessing an associative addition denoted by $+$ and let $C_n(x, y)$ be the circuit $x \Rightarrow + \Rightarrow + \ldots \Rightarrow + \rightarrow = \leftarrow y$ containing n addition-gates and the gate $=$ that checks the equality. $C_n(x, y) = 1$ is equivalent with the formula $x + x + \ldots + x = y$ with $2^n - 1$ additions.

In the case of this circuit,

$$C(x, y) = 1 \leftrightarrow \exists \boldsymbol{z} \quad z_1 = x + x \wedge z_2 = z_1 + z_1 \wedge \ldots \wedge y = z_n + z_n.$$

This existential formula has a length which is linear in n. Even if we use more symbols for the indices in order to write them using some finite alphabet, it will have at most a quadratic length. If we forget the quantifiers and we look at the quantifier-free conjunction with parameters x and y, this quantifier-free formula is satisfied if and only if the circuit was satisfied by x and y.

Following this idea, we modify the definition for the satisfiability problem for quantifier-free formulae with parameters in S. This is just an equivalent definition, and not a new problem.

Definition 8. The satisfiability problem for quantifier-free L-formulae with parameters in S:

Given a string $\boldsymbol{wa} \in S^*$, such that the subword $\boldsymbol{w}$ is a binary word made up by the special boolean constants $0, 1 \in S$ and encoding a quantifier-free L-formula $\varphi(\boldsymbol{x}, \boldsymbol{y})$;

It is asked whether

$$(S, \boldsymbol{a}) \models \exists \, \boldsymbol{y} \, \varphi(\boldsymbol{a}, \boldsymbol{y}).$$

The satisfaction of quantifier-free formulae with parameters in S is the same thing as the truth of existential formulae with parameters in S.

Theorem 2. *The satisfiability problem for quantifier-free formulae with parameters in S is complete for the class $\mathrm{NP}(S)$. Consequently, this problem belongs to $\mathrm{P}(S)$ if and only if S has the property $\mathrm{P} = \mathrm{NP}$.*

Proof. The problem is evidently in NP by guess and check. To prove the NP-completeness, we interpret the satisfiability problem of L-circuits with parameters in S in the satisfiability problem for formulae. Let $(C(\boldsymbol{x}, \boldsymbol{y}), \boldsymbol{a})$ be an instance for the circuit-problem. For each gate in C which is not an input-gate, a constant-gate or the output-gate, consider a new variable z_{gate}. The following holds:

$$\exists \boldsymbol{y} \, C(\boldsymbol{a}, \boldsymbol{y}) = 1 \leftrightarrow \exists \boldsymbol{y} \, \exists z_{\text{gate}_1} \ldots \exists z_{\text{gate}_k}$$

$$\bigwedge_{\text{all gates}} z_{\text{gate}} = \text{gate}(\text{predecessor gates}) \wedge \text{output} = 1.$$

This gives a quantifier-free formula of a length which is polynomially bounded in the length of the circuit and which is satisfiable if and only if the circuit is satisfiable. $\qquad \square$

Definition 9. The structure S is said to allow elimination of quantifiers if for every formula $\varphi(\boldsymbol{x})$ with quantifiers and no other free variables as in the tuple $\boldsymbol{x}$ there is a quantifier-free formula $\psi(\boldsymbol{x})$ such that:

$$S \models \forall \boldsymbol{x} \, (\varphi(\boldsymbol{x}) \leftrightarrow \psi(\boldsymbol{x})).$$

For equivalent definitions, we may require this only for formulae that are logically equivalent with prenex existential formulae, or even for formulae that are logically equivalent with formulae containing only one existential quantifier. Summing up all results got so far, we conclude:

Theorem 3. *The L-structure S has the property $\mathrm{P} = \mathrm{NP}$ if and only if there is a polynomial-time machine over S which transforms all formulae $\exists \, \boldsymbol{y} \, \varphi(\boldsymbol{x}, \boldsymbol{y})$ in a circuit $C(\boldsymbol{x})$ such that:*

$$\forall \, \boldsymbol{x} \, (\exists \, \boldsymbol{y} \, \varphi(\boldsymbol{x}, \boldsymbol{y}) \leftrightarrow C(\boldsymbol{x}) = 1).$$

Proof. The structure has $\mathrm{P} = \mathrm{NP}$ if and only if the decision problem for existential formulae with parameters in S is in $\mathrm{P}(S)$. Using Poizat's Theorem 1, there is a polynomial-time constructible sequence of circuits (C_n) such that for inputs $\boldsymbol{wa}$ of length n,

$$\exists\, \boldsymbol{y}\, \varphi(\boldsymbol{a}, \boldsymbol{y}) \leftrightarrow C_n(\boldsymbol{wa}) = 1.$$

and the binary word $\boldsymbol{w}$ encodes the existential formula $\exists\, \boldsymbol{y}\, \varphi(\boldsymbol{x}, \boldsymbol{y})$. Now fix the existential formula and let $C(\boldsymbol{x})$ be the circuit $C_n(\boldsymbol{w}, \boldsymbol{x})$. This means that the input gates corresponding to the subword $\boldsymbol{w}$ are replaced with constant-gates giving the corresponding booleans, and the input gates for the subword $\boldsymbol{a}$ remain free input gates.

For the other direction, recall that the satifiability of existential formulas with parameters in the structure is an NP-complete problem for this model of computation. If this problem lies in P, then P = NP. □

Definition 10. We say that the L-structure S has fast quantifier-elimination if it satisfies the condition occurring in Theorem 3.

In particular, all structures with P = NP allow quantifier-elimination.

Remarks. There are maybe some points which need a special emphasis:

- The model of computation is very different from the classical one. In particular, the computational devices are ideal, working with arbitrary structure elements, as for example real or complex numbers, and the structures are in general infinite. The property P = NP has to be consequently understood.
- On the other hand, if this theory is applied for some finite structure, one gets back the familiar classes P and NP from the classical theory of complexity.
- The notion "fast quantifier-elimination" is also slightly different from the similar notions used in the literature. Peoples tend to understand that the equivalent quantifier-free formula has to be short. This condition would be too strong for our purpose. Here the equivalent quantifier-free circuit has to be small, although the equivalent quantifier-free formula might be long.
- There are several results giving exponential lower bounds for the quantifier-elimination for the field of complex numbers or for the ordered field of real numbers. The known results are not sufficient for proving that those structures satisfy P ≠ NP! Even the exponential lower bound for purely existential formulae over the complex numbers is too weak: although the equivalent quantifier-free formula has exponential length, it is still not proved that there is no circuit of polynomial length which is equivalent with that formula!
- Last but not least: Theorems 1, 2 and 3 have been proved by Poizat in [6] at the pages 109, 149 and 156. The reader has observed that I have slightly modified the statements, because I have deleted all about supplementary parameters in machines or circuits. For this I have introduced the constant-gates.

3   A Structure with Fast Quantifier-Elimination

In the past section we shortly presented Poizat's theory about arbitrary algebraic structures that satisfy the condition P = NP. In this section we will come in contact with the structure with fast quantifier-elimination constructed by the author in [9], proving that we don't deal with the empty class.

For a short historical account: in [7] Poizat discussed the possibility of constructing a structure with P = NP and proposed some approaches; the most concrete proposal was to define a consistent truth-predicate over Malcev's freely generated tree-algebra. In [6] he produced a truth-predicate over an algebra with unary operations only, that will be presented below. His predicate V encodes the truth of existential formulae with only one free variable, in a way that to all formula $\exists y\, \varphi(x, y)$ there is a term $\tau_\varphi(x)$ such that $\forall x\, (\exists y\, \varphi(x, y) \leftrightarrow V(\tau_\varphi(x)))$. Because of the unary functions, we cannot have more than one free variable in a term. The construction of his predicate is rather difficult. In [4] Hemmerling, working with a similar underlying algebra, doubled the length of the binary words satisfying some PSPACE-complete predicate in order to make it sparse (see definition below). The doubling technic is used also in the approach of Gaßner, [3]. She doesn't use the classical properties of PSPACE-complete predicates (to manifest P = NP for computations with oracles) and tried a direct construction of a structure with P = NP by encoding machine instances. The machine-oriented approaches look however very difficult; it is always quite hard to write down all conditions to be checked for a such construction. In [9] the author combined Poizat's truth-predicate, the model-theoretic view about effective quantifier elimination as described above and the doubling technic. In fact the construction is based on the following rules:

1. As in Poizat's case, a general Elimination Lemma for unary structures with generic predicates.
2. By the Elimination Lemma, the satisfaction of $\exists y\, \varphi(x, y)$ depends only of some local information on x, which is encoded by a quantifier-free formula $\beta(x)$ of polynomial length.
3. The predicate V will encode the truth value of a special kind of $\forall\exists$-sentences.

3.1   Preliminaries

The Elimination Lemma is used in [9] without proof, so it shall be proved here. The other lemmas are quoted from [9] with some hints of proof.

Definition 11. Let $\mathcal{R}$ be an infinite set of elements called roots. The set M is the algebra freely generated by $\mathcal{R}$ with two independent unary successor operations, s_0 and s_1 such that: all elements $x \in M$ are terms in some $r \in \mathcal{R}$ and two elements x and y are equal if and only if they are the same term of the same root. The set of elements generated by a given $r \in \mathcal{R}$ is called a block. We add a unary predecessor operation p such that $p(x) = x$ defines the set of all roots and for all x, $p(s_0(x)) = p(s_1(x)) = x$. We add also a constant a to be interpreted by a fixed root and a unary predicate V called also colour, which will be constructed later. Elements x with $V(x)$ are called black, the other are called white. Our structure is (M, s_0, s_1, p, V, a) but shall be refered to as (M, V).

Definition 12. A triangle of height n is a conjunctive formula $T(x)$ as follows: For all $2^{n+1} - 1$ terms (using only the successors s_0 and s_1) $t(x)$ of length $\leq n$ exactly one of the atomic formulas $V(t(x))$ or $\neg V(t(x))$ occurs in the conjunction. No other atomic formula does occur in the conjunction $T(x)$. There are exactly $2^{2^{n+1}-1}$ possible triangles of height n.

Definition 13. For a tuple $z \in M$ we call m-neighborhood of z a conjunction of the following formulas: the formulas $T_{2m}(p^m(z_i))$ with $i = 1, \ldots, k$; and the formulae $p(y) = y$, $p(y) \neq y$, $y = y'$, $y \neq y'$, for all terms y, y' occurring in the triangles above, and exactly those equalities and negated equalities that are realized by the tuple z in M. If the tuple consists of only one element, we speak about an individual neighborhood.

Definition 14. The predicate V is called generic if it satisfies the following condition G:

$\quad G:$ if (M, V) realizes some finite individual neighborhood $\mathcal{N}(x)$

$\quad\quad\quad$ then (M, V) realizes $\mathcal{N}(x)$ infinitely many times.

A structure (M, V) that is an infinite disjoint union of identically coloured blocks has always a generic predicate.

Definition 15. Let us use the alphabet of 15 letters $\forall$, $\exists$, x, $'$, $)$, $($, $\neg$, $\vee$, $\wedge$, s_0, s_1, p, $=$, V, a for writing down formulae. Different variables are built by x and $'$ like: x, x', x'', $\ldots$ We denote by $|\varphi(x)|$ the length of a formula $\varphi(x)$ as word over this alphabet.

Lemma 1. *Let (M, V) be a structure consisting of a disjoint union of (not necessarily identic) blocks such that V is a generic predicate. Consider a formula $\psi(x)$ which is logically equivalent with a prenex $\exists$-formula. Let $|\psi(x)| = n$. Then there is a quantifier-free formula $\lambda(x)$ such that $M \models \forall x\, \psi(x) \leftrightarrow \lambda(x)$. Moreover, all the terms in x and a occurring in $\lambda(x)$ have length smaller than $2n$, and the formula $\lambda(x)$ depends only on the list of all isomorphism-types of individual $2n$-neighborhoods occurring in M.*

Consequently, in order to decide if a tuple $z \in M$ satisfies this $\psi(x)$, we must know the $2n$-neighborhood of the tuple (z, a) and the list of isomorphism-types of individual $2n$-neighborhoods which are realized in M.

Proof. This is Poizat's "Lemme d'élimination" proved in [7] for one free variable. Let $\psi(z)$ be logically equivalent with $\exists y\, \varphi(y, z)$. The quantifier-free formula φ is put in disjunctive normal form. All conjunctions are shorter than n and the existential block commutes with the big disjunction. Working with equations, we write a conjunction in the form:

$$\exists y\, \varphi_0(z) \wedge \varphi_1(z, y) \wedge \lambda_1(y_1) \wedge \ldots \wedge \lambda_k(y_k).$$

Here, $\varphi_1(z, y)$ is a conjunction of negated equalities of the form $t_1(x_i) \neq t_2(y_i)$ and $t_1(y_i) \neq t_2(y_j)$, and all terms appearing in the whole formula have lengths

$\leq 2n$. Because of the genericity of V we can always satisfy the inequalities, provided that the formulas $\lambda_i(y_i)$ are realizable in M. This can be decided if we know the list of isomorphism-types of individual $2n$-neighborhoods realized in M. The conjuction is then equivalent over M with:

$$\varphi_0(z) \wedge \exists y_1 \, \lambda_1(y_1) \wedge \ldots \wedge \exists y_k \, \lambda_k(y_k).$$

In the case that some $\exists y_l \lambda_l(y_l)$ is not consistent, or just not compatible with the list of individual $2n$-neighborhoods realized in the structure, all the conjuction disappears. In the contrary case, the whole conjunction is equivalent with the quantifier free formula $\varphi_0(z)$. □

Definition 16. The predicate V is called sparse if it satisfies the following condition:

$$\forall x \; [V(x) \quad \rightarrow \quad \exists n \in \mathbb{N} \; \exists \varepsilon \in \{0,1\}^n \; \exists r \;\; x = s_1^n s_0 s_{\varepsilon_1} \ldots s_{\varepsilon_n}(r) \wedge p(r) = r].$$

The sparse predicates are very useful: they allow a small list of isomorphism types of individual neighborhoods and for all elements, the corresponding individual neighborhood has a succint description.

Lemma 2. *Suppose that the predicate V is sparse. For all $x \in M$ the following holds: if x is at distance $> 3m$ from its root, then the individual m-neighborhood of x contains at most one black point, which is of the form $s_1^n p^m(x)$ with $0 \leq n \leq 2m$.*

Consequently, there is a unit-cost algorithm such that for input $x \in M$ and $m \in \mathbb{N}$ it constructs a quantifier-free formula $\beta(x)$ which determines the individual m-neighborhood of x up to isomorphism. The algorithm works in time $O(m)$.

Proof. A remark on the first part: it is easy to see that if an m-neighborhood of x contains two black points ore more, then x is at a distance $\leq 3m$ from the root.

The algorithm starts by exploring the $3m$ ancestors of x. If one finds a root, the formula $\beta(x)$ gives x as an s_i-term of the root and the information if this root is the constant a or not. If one doesn't find the root and there is no black point in the m-neighborhood, the algorithm gives x as an s_i-term of his $3m$-th ancestor, the information that this ancestor is not a root, and a new symbol Σ meaning "white neighborhood". If one doesn't find the root and there is a unique black point in the m-neighborhood of x, instead of Σ write down the formula $V(b)$ where the black point b is given as term in x. □

Lemma 3. *If V is sparse, there is a unit-cost algorithm such that for input consisting of a tuple $x \in M$ of length k and $m \in \mathbb{N}$ it constructs a quantifier-free formula $\beta(x)$ which determines the m-neighborhood of x up to isomorphism. The algorithm works in time $O(mk^2)$ and the length of $\beta(x)$ is $O(mk)$.*

Proof. First one writes down the conjunction of the individual formulae β as computed in Lemma 2. Then we observe that:

$$\mathcal{N}_m(x_i) \cap \mathcal{N}_m(x_j) \neq \emptyset \leftrightarrow$$

$$\leftrightarrow p^m(x_i) \in \{x_j, p(x_j), \ldots, p^{3m}(x_j)\} \vee p^m(x_j) \in \{x_i, p(x_i), \ldots, p^{3m}(x_i)\}.$$

For each pair (i, j) in this situation, write down x_i as a minimal term in x_j. If this doesn't happen, don't write anything. The new symbol Σ may play the role of conjunction of all negated equalities which are true instead. $\Box$

Lemma 4. *The number k of different free variables occurring in the formula $\exists y\, \varphi(x, y)$ as a word of length n in the 15-letter alphabet satisfies $k(k + 1) < 2n$. Consequently, the algorithm given by Lemma 3 for constructing the succint description $\beta(x)$ for the neighborhood $\mathcal{N}_{2n}(u)$ works in time $O(n^2)$. Moreover, $\beta(x)$ as a word in the 15-letter alphabet extended with Σ is shorter than $24n^2$.*

3.2   Construction and Main Result

In order to construct the predicate V we extend the 15-letter alphabet with the symbols Σ and $\rightarrow$ and we fix a coding of these symbols as binary words $\varepsilon_1 \ldots \varepsilon_5 \in \{0, 1\}^5$.

We consider all pairs of formulas $(\beta(x), \psi(x))$ in the language (s_0, s_1, p, a, V) such that:

- $\psi(x)$ is logically equivalent with an existential formula $\exists y\, \varphi(x, y)$ where $\varphi(x, y)$ is quantifier-free. Let n be the length of $\psi(x)$ in the 15-letter alphabet.
- $\beta(x)$ is a formula produced by Lemma 3 to describe the $2n$-neighborhood $\mathcal{N}_{2n}(x)$ for some tuple x of elements in some structure (M, V) consisting of an infinite union of identical blocks, with a root interpreting a and such that V is sparse.

We consider all $\forall\exists$-sentences θ of the form:

$$\forall x\, [\beta(x) \rightarrow \exists y\, \varphi(x, y)],$$

together with the existential sentences $\exists y\, \varphi(y)$.

Such a sentence θ of length l is encoded by the sequence of letters $\varepsilon_1 \ldots \varepsilon_{5l}$. We define the code:

$$[\theta] := s_1^{t+5l} \circ s_0 \circ s_1^t \circ s_{\varepsilon_{5l}} \circ \ldots \circ s_{\varepsilon_1}(a).$$

Here is t is a natural number such that $t + 5l = 121n^2$. (We use 121 because $121 = 24 \times 5 + 1$.) The elements $[\theta]$ defined here form the set of all codes.

The following Lemma follows by applying Lemma 1 two times successively.

Lemma 5. *Let (M, V) be a structure consisting of an infinite union of copies of a block, so that V is generic and sparse. Consider a sentence $\theta = \forall x\, [\beta(x) \rightarrow \exists y\, \varphi(x, y)]$ such that the existential sub-formula has length n. In order to know if θ is true in M it is enough to know the colour of terms $t(a)$ with $|t| < 2n^2$ and the list of isomorphism-types of individual $4n^2$-neighborhoods realized in M.*

We construct V inductively. The structure (M, V) will consist of an infinite union of identical blocks. We call an element "a black point" if it satisfies the predicate V as constructed up to the given point; the other elements are called white. Let M_0 be the structure that has the following set of black points: $\{ s_1^n s_0^{n+1}(r) \mid r \text{ root} \}$. All this poins are non-codes. M_0 is already a structure with a sparse generic predicate. It ensures the existence of sufficiently many types of individual neighborhoods, even before the construction starts.

We order the codes in a sequence $[\theta_s]$ according to their length and lexicographically ($s \geq 1$).

Construction step: If the structure M_{s-1} has been constructed, the structure M_s is defined in the following way: the code $[\theta_s]$ is painted in black if and only if $M_{s-1} \models \theta_s$. If this is the case, all the corresponding points in the other blocks become also black. ⊓

Then $M = \lim_{s \to \infty} M_s$.

Lemma 6. *The L-structure (M, V) constructed here has the following properties: the predicate V is generic and sparse, and for all encoded $\forall \exists$ formal L-sentences θ:*

$$(M, V) \models \quad \theta \leftrightarrow V([\theta]).$$

Proof. All structures M_s are generic and sparse, so we can apply Lemma 5 at every step. Consider some step s. The quantifier-free sentence which is equivalent with the encoded sentence depends on terms which are strictly shorter than the code to paint (and so their colour has been already decided). It depends also on the list of isomorphism-types of individual neighborhoods of a relatively small radius. This list does not change anymore, either by painting the new code, nor in the future of the construction. □

Theorem 4. *There is a deterministic unit-cost algorithm able to solve the satisfaction problem for quantifier-free formulae over (M, V) in uniform polynomial time $O(n^2)$ for formulae of length n. Consequently, the structure (M, V) satisfies $P = NP$ for the unit-cost model of computation and has fast quantifier-elimination.*

Proof. Consider an input of the form ψu with $\psi(x) = \exists y \, \varphi(x, y)$ pure existential formula of length n and $u \in M$ a tuple of the same length k as the tuple of different free variables x. The formula can be encoded using the elements $0 := s_0(a)$ and $1 := s_1(a)$.

Using Lemma 3 we get a quantifier-free formula $\beta(u)$ that determines up to isomorphism the $2n$-neighborhood of the tuple (u, a). The algorithm takes time $O(n^2)$ according to Lemma 4. Now construct the following sentence θ:

$$\forall x \, [\beta(x) \to \exists y \, \varphi(x, y)].$$

Compute the code $[\theta]$ in M and check if $V([\theta])$ does hold. Recall that in (M, V) the sentence θ does hold if and only if $V([\theta])$ holds.

If θ holds, then $\exists\, y\ \varphi(u, y)$. If θ does not hold, then there cannot be any tuple x with $2n$-neighborhood isomorphic with the corresponding neighborhood of u that satisfies $\exists\, y\ \varphi(x, y)$. In particular $\exists\, y\ \varphi(u, y)$ doesn't hold in M.   □

References

1. Lenore Blum, Felipe Cucker, Michael Shub and Steven Smale: Complexity and real computation. Springer Verlag, New York, 1998.
2. Lenore Blum, Michael Shub and Steven Smale: On a theory of computation of complexity over the real numbers. American Mathematical Society Bulletin, 21, 1989.
3. Christine Gaßner: A structure of finite signature with identity relation and with P = NP. Preprints of the University Greifswald number 1, 2, 13, 14 / 2004; 1, 9, 17 / 2005; 1 / 2006 (different versions and expositions).
4. Armin Hemmerling: P = NP for some structures over the binary words. Journal of Complexity, 21, 4, 557 - 578, 2005.
5. Saul Kripke: Outline of a theory of truth. The Journal of Philosophy 72, 19, 690 - 716, 1975.
6. Bruno Poizat: Les petits cailloux. ALEAS, Lyon, 1995.
7. Bruno Poizat: Une tentative malheureuse de construire une structure éliminant rapidement les quanteurs. Lecture Notes in Computer Science 1862, 61 - 70, 2000.
8. Mihai Prunescu: Non-effective quantifier elimination. Mathematical Logic Quarterly 47, (4), 557 - 561, 2001.
9. Mihai Prunescu: Structure with fast elimination of quantifiers. The Journal of Symbolic Logic 71, (1), 321 - 328, 2006.

Admissible Representations
in Computable Analysis

Matthias Schröder

LFCS, School of Informatics
University of Edinburgh, Edinburgh, UK
mschrode@inf.ed.ac.uk

Abstract. Computable Analysis investigates computability on real numbers and related spaces. One approach to Computable Analysis is Type Two Theory of Effectivity (TTE). TTE provides a computational framework for non-discrete spaces with cardinality of the continuum. Its basic tool are representations. A representation equips the objects of a given space with "names", which are infinite words. Computations are performed on these names.

We discuss the property of admissibility as a well-behavedness criterion for representations. Moreover we investigate and characterise the class of spaces which have such an admissible representation. This category turns out to have a remarkably rich structure.

Keywords: Computable Analysis, TTE, Admissibility, Topological Spaces, Cartesian-Closed Categories.

1 Introduction

Computable Analysis is a theory that investigates computability and complexity on spaces occuring in functional analysis like the real numbers or vector spaces. The traditional model for real number computation is floating point arithmetic: a real number is represented by a finite word of fixed length consisting of a mantissa and an exponent. Unfortunately, floating point arithmetic does not provide a reliable computational model for the real numbers. We illustrate this by giving two simple examples of arithmetical problems, for which floating point arithmetic fails spectacularly.

The first example is the following system of linear equations:

$$40157959.0 \cdot x + 67108865.0 \cdot y = 1.0$$

$$67108864.5 \cdot x + 112147127.0 \cdot y = 0.0$$

Applying the solution formula $x = \frac{b_1 \cdot a_{2,2} - b_2 \cdot a_{1,2}}{a_{1,1} \cdot a_{2,2} - a_{2,1} \cdot a_{1,2}}$, $y = \frac{a_{1,1} \cdot b_2 - a_{2,1} \cdot b_1}{a_{1,1} \cdot a_{2,2} - a_{2,1} \cdot a_{1,2}}$, floating point arithmetic using "double precision" variables (53bit mantissa) computes the pair $(112147127.0, -67108864.5)$ as the solution. However, the correct solution is the double of this pair. The reason for this flaw is that $a_{1,1} \cdot a_{2,2}$ and $a_{2,1} \cdot a_{1,2}$ in this example are very close to each other and that unfavourable

A. Beckmann et al. (Eds.): CiE 2006, LNCS 3988, pp. 471–480, 2006.

rounding of these products produces 1.0 as the determinant $a_{1,1} \cdot a_{2,2} - a_{2,1} \cdot a_{1,2}$, whereras the correct value is 0.5.

The second example, the so-called *logistic equation*, shows that floating point arithmetic even infringes the fundamental law of associativity of multiplication. The logistic function defines a non-converging sequence of reals (actually rationals) in the unit interval $[0, 1]$ by

$$x_0 := 0.5, \ x_{i+1} := 3.75 \cdot x_i \cdot (1.0 - x_i),$$

cf. [10, 12]. Depending on the order of evaluation of the product $3.75 \cdot x_i \cdot (1.0 - x_i)$, double precision floating point arithmetic produces the rather arbitrary values $0.8358\ldots$, $0.5887\ldots$, $0.3097\ldots$ for x_{100}. All three "results" are nowhere near the correct value (roughly 0.8882).

An alternative to floating point arithmetic would be the use of *rational arithmetic*. By forming a countable set, rational numbers can be represented unambiguously by finite words, so that reliable computation on $\mathbb{Q}$ is possible. However, the example of the logistic function shows that rational arithmetic is unfeasible: the space necessary to store the rational x_i roughly doubles in each iteration step. Hence even the computation of this simple sequence with rational arithmetic requires exponential time.

Thus we need an approximative computational model which is capable of approximating the reals with arbitrary precision. K. Weihrauch's Type Two Theory of Effectivity (TTE) provides such a computational model (cf. [22, 23]). The basic idea of TTE is to represent the objects of a given space by infinite sequence of symbols of an alphabet Σ. Such a naming function is a called a *representation* of that space. The actual computation is performed by a digital computer on these names. This means that a function f is computable if, for any name p of an argument x, every finite prefix of a name of $f(x)$ can be computed from some finite prefix of the input name p. N. Müller's iRRAM (cf. [12]), an efficient implementation of exact real arithmetic, originates from TTE.

A familiar example of a representation of the set of reals is the *decimal representation* $\varrho_{\text{dec}} \colon \{0, \ldots, 9, -, .\} \rightharpoonup \mathbb{R}$, defined by $\varrho_{\text{dec}}(a_{-k} \ldots a_0 . a_1 a_2 \ldots) := \sum_{i=-k}^{\infty} a_i / 10^i$ and $\varrho_{\text{dec}}(-a_{-k} \ldots a_0 . a_1 a_2 \ldots) := -\sum_{i=-k}^{\infty} a_i / 10^i$. However, the decimal representation fails to induce a reasonable notion of a computable real function, a fact that was already implicitly noticed by A. Turing in [21]. The surprising reason is that real multiplication by 3 is not computable, if the decimal representation is used. The reason for this unexpected flaw is of topological rather than recursion-theoretical nature, see Example 1. This misbehaviour of the decimal representation demonstrates the need of a well-behavedness criterion for representations. In this paper, we will investigate the property of *admissibility* which ensures that a representation behaves suitably in topological terms.

In Section 2 we summarise some basics of Type Two Theory of Effectivity. Section 3 introduces the notion of an admissible representation. In Section 4 we characterise the class of sequential topological spaces which have an admissible representation. It turns out that this is the category $\mathbf{QCB_0}$ of T_0-quotients of countably based spaces. Furthermore we investigate the structural properties of this category. Proofs are not included, as the paper summarises published results.

2   Basics of Type Two Theory of Effectivity

This section gives a short introduction to basic concepts of Type Two Theory of Effectivity (TTE). More details can be found e.g. in [22, 23, 4, 3].

Representations

The basic idea of TTE is to represent infinite objects like real numbers, functions or sets by infinite strings over some finite or countably infinite alphabet Σ and to compute on these names. The corresponding naming function δ mapping a name p to the encoded object $\delta(p)$ is called a *representation*. More precisely:

Definition 1 A *representation* of a set X is a partial surjection $\delta \colon \Sigma^\omega \to X$, where $\Sigma^\omega := \{p \mid p \colon \mathbb{N} \to \Sigma\}$.

We denote by $\mathrm{dom}(f) \subseteq Y$ the domain of a partial function $f \colon Y \to Z$. Hence $\mathrm{dom}(\delta)$ is referred to as the set of *names* of a representation δ.

Computability of Functions on Σ^ω

For functions on Σ^ω, computability is defined via computable monotone word functions or, equivalently, via Type-2 machines. A word function $\lambda \colon \Sigma^* \to \Sigma^*$ is called *monotone*, if $u \sqsubseteq v$ implies $\lambda(u) \sqsubseteq \lambda(v)$, where $\sqsubseteq$ denotes the prefix-relation on $\Sigma^* \cup \Sigma^\omega$. A partial function $g \colon \Sigma^\omega \to \Sigma^\omega$ is *computable*, if there is a computable monotone word-function $\lambda \colon \Sigma^* \to \Sigma^*$ such that

$$g(p) = \sup_{\sqsubseteq} \left\{ \lambda(p^{<n}) \mid n \in \mathbb{N} \right\}$$

holds for all $p \in \mathrm{dom}(g)$ and $\{p \in \Sigma^\omega \mid \{\lambda(p^{<n}) \mid n \in \mathbb{N}\}$ is infinite$\} = \mathrm{dom}(g)$. Here $p^{<n}$ denotes the prefix of length n of p, i.e. the word $p(0) \ldots p(n-1)$. Computability of multivariate functions $h \colon \Sigma^\omega \times \ldots \times \Sigma^\omega \to \Sigma^\omega$ is defined in a similar way. Computable functions on Σ^ω are topologically continuous w.r.t. the countably based[1] Cantor topology $\mathcal{O}(\Sigma^\omega)$ on Σ^ω. Any open set of the Cantor space has the form $\{p \in \Sigma^\omega \mid \exists w \in W. w \sqsubseteq p\}$, where $W \subseteq \Sigma^*$. Remember that a partial function f between two topological spaces $\mathfrak{X}$ and $\mathfrak{Y}$ is *topologically continuous*, if for every open set $V \in \mathcal{O}(\mathfrak{Y})$ there is an open set[2] $U \in \mathcal{O}(\mathfrak{X})$ satisfying $U \cap \mathrm{dom}(f) = f^{-1}[V]$.

Relative Computability

Given two representations $\delta \colon \Sigma^\omega \to X$ and $\gamma \colon \Sigma^\omega \to Y$, a partial function $f \colon X \to Y$ is called (δ, γ)-*computable*, if there is a computable function $g \colon \Sigma^\omega \to \Sigma^\omega$ *realising* f, which means that

$$\gamma(g(p)) = f(\delta(p)) \tag{1}$$

---

[1] A topological space is *countably based*, if it has a countable base (cf. [6, 20, 24]).
[2] We write $\mathcal{O}(\mathfrak{X})$ for the topology of a topological space $\mathfrak{X}$.

holds for all $p \in \mathrm{dom}(f\delta)$. In other words, the diagram

$$
\begin{array}{ccc}
\Sigma^\omega & \xrightarrow{\ g\ } & \Sigma^\omega \\
{\scriptstyle\delta}\downarrow & & \downarrow{\scriptstyle\gamma} \\
X & \xrightarrow{\ f\ } & Y
\end{array}
$$

has to commute. In this case we also say that f is *computable w.r.t.* the representations δ and γ. If δ and γ are ambient representations of X and Y, then we simply say that f is *computable* rather than f is (δ, γ)-computable. Relative computability of multivariate functions is defined in an analogous way.

Relative Continuity

Since any computable function on Σ^ω is continuous, the notion of relative continuous functions plays an important role in investigating which representations of a space induce a reasonable computability notion on that space. We say that a function $f \colon X \to Y$ is (δ, γ)-*continuous*, if there is a continuous function $g \colon \Sigma^\omega \to \Sigma^\omega$ satisfying Equation (1). In this situation we also say that f is relatively continuous w.r.t. δ and γ.

Proposition 1 motivates the concept of relative continuity.

Proposition 1. *Relative computability implies relative continuity (i.e. every* $(\delta_1, \ldots, \delta_k,\ \gamma)$-*computable function is* $(\delta_1, \ldots, \delta_k,\ \gamma)$-*continuous).*

We have already mentionned that for purely topological reasons multiplication by 3 is not computable w.r.t. the decimal representation.

Example 1. Real multiplication by 3 is not $(\varrho_{\mathrm{dec}}, \varrho_{\mathrm{dec}})$-continuous. For a proof, consider the name $p = 0.3333\ldots$ representing $\frac{1}{3}$. No finite prefix w of p provides the information whether the number represented by p is in $(-\infty, \frac{1}{3}]$ or in $[\frac{1}{3}, \infty)$, because w can be extended to a name of a number $> \frac{1}{3}$ as well as to a name of a number $< \frac{1}{3}$. But this information is necessary to determine the first symbol of any name of the result 1.

The Categories $\mathbf{Rep_t}$ and $\mathbf{Rep_c}$

We define $\mathbf{Rep_t}$ to be the category whose objects are the representations over the alphabet $\Sigma := \mathbb{N}$ and whose morphisms between objects δ and γ are the total (δ, γ)-continuous functions. By $\mathbf{Rep_c}$ we denote the subcategory whose morphisms are the relatively computable functions. Both categories are cartesian-closed (cf. [1, 23]). There is a canonical way to construct a representation $[\delta, \gamma]$ of the cartesian product $X \times Y$ and a representation $[\delta \to \gamma]$ of the (δ, γ)-continuous total functions (cf. [23]). The representations $[\delta, \gamma]$ and $[\delta \to \gamma]$ form, respectively, the product and the exponential of the objects δ and γ in both categories $\mathbf{Rep_t}$ and $\mathbf{Rep_c}$.

Computable Real Functions

The usual computability notion on the reals is induced by the binary signed-digit representation $\varrho_{sd} : \{-1, 0, 1, .\}^\omega \to \mathbb{R}$. Its main feature is the use of the negative digit -1. The signed-digit representation is defined by

$$\varrho_{sd}(a_{-k} \ldots a_0 . a_1 a_2 \ldots) := \sum_{i=-k}^{\infty} a_i \cdot 2^{-i}$$

for all $k \in \mathbb{N}$ and $a_{-k}, a_{-k+1}, \ldots \in \{-1, 0, 1\}$.

Definition 2. A function $f \colon \mathbb{R}^k \to \mathbb{R}$ is called *computable*, if it is computable w.r.t. to the binary signed-digit representation.

Definition 2 establishes a computational model for the real numbers which is essentially equivalent[3] to the ones considered by other authors like A. Grzegor-czyk [8], M. Pour-El and J. Richards [13], K.I. Ko [9], V. Stoltenberg-Hansen and J.V. Tucker [17], K. Weihrauch [23].

Example 2. Real addition, subtraction, multiplication, division as well as $\sqrt[n]{x}$, exp, log, sin, cos, tan, arcsin, arccos, arctan are computable real functions. Note that none of these functions is computable w.r.t. the decimal representation by not being relatively continuous.

An important observation is that any computable real function is continuous w.r.t. the familiar Euclidean topology, which is generated by the family of all rational open intervals (q_1, q_2) as a base.

Theorem 1. ([22, 23]) *Any computable function $f \colon \mathbb{R}^k \to \mathbb{R}$ is topologically continuous with respect to the Euclidean topology.*

A consequence of this theorem is that tests like "$x = 0$?" and "$x < y$?" are not computable by being discontinuous. However, in many cases these tests can be replaced by computable multivalued tests (cf. [3, 23]).

3   Admissible Representations

In this section we introduce and motivate the notion of an admissible representation. Admissibility ensures that representations are topologically well-behaved. More details can be found in [16, 14].

The Final Topology of a Representation

The definition of relative computability w.r.t. representations establishes an approximative computability model. Thus it comes as no surprise that any relatively computable function is continuous in some sense.

---

[3] For more details see [18, 23].

The *final topology* induced by a representation $\delta\colon \Sigma^\omega \to X$ is one tool to formalise this continuity. It is defined by

$$\mathcal{O}(\delta) := \left\{ U \subseteq X \mid \exists V \in \mathcal{O}(\Sigma^\omega)\,.\,\delta^{-1}[U] = V \cap \mathrm{dom}(\delta) \right\}.$$

The final topology contains every *finitely observable* property (cf. [20]), i.e. every property $U \subseteq X$ that can be obtained from each name of each element satisfying U by observing some finite prefix of that name. Another tool to describe the approximation structure provided by δ is given by the **Lim**-quotient induced by δ in the category **Lim** of limit space (cf. [15, 16]).

We call $\delta\colon \Sigma^\omega \to \mathfrak{X}$ a *quotient representation* of a topological space $\mathfrak{X}$, if the final topology $\mathcal{O}(\delta)$ is equal to the topology $\mathcal{O}(\mathfrak{X})$ of $\mathfrak{X}$. By being a topological quotient[4] of a countably based space, any space $\mathfrak{X}$ having a quotient representation is *sequential*, i.e. every sequentially open[5] set of $\mathfrak{X}$ is open (cf. [6, 7]).

Proposition 2. ([16]) *Let* $\delta\colon \Sigma^\omega \to X$ *and* $\gamma\colon \Sigma^\omega \to Y$ *be representations. Then any total* (δ, γ)*-continuous function* $f\colon X \to Y$ *is topologically continuous w.r.t. the final topologies* $\mathcal{O}(\delta)$ *and* $\mathcal{O}(\gamma)$.

A relatively continuous multivariate function is not necessarily topologically continuous w.r.t. the product topology of the respective final topologies, but at least it is sequentially continuous[6].

Proposition 3. ([16]) *For every* i *let* δ_i *be a quotient representation of a topological space* $\mathfrak{X}_i$. *Let* $f\colon \mathfrak{X}_1 \times \ldots \times \mathfrak{X}_k \to \mathfrak{X}_{k+1}$ *be* $(\delta_1, \ldots, \delta_k, \delta_{k+1})$*-continuous. Then* f *is sequentially continuous.*

As a consequence, f in Proposition 3 is topologically continuous w.r.t. the topology on $\mathrm{dom}(f)$ formed in the category **Seq** of sequential topological spaces. This is an important reason to work in the category **Seq** rather than in the category **Top** of topological spaces.

The Definition of Admissible Representations

The converse of Proposition 2 does not hold: Multiplication by 3 is continuous w.r.t. the final topology of the decimal representation ϱ_{dec} (which is the Euclidean topology), yet it is not $(\varrho_{\mathrm{dec}}, \varrho_{\mathrm{dec}})$-continuous.

The property of admissibility is defined in order to reconcile relative continuity with mathematical continuity. Thus an "admissible" representation $\delta\colon \Sigma^\omega \to X$ should at least have the property that the identity function id_X is (ϕ, δ)-continuous for every representation ϕ of X having the same final topology as δ.

---

[4] A topological space $\mathfrak{Z}$ is called a *topological quotient* of a topological space $\mathfrak{Y}$, if there is a surjection $q\colon \mathfrak{Y} \to \mathfrak{Z}$ such that the final topology $\{V \subseteq \mathfrak{Z} \mid q^{-1}[V] \in \mathcal{O}(\mathfrak{Y})\}$ is equal to topology $\mathcal{O}(\mathfrak{Z})$ of $\mathfrak{Z}$.

[5] A set $V \subseteq \mathfrak{X}$ is *sequentially open*, if every sequence $(x_n)_n$ that converges to some element in V is eventually in V. Clearly, every open set is sequentially open.

[6] A function between topological spaces is *sequentially continuous*, if it maps convergent sequences to convergent sequences (cf. [6, 7]). Topological continuity implies sequential continuity, but not conversely, unless the domain space is sequential.

Definition 3. A representation $\delta\colon \Sigma^\omega \rightharpoonup X$ is called *admissible*, if id_X is (ϕ,δ)-continuous for every representation $\phi\colon \Sigma^\omega \rightharpoonup X$ with $\mathcal{O}(\phi) = \mathcal{O}(\delta)$.

It turns out that this property is sufficient to entail:

Theorem 2. ([14, 16]) *Let $\delta\colon \Sigma^\omega \rightharpoonup X$ and $\gamma\colon \Sigma^\omega \rightharpoonup Y$ be admissible representations. Then a total function $f\colon X \to Y$ is (δ,γ)-continuous if and only if f is topologically continuous w.r.t. the final topologies $\mathcal{O}(\delta)$ and $\mathcal{O}(\gamma)$.*

For multivariate functions we have equivalence between relative continuity (w.r.t. admissible representations) and sequential continuity.

Theorem 3. ([14, 16]) *For $i \in \{1,\dots,k+1\}$ let δ_i be an admissible quotient representation of a topological space $\mathfrak{X}_i$. Then a function $f\colon \mathfrak{X}_1 \times \dots \times \mathfrak{X}_k \rightharpoonup \mathfrak{X}_{k+1}$ is $(\delta_1,\dots,\delta_k,\ \delta_{k+1})$-continuous if and only if f is sequentially continuous.*

Example 3 gives a justification for choosing the signed-digit representation rather than the decimal representation for defining real computability.

Example 3. The signed-digit representation ϱ_{sd} is admissible, whereas the decimal representation ϱ_{dec} is not.

If δ, δ' are continuously equivalent representations of X, i.e. id_X is (δ,δ')-continuous and (δ',δ)-continuous, then either both are admissible or neither is.

4   The Category of QCB-Spaces

In this section we introduce qcb-spaces and show their relationship to admissible representations. Moreover we describe their rich structure: they are closed under many useful constructions for modelling computation.

Any space having an admissible quotient representation is a topological quotient[4] of a countably based space. We name spaces with the latter property:

Definition 4. A topological space $\mathfrak{X}$ is a called a *qcb-space*, if $\mathfrak{X}$ is a topological quotient of a countably based space.

By **QCB** we denote the category of qcb-spaces and of topologically continuous functions, and by **QCB₀** the full subcategory consisting of those qcb-spaces that are T_0-spaces[7].

The incentive for this definition is the following surprising characterisation of spaces having an admissible quotient representation.

Theorem 4. ([16]) *A topological space has an admissible quotient representation if and only if it is a qcb-space and satisfies the T_0-property.*

---

[7] A topological space $\mathfrak{X}$ is called a T_0-*space*, if its *specialization order* $\sqsubseteq_{\mathfrak{X}}$ is a partial order. It is defined by: $x \sqsubseteq_{\mathfrak{X}} y$ if, for all $U \in \mathcal{O}(\mathfrak{X})$, $x \in U$ implies $y \in U$.

One can drop the T_0-property, if one considers "multivalued" representations.

Another useful characterisation uses the notion of a pseudobase. A family $\mathcal{B}$ of (not necessarily open) subsets of $\mathfrak{X}$ is called a *pseudobase* for a topological space $\mathfrak{X}$, if for every open set U and every sequence $(x_n)_n$ converging to some element x_∞ in U there is some $B \in \mathcal{B}$ and $n_0 \in \mathbb{N}$ with $\{x_\infty, x_n \mid n \geq n_0\} \subseteq B \subseteq U$.

Example 4. The family of all rational closed intervals $[q_1, q_2]$ forms a pseudobase of the Euclidean space consisting of non-open sets.

In contrast to bases, pseudobases do not characterise topological spaces unambiguously. For example, the powerset of X is a pseudobase of all topological spaces with underlying set X. However, pseudobases do become interesting, when they are countable. The following theorem is used in [16] to prove Theorem 4.

Theorem 5. ([16,5]) *A sequential topological space has an admissible quotient representation if and only if it has a countable pseudobase and satisfies the T_0-property. A sequential topological space is a qcb-space if and only if it has a countable pseudobase.*

The next proposition yields some useful properties of countably based spaces that remain valid in the more general class of qcb-spaces.

Proposition 4. ([16,2]) *Let $\mathfrak{X}$ be a qcb-space. Then:*

1. *$\mathfrak{X}$ is a sequential topological space.*
2. *$\mathfrak{X}$ is sequentially separable, i.e. there is a countable subset $A \subseteq \mathfrak{X}$ such that any element in $\mathfrak{X}$ is the limit of a sequence in A.*
3. *$\mathfrak{X}$ is hereditarily Lindelöf, i.e. every family $(U_i)_{i \in I}$ of open sets contains a countable subfamily $(U_j)_{j \in J}$ with $\bigcup_{j \in J} U_j = \bigcup_{i \in I} U_i$.*
4. *If $\mathfrak{X}$ is regular ("T_3"), then $\mathfrak{X}$ is perfectly normal ("T_6").*
5. *If $\mathfrak{X}$ is locally compact or metrisable, then it is countably based.*
6. *Compactness and sequential compactness are equivalent for subsets of $\mathfrak{X}$.[8]*

The category **QCB** has a very rich categorical structure. In particular it is cartesian-closed (i.e. all product spaces and function spaces exist), in contrast to its supercategory **Top** of topological spaces and its subcategory ω**Top** of countably based spaces.

Theorem 6. ([16,11,5]) *The categories **QCB** and **QCB$_0$** are cartesian-closed. Moreover both categories have all countable limits and countable colimits.*

The category **QCB** inherits this structure from its cartesian-closed supercategory **Seq** of sequential topological spaces (and **QCB$_0$** from the category **Seq$_0$** of sequential T_0-spaces). Theorem 6 can be proven by constructing corresponding countable pseudobases. For example, the exponential $\mathfrak{Y}^{\mathfrak{X}}$ of two sequential spaces $\mathfrak{X}$ and $\mathfrak{Y}$ is the set $\mathcal{C}(\mathfrak{X}, \mathfrak{Y})$ of all continuous functions $f : \mathfrak{X} \to \mathfrak{Y}$ endowed

---

[8] This result is due to Peter Nyikos (private communication).

with the sequentialization[9] of the compact-open topology on $\mathcal{C}(\mathfrak{X}, \mathfrak{Y})$. For given countable pseudobases $\mathcal{A}$ for $\mathfrak{X}$ and $\mathcal{B}$ for $\mathfrak{Y}$, the closure under finite intersection of the family of sets of the form

$$\{f \in \mathcal{C}(\mathfrak{X}, \mathfrak{Y}) \mid f[A_1 \cap \ldots \cap A_k] \subseteq B\},$$

where $A_1, \ldots, A_k \in \mathcal{A}$ and $B \in \mathcal{B}$, constitutes a countable pseudobase for $\mathfrak{Y}^{\mathfrak{X}}$.

Importantly, $\mathbf{QCB_0}$ inherits its cartesian-closed structure also from the category $\mathbf{Rep_t}$: the standard constructions of product representations and function space representations preserve admissibility. A third supercategory from which $\mathbf{QCB_0}$ inherits its product space and function space constructions is the category $\omega\mathbf{Equ}$ of countably based equilogical spaces (cf. [11]). The category $\omega\mathbf{Equ}$ can be considered as an exemplification of the domain-theoretic approach to Computable Analysis (cf. [1]). The fact that $\mathbf{QCB_0}$ is a common subcategory of $\mathbf{Rep_t}$ and $\omega\mathbf{Equ}$ via structure-preserving inclusion functors explains why the TTE computational model agrees to a large extent with domain-theoretic computational models, namely on spaces that can be built from countably based T_0-spaces by forming: finite or countable products, exponentials, (sequentialized) subspaces, finite or countable coproducts, T_0-quotients.

An interesting subcategory of $\mathbf{QCB_0}$ is the category $\omega\mathbf{P}$ of *topological predomains* (cf. [19,2]). It consists of those qcb-spaces that are monotone convergence spaces. A *monotone convergence space* is a topological space $\mathfrak{X}$ such that its specialization order[7] is a dcpo and every open of $\mathfrak{X}$ is Scott-open w.r.t. the specialization order. The category $\omega\mathbf{P}$ enjoys similar closure properties to $\mathbf{QCB_0}$.

Theorem 7. ([2]) *The category $\omega\mathbf{P}$ is cartesian-closed with countable limits and colimits.*

Topological predomains are a topological generalisation of dcpos which offer certain advantages over traditional dcpo-based domain theory: they can be used to model combinations of features such as effectivity, computational effects and polymorphism that traditional domain theory is unable to handle simultaneously.

5   Conclusion

We have introduced admissibility as a property guaranteeing well-behavedness of representations. The category $\mathbf{QCB_0}$ of spaces equipped with admissible representations has nice characterisations. Moreover, $\mathbf{QCB_0}$ enjoys very good structural properties allowing to model various aspects of computation.

References

1. Bauer, A.: A Relationship between Equilogical Spaces and Type Two Effectivity. Electronic Notes in Theoretical Computer Science **45** (2001)
2. Battenfeld, I, Schröder, M., Simpson, A.K.: Compactly Generated Domain Theory. Math. Struct. in Comp. Sci. **16** (2006)

---

[9] The *sequentialization* of a topology is the topology of all sequentially open[5] sets.

3. Brattka, V.: Computability over Topological Structures. In: Computability and Models. Kluwer Acadamic Publishers, New York (2003) 93–136
4. Brattka, V., Hertling, P.: Topological Properties of Real Number Representations. Theoretical Computer Science **284** (2002) 241–257
5. Escardó, M.H., Lawson, J.D., Simpson, A.K.: Comparing Cartesian-closed Categories of Core Compactly Generated Spaces. Topology and its Applications **143** (2004) 105–145
6. Engelking, R.: General Topology. Heldermann, Berlin (1989)
7. Franklin, S.P.: Spaces in which Sequences Suffice. Fundamenta Mathematicae **57** (1965) 107–115
8. Grzegorczyk, A.: On the Definitions of Computable Real Continuous Functions. Fundamenta Mathematicae **44** (1957) 61–71
9. Ko, K.I.: Complexity Theory of Real Functions. Progress in Theoretical Computer Science, Birkhäuser, Boston (1991)
10. Kulisch, U.: Numerical Algorithms with Automatic Result Verification. Lectures in Applied Mathematics **32** (1996) 471–502
11. Menni, M., Simpson, A.: Topological and Limit-space Subcategories of Countably-based Equilogical Spaces. Mathematical Structures in Computer Science **12** (2002)
12. Müller, N.: The iRRAM: Exact Arithmetic in C++. Lecture Notes in Computer Science **2064** (2001) 222–252
13. Pour-El, M.B., Richards, J.I.: Computability in Analysis and Physics. Perspectives in Mathematical Logic. Springer, Berlin (1989)
14. Schröder, M.: Extended Admissibility. Theoretical Computer Science **284** (2002) 519–538
15. Schröder, M.: Admissible Representations of Limit Spaces. Lecture Notes in Computer Science **2064** (2001) 273–295
16. Schröder, M.: Admissible Representations for Continuous Computations. PhD Thesis, Fachbereich Informatik, FernUniversität Hagen (2002)
17. Stoltenberg-Hansen, V., Tucker, J.V.: Effective Algebras: In: Handbook of Logic in Computer Science 4. Oxford Science Publications (1995) 357–526
18. Stoltenberg-Hansen, V., Tucker, J.V.: Concrete Models of Computation for Topological Algebras. Theoretical Computer Science **219** (1999) 347–378
19. Simpson, A.: Towards a Convenient Category of Topological Domains. In: Proceedings of thirteenth ALGI Workshop, RIMS, Kyoto University (2003)
20. Smyth, M.B.: Topology. In: Handbook of Logic in Computer Science 1. Oxford Science Publications (1992) 641–761
21. Turing, A.M.: On Computable Numbers, with an Application to the "Entscheidungsproblem". A Correction. Proceedings of the London Mathematical Society **43(2)** (1937) 544–546
22. Weihrauch, K.: Computability. Springer, Berlin (1987)
23. Weihrauch, K.: Computable Analysis. Springer, Berlin (2000)
24. Willard, S.: General Topology. Addison-Wesley, Reading (1970)

Do Noetherian Modules Have Noetherian Basis Functions?

Peter Schuster and Júlia Zappe

Mathematisches Institut, Universität München
{pschust, zappe}@math.lmu.de

Abstract. In Bishop-style constructive algebra it is known that if a module over a commutative ring has a Noetherian basis function, then it is Noetherian. Using countable choice we prove the reverse implication for countable and strongly discrete modules. The Hilbert basis theorem for this specific class of Noetherian modules, and polynomials in a single variable, follows with Tennenbaum's celebrated version for modules with a Noetherian basis function. In particular, the usual hypothesis that the modules under consideration are coherent need not be made. We further identify situations in which countable choice is dispensable.

MSC (2000). Primary 13E05; Secondary 13E15, 03E25, 03F65.

Keywords: Noetherian modules, commutative rings, Hilbert basis theorem, countable choice, constructive algebra.

1   Introduction

This article is conceived in the realm of constructive algebra [1] which parallels Bishop's constructive analysis [2, 3]. As compared with the – then dubbed classical – customary way of doing mathematics, the principal characteristic of the framework created by Bishop is the exclusive use of intuitionistic logic. According to Richman [4], this allows to view Bishop's theory as a generalisation of classical mathematics. Given Richman's more recent proposal to do Bishop-style constructive mathematics even without countable choice [5], we will indicate throughout where this principle is used and under which additional hypotheses its exertion can be avoided.[1]

One of the first constructively interesting proofs of the Hilbert basis theorem for modules over a commutative ring was given by J. Tennenbaum in

---

[1] As laid out in [2], p. 9, Bishop's clear-cut position was that "the axiom of choice ... is not a real source of nonconstructivity in classical mathematics. A choice function exists in constructive mathematics, because a choice is implied by the very meaning of existence." This argument was considered to apply at least to countable and dependent choice, which principles still are widely used in Bishop-style constructive mathematics (for a discussion we refer to [6]). These choice principles were considered to be particularly indispensable for the theory of Noetherian rings and modules ([1], p. 107).

A. Beckmann et al. (Eds.): CiE 2006, LNCS 3988, pp. 481–489, 2006.

his thesis [7], with the existence of a Noetherian basis operation as his particular notion of Noetherianity. The customary definition of the module being Noetherian is classically equivalent to the existence of a Noetherian basis operation, and with the appropriate form of the axiom of choice at hand any such operation gives rise to a function. While every discrete module that admits a Noetherian basis function is known to be Noetherian in the sense of Richman and Seidenberg ([1], VIII.4.1), it has not been clarified yet whether – or under which circumstances – Noetherian modules have Noetherian basis functions.

In the present paper we investigate the connection between Noetherianity and the existence of a Noetherian basis function, and show – making use of countable choice – that for countable and strongly discrete modules the two notions are equivalent (Theorem 1). We thus partially refute the conjecture, made in [1], VIII.4, Exercise 4, that Noetherianity is "probably not" sufficient for the existence of a Noetherian basis function on a strongly discrete module. In combination with Tennenbaum's result in the form of [1], VIII.4.4, we eventually obtain the Hilbert basis theorem for countable and strongly discrete Noetherian modules (Corollary 1). This approach enables one to do without the usually made additional assumption that the modules are coherent as long as one is content with the case of univariate polynomials.

We further refer to Perdry's article [8] for a comparison of several concepts of Noetherianity (exclusive of Tennenbaum's but inclusive of an inductive variant due to Martin-Löf [9]), and of the corresponding versions of the Hilbert basis theorem. A choice-free constructive approach to the Hilbert basis theorem for coherent modules was given in [10]; the ascending tree condition Richman used in that approach is equivalent to his and Seidenberg's ascending chain condition whenever countable choice is assumed.

Notation. If S is a set, we write S^* for the set of finite sequences of elements of S, and $S^{\mathbb{N}}$ for the set of infinite sequences. For a mapping $f : A \to B$ and a subset S of A, let $f[S]$ denote the image of S under f. We suppose that $\mathbb{N} = \{1, 2, 3, \dots\}$ and $\mathbb{N}_0 = \{0, 1, 2, \dots\}$.

A subset T of a set S is said to be *detachable* if for every $x \in S$ one can decide whether $x \in T$ or $x \notin T$, where the latter stands for the negation of the former. A set S is called *discrete* if every singleton subset of S is detachable: that is, if for all $x, y \in S$ one can decide whether $x = y$ or $x \neq y$, where again the latter stands for the negation of the former.

A set S is called *finitely enumerable* if for some $n \in \mathbb{N}_0$ there is a surjective mapping from $\{1, \dots, n\}$ to S, which encompasses the case $n = 0$ of $S = \emptyset$. If a discrete set S is finitely enumerable, then it is even *finite*: that is, there is a bijective mapping from $\{1, \dots, n\}$ to S for some $n \in \mathbb{N}_0$.

Throughout this paper, let R be a commutative ring with unit. For an R-module M and a finitely enumerable subset $S = \{x_1, \dots, x_n\}$ thereof, the submodule of M that is generated by S will be denoted by $(x_1, \dots, x_n)_R$. Any submodule of this kind is called *finitely generated*.

2   Noetherian Modules

The probably best known constructively relevant definition of a Noetherian module is due to Richman [11] and Seidenberg [12]. It requires the set of finitely generated submodules to satisfy the following ascending chain condition:

Definition 1. *An R-module M is called* Noetherian *if for every sequence $A_1 \subseteq A_2 \subseteq \ldots$ of finitely generated submodules of M one can find an index $n \in \mathbb{N}$ such that $A_n = A_{n+1}$.*

Note that this definition only refers to finitely generated submodules, and does not require the ascending chain of submodules to eventually stop. It is clear from the definition, however, that for any such sequence there exist arbitrarily large indexes $n \in \mathbb{N}$ with $A_n = A_{n+1}$. One easily proves that Noetherianity à la Richman and Seidenberg is classically equivalent to the usual definition of a Noetherian module.

Examples of Noetherian modules are the following: finite modules; the ring of integers $\mathbb{Z}$; every discrete field; and – by the Hilbert basis theorem from [1], VIII.1 – the polynomial ring $R[X_1, \ldots, X_n]$ over any coherent Noetherian ring R, such as $\mathbb{Z}$ or a discrete field ([1], p. 196). The algebraic numbers, for instance, form a discrete field ([1], p. 189).

Definition 2. *An R-module M is called* strongly discrete *if every finitely generated submodule of M is detachable.*

Needless to say, if $A = (x_1, \ldots, x_n)_R$ is a finitely generated submodule of M, then $b \in A$ means that one can find $r_1, \ldots, r_n \in R$ such that $b = \sum_{i=1}^n r_i x_i$, whereas $b \notin A$ stands for the negation of $b \in A$: that is, $b \neq \sum_{i=1}^n r_i x_i$ for all $r_1, \ldots, r_n \in R$.

The examples of Noetherian modules given above are all strongly discrete. A strongly discrete module is often said "to have detachable (finitely generated) submodules". Note that an R-module M is strongly discrete if and only if M/A is discrete for every finitely generated submodule A of M; in particular, if M is strongly discrete, then it is discrete.

The following characterisation of Noetherianity for strongly discrete modules makes it easier to compare this notion with the existence of a Noetherian basis function.

Proposition 1. *Let M be a strongly discrete R-module.*

$$M \text{ is Noetherian} \iff \forall (x_i) \in M^{\mathbb{N}} \exists n \in \mathbb{N} : x_n \in (x_1, \ldots, x_{n-1})_R.$$

Proof. Assume first that M is a Noetherian R-module. Let $x_1, x_2, \ldots$ be an infinite sequence in M, and set $A_i := (x_1, \ldots, x_i)_R$ for every $i \in \mathbb{N}$. Obviously, $A_1, A_2, \ldots$ is an ascending chain of finitely generated submodules of M. As M is Noetherian, there exists an index $n \in \mathbb{N}$ such that $A_n = A_{n+1}$, which is equivalent to $x_{n+1} \in (x_1, \ldots, x_n)_R$.

For the converse, let $A_1, A_2, \ldots$ be an ascending chain of finitely generated submodules of M. Construct a sequence $x_1, x_2, \ldots$ of elements of M, and a strictly increasing sequence $\alpha(1), \alpha(2), \ldots$ of positive integers such that

$$A_j = (x_{\alpha(j)}, \ldots, x_{\alpha(j+1)-1})R$$

for every $j \in \mathbb{N}$. Now construct a sequence $\beta(1), \beta(2), \ldots$ of positive integers such that for every $j \in \mathbb{N}$ the following two conditions hold:

(i) $\alpha(j) \leq \beta(j) \leq \alpha(j+1) - 1$, and
(ii) if $x_{\beta(j)} \in (x_1, \ldots, x_{\alpha(j)-1})R$, then

$$x_i \in (x_1, \ldots, x_{\alpha(j)-1})R \text{ for all } \alpha(j) \leq i \leq \alpha(j+1) - 1.$$

Note that we always can find such $\beta(j)$'s: in the case that there exists an $i \in \{\alpha(j), \ldots, \alpha(j+1) - 1\}$ such that $x_i \notin (x_1, \ldots, x_{\alpha(j)-1})R$, we let $\beta(j)$ be the smallest i of this kind; in the other case, we could take any index between $\alpha(j)$ and $\alpha(j+1) - 1$, but to pin down one of them, we set $\beta(j) = \alpha(j)$.

Now consider the sequence $x_{\beta(1)}, x_{\beta(2)}, \ldots$ in M. By hypothesis, there exists $n \in \mathbb{N}$ such that

$$x_{\beta(n)} \in (x_{\beta(1)}, \ldots, x_{\beta(n-1)})R \ .$$

Since $\beta(1) < \ldots < \beta(n-1) \leq \alpha(n) - 1$, it follows that

$$x_{\beta(n)} \in (x_1, \ldots, x_{\alpha(n)-1})R \ .$$

By the construction of the sequence $\beta(1), \beta(2), \ldots$, we thus have

$$x_i \in (x_1, \ldots, x_{\alpha(n)-1})R$$

for every $i \in \{\alpha(n), \ldots, \alpha(n+1) - 1\}$, which is to say that $A_n = A_{n+1}$.   □

In [11], p. 441 the right-hand equivalent of Proposition 1 was seen, also for a strongly discrete module, to be equivalent to the existence of a Noetherian basis operation à la Tennenbaum, which thus is a further equivalent of the module being Noetherian.

3   Noetherian Basis Functions

We suppose that every mapping $\varphi : M^* \to R^*$ which occurs in the sequel has the property that $\varphi[M^{n+1}] \subseteq R^n$ for every $n \in \mathbb{N}$.

Definition 3. *Let M be a discrete R-module. A Noetherian basis function for M is a mapping $\varphi : M^* \to R^*$ such that for every infinite sequence $x_1, x_2, \ldots$ of elements of M there exists an index $N \geq 2$ such that $x_N = \sum_{i=1}^{N-1} r_i x_i$, where $(r_1, \ldots, r_{N-1}) = \varphi(x_1, \ldots, x_N)$.*

In the definition of a Noetherian basis function given in [1], VIII.4, the existence of arbitrary large indexes $N \geq 2$ with the required property is asserted, which clearly follows from the definition we use ([1], VIII.4, Exercise 3).

To see that every finite R-module M admits a Noetherian basis function ([1], VIII.4, Exercise 1), define the mapping $\varphi : M^* \to R^*$ by setting

$$\varphi(x_1, \ldots, x_k) := (r_1, \ldots, r_{k-1})$$

where

$$r_i := \begin{cases} 1 & \text{if } x_i = x_k \text{ and } x_j \neq x_k \text{ for all } j < i \\ 0 & \text{otherwise} \end{cases}$$

for $i \in \{1, \ldots, k-1\}$. This φ obviously is a Noetherian basis function for M.

A perhaps more interesting example of a Noetherian basis function is one for the ring $\mathbb{Z}$ ([1], VIII.4.2). For integers $x_1, \ldots, x_m$, compute the nonnegative greatest common divisor d of $x_1, \ldots, x_{n-1}$. If $d = 0$ (that is, if $x_1 = \ldots = x_{n-1} = 0$), set $\varphi(x_1, \ldots, x_n) := (0, \ldots, 0)$; otherwise let r be the nonnegative remainder when x_n is divided by d. We choose integers $z_1, \ldots, z_{n-1}$ such that $x_n - r = \sum_{i=1}^{n-1} z_i x_i$, and set $\varphi(x_1, \ldots, x_n) := (z_1, \ldots, z_{n-1})$ to obtain a Noetherian basis function for $\mathbb{Z}$.

To avoid the seeming invocation of countable choice in the foregoing construction, it suffices to fix from the outset any variant of the Euclidean algorithm for $\mathbb{Z}$. This method can also be applied to other rings with a Euclidean algorithm, as there is $k[X]$ with k a discrete field ([1], II.5.7).

As an immediate consequence of Proposition 1, if a strongly discrete R-module has a Noetherian basis function, then it is Noetherian. This was proved for discrete modules in [1], VIII.4.1 by a method similar to our proof of Proposition 1. In [1], VIII.4, Exercise 4, the authors raise the question whether the reverse implication – that every Noetherian module has a Noetherian basis function – holds for strongly discrete modules, and say that this is "probably not" the case. Although this question is still open in general, in the particular case of countable modules, and generically with countable choice, the answer is on the positive as we will show in the sequel.

In [1], p. 11 a set S is called *countable* if for some detachable subset D of $\mathbb{N}$ there is a surjective mapping f from D to S. Note that this notion of a countable set subsumes the one of a finitely enumerable set. If a discrete set S is countable, then one can even obtain a bijection between S and a detachable subset of $\mathbb{N}$ by redefining f wherever necessary to avoid all repetitions ([1], I.2, Exercise 8). Since this fact will be crucial for the remainder of this note, we provide the argument in some detail. Pick $\infty \notin S$ such that $S \cup \{\infty\}$ is a discrete set, and define a mapping g from $\mathbb{N}$ to $S \cup \{\infty\}$ by setting

$$g(n) := \begin{cases} f(n) & \text{if } n \in D \text{ and } f(n) \notin \{g(1), \ldots, g(n-1)\} \\ \infty & \text{otherwise} \end{cases}$$

for every $n \in \mathbb{N}$. This g induces a bijection from $g^{-1}(S)$ to S, and $g^{-1}(S)$ is a detachable subset of $\mathbb{N}$. Hence a set is countable and discrete precisely when it is in a one-to-one correspondence with a detachable subset of $\mathbb{N}$. From now on we will exclusively use this characterisation of the countable and discrete sets.

Unlike the general proof that a countable union of countable sets is countable, to obtain the following more specific case there is no need to invoke countable choice.

Lemma 1. *If S is a countable and discrete set, then so is S^*.*

Proof. If S is a detachable subset of $\mathbb{N}$, then S^* is a detachable subset of $\mathbb{N}^*$. It is an established fact that one can construct a bijection between $\mathbb{N}$ and $\mathbb{N}^*$ even in a recursive manner.   □

In view of the peculiar notion of countable set that is in vigour, we need to rephrase the principle of countable choice in the following way:

> Let A and B be sets, and $F(x, y)$ any formula. Assume that A is a detachable subset of $\mathbb{N}$. If for every $x \in A$ there is $y \in B$ with $F(x, y)$, then there is a mapping φ from A to B such that $F(x, \varphi(x))$ for all $x \in A$.

The particular case $A = \mathbb{N}$ of this statement is a common way to put countable choice, and is readily seen to imply the general case of an arbitrary detachable subset A of $\mathbb{N}$. Whenever we speak of countable choice in the sequel, we mean this principle in the formulation given above.

To prove the next result in its general form we seem to need countable choice; some specific cases for which countable choice is definitely not required will be discussed after the proof.

Theorem 1. *Let M be a countable and strongly discrete R-module. If M is Noetherian, then M admits a Noetherian basis function.*

Proof. Since M is strongly discrete, for every finite sequence $x_1, \ldots, x_n$ in M with $n \geq 2$ there exist $r_1, \ldots, r_{n-1} \in R$ such that if $x_n \in (x_1, \ldots, x_{n-1})_R$, then $x_n = \sum_{i=1}^{n-1} r_i x_i$. Applying countable choice together with Lemma 1, we obtain a mapping $\varphi : M^* \to R^*$ such that if $x_n \in (x_1, \ldots, x_{n-1})_R$, then

$$x_n = \varphi(x_1, \ldots, x_n) \begin{pmatrix} x_1 \\ \vdots \\ x_{n-1} \end{pmatrix},$$

that is, $x_n = \sum_{i=1}^{n-1} r_i x_i$ whenever $\varphi(x_1, \ldots, x_n) = (r_1, \ldots, r_{n-1})$.

To show that φ is a Noetherian basis function, let $x_1, x_2, \ldots$ be an infinite sequence of elements of M. Since M is Noetherian, by Proposition 1 there is an index $N \geq 2$ such that $x_N \in (x_1, \ldots, x_{N-1})_R$. By the construction of φ it follows that

$$x_N = \varphi(x_1, \ldots, x_N) \begin{pmatrix} x_1 \\ \vdots \\ x_{N-1} \end{pmatrix}.$$

Hence φ is a Noetherian basis function for M.   □

Observe that in order to obtain φ we have applied countable choice only to the definition of M being a strongly discrete module. One can see this even more clearly if one restates this definition as follows. An R-module M is strongly discrete if and only if

$$\forall(x_1,\ldots,x_n) \in M^* \, \exists(r_1,\ldots,r_{n-1}) \in R^*$$

$$x_n \notin (x_1,\ldots,x_{n-1})_R \lor x_n = (r_1,\ldots,r_{n-1}) \begin{pmatrix} x_1 \\ \vdots \\ x_{n-1} \end{pmatrix}. \tag{1}$$

If, in addition, M is countable, then – with countable choice – from (1) we arrive at

$$\exists \varphi : M^* \to R^* \, \forall(x_1,\ldots,x_n) \in M^*$$

$$x_n \notin (x_1,\ldots,x_{n-1})_R \lor x_n = \varphi(x_1,\ldots,x_n) \begin{pmatrix} x_1 \\ \vdots \\ x_{n-1} \end{pmatrix}. \tag{2}$$

Any φ as granted by (2) is a Noetherian basis function for M whenever M is Noetherian.

We thus have invoked countable choice for $A = M^*$, $B = R^*$, and the formula $F(x,r)$ which is the conjunction of $|x| = |r| + 1$ and

$$x_{|x|} \notin (x_1,\ldots,x_{|x|-1})_R \lor x_n = (r_1,\ldots,r_{|r|}) \begin{pmatrix} x_1 \\ \vdots \\ x_{|x|-1} \end{pmatrix},$$

where $x \in M^*$, $r \in R^*$, and $|u|$ denotes the length k of any finite sequence $u = (u_1,\ldots,u_k)$. This $F(x,r)$ is decidable (because M is strongly discrete) and quantifier-free, which prompts the question whether the specific form of countable choice for this kind of formulas is necessary for proving Theorem 1.

The answer to this question is on the negative if, in addition, the commutative ring R is assumed to be countable and discrete. In this case, namely, also R^* is countable and discrete (Lemma 1). Hence we would only need countable choice, for $F(x,y)$ decidable and quantifier-free, in the particular form in which both A and B are detachable subsets of $\mathbb{N}$. This form, however, is trivially valid without any choice: a mapping φ as required can simply be obtained by defining $\varphi(x)$ as the least $y \in B$ with $F(x,y)$.

The additional condition that R is countable and discrete is satisfied automatically whenever M has a free and cyclic submodule: that is, M contains an isomorphic copy of R. In fact, if any M of this sort is countable and strongly discrete, then R is countable and (even strongly) discrete. The reason for this is that a detachable subset of a detachable subset of $\mathbb{N}$ is again a detachable subset of $\mathbb{N}$; whence every finitely generated submodule of a countable and strongly discrete module is again countable and (even strongly) discrete.

We conclude this note with a particular version of the Hilbert basis theorem. For Corollary 1, countable choice is involved in equal measure as for Theorem 1, of which it is a consequence.

Corollary 1. *Let M be a countable and strongly discrete R-module. If M is Noetherian, then the $R[X]$-module $M[X]$ is Noetherian.*

Proof. If M is Noetherian, then it has a Noetherian basis function (Theorem 1); whence $M[X]$ has a Noetherian basis function according to [1], VIII.4.4. Since $M[X]$, too, is discrete, we obtain with [1], VIII.4.1 that $M[X]$ is Noetherian.   □

As we have seen before, countable choice is not needed to prove Theorem 1 whenever R is countable and discrete; nor do the other ingredients of the proof of Corollary 1 require countable choice. Hence the particular case $M = R$ can be obtained without countable choice: if R is countable, strongly discrete, and Noetherian, then $R[X]$ is Noetherian.

Unlike the treatment of the Hilbert basis theorem for $M = R$ in [1], VIII.1, to obtain Corollary 1 we did not have to suppose that the module under consideration is coherent. Among other things, this additional hypothesis appears to be necessary to ensure that if R is strongly discrete, then so is $R[X]$. We have only needed that $M[X]$ is discrete, which is a trivial consequence of M being (strongly) discrete. Without R being coherent, however, it is unclear whether $R[X]$ is strongly discrete whenever so is R; whence our approach seems to be inapplicable to rings and modules of polynomials in more than one indeterminate.

Acknowledgements. Apart from being grateful to Florian Ranzi for a first introduction into this subject, the authors are indebted to Henri Lombardi, Erik Palmgren, Viggo Stoltenberg-Hansen, and the anonymous referees for corrections, suggestions, and questions. During parts of the preparation of the final version of this paper, the first author was hosted by the Matematiska institutionen of Uppsala universitet, while the second's visit to the Département de Mathématiques of the Université de Franche-Comté in Besançon was supported by the Bayerisch-Französisches Hochschulzentrum.

References

1. Mines, R., Richman, F., Ruitenburg, W.: A Course in Constructive Algebra. Springer-Verlag, New York (1988)
2. Bishop, E.: Foundations of Constructive Analysis. McGraw-Hill Book Co., New York (1967)
3. Bishop, E., Bridges, D.: Constructive Analysis. Springer-Verlag, Berlin (1985)
4. Richman, F.: Intuitionism as generalization. Philos. Math. (2) **5** (1990) 124–128
5. Richman, F.: The fundamental theorem of algebra: a constructive development without choice. Pacific J. Math. **196** (2000) 213–230
6. Schuster, P.M.: Countable choice as a questionable uniformity principle. Philos. Math. (3) **12** (2004) 106–134
7. Tennenbaum, J.: A Constructive Version of Hilbert's Basis Theorem. PhD thesis, University of California San Diego (1973)

8. Perdry, H.: Strongly Noetherian rings and constructive ideal theory. J. Symbolic Comput. **37** (2004) 511–535
9. Jacobsson, C., Löfwall, C.: Standard bases for general coefficient rings and a new constructive proof of Hilbert's basis theorem. J. Symbolic Comput. **12** (1991) 337–371
10. Richman, F.: The ascending tree condition: constructive algebra without countable choice. Comm. Algebra **31** (2003) 1993–2002
11. Richman, F.: Constructive aspects of Noetherian rings. Proc. Amer. Math. Soc. **44** (1974) 436–441
12. Seidenberg, A.: What is Noetherian? Rend. Sem. Mat. Fis. Milano **44** (1974) 55–61

Inverting Monotone Continuous Functions in Constructive Analysis

Helmut Schwichtenberg

Mathematisches Institut der Ludwig–Maximilians–Universität,
Theresienstr. 39, D-80333 München, Germany

Abstract. We prove constructively (in the style of Bishop) that every monotone continuous function with a uniform modulus of increase has a continuous inverse. The proof is formalized, and a realizing term extracted. This term can be applied to concrete continuous functions and arguments, and then normalized to a rational approximation of say a zero of a given function. It turns out that even in the logical term language "normalization by evaluation" is reasonably efficient.

1   Introduction

There have been many attempts to formalize constructive analysis as presented in Bishop's classic [1]. One reason to do this is to uncover the computational content of constructive proofs, an aspect that has been one of the motivations of the FTA project [2], where Kneser's constructive proof of the fundamental theorem of algebra was formalized in Coq. This work has recently been extended to build a "Constructive Coq Repository (C-CoRN)" at Nijmegen (Barendregt, Geuvers, Wiedijk, Cruz-Filipe [3]). However, extraction of reasonable program from proofs in this setup turned out to be problematic. One reason is that witnesses were missing from computational meaningful axioms (e.g., strong extensionality $\forall_{x,y}.f(x)\#f(y) \to x\#y$), another one that the Set, Prop distinction in Coq was found to be insufficient (cf. [3]). Here we desribe a different formalization of constructive analysis, from the point of view of later term extraction. In particular, we deal with the existence of a continuous inverse to a monotonically increasing continuous function. The proof uses the Intermediate Value Theorem IVT.

Some optimizations in definitions and proofs are necessary to produce extracted terms that can be evaluated efficiently. These are (a) addition of external code to the definitions of arithmetical operations, which is used (based on the corresponding function of the programming language) when the arguments are numerals; (b) introduction of the let-construct in extracted terms; (c) "non-computational" quantifiers [4].

The paper extends [5] by a formalization of and term extraction from the theorem on the existence of inverse functions. It turns out that in spite of the harder theorem (compared with IVT) one obtains even better extracted terms. – I have tried to make this paper readable independently of [5].

A. Beckmann et al. (Eds.): CiE 2006, LNCS 3988, pp. 490–504, 2006.

2 Inverse Functions

We prove that every continuous function with a uniform modulus of increase
has a continuous inverse. A constructive proof of this fact has been given by
Mandelkern [6]. More recently, J. Berger [7] introduced a concept he called "exact
representation of continuous functions", and based on this gave a construction
converting one such representation of an increasing function into another one
of its inverse. The proof below is based on a particular concept of a continuous
function, as a type-1 object (using separability of the real numbers).

The setup of constructive analysis is essentially the one of Bishop [1], and is
only sketched here. More detailed elaborations can be found in [8, 5].

We view a *real* x as a Cauchy sequence $(a_n)_n$ of rationals with a separately
given modulus M. When comparing two reals, $x < y$ needs a witness, but $x \leq y$
doesn't; in fact, we can prove $x \not< y \leftrightarrow y \leq x$. For reals $x = ((a_n)_n, M)$ and
$y := ((b_n)_n, N)$, define $x <_k y$ to mean $1/2^k \leq b_p - a_p$, for $p := \max(M(k + 2), N(k + 2))$.

Constructively we cannot compare two reals, but we can compare a real with
a proper interval:

Lemma 1 (ApproxSplit). *Let x, y, z be given and assume $x < y$. Then either
$z \leq y$ or $x \leq z$.*

Proof. Let $x := ((a_n)_n, M)$, $y := ((b_n)_n, N)$, $z := ((c_n)_n, L)$. Assume $x <_k y$,
and let $q := \max(p, L(k + 2))$ and $d := (b_p - a_p)/4$.

Case $c_q \leq \frac{a_p + b_p}{2}$. We show $z \leq y$. It suffices to prove $c_n \leq b_n$ for $n \geq q$. To
see this, observe

$$c_n \leq c_q + \frac{1}{2^{k+2}} \leq \frac{a_p + b_p}{2} + \frac{b_p - a_p}{4} = b_p - \frac{b_p - a_p}{4} \leq b_p - \frac{1}{2^{k+2}} \leq b_n$$

Case $c_q \not\leq \frac{a_p + b_p}{2}$. We show $x \leq z$, via $a_n \leq c_n$ for $n \geq q$.

$$a_n \leq a_p + \frac{1}{2^{k+2}} \leq a_p + \frac{b_p - a_p}{4} \leq \frac{a_p + b_p}{2} - \frac{b_p - a_p}{4} \leq c_q - \frac{1}{2^{k+2}} \leq c_n.$$

This concludes the proof.   □

A *continuous function* $f \colon I \to \mathbb{R}$ on a compact interval I with rational end
points is given by

(a) an approximating map $h_f \colon (I \cap \mathbb{Q}) \times \mathbb{N} \to \mathbb{Q}$ and a map $\alpha_f \colon \mathbb{N} \to \mathbb{N}$ such
   that $(h_f(a, n))_n$ is a Cauchy sequence with (uniform) modulus α_f;
(b) a modulus $\omega_f \colon \mathbb{N} \to \mathbb{N}$ of (uniform) continuity, which satisfies

$$|a - b| \leq 2^{-\omega_f(k)+1} \to |h_f(a, n) - h_f(b, n)| \leq 2^{-k} \quad \text{for } n \geq \alpha_f(k);$$

α_f and ω_f are required to be weakly increasing. One may also add a lower bound
N_f and an upper bound M_f for all $h_f(a, n)$.

Notice that a continuous function is given by objects of type level ≤ 1 only. This is due to the fact that it suffices to define its values on rational numbers.

To prove the Intermediate Value Theorem, we begin with an auxiliary lemma, which from a "correct" interval $c < d$ (that is, $f(c) \leq 0 \leq f(d)$ and $2^{-n} \leq d - c$) constructs a new one $c_1 < d_1$ with $d_1 - c_1 = \frac{2}{3}(d - c)$.

We say that $l \in \mathbb{N}$ is a *uniform modulus of increase* for $f\colon [a,b] \to \mathbb{R}$ if for all $c, d \in [a, b]$ and all $m \in \mathbb{N}$

$$2^{-m} \leq d - c \to f(c) <_{m+l} f(d).$$

Lemma 2 (IVTAux). *Let* $f\colon [a,b] \to \mathbb{R}$ *be continuous, and with a uniform modulus* l *of increase. Assume* $a \leq c < d \leq b$, *say* $2^{-n} < d - c$, *and* $f(c) \leq 0 \leq f(d)$. *Then we can construct* c_1, d_1 *with* $d_1 - c_1 = \frac{2}{3}(d - c)$, *such that again* $a \leq c \leq c_1 < d_1 \leq d \leq b$ *and* $f(c_1) \leq 0 \leq f(d_1)$.

Proof. Let $c_0 = c + \frac{d-c}{3} = \frac{2c+d}{3}$ and $d_0 = c + \frac{2(d-c)}{3} = \frac{c+2d}{3}$. From $2^{-n} < d - c$ we obtain $2^{-n-2} \leq d_0 - c_0$, so $f(c_0) <_{n+2+l} f(d_0)$. Now compare 0 with this proper interval, using ApproxSplit. In the first case we have $0 \leq f(d_0)$; then let $c_1 = c$ and $d_1 = d_0$. In the second case we have $f(c_0) \leq 0$; then let $c_1 = c_0$ and $d_1 = d$. □

Theorem 1 (IVT). *If* $f\colon [a, b] \to \mathbb{R}$ *is continuous with* $f(a) \leq 0 \leq f(b)$, *and with a uniform modulus of increase, then we can find* $x \in [a, b]$ *such that* $f(x) = 0$.

Proof. Iterating the construction in the auxiliary lemma IVTAux above, we construct two sequences $(c_n)_n$ and $(d_n)_n$ of rationals such that for all n

$$a = c_0 \leq c_1 \leq \cdots \leq c_n < d_n \leq \cdots \leq d_1 \leq d_0 = b,$$
$$f(c_n) \leq 0 \leq f(d_n),$$
$$d_n - c_n = (2/3)^n (b - a).$$

Let x, y be given by the Cauchy sequences $(c_n)_n$ and $(d_n)_n$ with the obvious modulus. As f is continuous, $f(x) = 0 = f(y)$ for the real number $x = y$. □

From the Intermediate Value Theorem we obtain

Theorem 2 (Inv). *Let* $f\colon [a, b] \to \mathbb{R}$ *be continuous with a uniform modulus of increase, and assume* $f(a) \leq a' < b' \leq f(b)$. *We can find a continuous* $g\colon [a', b'] \to \mathbb{R}$ *such that* $f(g(y)) = y$ *for every* $y \in [a', b']$ *and* $g(f(x)) = x$ *for every* $x \in [a, b]$ *such that* $a' \leq f(x) \leq b'$.

Proof. Let $f\colon [a, b] \to \mathbb{R}$ be continuous with a uniform modulus of increase, that is, some $l \in \mathbb{N}$ such that for all $c, d \in [a, b]$ and all $m \in \mathbb{N}$

$$2^{-m} \leq d - c \to f(c) <_{m+l} f(d).$$

Let $f(a) \leq a' < b' \leq f(b)$. We construct a continuous $g\colon [a', b'] \to \mathbb{R}$.

Let $u \in [a', b']$ be rational. Using $f(a) - u \leq a' - u \leq 0$ and $0 \leq b' - u \leq f(b) - u$, the IVT gives us an x such that $f(x) - u = 0$, as a Cauchy sequence (c_n). Let $h_g(u, n) := c_n$. Define the modulus α_g such that for $n \geq \alpha_g(k)$, $(2/3)^n(b - a) \leq 2^{-\omega_f(k+l+2)}$. For the uniform modulus ω_g of continuity assume $a' \leq u < v \leq b'$ and $k \in \mathbb{N}$. We claim that with $\omega_g(k) := k + l + 2$ (l from the hypothesis on the slope) we can prove the required property

$$|u - v| \leq 2^{-\omega_g(k)+1} \rightarrow |h_g(u, n) - h_g(v, n)| \leq 2^{-k} \quad (n \geq \alpha_g(k)).$$

Let $u' \leq u < v \leq b'$ and $n \geq \alpha_g(k)$. For $c_n^{(u)} := h_g(u, n)$ and $c_n^{(v)} := h_g(v, n)$ assume that $|c_n^{(u)} - c_n^{(v)}| > 2^{-k}$; we must show $|u - v| > 2^{-\omega_g(k)+1}$.

By the proof of the Intermediate Value Theorem we have

$$d_n^{(u)} - c_n^{(u)} \leq (2/3)^n(b - a) < 2^{-\omega_f(k+l+2)} \quad \text{for } n \geq \alpha_g(k).$$

Using $f(c_n^{(u)}) - u \leq 0 \leq f(d_n^{(u)}) - u$, the fact that a continuous function f has ω_f as a modulus of uniform continuity gives us

$$|f(c_n^{(u)}) - u| \leq |(f(d_n^{(u)}) - u) - (f(c_n^{(u)}) - u)| = |f(d_n^{(u)}) - f(c_n^{(u)})| \leq 2^{-k-l-2}$$

and similarly $|f(c_n^{(v)}) - v| \leq 2^{-k-l-2}$. Hence, using $|f(c_n^{(u)}) - f(c_n^{(v)})| \geq 2^{-k-l}$ (which follows from $|c_n^{(u)} - c_n^{(v)}| > 2^{-k}$ by the hypothesis on the slope),

$$|u - v| \geq |f(c_n^{(u)}) - f(c_n^{(v)})| - |f(c_n^{(u)}) - u| - |f(c_n^{(v)}) - v| \geq 2^{-k-l-1}.$$

Now $f(g(u)) = u$ follows from

$$|f(g(u)) - u| = |h_f(c_n, n) - u| \leq |h_f(c_n, n) - h_f(c_n, m)| + |h_f(c_n, m) - u|,$$

which is $\leq 2^{-k}$ for $n, m \geq \alpha_f(k + 1)$. Since continuous functions are determined by their values on the rationals, we have $f(g(y)) = y$ for $y \in [a', b']$.

For all $x \in [a, b]$ with $a' \leq f(x) \leq b'$, from $g(f(x)) < x$ we obtain the contradiction $f(x) = f(g(f(x))) < f(x)$ by the hypothesis on the slope, and similarly for $>$. Using $u \not< v \leftrightarrow v \leq u$ we obtain $g(f(x)) = x$. $\qquad\square$

As an example, consider the squaring function $f \colon [1, 2] \to [1, 4]$, given by the approximating map $h_f(a, n) := a^2$, constant Cauchy modulus $\alpha_f(k) := 1$, and modulus $\omega_f(k) := k + 1$ of uniform continuity. The modulus of oncrease is $l := 0$, because for all $c, d \in [1, 2]$

$$2^{-m} \leq d - c \rightarrow c^2 <_m d^2.$$

Then $h_g(u, n) := c_n^{(u)}$, as constructed in the IVT for $x^2 - u$, iterating IVTAux. The Cauchy modulus α_g is such that $(2/3)^n \leq 2^{-k+3}$ for $n \geq \alpha_g(k)$, and the modulus of uniform continuity is $\omega_f(k) := k + 2$.

3   Formalization

We now aim at formalizing the proof above, with the planned extraction of realizing terms in mind. For this purpose it is clearly important to represent the underlying mathematical objects in an appropriate way.

It is tempting to start with groups, rings, fields etc. (as in [2,3]). However, it turned out that in such a general approach it is hard to control the computational content of the proofs, and hence its extracted terms. This does not mean that an abstract approach is impossible for our task, but for the moment we prefer the more "concrete" setup, with explicit constructions of the objects.

- *Positive* natural numbers are written in binary; we take them as generated from 1 by two successors $n \mapsto 2n$ and $n \mapsto 2n + 1$. In the corresponding free algebra we have the constructors One, SZero and SOne.
- An *integer* is either a positive number, or zero, or a negative number.
- A *rational* is a pair of an integer and a positive, written i#n. Notice that equality of rationals is not the literal one, but given by the usual equivalence relation.
- A *real* is a pair of a Cauchy sequence of rationals and a modulus. We view the reals as a data type (i.e., no properties), with constructor RealConstr as M, whose components are written x seq and x mod. Within this data type we inductively define the predicate Real x, meaning that x is a (proper) real.
- A *continuous function* is viewed as an element of a data type with constructor ContConstr, whose fields are written f doml, f domr (for the left and right end point of its domain), f approx (for the approximating function), f uMod (for the uniform Cauchy modulus) and f uModCont (for the modulus of uniform continuity). Within this data type we have an inductively defined predicate Cont f, meaning that f is a (proper) continuous function.

From this material we can now build typed lambda terms, as usual. They are terms in the sense of Gödel's T [9], that is, contain (structural) recursion operators for every data type (i.e., free algebra), with arbitrary value types. These terms are the basis of our logical (better: arithmetical) system, which contains an induction scheme (w.r.t. arbitrary formulas) for every data type.

The formalization itself is (some¡what tedious but) straightforward; proof scripts are available at ¦www.minlog-system.de—, in the directory ¦examples/analysis—.

4   Terms and Their Evaluation

4.1   Computation Rules

Computable functionals are defined by "computation rules" [10, 11]; these rules are added to the standard conversion rules of typed λ-calculus. To simplify equational reasoning, terms with the same normal form are identified.

A *system of computation rules* for a defined constant D consists of finitely many equations $DP_i = Q_i$ $(i = 1, \ldots, n)$ with constructor patterns P_i, such that

P_i and P_j ($i \neq j$) are non-unifiable. *Constructor patterns* are lists of applicative terms with distinct variables, defined inductively as follows (we write $P(x)$ to indicate all variables in P; all expressions must be type-correct):

- $x(x)$ is a constructor pattern.
- If C is a constructor and $P(x)$ a constructor pattern, then $(CP)(x)$ is a constructor pattern.
- If $P(x)$ and $Q(y)$ are constructor patterns whose variables x and y are disjoint, then $(P, Q)(x, y)$ is a constructor pattern.

One instance of such rules is the definition of the fixed point operator $\mathcal{Y}_\rho$ of type $(\rho \Rightarrow \rho) \Rightarrow \rho$, by $\mathcal{Y}_\rho f = f(\mathcal{Y}_\rho f)$, which clearly defines a partial functional. Another important example are the (Gödel) structural recursion operators.

However, in practice one wants to define computable functionals by recursion equations, and if possible consider total functionals only. This can be achieved if the patterns on the lhs are "complete" (as for the structural recusion operator) and moreover the rules terminate (as for Gödel's T [9]). Then every closed term of ground type reduces to a "numeral" (or a "canonical term"), that is, a term built from constructors only.

For example, addition for rational numbers is defined by the computation rule converting (i1#k1)+(i2#k2) into i1*k2+i2*k1#k1*k2.

4.2   External Code as Part of Arithmetical Constants

A problem when computing on rationals with the rule above is that the gcd is not cancelled out automatically. Therefore we add "external code" to the internal representation of the function. It works as follows: whenever addition for rationals is called with numerical arguments, these arguments are converted into Scheme rationals, then added with the rational addition function of Scheme, and the result is converted back into the internal representation (using the #-constructor) of a rational.

4.3   Cleaning of Reals

After some computations involving real numbers it is to be expected that the rational numbers occurring in the Cauchy sequences may become rather complex. Hence under computational aspects it is necessary to be able to *clean up* a real, as follows.

Lemma 3. *For every real $x = ((a_n)_n, M)$ we can construct an equivalent real $y = ((b_n)_n, N)$ where the rationals b_n are of the form $c_n/2^n$ with integers c_n, and with modulus $N(k) = k + 2$.*

Proof. Let $c_n := \lfloor a_{M(n)} \cdot 2^n \rfloor$ and $b_n := c_n \cdot 2^{-n}$, hence

$$\frac{c_n}{2^n} \leq a_{M(n)} < \frac{c_n}{2^n} + \frac{1}{2^n} \quad \text{with } c_n \in \mathbb{Z}.$$

Then for $m \leq n$

$$
\begin{aligned}
|b_m - b_n| &= |c_m \cdot 2^{-m} - c_n \cdot 2^{-n}| \\
&\leq |c_m \cdot 2^{-m} - a_{M(m)}| + |a_{M(m)} - a_{M(n)}| + |a_{M(n)} - c_n \cdot 2^{-n}| \\
&\leq 2^{-m} + 2^{-m} + 2^{-n} \\
&< 2^{-m+2},
\end{aligned}
$$

hence $|b_m - b_n| \leq 2^{-k}$ for $n \geq m \geq k+2 =: N(k)$, so $(b_n)_n$ is a Cauchy sequence with modulus N.

To prove that x is equivalent to $y := ((b_n)_n, N)$, observe

$$
\begin{aligned}
|a_n - b_n| &\leq |a_n - a_{M(n)}| + |a_{M(n)} - c_n \cdot 2^{-n}| \\
&\leq 2^{-k-1} + 2^{-n} \quad \text{for } n, M(n) \geq M(k+1) \\
&\leq 2^{-k} \quad \text{if in addition } n \geq k+1.
\end{aligned}
$$

Hence $|a_n - b_n| \leq 2^{-k}$ for $n \geq \max(k+1, M(k+1))$, and therefore $x = y$. $\qquad \square$

5 Extracted Terms

5.1 Realizability

We first describe some proof-theoretic background on term extraction, as it is implemented in the Minlog proof assistant (www.minlog-system.de). It is based on *modified realizability* as introduced by Kreisel [12]: from every constructive proof M (in natural deduction) of a formula A with computational content one extracts a term $[\![M]\!]$ "realizing" A. This term usually is much shorter than the proof it came from, because in the process all subproofs of formulas without computational content can be ignored. The extracted term has a type $\tau(A)$ which depends on the logical shape of the proven formula A only.

An important aspect of this "internal" term extraction (compared with say the extraction of OCaml programs in Coq [13]) is that one stays within the language of the logical theory, and hence – for a particular proof M – can *prove* within the system that the extracted term indeed realizes the formula A (the "Soundness Theorem").

Of course, there is a good reason to extract programs rather than terms: running programs is much faster than evaluating (closed) terms. However, the point made in the previous paragraph is a strong argument for term extraction, particularly in safety critical applications. Moreover, as should become clear from what is done in the present paper, with some care one may well design proofs (and the underlying data types) in such a way that the extracted terms are short and easy to read and evaluate. One can then go on and (automatically) translate these terms into code of a functional programming language, for faster evaluation (cf. [5] for an example).

5.2   Quantifiers Without Computational Content

Besides the usual quantifiers, $\forall$ and $\exists$, Minlog has so-called *non-computational quantifiers*, $\forall^{nc}$ and $\exists^{nc}$, which allow for the extraction of simpler terms. The nc-quantifiers, which were first introduced in [4], can be viewed as a refinement of the Set/Prop distinction in constructive type systems like Coq or Agda. Intuitively, a proof of $\forall^{nc}_x A(x)$ ($A(x)$ non-Harrop, i.e., with a strictly positive occurrence of an existential quantifier) represents a procedure that assigns to every x a proof $M(x)$ of $A(x)$ where $M(x)$ does not make "computational use" of x, i.e., the extracted term $[\![M(x)]\!]$ does not depend on x. Dually, a proof of $\exists^{nc}_x A(x)$ is a proof of $M(x)$ for some x where the witness x is "hidden", that is, not available for computational use. Consequently, the types of extracted terms for nc-quantifiers are $\tau(\forall^{nc}_{x\rho} A) = \tau(\exists^{nc}_{x\rho} A) = \tau(A)$ as opposed to $\tau(\forall_{x\rho} A) = \rho \Rightarrow \tau(A)$ and $\tau(\exists_{x\rho} A) = \rho \times \tau(A)$. The extraction rules are, for example in the case of $\forall^{nc}$-introduction and -elimination, $[\![(\lambda x.M^{A(x)})^{\forall^{nc}_x A(x)}]\!] = [\![M]\!]$ and $[\![(M^{\forall^{nc}_x A(x)}t)^{A(t)}]\!] = [\![M]\!]$ as opposed to $[\![(\lambda x.M^{A(x)})^{\forall_x A(x)}]\!] = [\![\lambda x M]\!]$ and $[\![(M^{\forall_x A(x)}t)^{A(t)}]\!] = [\![Mt]\!]$. In order for the extracted terms to be correct the variable condition for $\forall^{nc}$-introduction needs to be strengthened by requiring in addition the abstracted variable x not to occur in the extracted term $[\![M]\!]$. Note that for a Harrop formula A the formulas $\forall^{nc}_x A$ and $\forall_x A$ are equivalent; similarly, $\exists^{nc}_x A$ and $\exists_x A$ are equivalent.

5.3   Animation

Suppose a proof of a theorem uses a lemma. Then the proof term contains just the name of the lemma, say L. In the term extracted from this proof we want to preserve the structure of the original proof as much as possible, and hence we use a new constant cL at those places where the computational content of the lemma is needed. When we want to execute the program, we have to replace the constant cL corresponding to a lemma L by the extracted program of its proof. This can be achieved by adding computation rules for cL and cGA. We can be rather flexible here and enable/block rewriting by using `animate`/`deanimate` as desired.

5.4   Removal of Duplicated Parts in Terms

In machine generated terms (e.g., those obtained by term extraction) it often happens that a subterm has many occurrences in a term, which leads to unwanted recomputations when evaluating it. A possible cure is to "optimize" the term after extraction, and replace for instance $M[x := N]$ with many occurrences of x in M by $(\lambda x M)N$ (or a corresponding "let"-expression). However, this can already be done at the proof level: When an object (value of a variable or realizer of a premise) might be used more than once, make sure (if necessary by a cut) that the goal has the form $A \to B$ or $\forall_x A$. Now use the "identity lemma" Id: $\hat{P} \to \hat{P}$, whose predicate variable $\hat{P}$ is then instantiated with $A \to B$ or $\forall_x A$; its realizer has the form $\lambda f, x.fx$. However, if Id is not animated, the extracted term has the form $\mathrm{cId}(\lambda x M)N$, which is printed as $[\texttt{let } x\ N\ M]$.

5.5 Extracted Terms

The term extracted from the proof of `ApproxSplit` is

```
(Rec real=>real=>real=>pos=>boole)
([as4,M5]
  (Rec real=>real=>pos=>boole)
  ([as9,M10]
    (Rec real=>pos=>boole)
    ([as13,M14,n15]
      as13(M5(S(S n15))max M10(S(S n15))max M14(S(S n15)))<=
      (as4(M5(S(S n15))max M10(S(S n15)))+
      as9(M5(S(S n15))max M10(S(S n15))))/2)))
```

of type `real=>real=>real=>pos=>boole`. It takes three reals x, y, z with moduli M, N, K (here given by their Cauchy sequences `as4`, `as9`, `as13` and moduli `M5`, `M10`, `M14`) and a positive number k (here `n15`), and computes $p := \max(M(k + 2), N(k + 2))$ and $q := \max(p, L(k + 2))$. Then the choice whether to go right or left is by computing the boolean value $c_q \leq \frac{a_p + b_p}{2}$.

For the auxiliary lemma `IVTAux` we obtain the extracted term

```
[f0,n1,n2]
 (cId rat@@rat=>rat@@rat)
 ([cd4]
   [let cd5
     ((2#3)*left cd4+(1#3)*right cd4@
      (1#3)*left cd4+(2#3)*right cd4)
     [if (cApproxSplit(RealConstr(f0 approx left cd5)
                                 ([n6]f0 uMod(S(S n6))))
                       (RealConstr(f0 approx right cd5)
                                 ([n6]f0 uMod(S(S n6))))
          0
          (S(S(n2+n1))))
      (left cd4@right cd5)
      (left cd5@right cd4)]])
```

of type `cont=>pos=>pos=>rat@@rat=>rat@@rat`. As in the proof above, it takes a continuous f (here `f0`), a uniform modulus l of increase (here `n1`), a positive number n (here `n2`) and two rationals c, d (here the pair `cd4`) such that $2^{-n} < d - c$. Let $c_0 := \frac{2c+d}{3}$ and $d_0 := \frac{c+2d}{3}$ (here the pair `cd5`, introduced via `let` because it is used four times). Then `ApproxSplit` is applied to $f(c_0)$, $f(d_0)$, 0 and the witness $n + 2 + l$ (here `S(S(n2+n1))`) for $f(c_0) < f(d_0)$. In the first case we go left, that is $c_1 := c$ and $d_1 := d_0$, and in the second case we go right, that is $c_1 := c_0$ and $d_1 := d$.

In the proof of the Intermediate Value Theorem, the construction step in `IVTAux` (from a pair c, d to the "better" pair c_0, d_0) had to be iterated, to produce two sequences $(c_n)_n$ and $(d_n)_n$ of rationals. This is the content of a separate lemma `IVTcds`, whose extracted term is

```
[f0,n1,n2](cDC rat@@rat)(f0 doml@f0 domr)
                     ([n4]cIVTAux f0 n1(n2+n4))
```

of type `cont=>pos=>pos=>pos=>rat@@rat`. It takes a continuous function
$f: [a, b] \to \mathbb{R}$ (here `f0`), a uniform modulus l of increase (here `n1`), and a positive
number k_0 (here `n2`) such that $2^{-k_0} < b - a$. Then the axiom of dependent choice
DC is used, to construct from an initial pair $(c_0, d_0) = (a, b)$ of rationals (here `f0`
`doml@f0 domr`) a sequence of pairs of rationals, by iterating the computational
content `cIVTAux` of the lemma `IVTAux`.

The proof of the Inversion Theorem does not use the Intermediate Value
theorem directly, but its essential ingredient `IVTcds`. Its extracted term is

```
[f0,n1,n2,n3,a4,a5]
 ContConstr a4 a5
 ([a6,n7]
   left((cACT rat pos=>rat@@rat)
        ([a8]
          (cIPT pos=>rat@@rat)
          ((cIPT pos=>rat@@rat)
           (cIVTcds
            (ContConstr f0 doml f0 domr
             ([a12,n13]f0 approx a12 n13-a8)
             f0 uMod
             f0 uModCont)
            n1
            n2)))
        a6
        n7))
 ([n6]n3+f0 uModCont(S(S(n6+n1))))
 ([n6]S(S(n6+n1)))
```

It takes a continuous function f (here `f0`), a uniform modulus l of increase (here
`n1`), positive numbers k_0, k_1 (here `n2`, `n3`) such that $2^{-k_0-1} < b - a < 2^{k_1}$ and
two rationals $a_1 < a_2$ (here `a4`, `a5`) in the range of f. Then the continuous inverse
g is constructed (via `ContConstr`) from

- an approximating map,
- a uniform Cauchy modulus (involving the one from f), and
- an easy and explicit modulus of uniform continuity.

The approximating map takes a, u (here `a6`, `n7`). Ignoring the computational
content `cACT`, `cIPT` of `ACT`, `IPT` (which are identities), it yields the left component
(i.e., the Cauchy sequence) of the result of applying `cIVTcds` to a continuous
function close to the original f.

To compute numerical approximations of values of an inverted function we
need `RealApprox`, stating that every real can be approximated by a rational. Its
extracted term is

```
(Rec real=>pos=>rat)([as2,M3,n4]as2(M3 n4))
```

of type `real=>pos=>rat`. It takes a real x (here given by the Cauchy sequence as2 and modulus M3) and a positive number k (here n4), and computes a rational a such that $|x - a| \leq 2^{-k}$. Notice that the `Rec`-operator is somewhat trivial here: it just takes the given real apart. This is because the data type of the reals has no inductive constructor.

To compose `Inv` with `RealApprox`, we prove a proposition `InvApprox` stating that given an error bound, we can find a rational approximating the value of the inverted function g up to this bound. Clearly we need to refer to this value and hence the inverted function g in the statement of the theorem, but on the other hand we do not want to see a representation of g in the extracted term, but only the construction of the rational approximation from the error bound. Therefore in the statement of `InvApprox` we use the non-computational quantifier $\exists^{nc}$ (see Sect.5.2), for the inversion g of the given continuous f. The extracted term of `InvApprox` then simply is

```
[f0,n1,n2,n3,a4,a5,a6]
 cRealApprox
 (RealConstr((cInv f0 n1 n2 n3 a4 a5)approx a6)
  ([n8](cInv f0 n1 n2 n3 a4 a5)uMod(S(S n8))))
```

of type `cont=>pos=>pos=>pos=>rat=>rat=>rat=>pos=>rat`.

Now we "animate" the auxiliary lemmas, that is, add computation rules for all constants with "c" in front of name of the lemma. For `InvApprox` this gives

```
[f0,n1,n2,n3,a4,a5,a6,n7]
 left((cDC rat@@rat)(f0 doml@f0 domr)
     ([n8]
       (cId rat@@rat=>rat@@rat)
       ([cd10]
         [let cd11
           ((2#3)*left cd10+(1#3)*right cd10@
           (1#3)*left cd10+(2#3)*right cd10)
           [if (0<=(f0 approx left cd11
                   (f0 uMod(S(S(S(S(S(S(S(n2+n8+n1)))))))))-
                a6+
                (f0 approx right cd11
                   (f0 uMod(S(S(S(S(S(S(S(n2+n8+n1)))))))))-
                a6))/2)
             (left cd10@right cd11)
             (left cd11@right cd10)]]))
     (n3+f0 uModCont(S(S(S(S(S(n7+n1)))))))))
```

Let us now use this term to compute numerical approximations of values of an inverted function. First we construct the continuous function $x \mapsto x^2$ on $[1, 2]$, with its (trivial) uniform Cauchy modulus and modulus of uniform continuity, and give it the name sq:

```
(define sq (pt "contConstr 1 2([a0,n1]a0*a0)([n0]1)S"))
```

We now apply the extracted term of theorem `InvApprox` to

- the continuous `sq` to be inverted,
- a uniform modulus l of increase,
- a positive number k_0 such that $2^{-k_0-1} < b - a$, and a positive number k_1 such that $b - a < 2^{k_1}$ (which all happen to be 1 in this case),
- two rational bounds a_1, b_1 for an interval in the range,

and normalize the result:

```
(define inv-sq-approx
  (normalize-term
    (apply mk-term-in-app-form
      (list (proof-to-extracted-term
              (theorem-name-to-proof."InvApprox"))
            sq ;continuous function to be inverted
            (pt "1") ;uniform modulus of increase
            (pt "1") (pt "1") ;bounds for b-a
            (pt "1") (pt "4") ;interval in range
            )))))
```

which prints as

```
[a0,n1]
 left(((cDC rat@@rat)(1@2)
      ([n2]
        (cId rat@@rat=>rat@@rat)
        ([cd4]
          [let cd5
            ((2#3)*left cd4+(1#3)*right cd4@
             (1#3)*left cd4+(2#3)*right cd4)
            [if (0<=(left cd5*left cd5-a0+
                    (right cd5*right cd5-a0))/2)
              (left cd4@right cd5)
              (left cd5@right cd4)]]))
      (S(S(S(S(S(S(S(S n1)))))))))
```

The term `sqrt-two-approx` has type `rat=>pos=>rat`, where the first argument is for the rational to be inverted and the second argument k is for the error bound 2^{-k}. We can now directly (that is, without first translating into a programming language) use it to compute an approximation of say $\sqrt{3}$ to 20 binary digits. To do this, we need to "animate" `Id` and then normalize the result of applying `inv-sq-approx` to 3 and 20 (we use normalization by evaluation here, for efficiency reasons):

```
(animate "Id")
(pp (nbe-normalize-term-without-eta
      (make-term-in-app-form sqrt-two-approx (pt "20"))))
```

The result (returned in .7 seconds) is the rational

4402608752054#2541865828329

or 1.7320382149943123, which differs from $\sqrt{3} = 1.7320508075688772$ at the fifth (decimal) digit.

5.6   Translation into Scheme Expressions

For a further speed-up (beyond the use of external code; cf. Sect. 4.2), we can also translate this internal term (where "internal" means "in our underlying logical language", hence usable in formal proofs) into an expression of a programming language (Scheme in our case), by evaluating (term-to-expr inv-sq-approx):

```
(lambda (a0)
  (lambda (n1)
    (car
      (((cdc (cons 1 2))
        (lambda (n2)
          (lambda (cd4)
            (let ([cd5
                    (cons (+ (* 2/3 (car cd4))
                             (* 1/3 (cdr cd4)))
                          (+ (* 1/3 (car cd4))
                             (* 2/3 (cdr cd4))))])
              (if (<= 0
                      (/ (+ (- (* (car cd5) (car cd5)) a0)
                            (- (* (cdr cd5) (cdr cd5)) a0))
                         2))
                  (cons (car cd4) (cdr cd5))
                  (cons (car cd5) (cdr cd4)))))))
        (+ (+ (+ (+ (+ (+ (+ n1 1) 1) 1) 1) 1) 1) 1)))))
```

This Scheme program is very close to the internal term displayed above; we have replaced the internal constant cDC (computational content of the axiom of dependent choice) by the corresponding Scheme function (a curried form of iteration):

```
(define cdc
  (lambda (init)
    (lambda (step)
      (lambda (n)
        (if (= 1 n)
            init
            ((step n) (((cdc init) step) (- n 1)))))))),
```

the internal arithmetical functions +, *, /, <= by the ones from the programming language and the internal pairing and unpairing functions by cons, car and cdr. – It turns out that this code is reasonably fast: evaluating

```
(((ev (term-to-expr inv-sq-approx)) 3) 200)
```

gives the result in .5 seconds, with an accuracy of 200 binary digits.

6   Conclusion, Future Work

The present case study shows that it is possible – albeit after some formalization effort – to machine extract reasonable terms from proofs in constructive analysis, and that ordinary evaluation of these terms can be used to numerically compute approximations to say reals whose existence is claimed by the theorems, with a prescribed precision.

As for future work, an obvious canditate is to do the same for the Cauchy-Euler construction of approximate solutions to ordinary differential equations. A particularly promising candiate is the treatment of ordinary differential equations in Chapt. 1 of Hurewicz's textbook [14], which can easily be adapted to our constructive setting. It should also be possible to compare estimates for solutions of ordinary differential equations with the treatment of the same problem in the interval analysis setting of Moore [15].

References

1. Bishop, E.: Foundations of Constructive Analysis. McGraw-Hill, New York (1967)
2. Geuvers, H., Wiedijk, F., Zwanenburg, J.: A constructive proof of the fundamental theorem of algebra without using the rationals. In Callaghan, P., Luo, Z., McKinna, J., Pollack, R., eds.: Proc. Types 2000. Volume 2277 of LNCS., Springer Verlag, Berlin, Heidelberg, New York (2000) 96–111
3. Cruz-Filipe, L.: Constructive Real Analysis: a Type-Theoretical Formalization and Applications. PhD thesis, Nijmegen University (2004)
4. Berger, U.: Program extraction from normalization proofs. In Bezem, M., Groote, J., eds.: Typed Lambda Calculi and Applications. Volume 664 of LNCS., Springer Verlag, Berlin, Heidelberg, New York (1993) 91–106
5. Schwichtenberg, H.: Program extraction in constructive analysis. Submitted to: Logicism, Intuitionism, and Formalism – What has become of them? (eds. S. Lindström, E.Palmgren, K. Segerberg, V. Stoltenberg-Hansen) (2006)
6. Mandelkern, M.: Continuity of monotone functions. Pacific J. of Math. **99**(2) (1982) 413–418
7. Berger, J.: Exact calculation of inverse functions. Math. Log. Quart. **51**(2) (2005) 201–205
8. Andersson, P.: Exact real arithmetic with automatic error estimates in a computer algebra system. Master's thesis, Mathematics department, Uppsala University (2001)
9. Gödel, K.: Über eine bisher noch nicht benützte Erweiterung des finiten Standpunkts. Dialectica **12** (1958) 280–287
10. Berger, U., Eberl, M., Schwichtenberg, H.: Term rewriting for normalization by evaluation. Information and Computation **183** (2003) 19–42
11. Berger, U.: Uniform Heyting Arithmetic. Annals Pure Applied Logic **133** (2005) 125–148

12. Kreisel, G.: Interpretation of analysis by means of constructive functionals of finite types. In Heyting, A., ed.: Constructivity in Mathematics. North–Holland, Amsterdam (1959) 101–128
13. Letouzey, P.: A New Extraction for Coq. In Geuvers, H., Wiedijk, F., eds.: Types for Proofs and Programs, Second International Workshop, TYPES 2002. Volume 2646 of Lecture Notes in Computer Science., Springer-Verlag (2003)
14. Hurewicz, W.: Lectures on Ordinary Differential Equations. MIT Press, Cambridge, Mass. (1958)
15. Moore, R.E.: Interval Analysis. Prentice-Hall (1966)

Partial Recursive Functions in Martin-Löf Type Theory

Anton Setzer*

Dept. of Computing Science, University of Wales Swansea,
Singleton Park, Swansea SA2 8PP, UK
Tel.: +44 1792 513368; Fax: +44 1792 295651
a.g.setzer@swan.ac.uk
http:www.cs.swan.ac.uk/~csetzer/

Abstract. In this article we revisit the approach by Bove and Capretta for formulating partial recursive functions in Martin-Löf Type Theory by indexed inductive-recursive definitions. We will show that all inductive-recursive definitions used there can be replaced by inductive definitions. However, this encoding results in an additional technical overhead. In order to obtain directly executable partial recursive functions, we introduce restrictions on the indexed inductive-recursive definitions used. Then we introduce a data type of partial recursive functions. This allows to define higher order partial recursive functions like the map functional, which depend on other partial recursive functions. This data type will be based on the closed formalisation of indexed inductive-recursive definitions introduced by Dybjer and the author. All elements of this data type will represent partial recursive functions, and the set of partial recursive functions will be closed under the standard operations for forming partial recursive functions, and under the total functions.

Keywords: Martin-Löf type theory, computability theory, recursion theory, Kleene index, Kleene brackets, partial recursive functions, inductive-recursive definitions, indexed induction-recursion.

1 Introduction

A problem when developing computability theory in Martin-Löf type theory is that the function types only contain total functions, therefore partial recursive functions are not first class objects. One approach to overcome this problem has been taken by Bove and Capretta (e.g. [BC05a, BC05b]), who have shown how to represent partial recursive functions by indexed inductive-recursive definitions (IIRD), and in this article we will investigate their approach. In order to illustrate it, we make use of a toy example. We choose a notation which is closer to that used in computability theory.

Assume the partial recursive function $f : \mathbb{N} \rightharpoonup \mathbb{N}$ defined by

$$f(0) :\simeq 0 , \qquad f(n+1) :\simeq f(f(n)) .$$

* Supported by EPSRC grant GR/S30450/01.

A. Beckmann et al. (Eds.): CiE 2006, LNCS 3988, pp. 505–515, 2006.

This function is constantly zero, but we want to represent it directly in Martin-Löf type theory, so that we can prove for instance, that it is in fact constantly zero. In order to do this, Bove and Capretta introduce

$$f(\cdot)\downarrow : \mathbb{N} \to \text{Set} , \quad \text{eval}_f : (n : \mathbb{N}, p : f(n)\downarrow) \to \mathbb{N} .$$

Here $f(n)\downarrow$ expresses that $f(n)$ is defined and $\text{eval}_f(n, p)$ computes, depending on $n : \mathbb{N}$ and a proof $p : f(n)\downarrow$, the value $f(n)$.

In the literature, $f(\cdot)\downarrow$ is often referred to as the accessibility predicate for f. If we define for arguments a, b of f that $a \prec b$ if and only if the call of $f(b)$ recursively calls $f(a)$, then $f(a)\downarrow$ if and only if a is in the accessible part of $\prec$. The approach by Bove/Capretta can be seen as a general method of determining the accessible part of $\prec$ for a large class of recursively defined functions.

If we take the definition of f as it stands, we see that the definitions of $f(\cdot)\downarrow$ and eval_f refer to each other. $f(\cdot)\downarrow$ has two constructors defined$_f^0$, defined$_f^S$ corresponding to the two rewrite rules, and we obtain the following introduction and equality rules:

$$\text{defined}_f^0 : f(0)\downarrow , \quad \text{eval}_f(0, \text{defined}_f^0) = 0 ,$$
$$\text{defined}_f^S : (n : \mathbb{N}, p : f(n)\downarrow, q : f(\text{eval}_f(n, p))\downarrow) \to f(n+1)\downarrow ,$$
$$\text{eval}_f(n + 1, \text{defined}_f^S(n, p, q)) = \text{eval}_f(\text{eval}_f(n, p), q) .$$

The constructor defined$_f^S$ has arguments $n : \mathbb{N}$, $p : f(n)\downarrow$, and if $f(n) \simeq m$, a proof $q : f(m)\downarrow$. Then $p' := \text{defined}_f^S(n, p, q)$ proves $f(n + 1)\downarrow$ and we have $\text{eval}_f(n+1, p') = \text{eval}_f(m, q)$. We observe that defined$_f^S$ refers to $\text{eval}_f(n, p)$, so we have to define simultaneously $f(\cdot)\downarrow$ inductively, while defining eval_f recursively. This is an instance of an IIRD, as introduced by Dybjer [Dyb00, Dyb94]. We will see below that such kind of IIRD can be reduced to inductive definitions.

Bove and Capretta face the problem that they cannot define a data type of partial recursive functions (unless using impredicative type theory) and therefore cannot deal with partial recursive functions depending on other partial recursive functions as an argument. A simple example would be to define depending on a partial recursive function $f : \mathbb{N} \to \mathbb{N}$ (e.g. f as above)

$$g : \text{List}(\mathbb{N}) \to \text{List}(\mathbb{N}) , \quad g(l) :\simeq \text{map}(f, l) .$$

Here $\text{map}(f, [n_0, \ldots, n_k]) :\simeq [f(n_0), \ldots, f(n_k)]$. In order to define the above directly, we need to define map, depending on an arbitrary partial recursive function f. More complex examples of this kind are discussed in [BC05a].

In this article we will show how to overcome this restriction by introducing a data type of partial recursive functions. This will be based on the closed formulation of IIRD, as developed by P. Dybjer and the author. In order to have that all functions represented by an IIRD correspond directly to a partial recursive function, without using search functions, we will impose restrictions on the set of IIRD used. The data type given in this article will define exactly those restricted IIRD. We will then show that the functions given by those indices are all partial recursive, and that they are closed under the standard constructions for defining partial recursive functions, and under the total functions.

Future work. With the above research we will be able to define functions referring to indices of other partial recursive functions. However, we will not yet be able to deal with mutually recursive functions, in which one function refers to another as a whole. In order to deal with this, we will in a follow-up paper introduce partial recursive functions with function arguments represented as oracles. We will then obtain a recursion theorem stating that recursion equations defined using this principle can always be solved.

2   Inductive-Recursive Definitions

Before formulating the data type of partial recursive definitions as a data type of IIRD let us sketch briefly Dybjer's notion of IIRD.

Dybjer introduced first the simpler notion of inductive-recursive definitions (IRD). In IRD one defines a set U : Set together with a function T : U $\to$ D where D : Type. Inductive-recursive definitions emerged first as universes, i.e. D = Set. Universes are sets of sets, which are given by a set U of codes for sets, and a decoding function T : U $\to$ Set, which determines for every code a : U the set T(a) : Set it denotes.

Strictly positive inductive definitions are given by a set U together with con structors $C : A_1 \to \cdots A_n \to$ U, where A_i can refer to U strictly positively: (1) Either A_i is a set, which were defined before one started to introduce U. Arguments of C referring to such sets are called *non-inductive arguments*. (2) Or A_i is of the form $(B_1 \to \cdots \to B_m \to$ U) for some sets B_i defined before U was introduced. Arguments referring to such sets are called *inductive arguments*. In dependent type theory, we can extend inductive definitions by allowing A_i to refer to previous arguments, i.e. we get $C : (a_1 : A_1, \ldots, a_n : A_n) \to$ U, and A_i might depend on a_j for $j < i$. However, closer examination reveals that only dependencies on non-inductive arguments are possible: When introducing the constructor C, U has not been defined yet, so we are not able to define any sets depending on it.

In an IRD of U : Set and T : U $\to$ D, A_i might refer to previous inductive arguments via T: if $a_j : A_j = (B_1 \to \cdots \to B_k \to$ U), and $j < i$, then A_i might make use of T($a_j(b_1, \ldots, b_k)$). An example is the constructor $\widehat{\Pi}$ expressing the closure of a universe under the dependent function type (which is often written as $\Pi(A, B)$): $\widehat{\Pi}$ has type $(a : $ U$, b : $ T$(a) \to$ U) $\to$ U and the type of the second argument b of $\widehat{\Pi}$ depends on T(a).

The intuition why this is a good predicative definition is that one defines the elements of U inductively. Whenever one introduces a new element of U, one computes recursively T applied to it. Therefore, when referring to a previous inductive argument, we can make use of T applied to it.

When applying this principle, one notices that one often needs to define several universes (U_i, T_i) simultaneously. Indexed inductive-recursive definitions IIRD extend the principle of IRD so that it allows to define U : $I \to$ Set and

T : $(i : I, u : U(i)) \rightarrow D[i]$ simultaneously for all $i : I$. Here I : Set, and $i : I \Rightarrow D[i]$: Type. The constructors of U(i) might refer to U(j) for any j in a strictly positive way, and make use of T applied to previous inductive arguments.

Reduction to inductive definitions. When formalising the representation of partial recursive functions as IIRD in general, one wants to represent partial recursive functions $f : (x : A) \rightarrow B(x)$ for arbitrary A : Set, $B : A \rightarrow$ Set. Such functions will be translated into IIRD with index set A and $D[x] := B(x)$. Note that $D[x]$: Set. In [DS06] Dybjer and the author have shown that IIRD with $D[x]$: Set can always be reduced to indexed inductive definitions. We sketch the idea briefly by taking the example from the introduction. Remember that defined$^{\text{S}}_f$ had type

$$\text{defined}^{\text{S}}_f : (n : \mathbb{N}, p : f(n)\!\downarrow, q : f(\text{eval}_f(n, p))\!\downarrow) \rightarrow f(n+1)\!\downarrow \ .$$

In order to avoid the use of $\text{eval}_f(n, p)$ in the type of defined$^{\text{S}}_f$, one introduces first an inductive definition of a set $f(n)\!\downarrow^{\text{aux}} : \mathbb{N} \rightarrow$ Set, with constructors

$$\text{defined}^{0,\text{aux}}_f : f(0)\!\downarrow^{\text{aux}} \ ,$$
$$\text{defined}^{\text{S},\text{aux}}_f : (n : \mathbb{N}, p : f(n)\!\downarrow^{\text{aux}}, m : \mathbb{N}, q : f(m)\!\downarrow^{\text{aux}}) \rightarrow f(n+1)\!\downarrow^{\text{aux}} \ .$$

Then we compute recursively $\text{eval}^{\text{aux}}_f : (n : \mathbb{N}, p : f(n)\!\downarrow^{\text{aux}}) \rightarrow \mathbb{N}$ (this definition is now separated from the inductive definition of $f(\cdot)\!\downarrow$) by

$$\text{eval}^{\text{aux}}_f(0, \text{defined}^{0,\text{aux}}_f) := 0 \ ,$$
$$\text{eval}^{\text{aux}}_f(n+1, \text{defined}^{\text{S},\text{aux}}_f(n, p, m, q)) := \text{eval}^{\text{aux}}_f(m, q) \ .$$

$p : f(n)\!\downarrow^{\text{aux}}$ proves that $f(n)$ is defined, provided that, whenever we made use of $\text{defined}^{\text{S},\text{aux}}_f(n, p, m, q)$, we had $m = \text{eval}^{\text{aux}}_f(n, p)$. We introduce a correctness predicate expressing this:

$$\text{Corr}_f : (n : \mathbb{N}, p : f(n)\!\downarrow^{\text{aux}}) \rightarrow \text{Set} \ , \qquad \text{Corr}_f(0, \text{defined}^{0,\text{aux}}_f) := \text{True} \ ,$$
$$\text{Corr}_f(n+1, \text{defined}^{\text{S},\text{aux}}_f(n, p, m, q)) :=$$
$$\text{Corr}_f(n, p) \wedge \text{Corr}_f(m, q) \wedge m =_{\mathbb{N}} \text{eval}^{\text{aux}}_f(n, p) \ .$$

One can now simulate $f(\cdot)\!\downarrow$ by

$$f(\cdot)\!\downarrow' : \mathbb{N} \rightarrow \text{Set} \ , \ f(n)\!\downarrow' := (p : f(n)\!\downarrow^{\text{aux}}) \times \text{Corr}_f(n, p) \ ,$$

and simulate eval_f by

$$\text{eval}'_f : (n : \mathbb{N}, p : f(n)\!\downarrow') \rightarrow \mathbb{N} \ , \ \text{eval}'_f \ (n, \langle p, q \rangle) := \text{eval}^{\text{aux}}_f(n, p) \ .$$

In [DS06] this reduction has been carried out in detail and there we were able to show that indeed all IIRD with target type of T being a set can be simulated by indexed inductive definitions.

Note that this reduction adds an additional overhead. So we assume when verifying the correctness of partial recursive functions on the machine one probably prefers to use the original IIRD.

Restrictions to the class of IIRD. Bove and Capretta have shown that the set of partial recursive functions definable this way is Turing-complete – it contains all partial recursive functions on $\mathbb{N}$ – and that a large class of recursion schemes can be represented this way. However, not all IIRD having the above types correspond to directly executable partial recursive functions.

1. We need to replace general IIRD by restricted IIRD. The concepts of general and restricted IIRD were investigated in [DS01, DS06]. In general IIRD, one introduces constructors for the inductively defined set U and then determines for each constructor C depending on its arguments $\vec{a}$ the i s.t. $C(\vec{a}) : U(i)$. In the example used in the introduction we used in fact such kind of general IIRD. We have a constructor defined$_f^S$, and we state that defined$_f^S(n, p, q) :$ $f(n+1)\downarrow$, so the index $n+1$ depends on the arguments n, p, q. The problem with this is that it allows to define multivalued functions: Nothing prevents us from adding a second constructor defined$_f^{S,'}(n) : f(n+1)\downarrow$, s.t.

   eval$_f(n+1, $defined$_f^{S,'}(n))$ returns a different value, e.g. 5. This corresponds to adding contradictory rewrite rules, such as $f(n+1) \longrightarrow 5$.

   Furthermore this principle does not mean that the functions are directly executable, unless one has a proof of $f(n)\downarrow$ (in which case eval$_f$ allows of course to compute the value of $f(n)$). In order to evaluate $f(n)$, one has to guess the arguments of the constructor in such a way that we obtain an element of $f(n)\downarrow$, which requires in general a search process. $f(n)$ is still partially recursive (the $f(n)$ are always computable since one can always search for a proof $p : f(n)\downarrow$, and then compute eval$_f(n, p)$), but we do not regard the search for arguments as a means of directly executing a function. In order to obtain directly executable partial recursive functions we need to determine for each index its constructor. This corresponds to restricted IIRD as introduced in [DS06]. If one defines in restricted IIRD U : $I \to$ Set, one needs to determine, depending on i the set of constructors having result type U(i). The initial example can be represented as a restricted IIRD by defining it as follows:

   $$f(n)\downarrow := \text{case } n \text{ of } 0 \quad \to \text{data } C_f^0$$
   $$S(n') \to \text{data } C_f^S(p : f(n')\downarrow, \quad q : f(\text{eval}_f(n', p))\downarrow)$$

   So $f(0) \downarrow$ has constructor C_f^0, and $f(S(n')) \downarrow$ has constructor C_f^S with arguments $p : f(n')\downarrow$ and $q : f(\text{eval}_f(n', p))\downarrow$.

2. In order to avoid multivalued functions, for each argument $a : A$ there should be at most one constructor of type $f(a)\downarrow$. One can easily achieve that there is always exactly one constructor – if there is none, one can always add the constructor $C : (p : f(a) \downarrow) \to f(a)\downarrow$ corresponding to black hole recursion (i.e. rewrite rule $f(a) \longrightarrow f(a)$).

3. We disallow non-inductive arguments. Non-inductive arguments, except for the empty set and the one-element set might result in multivalued functions (different choices for one argument of the constructor might yield different proofs of $f(n){\downarrow}$ and therefore different values of $\text{eval}_f(n, p)$). Another problem with non-trivial non-inductive arguments is that when evaluating the partial recursive function, one needs to search for instances of these non-inductive argument. Therefore one does not obtain directly executable functions. We could allow non-indexed arguments indexed over the one-element set 1 and the empty set $\emptyset$. But arguments indexed over 1 can be ignored, and arguments indexed over $\emptyset$ have the effect that $f(a){\uparrow}$, which can alternatively be obtained by using black-hole recursion as above.

4. Inductive arguments should be single ones. In general IIRD of a set U, the constructor might have an inductive argument of the form $(x : A) \to \text{U}(i(x))$. In our setting such an inductive argument would be of the form $(x : A) \to f(i(x)){\downarrow}$, which expresses $f(i(x))$ is defined for all $x : A$. Such an argument requires the evaluation of f for possibly infinitely many values $i(x)$ $(x : A)$, which is non-computable. One can search for a proof of $(x : A) \to f(i(x))$ and use this search process as a means of evaluating f. However, such a search will miss the situation where $(x : A) \to f(i(x))$ is true (even constructively), but unprovable in the type theory in question. Furthermore, we do not regard such a search process as a means of directly executing a function.

So in order to obtain directly executable functions, we need to restrict inductive arguments to single valued ones, i.e. in the above situation to arguments of the form $p : f(i){\downarrow}$.

3   A Data Type of Partial Recursive Functions

In [BC05a] it was pointed out that one of the limitations of their approach is that they cannot define partial recursive functions referring to other partial recursive functions as a whole. We have given a toy example in the introduction (the function $g : \text{List}(\mathbb{N}) \to \text{List}(\mathbb{N})$). In computability theory one overcomes this problem by introducing Kleene-indices for partial recursive functions. Then one can define for instance map by having two natural numbers as arguments, one which is a Kleene-index for a partial recursive function, and the second one a code for a list. In order to do the same using the approach by Bove/Capretta, we will introduce a data type of codes for the IIRD we were referring to above. This data type is a subtype of the data type of IIRD introduced in [DS06] (see as well [DS01, DS03, DS99]). This shows the consistency of the new rules introduced in this article.

Assume $A : \text{Set}, B : A \to \text{Set}$ fixed. Unless explicitly needed, we suppress in the following dependencies on A, B. We regard a partial recursive function $f : (a : A) \to B(a)$ as being given by an IIRD, and introduce the data type Rec of codes for those IIRD, which correspond according to the previous section to partial recursive functions. So the set Rec will as well be the set of codes for partial recursive functions. We have formation rule:

$$\mathrm{Rec} : \mathrm{Set} \ .$$

The set defined inductively by a code $e : \mathrm{Rec}$ is given as

$$f_e(\cdot)\!\downarrow \ : A \to \mathrm{Set}$$

If we understand each IIRD as defining a partial recursive function, $f_e(a)\!\downarrow$ means that the function with index e is defined for argument a. The function defined recursively is

$$\mathrm{eval}_e : (a : A, p : f_e(a)\!\downarrow) \to B(a) \ ,$$

which, if the IIRD is interpreted as the definition of a partial recursive function, computes the result of this function.

Rec is a restricted IIRD, which means that for each $a : A$ we can determine the type of arguments for the constructor with result $f_e(a)\!\downarrow$. Let Rec'_a be the type of codes for possible arguments for the constructor of an IIRD e with result $f_e(a)\!\downarrow$. Then an element of Rec is given by an element of Rec'_a for each $a : A$. So we have the following formation and equality rule:

$$\mathrm{Rec}' : A \to \mathrm{Set} \ , \qquad \mathrm{Rec} = (a : A) \to \mathrm{Rec}'_a \ .$$

The type of the arguments of the constructor of $f_e(a)\!\downarrow$ and the result of $\mathrm{eval}_e(a, p)$ for the constructed element p will depend on $f_e(\cdot)\!\downarrow$ and eval_e. Since, when introducing Rec'_a, $f_e(\cdot)\!\downarrow$ and eval_e are not available, we define more generally, depending on $a : A$, $e : \mathrm{Rec}'_a$, for general X and Y, having the types of $f_e(\cdot)\!\downarrow$, eval_e respectively, the following operations:

$$\mathrm{Arg}_{a,e} : (X : A \to \mathrm{Set}, Y : (a' : A, x : X(a')) \to B(a')) \to \mathrm{Set}$$
$$\mathrm{Eval}_{a,e} : (X : A \to \mathrm{Set}, Y : (a' : A, x : X(a')) \to B(a'), \mathrm{Arg}_{a,e}(X, Y)) \to B(a)$$

$\mathrm{Arg}_{a,e(a)}(f_e(\cdot)\!\downarrow, \mathrm{eval}_e)$ will be the type of the arguments of the constructor of $f_e(a)\!\downarrow$. If using arguments p we have constructed $q : f_e(a)\!\downarrow$, then $\mathrm{eval}_e(a, q) = \mathrm{Eval}_{a,e(a)}(f_e(\cdot)\!\downarrow, \mathrm{eval}_e, p)$. If we call the constructor for $f_e(a)\!\downarrow$ $\mathrm{tot}_{e,a}$, then the introduction and equality rules for $f_e(\cdot)\!\downarrow$ and eval_e are as follows:

$$\mathrm{tot}_{e,a} : \mathrm{Arg}_{a,e(a)}(f_e(\cdot)\!\downarrow, \mathrm{eval}_e) \to f_e(a)\!\downarrow \ ,$$
$$\mathrm{eval}_e(a, \mathrm{tot}_{e,a}(p)) = \mathrm{Eval}_{a,e(a)}(f_e(\cdot)\!\downarrow, \mathrm{eval}_e, p) \ .$$

We define additionally outside type theory for closed $e : \mathrm{Rec}$ the partial recursive function $\{e\} : (a : A) \to B(a)$. $\{e\}$ will be defined in such a way that we can prove outside type theory $\{e\}(a)\!\downarrow \Leftrightarrow \exists p.p : f_e(a)\!\downarrow$ and that if $p : f_e(a)\!\downarrow$, then $\{e\}(a) \simeq \mathrm{eval}_e(a, p)$.

In order to define $\{e\}(a)$, we define recursively (outside type theory) an auxiliary partial recursive function

$$\mathrm{compute}^{\mathrm{aux}} : (e : \mathrm{Rec}, a : A, e' : \mathrm{Rec}'_a) \to B(a)$$

compute$^{\mathrm{aux}}(e, a, e')$ roughly speaking computes the subcomputation of $\{e\}(a)$, where we consider the subcode e' of $e(a)$. However, in the definition we do not assume e' to be a subcode of $e(a)$. Then we define for $e : \mathrm{Rec}$

$$\{e\}(a) \simeq \mathrm{compute}^{\mathrm{aux}}(e, a, e(a))$$

Rec'_a has 2 constructors:

1. Initial (constant) case: the constructor for $f_e(a)\!\downarrow$ has no arguments (or more precisely the trivial argument $x : \{*\}$ for the one-element set $\{*\}$). $\mathrm{eval}_e(a, p)$ returns, independently of p, a fixed element $b : B(a)$:

$$\mathrm{const}_a : B(a) \to \mathrm{Rec}'_a \qquad \mathrm{Arg}_{a, \mathrm{const}_a(b)}(X, Y) = \{*\}$$
$$\mathrm{Eval}_{a, \mathrm{const}_a(b)}(X, Y, *) = b$$
$$\mathrm{compute}^{\mathrm{aux}}(e, a, \mathrm{const}_a(b)) \simeq b$$

2. A single inductive argument. The constructor of $f_e(a)\!\downarrow$ has as an inductive argument $p : f_e(a')\!\downarrow$, and depending on $m : \mathrm{eval}_e(a', p)$ later arguments. As a partial recursive function this means that we make a recursive call to $f_e(a')\!\downarrow$. Depending on the result m, we choose further steps. So the constructor rec_a of Rec'_a needs to have as arguments a' and a function $e' : B(a') \to \mathrm{Rec}'_a$, which determines depending on the result $b : B(a')$ of $\mathrm{eval}_e(a', p)$ the later arguments of the constructor. We obtain

$$\mathrm{rec}_a : (a' : A, e' : B(a') \to \mathrm{Rec}'_a) \to \mathrm{Rec}'_a$$
$$\mathrm{Arg}_{a, \mathrm{rec}_a(a', e')}(X, Y) = (x : X(a')) \times \mathrm{Arg}_{a, e'(Y(a', x))}(X, Y)$$
$$\mathrm{Eval}_{a, \mathrm{rec}_a(a', e')}(X, Y, \langle x, y \rangle) = \mathrm{Eval}_{a, e'(Y(a', x))}(X, Y, y)$$
$$\mathrm{compute}^{\mathrm{aux}}(e, a, \mathrm{rec}_a(a', g)) \simeq \begin{cases} \mathrm{compute}^{\mathrm{aux}}(e, a, g(b)), \\ \qquad \text{if } \mathrm{compute}^{\mathrm{aux}}(e, a', e(a)) \simeq b, \\ \perp, \quad \text{if } \mathrm{compute}^{\mathrm{aux}}(e, a', e(a)) \uparrow. \end{cases}$$

We usually omit the parameter a in const_a, rec_a. We observe that Rec'_a is an inductively defined set: it is like a W-type with branching degrees $(B(a))_{a:A}$, but with additional leaves $\mathrm{const}(b)$. Arg and Eval are then defined by recursion on Rec'_a. $f_e(\cdot)\!\downarrow$ and eval_e are given by an IIRD which is determined by e.

Reference to other partial recursive functions. If one wants to show the closure of the resulting set of partial recursive functions under operations like composition, one sees that one needs the possibility to refer to other partial recursive functions, which is not available in the above calculus. In the context of dependent type theory, allowing this will cause one problem: we want to refer in the definition of partial recursive functions to other partial recursive functions g of any type $(c : C) \to D(c)$ where $\langle C, D \rangle : \mathrm{Fam}(\mathrm{Set})$. Here $\mathrm{Fam}(\mathrm{Set}) :=$ $(X : \mathrm{Set}) \times (X \to \mathrm{Set}) : \mathrm{Type}$ is the type of families of sets. If we want to allow reference to arbitrary such functions, we will end up with $\mathrm{Rec}_{A,B} : \mathrm{Type}$ instead of $\mathrm{Rec}_{A,B} : \mathrm{Set}$. (We will no longer suppress the arguments A, B of Rec.) This causes problems when defining partial recursive functions having elements of $\text{Rec}_{A,B}$ as arguments. However, one can usually restrict the domain and codomain of partial recursive functions used to elements of a universe, and therefore obtain $\text{Rec}_{A,B}$ to be a set. If one is for instance interested in functions occurring in traditional computability theory only, one can restrict oneself to a universe $\{\mathbb{N}^k \mid k \in \mathbb{N}\}$, i.e. $\text{U} := \mathbb{N} : \text{Set}$ and for $n : \mathbb{N}$ $\text{T}(n) := \mathbb{N}^n : \text{Set}$.

In order to keep the notations simple we will in the following extend $\text{Rec}_{A,B}$ to $\text{Rec}^+_{A,B}$ in such a way that it refers to partial recursive functions of arbitrary $\langle C, D \rangle : \text{Fam}(\text{Set})$, which means it is a true type, and keep in mind that if $\langle C, D \rangle$ are restricted to elements of a universe, we obtain $\text{Rec}^+_{A,B} : \text{Set}$.

There are two alternative ways of dealing with reference to other partial recursive functions:

1. The conceptually easier one is to treat such references as recursive calls of simultaneously defined functions. In order to represent $f : (a : A) \to C(a)$ which makes use of $g : (a : A') \to C'(a)$, we combine f, g into one function $fg : (a . A'') \to C''(a)$ as follows:
   We have $\langle A, C \rangle, \langle A', C' \rangle : \text{Fam}(\text{Set})$. Define

$$\langle A, C \rangle \oplus \langle A', C' \rangle := \langle A + A', [C, C'] \rangle$$

where $A + A'$ is the disjoint union of A and A' and

$$[C, C'] : (A + A') \to \text{Set} ,$$
$$[C, C'](\text{inl}(x)) := C(x) , \quad [C, C'](\text{inr}(x')) := C'(x') .$$

Let $\langle A'', C'' \rangle := \langle A, C \rangle \oplus \langle A', C' \rangle$. Then define

$$f \oplus g : (a : A'') \to C''(a) ,$$
$$(f \oplus g)(\text{inl}(a)) \simeq f(a) , \quad (f \oplus g)(\text{inr}(b)) \simeq g(b) .$$

Define $\text{emb}_{\text{inl}} : (a : A, \text{Rec}'_{A,C,a}) \to \text{Rec}'_{A'',C'',\text{inl}(a)}$ and
$\text{emb}_{\text{inr}} : (a' : A', \text{Rec}'_{A',C',a'}) \to \text{Rec}'_{A'',C'',\text{inr}(a')}$, by replacing all occurrences of $\text{rec}(x, g)$ by $\text{rec}(\text{inl}(x), g)$ or $\text{rec}(\text{inr}(x), g)$, respectively.
Let $fg := f \oplus g$. Assume we can define f recursively by making use of g. We obtain an index e_{fg} of fg by setting $e_{fg}(\text{inr}(b)) := e_g(b)$ and defining $e_{fg}(\text{inl}(a))$ like the recursive definition of f, but replacing recursive calls to $f(a')$ by recursive calls to $fg(\text{inl}(a'))$ and calls of $g(b)$ by recursive calls to $fg(\text{inr}(b))$.
As an example we show how to define,
assuming $g : (a : A) \to C(a)$, $h : (a : A) \to C(a) \to B(a)$,
the function $\qquad f : (a : A) \to B(a)$, $\quad f(a) \simeq h(a, g(a))$.
Let $C' := (a : A) \times C(a)$, $B' : C' \to \text{Set}$, $B'(\langle a, c \rangle) := C(a)$.
Let $h' : (c : C') \to B'$, $h'(\langle a, c \rangle) := h(a, c)$ be the uncurried form of h.
Let $e_g : \text{Rec}_{A,C}$, $e_{h'} : \text{Rec}_{C',B'}$ be indices for g and h'.
Let $\langle A^+, B^+ \rangle := \langle A, B \rangle \oplus \langle A, C \rangle \oplus \langle C', B \rangle$.
Let $fgh := f \oplus g \oplus h'$.

Let in_f, in_g, in_h be the left, middle and right injection from A, A, C', respectively, into A^+.

Let $e'_g := \lambda b.\text{emb}_{\text{in}_g}(e_g(b)) : (b : B) \to \text{Rec}'_{A^+,B^+,\text{in}_g(b)}$,

$\quad e'_h := \lambda c.\text{emb}_{\text{in}_h}(e_h(c)) : (c : C') \to \text{Rec}'_{A^+,B^+,\text{in}_h(c)}$.

We introduce an index $e_{fgh} : \text{Rec}_{A^+,B^+}$ as follows:

$$e_{fgh}(\text{in}_g(b)) := e'_g(b) , \qquad e_{fgh}(\text{in}_h(c)) := e'_h(c)$$
$$e_{fgh}(\text{in}_f(a)) := \text{rec}(\text{in}_g(a), \lambda c.\text{rec}(\text{in}_h(\langle a, c \rangle), \lambda b.\text{const}(b)))$$

Using extensional equality one can see that there is a bijection
$g^{\cong}(a) : f_{e_{fgh}}(\text{in}_g(a))\downarrow \cong f_{e_g}(a)\downarrow$, and we have
$\text{eval}_{e_{fgh}}(\text{in}_g(a), p) = \text{eval}_{e_g}(a, g^{\cong}(a, p))$, similarly for $f_{e_{fgh}}(\text{in}_h(a))\downarrow$ and
$f_{e_h}(a)\downarrow$. Furthermore, the argument of the constructor $\text{tot}_{e_{fgh},\text{in}_f(a)}$ for
$f_{e_{fgh}}(\text{in}_f(a))\downarrow$ has type

$$(p : f_{e_{fgh}}(\text{in}_g(a))\downarrow) \times (f_{e_{fgh}}(\text{in}_h(\langle a, \text{eval}_{e_{fgh}}(\text{in}_g(a), p)\rangle))\downarrow \times \{*\}) ,$$

and we have

$$\text{eval}_{e_{fgh}}(\text{in}_f(a), \text{tot}_{e_{fgh},\text{in}_f(a)}(\langle p, \langle q, * \rangle \rangle))$$
$$= \text{eval}_{e_{fgh}}(\text{in}_h(\langle a, \text{eval}_{e_{fgh}}(\text{in}_g(a), p)\rangle), q)$$

Modulo the aforementioned isomorphisms, this means that $f_{e_{fgh}}(\text{in}_f(a))\downarrow$ iff
there exists $p : f_{e_g}(a)\downarrow$ and $q : f_{e_h}(\langle a, \text{eval}_{e_g}(a, p)\rangle)\downarrow$, and that
$\text{eval}_{e_{fgh}}(\text{in}_f(a), \_) = \text{eval}_{e_h}(\langle a, \text{eval}_{e_g}(a, p)\rangle, q)$, i.e. the function defined by
e_{fgh} composed with in_f is the composition of the functions given by e_g and
e_h.

In general we define

$$\text{Rec}^+_{A,B} := (C : \text{Set}) \times (D : C \to \text{Set}) \times \text{Rec}_{C+A,[D,B]} ,$$

and for $e = \langle C, D, e' \rangle : \text{Rec}^+_{A,B}$ we define $f^+_e(\cdot)\downarrow : A \to \text{Set}$,
$f^+_e(a)\downarrow := f_{e'}(\text{inr}(a))\downarrow$ and $\text{eval}^+_e : (a : A, p : f^+_e(a)\downarrow) \to B(a)$,
$\text{eval}^+_e(a, p) := \text{eval}^+_{e'}(\text{inr}(a), p)$.

2. The approach which is easier for implementing proofs is to extend Rec by a
constructor which calls a partial recursive function having an arbitrary type.
So we extend $\text{Rec}_{A,B}$, $\text{Rec}'_{A,B,a}$ to types $\text{Rec}^+_{A,B}$, $\text{Rec}^{+,\prime}_{A,B,a}$ with an additional
constructor

$$\text{call}_a : (C : \text{Set}, D : C \to \text{Set}, e : \text{Rec}^+_{C,D}, c : C, g : D(c) \to \text{Rec}^{+,\prime}_{A,B,a})$$
$$\to \text{Rec}^{+,\prime}_{A,B,a}$$
$$\text{Arg}_{a,\text{call}_a(C,D,e,c,g)}(X, Y) = (p : f_{C,D,e}(c)\downarrow) \times \text{Arg}_{a,g(\text{eval}_{C,D,e}(c,p))}(X, Y)$$
$$\text{Eval}_{a,\text{call}_a(C,D,e,c,g)}(X, Y, \langle p, q \rangle) = \text{Eval}_{a,g(\text{eval}_{C,D,e}(c,p))}(X, Y, q)$$

Note that with this approach $\text{Rec}^+_{C,D} : \text{Type}$ are defined simultaneously for
all $\langle C, D \rangle : \text{Fam}(\text{Set})$ and simultaneously with $f_{C,D}(\cdot)\downarrow$, $\text{eval}_{C,D}$, Arg, Eval.

With both approaches we can show the following theorem:

Theorem 1. *(a) The type of partial recursive functions represented by Rec^+ contains all total functions and is closed under composition, primitive recursion into higher types, and the μ-operator for partial recursive functions.*
(b) All partial recursive functions $\mathbb{N}^n \to \mathbb{N}$ are represented in $\mathrm{Rec}^+_{\mathbb{N}^n,\mathbb{N}}$. For this the restriction of calls of other partial recursive functions to types being elements of a universe containing $\{\mathbb{N}^k \mid k \in \mathbb{N}\}$ suffices.
(c) If $e : \mathrm{Rec}^+_{A,B}$ is derived without a context, then we have $\{e\}(a)\!\downarrow$ iff $p : f_e(a)\!\downarrow$ for some p. Furthermore, if $p : f_e(a)\!\downarrow$, then $\{e\}(a) \simeq \mathrm{eval}_e(a,p)$, and $\{e\}$ is partial recursive.

References

[BC05a]   A. Bove and V. Capretta. Modelling general recursion in type theory. *Mathematical Structures in Computer Science*, 15(4):671–708, August 2005.

[BC05b]   A. Bove and V. Capretta. Recursive functions with higher order domains. In P. Urzyczyn, editor, *Typed Lambda Calculi and Applications.*, volume 3461 of *LNCS*, pages 116–130. Springer, 2005.

[Cap05]   Venanzio Capretta. General recursion via coinductive types. *Logical Methods in Computer Science*, 1(2):1–18, 2005.

[DS99]   Peter Dybjer and Anton Setzer. A finite axiomatization of inductive-recursive definitions. In J.-Y. Girard, editor, *Typed Lambda Calculi and Applications*, volume 1581 of *Lecture Notes in Computer Science*, pages 129–146, 1999.

[DS01]   Peter Dybjer and Anton Setzer. Indexed induction-recursion. In R. Kahle, P. Schroeder-Heister, and R. Stärk, editors, *Proof Theory in Computer Science*, pages 93 – 113. LNCS 2183, 2001.

[DS03]   Peter Dybjer and Anton Setzer. Induction-recursion and initial algebras. *Annals of Pure and Applied Logic*, 124:1 – 47, 2003.

[DS06]   Peter Dybjer and Anton Setzer. Indexed induction-recursion. *Journal of Logic and Algebraic Programming*, 66:1 – 49, 2006.

[Dyb94]   Peter Dybjer. Inductive families. *Formal Aspects of Comp.*, 6:440–465, 1994.

[Dyb00]   Peter Dybjer. A general formulation of simultaneous inductive-recursive definitions in type theory. *Journal of Symbolic Logic*, 65(2):525–549, June 2000.

Partially Ordered Connectives and Σ_1^1 on Finite Models[*]

Merlijn Sevenster and Tero Tulenheimo

[1] Institute for Logic, Language and Computation, University of Amsterdam, Plantage
Muidergracht 24, 1018 TV, Amsterdam, The Netherlands
sevenstr@science.uva.nl
[2] Academy of Finland; Department of Philosophy, University of Helsinki,
P.O. Box 9 (Siltavuorenpenger 20 A), 00014 University of Helsinki, Finland
tero.tulenheimo@helsinki.fi

Abstract. In this paper we take up the study of Henkin quantifiers with
boolean variables [4], also known as partially ordered connectives [19].
We consider first-order formulae prefixed by partially ordered connectives, denoted **D**, on finite structures. **D** is characterized as a fragment
of second-order existential logic $\Sigma_1^1 \heartsuit$, whose formulae do not allow existential variables as arguments of predicate variables. By means of a game
theoretic argument, it is shown that $\Sigma_1^1 \heartsuit$ harbors a strict hierarchy induced by the arity of predicate variables, and that it is not closed under
complementation. It is further shown that allowing at most one existential variable to appear as an argument of a predicate variable, already
yields a logic coinciding with full Σ_1^1.

Keywords: Henkin quantifiers, partially ordered connectives, NP vs.
coNP, finite model theory.

1 Introduction

Fagin's Theorem [9]—characterizing NP in terms of the expressive power of
Σ_1^1 over finite models—reveals the intimate connection between finite model
theory and complexity theory. As a methodological consequence it appears that
questions and results regarding a complexity class may bear relevance for logic
and vice versa. For instance, the complexity theorist's headache caused by the
NP = coNP-problem can now be shared by the logician working on the $\Sigma_1^1 = \Pi_1^1$-
problem.[1] Indeed, logicians working in finite model theory address this problem.
By and large they go about by mapping out *fragments* of various logics. A case
in point is Fagin's [10] study of the *monadic* fragments of Σ_1^1 and Π_1^1, showing
that they do not coincide.

[*] The authors gratefully acknowledge Peter van Emde Boas, Lauri Hella, Benedikt
Löwe and the referees for their contributions to the paper, and NWO for financial
support (visitor's grant B 62-608).
[1] Solving the NP=coNP-problem is worth a headache: if NP$\neq$coNP, then P$\neq$NP.

A. Beckmann et al. (Eds.): CiE 2006, LNCS 3988, pp. 516–525, 2006.
© Springer-Verlag Berlin Heidelberg 2006

The results in [10] aroused a lot of interest in monadic languages [2, 3, 20], but we are still waiting for methods to separate binary, existential, second-order logic from 3-ary, existential, second-order logic, see [5], or even from binary, universal, second-order logic.

The present paper will be concerned with the finite model theory of languages involving (what we will call) *Henkin quantifiers with restricted quantifiers*, also known as *partially ordered connectives*. Henkin quantifiers $H^n_k xy$ are objects of the form

$$\begin{pmatrix} \forall x_{11} \ldots \forall x_{1k} \; \exists y_1 \\ \vdots \quad \ddots \quad \vdots \quad \vdots \\ \forall x_{n1} \ldots \forall x_{nk} \; \exists y_n \end{pmatrix} \tag{1}$$

that prefix first-order formulae ϕ. Here and henceforth, a series of variables as in $x_{11}, \ldots, x_{nk}$ is abbreviated by $\boldsymbol{x}$. On suitable structures $\mathfrak{A}$, the formula $H^n_k \boldsymbol{xy} \; \phi(\boldsymbol{x}, \boldsymbol{y})$ is defined to be true iff there are k-ary functions $f_1, \ldots, f_n$ on the universe of $\mathfrak{A}$ such that

$$\mathfrak{A} \models \forall \boldsymbol{x} \; \phi(\boldsymbol{x}, f_1(\boldsymbol{x}_1), \ldots, f_n(\boldsymbol{x}_n)) \tag{2}$$

where $\boldsymbol{x}_i = x_{i1}, \ldots, x_{ik}$. It is a milestone result in the theory of Henkin quantication that the logic obtained by applying Henkin quantifiers to first-order formulae, denoted **H**, coincides with Σ^1_1, cf. [8, 21]. Referring to Fagin's Theorem, Blass and Gurevich [4, Theorem 1] draw the conclusion that NP can be characterized in terms of **H** as well. In the same publication the authors study what constraints can be imposed on the existentially quantified variables in a Henkin quantifier, such as $\boldsymbol{y}$ in (1), without the quantifier losing its power to express NP-complete problems. It turns out that Henkin quantifiers of the form

$$\begin{pmatrix} \forall x_{11} \ldots \forall x_{1k} \; \exists \alpha_1 \\ \forall x_{21} \ldots \forall x_{2k} \; \exists \alpha_2 \end{pmatrix} \tag{3}$$

cannot express NP-complete problems, unless NL = NP. The variables α_1 and α_2 appearing in (3) are *boolean variables* that range over a fixed two-element domain. In this sense $\exists \alpha_i$ is a 'restricted quantifier', whence the term 'Henkin quantifier with restricted quantifiers'.

The model theory for Henkin quantifiers with restricted quantifiers was taken up in [19], be it under the name of 'partially ordered connectives' and written in the following format:

$$\begin{pmatrix} \forall x_{11} \ldots \forall x_{1k} \; \bigvee i_1 \\ \vdots \quad \ddots \quad \vdots \quad \vdots \\ \forall x_{n1} \ldots \forall x_{nk} \; \bigvee i_n \end{pmatrix} \tag{4}$$

denoted $D^n_k \boldsymbol{xi}$. The usage of the symbol $\bigvee$ reflects the fact that the variables i_j range over a fixed finite domain. Sandu and Väänänen [19, Proposition 2] show that any first-order formula ϕ prefixed by the partially ordered connective $D^2_1 \boldsymbol{xi}$ can be translated into $H^2_1 \boldsymbol{xy} \; \phi'$, for some first-order ϕ'. Furthermore, they provide an *Ehrenfeucht-Fraïssé game* for partially ordered connectives and use it to prove non-definability results. Note that there are first-order formulae ϕ that can express NP-complete problems, when prefixed with the partially ordered connective $\mathsf{D}_1^3 xi$, in virtue of Blass and Gurevich's result; 3-colorability of graphs is a case in point. Other publications on Henkin quantifiers and partially ordered connectives in relation to complexity theory include [13, 14, 16, 17, 18].

In this paper we characterize the logic **D**—the result of applying (4) to first-order formulae for arbitrary k, n—as a fragment of Σ_1^1. The relevant fragment only allows universally quantified variables to appear as arguments of (existentially quantified) predicate variables. As this fragment is rather natural, it may be of interest to the descriptive complexity community to observe that (a) **D** can express a property expressible in $(k+1)$-ary, existential, second-order logic that cannot be expressed in k-ary, existential, second-order logic; and that (b) **D** is not closed under complementation, as it can express 2-COLORABILITY but not its complement. Along the way we prove that the Henkin quantifier $\mathsf{H}_1^2 x$ is not definable in **D** and that **D** is strictly contained in NP.

In Section 2, we introduce the apparatus necessary to get going. In Section 3, **D** is characterized as a fragment of Σ_1^1. Using this characterization, we prove result (a). In Section 4, an Ehrenfeucht-Fraïssé game for **D** is given, and it is used to show that **D** is not closed under complementation, cf. (b). In Section 5, we show that if $\Sigma_1^1 \heartsuit$ is extended so as to allow predicate variables to have at most one existential variable among their arguments, the resulting logic coincides with full Σ_1^1.

2   Preliminaries

A *vocabulary* τ is a finite set of relation symbols, rigidly including the equality symbol. Vocabularies do not contain constant or function symbols. Results can easily be extended to vocabularies with constant symbols, though. A *finite τ-structure* $\mathfrak{A} = \langle A, \langle R^{\mathfrak{A}} \rangle_{R \in \tau} \rangle$ consists of a finite set A, referred to as the *universe of* $\mathfrak{A}$, and interpretations of the relation symbols of τ on A. Here and henceforth, every structure is finite and for this reason we omit mentioning this. The equality symbol is interpreted as the identity relation. If τ only contains one binary relation symbol, other than the equality symbol, then any τ-structure is called a *digraph (directed graph)*. If $\mathfrak{G} = \langle G, R^{\mathfrak{G}} \rangle$ is a digraph and $R^{\mathfrak{G}}$ is symmetric, then $\mathfrak{G}$ is a *graph*. A class relevant to this paper is n-COLORABILITY holding of those finite graphs whose chromatic number is $\leq n$. Conversely, let $\overline{n\text{-COLORABILITY}}$ denote the complement of n-COLORABILITY with respect to the class of graphs.

Define an *implicit matrix τ-formula* γ as a function of type $\{0, 1\}^k \to \mathbf{FO}(\tau)$, where k is an integer and $\mathbf{FO}(\tau)$ is first-order logic over τ. Let $\mathbf{D}_k(\tau)$ be the logic with formulae of the form $\mathsf{D}_k^n xi \ \gamma(i)(x)$, for arbitrary n. The notions of *bound* and *free variable* are canonically extended from first-order logic so as to apply to the variables i as well. A *sentence* is a formula without free variables. We shall usually omit explicit indication of as many variables from the formulae as possible without losing readability. In this manner we may write $\mathsf{D}_k^n \gamma$ instead of $\mathsf{D}_k^n xi \ \gamma(i)(x)$. Put $\mathbf{D}(\tau) = \bigcup_k \mathbf{D}_k(\tau)$.

Let $\mathfrak{A}$ be a τ-structure and let $\Gamma = D_k^n \boldsymbol{x} \boldsymbol{i} \; \gamma(\boldsymbol{i})(\boldsymbol{x}) \in \mathbf{D}$. Then, Γ is true on $\mathfrak{A}$ iff there exist functions $f_1, \ldots, f_n : A^k \to \{0, 1\}$ such that

$$\mathfrak{A} \models \forall \boldsymbol{x} \; \gamma(f_1(\boldsymbol{x}_1), \ldots, f_n(\boldsymbol{x}_n))(\boldsymbol{x}) \tag{5}$$

Let $\Sigma_{1,k}^1(\tau)$ be the fragment of $\Sigma_1^1(\tau)$ whose predicate variables have arity $\leq k$. Our particular interest will pertain to the fragments $\Sigma_{1,k}^1(\tau)$, called k-ary, *existential, second-order logic*. If k equals 1 or 2, we arrive at *monadic* and *binary*, existential, second-order logic: $\Sigma_{1,1}^1(\tau) = M\Sigma_1^1(\tau)$ and $\Sigma_{1,2}^1(\tau) = B\Sigma_1^1(\tau)$. For the semantics of first and second-order logic, we refer the reader to [6].

If Φ and Ψ are τ-sentences for which the satisfaction relation $\models$ is defined, and for every τ-structure we have that $\mathfrak{A} \models \Phi$ iff $\mathfrak{A} \models \Psi$, then we say that they are *equivalent*.

Let $\mathbf{L}(\tau)$ and $\mathbf{L}'(\tau)$ be logics for which $\models$ is defined and let C be a class of (finite) τ-structures. Then C is *characterized* by $\Phi \in \mathbf{L}(\tau)$ if for every τ-structure $\mathfrak{A}$ it is the case that $\mathfrak{A} \in$ C iff $\mathfrak{A} \models \Phi$. If some of its formulae characterize the class C, then $\mathbf{L}(\tau)$ is said to characterize or *express* C as well. We write $\mathbf{L}(\tau) \leq \mathbf{L}'(\tau)$ to denote that for every formula $\Phi \in \mathbf{L}(\tau)$ there is an equivalent $\Psi \in \mathbf{L}'(\tau)$. We write $\mathbf{L}(\tau) = \mathbf{L}'(\tau)$, if $\mathbf{L}(\tau) \leq \mathbf{L}'(\tau)$ and $\mathbf{L}(\tau) \geq \mathbf{L}'(\tau)$. If $\mathbf{L}(\tau) \leq \mathbf{L}'(\tau)$ and there is a class characterizable in $\mathbf{L}'(\tau)$ that is not characterizable in $\mathbf{L}(\tau)$, we write $\mathbf{L}(\tau) < \mathbf{L}'(\tau)$.

By means of a game theoretical argument we show that $\mathbf{D}$ cannot characterize the class of structures with a universe of even cardinality, EVEN. The latter class, however, is definable by a Henkin quantifier with unrestricted quantifiers.

Proposition 1. *There exists a first-order formula* ϕ, *such that* $\mathsf{H}_1^2 \; \phi$ *characterizes* EVEN.

Proof. Structure $\mathfrak{A}$ has a universe A with even cardinality iff there exists a function $f : A \to A$ such that for every $a \in A$, $f(f(a)) = a$ and $f(a) \neq a$. The latter condition is expressed by the following formula: $\mathsf{H}_1^2 x_1 x_2 y_1 y_2 \; \phi(x_1, x_2, y_1, y_2)$, where $\phi(x_1, x_2, y_1, y_2) = (x_1{=}x_2 \to y_1{=}y_2) \land (y_1{=}x_2 \to y_2{=}x_1) \land (x_1{\neq}y_1)$. $\square$

3   A Characterization of $\mathbf{D}_k$

In this section $\mathbf{D}_k(\tau)$ is characterized as a fragment of $\Sigma_{1,k}^1(\tau)$. First we lay down a translation result. To this end, let $\Gamma = D_k^n \; \gamma$ be a $\mathbf{D}_k(\tau)$-formula, where

$$\Gamma = D_k^n \; \gamma = \begin{pmatrix} \forall x_{11} \ldots \forall x_{1n} & \bigvee i_1 \\ \vdots \quad \ddots \quad \vdots & \vdots \\ \forall x_{k1} \ldots \forall x_{kn} & \bigvee i_k \end{pmatrix} \gamma \tag{6}$$

Define the translation of Γ into $\Sigma_{1,k}^1(\tau)$, written $T(\Gamma)$, as follows

$$\exists X_1 \ldots \exists X_k \forall \boldsymbol{x} \begin{bmatrix} X_1(\boldsymbol{x}_1) \land \ldots \land \; X_k(\boldsymbol{x}_k) \to \gamma(1, \ldots, 1)(\boldsymbol{x}) \\ \vdots \\ \neg X_1(\boldsymbol{x}_1) \land \ldots \land \neg X_k(\boldsymbol{x}_k) \to \gamma(0, \ldots, 0)(\boldsymbol{x}) \end{bmatrix} \tag{7}$$

where the X_i are k-ary predicate variables. The square brackets enclosing the implications should be read as their conjunction; using them reflects the matrix-style of presenting γ. The block of implications is referred to as γ's *explication*. The translation hinges on the insight that every function $f : A^k \to \{0,1\}$ can be mimicked by the set $X = \{a \in A^k \mid f(a) = 1\}$.

Proposition 2. *For every sentence* $\Gamma \in \mathbf{D}_k(\tau)$, Γ *and* $T(\Gamma)$ *are equivalent.*

We proceed by giving a characterization of $\mathbf{D}_k$ as a fragment of $\Sigma^1_{1,k}$.

Definition 1. *Let* Φ *be a second-order* τ-*formula. Call* Φ *sober if for every predicate variable* X *in* Φ, *it is the case that (i)* X *is not bound in* Φ *and (ii)* $X(\mathbf{x})$ *occurring in* Φ *implies that all variables in* $\mathbf{x}$ *are free in* Φ. *Let* $\Sigma^1_1 \heartsuit k(\tau)$ *be the fragment of* $\Sigma^1_{1,k}(\tau)$ *containing all formulae of the form*

$$\exists X_1 \ldots \exists X_m \forall x_1 \ldots \forall x_n \ \Phi \tag{8}$$

where Φ *is sober. Put* $\Sigma^1_1 \heartsuit(\tau) = \bigcup_k \Sigma^1_{1,k} \heartsuit(\tau)$.

So any sober formula is a second-order formula, but only in virtue of the fact that it contains predicate variables. If Φ is a sober formula occurring in a $\Sigma^1_{1,k} \heartsuit(\tau)$-formula as in (8), then there are no existentially quantified variables among the arguments of a predicate variable. In Section 5 we see that the slightest extension in this respect results in a logic that enjoys the expressive power of full NP.

As an example, consider the $\Sigma^1_1 \heartsuit$-formula $\exists X_1 \exists X_2 \exists X_3 \forall x_1 \forall x_2 \ (\Phi \wedge \Phi')$ that characterizes 3-COLORABILITY, where $(\Phi \wedge \Phi')$ is a sober formula:

$$\Phi = \left(\bigvee_{i \in \{1,2,3\}} X_i(x_1) \right) \wedge \left(\bigwedge_{i \in \{1,2,3\}} \bigwedge_{j \in \{1,2,3\} - \{i\}} \neg(X_i(x_1) \wedge X_j(x_1)) \right) \tag{9}$$

$$\Phi' = \left(\bigwedge_{i \in \{1,2,3\}} (X_i(x_1) \wedge X_i(x_2) \to \neg R(x_1, x_2)) \right) \tag{10}$$

Theorem 1. $\mathbf{D}_k(\tau) = \Sigma^1_{1,k} \heartsuit(\tau)$. *Hence,* $\mathbf{D}(\tau) = \Sigma^1_1 \heartsuit(\tau)$.

Proof. The inclusion from-left-to-right is accounted for by the translation $T(\cdot)$. The converse inclusion is more involved, hinging on the claim that every sober formula is equivalent to the explication of an implicit matrix formula. $\square$

The characterization of $\mathbf{D}$ may speed up the finding of interesting properties it enjoys, for second-order logic happens to be more intensively studied than partially ordered connectives. Finding formulae with partially ordered connectives expressing a particular property on structures can be hard labor. Now that we have characterized $\mathbf{D}_k$, we can safely conclude that any property expressible in $\Sigma^1_{1,k} \heartsuit(\tau)$ is expressible in $\mathbf{D}_k(\tau)$ as well. A concrete—and relevant!—example of this mode of research can be found in the following theorem.

Theorem 2. *Let* $k \geq 2$ *be an integer and let* τ_k *be a vocabulary with at least one k-ary relation symbol and the linear order symbol[2]* $<$. *Then,* $\mathbf{D}_{k-1}(\tau_k) < \mathbf{D}_k(\tau_k)$.

---

[2] That is, on a τ_k-structure $\mathfrak{A}$, the extension of $<$ is a linear order on A.

Proof. Ajtai [1] proved the following result: Let $k \geq 2$ and let $\tau_k = \{P, <\}$, where P is a k-ary relation symbol. Then, the class $\mathbf{C}_k$ of τ_k-structures $\mathfrak{A}$ such that $P^{\mathfrak{A}}$ has even cardinality is not characterizable in $\Sigma^1_{1,k-1}(\tau_k)$, but it is characterizable in $\Sigma^1_{1,k}(\tau_k)$.[3]

To separate $\mathbf{D}_k$ from $\mathbf{D}_{k-1}$, we show that $\mathbf{C}_k$ is expressible by a formula in $\mathbf{D}_k(\tau_k)$. This suffices to prove the statement, since by Theorem 1, $\Sigma^1_{1,k-1}\heartsuit(\tau_k)$ is a fragment of $\Sigma^1_{1,k-1}(\tau_k)$, and for this reason cannot express $\mathbf{C}_k$.

Intuitively, the $\Sigma^1_{1,k}\heartsuit(\{P, <\})$-formula Υ_k that characterizes $\mathbf{C}_k$ over τ_k-structures lifts the linear order $<$ to a linear order ψ_k of k-tuples. With respect to this lifted linear order, Υ_k expresses that there exists a subset of k-tuples of objects from the domain Q such that

1. Q is a subset of $P^{\mathfrak{A}}$;
2. the ψ_k-minimal k-tuple that is in $P^{\mathfrak{A}}$ is also in Q, and the ψ_k-maximal k-tuple that is in $P^{\mathfrak{A}}$ is not in Q;
3. if two k-tuples are in $P^{\mathfrak{A}}$ and there is no k-tuple in between them (in the ordering constituted by ψ_k) that is in $P^{\mathfrak{A}}$, then exactly one of the k-tuples is in Q.

We omit further details in the interest of space.   □

4   Ehrenfeucht-Fraïssé Game for D

Ehrenfeucht-Fraïssé games or *model comparison games* are usually employed to prove that some property is not definable in a certain logic. These games were first introduced for first-order logic in [7, 11].

Let the *quantifier rank* of a first-order formula be its maximum number of nested quantifiers. Let m be an integer. If $\mathfrak{A}, \mathfrak{B}$ are τ-structures, $\boldsymbol{x}^{\mathfrak{A}} = \langle x^{\mathfrak{A}}_1, \ldots, x^{\mathfrak{A}}_r \rangle \in A^r$, and $\boldsymbol{x}^{\mathfrak{B}} = \langle x^{\mathfrak{B}}_1, \ldots, x^{\mathfrak{B}}_r \rangle \in B^r$, then the *m-round Ehrenfeucht-Fraïssé game on the structures $\mathfrak{A}$ and $\mathfrak{B}$*, denoted by

$$EF^{\mathbf{FO}}_m(\langle \mathfrak{A}, \boldsymbol{x}^{\mathfrak{A}} \rangle, \langle \mathfrak{B}, \boldsymbol{x}^{\mathfrak{B}} \rangle),$$

is an m-round game proceeding as follows: There are two players, Spoiler and Duplicator. During the ith round, Spoiler first chooses a structure $\mathfrak{A}$ (or $\mathfrak{B}$) and an element called c_i (or d_i) from the domain of the chosen structure. Duplicator replies by choosing an element d_i (or c_i) from the domain of the other structure $\mathfrak{B}$ (or $\mathfrak{A}$). Duplicator wins the play $\langle \langle c_1, d_1 \rangle, \ldots, \langle c_m, d_m \rangle \rangle$, if the relation

$$\{\langle x^{\mathfrak{A}}_i, x^{\mathfrak{B}}_i \rangle \mid 1 \leq i \leq r\} \cup \{\langle c_i, d_i \rangle \mid 1 \leq i \leq m\} \tag{11}$$

is a *partial isomorphism* between $\mathfrak{A}$ and $\mathfrak{B}$; otherwise, Spoiler wins the play. If against any sequence of moves by Spoiler, Duplicator is able to make her moves

---

[3] The result uses *hypergraphs*, that is, structures interpreting relation symbols of unbounded arity. As a consequence, the result does not imply that $\Sigma^1_{1,2}(\tau)$ is strictly weaker than $\Sigma^1_{1,3}(\tau)$, where τ is a vocabulary that contains only unary and binary predicates, cf. [5].

so as to win the resulting play, we say that Duplicator has a *winning strategy in* $EF_m^{\mathbf{FO}}(\langle \mathfrak{A}, x^{\mathfrak{A}} \rangle, \langle \mathfrak{B}, x^{\mathfrak{B}} \rangle)$. The notion of winning strategy for Spoiler is defined analogously. By the Gale-Stewart Theorem [12], Ehrenfeucht-Fraïssé games are determined; that is, precisely one of the players has a winning strategy. The effectiveness of these games is established in the following seminal result.

Theorem 3 ([7, 11]). *For every integer m, the following are equivalent:*

- $\langle \mathfrak{A}, x^{\mathfrak{A}} \rangle$ *and* $\langle \mathfrak{B}, x^{\mathfrak{B}} \rangle$ *satisfy the same first-order formulae (possibly with free variables from* x*) of quantifier rank* $\leq m$
- *Duplicator has a winning strategy in* $EF_m^{\mathbf{FO}}(\langle \mathfrak{A}, x^{\mathfrak{A}} \rangle, \langle \mathfrak{B}, x^{\mathfrak{B}} \rangle)$.

Readers unfamiliar with these games may find it helpful to consult [6], and [10, 15] for similar games for $M\Sigma_1^1$.

The notion of quantifier rank is extended to implicit matrix formulae as follows: $qr(\gamma) = \max\{qr(\gamma(i)) \mid i \in \{0,1\}^k\}$, for γ of type $\{0,1\}^k \to \mathbf{FO}$.

The model comparison game for $\mathbf{D}$ has two phases: a *watercoloring phase* and a *first-order phase*. Let $\mathfrak{A}$ and $\mathfrak{B}$ be τ-structures and let m be an integer. Then, the m-round, watercolor $\mathbf{D}_k^n$-*Ehrenfeucht-Fraïssé game on the structures* $\mathfrak{A}$ *and* $\mathfrak{B}$, denoted as

$$EF_m^{\mathbf{D}_k^n}(\mathfrak{A}, \mathfrak{B}) \, ,$$

is an $(m+1)$-round game proceeding as follows: First we have the watercoloring phase. Spoiler picks out for every $1 \leq i \leq n$ a subset A_i from A^k. Duplicator picks out a subset B_i of B^k, for every $1 \leq i \leq n$. Next, Spoiler chooses a tuple $x_i^{\mathfrak{B}} \in B^k$, for every $1 \leq i \leq n$, and Duplicator replies by choosing a tuple $x_i^{\mathfrak{A}} \in A^k$. If for every $1 \leq i \leq n$ the selected tuples satisfy $x_i^{\mathfrak{A}} \in A_i$ iff $x_i^{\mathfrak{B}} \in B_i$, then the game proceeds to the first-order phase as $EF_m^{\mathbf{FO}}(\langle \mathfrak{A}, x^{\mathfrak{A}} \rangle, \langle \mathfrak{B}, x^{\mathfrak{B}} \rangle)$; otherwise, Duplicator loses right away.

It is interesting to note that in the first-order Ehrenfeucht-Fraïssé game that is started up after the watercolor phase, the actual colorings are immaterial. The watercolors fade away quickly, so to say.

Proposition 3. *Let* $\mathfrak{A}$ *and* $\mathfrak{B}$ *be* τ-*structures, and let* k, n *be integers. Let* $\Gamma = \mathbf{D}_k^n \, \gamma$ *be any* $\mathbf{D}_k$-*sentence with* $qr(\gamma) \leq m$. *Then, the first assertion implies the second:*

- *Duplicator has a winning strategy in* $EF_m^{\mathbf{D}_k^n}(\mathfrak{A}, \mathfrak{B})$
- $\mathfrak{A} \models \Gamma$ *implies* $\mathfrak{B} \models \Gamma$.

Hence, if the first assertion holds for arbitrary k, n, *then the second assertion holds for every* $\Gamma \in \mathbf{D}$, *where* $qr(\Gamma) \leq m$.

Proof. The game is a simple adaptation of the one presented in [19].   □

Fagin [10] showed that the monadic fragments of Σ_1^1 and Π_1^1 do not coincide, as the latter harbors CONNECTED but the former does not. Thus we say that $M\Sigma_1^1$ *is not closed under complementation.*

Using the model comparison games for $\mathbf{D}$, we show that $\mathbf{D}$ is not closed under complementation either. This result may be interesting, because $\mathbf{D} = \Sigma_1^1 \heartsuit$ is a fragment of Σ_1^1 that is not bounded by the arity of the predicate variables and has a non-empty intersection with k-ary, existential, second-order logic, for arbitrary k, see Theorem 2. Clearly, these properties are not enjoyed by $M\Sigma_1^1$.

For any two τ-structures $\mathfrak{A}$ and $\mathfrak{B}$ with non-intersecting universes, let $\mathfrak{A} \cup \mathfrak{B}$ denote the τ-structure with universe $A \cup B$ and $R^{\mathfrak{A}\cup\mathfrak{B}} = R^{\mathfrak{A}} \cup R^{\mathfrak{B}}$, for any $R \in \tau$.

Theorem 4. $\overline{\text{2-COLORABILITY}}$ *cannot be expressed in* $\mathbf{D}$. *Hence,* $\mathbf{D}$ *is not closed under complementation.*

Proof. For contradiction, suppose $\overline{\text{2-COLORABILITY}}$ were characterizable in $\mathbf{D}$. So there would be a particular sentence in $\mathbf{D}$ that characterizes $\overline{\text{2-COLORABILITY}}$, say Γ. This sentence Γ would have a partially ordered connective with dimensions k, n prefixing an implicit matrix τ-formula of quantifier rank m. Now if we are able to find structures $\mathfrak{A}$ and $\mathfrak{B}$ such that (i) $\mathfrak{A}$ is not 2-colorable but $\mathfrak{B}$ is 2-colorable, and (ii) Duplicator has a winning strategy in $EF_m^{\mathbf{D}_k^n}(\mathfrak{A}, \mathfrak{B})$, we may reason as follows: Since Γ is supposed to characterize $\overline{\text{2-COLORABILITY}}$, we derive from (i) that $\mathfrak{A} \models \Gamma$ and $\mathfrak{B} \not\models \Gamma$. But from (ii) and $\mathfrak{A} \models \Gamma$ it follows by Proposition 3, that $\mathfrak{B} \models \Gamma$. A contradiction. So if such structures $\mathfrak{A}$ and $\mathfrak{B}$ are found for all m, k, n, we may conclude that no sentence Γ exists in $\mathbf{D}$ that expresses $\overline{\text{2-COLORABILITY}}$.

It remains to be shown that for arbitrary m, k, n, there indeed exist graphs $\mathfrak{A}$ and $\mathfrak{B}$ meeting (i) and (ii). To this end, fix integers m, k, n and consider the graphs $\mathfrak{C}$ and $\mathfrak{D}$, where

$$C = \{c_1, \ldots, c_N\}$$
$$R^{\mathfrak{C}} = \text{the symmetric closure of } \{\langle c_i, c_{i+1}\rangle \mid 1 \le i \le N - 1\} \cup \{\langle c_N, c_1\rangle\}$$
$$D = \{d_1, \ldots, d_{N+1}\}$$
$$R^{\mathfrak{D}} = \text{the symmetric closure of } \{\langle d_i, d_{i+1}\rangle \mid 1 \le i \le N\} \cup \{\langle d_{N+1}, d_1\rangle\}$$

and $N = 2^{m+k\cdot n}$. So $\mathfrak{C}$ and $\mathfrak{D}$ are cycles of even and odd length, respectively. A cycle is 2-colorable iff it is of even length, hence $\mathfrak{D}$ is not 2-colorable whereas $\mathfrak{C}$ is. Obviously, the structure $\mathfrak{C} \cup \mathfrak{D}$ is not 2-colorable either.

Let us proceed to show that Duplicator has a winning strategy in $EF_m^{\mathbf{D}_k^n}(\mathfrak{C}\cup\mathfrak{D}, \mathfrak{C})$. Suppose Spoiler selects, for every $1 \le i \le n$, a set $X_i \subseteq (C\cup D)^k$. Let Duplicator respond with X_i restricted to $\mathfrak{C}$, that is, with $Y_i = X_i \cap C^k$, for every $1 \le i \le n$. Suppose Spoiler selects the tuple $\boldsymbol{x}_i^{\mathfrak{C}} \in C^k$, for every $1 \le i \le n$. Let Duplicator respond by simply copying these tuples on $(C \cup D)^k$, that is, setting $\boldsymbol{x}_i^{\mathfrak{C}\cup\mathfrak{D}} = \boldsymbol{x}_i^{\mathfrak{C}}$. The game advances to the first-order phase, since obviously $\boldsymbol{x}^i \in X_i$ iff $\boldsymbol{x}^i \in Y_i$. A standard argument suffices to show that Duplicator has a winning strategy in

$$EF_m^{\mathbf{FO}}(\langle \mathfrak{C} \cup \mathfrak{D}, \boldsymbol{x}_1^{\mathfrak{C}\cup\mathfrak{D}}, \ldots, \boldsymbol{x}_n^{\mathfrak{C}\cup\mathfrak{D}}\rangle, \langle \mathfrak{C}, \boldsymbol{x}_1^{\mathfrak{C}}, \ldots, \boldsymbol{x}_n^{\mathfrak{C}}\rangle),$$

compare [6, p. 23].

As noted in Introduction, Blass and Gurevich have shown that $\mathbf{D}$ can characterize the class of 3-colorable graphs. In the same way it is capable of characterizing 2-COLORABILITY. We just showed that the complement of this class is not expressible in $\mathbf{D}$. Therefore, $\mathbf{D}$ is not closed under complementation.   □

Since $\mathfrak{C}$'s universe has even cardinality but $\mathfrak{D}$'s has not, we conclude that also the class EVEN is not characterizable in $\mathbf{D}$. By contrast, in Proposition 1 we showed that this class is characterizable by a sentence of the form $\mathsf{H}_1^2 \, \phi$. So already the simplest Henkin quantifier not definable in first-order logic, cannot be defined in $\mathbf{D}$. Since EVEN is obviously characterizable in binary Σ_1^1, $\mathbf{D} < \Sigma_1^1$.

5   Revisiting $\Sigma_1^1 \heartsuit$

We mapped out some finite model theory for $\mathbf{D}$ and observed that it is not closed under complementation, and not bounded by an arity constraint. We saw that $\mathbf{D}$ comprises a fragment of Σ_1^1 whose formulae do not allow for a single existential variable to appear as the argument of a predicate variable. Amusingly, this boundary is rather sharp: already the slightest extension yields a logic coinciding with Σ_1^1. Let us write $\Sigma_1^1\clubsuit$ for the fragment of Σ_1^1 that has formulae of the form

$$\exists X_1 \ldots \exists X_m \mathsf{Q}_1 x_1 \ldots \mathsf{Q}_n x_n \; \varPhi \tag{12}$$

where $\varPhi$ is sober as before, and for at most one $i \in \{1, \ldots, n\}$, we have that $\mathsf{Q}_i = \exists$; all other quantifiers are universal quantifiers. Using a result by Krynicki [16], it is not hard to see that $\Sigma_1^1\clubsuit = \mathrm{NP}$ on finite structures. Krynicki showed, namely, that first-order logic prefixed by the quantifier below (with unbounded k) coincides with full Σ_1^1:

$$\begin{pmatrix} \forall x_{11} \ldots \forall x_{1k} \bigvee i \\ \forall x_{21} \ldots \forall x_{2k} \; \exists y \end{pmatrix} \tag{13}$$

The semantics of (13) is readily defined in view of the semantics of (1) and (4), involving one function variable of type $A^k \to \{0,1\}$ and one function variable of type $A^k \to A$. The former function variable can be mimicked by a k-ary predicate variable as in the translation $T(\cdot)$. The latter k-ary function variable can be mimicked by a $(k+1)$-ary predicate variable along the obvious path, be it at the cost of introducing one existential quantifier.

References

1. M. Ajtai. Σ_1^1-formulae on finite structures. *Annals of Pure and Applied Logic*, 24:1–48, 1983.
2. M. Ajtai and R. Fagin. Reachability is harder for directed than for undirected graphs. *Journal of Symbolic Logic*, 55:113–150, 1990.
3. M. Ajtai, R. Fagin, and L. Stockmeyer. The closure of monadic NP. *Journal of Computer and System Sciences*, 60(3):660–716, 2000.

4. A. Blass and Y. Gurevich. Henkin quantifiers and complete problems. *Annals of Pure and Applied Logic*, 32:1–16, 1986.
5. A. Durand, C. Lautemann, and T. Schwentick. Subclasses of Binary-NP. *Journal of Logic and Computation*, 8(2):189–207, 1998.
6. H.-D. Ebbinghaus and J. Flum. *Finite Model Theory*. Springer-Verlag, Berlin, 1999.
7. A. Ehrenfeucht. An application of games to the completeness problem for formalized theories. *Fundamenta Mathematicae*, 49:129–141, 1961.
8. H. B. Enderton. Finite partially ordered quantifiers. *Zeitschrift für Mathematische Logik und Grundlagen der Mathematik*, 16:393–397, 1970.
9. R. Fagin. Generalized first-order spectra and polynomial-time recognizable sets. In R. M. Karp, editor, *SIAM-AMS Proceedings, Complexity of Computation*, volume 7, pages 43–73, 1974.
10. R. Fagin. Monadic generalized spectra. *Zeitschrift für Mathematische Logik und Grundlagen der Mathematik*, 21:89–96, 1975.
11. R. Fraïssé. Sur quelques classifications des systèmes de relations. *Publications Scientifiques*, Série A, 35–182 1, Université d'Alger, 1954.
12. D. Gale and F. Stewart. Infinite games with perfect information. In H. W. Kuhn and A. W. Tucker, editors, *Contributions to the Theory of Games II*, volume 28 of *Annals of Mathematics Studies*, pages 245–266. Princeton University Press, Princeton, 1953.
13. G. Gottlob. Relativized logspace and generalized quantifiers over finite ordered structures. *The Journal of Symbolic Logic*, 62(2):545–574, 1997.
14. L. Hella and G. Sandu. Partially ordered connectives and finite graphs. In M. Krynicki, M. Mostowski, and L. W. Szczerba, editors, *Quantifiers: Logics, Models and Computation*, volume II of *Synthese library: studies in epistemology, logic, methodology, and philosophy of science*, pages 79–88. Kluwer Academic Publishers, Dordrecht, 1995.
15. N. Immerman. *Descriptive Complexity*. Graduate texts in computer science. Springer, New York, 1999.
16. M. Krynicki. Hierarchies of finite partially ordered connectives and quantifiers. *Mathematical Logic Quarterly*, 39:287–294, 1993.
17. M. Krynicki and M. Mostowski. Henkin quantifiers. In M. Krynicki, M. Mostowski, and L.W. Szczerba, editors, *Quantifiers: Logics, Models and Computation*, volume I of *Synthese library: studies in epistemology, logic, methodology, and philosophy of science*, pages 193–262. Kluwer Academic Publishers, The Netherlands, 1995.
18. G. Sandu. The logic of informational independence and finite models. *Logic Journal of the IGPL*, 5(1):79–95, 1997.
19. G. Sandu and J. Väänänen. Partially ordered connectives. *Zeitschrift für Mathematische Logik und Grundlagen der Mathematik*, 38:361–372, 1992.
20. G. Turán. On the definability of properties of finite graphs. *Discrete Mathematics*, 49:291–302, 1984.
21. W. Walkoe. Finite partially-ordered quantification. *Journal of Symbolic Logic*, 35:535–555, 1970.

Upper and Lower Bounds for the Computational Power of P Systems with Mobile Membranes

Shankara Narayanan Krishna

Department of Computer Science and Engineering,
Indian Institute of Technology, Bombay,
Powai, Mumbai, India 400 076
krishnas@cse.iitb.ac.in

Abstract. We continue the study of P systems with mobile membranes introduced in [7], which is a variant of P systems with active membranes, but having none of the features like polarizations, label change and division of non-elementary membranes. This variant was shown to be computationally universal (RE) using only the simple operations of endocytosis and exocytosis; moreover, if elementary membrane division is allowed, it is capable of solving NP-complete problems. It was shown in [5] that 4 membranes are sufficient for universality while using only endo/exo operations. In this paper, we study the computational power of these systems more systematically: we examine not only the power due to the number of membranes, but also with respect to the kind of rules used, thereby trying to find out the border line between universality and non-universality. We show that 3 membranes are sufficient for computational universality, whereas two membranes are not, if λ-free rules are used.

1 Introduction

P systems are a class of distributed parallel computing models inspired from the way the living cells process chemical compounds, energy, and information. One of the central operations in cell biology is cell division, and with this inspiration, P systems with active membranes were introduced in [12]. This variant was shown to be computationally universal as well as to be able to solve hard problems. The features used by this variant include the use of polarizations ($+$, $-$, 0) and division of non-elementary as well as elementary membranes, giving rise to an exponential workspace. These features are quite powerful, thus making the system powerful. Many attempts have been made to define equivalent systems having none of the above features, but in general, removal of one feature has requested the introduction of other powerful operations [13]. [7] is an attempt in this direction, wherein a variant of P systems with none of the above mentioned features was introduced, but instead use two simple operations : *endocytosis* and *exocytosis*. These operations are different and simpler than the operations considered in [1], [2], [4] and [3]. In [7], computational universality of these systems was obtained using 9 membranes. This was then improved to 4 membranes in [5]. However, the

A. Beckmann et al. (Eds.): CiE 2006, LNCS 3988, pp. 526–535, 2006.

power of 3 and 2 membranes was not clear. It was also not clear whether the kind of evolution rules used, could have any significant effect on the computational power.

In this paper, we improve the universality result of [5] obtaining a universality result with 3 membranes. Further, we try to understand and analyze where the borderline between universality and non-universality lies. For instance, what happens when two membranes are used instead of three? We obtain a very subtle borderline here with two membranes : if λ-free rules are used, we show that it is impossible to obtain universality, whereas if general rules are used with two membranes, the power remains open. This leaves a very interesting question open about the power of two kinds of systems : those with two membranes having λ-rules and those with three membranes having λ-free rules. To the best of our knowledge, this is the first time a characterization of the computational power of P systems has been made, based on λ-free and λ-rules. In formal language theory, inclusion of λ-rules is known to increase the computational power of certain systems (eg. MAT_{ac} and MAT_{ac}^{λ}) from non-RE to RE [14], but we do not know the effect of this as far as P systems are concerned. However, we conjecture that two membranes with λ-rules cannot be equivalent to RE and hence that the universality result with three membranes is optimal with respect to the number of membranes.

The next section is devoted towards some formal language theory prerequisites. In Section 3, we recall the basics of P systems with mobile membranes, whose power we investigate, in this paper. Section 4 describes the universality of 3 membranes, and states the upper and lower computational bounds of systems with two membranes, the proofs of which can be found in the full version of the paper [6].

2   Some Prerequisites

We refer to [14] for the elements of formal language theory we use here. We list a few notions and notations: N denotes the set of natural numbers; V denotes a finite alphabet; V^* is the is the free monoid generated by V under the operation of concatenation and the empty string denoted by λ, as unit element; by $NFIN, NREG, NCF, N0L, NRC_{p,f}, NCS$ and NRE we denote the family of finite sets, regular sets, context-free sets, zero-interaction Lindenmayer sets, random-context sets, context-sensitive sets and recursively enumerable sets of natural numbers, respectively. These can also be looked at as the family of sets of numbers recognized by these languages. For $k \geq 1$ and a family of languages FL, by $N_k FL$ we denote the length sets of FL excluding the initial segment upto $k-1$. Equivalently, $N_k FL = \{k+L \mid L \in NFL\}$, where $k+L = \{k+n \mid n \in L\}$. A multiset over an alphabet V is represented by a string over V (and by all its permutations) and each string precisely identifies a multiset. It is known that $NFIN \subset NREG = NCF \subseteq N0L \subset NRC_{p,f} \subseteq NCS \subset NRE$.

We briefly mention the definition of random context grammars and 0L systems here, since we use them in Theorem 1. A random context grammar is a construct

$G = (N, T, S, R)$ where N is the set of non-terminals, T is the set of terminals, S is the start symbol and R is a set of rules of the form $p : (A \to w, E_1, E_2)$ where $A \to w$ is a context-free production over $N \cup T$ and E_1, E_2 are subsets of N. Then, p can be applied to a string $x \in (N \cup T)^*$ only if $x = x_1 A x_2, E_1 \subseteq alph(x_1 x_2)$, and $E_2 \cap alph(x_1 x_2) = \emptyset$. $alph(x_1 x_2)$ stands for the set of symbols occurring in $x_1 x_2$. If E_1 or E_2 is the empty set, then no condition is imposed by E_1 or E_2 respectively. E_1 is said to be permitting and E_2 is said to be the set of forbidding context conditions of p. We denote by $RC_{p,f}$ the family of languages generated by random context grammars with permitting and forbidding contexts and λ-free rules. It is known that the family $RC_{p,f}$ is closed with respect to left quotient by letters, where the left quotient of a family L of languages with respect to a symbol a is defined as $\partial_a^l(L) = \{x \mid ax \in L\}$.

A 0L system is a construct $G = (V, w, R)$ where V is an alphabet, $w \in V^*$ is the axiom, and R is a finite set of rules of the form $a \to v$ with $a \in V, v \in V^*$ such that for each $a \in V$, there is atleast one rule $a \in v$ in R. For $w_1, w_2 \in V^*$, we write $w_1 \Rightarrow w_2$ if $w_1 = a_1 a_2 \ldots a_n, w_2 = v_1 v_2 \ldots v_n$, for $a_i \to v_i \in R, 1 \le i \le n$. The language generated by G is $L(G) = \{x \in V^* \mid w \Rightarrow^* x\}$.

For basic elements of membrane computing we refer to [13]; for the state-of-the art of the domain, the reader may consult the bibliography from the web address http://psystems.disco.unimib.it. For proving computational universality, we use the concept of Minsky's register machine [11].

The proofs about membrane systems in this paper are based on the concept of Minsky's register machine [11]. Such a machine runs a program consisting of numbered instructions of several simple types. Several variants of register machines with different number of registers and different instruction sets were shown to be computationally universal (e.g., see [10], [11] for some original definitions and [8] for the definitions we use in this paper).

An *n-register machine* is a construct $M = (n, P, i, E)$, where: (i) n is the number of registers, (ii) P is a set of labeled instructions of the form $j : (op(r), k, l)$, where $op(r)$ is an operation on register r of M, j, k, l are labels from the set $Lab(M)$ (which numbers the instructions in a one-to-one manner), (iii) i is the initial label, and (iv) E is the final label.
The machine is capable of the following instructions:

$(add(r), k, l)$: Add one to the contents of register r and proceed to instruction k or to instruction l; in the deterministic variants usually considered in the literature we demand $k = l$.

$(sub(r), k, l)$: If register r is not empty, then subtract one from its contents and go to instruction k, otherwise proceed to instruction l.

halt: Stop the machine. This additional instruction can only be assigned to the final label E.

In their *deterministic variant*, such n-register machines can be used to compute any partial recursive function $f : \mathbf{N}^\alpha \to \mathbf{N}^\beta, \alpha, \beta > 0$; starting with $(n_1, \ldots, n_\alpha) \in \mathbf{N}^\alpha$ in registers 1 to α, M has computed $f(n_1, \ldots, n_\alpha) = (r_1, \ldots, r_\beta)$ if it halts in the final label E with registers 1 to β containing

r_1 to r_β. If the final label cannot be reached, then $f(n_1, \ldots, n_\alpha)$ remains undefined.

A deterministic m-register machine can also analyze an input $(n_1, \ldots, n_\alpha) \in \mathbf{N}_0^\alpha$ in registers 1 to α, which is recognized if the register machine finally stops by the halt instruction with all its registers being empty. If the machine does not halt, the analysis was not successful. In their *non-deterministic variant*, n-register machines can compute any recursively enumerable set of non-negative integers (or of vectors of non-negative integers). Starting with all registers being empty, we consider a computation of the n-register machine to be successful, if it halts with the result being contained in the first (β) register(s) and with all other registers being empty. In fact, [11] has shown that 3 registers are enough for computing any recursively enumerable set of numbers, such that the input is in register 1, register 3 is never decremented, and the machine, when it halts, has the output in register 3.

3   P Systems with Mobile Membranes

We now briefly recall P systems with mobile membranes introduced in [7]. A *P system with mobile membranes* is a construct $\Pi = (V, H, \mu, w_1, \ldots, w_n, R)$, where: $n \geq 1$ (the initial *degree* of the system); V is an alphabet (its elements are called *objects*); H is a finite set of *labels* for membranes; μ is a *membrane structure*, consisting of n membranes, labeled (not necessarily in a one-to-one manner) with elements of H; $w_1, w_2, \ldots, w_n$ are strings over V, describing the *multisets of objects* placed in the n regions of μ, and R is a finite set of *developmental rules*, of the following forms:

a. $[_m a \to v]_m$, for $m \in H, a \in V, v \in V^*$; object evolution rules.

b. $[_h a]_h [_m \ ]_m \to [_m [_h b]_h]_m$, for $h, m \in H, a, b \in V$; endocytosis rules: an elementary membrane labeled h enters the adjacent membrane labeled m, under the control of object a; the labels h and m remain unchanged during this process, however, the object a may be modified to b during the operation; m is not necessarily an elementary membrane.

c. $[_m [_h a]_h]_m \to [_m \ ]_m [_h b]_h$, for $h, m \in H, a, b \in V$; exocytosis: an elementary membrane labeled h is sent out of a membrane labeled m, under the control of object a; the labels of the two membranes remain unchanged, but the object a from membrane h may be modified during this operation; membrane m is not necessarily elementary.

d. $[_h a]_h \to [_h b]_h [_h c]_h$, for $h \in H, a, b, c \in V$; division rules for elementary membranes: in reaction with an object a, the membrane labeled h is divided into two membranes labeled h, with the object a replaced in the two new membranes by possibly new objects. *Note that we do not use division rules for our investigations in this paper.* The rules are applied according to the following principles:

1. All rules are applied in parallel, non-deterministically choosing the rules, the membranes, and the objects, but in such a way that the parallelism is maximal; this means that in each step we apply a set of rules such that no

further rule can be added to the set, no further membranes and objects can evolve at the same time.

2. The membrane m from each type (a) – (c) of rules as above is said to be *passive*, while the membrane h is said to be *active*. In any step of a computation, any object and any active membrane can be involved in at most one rule, but the passive membranes are not considered involved in the use of rules (hence they can be used by several rules at the same time as passive membranes); for instance, a rule $[_m a \rightarrow v]_m$, of type (a), is considered to involve only the object a, not also the membrane m.

3. The evolution of objects and membranes takes place in a bottom-up manner. After having a (maximal) set of rules chosen, they are applied starting from the innermost membranes, level by level, up to the skin membrane (all these sub-steps form a unique evolution step, called a *transition* step).

4. When a membrane is moved across another membrane, by endocytosis or exocytosis, its whole contents (its objects) are moved; because of the bottom-up way of using the rules, the inner objects first evolve (if there are rules applicable for them), and then any membrane is moved with the contents as obtained after this inner evolution.

5. If a membrane exits the system (by exocytosis), then its evolution stops, even if there are rules of type (a) which would be applicable to it provided that the membrane would be in the system.

6. All objects and membranes which do not evolve at a given step (for a given choice of rules which is maximal) are passed unchanged to the next configuration of the system.

By using the rules in this way, we get transitions among the configurations of the system. A sequence of transitions is a computation, and a computation is successful if it halts (it reaches a configuration where no rule can be applied). During a computation, membranes can leave the skin membrane (by means of rules of type (c)). The number of objects present in each membrane that is sent out of the system contributes to the output. If only one membrane is sent out at the end of a halting configuration, then the number of objects present in that membrane is the output of the system. If multiple membranes are sent out, then each membrane contributes a number. The set of all such numbers is the output of a successful computation; a non-halting computation provides no output.

The set of all numbers computed in this way by Π is denoted $\mathbf{N}(\Pi)$. The family of all sets of numbers $\mathbf{N}(\Pi)$ generated by systems Π having atmost n membranes, using endocytosis and exocytosis rules, is denoted by $\mathbf{NMP}_n^\lambda(endo, exo))$. If no rules of the from $a \rightarrow \lambda$ are used in Π, then we omit the superscript λ. If a type of rules is not used, then we omit its "name" from the list. For instance, if only exocytosis rules are used and if all rules are λ-free, we write or $\mathbf{NMP}_n(exo)$. Note that the exocytosis rules allow only replacing an object by another object and hence, any membrane leaving the skin membrane has at least one object. Thus, $\mathbf{NMP}_n(endo, exo) \geq 1, n \geq 2$.

Lemma 1. $\mathbf{NMP}_n(endo, exo) \subseteq \mathbf{NMP}_n^\lambda(endo, exo)$.

4 The Power of Endocytosis and Exocytosis

We first examine the power of systems with 2 membranes. We show that with 2 membranes one can compute all numbers (≥ 1) that are computable by a $0L$ system (lower bound). We also show that we can obtain an sub Turing ($< RE$) upper bound for systems of degree two, with λ-free rules.

Theorem 1. *1.* $NMP_2(exo) \subseteq N_1 RC_{p,f} \subset N_1 RE$, *2.* $N_1 0L \subseteq NMP_2^\lambda(exo)$.

Proof. In systems with two membranes $\mu = [_1[_2 \ ]_2]_1$, $[_2 \ ]_2$ is the output membrane and there are no endo operations. Thus, there is only one exo operation (when membrane 2 is sent out of the skin) and object evolution rules for membranes 1,2. Further, none of the objects in membrane 1 contribute to the output. Therefore, without loss of generality, to assume that membrane 1 is empty, while considering systems with 2 membranes. For a detailed explanation of this claim, see [6].

Table 1. Rules for the random context grammar

1. $(S \rightarrow Ow, \emptyset, \emptyset)$
2. $(O \rightarrow F, \emptyset, V' \cup V'' \cup \{\eta_a \mid a \in V\})$
3. $(a \rightarrow v', \{F\}, \{O, E\})$, if $a \rightarrow v \in R$
4. $(a \rightarrow a', \{F\}, \{O, E\})$, if $a \in U_2$
5. $(F \rightarrow O, \emptyset, V \cup V'' \cup \{\eta_a \mid a \in V\})$
6. $(a' \rightarrow a, \{O\}, \{F, E\})$
7. $(a \rightarrow \eta_b, \{F\}, \{O, E\} \cup \{\eta_e \mid e \in V\})$, if $[_1[_2 a]_2]_1 \rightarrow [_2 b]_2[_1 \ ]_1$
8. $(a' \rightarrow a'', \{F\} \cup \{\eta_e\}, \emptyset)$, for $a, e \in V$
9. $(a \rightarrow a'', \{F\} \cup \{\eta_e\}, \emptyset)$, where $a \in U_1, e \in V$
10. $(a \rightarrow a'', \{F\} \cup \{e''\}, \emptyset)$, where $a \in U_1, e \in V$
11. $(a' \rightarrow a'', \{F\} \cup \{e''\}, \emptyset)$, for $a, e \in V$
12. $(F \rightarrow E, \emptyset, V \cup V')$
13. $(\eta_a \rightarrow \hat{a}, \{E\}, V \cup V' \cup \{\eta_e \mid e \in V\})$, $a \in V$
14. $(a'' \rightarrow \hat{a}, \{E\}, V \cup V' \cup \{\eta_e \mid e \in V\})$, $a \in V$
15. $(E \rightarrow H, \emptyset, V \cup V' \cup V'' \cup \{\eta_a \mid a \in V\})$, $a \in V$.

We give the idea behind 1 as well as the construction of a $RC_{p,f}$ system. The detailed explanations of correctness can be found in the full version of the paper [6]. We show that $\mathbf{N_1} MP_2(exo) \subseteq \mathbf{N_1} RC_{p,f}$, which implies $\mathbf{N_1} MP_2(exo) \subset \mathbf{N_1} RE$ since $\mathbf{N_1} RC_{p,f} \subseteq \mathbf{N_1} CS \subset \mathbf{N_1} RE$.

Let $\Pi = (V, \{1, 2\}, [_1[_2 \ ]_2]_1, \emptyset, w, R)$ be a P system with mobile membranes of degree two. Let U_1, U_2 be two subsets of V such that

$$U_1 = \{a \in V \mid \text{all rules for } a \text{ are of the form } [_1[_2 a]_2]_1 \rightarrow [_2 b]_2[_1 \ ]_1\},$$
$$U_2 = \{a \in V \mid a \text{ has no rules in } R\}$$

For the alphabet V, let $V' = \{a' \mid a \in V\}$, $V'' = \{a'' \mid a \in V\}$, $\hat{V} = \{\hat{a} \mid a \in V\}$. Let us construct a random context grammar $G = (N, T, S, R')$ with $N = V \cup V' \cup V'' \cup \{\eta_a \mid a \in V\} \cup \{F, O, E\}$, $T = \{H\} \cup \hat{V}$. The rules R' are given in Table 1. We obtain a string $Hw, w \in \widehat{V}^*$ in $L(G)$, where, the length of w would be same as the number of symbols in the membrane that is sent out. Since we have an extra symbol H here, we can take a left quotient of the string Hw with respect to the letter H, and obtain w. Since the family $RC_{p,f}$ is closed with respect to left quotient by letters, w is also in $L(G)$. This establishes the claim made by 1.

The proof of 2 is as follows: Consider any $0L$ system $G = (V, w, P)$ and construct the P system $\Pi_1 = (V \cup \{d\}, \{1, 2\}, [_1[_2\ ]_2]_1, w_1, wd, R)$ such that $d \notin V$ and having rules $R = P \cup \{[_2d \rightarrow d]_2, [_1[_2d]_2]_1 \rightarrow [_2d]_2[_1]_1\}$. Note that w_1 can be any arbitrary multiset. Clearly, the rules in P can be applied as long as we want, along with $d \rightarrow d$, and at any point, we can use the exo rule. Clearly, membrane 2 has an extra d and the multiplicities of objects in V will be same as that computed by the $0L$ system. $\qquad \square$

Next we show computational universality can be obtained with 3 membranes.

Theorem 2. $\mathbf{N}_1 RE = NMP_3^\lambda(endo, exo)$.

Proof. We only prove the assertion $\mathbf{N}_1 RE \subseteq NMP_3^\lambda(endo, exo)$, and infer the other inclusion from the Church-Turing thesis. The proof is based on the observation that each set from $\mathbf{N}_1 RE$ is the range of a recursive function. Thus, we will prove the following assertion. For each recursively enumerable function $f : \mathbf{N} \rightarrow \mathbf{N}$, there is a mobile P System Π with 3 membranes satisfying the following condition: For any arbitrary $x \in \mathbf{N}$, the system Π first "generates" a multiset of the form o_1^x and halts if and only if $f(x)$ is defined, and, if so, the result of the computation is $f(x)$.

In order to prove this assertion, we consider a register machine with 3 registers, the last one being a special output register which is never decremented. Let there be a program P consisting of n instructions $P_1, \ldots, P_n$ which computes f. Let P_n correspond to the instruction HALT and P_1 be the first instruction. The input value x is expected to be in register 1 and the output value in register 3. Without loss of generality, we can assume that all registers other than the first one are empty at the beginning of a computation.

We construct the mobile P system $\Pi = (V, H, \mu, w_1, w_2, w_3, R)$ where

$$V = \{s, s_1, s_2, s_1', s_2', s_1'', s_2'', s_2'''\} \cup \{a, d, d', e, e', N\} \cup \{o_1, o_2, o_3\}$$

$$\cup \{P_i \mid 1 \le i \le n\} \cup \{P_{kj}^i, P_l^m \mid 1 \le i \le 6, m = 1, 2, j = (sub(1), k, l)\}$$

$$\cup \{Q_l^m, Q_{kj}^i \mid 1 \le i \le 6, m = 1, 2, \text{ and } j = (sub(2), k, l)\}$$

$$\cup \{E, E', E'', E''', E^4, E^5, E^6, \dagger\},$$

$$H = \{1, 2, 3\}, \ \mu = [_3[_1\ [_2\ ]_2\ ]_1]_3, \ w_1 = sa, w_2 = s_2, w_3 = \emptyset.$$

Proof Idea: Membrane 1 contains the current contents of registers 1 and 3 (in the form of o_1, o_3), and membrane 2, the current contents of register 2 (as o_2). A copy of the current instruction is always kept in both membranes 1 and 2. This helps to decrement the registers 1 and 2, separately in the two membranes, and to update the next instruction. When the halting instruction E is obtained, the contents of register 3, present in membrane 1 is the output. We have the following rules:

1.*Creation of $o_1^x, x \geq 0$ in membrane 1, representing input x in register 1*

$$1. \, [_2 s_2 \rightarrow s_2]_2, [_1 s \rightarrow s_1]_1, \quad 2. \, [_1 s_1 \rightarrow o_1 s_1]_1, [_1 s_1 \rightarrow s_1]_1,$$
$$3. \, [_1 \, [_2 s_2]_2 \,]_1 \rightarrow [_2 s_2']_2 [_1 \,]_1,$$
$$4. \, [_1 a]_1 [_2 \,]_2 \rightarrow [_2 \, [_1 \dagger]_1]_2, \, [_1 s_1]_1 [_2 \,]_2 \rightarrow [_2 \, [_1 s_1']_1]_2, \, [_2 s_2' \rightarrow s_2'']_2,$$
$$5. \, [_2 \, [_1 a]_1 \,]_2 \rightarrow [_1 N \,]_1 [_2 \,]_2, [_1 s_1' \rightarrow s_1'']_1, [_2 s_2'' \rightarrow s_2''']_2,$$
$$6. \, [_1 s_1'' \rightarrow P_1]_1, [_2 s_2''' \rightarrow P_1]_2, [_i N \rightarrow \lambda]_i, i = 1, 2.$$

We start with the initial configuration $[_3 \, [_1 \, [_2 s_2]_2 \, sa]_1 \,]_3$. Rule 1 is first applied to membranes 1 and 2, replacing s by s_1 in membrane 1. In the next few steps, rule 2 can be used, creating as many o_1's as required, and assuming that $s_2 \rightarrow s_2$ is used in membrane 2. At some point, rule 3 is used, by which membranes 1 and 2 become adjacent to each other, and s_2 becomes s_2'. Next, we have two choices for membrane 1 : either the endo rule (4) involving a or the endo rule (4) involving s_1 has to be applied. In the former case, we get an infinite computation (rule 29), so let us assume we replace s_1 by s_1' using the endo rule for s_1. In parallel, s_2' evolves to s_2'' in membrane 2. We have now reached the configuration $[_3 \, [_2 \, [_1 as_1' o_1^i]_1 \, s_2'']_2 \,]_3$. Rules 5, 6 now follow, leading to the configuration $[_3 \, [_2 P_1]_2 [_1 P_1 o_1^i]_1 \,]_3$.

2.*Simulation of instructions P_j, where $j = (add(r), k), 1 \leq j, k \leq n, 1 \leq r \leq 3$*

$$7. \, [_1 P_j \rightarrow o_r P_k]_1, [_2 P_j \rightarrow P_k]_2, \, j = (add(r), k), r = 1, 3,$$
$$8. \, [_2 P_j \rightarrow o_2 P_k]_2, [_1 P_j \rightarrow P_k]_1, \, j = (add(2), k).$$

Assume we are in a configuration $[_3 \, [_1 P_i o_1^q o_3^t]_1 [_2 P_i o_2^u]_2 \,]_3, q, t, u \geq 0$. If $P_i : (add(r), k), r \in \{1, 3\}$, then, in membrane 1, we replace P_i by $o_1 P_k$ or $o_3 P_k$ and in membrane 2, we replace P_i by P_k. A similar case holds for incrementing register 2. Thus, membrane 1 contains the current contents of registers 1 and 3, and membrane 2 contains the contents of register 2.

3.*Simulation of instructions P_j, where $j = (sub(1), k, l), 1 \leq j, l, k \leq n$*

We look at the simulation of an instruction decrementing register 1. We need to ensure that after simulation, we either have P_l in membranes 1 and 2 (if register 1 was zero) or have P_k in both membranes (if register 1 was non-zero). We look at both the cases here.

Case 1: Register 1 is non-zero Let $[_3 \, [_2 o_2^u P_j]_2 [_1 P_j o_1^s o_3^t]_1 \,]_3, s > 0$ be the current configuration. We start with rule 9, by which membrane 1 enters membrane 2 replacing P_j by P_{kj}^1, while, in membrane 2, P_j evolves to P_{kj}^1. In the next step, since o_1's are present in membrane 1, rule 10 is applied, by which membrane 1 comes out of membrane 2, replacing an o_1 by d, while, in parallel, P_{kj}^1 evolves into P_{kj}^2 in both membrane 1 and 2. In the next step, we have two choices of applicable rules : rules 11,19 in parallel, or rule 12. Rule 11 is a trap rule, and the symbol $\dagger$ can never be removed. By rule 12, P_{kj}^2 evolves into P_{kj}^3 in both membranes, and membrane 2 enters membrane 1. Rule 13 comes next,

Table 2. Rules for decrementing register 1

$9.[_1 P_j]_1[_2]_2 \to [_2 [_1 P^1_{kj}]_1]_2, [_2 P_j \to P^1_{kj}]_2$	
$10. [_2 [_1 o_1]_1]_2 \to [_1 d]_1[_2]_2, [_i P^1_{kj} \to P^2_{kj}]_i, i = 1,2$	

	Register 1 non-zero		Register 1 is zero
11	$[_1 d]_1[_2]_2 \to [_2 [_1 \dagger]_1]_2$	18	$[_2 [_1 P^2_{kj}]_1]_2 \to [_1 P^1_l]_1[_2]_2$
12	$[_1 P^2_{kj} \to P^3_{kj}]_1, [_2 P^2_{kj}]_2[_1]_1 \to [_1 [_2 P^3_{kj}]_2]_1$	19	$[_2 P^2_{kj} \to eP^1_l]_2$
13	$[_1 P^3_{kj} \to P^4_{kj}]_1, [_1 [_2 P^3_{kj}]_2]_1 \to [_2 P^4_{kj}]_2[_1]_1$	20	$[_2 e]_2[_1]_1 \to [_1 [_2 N]_2]_1,$
14	$[_2 P^3_{kj} \to \dagger]_2$		$[_i P^1_l \to P^2_l]_i, i = 1,2$
15	$[_2 P^4_{kj} \to P^5_{kj}]_2, [_1 P^4_{kj}]_1[_2]_2 \to [_2 [_1 P^5_{kj}]_1]_2$	21	$[_1 [_2 P^2_l]_2]_1 \to [_2 P_l]_2[_1]_1,$
16	$[_2 [_1 d]_1]_2 \to [_1 N]_1[_2]_2,$		$[_1 P^1_l \to P_l]_1$
	$[_i P^5_{kj} \to P^6_{kj}]_i, i = 1,2$	22	$[_2 P^2_l \to \dagger]_2$
17	$[_i P^6_{kj} \to P_k]_i, i = 1,2$		

P^3_{kj} evolves into P^4_{kj}, and membrane 2 comes out of membrane 1. Now we have $[_3 [_2 P^4_{kj} o^u_2]_2 [_1 P^4_{kj} o^{s-1}_1 o^t_3]_1]_3$. In the next step, 15 is the only applicable rule, by which P^4_{kj} is replaced by P^5_{kj} in both membranes, and membrane 1 enters membrane 2. We now use rule 16 to remove the d introduced by rule 10, obtaining $[_3 [_2 P^6_{kj} o^u_2]_2 [_1 N P^6_{kj} o^m_1 o^q_3]_1]_3$. Rules 17 and 6 are used next, replacing both P^6_{kj}'s by P_k, the next instruction, and erasing N.

Case 2 : Register 1 is zero Let $[_3 [_2 o^u_2 P_j]_2 [_1 P_j o^s_3]_1]_3$ be the current configuration. The initial two steps in both membranes is same as in Case 1. The difference is that we do not have an o_1 in membrane 1. To proceed, we use rule 18 to membrane 1 and rule 19, to P^2_{kj} in membrane 2. This gives $[_3 [_2 o^u_2 e P^1_l]_2 [_1 P^1_l o^s_3]_1]_3$. In the next step, 20 is the only applicable rule, which removes e in the process of membrane 2 entering membrane 1, while P^1_l becomes P^2_l in both membranes. Next, P^2_l is replaced by P_l in both membranes by rule 21 and N is erased. Note that not using the rules as mentioned here would give no result.

4.Simulation of instructions P_j, where $j = (sub(2), l, k), 1 \le j, l, k \le n$
The simulation of an instruction decrementing register 2 is analogous to that of register 1, the only difference being that we use as intermediate symbols Q_{jk} instead of P_{jk}, and d', e' instead of d, e.

5. Halting and Handling Exception: Let $P_n = E$ be the halting instruction

23. $[_1 E]_1 [_2]_2 \to [_2 [_1 E']_1]_2,$ 24. $[_1 E' \to E'', E^4 \to E'']_1,$

25. $[_2 [_1 E'']_1]_2 \to [_1 E''']_1 [_2]_2,$ 26. $[_1 E'']_1 [_2]_2 \to [_2 [_1 E^4]_1]_2,$

27. $[_1 E''' \to E^5, E^5 \to E^6]_1,$ 28. $[_3 [_1 E^6]_1]_3 \to [_1 o_3]_1 [_3]_3,$

29. $[_i \dagger \to \dagger]_i, i = 1,2,$ 30. $[_1 \dagger]_1 [_2]_2 \to [_2 [_1 \dagger]_1]_2,$ 31. $[_1 E'' \to \dagger]_1.$

Assume that we reach the halting instruction P_n. This means, we are in a configuration $[_3 [_1 E o^i_1 o^j_3]_1 [_2 E o^k_2]_2]_3$, $i, j, k \ge 0$. Clearly, there are no instructions to be simulated after E. Rule 23 is used, by which membrane 1 enters membrane 2, replacing E by E'. Nothing happens to the contents of membrane 2 from this point onward. If there are o_1's in membrane 1, they will be removed and membrane 1 will come out of membrane 2, by rule 10. In parallel, E' evolves to E''

by rule 24. Next, rule 26 is used, by which membrane 1 again enters membrane 2, replacing E'' by E^4. The o_1's, if any, are removed and E^4 is replaced by E''. This continues as long as there are o_1's in membrane 1. When all the o_1's are exhausted, we obtain the configuration $[_3 \ [_2 \ Eo_2^k[_1E''o_3^j]_1]_2 \ ]_3, j, k \geq 0$.

The only applicable rule next, is 25, by which, membrane 1 comes out of membrane 2, replacing E'' by E''', giving the configuration $[_3 \ [_1E'''o_3^j]_1[_2Eo_2^k]_2 \ ]_3$. E''' next evolves into E^5 by rule 27. In parallel, rule 30 (if applicable) will be used, provided a trap symbol † was introduced in membrane one in the past. If this happens, we get no output, since there is no way to send membrane 1 out of membrane 2. Otherwise, we replace in the next step, E^5 by E^6 and subsequently by o_3 (rule 28) while expelling membrane 1 out of the system. Membrane 1 will contain $j + 1$ o_3's, $j \geq 0$, provided j was the output of the register machine.   □

References

1. A. Alhazov, T. O. Ishdorj, Membrane operations in P systems with active membranes, *Proc. Second BWMC*, TR 01/04, Sevilla University, 2004, 37–44.
2. G. Bel Enguix, M.D. Jiménez-Lopez, Linguistic membrane systems and applications, in *Applications of Membrane Computing*, Springer, 2005, 347-388.
3. D. Besozzi, C. Zandron, G. Mauri, N. Sabadini, P systems with gemmation of mobile membranes, *Proc. ICTCS 2001*, LNCS 2202, Springer, 2001, 136–153.
4. E. Csuhaj-Varjú, A. Di Nola, Gh. Păun, M.J. Perez-Jiménez, G. Vaszil, Editing configurations of P systems, *Proc. 3rd BWMC*, Jan 31- Feb 4, 2005, 131-154.
5. S. N. Krishna, The Power of Mobility : Four Membranes Suffice, *Proc. CiE 2005*, Amsterdam, LNCS 3526, 242-251.
6. S. N. Krishna, Upper and Lower Bounds for the Computational Power of P Systems with Mobile Membranes, Technical Report, IIT Bombay, (2006). (www.cse.iitb.ac.in/~krishnas/cie06tr.ps).
7. S. N. Krishna, Gh. Păun, P systems with mobile membranes *Natural Computing*, 4(**3**), 2005, 255-274.
8. R. Freund and M. Oswald, GP Systems with Forbidding Context, *Fundamenta Informaticae*, 49(**1–3**), 2002, 81–102.
9. R. Freund and M. Oswald, P systems with activated/prohibited membrane channels, *Proc. of WMC 2002*, LNCS 2597, Springer-Verlag, 2003, 261–269.
10. J. Lambek, How to program an infinite abacus, *Canad. Math. Bull.*,4, 295-302, 1961.
11. M.L. Minsky, Finite and Infinite Machines, *Prentice Hall, EngleWood Cliffs*, New Jersey, 1967.
12. Gh. Păun, P systems with active membranes: Attacking NP-complete problems, *Journal of Automata, Languages and Combinatorics*, 6(**1**), 2001, 75–90.
13. Gh. Păun, *Computing with Membranes. An Introduction*, Springer, 2002.
14. G. Rozenberg, A. Salomaa, eds., *Handbook of Formal Languages*, Springer, 1997.

Gödel's Conflicting Approaches
to Effective Calculability

Wilfried Sieg

Department of Philosophy
Carnegie Mellon University

Identifying the informal concept of effective calculability with a rigorous mathematical notion like general recursiveness or Turing computability is still viewed as problematic, and rightly so. In a 1934 conversation with Church, Gödel suggested finding axioms for the notion of effective calculability and "doing something on that basis" instead of identifying effective calculability with λ-definability; that identification he found "thoroughly unsatisfactory". He introduced in his contemporaneous Princeton lectures (Gödel 1934) the class of general recursive functions through the equational calculus, but was not convinced at the time that this mathematical notion encompassed all effectively calculable functions. (See (Davis 1982) and (Sieg 1997).)

Gödel articulated different and conflicting approaches to the underlying methodological issues during the three decades from 1934 to 1964. The significant shifts in his position underline the difficulty of the problems surrounding the Church-Turing Thesis. In (1936) and (1946) he emphasized that the importance of the notion of general recursive function is largely due to its absoluteness. Yet he also claimed in (193?) that the analysis of the manner in which the calculation of number theoretic functions proceeds leads to the characteristic features of the equational calculus; thus, it provides a "correct definition" of effectively calculable function. In (1951) he calls Turing's reduction of the "concept of finite procedure to that of a machine with a finite number of parts" the most satisfactory way of arriving at a precise definition of the former concept. Finally, in (1964) Gödel saw, quite emphatically, Turing's work as providing a correct analysis of mechanical procedures (thus also of effective calculability) and a proof of the fact that the analyzed notion is equivalent to that of a Turing machine.

Eight years later Gödel detected "a philosophical error in Turing's work" (of 1936) and attributed to Turing the claim that "mental procedures cannot go beyond mechanical procedures". Turing, however, did not maintain such a claim when reducing mechanical procedures (carried out by a human computer) to machine computations. A deepened analysis of Turing's reduction can serve, ironically, as a springboard for the methodological approach Gödel had recommended in 1934, but never followed up, namely an axiomatic characterization of computability.

References

Davis, Martin (1982) Why Gödel didn't have Church's Thesis; Information and Control 54, 3-24.

A. Beckmann et al. (Eds.): CiE 2006, LNCS 3988, pp. 536–537, 2006.

Gödel, Kurt (1934) On undecidable propositions of formal mathematical systems; in: *Collected Works I*, 346-369.
— (1936) Über die Länge von Beweisen; in: *Collected Works I*, 396-399.
— (193?) Undecidable Diophantine propositions; in: *Collected Works III*, 164-175.
— (1946) Remarks before the Princeton bicentennial conference on problems in mathematics; in: *Collected Works II*, 150-153.
— (1951) Some basic theorems on the foundations of mathematics and their implications; in: *Collected Works III*, 304-323.
— (1964) Postscriptum for 1934; in: *Collected Works I*, 369-371.
— (1972) Some remarks on the undecidability results; *Collected Works II*, 305-306.
Sieg, Wilfried (1997) Step by recursive step: Church's analysis of effective calculability; The Bulletin of Symbolic Logic 3 (2), 154-180.
Turing, Alan (1936) On computable numbers, with an application to the *Entscheidungsproblem*; Proceedings of the London Mathematical Society (Series 2) 42, 230-265.

Co-total Enumeration Degrees

Boris Solon

Ivanovo State University of Chemistry and Technology,
Department of Mathematics

Abstract. This paper is dedicated to the study of the enumeration degrees which contain sets the complements of which are the graphs of some total functions. Such e-degrees are called co-total. That every total e-degree $\mathbf{a} \geq \mathbf{0}'_e$ contains such total function f that $\deg_e(\overline{\mathrm{graph}(f)})$ is a quasi-minimal e-degree has been proved. Some known results of McEvoy and Gutteridge with the aid of co-total e-degrees become stronger as well.

The notations and terminology similar to those of the monograph [10] are used. Let ω denote the set of positive integers; $A, B, \ldots, X, Y$ (with or without indices) are used to denote the subsets of ω; $\overline{A} = \omega - A$; $c_A(x) = \{(x,1) : x \in A\} \cup \{(x,0) : x \notin A\}$ is a characteristic function of A. Let D_u as usual be a finite set with canonical index u; $\langle x, y \rangle$ be the Cantor number of an ordered pair (x, y). If z is a Cantor number of (x, y) then let $\langle z \rangle_1 = x$ and $\langle z \rangle_2 = y$. Let also $\langle A \rangle_1 = \{x : \exists y(\langle x, y \rangle \in A)\}$ and $\langle A \rangle_2 = \{y : \exists x(\langle x, y \rangle \in A)\}$. Let W_t be a computably enumerable (c.e.) set with c.e. index t, $K = \{t : t \in W_t\}$ and $K_0 = \{\langle x, t \rangle : x \in W_t\}$. Below symbol D is used as a variable which ranges over a set of all finite sets.

Given a partial function $\alpha : \omega \to \omega$ let $\mathrm{dom}(\alpha)$, $\mathrm{rang}(\alpha)$ and $\mathrm{graph}(\alpha) = \{\langle x, \alpha(x) \rangle : x \in \mathrm{dom}(\alpha)\}$ be the domain, the range and the graph of α respectively. We restrict the use of the symbols f, g only to denote *total* functions, i.e. $\mathrm{dom}(f) = \mathrm{dom}(g) = \omega$. If $\mathrm{graph}(\alpha) \subseteq \mathrm{graph}(\beta)$ then we shall write $\alpha \subseteq \beta$ for brief. A set A is said to be *single-valued*, if $A = \tau\alpha$ for some partial function α.

We recall, [3], that $A \leq_e B$ (*A is enumeration reducible to B or A is e-reducible to B*), if there is a uniform algorithm for enumerating A given any enumeration of B. Formally,

$$A \leq_e B \iff (\exists t)(\forall x)[x \in A \iff (\exists u)[\langle x, u \rangle \in W_t \;\&\; D_u \subseteq B]].$$

Let $\Phi_t : 2^\omega \to 2^\omega : \Phi_t(X) = \{x : (\exists u)[\langle x, u \rangle \in W_t \;\&\; D_u \subseteq X]\}$. Then $A \leq_e B \iff (\exists t)[A = \Phi_t(B)]$. Φ_t is called *the enumeration operator* or *e-operator with c.e. index t*. Let as usual $A \equiv_e B \iff A \leq_e B \;\&\; B \leq_e A$, let $\deg_e(A) = \{B : B \equiv_e A\}$ be *the e-degree of A* and finally let $\deg_e(A) \leq \deg_e(B) \iff A \leq_e B$. It is easy to see that this relation defines a partial ordering relation on the e-degrees. Bold Latin letters range over e-degrees and corresponding light capital letters automatically denote a representative set of the same degree. We will write for partial function α, β $\alpha \leq_e A$ or $\alpha \leq_e \beta$ if $\tau\alpha \leq_e A$ or $\tau\alpha \leq_e \tau\beta$, respectively. Denote by $\mathbf{D}_e$ a set of the e-degrees partially ordered by $\leq$. It is well known

A. Beckmann et al. (Eds.): CiE 2006, LNCS 3988, pp. 538–545, 2006.

that $\mathbf{D}_e$ forms an upper semilattice with the least element $\mathbf{0}_e = \{W_t : t \in \omega\}$ in which the least upper bound of the e-degrees $\mathbf{a}$ and $\mathbf{b}$ is $\mathbf{a} \cup \mathbf{b} = \deg_e(A \oplus B)$ where $A \oplus B = \{2x : x \in A\} \cup \{2x + 1 : x \in B\}$.

Following K. McEvoy [5] we define a jump operator $'$ on $\mathbf{D}_e$. Let $K_A = \{x : x \in \Phi_x(A)\}$ and $\mathbf{J}(A) = K_A \oplus \bar{K}_A$. It is clear that $\mathbf{J}(A) \equiv_e A \oplus \bar{K}_A$. Let $\mathbf{a}' = (\deg_e(A))' = \deg_e(\mathbf{J}(A))$.

An e-degree is said to be *total* if it contains a graph of some total function. It is clear that an e-degree $\mathbf{a}$ is total iff it contains the set A such that $A \equiv_e A \oplus \bar{A}$. We denote by $\mathbf{T}$ a partial ordering set of all total e-degrees. As for any A and B

$$A \leq_T B \iff A \oplus \bar{A} \leq_e B \oplus \bar{B},$$

then there is the isomorphism between $\mathbf{D}_T$ and $\mathbf{T}$.

Yu. Medvedev announced in [6] that there is a non-c.e. set A such that

$$(\forall f)[f \leq_e A \Rightarrow f \text{ is computable}].$$

In Rogers's monograph ([8],p.280) this result was proved in the following way:

$$(\exists \alpha)[\alpha \text{ is not partial computable } \& \ (\forall f)[f \leq_e \alpha \rightarrow f \text{ is computable}]].$$

It is clear that $\deg_e(\text{graph}(\alpha))$ is not total, i.e. it is *non-total*. Thus $\mathbf{D}_e - \mathbf{T} \neq \emptyset$. J. Case [2] called Medvedev's sets as *quasi-minimal* and their degrees *quasi-minimal e-degrees*.

One of the relativizations of the notion of quasi-minimality is known as c-quasi-minimality. A set A is called C-*quasi-minimal* (and the e-degree $\mathbf{a}$ is called $\mathbf{c}$ -*quasi-minimal*) if $C <_e A$ and $(\forall f)[f \leq_e A \rightarrow f \leq_e C]$. The existence of $\mathbf{c}$-quasi-minimal e-degrees for any $\mathbf{c} \in \mathbf{D}_e$ can be received from the proof of Medvedev's theorem in [8].

In [9] L. Sasso studies three reducibilities on partial functions which are near to e-reducibility (and agree with e-reducibility on total functions). For these reducibilities he introduced the notion of quasi-minimal cover for an ideal. We give this notion for e-degrees:

Definition 1. *Let* A *be an ideal in* $\mathbf{D}_e$, *e-degree* $\mathbf{b}$ *is called a quasi-minimal cover for* A *if* $(\forall \mathbf{x})[\mathbf{x} \in \mathsf{A} \Rightarrow \mathbf{x} \leq \mathbf{b}]$ *and for all* $\mathbf{f} \in \mathbf{T}$

$$\mathbf{f} \leq \mathbf{b} \Rightarrow (\exists \mathbf{x})[\mathbf{x} \in \mathsf{A} \ \& \ \mathbf{f} \leq \mathbf{x}].$$

It is obvious that c-quasi-minimality is a special case of the quasi-minimality in the sense of Sasso, i.e. e-degree $\mathbf{a}$ is $\mathbf{c}$-quasi-minimal iff $\mathbf{a}$ is a quasi-minimal cover for $(\mathbf{c}) = \{\mathbf{x} : \mathbf{x} \leq \mathbf{c}\}$.

Now we introduce the notion of co-total e-degree.

Definition 2. *An e-degree* $\deg_e(A)$ *is said to be co-total if* $\bar{A} = \text{graph}(f)$ *for some total function* f.

Denote by $\mathbf{CT}$ a set of all co-total e-degrees. As every total e-degree $\mathbf{a}$ contains a set A such that $A \equiv_e A \oplus \bar{A}$ and $A \oplus \bar{A} \equiv_e \overline{A \oplus \bar{A}}$ thusevery total e-degree is co-total, i.e. $\mathbf{T} \subseteq \mathbf{CT}$. It is easy to see that $\Pi_1^0 \subseteq \mathbf{CT}$ and $\mathbf{CT} \cap \Pi_2^0 \subseteq \Delta_2^0$. L. Gutteridge showed in [4] that there are quasi-minimal co-total e-degrees, i.e. $\mathbf{T} \subset \mathbf{CT}$. In [1] it was shown that under every non-zero total e-degree there are quasi-minimal incomparable e-degrees $\deg_e(A)$ and $\deg_e(\overline{A})$ thus $\Delta_2^0 - \mathbf{CT} \neq \emptyset$. In particular, not every non-total e-degree is co-total. In [7] it was announced that there is non-zero co-total e-degree $\mathbf{a} \leq \mathbf{0}'_e$ which forms a minimal pair with every e-degree belonging to Π_1^0. In particular, from here it follows that there are co-total e-degrees below $\mathbf{0}'_e$ not belonging to Π_1^0.

Theorem 1. *Every total e-degree* $\mathbf{a} \geq \mathbf{0}'_e$ *contains a total function* f *such that* $\deg_e(\mathrm{graph}(f))$ *is the quasi-minimal e-degree.*

Proof. Let $\mathbf{a} \geq \mathbf{0}'_e$, A be a retraceable set and $\{a_s\}_{s\in\omega}$ be the direct enumeration of A. We construct step by step function f with the help of the construction which is computable in A in such way that $\mathrm{rang} f = A$ and $\deg_e(\mathrm{graph}(f))$ is a quasi-minimal e-degree. At the step $t + 1$ we denote by f_t the finite initial segments which were constructed at the end of step t. Let $l_t = 1 + \max \mathrm{dom}(f_t)$. In the following the symbol σ is used as a variable which ranges over a set of all finite initial A-segments (i.e. such that $\mathrm{rang}(\sigma) \subset A$).

The start of the construction.
  Step 0. Set $f_0 = \emptyset$ and $l_0 = 0$.
  Step 2s+1. Let $t = 2s$. See whether

$$(\exists D)[\Phi_s(D) \text{ is not single} - valued]. \tag{1}$$

If (1) is true then let D^* be D which satisfies (1) and has the least canonical index. In this case we have two subcases:

$$(\exists \sigma)[f_t \subset \sigma \ \& \ \langle D^* \rangle_1 \subseteq \mathrm{dom}(\sigma) \ \& \ \Phi_s(\overline{\mathrm{graph}(\sigma)}) \text{ is single} - valued]. \tag{2}$$

If (2) is true then let σ^* be σ such that it satisfies (1.1) and its graph has the least canonical index. Set $f_{t+1} = \sigma^*$.
  If (2) is not true then we have

$$(\forall \sigma)[f_t \subset \sigma \ \& \ \langle D^* \rangle_1 \subseteq \mathrm{dom}(\sigma) \Rightarrow \Phi_s(\overline{\mathrm{graph}(\sigma)}) \text{ is not single} - valued]. \tag{3}$$

Let σ^* be such σ that its graph has the least canonical index and it satisfies the following condition

$$f_t \subset \sigma \ \& \ \langle D^* \rangle_1 \subseteq \mathrm{dom}(\sigma) \ \& \ D^* \subseteq \mathrm{dom}(\sigma) \times \omega - \mathrm{graph}(\sigma).$$

Set $f_{t+1} = \sigma^*$.
  If (1) is not true then set $f_{t+1} = f_t$ and we pass to the next step.
  Step 2s+2. Let $t = 2s + 1$. Set $f_{t+1} = f_t \cup \{(l_t, a_s)\}$.
  The end of the construction.
  Let $f = \bigcup_{t\in\omega} f_t$. We shall prove that function f resulting from the construction satisfies the theorem. The construction is such that all steps $2s + 1$ are computable in $\overline{K_0} \oplus A$, and all steps $2s + 2$, $s \in \omega$, are computable in A.

As $\mathbf{a} \geq \mathbf{0}'_e$ then our construction as a whole is computable in A, hence $f \leq_e A$. From the construction we see that $\mathrm{rang}(f) = A$, hence $A \leq_e f$.

Let total function $g \leq_e \overline{\mathrm{graph}(f)}$ and $\mathrm{graph}(g) = \Phi_{s_0}(\overline{\mathrm{graph}(f)})$ for some s_0. We shall consider the step $2s_0 + 1$, let $t_0 = 2s_0$. If at this step the condition (1) is not true then we have

$$(\forall D)[\Phi_{s_0}(D) \ is \ single - valued],$$

then $\Phi_{s_0}(\omega)$ is a single-valued set. It is clear that $\mathrm{graph}(g) = \Phi_{s_0}(\overline{\mathrm{graph}(f)}) \subset \Phi_{s_0}(\omega)$ and $\mathrm{graph}(g) = \Phi_{s_0}(\omega)$ by the totality of g. Hence $\mathrm{graph}(g)$ is c.e. and g is a computable function.

If the condition (1) is true then for the subcase (2) we have $\Phi_{s_0}(\overline{\mathrm{graph}(f_{t_0+1})})$ is single-valued. As $\overline{\mathrm{graph}(f)} \subseteq \overline{\mathrm{graph}(f_{t_0+1})}$ then

$$\mathrm{graph}(g) = \Phi_{s_0}(\overline{\mathrm{graph}(f)}) \subseteq \Phi_{s_0}(\overline{\mathrm{graph}(f_{t_0+1})}),$$

and then $\mathrm{graph}(g) = \Phi_{s_0}(\overline{\mathrm{graph}(f_{t_0+1})})$. Hence g is a computable function.

Assume that the subcase (3) holds. Then we obtain that $\Phi_{s_0}(D^*)$ is not single-valued where $D^* \subseteq \overline{\mathrm{graph}(f_{t_0+1})}$ and $\langle D^* \rangle_1 \subseteq \mathrm{dom}(\mathrm{graph}(f_{t_0+1}))$. Then $\mathrm{graph}(g) = \Phi_{s_0}(\overline{\mathrm{graph}(f)})$ is not single-valued what contradicts the premise. Thus $\deg_e(\mathrm{graph}(f))$ is a quasi-minimal e-degree and the theorem is proved completely.

The following theorem is a stronger analogue of McEvoy's theorem [5] which states that for every total e-degree $\mathbf{b} \geq \mathbf{0}'_e$ there is such quasi-minimal e-degree $\mathbf{a}$ that $\mathbf{a}' = \mathbf{b}$.

Theorem 2. *For every total e-degree* $\mathbf{b} \geq \mathbf{0}'_e$ *there is a co-total quasi-minimal e-degree* $\mathbf{a}$ *such that* $\mathbf{a}' = \mathbf{b}$.

Proof. Let $B \in \mathbf{b}$ such that $B \equiv_e c_B$. We construct step by step function f which satisfies the requirements:

(Q): $(\forall z)[\overline{\mathrm{graph}(f)} \neq W_z]$ & $(\forall g)[g \leq_e \overline{\mathrm{graph}(f)} \Rightarrow g \ is \ computable]$,
(J): $\mathbf{J}(\mathrm{graph}(f)) \equiv_e B$.

We note that a total function which satisfies the requirement (Q) was constructed in [4] with the help of the complex priority construction. Here we offer a simple interval construction with the help of which we shall construct a total function satisfying the requirements (Q) and (J).

At step $t + 1$ we denote by f_t the finite initial segments of f which was constructed at the end of step t. Let $l_t = 1 + \max \mathrm{dom}(f_t)$. In the following the symbol σ is used as a variable which ranges over a set of all finite initial segments.

The start of the construction.

Step 0. Set $f_0 = \emptyset$ and $l_0 = 0$.

Step 4s+1. Let $t = 4s$. See whether

$$(\exists y)[\langle l_t, y \rangle \in W_s]. \tag{4}$$

If (4) is true then set $f_{t+1} = f_t \cup \{(l_t, y^*)\}$ where y^* is the least y satisfying (4). If (4) is not true then set $f_{t+1} = f_t \cup \{(l_t, 0)\}$ and passes to the next step.

Step 4s+2. Let $t = 4s + 1$. See whether

$$(\exists D)[\Phi_s(D) \text{ is not single} - \text{valued}]. \tag{5}$$

If (5) is true then let D^* be D which satisfies (5) and has the least canonical index. In this case we have two subcases:

$$(\exists \sigma)[f_t \subset \sigma \ \& \ \langle D^* \rangle_1 \subseteq \text{dom}(\sigma) \ \& \ \Phi_s(\overline{\text{graph}(\sigma)}) \text{ is single} - \text{valued}]. \tag{6}$$

If (6) is true then let σ^* be σ such that it satisfies (6) and its graph has the least canonical index. Set $f_{t+1} = \sigma^*$.

If (6) is not true then we have

$$(\forall \sigma)[f_t \subset \sigma \ \& \ \langle D^* \rangle_1 \subseteq \text{dom}(\sigma) \Rightarrow \Phi_s(\overline{\text{graph}(\sigma)}) \text{ is not single} - \text{valued}]. \tag{7}$$

Let σ^* be such σ that its graph has the least canonical index and it satisfies the following condition

$$f_t \subset \sigma \ \& \ \langle D^* \rangle_1 \subseteq \text{dom}(\sigma) \ \& \ D^* \subseteq \text{dom}(\sigma) \times \omega - \text{graph}(\sigma).$$

Set $f_{t+1} = \sigma^*$.

If (5) is not true then set $f_{t+1} = f_t$ and we pass to the next step.

Step 4s+3. Let $t = 4s + 2$. See whether

$$(\exists \sigma)[f_t \subset \sigma \ \& \ s \in \Phi_s(\overline{\text{graph}(\sigma)})]. \tag{8}$$

If (8) is true then let $D^* \subset \overline{\text{graph}(\sigma)}$ be a finite set such that it has the least canonical index and $s \in \Phi_s(D^*)$. Set $f_{t+1} = \sigma^*$ where σ^* such that its graph has the least canonical index, it satisfies the condition (8) and $\langle D^* \rangle_1 \subset \text{dom}(\sigma^*)$. If (8) is not true then set $f_{t+1} = f_t$ and passes to the next step.

Step 4s+4. Let $t = 4s + 3$, set

$$f_{t+1} = f_t \cup \{(l_t, 1 - c_B(s))\}$$

and passes to the next step.

The end of the construction. Let $f = \bigcup_{t \in \omega} f_t$. Now we shall prove that function f resulting from the construction satisfies the requirements (Q) and (J).

The steps $4s + 1$, $s \in \omega$ provide $\overline{\text{graph}(f)} \neq W_s$. Let a total function $g \leq_e \overline{\text{graph}(f)}$ and $\text{graph}(g) = \Phi_{s_0}(\overline{\text{graph}(f)})$ for some s_0. We shall consider the step $4s_0+2$, let $t_0 = 4s_0+1$. Repeating the corresponding part of the proof of theorem 1 we shall prove that g is a computable function. Hence the requirement (Q) is satisfied.

Our construction provides that all steps $4s + 1$, $4s + 2$, $4s + 3$, $s \in \omega$ are computable in $\mathbf{0}'_e$ and the steps $4s + 4$, $s \in \omega$ are computable in B. As $\mathbf{0}'_e \leq \mathbf{b}$ then our construction as a whole is computable in B, hence $f \leq_e B$. From steps $4s+3$ it follows

$$(\forall x)[x \in \Phi_x(\overline{\text{graph}(f)}) \iff f_{4x+3} \neq f_{4x+2}],$$

from which $\mathbf{J}(\overline{\text{graph}(f)}) \leq_e B$.

To check $B \leq_e \mathbf{J}(\overline{\mathrm{graph}(f)})$ we shall show that the sequence of initial segments $\{f_t\}_{t \in \omega}$ and hence the sequence of computable sets $\{\overline{\mathrm{graph}(f)_t}\}_{t \in \omega}$ is computable in $\mathbf{J}(\overline{\mathrm{graph}(f)})$. Then $\lambda x.c_B(x) = 1 - f_{4x+4}(l_x)$, therefore $B \leq_e \mathbf{J}(\overline{\mathrm{graph}(f)})$. It is clear that $\mathrm{graph}(f) \leq_e \mathbf{J}(\overline{\mathrm{graph}(f)})$. All steps except for $4s + 4$, $s \in \omega$ are computable in $\mathbf{0}'_e$, and at steps $4s + 4$, $s \in \omega$ we made the structure

$$f_{4s+4} = f_{4s+3} \cup \{(l_{4s+3}, 1 - c_{\overline{\mathrm{graph}(f)}}(l_{4s+3})\},$$

which is computable in $\overline{\mathrm{graph}(f)}$. Hence $\mathbf{J}(\overline{\mathrm{graph}(f)}) \equiv_e B$ and the requirement (J) is satisfied.

Let $\mathbf{a} = \deg_e(\mathrm{graph}(f))$. Our construction and the satisfiability of (Q) provide that $\mathbf{a}$ is a co-total quasi-minimal e-degree and the satisfiability of (J) provides that $\mathbf{a}' = \mathbf{b}$. The following theorem is a generalization of Guttteridge's theorem [4] which states that there is a quasi-minimal co-total e-degree $\mathbf{a}$.

Theorem 3 *For every total e-degree* $\mathbf{b}$ *there is* $\mathbf{b}$-*quasi-minimal co-total e-degree* $\mathbf{a}$.

Proof. Let $B \in \mathbf{b}$ such that $B \equiv_e c_B$. We construct step by step a function f which satisfies the requirement:

(BQ): $(\forall s)[\overline{\mathrm{graph}(f)} \neq \Phi_s(B)]$ & $(\forall g)[y \leq_e \overline{\mathrm{graph}(f)} \Rightarrow g \leq_e B]$.

At the step $t + 1$ we denote by f_t the finite initial segment of f which was constructed at the end of step t. f_t has a form $c_B \oplus \sigma_t$ where σ_t is an initial segment which we choose at step t. Let $l_t = 1 + \max \mathrm{dom}(\sigma_t)$. In the following the symbol σ is used as a variable which ranges over a set of all finite initial segments. Thus $f = \bigcup_{t \in \omega} f_t = c_B \oplus \bigcup_{t \in \omega} \sigma_t = c_B \oplus \alpha$.

The start of the construction.

Step 0. Set $f_0 = c_B \oplus \emptyset$ and $l_0 = 0$.

Step 2s+1. Let $t = 2s$. See whether

$$(\exists y)[\langle 2l_t + 1, y \rangle \in \Phi_s(B)]. \tag{9}$$

If (9) is true then set $f_{t+1} = f_t \cup \{(2l_t + 1, y^*)\}$ where y^* is the least y satisfying (9). If (9) is not true then set $f_{t+1} = f_t \cup \{(2l_t + 1, 0)\}$ and passes to the next step.

Step 2s+2. Let $t = 2s + 1$. See whether

$$(\exists D)[\Phi_s(D) \; is \; not \; single - valued]. \tag{10}$$

If (10) is true then let D^* be D which satisfies (10) and has the least canonical index. In this case we have two subcases:

$$(\exists \sigma)[f_t \subset \sigma \; \& \; \langle D^* \rangle_1 \subseteq \mathrm{dom}(c_B \oplus \sigma) \; \&$$
$$\& \; \Phi_s(\overline{\mathrm{graph}(c_B \oplus \sigma)}) \; is \; single - valued]. \tag{11}$$

If (11) is true then let σ_{t+1} be σ such that it satisfies (11) and its graph has the least canonical index. Set $f_{t+1} = c_B \oplus \sigma_{t+1}$.

If (11) is not true then we have

$$(\forall \sigma)[f_t \subset \sigma \ \& \ \langle D^* \rangle_1 \subseteq \mathrm{dom}(c_B \oplus \sigma) \Rightarrow$$
$$\Rightarrow \Phi_s(\overline{\mathrm{graph}(c_B \oplus \sigma)}) \ is \ not \ single-valued]. \qquad (12)$$

Let σ_{t+1} be such σ that its graph has the least canonical index and it satisfies the following condition

$$f_t \subset \sigma \ \& \ \langle D^* \rangle_1 \subseteq \mathrm{dom}(c_B \oplus \sigma) \ \& \ D^* \subseteq \mathrm{dom}(c_B \oplus \sigma) \times \omega - \mathrm{graph}(c_B \oplus \sigma).$$

Set $f_{t+1} = c_B \oplus \sigma_{t+1}$.

If (10) is not true then set $f_{t+1} = f_t$ and we pass to the next step.

The end of the construction.

First we shall prove that the requirement (BQ) is satisfied. Let a total function $g \leq_e \overline{\mathrm{graph}(f)}$ and $\mathrm{graph}(g) = \Phi_{s_0}(\overline{\mathrm{graph}(f)})$ for some s_0. We shall consider the step $2s_0 + 2$, let $t_0 = 2s_0 + 1$. If at this step the condition (2) is not true then we have

$$(\forall D)[\Phi_{s_0}(D) \ is \ single-valued],$$

then $\Phi_{s_0}(\omega)$ is a single-valued set. Then it is clear that g is a computable function.

If the condition (10) is true then for the subcase (11) we have that

$$\Phi_{s_0}(\overline{\mathrm{graph}(f_{t_0+1})}) = \Phi_{s_0}(\overline{\mathrm{graph}(c_B \oplus \sigma_{t_0+1})})$$

and both are single-valued. As $\overline{\mathrm{graph}(f)} \subseteq \overline{\mathrm{graph}(f_{t_0+1})}$ then

$$\mathrm{graph}(g) = \Phi_{s_0}(\overline{\mathrm{graph}(f)}) = \Phi_{s_0}(\overline{\mathrm{graph}(c_B \oplus \sigma_{t_0+1})}).$$

Hence $g \leq_e B$.

Assume that the subcase (12) holds. Then we obtain that $\Phi_{s_0}(D^*)$ is not single-valued where $D^* \subseteq \overline{\mathrm{graph}(f_{t_0+1})}$ and $\langle D^* \rangle_1 \subseteq \mathrm{dom}(f_{t_0+1})$. Then $\mathrm{graph}(g) = \Phi_{s_0}(\overline{\mathrm{graph}(f)})$ is not single-valued what contradicts the premise. Thus the requirement (BQ) is satisfied.

Finally let $\mathbf{a} = \deg_e(\overline{\mathrm{graph}(f)})$ where f is the result of our construction. The construction and the steps $2s + 1$, $s \in \omega$ guarantee $\mathbf{b} < \mathbf{a}$. The satisfiability of (BQ) guarantees that $\mathbf{a}$ is $\mathbf{b}$-quasi-minimal e-degree. The theorem is proved completely.

References

1. Arslanov M.M., Cooper S.B., Kalimullin I.S.: The properties of the splitting of total e-degrees. Algebra i Logica **43**(2003) 1–25
2. Case J.: Enumeration reducibility and partial degreees. Annals Math. Logic **2**(1971) 419–439
3. Fridberg R., Rogers H.: Reducibility and completness for sets of integers. Z. math. Logic Grundl. Math. **5**(1959) 117–125
4. Gutteridge L.: Some results on e-reducibility. Ph.D.Diss (1971)

5. McEvoy K.:   Jumps of quasi-minimal enumeration degrees. J. Symb. Logic **50**(1985) 839–848
6. Medvedev Yu.T.: Degrees of difficulty of the mass problem. Dokl. Acad. Nauk SSSR **104**(1955) 501–504
7. Pankratov A.: The research of some properties of co-total e-degrees. Intern. conf. "Logic and applications", Proceedings, Novosibirsk (2000) 79
8. Rogers H.,Jr.: Theory of Recursive Functions and Effective Computability. McGraw-Hill. New York. 1967
9. Sasso L. P.: A survey of partial degrees. J. Symb. Logic **40** (1975) 130–140
10. Soar Robert I.: Recursively Enumerable Sets and Degrees. Springer-Verlag. Berlin, Heidelberg, New York, London. 1987

Relativized Degree Spectra

Alexandra A. Soskova*

Faculty of Mathematics and Computer Science, Sofia University,
5 James Bourchier Blvd., 1164 Sofia, Bulgaria
asoskova@fmi.uni-sofia.bg

Abstract. A relativized version of the notion of Degree spectrum of a structure with respect to finitely many abstract structures is presented, inspired by the notion of relatively intrinsic sets. The connection with the notion of Joint spectrum is studied. Some specific properties like Minimal Pair type theorem and the existence of Quasi-Minimal degree with respect to the Relative spectrum are shown.

1 Introduction

Let $\mathfrak{A}$ be a countable partial structure. The Degree spectrum $DS(\mathfrak{A})$ of the structure $\mathfrak{A}$ is the set of all enumeration degrees generated by all enumerations of $\mathfrak{A}$. The notion is introduced by *Richter* in [8] and studied by *Knight, Ash, Jockush, Downey* and *Soskov* in [7, 3, 6, 10]. It is a kind of a measure of complexity of the structure. The Co-spectrum $CS(\mathfrak{A})$ of the structure $\mathfrak{A}$ is the set of all enumeration degrees which are lower bounds of the $DS(\mathfrak{A})$. A typical example of a Degree spectrum is the cone of all total enumeration degrees, greater than or equal to some enumeration degree $\mathbf{a}$ and the respective Co-spectrum is equal to the set of all degrees less than or equal to $\mathbf{a}$. In [10] Soskov shows that the Degree spectra behave with respect to their Co-spectra very much like the cones of enumeration degrees. The Degree spectra have some general and specific properties. For example each Degree spectrum is closed upwards, i.e. if $\mathbf{a} \in DS(\mathfrak{A})$ then each total enumeration degree $\mathbf{b}$ greater than or equal to $\mathbf{a}$ is in $DS(\mathfrak{A})$. But not every upwards closed set of enumeration degrees is a spectrum of a structure. Some typical specific properties of the Degree spectra and their Co-spectra are the Minimal Pair type theorem and the existence of Quasi-Minimal degree. For every Degree spectrum $DS(\mathfrak{A})$ there exist total enumeration degrees $\mathbf{f_0}$ and $\mathbf{f_1}$, elements of $DS(\mathfrak{A})$, which determine completely the elements of the Co-spectrum $CS(\mathfrak{A})$, i.e. the set of all enumeration degrees less than or equal to both $\mathbf{f_0}$ and $\mathbf{f_1}$ is exactly $CS(\mathfrak{A})$. The degrees $\mathbf{f_0}$ and $\mathbf{f_1}$ are called Minimal Pair for $DS(\mathfrak{A})$. For each Degree spectrum $DS(\mathfrak{A})$ there is an enumeration degree $\mathbf{q} \notin CS(\mathfrak{A})$, called Quasi-Minimal for $DS(\mathfrak{A})$, such that for each total degree $\mathbf{a}$ if $\mathbf{a} \geq \mathbf{q}$, then $\mathbf{a} \in DS(\mathfrak{A})$ and if $\mathbf{a} \leq \mathbf{q}$, then $\mathbf{a} \in CS(\mathfrak{A})$.

In this paper we introduce and study a generalized notion of Degree spectrum of the structure $\mathfrak{A}$, relatively given structures $\mathfrak{A}_1, \ldots, \mathfrak{A}_n$, inspired by the notion

---

* This work was partially supported by Sofia University Science Fund. The author thanks the anonymous referees for helpful comments.

A. Beckmann et al. (Eds.): CiE 2006, LNCS 3988, pp. 546–555, 2006.

of relatively intrinsic on $\mathfrak{A}$ sets. An internal characterization of the relatively intrinsic on $\mathfrak{A}$ sets is presented in [2], [4] and in [11] with respect to the infinite sequence of sets.

The *Relative spectrum* $RS(\mathfrak{A}, \mathfrak{A}_1, \ldots, \mathfrak{A}_n)$ of $\mathfrak{A}$ with respect to the structures $\mathfrak{A}_1, \ldots, \mathfrak{A}_n$ is the set of all enumeration degrees generated by all enumerations of $\mathfrak{A}$, such that the structure $\mathfrak{A}_k$ is relatively k-intrinsic on $\mathfrak{A}$, i.e. $\mathfrak{A}_k$ is admissible in the kth jump of $\mathfrak{A}$. In other words we consider the set of all enumeration degrees of the presentations of the structure $\mathfrak{A}$ in which the degrees of $\mathfrak{A}_k$ fall below the kth jump of the degrees of $\mathfrak{A}$, $k \leq n$. We will show that this generalized notion of Degree spectra posses all general and specific properties of the Degree spectra of a structure. And we will compare this notion with the notion of Joint Spectrum of $\mathfrak{A}$ with respect to the structures $\mathfrak{A}_1, \ldots, \mathfrak{A}_n$, considered in [12], [13].

2 Preliminaries

Let $\mathfrak{A} = (\mathbb{N}; R_1, \ldots, R_s)$ be a partial structure over the set of all natural numbers $\mathbb{N}$, where each R_i is a subset of $\mathbb{N}^{r_i}$ and $=, \neq$ are among $R_1, \ldots, R_s$.

An *enumeration* f *of* $\mathfrak{A}$ is a total mapping from $\mathbb{N}$ onto $\mathbb{N}$.

For $A \subseteq \mathbb{N}^a$ define $f^{-1}(A) = \{\langle x_1 \ldots x_a \rangle : (f(x_1), \ldots, f(x_a)) \in A\}$. Denote by $f^{-1}(\mathfrak{A}) = f^{-1}(R_1) \oplus \ldots \oplus f^{-1}(R_s)$.

For any sets of natural numbers A and B the set A is enumeration reducible to B ($A \leq_e B$) if there is an enumeration operator Γ_z such that $A = \Gamma_z(B)$. By $d_e(A)$ we denote the enumeration degree of the set A and by $\mathcal{D}_e$ the set of all enumeration degrees. The set A is total if $A \equiv_e A^+$, where $A^+ = A \oplus (\mathbb{N} \backslash A)$. A degree $\mathbf{a}$ is total if $\mathbf{a}$ contains the e-degree of a total set. The jump operation "$'$" denotes here the enumeration jump introduced by *Cooper* in [5].

Let $B_0, \ldots, B_n$ be arbitrary subsets of $\mathbb{N}$. Define the set $\mathcal{P}(B_0, \ldots, B_i)$ by induction on $i \leq n$, as follows:

1. $\mathcal{P}(B_0) = B_0$;
2. If $i < n$, then $\mathcal{P}(B_0, \ldots, B_{i+1}) = (\mathcal{P}(B_0, \ldots, B_i))' \oplus B_{i+1}$.

We will use the following modification of Jump Inversion Theorem from [9]:

Theorem 1 ([9]). *Let* $\{A_r^k\}_{r \in \mathbb{N}}$*,* $k = 0, \ldots, n-1$ *be* n *sequences of subsets of* $\mathbb{N}$*, such that for every* r *and for all* k*,* $0 \leq k < n$*,* $A_r^k \nleq_e \mathcal{P}(B_0, \ldots, B_k)$ *and let* Q *be a total set, such that* $\mathcal{P}(B_0, \ldots, B_n) \leq_e Q$*. Then there exists a total set* F *having the following properties:*

1. $B_i \leq_e F^{(i)}$*, for all* $i \leq n$*;*
2. $A_r^k \nleq_e F^{(k)}$*, for all* r *and all* $k < n$*;*
3. $F^{(n)} \equiv_e Q$*.*

3 Relative Spectra of Structures

Definition 2. *The Degree spectrum of* $\mathfrak{A}$ *is the set*

$$DS(\mathfrak{A}) = \{d_e(f^{-1}(\mathfrak{A})) : f \text{ is an enumeration of } \mathfrak{A}\}.$$

Let $\mathfrak{A}_1, \ldots, \mathfrak{A}_n$ be arbitrary abstract structures on $\mathbb{N}$.

Definition 3. *The Relative spectrum of the structure* $\mathfrak{A}$ *with respect to* $\mathfrak{A}_1, \ldots,$ $\mathfrak{A}_n$ *is the set*

$$\mathrm{RS}(\mathfrak{A}, \mathfrak{A}_1, \ldots, \mathfrak{A}_n) = \{d_e(f^{-1}(\mathfrak{A})) : f \text{ is an enumeration of } \mathfrak{A} \text{ such that:}$$
$$(\forall k \leq n)(f^{-1}(\mathfrak{A}_k) \leq_e (f^{-1}(\mathfrak{A}))^{(k)})\}.$$

Definition 4. *Let* $k \leq n$. *An enumeration* f *of* $\mathfrak{A}$ *is* k-*acceptable with respect to the structures* $\mathfrak{A}_1, \ldots, \mathfrak{A}_k$, *if* $f^{-1}(\mathfrak{A}_i) \leq_e (f^{-1}(\mathfrak{A}))^{(i)}$, *for each* $i \leq k$.

In fact the Relative spectrum of $\mathfrak{A}$ is the set, generated by all n-acceptable enumerations of $\mathfrak{A}$ with respect to $\mathfrak{A}_1, \ldots, \mathfrak{A}_n$. First we show that the Relative spectra are closed upwards.

Lemma 5. *If* F *is a total set,* f *is a* n-*acceptable enumeration of* $\mathfrak{A}$ *with respect to* $\mathfrak{A}_1, \ldots, \mathfrak{A}_n$ *and* $f^{-1}(\mathfrak{A}) \leq_e F$, *then there exists a* n-*acceptable enumeration* g *of* $\mathfrak{A}$ *with respect to* $\mathfrak{A}_1, \ldots, \mathfrak{A}_n$, *such that*

1. $g^{-1}(\mathfrak{A}) \equiv_e F \oplus f^{-1}(\mathfrak{A}) \equiv_e F$;
2. $g^{-1}(B) \leq_e F \oplus f^{-1}(B)$, *for every* $B \subseteq \mathbb{N}$.

Proof (sketch). Let $s \neq t \in \mathbb{N}$, $f(x_s) \simeq s$ and $f(x_t) \simeq t$. Define

$$g(x) \simeq \begin{cases} f(x/2) & \text{if } x \text{ is even,} \\ s & \text{if } x = 2z+1 \text{ and } z \in F, \\ t & \text{if } x = 2z+1 \text{ and } z \notin F. \end{cases}$$

It is clear that $f^{-1}(\mathfrak{A}) \leq_e g^{-1}(\mathfrak{A})$. Since "=" and "$\neq$" are among the underlined predicates of $\mathfrak{A}$, $F \leq_e g^{-1}(\mathfrak{A})$.

Consider the predicate R_i of $\mathfrak{A}$. Let $x_1, \ldots, x_{r_i}$ be arbitrary natural numbers. Define the natural numbers $y_1, \ldots, y_{r_i}$ by means of the following recursive in F procedure. Let $1 \leq j \leq r_i$. If x_j is even then let $y_j = x_j/2$. If $x_j = 2z+1$ and $z \in F$, then let $y_j = x_s$. If $x_j = 2z+1$ and $z \notin F$, then let $y_j = x_t$. Clearly

$$\langle x_1, \ldots, x_{r_i} \rangle \in g^{-1}(R_i) \iff \langle y_1, \ldots, y_{r_i} \rangle \in f^{-1}(R_i).$$

Thus $g^{-1}(R_i) \leq_e F \oplus f^{-1}(\mathfrak{A})$. So, we obtain that $g^{-1}(\mathfrak{A}) \equiv_e F \oplus f^{-1}(\mathfrak{A}) \equiv_e F$.

From the definition of g it follows that $g^{-1}(B) \leq_e F \oplus f^{-1}(B)$, for any $B \subseteq \mathbb{N}$. Then, for each $i \leq n$, $g^{-1}(\mathfrak{A}_i) \leq_e F \oplus f^{-1}(\mathfrak{A}_i) \leq_e F \oplus (f^{-1}(\mathfrak{A}))^{(i)} \leq_e F \oplus F^{(i)} \equiv_e F^{(i)} \equiv_e (g^{-1}(\mathfrak{A}))^{(i)}$.

Corollary 6. *If* $\mathbf{b}$ *is a total e-degree,* $\mathbf{a} \in \mathrm{RS}(\mathfrak{A}, \mathfrak{A}_1, \ldots, \mathfrak{A}_n)$, *and* $\mathbf{a} \leq \mathbf{b}$, *then* $\mathbf{b} \in \mathrm{RS}(\mathfrak{A}, \mathfrak{A}_1, \ldots, \mathfrak{A}_n)$.

Denote by $\mathcal{P}_k^f = \mathcal{P}(f^{-1}(\mathfrak{A}), f^{-1}(\mathfrak{A}_1), \ldots, f^{-1}(\mathfrak{A}_k))$, for every enumeration f of $\mathfrak{A}$ and $k \leq n$.

Lemma 7. *Let* f *be an arbitrary enumeration of* $\mathfrak{A}$, *then there exists a* n-*acceptable enumeration* g *of* $\mathfrak{A}$ *with respect to* $\mathfrak{A}_1, \ldots, \mathfrak{A}_n$, *such that* $f^{-1}(\mathfrak{A}) \leq_e g^{-1}(\mathfrak{A})$ *and* $g^{-1}(\mathfrak{A})$ *is a total set.*

Let Q be a total set such that $\mathcal{P}_n^f \leq_e Q$. Apply Theorem 1 and the construction of Lemma 5.

Definition 8. Let $k \leq n$. The kth Jump Relative spectrum of $\mathfrak{A}$ with respect to $\mathfrak{A}_1, \ldots, \mathfrak{A}_n$ is the set

$$\mathrm{RS}_k(\mathfrak{A}, \mathfrak{A}_1, \ldots, \mathfrak{A}_n) = \{\mathbf{a}^{(k)} : \mathbf{a} \in \mathrm{RS}(\mathfrak{A}, \mathfrak{A}_1, \ldots, \mathfrak{A}_n)\}.$$

Proposition 9. Let $k \leq n$. $\mathrm{RS}_k(\mathfrak{A}, \mathfrak{A}_1, \ldots \mathfrak{A}_n)$ is closed upwards, i.e. if $\mathbf{b}$ is a total e-degree, $\mathbf{a} \in \mathrm{RS}(\mathfrak{A}, \mathfrak{A}_1, \ldots \mathfrak{A}_n)$ and $\mathbf{a}^{(k)} \leq \mathbf{b}$, then $\mathbf{b} \in \mathrm{RS}_k(\mathfrak{A}, \mathfrak{A}_1, \ldots, \mathfrak{A}_n)$.

Proof. Let G be a total set, $G \in \mathbf{b}$, and $(f^{-1}(\mathfrak{A}))^{(k)} \leq_e G$, for some n-acceptable enumeration f of $\mathfrak{A}$, with respect to $\mathfrak{A}_1, \ldots, \mathfrak{A}_n$. Then $\mathcal{P}_k^f \leq_e (f^{-1}(\mathfrak{A}))^{(k)} \leq_e G$. By Theorem 1 there exists a total set F, such that $f^{-1}(\mathfrak{A}) \leq_e F$, $f^{-1}(\mathfrak{A}_i) \leq_e F^{(i)}$, for $i \leq k$ and $F^{(k)} \equiv_e G$. As in Lemma 5, we construct a k-acceptable enumeration g of $\mathfrak{A}$, with respect to $\mathfrak{A}_1, \ldots, \mathfrak{A}_k$, so that $g^{-1}(\mathfrak{A}) \equiv_e F$, So, $g^{-1}(\mathfrak{A}_i) \leq_e (g^{-1}(\mathfrak{A}))^{(i)}$, for $i \leq k$. But for $k \leq j \leq n$ we have $g^{-1}(\mathfrak{A}_j) \leq_e F \oplus f^{-1}(\mathfrak{A}_j) \leq_e F \oplus (f^{-1}(\mathfrak{A}))^{(j)} \leq_e F \oplus F^{(j)} \equiv_e F^{(j)} \equiv_e (g^{-1}(\mathfrak{A}))^{(j)}$. Thus $G \equiv_e (g^{-1}(\mathfrak{A}))^{(k)}$, $d_e(g^{-1}(\mathfrak{A})) \in \mathrm{RS}(\mathfrak{A}, \mathfrak{A}_1, \ldots, \mathfrak{A}_n)$ and hence $d_e(G) \in \mathrm{RS}_k(\mathfrak{A}, \mathfrak{A}_1, \ldots, \mathfrak{A}_n)$.

4 Relative Co-spectra of Structures

Let $\mathcal{A}$ be a set of enumeration degrees. The co-set of $\mathcal{A}$ is the set of all lower bounds of $\mathcal{A}$.

Definition 10. The Relative co-spectrum of $\mathfrak{A}$ with respect to $\mathfrak{A}_1, \ldots, \mathfrak{A}_n$, is the co-set of $\mathrm{RS}(\mathfrak{A}, \mathfrak{A}_1, \ldots, \mathfrak{A}_n)$, i.e.

$$\mathrm{CRS}(\mathfrak{A}, \mathfrak{A}_1, \ldots, \mathfrak{A}_n) = \{\mathbf{b} : \mathbf{b} \in \mathcal{D}_e \& (\forall \mathbf{a} \in \mathrm{RS}(\mathfrak{A}, \mathfrak{A}_1, \ldots, \mathfrak{A}_n))(\mathbf{b} \leq \mathbf{a})\}.$$

Definition 11. Let $k \leq n$. The Relative kth co-spectrum of $\mathfrak{A}$ with respect to $\mathfrak{A}_1, \ldots, \mathfrak{A}_n$, is the co-set of $\mathrm{RS}_k(\mathfrak{A}, \mathfrak{A}_1, \ldots, \mathfrak{A}_n)$, i.e.

$$\mathrm{CRS}_k(\mathfrak{A}, \mathfrak{A}_1, \ldots, \mathfrak{A}_n) = \{\mathbf{b} : \mathbf{b} \in \mathcal{D}_e \& (\forall \mathbf{a} \in \mathrm{RS}_k(\mathfrak{A}, \mathfrak{A}_1, \ldots, \mathfrak{A}_n))(\mathbf{b} \leq \mathbf{a})\}.$$

Proposition 12. $\mathrm{CRS}_k(\mathfrak{A}, \mathfrak{A}_1, \ldots, \mathfrak{A}_k, \ldots, \mathfrak{A}_n) = \mathrm{CRS}_k(\mathfrak{A}, \mathfrak{A}_1, \ldots, \mathfrak{A}_k)$.

Proof. It is clear that $\mathrm{RS}_k(\mathfrak{A}, \mathfrak{A}_1, \ldots, \mathfrak{A}_k, \ldots, \mathfrak{A}_n) \subseteq \mathrm{RS}_k(\mathfrak{A}, \mathfrak{A}_1, \ldots, \mathfrak{A}_k)$. Thus $\mathrm{CRS}_k(\mathfrak{A}, \mathfrak{A}_1, \ldots, \mathfrak{A}_k) \subseteq \mathrm{CRS}_k(\mathfrak{A}, \mathfrak{A}_1, \ldots, \mathfrak{A}_k, \ldots, \mathfrak{A}_n)$.

Let $\mathbf{a} \in \mathrm{CRS}_k(\mathfrak{A}, \mathfrak{A}_1 \ldots \mathfrak{A}_k \ldots, \mathfrak{A}_n)$, $A \in \mathbf{a}$ and assume that $A \not\leq_e (f^{-1}(\mathfrak{A}))^{(k)}$ for some k-acceptable enumeration f of $\mathfrak{A}$ with respect to $\mathfrak{A}_1, \ldots, \mathfrak{A}_k$. Then $A \not\leq_e \mathcal{P}_k^f$. Hence by Theorem 1 for $B_0 = f^{-1}(\mathfrak{A}), B_1 = f^{-1}(\mathfrak{A}_1), \ldots, B_n = f^{-1}(\mathfrak{A}_n)$, $B_{n+1} = \mathbb{N}$, there exists a total set F, such that $f^{-1}(\mathfrak{A}) \leq_e F$, for each $i \leq n$ $f^{-1}(\mathfrak{A}_i) \leq_e F^{(i)}$, and $A \not\leq_e F^{(k)}$. As in Lemma 5, we construct a k-acceptable enumeration g of $\mathfrak{A}$ with respect to $\mathfrak{A}_1, \ldots, \mathfrak{A}_k$, such that $g^{-1}(\mathfrak{A}) \equiv_e F$. Then $A \not\leq_e (g^{-1}(\mathfrak{A}))^{(k)}$ and $g^{-1}(\mathfrak{A}_i) \leq_e (g^{-1}(\mathfrak{A}))^{(i)}$, for $i \leq k$. But for $k \leq j \leq n$, $g^{-1}(\mathfrak{A}_j) \leq_e F \oplus f^{-1}(\mathfrak{A}_j) \leq_e F \oplus F^{(j)} \equiv_e F^{(j)} \equiv_e (g^{-1}(\mathfrak{A}))^{(j)}$, i.e. g is a n-acceptable enumeration of $\mathfrak{A}$ with respect to $\mathfrak{A}_1, \ldots, \mathfrak{A}_n$ and $A \not\leq_e (g^{-1}(\mathfrak{A}))^{(k)}$, which contradicts with the choice of A.

In order to obtain a forcing normal form of the sets with enumeration degrees in $\mathrm{CRS}_k(\mathfrak{A}, \mathfrak{A}_1, \ldots, \mathfrak{A}_n)$ we shall define the notion of forcing relation $\tau \Vdash_k F_e(x)$ and the relations $f \models_k F_e(x)$, for $k \leq n$, as in [11].

Let $W_0, \ldots, W_z, \ldots$ be a Gödel's enumeration of the c.e. sets and D_v be the finite set having the canonical code v. Let f be an enumeration of $\mathfrak{A}$.

For every $i \leq n$, e and x in $\mathbb{N}$ define the relations $f \models_i F_e(x)$ and $f \models_i \neg F_e(x)$ by induction on i:

1. $f \models_0 F_e(x) \iff (\exists v)(\langle v, x \rangle \in W_e \,\&\, D_v \subseteq f^{-1}(\mathfrak{A}))$;
2. $f \models_{i+1} F_e(x) \iff (\exists v)(\langle v, x \rangle \in W_e \,\&\, (\forall u \in D_v)(u = \langle 0, e_u, x_u \rangle \,\&$
   $f \models_i F_{e_u}(x_u) \vee u = \langle 1, e_u, x_u \rangle \,\&\, f \models_i \neg F_{e_u}(x_u) \vee u = \langle 2, x_u \rangle \,\&$
   $x_u \in f^{-1}(\mathfrak{A}_{i+1})))$;
3. $f \models_i \neg F_e(x) \iff f \not\models_i F_e(x)$.

From the definition it follows that for any $A \subseteq \mathbb{N}$ and $k \leq n$

$$A \leq_e \mathcal{P}_k^f \iff (\exists e)(A = \{x : f \models_k F_e(x)\}).$$

The forcing conditions, called *finite parts*, are finite mappings τ of $\mathbb{N}$ in $\mathbb{N}$.

For any $i \leq n$, e and x in $\mathbb{N}$ and every finite part τ define the forcing relations $\tau \Vdash_i F_e(x)$ and $\tau \Vdash_i \neg F_e(x)$ following the definition of relation "$\models_i$".

1. $\tau \Vdash_0 F_e(x) \iff (\exists v)(\langle v, x \rangle \in W_e \,\&\, D_v \subseteq \tau^{-1}(\mathfrak{A}))$;
2. $\tau \Vdash_{i+1} F_e(x) \iff \exists v(\langle v, x \rangle \in W_e \,\&\, (\forall u \in D_v)(u = \langle 0, e_u, x_u \rangle \,\&$
   $\tau \Vdash_i F_{e_u}(x_u) \vee u = \langle 1, e_u, x_u \rangle \,\&\, \tau \Vdash_i \neg F_{e_u}(x_u) \vee u = \langle 2, x_u \rangle \,\&$
   $x_u \in \tau^{-1}(\mathfrak{A}_{i+1})))$;
3. $\tau \Vdash_i \neg F_e(x) \iff (\forall \rho \supseteq \tau)(\rho \not\Vdash_i F_e(x))$.

For any $i \leq n, e, x \in \mathbb{N}$ denote by $X^i_{\langle e, x \rangle} = \{\rho : \rho \Vdash_i F_e(x)\}$.

Definition 13. Let $k \leq n+1$. An enumeration f of $\mathfrak{A}$ is k-*generic* with respect to $\mathfrak{A}_1, \ldots, \mathfrak{A}_n$, if for every $j < k$, $e, x \in \mathbb{N}$

$$(\forall \tau \subseteq f)(\exists \rho \in X^j_{\langle e, x \rangle})(\tau \subseteq \rho) \implies (\exists \tau \subseteq f)(\tau \in X^j_{\langle e, x \rangle}).$$

In [11] the following properties of the k-generic enumerations are shown:

1. The forcing relation is monotone.
2. If f is a $(k+1)$-generic enumeration of $\mathfrak{A}$, with respect to $\mathfrak{A}_1, \ldots, \mathfrak{A}_n$, then

$$f \models_k (\neg) F_e(x) \iff (\exists \tau \subseteq f)(\tau \Vdash_k (\neg) F_e(x)).$$

Definition 14. Let $A \subseteq \mathbb{N}$ and $k \leq n$. The set A is *forcing k-definable* on $\mathfrak{A}$ with respect to $\mathfrak{A}_1, \ldots, \mathfrak{A}_n$ if there exist a finite part δ and $e \in \mathbb{N}$ such that

$$x \in A \iff (\exists \tau \supseteq \delta)(\tau \Vdash_k F_e(x)).$$

Proposition 15. *Let $\{A_r^k\}_{r \in \mathbb{N}}$, $k = 0, \ldots, n$ be $n+1$ sequences of subsets of $\mathbb{N}$, such that for every r and for all k, $0 \leq k \leq n$, the set A_r^k be not forcing k-definable on $\mathfrak{A}$ with respect to $\mathfrak{A}_1, \ldots, \mathfrak{A}_n$. Then there exists a $(n+1)$-generic enumeration f of $\mathfrak{A}$ such that $A_r^k \not\leq_e \mathcal{P}_k^f$ for all r and $k \leq n$.*

Corollary 16. *Let* $\{A_r^k\}_{r\in\mathbb{N}}$, $k = 0,\ldots,n$ *be* $n + 1$ *sequences of subsets of* $\mathbb{N}$, *such that for every* r *and for all* k, $0 \le k \le n$, *the set* A_r^k *be not forcing* k-*definable on* $\mathfrak{A}$ *with respect to* $\mathfrak{A}_1,\ldots,\mathfrak{A}_n$. *Then there exists a* n-*acceptable enumeration* f *of* $\mathfrak{A}$ *with respect to* $\mathfrak{A}_1,\ldots,\mathfrak{A}_n$, *such that the enumeration degree of* $f^{-1}(\mathfrak{A})$ *is total and* $A_r^k \not\le_e (f^{-1}(\mathfrak{A}))^{(k)}$ *for all* r *and* $k \le n$.

This follows from the previous proposition, Theorem 1 and Lemma 5.

Theorem 17. *For every* $A \subseteq \mathbb{N}$ *and* $k \le n$, *the following are equivalent:*

1. $d_e(A) \in \mathrm{CRS}_k(\mathfrak{A}, \mathfrak{A}_1,\ldots,\mathfrak{A}_n)$.
2. $A \le_e \mathcal{P}_k^f$, *for every* k-*acceptable enumeration* f *of* $\mathfrak{A}$ *with respect to* $\mathfrak{A}_1,\ldots,$ $\mathfrak{A}_k$.
3. A *is forcing* k-*definable on* $\mathfrak{A}$ *with respect to* $\mathfrak{A}_1,\ldots,\mathfrak{A}_n$.

5   Normal Form Theorem

In this section a normal form of the forcing k-definable sets on the structure $\mathfrak{A}$ with respect to $\mathfrak{A}_1,\ldots,\mathfrak{A}_n$ is presented. According to [11], these sets coincide with the sets which are definable on $\mathfrak{A}$ by means of *positive* recursive Σ_k^0 formulae [1].

Let $\mathcal{L} = \{T_1,\ldots,T_s\}$ be the first order language corresponding to the structure $\mathfrak{A}$. Let $\mathcal{L}_1,\ldots,\mathcal{L}_n$ be the languages of $\mathfrak{A}_1,\ldots,\mathfrak{A}_n$. Assume that the languages $\mathcal{L},\mathcal{L}_1,\ldots,\mathcal{L}_n$ are disjoined.

For each $i \le n$, define the elementary Σ_i^+ formulae and the Σ_i^+ formulae by induction on i, as follows.

Definition 18. (1) The elementary Σ_0^+ formulae are formulae in prenex normal form with a finite number of existential quantifiers and a matrix which is a finite conjunction of atomic predicates built up from the variables and the predicate symbols $T_1,\ldots,T_s$.

(2) An elementary Σ_{i+1}^+ formula is in the form

$$\exists Y_1 \ldots \exists Y_m \Phi(X_1,\ldots,X_l,Y_1,\ldots,Y_m),$$

where Φ is a finite conjunction of atoms built up from the variables $X_1,\ldots,$ $X_l, Y_1,\ldots,Y_m$ and the predicate symbols from $\mathcal{L}_{i+1}$, Σ_i^+ formulae and negations of Σ_i^+ formulae with free variables among $X_1,\ldots,X_l, Y_1,\ldots,Y_m$.

(3) A Σ_i^+ formula with free variables among $X_1,\ldots,X_l$ is an c.e. infinitary disjunction of elementary Σ_i^+ formulae with free variables among $X_1,\ldots,X_l$.

Let Φ be a Σ_i^+ formula with free variables among $W_1,\ldots,W_r$ and let $t_1,\ldots,t_r$ be elements of $\mathbb{N}$. Then by $(\mathfrak{A},\mathfrak{A}_1,\ldots,\mathfrak{A}_n) \models \Phi(W_1/t_1,\ldots,W_r/t_r)$ we denote that Φ is true on a structure, obtained from $\mathfrak{A}$ by adding the predicates from $\mathfrak{A}_1,\ldots,\mathfrak{A}_n$, under the variable assignment v such that $v(W_1) = t_1,\ldots,v(W_n) = t_n$.

Definition 19. Let $A \subseteq \mathbb{N}$ and let $k \leq n$. The set A is *formally k-definable* on $\mathfrak{A}$ with respect to $\mathfrak{A}_1, \ldots, \mathfrak{A}_n$ if there exists a recursive sequence $\{\Phi^{\gamma(x)}\}$ of Σ_k^+ formulae with free variables among $W_1, \ldots, W_r$ and elements $t_1, \ldots, t_r$ of $\mathbb{N}$ such that for every $x \in \mathbb{N}$ the following equivalence holds:

$$x \in A \iff (\mathfrak{A}, \mathfrak{A}_1, \ldots, \mathfrak{A}_n) \models \Phi^{\gamma(x)}(W_1/t_1, \ldots, W_r/t_r).$$

The next theorem is proved, following the construction from [11].

Theorem 20. *A set $A \subseteq \mathbb{N}$ is forcing k-definable on $\mathfrak{A}$ with respect to $\mathfrak{A}_1, \ldots, \mathfrak{A}_n$ if and only if A is formally k-definable on $\mathfrak{A}$ with respect to $\mathfrak{A}_1, \ldots, \mathfrak{A}_n$.*

6 The Connection with the Joint Spectra

In [12] another generalization of the notion of Degree spectra is considered.

Definition 21. *The Joint spectrum of $\mathfrak{A}, \mathfrak{A}_1, \ldots, \mathfrak{A}_n$ is the set*

$$\mathrm{DS}(\mathfrak{A}, \mathfrak{A}_1, \ldots, \mathfrak{A}_n) = \{\mathbf{a} : \mathbf{a} \in \mathrm{DS}(\mathfrak{A}), \mathbf{a}' \in \mathrm{DS}(\mathfrak{A}_1), \ldots, \mathbf{a}^{(n)} \in \mathrm{DS}(\mathfrak{A}_n)\}.$$

The co-set of $\mathrm{DS}(\mathfrak{A}, \mathfrak{A}_1, \ldots, \mathfrak{A}_n)$ is denoted by $\mathrm{CS}(\mathfrak{A}, \mathfrak{A}_1, \ldots, \mathfrak{A}_n)$. *The kth Jump spectrum of $\mathfrak{A}, \mathfrak{A}_1, \ldots, \mathfrak{A}_n$* is the set $\mathrm{DS}_k(\mathfrak{A}, \mathfrak{A}_1, \ldots, \mathfrak{A}_n)$ of all kth jumps of the elements of the Joint spectrum $\mathrm{DS}(\mathfrak{A}, \mathfrak{A}_1, \ldots, \mathfrak{A}_n)$. The co-set of $\mathrm{DS}_k(\mathfrak{A}, \mathfrak{A}_1, \ldots, \mathfrak{A}_n)$ is denoted by $\mathrm{CS}_k(\mathfrak{A}, \mathfrak{A}_1, \ldots, \mathfrak{A}_n)$.

The properties of both notions of spectra are very similar, for example the Joint Spectra are closed upwards, the kth Co-spectrum depends only on the first k structures.

Proposition 22. $\mathrm{CS}(\mathfrak{A}, \mathfrak{A}_1, \ldots, \mathfrak{A}_n) = \mathrm{CRS}(\mathfrak{A}, \mathfrak{A}_1, \ldots, \mathfrak{A}_n)$.

This follows from the fact that $\mathrm{CS}(\mathfrak{A}, \mathfrak{A}_1, \ldots, \mathfrak{A}_n) = \mathrm{CS}(\mathfrak{A})$ by [12], and $\mathrm{CRS}(\mathfrak{A}, \mathfrak{A}_1, \ldots, \mathfrak{A}_n) = \mathrm{CSR}(\mathfrak{A}) = \mathrm{CS}(\mathfrak{A})$ by Proposition 12.

The difference between the co-sets of these spectra we can see first from the forcing normal form of both sets. In [12] is shown that for any set $A \subseteq \mathbb{N}$:

$$d_e(A) \in \mathrm{CS}_k(\mathfrak{A}, \mathfrak{A}_1, \ldots, \mathfrak{A}_n) \iff A \leq_e \mathcal{P}(f^{-1}(\mathfrak{A}), f_1^{-1}(\mathfrak{A}_1) \ldots, f_k^{-1}(\mathfrak{A}_k)),$$

for every enumerations f of $\mathfrak{A}$, f_1 of $\mathfrak{A}_1$, $\ldots$, f_k of $\mathfrak{A}_k$. While by Theorem 17:

$$d_e(A) \in \mathrm{CRS}_k(\mathfrak{A}, \mathfrak{A}_1, \ldots, \mathfrak{A}_n) \iff A \leq_e \mathcal{P}(f^{-1}(\mathfrak{A}), f^{-1}(\mathfrak{A}_1) \ldots, f^{-1}(\mathfrak{A}_k)),$$

for any k-acceptable enumeration f of $\mathfrak{A}$ with respect to $\mathfrak{A}_1, \ldots, \mathfrak{A}_k$.

Second, from the normal form of forcing k-definable sets from [12] we know that these sets are definable on $\mathfrak{A}, \mathfrak{A}_1, \ldots, \mathfrak{A}_n$ by a recursive sequence of Σ_k^+ formulae, which differ from these considered here only by the induction step 2, where the existential quantifiers for the structure $\mathfrak{A}_{i+1}$ are different from the others. More precisely, in [12]:

(2) *An elementary Σ_{i+1}^+ formula* with free variables among $\bar{X}^0 \ldots \bar{X}^{i+1}$ is in the form

$$\exists \bar{Y}^0 \ldots \exists \bar{Y}^{i+1} \Phi(\bar{X}^0 \ldots \bar{X}^{i+1}, \bar{Y}^0, \ldots, \bar{Y}^{i+1})$$

where Φ is a finite conjunction of Σ_i^+ formulae and negations of Σ_i^+ formulae with free variables among $\bar{Y}^0 \ldots \bar{Y}^i, \bar{X}^0 \ldots \bar{X}^i$ and atoms of $\mathcal{L}_{i+1}$ with variables among $\bar{X}^{i+1}, \bar{Y}^{i+1}$;

Notice that, the variables for each structure are different. Moreover, when we get the value of a Σ_i^+ formula in $(\mathfrak{A}, \mathfrak{A}_1 \ldots, \mathfrak{A}_n)$ under an assignment then we treat the structure $(\mathfrak{A}, \mathfrak{A}_1 \ldots, \mathfrak{A}_n)$ as a many-sorted structure with separated sorts.

From this point we will prove that there are structures $\mathfrak{A}$ and $\mathfrak{A}_1$, for which $\mathrm{CS}_1(\mathfrak{A}, \mathfrak{A}_1) \neq \mathrm{CRS}_1(\mathfrak{A}, \mathfrak{A}_1)$.

Example 23. Fix an effective bijective coding of the pairs of natural numbers. Denote by $\langle i, j \rangle$ the code of the ordered pair (i, j). Let R and S be binary predicates defined as follows: for every $i, j \in \mathbb{N}$, $R(\langle i, j \rangle, \langle i+1, j \rangle)$, i.e. R is the graph of the successor function for the first coordinate. For every $i, j \in \mathbb{N}$, $S(\langle i, j \rangle, \langle i, j+1 \rangle)$, i.e. S is the graph of the successor function for the second coordinate. Let $\mathfrak{A} = (\mathbb{N}, R, S, =, \neq)$ and let the language of $\mathfrak{A}$ be $\mathcal{L} = (R, S, =, \neq)$.

Consider a set M which is Σ_3^0, but not Σ_2^0 in the arithmetical hierarchy, and let $M = \{j_0, \ldots, j_i, \ldots\}$ be a fixed enumeration of the elements of M.
Define $\mathfrak{A}_1 = (\mathbb{N}, P, =, \neq)$, where $P(\langle i, j_i \rangle) \iff j_i \in M$. Let $\mathcal{L}_1 = (P, =, \neq)$.

Claim: $d_e(M) \in \mathrm{CRS}_1(\mathfrak{A}, \mathfrak{A}_1)$ and $d_e(M) \notin \mathrm{CS}_1(\mathfrak{A}, \mathfrak{A}_1)$.
Let $t_0 = \langle 0, 0 \rangle$. Then $d_e(M) \in \mathrm{CRS}_1(\mathfrak{A}, \mathfrak{A}_1)$, since

$$j \in M \iff \exists Y_0 \ldots \exists Y_i \exists Z_0 \ldots \exists Z_j (Y_0 = t_0 \ \& \ R(Y_0, Y_1) \ \& \ldots \& \ R(Y_{i-1}, Y_i)$$
$$\& \ Y_i = Z_0 \ \& \ S(Z_0, Z_1) \ \& \ldots \& \ S(Z_{j-1}, Z_j) \ \& \ P(Z_j)).$$

On the other hand if $A \subseteq \mathbb{N}$ and $d_e(A) \in \mathrm{CS}_1(\mathfrak{A}, \mathfrak{A}_1)$, then A is Σ_2^0 set in the arithmetical hierarchy. This follows from the fact that for any elementary Σ_1^+ formula $\Phi(W_1, \ldots, W_r)$ we can effectively find an elementary Σ_1^+ formula $\Psi(W_1, \ldots, W_r)$, where the predicate symbol P does not occur in Ψ, such that for any fixed $t_1, \ldots, t_r \in \mathbb{N}$

$$(\mathfrak{A}, \mathfrak{A}_1) \models \Phi(W_1/t_1, \ldots, W_r/t_r) \iff (\mathfrak{A}, \mathfrak{A}_1) \models \Psi(W_1/t_1, \ldots, W_r/t_r).$$

7   Minimal Pair Theorem

In [10] a Minimal Pair Theorem for Degree spectrum of a structure $\mathfrak{A}$ is presented. There it is proved that for each constructive ordinal α there exist elements $\mathbf{f}$ and $\mathbf{g}$ of $\mathrm{DS}(\mathfrak{A})$ such that for any enumeration degree $\mathbf{a}$ and any $\beta + 1 < \alpha$

$$\mathbf{a} \leq \mathbf{f}^{(\beta)} \ \& \ \mathbf{a} \leq \mathbf{g}^{(\beta)} \Rightarrow \mathbf{a} \in \mathrm{CS}_\beta(\mathfrak{A}).$$

We shall prove an analogue of the Minimal Pair Theorem for the Relative spectrum.

Theorem 24. *For any structures* $\mathfrak{A}, \mathfrak{A}_1, \ldots, \mathfrak{A}_n$, *there exist enumeration degrees* $\mathbf{f}$ *and* $\mathbf{g}$ *in* $\mathrm{RS}(\mathfrak{A}, \mathfrak{A}_1, \ldots, \mathfrak{A}_n)$, *such that for any enumeration degree* $\mathbf{a}$ *and* $k \leq n$:

$$\mathbf{a} \leq \mathbf{f}^{(k)} \ \& \ \mathbf{a} \leq \mathbf{g}^{(k)} \Rightarrow \mathbf{a} \in \mathrm{CRS}_k(\mathfrak{A}, \mathfrak{A}_1, \ldots, \mathfrak{A}_n).$$

Proof. Let h be an arbitrary enumeration of $\mathfrak{A}$. By Lemma 7 there exists a n-acceptable enumeration f of $\mathfrak{A}$ with respect to $\mathfrak{A}_1, \ldots, \mathfrak{A}_n$, such that $h^{-1}(\mathfrak{A}) \leq_e f^{-1}(\mathfrak{A})$ and $F = f^{-1}(\mathfrak{A})$ is a total set. Hence $d_e(F) \in \mathrm{RS}(\mathfrak{A}, \mathfrak{A}_1, \ldots, \mathfrak{A}_n)$ and since f is n-acceptable enumeration of $\mathfrak{A}$ with respect to $\mathfrak{A}_1, \ldots, \mathfrak{A}_n$, $F^{(k)} \equiv_e \mathcal{P}_k^f$. For each $k \leq n$, denote by $\{X_r^k\}_{r \in \mathbb{N}}$ the sequence of all sets enumeration reducible to $\mathcal{P}_k^f$.

For each $k \leq n$ consider the sequence $\{A_r^k\}_{r \in \mathbb{N}}$ of these sets among the sets $\{X_r^k\}_{r \in \mathbb{N}}$, which are not forcing k-definable on $\mathfrak{A}$ with respect to $\mathfrak{A}_1, \ldots, \mathfrak{A}_n$. By Corollary 16 there is a n-acceptable enumeration g such that for all r, and all $k = 0, \ldots, n$, $A_r^k \not\leq_e (g^{-1}(\mathfrak{A}))^{(k)}$ and $g^{-1}(\mathfrak{A})$ is a total set. Let $G = g^{-1}(\mathfrak{A})$. It is clear that $d_e(G) \in \mathrm{RS}(\mathfrak{A}, \mathfrak{A}_1, \ldots, \mathfrak{A}_n)$.

Suppose now, that $k \leq n$ and a set X, $X \leq_e F^{(k)}$ and $X \leq_e G^{(k)}$. From $X \leq_e F^{(k)}$ and $F^{(k)} \equiv_e \mathcal{P}_k^f$, it follows that $X = X_r^k$ for some r. Assume for contradiction that X is not forcing k-definable on $\mathfrak{A}$ with respect to $\mathfrak{A}_1, \ldots, \mathfrak{A}_n$. Then $X = A_l^k$ for some l and then $X \not\leq_e G^{(k)}$. Hence X is forcing k-definable on $\mathfrak{A}$ with respect to $\mathfrak{A}_1, \ldots, \mathfrak{A}_n$. By Theorem 17, $d_e(X) \in \mathrm{CRS}_k(\mathfrak{A}, \mathfrak{A}_1, \ldots, \mathfrak{A}_n)$. Let $\mathbf{f} = d_e(F)$ and $\mathbf{g} = d_e(G)$.

8   Quasi-minimal Degree

Let $\mathcal{A}$ be a set of enumeration degrees and $co(\mathcal{A})$ be the co-set of $\mathcal{A}$. The degree $\mathbf{q}$ is *quasi-minimal with respect to* $\mathcal{A}$ if the following conditions hold ([10]):

1. $\mathbf{q} \notin co(\mathcal{A})$.
2. If $\mathbf{a}$ is a total degree and $\mathbf{a} \geq \mathbf{q}$, then $\mathbf{a} \in \mathcal{A}$.
3. If $\mathbf{a}$ is a total degree and $\mathbf{a} \leq \mathbf{q}$, then $\mathbf{a} \in co(\mathcal{A})$.

It is shown in [10] that for any structure $\mathfrak{A}$, there is a quasi-minimal degree $\mathbf{q}$ with respect to $\mathrm{DS}(\mathfrak{A})$, i.e. $\mathbf{q} \notin \mathrm{CS}(\mathfrak{A})$ and for every total degree $\mathbf{a}$: if $\mathbf{a} \geq \mathbf{q}$, then $\mathbf{a} \in \mathrm{DS}(\mathfrak{A})$ and if $\mathbf{a} \leq \mathbf{q}$, then $\mathbf{a} \in \mathrm{CS}(\mathfrak{A})$.

Theorem 25. *For any structures* $\mathfrak{A}, \mathfrak{A}_1, \ldots, \mathfrak{A}_n$ *there exists an enumeration degree* $\mathbf{q}$ *such that:*

1. $\mathbf{q} \notin \mathrm{CRS}(\mathfrak{A}, \mathfrak{A}_1, \ldots, \mathfrak{A}_n)$;
2. *If* $\mathbf{a}$ *is a total degree and* $\mathbf{a} \geq \mathbf{q}$, *then* $\mathbf{a} \in \mathrm{RS}(\mathfrak{A}, \mathfrak{A}_1, \ldots, \mathfrak{A}_n)$;
3. *If* $\mathbf{a}$ *is a total degree and* $\mathbf{a} \leq \mathbf{q}$, *then* $\mathbf{a} \in \mathrm{CRS}(\mathfrak{A}, \mathfrak{A}_1, \ldots, \mathfrak{A}_n)$.

Proof (sketch). Let f be a partial generic enumeration of $\mathfrak{A}$ constructed as in [10]. Then by [10], $d_e(f^{-1}(\mathfrak{A}))$ is quasi-minimal with respect to $\mathrm{DS}(\mathfrak{A})$. By Theorem 4. from [13] there is a quasi-minimal over $f^{-1}(\mathfrak{A})$ set F, such that $f^{-1}(\mathfrak{A}) <_e F$, $f^{-1}(\mathfrak{A}_i) \leq_e F^{(i)}$, for $i \leq n$, and for any total set A, if $A \leq_e F$, then $A \leq_e f^{-1}(\mathfrak{A})$.

The set F is constructed as a partial regular enumeration which is quasi-minimal over $f^{-1}(\mathfrak{A})$ with respect to $f^{-1}(\mathfrak{A}_i)$, $i \leq n$. Take $\mathbf{q} = d_e(F)$.

Since $d_e(f^{-1}(\mathfrak{A})) \notin \mathrm{CS}(\mathfrak{A})$ and $d_e(f^{-1}(\mathfrak{A})) < \mathbf{q}$ then $\mathbf{q} \notin \mathrm{CS}(\mathfrak{A})$. But $\mathrm{CS}(\mathfrak{A}) = \mathrm{CRS}(\mathfrak{A}, \mathfrak{A}_1, \ldots, \mathfrak{A}_n)$.

Let X be a total set.

If $X \leq_e F$, then by the choice of F, $X \leq_e f^{-1}(\mathfrak{A})$. Thus $d_e(X) \in \mathrm{CS}(\mathfrak{A}) = \mathrm{CRS}(\mathfrak{A}, \mathfrak{A}_1, \ldots, \mathfrak{A}_n)$ by the choice of $f^{-1}(\mathfrak{A})$.

If $X \geq_e F$, then $X \geq_e f^{-1}(\mathfrak{A})$. Since "$=$" is in $\mathfrak{A}$, $\mathrm{dom}(f) \leq_e X$ and since X is a total set, $\mathrm{dom}(f)$ is r.e. in X. Let ρ be a recursive in X enumeration of $\mathrm{dom}(f)$. Set $h = \lambda n.f(\rho(n))$. Thus $h^{-1}(\mathfrak{A}) \leq_e X$ and $h^{-1}(\mathfrak{A}_i) \leq_e X^{(i)}$, for $i \leq n$. Construct an enumeration g as in Lemma 5, $g^{-1}(\mathfrak{A}) \equiv_e X$, and for each $i \leq n$, $g^{-1}(\mathfrak{A}_i) \leq_e X \oplus h^{-1}(\mathfrak{A}_i) \leq_e X \oplus X^{(i)} \equiv_e (g^{-1}(\mathfrak{A}))^{(i)}$. And then $d_e(X) \in \mathrm{RS}(\mathfrak{A}, \mathfrak{A}_1, \ldots, \mathfrak{A}_n)$.

References

1. Ash, C. J. : Generalizations of enumeration reducibility using recursive infinitary propositional senetences. Ann. Pure Appl. Logic **58** (1992) 173–184.
2. Ash, C. J., Knight, J. F., Manasse, M., Slaman, T. : Generic copies of countable structures. Ann. Pure Appl. Logic **42** (1989) 195–205.
3. Ash, C. J., Jockush, C., Knight, J. F. : Jumps of orderings. Trans. Amer. Math. Soc. **319** (1990) 573–599.
4. Chisholm, J. : Effective model theory vs. recursive model theory. J. Symbolic Logic **55** (1990) 1168–1191.
5. Cooper, S. B. : Partial degrees and the density problem. Part 2: The enumeration degrees of the Σ_2 sets are dense. J. Symbolic Logic **49** (1984) 503–513.
6. Downey, R. G., Knight, J. F. : Orderings with αth jump degree $\mathbf{0}^{(\alpha)}$. Proc. Amer. Math. Soc. **114** (1992) 545–552.
7. Knight, J. F. : Degrees coded in jumps of orderings. J. Symbolic Logic **51** (1986) 1034–1042.
8. Richter, L. J. : Degrees of structures. J. Symbolic Logic **46** (1981) 723–731.
9. Soskov, I. N. : A jump inversion theorem for the enumeration jump. Arch. Math. Logic **39** (2000), 417–437.
10. Soskov, I. N. : Degree spectra and co-spectra of structures. Ann. Univ. Sofia **96** (2004), 45–68.
11. Soskov, I. N., Baleva, V. : Ash's theorem for abstract structures. Proceedings of Logic Colloquium'2002 (to appear).
12. Soskova, A. A., Soskov, I. N. : Co-spectra of joint spectra of structures. Ann. Univ. Sofia **96** (2004) 35–44.
13. Soskova, A. A. : Minimal Pairs and Quasi-minimal degrees for the Joint Spectra of Structures. New Computational Pradigms (S. B. Cooper; B. Loewe, eds.) Lecture Notes in Comp. Sci **3526** (2005) Springer-Verlag 451–460.

Phase Transition Thresholds for Some Natural Subclasses of the Computable Functions

Andreas Weiermann*

Fakulteit Bètawetenschappen
Departement Wiskunde, P.O. Box 80010, 3508 TA Utrecht
The Netherlands
weierman@math.uu.nl

Abstract. In this paper we first survey recent advances on phase transition phenomena which are related to natural subclasses of the recursive functions. Special emphasis is put on descent recursive functions, witness bounding functions for well-partial orders and Ramsey functions. In the last section we prove in addition some results which show how the asymptotic of the standard Ramsey function is affected by phase transitions for associated parameterized Ramsey functions.

1 Introduction

Phase transition is a type of behaviour wherein small changes of a parameter of a system cause dramatic shifts in some globally observed behaviour of the system, such shifts being usually marked by a sharp 'threshold point'. (An everyday life example of such thresholds are ice melting and water boiling temperatures.) This kind of phenomena nowadays occur throughout many mathematical and computational disciplines: statistical physics, evolutionary graph theory, percolation theory, computational complexity, artificial intelligence etc.

The last few years have seen an unexpected series of achievements that bring together independence results in logic, analytic combinatorics and Ramsey Theory. These achievements can be intuitively described as phase transitions from provability to unprovability of an assertion by varying a threshold parameter [23, 26]. Another face of this phenomenon is the transition from slow-growing to fast-growing computable functions [25, 28].

In this paper we survey recent advances on phase transition phenomena which are related to natural subclasses of the recursive functions.

For the purpose of motivation let us assume that we have given some algorithm A which performs computations on a given set D of data. We assume that D is equipped with a norm function $N : D \to \mathbb{N}$ such that for every $k \in \mathbb{N}$ the set $\{d \in D : N(d) \leq k\}$ is finite. Moreover let us assume that every computation tree

---

* Research supported by a Heisenberg-Fellowship of the Deutsche Forschungsgemeinschaft. The research has in addition been supported in part by NWO grant 613.080.000.

A. Beckmann et al. (Eds.): CiE 2006, LNCS 3988, pp. 556–570, 2006.

for A is finitely branching. A transition in the computation tree is denoted by $\rightarrow$. Thus $d \rightarrow d'$ indicates that the the algorithms performs a calculation to obtain d' out of d. If A is terminating then every sequence $d_0 \rightarrow d_1 \rightarrow d_2 \ldots$ must terminate after finitely many steps and by our assumption on N and the branching we obtain by König's Lemma the following: Given $k \in \mathbb{N}$ there exists $m \in \mathbb{N}$ such that every sequence $d_0 \rightarrow d_1 \rightarrow d_2 \ldots$ with $N(d_0) \leq k$ terminates after M steps. The minimal such m determines the computation lengths function $CompL_A$, i.e. $CompL_A(k)$ is the least m such that every sequence $d_0 \rightarrow d_1 \rightarrow d_2 \ldots$ with $N(d_0) \leq k$ terminates after m steps.

It is now quite obvious to look for methods M for proving termination of A. Moreover in this respect it is also very natural to classify $CompL_A$ which can be considered as some measure for the computational complexity of A.

From the logical point of view it would be nice to see whether there are general principles yielding classifications for $CompL_A$ depending on the method M which is used for proving termination of A.

To study phase transition phenomena in this context we equip the problem under investigation with a control function $f : \mathbb{N} \rightarrow \mathbb{N}$ and we demand that $N(d_i) \leq k + f(i)$ for any sequence of computations $d_0 \rightarrow d_1 \rightarrow d_2 \ldots \rightarrow d_i \ldots$. Hereby we assume that f is reasonably simple. Then classifying $CompL_A$ can be seen as a problem depending on parameters M and f. In particular when phase transitions for $CompL_A$ are studied the function f will play the role the order parameter plays in physics. The expectation is that for very slow growing functions f the function $CompL_A$ has moderate complexity but that the complexity of $CompL_A$ explodes as soon as f exceeds a certain threshold function. In analogy with physics it is natural to consider renormalization and universality in this context [27].

Investigations on this subject have given rise to rich and intriguing peaces of logic and mathematics where methods from Ramsey theory, analytic combinatorics and logic can be cross-fertilized.

In the following sections we consider different types of termination proof methods and consider the phase transition problem in each case separately. It is our aim to provide rules of thumb so that it is possible to guess the phase transition thresholds a priori. In the final section we study phase transitions in Ramsey theory.

2   Phase Transitions for Ordinal Sequences

In this and the following section we base our investigations on a principle suggested by Harvey Friedman. This turns out to be tailor made for our intended applications.

Obviously termination proofs can be carried out by using ordinals through mapping computation sequences into descending chains of ordinals. Typically such a mapping assigns an ordinal to a data element in an effective way so that resulting norms of ordinals from a descending ordinal sequence are also controlled by a function say $g : \mathbb{N} \rightarrow \mathbb{N}$.

After putting the problem into an abstract setting we arrive at Friedman's principle of combinatorial well-foundedness. For stating it let us fix a countable ordinal α and a norm function $N : \alpha \to \mathbb{N}$ such that for every $k \in \mathbb{N}$ the set $\{\beta < \alpha : N(\beta) \leq k\}$ is finite. Let

$$\mathrm{CWF}(\alpha, g) = (\forall k)(\exists M)$$
$$(\forall \alpha_0, \ldots, \alpha_M < \alpha)\big[(\forall i \leq M)[N\alpha_i \leq k + g(i)] \to (\exists i < M)[\alpha_i \leq \alpha_{i+1}]\big].$$

The associatec complexity function is

$$D(\alpha, g)(k) := \min\{M :$$
$$(\forall \alpha_0, \ldots, \alpha_M < \alpha)\big[(\forall i \leq M)[N\alpha_i \leq k + g(i)] \to (\exists i < M)[\alpha_i \leq \alpha_{i+1}]\big]\}.$$

By Friedman's results, it is well known that proof-theoretic ordinals α and natural associated norm functions (e.g. given by a term length function) the function $D(\alpha, g)$ grows rapidly even for rather small values of α.

To fix the context let as consider the ordinals segment of ordinals below ε_0 and define a length norm function using Cantor normal forms as follows. $N(0) := 0$ and $N(\alpha) := n + N(\alpha_1) + \cdots + N(\alpha_n)$ if $\alpha = \omega^{\alpha_1} + \cdots + \omega^{\alpha_n} > \alpha_1 \geq \ldots \geq \alpha_n$. A corresponding sup norm function lh can be defined as follows $|0| := 0$ and $|\alpha| = \max\{m_1, \ldots, m_n, |\alpha_1|, \ldots, |\alpha_n|\}$ if $\alpha = \omega^{\alpha_1} \cdot m_1 + \cdots + \omega^{\alpha_n} \cdot m_n > \alpha_1 > \ldots > \alpha_n$.

Theorem 1 (Friedman). *Let $g(i) = i$ be the identity function and let N be the length or sup norm.*

1. *$D(\omega^d, g)$ is primitive recursive.*
2. *$D(\omega^\omega, g)$ is Ackermannian.*
3. *$D(\omega^{\omega^d}, g)$ is multiple recursive.*
4. *$D(\omega^{\omega^\omega}, g)$ is not multiple recursive.*
5. *If $\alpha < \varepsilon_0$ then $D(\alpha, g)$ is provably recursive in PA.*
6. *$D(\varepsilon_0, g)$ is not provably recursive in PA.*

So pushing ordinals beyond certain thresholds is reflected by (perhaps expected) phase transitions of the resulting complexity functions.

Another perhaps even more intriguing phase transition occurs when the ordinal notation under consideration is fixed but the control function g is varied.

Further let us agree on the following assignment of fundamental sequences. If $\alpha = \omega^{\alpha_1} + \cdots + \omega^{\alpha_n + 1} > \alpha_1 \geq \ldots \geq \alpha_n + 1$ then $\alpha[x] := \omega^{\alpha_1} + \cdots + \omega^{\alpha_n} \cdot x$. If $\alpha = \omega^{\alpha_1} + \cdots + \omega^{\alpha_n} > \alpha_1 \geq \ldots \geq \alpha_n$ and α_n is a limit then $\alpha[x] := \omega^{\alpha_1} + \cdots + \omega^{\alpha_n[x]}$.

To be able to state the phase transition in terms of hierarchies of recursive functions let us recall the definition of the Hardy hierarchy.

$$H_0(x) := x$$
$$H_{\alpha+1}(x) := H_\alpha(x+1)$$
$$H_\lambda(x) := H_{\lambda[x]}(x)$$

For a given weakly increasing unbounded function $F : \mathbb{N} \to \mathbb{N}$ we define its inverse function F^{-1} as follows. $F^{-1}(x) := \min\{y : F(y) \geq x\}$. Typically such an inverse function F^{-1} grows rather slow when the original function F grows reasonably fast.

Theorem 2 (Weiermann [27]). *Let* $g_\alpha(i) := i^{\frac{1}{H_\alpha^{-1}(i)}}$ *and let* N *be the length or sup norm.*

1. *If* $\alpha < \omega^\omega$ *then* $D(\omega^\omega, g_\alpha)$ *is primitive recursive.*
2. $D(\omega^\omega, g_{\omega^\omega})$ *is Ackermannian.*

We are now going to formulate the general rationale behind this type of results. Of fundamental importance is here the study of count functions.

Definition 1

$$c_\alpha(n) := |\{\beta < \alpha : N(\beta) \le n\}|.$$

These count functions c_α come along with an intriguing mathematical theory which is based on generatingfunctionology [31]. To get good asymptotic bounds on $c_\alpha(n)$ for n large one applies Cauchy's integral formula to the generating function $C(z) := \sum_{i=0}^\infty c_\alpha(n) \cdot z^n$

Theorem 3. *Let* N *be the length norm and let the count functions be defined with respect to this norm.*

1. $c_{\omega^d}(n) \sim \frac{n^d}{(d!)^2}$
2. $\log(c_{\omega^\omega}(n)) \sim \pi \cdot \sqrt{\frac{2n}{3}}$
3. $\log(c_{\omega^{\omega^\omega}}(n)) \sim \frac{\pi^2}{6} \frac{n}{\log(n)}.$

To formulate the phase transition principle let us further define $\alpha[\![x]\!] := \max\{\beta < \alpha : N(\beta) \le N(\alpha) - 1 + x\}$.

Rule of thumb 1. *Let* $g_{\alpha,\beta}(x) := c_{\alpha[\![H_\beta^{-1}(i)]\!]}^{-1}(i)$. *If* $\beta \le \alpha$ *then* $D(\alpha, g_{\alpha,\beta})$ *is primitive recursive in* H_β *and* H_β *is primitive recursive in* $D(\alpha, g_{\alpha,\beta})$

This would imply the following Rule of thumb.

Rule of thumb 2. *Let* $g_{\alpha,\beta}(x) := c_{\alpha[\![H_\beta^{-1}(i)]\!]}^{-1}(i)$. *Let* T *be a fragment of* PA *with proof-theoretic* (Π_2^0-) *ordinal* α *and let* N *be the length or sup norm.*

1. *If* $\beta < \alpha$ *then* $D(\alpha, g_{\alpha,\beta})$ *is provably recursive in* T
2. $D(\alpha, g_{\alpha,\beta})$ *is not provably recursive in* T.

These general rules apply to larger segments of ordinals and lead to the following applications (which already have been proved rigorously). Let $|x| := \log_2(x+1)$ be the binary length of x where $|0| := 0$.

Theorem 4 (Weiermann). *Let* $n \ge 1$, T *be* $I\Sigma_n$ *and* $\alpha := \omega_{n+1}$ *and let* N *be the length norm. Let* $g_{\alpha,\beta}(n) := |i| \cdot {}^{H_\beta^{-1}(i)}\sqrt{\log_{n-1}(i)}$.

1. *If* $\beta < \alpha$ *then* $D(\alpha, g_{n,\beta})$ *is provably recursive in* T
2. $D(\alpha, g_{\alpha,\alpha})$ *is not provably recursive in* T.

In case of PA we obtain the following phase transition result.

Theorem 5 (Arai, Weiermann[1, 23]). *Let* $g_\beta(n) := |i| \cdot |i|_{H_\beta^{-1}(i)}$ *and let N be the length norm.*

1. *If $\beta < \varepsilon_0$ then $D(\varepsilon_0, g_\beta)$ is provably recursive in* PA
2. $D(\varepsilon_0, g_{\varepsilon_0})$ *is not provably recursive in* PA.

Remark: The phase transitions for CWF principles are in a sense continuous since they involve H_α^{-1} for varying α. In analogy with physics one might consider this as a second order phase transition.

3   Phase Transitions for Sequences in Well Partial Orderings

A partial ordering $\langle X, \leq_X \rangle$ is a well-partial ordering iff for all functions $F : \mathbb{N} \to X$ there exist natural numbers i, j such that $i < j$ and $F(i) \leq_X F(j)$. A sequence $F : \mathbb{N} \to X$ is called bad if there do not exist natural numbers i, j such that $i < j$ and $F(i) \leq_X F(j)$. So a partial order is a well-partial order iff there does not exist an infinite bad sequence for it.

Obviously termination proofs can be carried out by using well partial orders through mapping computation sequences into bad sequences. Typically such a mapping assigns an initial sequence of data elements to an element of the well-partial order in an effective way so that resulting sequences of elements in the well-partial-order are again also controlled in norm by some function $g : X \to \mathbb{N}$.

After putting the problem into an abstract setting we arrive at Friedman's principle of combinatorial well-partial-orderedness. For stating it let us fix a well partial order $\langle X, \leq_X \rangle$ and a norm function $N : X \to \mathbb{N}$ such that for every $k \in \mathbb{N}$ the set $\{\beta \in X : N(\beta) \leq k\}$ is always finite. Let

$\mathrm{CWP}(X, g) = (\forall k)(\exists M)$
$(\forall \alpha_0, \dots, \alpha_M < \alpha)\big[(\forall i \leq M)[N\alpha_i \leq k + g(i)] \to (\exists i \leq M)(\exists j \leq M)[i < j \wedge \alpha_i \leq \alpha_j]\big]$. The associated complexity function is
$D(X, g)(k) := \min\{M :$
$(\forall \alpha_0, \dots, \alpha_M < \alpha)\big[(\forall i \leq M)[N\alpha_i \leq k + g(i)] \to (\exists i \leq M)(\exists j \leq M)[i < j \wedge \alpha_i \leq \alpha_j]\big]\}.$

By Friedman's results it is well known that for several natural well-partial oders and associated norm functions (e.g. given by a length function) the function $D(\alpha, g)$ grows rapidly.

Basic examples are provided by Dickson's Lemma and Higman's Lemma. Assume that $\langle Y, \leq_Y \rangle$ is a partial ordering. Then we can induce a partial ordering $\leq_Y^k$ on Y^k the set of k-tuples of elements in X as follows $\langle x_0, \dots, x_{k-1} \rangle \leq_Y^k \langle y_0, \dots, y_{k-1} \rangle$ if $x_i \leq_Y y_i$ for $i = 0, \dots, k-1$. Moreover we can induce a partial ordering on Y^* the set of finite sequences of elements in X as follows $\langle x_0, \dots, x_{k-1} \rangle \leq_Y^* \langle y_0, \dots, y_{l-1} \rangle$ if there exist $i_0, \dots, i_{k-1}$ such that $0 \leq i_0 < i_1 < \dots < i_{k-1}$ and such that $x_m \leq_Y y_{i_m}$ for $m = 0, \dots, k-1$.

Theorem 6 (Dickson, Higman). *If $\langle Y, \leq_Y \rangle$ is a well partial ordering then so are $\langle Y^k, \leq_Y^k \rangle$ and $\langle Y^*, \leq_Y^* \rangle$*

If $Y \subseteq \mathbb{N}$ then for finite sequences with values in Y there are two obvious norm functions. Let $|\langle x_0, \ldots, x_{k-1}\rangle| := \max\{x_0, \ldots, x_{k-1}\}$ be the sup-norm and $N(\langle x_0, \ldots, x_{k-1}\rangle) := \sum_{l=0}^{k-1} x_l$ be the lengths norm. (In more general situations one defines the norms on sequences of course in terms of the norm given on the space Y.).

Theorem 7 (Friedman). *Let $d = \{0, \ldots, d-1\}$ and $\mathbb{N}$ be well-quasiordered by their natural orderings. Let $\mathbb{N}^d$ and d^* be ordered be ordered by the induced orderings. Let $g(i) = i$ be the identity function and N be either the sup norm or the length norm.*

1. *$D(\mathbb{N}^d, g)$ is primitive recursive.*
2. *$k \mapsto D(\mathbb{N}^k, g)(k)$ is Ackermannian.*
3. *$D(d^*, g)$ is multiple recursive.*
4. *$k \mapsto D(k^*, g)(k)$ is not multiple recursive.*

Theorem 8 (Weiermann [30]). *Let $k = \{0, \ldots, k-1\}$ be well-quasiordered by its natural ordering. Let d^* be ordered by the induced ordering. Let $g_r(i) = r \cdot |i|$ and N be the length norm.*

1. *If $r < 1$ then $k \mapsto D(k^*, g_r)$ is multiple recursive.*
2. *If $r > 1$ then $k \mapsto D(k^*, g)(k)$ is not multiple recursive.*

We conjecture that $k \mapsto D(k^*, g_r)$ is multiple recursive for $r = 1$.

Theorem 9 (Weiermann [27]). *Let $g_\alpha(i) := i^{\overline{H_\alpha^{-1}(i)}}$ and N be the sup or length norm.*

1. *If $\alpha < \omega^\omega$ then $k \mapsto D(\mathbb{N}^k, g_\alpha)(k)$ is primitive recursive.*
2. *$k \mapsto D(\mathbb{N}^k, g_{\omega^\omega})(k)$ is Ackermannian.*

To obtain farer reaching well-partial orders it is convenient to consider finite trees under homeomorphic embeddability. To stay within the realm of ε_0 it is convenient to restrict the consideration to the set $\mathcal{B}$ of binary trees. A convenient way to introduce $\mathcal{B}$ is as follows. Let 0 be a constant (a 0-ary function symbol) and let φ be a binary function symbol. Let $\mathcal{B}$ be the least set of terms such that

1. $0 \in \mathcal{B}$
2. If $\alpha, \beta \in \mathcal{B}$ then $\varphi(\alpha, \beta) \in \mathcal{B}$.

In the sequel we abbreviate $\varphi(\alpha, \beta)$ by $\varphi\alpha\beta$. The *homeomorphic embeddability relation* $\unlhd$ is the least binary relation on $\mathcal{B}$ such that

1. If $\alpha = 0$ then $\alpha \unlhd \beta$.
2. If $\alpha = \varphi\alpha_1\alpha_2$ and $\beta = \varphi\beta_1\beta_2$ and $\alpha \unlhd \beta_1$ or $\alpha \unlhd \beta_2$ then $\alpha \unlhd \beta$.
3. If $\alpha = \varphi\alpha_1\alpha_2$ and $\beta = \varphi\beta_1\beta_2$ and $\alpha_1 \unlhd \beta_1$ and $\alpha_2 \unlhd \beta_2$ then $\alpha \unlhd \beta$.

Theorem 10 (Higman, Kruskal [9]). *$\langle \mathcal{B}, \unlhd \rangle$ is a well partial order.*

Theorem 11 (Friedman). *Let $g(i) = i$. Then the function $D(\mathcal{B}, g)$ is not provably recursive in* PA.

The associated phase transition result runs as follows. Let $N(0) := 0$ and $N(\varphi\alpha\beta) := 1 + (\alpha) + N(\beta)$.

Theorem 12 (Weiermann [30]). *Let $g_r(i) = r \cdot |i|$ and let N be a length norm.*

1. *If $r \leq \frac{1}{2}$ then $k \mapsto D(\mathcal{B}, g_r)$ is elementary recursive.*
2. *If $r > \frac{1}{2}$ then $k \mapsto D(\mathcal{B}, g)(k)$ is not provably recursive in* PA.

Extraction of a general pattern from Theorem 8 and Theorem 12 leads to the following rule.

Rule of thumb 3. *Assume that $\langle Y, \leq_Y \rangle$ is a normed well partial ordering of maximal order type α such that $\log_2 |\{y \in Y : N(y) \leq n\}| \sim n \cdot c$ for some $c > 1$. Let $g_r(i) := r \cdot |i|$.*

1. *If $r \leq \frac{1}{\log_2(c)}$ then $D(Y, g_r)$ is elementary recursive.*
2. *If $r > \frac{1}{\log_2(c)}$ then $D(Y, g_r)$ eventually dominates H_β for all $\beta < \alpha$.*

The phase transitions for CWP principles in case of Higman's Lemma or the Higman-Kruskal theorem are in a sense discontinuous since they appear at a real number threshold. In analogy with physics one might consider this as a first order phase transition.

4   Phase Transitions in Ramsey Theory

In principle termination proofs can also be carried out by using Ramseyan theorems by providing appropriate partitions having only finite homogeneous sets but this connection seems us to be artificial at present. In this section we study therefore thresholds which are associated to Ramseyan statements as an investigation in its own right. We also indicate how classical open problems in Ramsey theory can be attacked via studying associated phase transitions.

4.1   Phase Transitions for Rapidly Growing Ramsey Functions

Let us recall the classical Kanamori-McAloon and the Paris-Harrington principles. If $X \subseteq \mathbb{N}, d \in \mathbb{N}$, let $[X]^d$ be the set of all subsets of X with d elements. As usual in Ramsey Theory, we identify a positive integer m with its set of predecessors $\{0, \ldots, m-1\}$. If C is a colouring defined on $[X]^d$ (with values in $\mathbb{N}$) we write $C(x_1, \ldots, x_d)$ for $C(\{x_1, \ldots, x_d\})$ where $x_1 < \cdots < x_d$. A subset H of X is called *homogeneous* or *monochromatic* for C if C is constant on $[H]^d$. We write

$$X \to (m)_k^d$$

if for all $C : [X]^d \to k$ there exists $H \subseteq X$ s.t. $card(H) = m$ and H is homogeneous for C. Ramsey [17] proved the following result, known as the Finite Ramsey Theorem.

$$(\forall d)(\forall k)(\forall m)(\exists \ell)[\ell \to (m)_k^d].$$

Let Let $R_k^d(m) := \min\{\ell : \ell \to (m)_k^d\}$. Erdös and Rado gave in [7] a primitive recursive upper bound on $R_k^d(m)$ as a function of d, k, m. The asymptotics of R_k^d is a main concern in Ramsey Theory [8] and we will come back to it later.

The Paris-Harrington principle is a seemingly innocent variant of the Finite Ramsey Theorem. Let f be a number-theoretic function. A set X is called f-relatively large if $card(X) \geq f(\min X)$. If $f = id$, the identity function, we call such a set relatively large or just large. We write

$$X \to_f^* (m)_k^d$$

if for all $C : [X]^d \to k$ there exists $H \subseteq X$ s.t. $card(H) = m$, H is homogeneous for C and H is relatively f-large. The Paris-Harrington principle is just the Finite Ramsey Theorem with the extra condition that the homogeneous set is also relatively large

$$(\text{PH}) := (\forall d)(\forall k)(\forall m)(\exists \ell)[\ell \to_{id}^* (m)_k^d].$$

Paris and Harrington showed by model-theoretic methods that (PH) is true but unprovable in PA.

Let $R_k^d(f)(m) := \min\{\ell : \ell \to_{id}^* (m)_k^d\}$.

The following phase transition result has been obtained for the parameterized Ramsey functions $R_k^d(f)$. Let $|\cdot|_d$ be the d-times iterated binary length function and $\log^*$ the inverse of the superexponential function: Recall that

$$|x| := \log_2(x+1), \quad |x|_{d+1} := ||x|_d|, \quad \text{and} \quad \log^* x := \min\{d : |x|_d \leq 2\}$$

Recall that for a weakly increasing and unbounded function $f : \mathbb{N} \longrightarrow \mathbb{N}$ we we denote by f^{-1} the functional inverse of f.

Theorem 13 (Weiermann [26]). *For $\alpha \leq \varepsilon_0$ let*

$$f_\alpha(i) = |i|_{H_\alpha^{-1}(i)}.$$

Then

1. *The function $d, k, m \mapsto R_k^d(\log^*)(m)$ is primitive recursive*
2. *For any fixed positive integer the function $d, k, m \mapsto R_k^d(|\cdot|_q)(m)$ is not provably recursive in PA.*
3. *The function $d, k, m \mapsto R_k^d(f_\alpha)(m)$ is provably recursive in PA iff $\alpha < \varepsilon_0$.*

The phase transition in case of fixed dimension d can be characterized as follows.

Theorem 14 (Weiermann [29]). *Let*

$$f_\alpha^d(i) = \left\lfloor \frac{|i|_d}{H_\alpha^{-1}(i)} \right\rfloor.$$

Then for d fixed the function
$k, m \mapsto R_k^{d+1}(f_\alpha^d)(m)$ is provably total in $I\Sigma_d$ iff $\alpha < \omega_{d+1}$.

Now let us consider the Kanamori McAloon Ramseyan theorem.

Fix a number-theoretic function $f : \mathbb{N} \to \mathbb{N}$. A function $C : [X]^d \to \mathbb{N}$ is called *f-regressive* if for all $s \in [X]^d$ such that $f(\min(s)) > 0$ we have $C(s) < f(\min(s))$. When f is the identity function we just say that C is regressive. A set H is *min-homogeneous* for C if for all $s, t \in [H]^d$ with $\min(s) = \min(t)$ we have $C(s) = C(t)$. We write

$$X \to (m)^d_{f\text{-}reg}$$

if for all f-regressive $C : [X]^d \to \mathbb{N}$ there exists $H \subseteq X$ s.t. $\mathrm{card}(H) = m$ and H is min-homogeneous for C. In [10] Kanamori and McAloon introduced the following statement and proved it for any choice of f.

$$(\mathrm{KM})_f :\equiv (\forall d)(\forall m)(\exists \ell)[\ell \to (m)^d_{f\text{-}reg}].$$

The main result of [10], proved by a model-theoretic argument, is that $(\mathrm{KM})_{id}$ is unprovable in PA. As a corollary one obtains the (provable in PA) equivalence of (KM) with (PH).

Let $R^d_{\min}(f)(m) := \min\{[\ell : \ell \to (m)^d_{f\text{-}reg}]\}$.

In his Ph.D. thesis [15], Lee showed that the situation of Theorem 13 occurs in the case of (KM). That is, the phase transition threshold is the same as the one for (PH) when unbounded dimensions are considered.

Theorem 15 (Lee [15]). *For $\alpha \leq \varepsilon_0$ let*

$$f_\alpha(i) = |i|_{H_\alpha^{-1}(i)}.$$

Then

1. *The function $d, k, m \mapsto R^d_{\min}(\log^*)(m)$ is primitive recursive*
2. *For any fixed positive integer the function $d, k, m \mapsto R^d_{\min}(|\cdot|_q)(m)$ is not provably recursive in* PA.
3. *The function $d, k, m \mapsto R^d_{\min}(f_\alpha)(m)$ is provably recursive in* PA *iff $\alpha < \varepsilon_0$.*

Carlucci, Lee and Weiermann obtained the following phase transition in case of fixed dimensions.

Theorem 16 (Carlucci, Lee, Weiermann[3]). *Let*

$$f_\alpha^d(i) = \lfloor {}^{H_\alpha^{-1}(i)}\sqrt{\log_d(i)} \rfloor.$$

Then the function $d, m \mapsto R^{d+1}_{\min}(f_\alpha)(m)$ is provably recursive in $I\Sigma_d$ iff $\alpha < \omega_{d+1}$.

Remarks:

1. The case $d = 1$ has been treated already by Kojman, Lee, Omri and Weiermann in [12] generalizing methods from Kojman and Shelah [13] and [5].
2. Related phase transition results can be shown for Friedman's Ramsey theorem and the canonical Ramsey theorem [Carlucci Weiermann, in preparation].

4.2   A Phase Transition for R_3^3

Let us recall that for given positive integers d and c the Ramsey function R_c^d is defined as follows: $R_c^d(m)$ is the least number R such that for every function $F : [R]^d \to c$ there exists a set $Y \subseteq [R]$ such that $F \restriction [Y]^d$ has a constant value and $card(Y) \geq m$. It is well known that there exists constants c_1, c_2, c_3, c_4 such that for all but finitely many m

$$2^{c_1 \cdot m^2} \leq R_2^3(m) \leq 2^{2^{c_2 \cdot m}} \tag{1}$$

and

$$2^{c_3 \cdot m^2 (\log(m))^2} \leq R_3^3(m) \leq 2^{2^{c_4 \cdot m}}. \tag{2}$$

For $c \geq 4$ it is known that there exists a double exponential lower bound for the function R_c^0. The asymptotics of R_2^3 and R_3^3 are not known. It is a longstanding open problem to prove or disprove that R_3^3 has a double exponential lower bound. (As far as we know the Erdös award offered for solving this problem is USD 500.)

For attacking this problem (and related problems) we propose to investigate the phase transition problem for the associated Paris Harrington function $R_3^3(f)$ (resp. other Paris Harrington functions in question). We show that a classification for the phase transition for $R_3^3(f)$ will yield advance on the asymptotic of R_3^3.

We now study the grwoth rate behaviour of the function $R_3^3(f)$ from the previous section when f varies from very slow growing functions to slow growing functions f.

Theorem 17. Let $f(i) = \frac{1}{c_4}\|i\|$. Then $R_3^3(f)(m) \leq 2^{2^{c_4 \cdot m}}$ for all but finitely many m.

Proof. By (2) there is a number K such that for $R_3^3(m) \leq 2^{2^{c_4 \cdot m}}$ for all $m \geq K$. We show that $R_3^3(f)(m) \leq R_3^3(m) =: R$. Let $F : [R]^3 \to 3$ be given. Then there exists $Y \subseteq R$ such that $F \restriction [Y]^3$ has constant value and $card(Y) \geq m$. We claim that even $card(Y) \geq f(\min(Y))$ is true. Indeed $f(\min(Y)) \leq f(R) \leq f(2^{2^{c_4 \cdot m}}) = m \leq card(Y)$. $\square$

Theorem 18. Let $\alpha > 0$ and $f(i) = |i|^\alpha$. If $R_3^3(m) \leq 2^{m^{\frac{1}{\alpha}}}$ for infinitely many m then $R_3^3(f)(m) \leq 2^{m^{\frac{1}{\alpha}}}$ for infinitely many m.

Proof. Pick an m such that $R_3^3(m) \leq 2^{m^{\frac{1}{\alpha}}}$. We show that $R_3^3(f)(m) \leq R_3^3(m)$ $=: R$. Let $F : [R]^3 \to 3$ be given. Then there exists $Y \subseteq [R]$ such that $F \restriction [Y]^3$ has constant value and $card(Y) \geq m$. We claim that even $card(Y) \geq f(\min(Y))$ is true. Indeed $f(\min(Y)) \leq f(R) \leq f(2^{m^{\frac{1}{\alpha}}}) = m \leq card(Y)$. $\square$

Corollary 1. Let $\alpha > 0$ and $f(i) = |i|^\alpha$. If $R_3^3(f)(m) > 2^{m^{\frac{1}{\alpha}}}$ for all but finitely many m then $R_3^3(m) > 2^{m^{\frac{1}{\alpha}}}$ holds for all but finitely many m.

Proof. If $R_3^3(f)(m) > 2^{m^{\frac{1}{\alpha}}}$ for all but finitely many m then there are not infinitely many m such that $R_3^3(f)(m) \leq 2^{m^{\frac{1}{\alpha}}}$ hence there are not infinitely many

m such that $R_3^3(m) \leq 2^{m^{\frac{1}{\alpha}}}$ and hence $R_3^3(m) > 2^{m^{\frac{1}{\alpha}}}$ holds for all but finitely many m.

Theorem 19. *Let $\varepsilon > 0$. If $f(i) = \varepsilon \cdot |i|$ then $R_3^3(f)$ eventually dominates every primitive recursive function.*

Theorem 20. *Let $\varepsilon > 0$, $\alpha := \frac{1}{2} + \varepsilon$ and $f(i) = |i|^\alpha$. Then $R_3^3(f)(m) \geq 2^{2^{(1+\varepsilon)m/2}}$ for all but finitely many m.*

Proof. Choose $\delta > 0$ sufficiently small so that

$$(2 - \delta)(\frac{1}{2} + \varepsilon) > 1 + \frac{3}{2}\varepsilon. \tag{3}$$

Choose K_0 such that $R_2^3(m) \geq 2^{m^{2-\delta}} + 1$ for $m \geq K_0$. We show that the asserted inequality holds for $m \geq 2^{K_0^{1+\frac{3}{2}\varepsilon}} + 1$. Define

$$v_0 := 1,$$
$$v_1 := R_2^3(m) - 1,$$
$$v_{i+1} := R_2^3(f(v_i)) - 1 \text{ for } i \geq 1,$$
$$v := v_{m'-1},$$

where m' is the least integer not greater then $\frac{m}{2}$.

Choose $G_1 : [v_0, v_1[^3 \rightarrow 2$ such that for all Y with $G_0 \upharpoonright [Y]^3$ having constant value we have $card(Y) < m$. Choose $G_{i+1} : [v_i, v_{i+1}[^3 \rightarrow 2$ such that for all Y with $G_0 \upharpoonright [Y]^3$ having constant value we have $card(Y) < f(v_i)$.

Define $G : [v_0, v[^3 \rightarrow 3$ as follows

$$G(x, y, z) := \begin{cases} G_i(x, y, z) & \text{if } v_i \leq x < y < z \leq v_{i+1} \text{ for some } i \\ 3 & \text{otherwise} \end{cases} \tag{4}$$

We claim that $v < R_3^3(f)(m)$. The counter example partition is provided by G. Assume that $Y \subseteq v$ and that $G \upharpoonright [Y]^3$ has constant value. We have to show that $card(Y) < \max\{m, f(\min(Y)\} =: m''$. We may assume that $card(Y) \geq 3$.

Case 1: $Y \subseteq [v_0, v_1[$. Then $G_0 \upharpoonright [Y]^3 = G \upharpoonright [Y]^3$ has constant value. Hence $card(Y) < m \leq m''$.

Case 2: $Y \subseteq [v_i, v_{i+1}[$ for some i with $0 < i < m' - 1$. Then $G_i \upharpoonright [Y]^3 = G \upharpoonright [Y]^3$ has constant value. Hence $card(Y) < f(v_i)) \leq f(\min(Y)) \leq m''$.

Case 3: For all $i < m' - 1$ the set Y is not contained in $[v_i, v_{i+1}[$. Then $G \upharpoonright [Y]^3$ has constant value 3. Moreover for all $i < m' - 1$ we have

$$card(Y \cap [v_i, v_{i+1}[) \leq 2. \tag{5}$$

Indeed if we would find three elements x, y, z in some $Y \cap [v_i, v_{i+1}[$ then

$$G(x, y, z) \neq 3.$$

By (5) we obtain that $card(Y) \leq (m' - 1) \cdot 2 < m \leq m''$.

We now claim that

$$v_i \geq 2^{K_0^{(1+\frac{3}{2}\varepsilon)^i}} \tag{6}$$

and

$$f(v_i) \geq K_0 \tag{7}$$

for $i \geq 1$. Proof of the claim by induction on i. Let us check the case $i = 1$. Then $v_1 = R_2^3(m) - 1 \geq m \geq 2^{K_0^{(1+\frac{3}{2})^1}}$ and $f(v_1) = |v_1|^{\frac{1}{2}+\varepsilon} \geq |2^{m^{2-\delta}}|^{\frac{1}{2}+\varepsilon} \geq m^{1+\frac{3}{2}\varepsilon} \geq K_0$. Now assume that the claim holds for i. Since $f(v_i) \geq K_0$ we can apply the asymptotic for R_2^3 to $f(v_i)$. We thus obtain

$$\begin{aligned} v_{i+1} &= R_2^3(f(v_i)) - 1 \\ &\geq 2^{(f(v_i))^{2-\delta}} \\ &\geq 2^{(K_0^{1+\frac{3}{2}\varepsilon)^i})^{\alpha \cdot (2-\delta)}} \\ &\geq 2^{K_0^{(1+\frac{3}{2}\cdot\varepsilon)^{i+1}}}. \end{aligned}$$

This moreover implies $f(v_{i+1}) \geq K_0$.

Summing up we have shown $R_3^3(f)(m) > 2^{K_0^{(1+\frac{3}{2}\cdot\varepsilon)^{m'-1}}}$. Thus for all but finitely many m we obtain $R_3^3(f)(m) > 2^{K_0^{(1+\varepsilon)^{m/2}}}$.

Similarly one shows the following result indicating the relevance of the threshold at $\frac{1}{2}$.

Theorem 21. Let $\delta \geq 2$, $\gamma > 0$. and $\varepsilon > 0$. Put $\alpha := \frac{1}{\delta} + \varepsilon$ and $f(i) = |i|^{\alpha}$. Assume that $R_2^3(n) \geq 2^{n^{\delta} \cdot \gamma}$ for all but finitely many n. Then $R_3^3(f)(m) \geq 2^{2^{(1+\varepsilon)^m}}$ for all but finitely many m.

Remark: Related phase transitions can be proved for all functions R_3^d for $d \geq 3$.

4.3   A Phase Transition for R_2^2

Finally we study the Ramsey function for pairs. Here we study a phase transition in terms of densities a concept which goes back to J. Paris [11]. Let f be a number theoretic function. We call a finite set X of natural numbers 0-dense(f, k, l) iff $card(X) \geq \max\{3, f(\min(X))\}$. We call X $n + 1$-dense(f,k,l) iff for any $F : [X]^k \to l$ there exists a $Y \subseteq X$ such that $F \upharpoonright [Y]^2$ is constant and Y is n-dense(f,k,l).

Recall that $R_2^2(k)$ is the least m such that for every $F : [m]^2 \to 2$ there exists a monochromatic $Y \subseteq m$ such that $card(Y) \geq k$. Then, by Erdös's probabilistic method, we know the classical lower bound $R_2^2(k) \geq 2^{\frac{k}{2}}$ for all k [8]. Elementary combinatorics yields further the well known upper bound $R_2^2(k) \leq 2^{2 \cdot k}$ for all k [8]. It is open (an Erdös USD 100 problem) whether the limit $\lim_{k \to \infty} (R_2^2(k))^{\frac{1}{k}}$ exists. (Determining the value is an Erdös USD 250 problem.)

Let $\log_4$ denote the logarithm with respect to base 4.

Theorem 22. *Let* $f(i) := \lceil \log_4(\log_4(i - \frac{1}{2})) \rceil$. *Assume that*

$$\rho := \lim_{k \to \infty} (R_2^2(k))^{\frac{1}{k}}$$

exists. Let

$$\mu := \inf\{r \in \mathbb{R} : (\exists K)(\forall m \geq K)\, [4^{4^m}, 4^{4^m} + \lfloor (r)^{\lfloor (r)^m \rfloor} \rfloor[\ is\ 2 - dense(f, 2, 2)\}.$$

Then $\rho = \mu$.

Proof. It is easy to see that $[4^{4^m}, 4^{4^m} + 4^{4^m}[$ is $2 - dense(f, 2, 2)\}$ using the well known upper bound on R_2^2. Thus $\mu \leq 4$ and $\rho \leq 4$. In the sequel we assume $\rho < 4$. In the case $\rho = 4$ we have $R_2^2(m) < 4^{4^m}$ for almost all m and the following argument shows that $\mu \leq \rho$.

To prove $\rho \geq \mu$ let $\varepsilon > 0$. We may assume $\rho + \varepsilon \leq 4$. Then for almost all m we have

$$R_2^2(m) < \lfloor (\rho + \varepsilon)^m \rfloor. \tag{8}$$

In particular for allmost all m we have

$$R_2^2(\lfloor (\rho + \varepsilon)^m \rfloor) < \lfloor (\rho + \varepsilon)^{\lfloor (\rho+\varepsilon)^m \rfloor} \rfloor. \tag{9}$$

For m large enough so that (8) and (9) hold let $I_m := [4^{4^m}, 4^{4^m} + \lfloor (\rho + \varepsilon)^{\lfloor (\rho+\varepsilon)^m \rfloor} \rfloor[$. We claim that I_m is $2 - dense(f, 2, 2)$. Let $P : [I_m]^2 \to 2$ be any partition. Then there exists by (9) a $Y \subset I_m$ such that $P \upharpoonright [Y]^2$ is constant and such that $card(Y) \geq \lfloor (\rho+\varepsilon)^m \rfloor$. We claim that Y is $1 - dense(f, 2, 2)$. For proving this let $Q : [Y]^2 \to 2$ be any partition. Then there exists by (8) a $Z \subseteq Y$ such that $Q \upharpoonright [Z]^2$ is constant and $card(Z) \geq m$. We claim that Z is $0 - dense(f, 2, 2)$. Indeed, $f(min(Z) \leq f(4^{4^m} + \lfloor (\rho + \varepsilon)^{\lfloor (\rho+\varepsilon)^m \rfloor} \rfloor) \leq f(4^{4^m} \cdot 2) \leq m \leq card(Z)$. Thus $\rho + \varepsilon \geq \mu$ for any $\varepsilon > 0$, hence $\rho \geq \mu$.

To prove $\mu \geq \rho$ let again $\varepsilon > 0$. Then for almost all m we have

$$R_2^2(m) > \lfloor (\rho - \varepsilon)^m \rfloor. \tag{10}$$

In particular for allmost all m we have

$$R_2^2(\lfloor (\rho - \varepsilon)^m \rfloor) > \lfloor (\rho - \varepsilon)^{\lfloor (\rho-\varepsilon)^m \rfloor} \rfloor. \tag{11}$$

Let for large enough m $J_m := [4^{4^m}, 4^{4^m} + \lfloor (\rho - \varepsilon)^{\lfloor (\rho-\varepsilon)^m \rfloor} \rfloor[$. We claim that J_m is not $2 - dense(f, 2, 2)$. Indeed, by (11) there exists a partition $P : [J_m]^2 \to 2$ such that $card(Y) < \lfloor (\rho + \varepsilon)^m \rfloor$ for all $Y \subseteq J_m$ such that $P \upharpoonright [Y]^2$ is constant. Pick any such Y. We claim that Y is not $1 - dense(f, 2, 2)$. Indeed by (11) there exists a partition $Q : [Y]^2 \to 2$ such that $card(Z) < m$ for all $Z \subseteq Y$ with $Q \upharpoonright [Z]^2$ is constant. We claim that any such Z is not $0 - dense(f, 2, 2)$. Indeed, $f(min(Z) \geq f(4^{4^m}) \geq m > card(Z)$. Thus $\rho - \varepsilon < \mu$ for any $\varepsilon > 0$, hence $\rho \leq \mu$.

Similar proofs yield the following results.

Theorem 23. *Let* $f(i) := \lceil \log_4(\log_4(i - \frac{1}{2})) \rceil$. *Assume that*

$$\rho := \lim_{k \to \infty} (R_2^2(k))^{\frac{1}{k}}$$

exists. Let

$$\mu := \sup\{r \in \mathbb{R} : (\exists K)(\forall m \geq K)[4^{4^m}, 4^{4^m} + \lfloor (r)^{\lfloor (r)^m \rfloor} \rfloor[\text{ is not } 2-\text{dense}(f, 2, 2)\}.$$

Then $\rho = \mu$.

Theorem 24. *Let* $f(i) := \lceil \log_4(\log_4(i - \frac{1}{2})) \rceil$. *Assume that*

$$\mu - \sup\{r \in \mathbb{R} : (\exists K)(\forall m \geq K)[4^{4^m}, 4^{4^m} + \lfloor (r)^{\lfloor (r)^m \rfloor} \rfloor[\text{ is not } 2-\text{dense}(f, 2, 2)\}$$

$$= \inf\{r \in \mathbb{R} : (\exists K)(\forall m \geq K)[4^{4^m}, 4^{4^m} + \lfloor (r)^{\lfloor (r)^m \rfloor} \rfloor[\text{ is } 2 - \text{dense}(f, 2, 2)\}.$$

Then

$$\rho := \lim_{k \to \infty} (R_2^2(k))^{\frac{1}{k}}$$

exists and $\rho = \mu$.

References

1. T. Arai, *On the slowly well orderedness of ε_0*, Math. Log. Q., 48 (2002), 125–130.
2. W. Buchholz, A. Cichon and A. Weiermann, *A uniform approach to fundamental sequences and hierarchies*, Math. Log. Q., 40 (1994), 273–286.
3. L. Carlucci, G. Lee and A. Weiermann *Classifying the phase transition threshold for regressive Ramsey functions*. Preprint 2006 (submitted to TAMS).
4. L.E. Dickson: Finiteness of the odd perfect and primitive abundant numbers with n distinct prime factors. American Journal of Mathematics, Vol. 35 (1913), 413–422.
5. P. Erdős and G. Mills, *Some bounds for the Ramsey-Paris-Harrington numbers*, J. Combin. Theory Ser. A, 30 (1981), no. 1, 53–70.
6. P. Erdős, A. Hajnal, A. Máté, and R. Rado. *Combinatorial set theory: Partition relations for cardinals*, Studies in Logic and the Foundations of Mathematics, 106. North-Holland, 1984.
7. P. Erdős and R. Rado, *Combinatorial theorems on classifications of subsets of a given set*, Proc. London Math. Soc. (3), 2 (1952), 417–439.
8. R. L. Graham, B. L. Rotschild and J. H. Spencer, *Ramsey Theory*, Wiley, 1980.
9. Graham Higman: Ordering by divisibility in abstract algebras. Proceedings of the London Mathematical Society 3(2) (1952) pp. 326–336.
10. A. Kanamori and K. McAloon, *On Gödel incompleteness and finite combinatorics*, Ann. Pure Appl. Logic, 33 (1987), no. 1, 23–41.
11. L. Kirby, J.B Paris. Initial segments of models of Peano's axioms. In: *Set theory and hierarchy theory, V (Proc. Third Conf., Bierutowice, 1976)*, pp. 211–226. Lecture Notes in Math., Vol. 619. Springer, Berlin (1977).
12. M. Kojman, G. Lee, E. Omri and A. Weiermann, *Sharp thresholds for the Phase Transition between Primitive Recursive and Ackermaniann Ramsey Numbers*, Preprint, 2005.

13. M. Kojman and S. Shelah, *Regressive Ramsey numbers are Ackermannian*, J. Comb. Theory Ser. A, 86 (1999), no.1, 177–181.

14. S. Kripke: The problem of entailment. Journal of Symbolic Logic. Vol. 24 (1959), 324 (abstract).

15. G. Lee, *Phase Transitions in Axiomatic Thought*, PhD thesis (written under the supervision of A. Weiermann), Münster 2005, 121 pages.

16. G. Moreno Socias: An Ackermannian polynomial ideal. Lecture Notes in Computer Science 539 (1991), 269-280.

17. F. P. Ramsey, *On a problem of Formal Logic*, Proc. London Math. Soc., ser. 2, 30 (1930), 338–384.

18. H. Schwichtenberg, *Eine Klassifikation der ε_0-rekursiven Funktionen*, Z. Math. Logik Grundlagen Math., 17 (1971), 61–74.

19. S.G. Simpson. Ordinal numbers and the Hilbert basis theorem. J. Symbolic Logic 53 (1988), no. 3, 961–974.

20. Rick L. Smith: The consistency strength of some finite forms of the Higman and Kruskal theorems. In *Harvey Friedman's Research on the Foundations of Mathematics*, L. A. Harrington et al. (editors), (1985), pp. 119–136.

21. S. S. Wainer, *A classification of the ordinal recursive functions*, Archiv für Mathematische Logik und Grundlagenforschung, 13 (1970), 136–153.

22. A. Weiermann, *How to characterize provably total functions by local predicativity*, J. Symbolic Logic, 61 (1996), no.1, 52–69.

23. A. Weiermann, *An application of graphical enumeration to* PA, J. Symbolic Logic 68 (2003), no. 1, 5–16.

24. A. Weiermann, *An application of results by Hardy, Ramanujan and Karamata to Ackermannian functions*, Discrete Mathematics and Computer Science, 6 (2003), 133–142.

25. A. Weiermann: A very slow growing hierarchy for Γ_0. Logic Colloquium '99, 182–199, Lect. Notes Log., 17, Assoc. Symbol. Logic, Urbana, IL, 2004.

26. A. Weiermann, *A classification of rapidly growing Ramsey functions*, Proc. Amer. Math. Soc., 132 (2004), no. 2, 553–561.

27. A. Weiermann, *Analytic Combinatorics, proof-theoretic ordinals and phase transitions for independence results*, Ann. Pure Appl. Logic, 136 (2005), 189–218.

28. A. Weiermann:An extremely sharp phase transition threshold for the slow growing hierarchy. Mathematical Structures in Computer Science (to appear).

29. A. Weiermann, *Classifying the phase transition for Paris Harrington numbers*, *Preprint*, 2005.

30. A. Weiermann: *Phase transitions for some Friedman style independence results* Preprint 2006, to appear in MLQ.

31. H.S. Wilf: generatingfunctionology. Second edition. Academic Press, Inc., Boston, MA, 1994. x+228 pp.

Non-deterministic Halting Times for Hamkins-Kidder Turing Machines

P.D. Welch

School of Mathematics, University of Bristol
p.welch@bristol.ac.uk

In this talk we consider some issues related to the Infinite Time Turing Machine (ITTM) model of Hamkins & Lewis [3]. In particular our main results (Propositions 1 & 2) relate to Bounding times of the lengths of certain computations, and their application to certain questions raised in [2] on "non-determinism" both in terms of non-deterministically halting ordinals (Theorem 2) and pointclasses defined by using such non-deterministic machines (Proposition 6).

In ITTM's, a standard Turing machine (with some inessential minor modifications) is allowed to run transfinitely in ordinal time. The machine's behaviour at limit stages of time λ is completely specified by requiring that (i) the machine enter a special limit state q_L; (ii) the read/write head return to the initial starting cell at the leftmost end of the tape; (iii) the cell values - which we shall assume are taken from the alphabet of $\{0,1\}$ - are the limsup of their previous values: that is if cell i on the tape has contents $C_i(\gamma) \in \{0,1\}$ at time γ, then $C_i(\lambda) = \limsup_{\gamma \to \lambda} \langle C_i(\gamma) | \gamma < \lambda \rangle$. The original machine specified three infinite tapes: input, scratch, and output, with a read/write head positioned over one cell from each tape simultaneously. The machine's actions at successor stages is determined by its (finite) program in the ordinary way.

A number of intriguing questions immediately spring to mind. The question of the identity of the "decidable" reals (for which $x \in 2^{\mathbb{N}}$ is there a program P_e which halts on all inputs, and so that on input x P_e halts with output 1: "$P_e(x) \downarrow$ 1" ?), and of the semi-decidable reals, is answered in Welch[7]. Hamkins & Lewis [3] had previously showed, *inter alia*, that Π_1^1 predicates of reals are decidable, and that the decidable, (and semi-decidable) pointclasses of reals are strictly between Π_1^1 and Δ_2^1 in the projective hierarchy.

We shall be concerned here rather with the question of *halting times*, or how long such a computation takes, if it is going to halt.

Definition 1. $P_e(x) \downarrow^\alpha$ *will denote that program* $P_e(x) \downarrow$ *in exactly* α *steps.* $P_e(x) \downarrow^{\leq \alpha}, P_e(x) \downarrow^{< \alpha}$ *are defined analogously.*

To clarify the above: $P_e(x) \downarrow^\alpha$ means that at ordinal time α the read/write head is in particular state q_s and is reading a triple of cells (one from each of the three tapes) so that it's program determines that it goes into a halting state q_h. Thus a machine may halt exactly at some limit stage of time α where then $q_s = q_L$.

Suppose x is simple: perhaps it is an integer (*i.e.* it is a binary code for $n \in \mathbb{N}$ followed by an infinite string of 0's), perhaps it is 0 (in the above sense) itself.

A. Beckmann et al. (Eds.): CiE 2006, LNCS 3988, pp. 571–574, 2006.
© Springer-Verlag Berlin Heidelberg 2006

What possible halting times as e varies are there for $P_e(x)$? [3] calls an ordinal *clockable* if it is the halting time of a computation with input 0.

Further, let us define:

Definition 2. *"$P_e(x)\downarrow y$" will denote that $P_e(x)\downarrow$ and that $y \in 2^{\mathbb{N}}$ is the contents of the output tape on halting. (Again $P_e(x)\downarrow^\alpha y$ etc. are defined analogously).*

Then we say that y is *writable* if it is the output of some program: $P_e(0)\downarrow y$. An ordinal β *is writable* if some $y \in WO$ is writable, and y codes a wellordering of rank β. What possible ordinals are writable? It is easy to adjust a program that demonstrates that β is writable, to one that shows $\beta' < \beta$ is writable for some particular β'. Thus the writable ordinals are an initial segment, λ, of all ordinals. Hamkins and Lewis [3] showed that there are gaps in the clockable ordinals, and also the following:

Theorem 1. *(Hamkins and Lewis [3]) If β is admissible then it is not clockable.*

(For notions of *admissible ordinal* and *admissible set* see [1].) Welch [8] shows that λ, the suprema of the writable ordinals, is also the supremum of the clockable ordinals.

One may generalise these questions to those involving arbitrary input x. The following is Definition 24 of Deolalikar, Hamkins & Schindler [2]:

Definition 3. *An ordinal α is* nondeterministically clockable *if there is an algorithm P_e which halts in time at most α for all input and in time exactly α for some input.*

Symbolically: α is nondeterministically clockable iff

$$\exists e \in \mathbb{N}[\forall x \in 2^{\mathbb{N}} P_e(x)\downarrow^{\leq\alpha} \wedge \exists x \in 2^{\mathbb{N}} P_e(x)\downarrow^\alpha].$$

This notion arises in the paper [2], which was concerned with various complexity pointclasses defined using halting times of computations on these machines, with or without existential 'non-determinacy' witnesses. A Turing machine is is non-deterministic if, in effect, tests all possible runs $P_e(x)$. An ordinal α is then non-deterministically clockable, if such a machine halts on all inputs in less than or equal to α steps, and halts on some input in exactly α steps. Hence the nomenclature in the last definition.

We show the following by applying the Barwise Compactness Theorem ([1]).

Theorem 2. *If β is admissible then it is not nondeterministically clockable.*

This is in fact a corollary of a more general *Bounding Lemma* (where we identify $\mathbb{R}$ with $2^{\mathbb{N}}$):

Proposition 1. *(Bounding Lemma) Suppose β be admissible. Let $F : \mathbb{R} \longrightarrow \mathbb{R}$ be an ITTM-computable total function, so that $\forall x P_e(x)\downarrow^{\leq\beta}$ where P_e computes F. Then $\exists \gamma < \beta \; \forall x P_e(x)\downarrow^{<\gamma}$.*

Let $x \in 2^{\mathbb{N}}$. Then, as is usual, we let ω_{1ck}^x denote the supremum of all ordinals that are recursive in x (that is, those ordinals α with a corresponding $y \in WO$ with rank of y equalling α, and the characteristic function of y is Turing recursive (in the ordinary sense of recursive) in x.

The following question is posed in [2]:

Question 6. *Suppose an algorithm halts on each input x in fewer than ω_{1ck}^x steps. Then does it halt uniformly before ω_{1ck}?*

As they say an affirmative answer explains some of the phenomena observed in their paper. It is perhaps somewhat remarkable that processes that uniformly are required to halt only by an ordinal recursive in the input, in fact must halt *uniformly* by some recursive ordinal, but the next proposition shows that this is indeed the case (we drop the subscript ck and write ω_1^x for the first ordinal not recursive in x etc.). We prove that we have *Uniform Bounding*:

Proposition 2. *Let $F : \mathbb{R} \longrightarrow \mathbb{R}$ be ITTM-computable and total as witnessed by the program P_e. If $\forall x P_e(x)\downarrow^{<\omega_1^x}$ then $\exists \gamma < \omega_{1ck} \ \forall x P_e(x)\downarrow^{<\gamma}$.*

What we are calling here the Uniform Bounding Lemma is in fact a straightforward application of the Bounding Lemma in Higher Recursion Theory. Although that is usually stated close to the form Spector gave it ([6], or see [4] 4A.5) concerning as it does Σ_1^1 sets of codes for *recursive* ordinals, the argument applies in general to sets of ordinal codes defined by Σ_1^1-formulae - and we use it as such.

We consider some further queries arising from the paper [2]. These concerned various complexity pointclasses defined using halting times of computations on Infinite Time Turing machines, with or without existential 'non-determinacy' witnesses. These classes were first explicitly introduced by Schindler in [5].

Definition 4. *Let $f : \mathbb{R} \longrightarrow On$. (i) $A \in P^f$ if there is an infinite time Turing machine deciding (with a $0,1$ output), whether or not each $x \in A$ in fewer than $f(x)$ many steps.*

(ii) $A \in NP^f$ when there is an infinite time Turing machine T such that $x \in A$ if and only if there is $y \in \mathbb{R}$ such that T accepts (x, y), and further T halts on any input (x, y) in fewer than $f(x)$ many steps.

We thus think of f as a bounding function on the number of steps needed to determine whether x is, or is not, in some pointclass A, by using some total (so always either accepting or rejecting) ITTM program. Here f may be a constant function, and in the case that it is, with constant value ω^ω then [2] call the pointclasses P and NP. They analyse these classes for a variety of f and show, for example:

Theorem 3. *[2] $P \neq NP \cap \text{co-}NP$.*

Concomitant with the classes P^f are the following pointclasses definable in a simple way over the $f(x)$ level of the constructible hierarchy over x :

Definition 5. $\Gamma^f = \{A \subseteq \mathbb{R} : \exists \Sigma_1 \varphi \forall x [x \in A \longleftrightarrow L_{f(x)}[x] \models \varphi[x]]\}$.

So, as in [2] Sect.6, call f *suitable* if for any x $f(x) \geq \omega + 1$ and for any x, y $x \leq_T y \Longrightarrow f(x) \leq_T f(y)$.

Now let f be suitable, such that for any $x \in \mathbb{R}$ $L_{f(x)}[x]$ is an admissible set that is a union of admissible sets. Then:

Proposition 3. $NP^f = \Gamma^f; P^f = \Gamma^f \cap \text{co-}\Gamma^f = NP^f \cap \text{co}NP^f$. *Thus in general* NP^f *does not equal the dual class* $\Gamma^f \cap \text{co-}\Gamma^f$.

This answers another of the queries of [2].

References

[1] K.J. Barwise. *Admissible Sets and Structures*. Perspectives in Mathematical Logic. Springer Verlag, 1975.

[2] V. Deolalikar, J.D. Hamkins, and R-D. Schindler. $P \neq NP \cap co - NP$ for the infinite time Turing machines. *Journal of Logic and Computation*, 15:577–592, October 2005.

[3] J. D. Hamkins and A. Lewis. Infinite time Turing machines. *Journal of Symbolic Logic*, 65(2):567–604, 2000.

[4] Y. Moschovakis. *Descriptive Set theory*. Studies in Logic series. North-Holland, Amsterdam, 1980.

[5] R-D. Schindler. $P \neq NP$ for infinite time Turing machines. *Monatsheft für Mathematik*, to appear.

[6] C. Spector. Recursive wellorderings. *Journal of Symbolic Logic*, 20:151–163, 1955.

[7] P. D. Welch. Eventually infinite time Turing degrees: infinite time decidable reals. *Journal for Symbolic Logic*, 65(3):1193–1203, 2000.

[8] P. D. Welch. The length of infinite time Turing machine computations. *Bulletin of the London Mathematical Society*, 32:129–136, 2000.

Kurt Gödel and Computability Theory

Richard Zach[*]

Department of Philosophy, University of Calgary
Calgary, AB T2N 1N4, Canada
rzach@ucalgary.ca

Abstract. Although Kurt Gödel does not figure prominently in the history of computabilty theory, he exerted a significant influence on some of the founders of the field, both through his published work and through personal interaction. In particular, Gödel's 1931 paper on incompleteness and the methods developed therein were important for the early development of recursive function theory and the lambda calculus at the hands of Church, Kleene, and Rosser. Church and his students studied Gödel 1931, and Gödel taught a seminar at Princeton in 1934. Seen in the historical context, Gödel was an important catalyst for the emergence of computability theory in the mid 1930s.

1   Introduction

Kurt Gödel's contributions to logic rank among the most important work in logic, and among the most important in 20th century mathematics. The theory of computability, and much of theoretical computer science more generally, has its roots, historically as well as conceptually, in the field of logic, and so it is a given that many of Gödel's results are also important in the field of theoretical computer science. However, it would be an exaggeration to say that Gödel was himself a pioneer of the field. That distinction belongs to those who lay the groundwork for a mathematical analysis of the concept of computation: Church, Kleene, Post, Rosser, and Turing, and those who followed in their footsteps. Nevertheless, the early work of Church, Kleene and Rosser was heavily influenced by Gödel, and it is perhaps not an exaggeration to say that their work was made possibly only by Gödel's earlier contributions.

The historical background both for Gödel's early work and that of Church, Rosser, and Kleene lies in the context of the foundational debate of the 1920s. Hilbert's program for the foundations of mathematics was the driving force behind many of the advances in logic during that time. His belief that all mathematical questions are in principle decidable underwrote his belief that the formal systems of mathematics considered then, such as arithmetic, analysis, and set theory, are complete in the sense that for any sentence A in the respective language, either A or $\neg A$ is derivable in the system. (Although Hilbert himself had

---

[*] Research supported by the Social Sciences and Humanities Research Council of Canada.

A. Beckmann et al. (Eds.): CiE 2006, LNCS 3988, pp. 575–583, 2006.

reservations whether this is the case in "higher domains", e.g., set theory, he did believe it was true for first and second-order arithmetic.) In a related sense, this was also the basis for Hilbert's conjecture that first-order logic is complete in the sense that any valid sentence is derivable from the axioms of the predicate calculus. (It was known by the mid 1920s that first-order logic is not complete in the first, syntactic sense described above—there are formulas A such that neither A nor $\neg A$ is derivable in first-order logic alone.) It was also the basis for his aim in the work on the decision problem for logic, i.e., that it should be possible to find a procedure to decide, for any given sentence of first-order logic, whether it is provable from the axioms of the predicate calculus or not. Hilbert's firm belief that classical mathematics is secure in the sense that the axioms of arithmetic and set theory do no lead to contradictions suggested that it should be possible to prove that these axioms are consistent, and since the statement of consistency is a purely combinatorial one about what sequences of formulas of certain sorts there are, that consistency could be proved using elementary, "finitary" methods. These methodologically motivated questions, then, guided the work of the Hilbert school: to solve the decision problem by giving a decision problem for predicate logic; to prove that arithmetic and logic are complete; and to find a finitary consistency proof of arithmetic and analysis.

In 1929 and 1930, Gödel solved the latter two problems. In his dissertation (1929; 1930), he showed that first-order logic is complete, and in his *Habilitationsschrift* (1930; 1931) he showed that arithmetic is incomplete. Very soon afterward he himself accepted the consequence of the second incompleteness theorem that no finitary consistency proof of arithmetic can be given, a consequence that others (e.g., von Neumann and Herbrand) accepted more readily. Although Church and Turing gave the definitive (negative) solution to the decision problem, Gödel also actively contributed to the literature on Hilbert's first task (Gödel, 1932, 1933).

Church's first publications on the λ-calculus were similarly concerned with foundational problems in mathematics: Church's stated aim was to develop a new axiomatization of logic which avoids the paradoxes, but in a manner different from Russell's theory of types or axiomatic set theory. Although we now think of the (simple) λ-calculus as a formalism for expressing computable functions, Church did not originally conceive of it in that way—for him, the system which evolved into the λ-calculus was a *logical* formalism which, he hoped, would be capable of serving as a contradiction-free formalization of mathematics. Unfortunately, Church's original system proved to be inconsistent (Kleene and Rosser, 1935). Kleene's and Rosser's proof that it was inconsistent made essential use of the method of Gödel coding introduced in (Gödel, 1931). Kleene's (1935) development of arithmetic and the representability of recursive functions within the λ-calculus was motivated, in part, by the aim of reproducing Gödel's incompleteness result in the context of the λ-calculus, and his important normal form theorem also relied on Gödel coding. It was in the context of this turn towards *meta*mathematical investigations of the λ-calculus along the lines of Gödel (1931) that the notion of λ-definability achieved pride of place in the work of Church, Kleene, and Rosser. The positive results obtained by Kleene to the effect that a great many recursive functions could be formalized in the λ-calculus led Church to formulate what now has come to be known as Church's Thesis, viz., that every effectively computable function is λ-definable. And again it was Gödel, who at the time (1934) was in Princeton, who led Church and his students to take a broader view: his skepticism about Church's thesis when first formulated regarding λ-definability and his proposal that general recursiveness might be a better candidate for a precise characterization of effective computability led Kleene to show that the two notions are coextensive: every λ-definable function is general recursive and conversely (Kleene, 1936b).

In what follows, I will give an outline of the early history of recursion theory, with special emphasis on the role Gödel and his results played in it. In my survey of these developments, I rely heavily on the recollections of Kleene (1981; 1987) and the analyses of Davis (1982) and Sieg (1997), as well as chapter V of Dawson's (1997) biography of Gödel.

2   Church's System and Gödel's Incompleteness Result

In the years 1929–1931, Church developed an alternative formulation of logic (Church, 1932, 1933), which he hoped would serve as a new foundation of mathematics which would avoid the paradoxes. Church taught a course on logic in the Fall of 1931, where Kleene, then a graduate student, took notes. During that time, Church and Kleene were first introduced to Gödel's work on incompleteness: the occasion was a talk by John von Neumann on Gödel's work. Church and Kleene immediately studied the paper in detail. At the time, it was not yet clear how general Gödel's results were. Church believed that the incompleteness of Gödel's system P (a type-theoretic higher-order formulation of Peano arithmetic) relies essentially on some feature of type theory, and that Gödel's result would not apply to Church's own system. It nevertheless seems like it became a pressing issue for Church to determine to what extent Gödel's results and methods could be carried out in his system. He set Kleene to work on the task of obtaining Peano arithmetic in the system. Kleene succeeded in carrying this out in the first half of 1932. It involved, in particular, showing that various number-theoretic functions are λ-definable. In July 1932, Gödel wrote to Church, asking if Church's system could be proved consistent relative to *Principia Mathematica*. Church was skeptical of the usefulness of such a relative consistency proof. He wrote,

> In fact, the only evidence for the freedom from contradiction of *Principia Mathematica* is the empirical evidence arising from the fact that the system has been in use for some time, many of its consequences have been drawn, and no one has found a contradiction. If my system be really free from contradiction, then an equal amount of work in deriving its consequences should provide an equal weight of empirical evidence for its freedom from contradiction. [...]

But it remains barely possible that a proof of freedom from contradiction for my system can be found somewhat along the lines suggested by Hilbert. I have, in fact, made several unsuccessful attempts to do this.

Dr. von Neumann called my attention last fall to your paper entitled "Über formal unentscheidbare sätze der Principia Mathematica." I have been unable to see, however, that your conclusions in §4 of this paper apply to my system. Possibly your argument can be modified so as to make it apply to my system, but I have not been able to find such a modification of your argument. (Church to Gödel, July 27, 1932. Gödel 2003a, 368–369).

Section §4 of Gödel (1931) which Church mentions here is the section in which Gödel sketched the *second* incompleteness theorem. Since Gödel did not provide a complete proof of the theorem—indeed, the first complete proof did not appear until Hilbert and Bernays (1939)—Church was surely justified in doubting that the result applies to his system. It leaves open the question, however, of whether Church believed, at the time, that the construction of the *first* incompleteness theorem do go through in his system.

Kleene reports (Crossley, 1975) that he carried out the development of Peano arithmetic in Church's system between January and June 1932, and then wrote up the results over the following year. The paper reporting these results (Kleene, 1935) was received by the *American Journal of Mathematics* on October 9, 1933, and in revised version on June 18, 1934. The paper also contains the arithmetization of syntax, making use of Gödel's methods and results, and a proof that all primitive recursive functions are λ-definable. Kleene also showed that for any formula in the formalism of *Principia Mathematica*, the question of whether it is provable is equivalent to the question of whether a certain expression of Church's system has a normal form. Only a few months after Kleene submitted the final version of his paper, in November 1934, Rosser and he submitted another paper to the *Annals of mathematics* (Kleene and Rosser, 1935). In it, they showed that Church's system, as well as Curry's combinatory logic (Curry, 1930), were inconsistent. In their proof, they again made extensive use of Gödel's arithmetization of syntax, and were able to derive a version of Richard's paradox within the system. The fragment of Church's system with the logical axioms removed is demonstrably consistent: it is the simple λ-calculus (see Barendregt 1997 for the impact of λ-calculus in computer science, and Seldin 2006 for a history of the λ-calculus).

Church, then, turned out to be right: Gödel's second incompleteness theorem does not apply to his system—because the theorem only applies to consistent formal systems. But in order to obtain this result, and many of the positive results due to Kleene which provided the foundation for Church's undecidability results a year later, Gödel's methods were of crucial importance, both because they motivated a certain line of inquiry and because Kleene, Rosser, and Church were able to build on them.

The methods introduced in Gödel (1931) and used by Kleene and Rosser to show that Church's system was inconsistent also figure prominently in Church's negative solution of the decision problem. Church (1935, 1936b) first showed that the question of whether a given expression of the λ-calculus has a normal form is not recursive. In the same paper, Church also stated what is now known as "Church's Thesis," viz., that the general recursive functions (and hence, the λ-definable ones) are just the "effectively computable" ones. The theorem and the thesis combine to yield the result that having a normal form is not an effectively decidable property. The genesis of Church's Thesis will be outlined in the next section. Here, I want to stress only that the result itself, and with it the negative solution of the decision problem for first-order logic (Church, 1936a), made essential use of Gödel's work.

Kleene (1987) himself emphasizes the importance of Gödel (1931) in the work that he and Rosser carried out in their seminal contributions to recursion theory in the early 1930s:

> After the colloquium [by von Neumann in the fall of 1931], Church's course continued uninterruptedly concentrating on his formal system; but on the side we all read Gödel's paper, which to me opened up a whole new world of fascinating ideas and perspectives.

3   Gödel and Church's Thesis

Gödel's (1931) had a dramatic and lasting influence on the pioneers of recursion theory and the development of the λ-calculus. Gödel had a more direct and personal influence in the formation of Church's thesis. He visited Princeton in the 1933/34 academic year and gave a series of lectures there between February and May 1934, which was attended by Church, Kleene, and Rosser. Kleene's work on defining various number-theoretic functions in the λ-calculus (1935) first prompted Church to put forward a tentative version of the thesis in late 1933 or early 1934, in the form: every effectively calculable function is λ-definable. In conversation, Gödel expressed skepticism about the thesis.

Towards the end of his Princeton lectures, Gödel introduced the notion of general recursive function. This notion was based on a suggestion by Herbrand in a letter to Gödel of April 7, 1931 (Gödel, 2003b, 14–21). In the lectures, Gödel (1934, 368–369), defined the general recursive functions as those which can be computed using a specific set of substitution rules from a set of defining equations, and for which the result of the computation is uniquely determined. (For a discussion of the connection between Herbrand's and Gödel's notions, see Sieg 2005.) Gödel did not at first propose the definition of general recursive function as an explication of the informal notion of "effectively computable," but only as an explication of the notion of "recursive function." In 1931, Gödel had introduced the primitive recursive functions (although he called them then just "recursive functions"). It was already known since the mid-1920s (Hilbert, 1926; Ackermann, 1928) that there are non-primitive recursive functions which can be defined by double recursion, and in the early 1930s, Péter (1934, 1935) studied such recursive functions in more detail.

Gödel was interested in a precise characterization of intuitively recursive functions. Kleene (1936b) soon succeeded in establishing that the general recursive functions are exactly the λ-definable ones, and this seems to have been a reason for Church to propose his thesis in print in 1935. Kleene (1936a) is a systematic study of Gödel's class of general recursive functions. It contains Kleene's normal form theorem, that every general recursive function can be written as $f(\mu x[g(x) = 0])$, with f, g primitive recursive, Kleene's T predicate, and examples of non-recursive functions and relations based on it.

For a more detailed historical discussion on the origin of Church's Thesis and Gödel's influence, see Davis (1982) and Sieg (1997).

4   Gödel and Complexity Theory

Another work of Gödel's played a role in the development and gradual acceptance of Church's Thesis—although Gödel himself apparently became convinced of the truth of the thesis only through Turing's work. That work was an abstract on length of proofs (Gödel, 1936). In stating his Thesis, Church (1936b, §7) had introduced the notion of functions *computable in a logic S*: f is is computable in S if there is some term ϕ so that for every numeral m there is a numeral n with $S \vdash \phi(m) = n$ iff $f(m) = n$ (following Kleene 1952, §59, such functions are also called *reckonable in S*). In a note added in proof, Gödel (1936) remarked that this notion of computability is absolute, in the sense that if a function is computable in a higher-order system S, it already is computable in first-order arithmetic—i.e., the general recursive functions are all the functions computable in any consistent system S containing arithmetic. The reason for this is, of course, that if the system is formal in the sense that its proofs are recursively enumerable, then then function is computable by searching through all proofs until one finds one of $\phi(m) = n$, and this procedure is insensitive to the logical strength of the theory S. This result served both Church and later also Gödel as evidence for the Church-Turing Thesis (see Gödel 1946 and Sieg 1997, 2006).

The main part of (Gödel, 1936), however, was not concerned with computability so much as with proof complexity. The result that Gödel announced concerned speed-up of proofs (measured as number of symbols) between n-th and $(n+1)$st-order arithmetic. (Buss 1994 contains a proof of the result.) 20 years later, Gödel was again thinking about proof complexity. In an intriguing letter to John von Neumann on March 20, 1956 (Gödel, 2003b, 372–377), Gödel discussed the complexity of deciding for a formula A of first-order logic, whether A has a proof with k symbols or less. Cook has shown that this problem is NP-complete (see Hartmanis 1989 and Buss 1995).

Unlike Gödel's earliest work, his thoughts on proof complexity and feasible computation in the letter to von Neumann had no impact on the historical development of computability and complexity theory. It nevertheless shows that questions of the nature of computability, even though they were not at the forefront of Gödel's thought or prominent in his publications, did occupy Gödel throughout his professional career.

Bibliography

Ackermann, Wilhelm. 1928.   Zum Hilbertschen Aufbau der reellen Zahlen. *Mathematische Annalen* 99: 118–133.

Barendregt, Henk. 1997. The impact of the lambda calculus in logic and computer science. *The Bulletin of Symbolic Logic* 3(2): 181–215.

Buss, Samuel R. 1994. On Gödel's theorems on lengths of proofs I: Number of lines and speedups for arithmetic. *Journal of Symbolic Logic* 39: 737–756.

Buss, Samuel R. 1995. On Gödel's theorems on lengths of proofs II: Lower bounds for recognizing *k* symbol provability. In *Feasible Mathematics II*, eds. P. Clote and J. Remmel, 57–90. Basel: Birkhäuser.

Church, Alonzo. 1932. A set of postulates for the foundation of logic. *Annals of Mathematics* 33: 346–366.

Church, Alonzo. 1933. A set of postulates for the foundation of logic (second paper). *Annals of Mathematics* 34: 839–864.

Church, Alonzo. 1935. An unsolvable problem of elementary number theory. *Bulletin of the American Mathematical Society* 41: 332–333.

Church, Alonzo. 1936a. A note on the Entscheidungsproblem. *Journal of Symbolic Logic* 1: 40–41.

Church, Alonzo. 1936b. An unsolvable problem of elementary number theory. *American Journal of Mathematics* 58: 345–363.

Crossley, J. N. 1975. Remniscences of logicians. In *Algebra and Logic. Papers from the 1974 Summer Research Institute of the Australian Mathematical Society*, ed. J. N. Crossley, LNM 450, 1–62. Berlin: Springer.

Curry, Haskell B. 1930. Grundlagen der kombinatorischen Logik. *American Journal of Mathematics* 52: 509–536, 789–834.

Davis, Martin. 1982. Why Gödel didn't have Church's thesis. *Information and Control* 54: 3–24.

John W. Dawson, Jr. 1997. *Logical Dilemmas. The Life and Work of Kurt Gödel.* Wellesley, Mass.: A K Peters.

Gödel, Kurt. 1929. Über die Vollständigkeit des Logikkalküls. Dissertation, Universität Wien. Reprinted and translated in (Gödel, 1986, 60–101).

Gödel, Kurt. 1930. Die Vollständigkeit der Axiome des logischen Funktionenkalküls. *Monatshefte für Mathematik und Physik* 37: 349–360. Reprinted and translated in (Gödel, 1986, 102–123).

Gödel, Kurt. 1930. Einige metamathematische Resultate über Entscheidungsdefinitheit und Widerspruchsfreiheit. *Anzeiger der Akademie der Wissenschaften in Wien* 67: 214–215. Reprinted and translated in (Gödel, 1986, 140–143).

Gödel, Kurt. 1931. Über formal unentscheidbare Sätze der *Principia Mathematica* und verwandter Systeme I. *Monatshefte für Mathematik und Physik* 38: 173–198. Reprinted and translated in (Gödel, 1986, 144–195).

Gödel, Kurt. 1932. Ein Spezialfall des Entscheidungsproblem der theoretischen Logik. *Ergebnisse eines mathematischen Kolloquiums* 2: 27–28. Reprinted and translated in (Gödel, 1986, 130–235).

Gödel, Kurt. 1933. Zum Entscheidungsproblem des logischen Funktionenkalküls. *Monatshefte für Mathematik und Physik* 40: 433–443. Reprinted and translated in (Gödel, 1986, 306–327).

Gödel, Kurt. 1934. On undecidable propositions of formal mathematical systems. Lecture notes by Stephen C. Kleene and J. Barkely Rosser, Princeton University. Reprinted in (Gödel, 1986, 338–371).

Gödel, Kurt. 1936. Über die Länge von Beweisen. *Ergebnisse eines mathematisches Kolloquiums* 7: 23–24. Reprinted and translated in (Gödel, 1986, 394–399).

Gödel, Kurt. 1946. Remarks before the Princeton bicentennial conference on problems in mathematics. In *Collected Works*, eds. Solomon Feferman et al., vol. 2, 144–153. Oxford: Oxford University Press.

Gödel, Kurt. 1986. *Collected Works*, vol. 1, eds. Solomon Feferman et al. Oxford: Oxford University Press.

Gödel, Kurt. 2003a. *Collected Works*, vol. 4, eds. Solomon Feferman et al. Oxford: Oxford University Press.

Gödel, Kurt. 2003b. *Collected Works*, vol. 5, eds. Solomon Feferman et al. Oxford: Oxford University Press.

Hartmanis, Juris. 1989. Gödel, von Neumann and the P=?NP problem. *Bulletin of the European Association for Theoretical Computer Science (EATCS)* 38: 101–107.

Hilbert, David. 1926. Über das Unendliche. *Mathematische Annalen* 95: 161–90. Lecture given Münster, 4 June 1925. English translation in (van Heijenoort, 1967, 367–392).

Hilbert, David and Paul Bernays. 1939. *Grundlagen der Mathematik*, vol. 2. Berlin: Springer.

Kleene, Stephen C. 1935. A theory of positive integers in formal logic. *American Journal of Mathematics* 57: 153–173, 219–244.

Kleene, Stephen C. 1936a. General recursive functions of natural numbers. *Mathematische Annalen* 112: 727–742.

Kleene, Stephen C. 1936b. λ-definability and recursiveness. *Duke Mathematical Journal* 2: 340–353.

Kleene, Stephen C. 1952. *Introduction to Metamathematics*. Amsterdam: North-Holland.

Kleene, Stephen C. 1981. Origins of recursive function theory. *Annals of the History of Computing* 3: 52–67.

Kleene, Stephen C. 1987. Gödel's impression on students of logic in the 1931s. In *Gödel Remembered*, eds. Paul Weingartner and Leopold Schmetterer, 49–64. Bibliopolis.

Kleene, Stephen C. and J. Barkley Rosser. 1935. The inconsistency of certain formal logics. *Annals of Mathematics* 36: 630–636.

Péter, Rósza. 1934. über den Zusammenhang der verschiedenen Begriffe der rekursiven Funktionen. *Mathematische Annalen* 110: 612–632.

Péter, Rósza. 1935. Konstruktion nichtrekursiver Funktionen. *Mathematische Annalen* 111: 42–60.

Seldin, Jonathan P. 2006. The logic of Curry and Church. In *Handbook of the History of Logic*, eds. Dov Gabbay and John Woods, vol. 5. Amsterdam: Elsevier. Forthcoming.

Sieg, Wilfried. 1997. Step by recursive step: Church's analysis of effective computability. *Bulletin of Symbolic Logic* 3: 154–180.

Sieg, Wilfried. 2005. Only two letters: The correspondence between Herbrand and Gödel. *Bulletin of Symbolic Logic* 11: 172–184.

Sieg, Wilfried. 2006. Gödel on computability. *Philosophia Mathematica* 14. Forthcoming.

van Heijenoort, Jean, ed. 1967. *From Frege to Gödel. A Source Book in Mathematical Logic, 1897–1931.* Cambridge, Mass.: Harvard University Press.

A Computability Theory of Real Numbers[*]

Xizhong Zheng[1,2]

[1] Department of Computer Science, Jiangsu University, Zhenjiang 212013, China
[2] Theoretische Informatik, BTU Cottbus, D-03044 Cottbus, Germany
zheng@informatik.tu-cottbus.de

Abstract. In mathematics, various representations of real numbers have been investigated. Their standard effectivizations lead to equivalent definitions of computable real numbers. For the primitive recursive level, however, these effectivizations are not equivalent any more. Similarly, if the weaker computability is considered, we usually obtain different weak computability notions of reals according to different representations of real number. In this paper we summarize several recent results about weak computability of real numbers and their hierarchies.

1 Introduction

The classic computability theory is a well developed theory which deals with the effectivity of subsets and functions on discrete domains like natural numbers. For the real numbers, we consider w.l.o.g. the reals $x \in [0, 1]$ which are corresponded naturally to their binary expansion set $A \subseteq \mathbb{N}$ in the way of $x = x_A := \sum_{n \in A} 2^{-(n+1)}$. If a set is identified with its characteristic sequence, then the real x_A of binary expansion A can also be denoted by $0.A$. In this way, the computability of subsets of natural numbers can be transferred straightforwardly to reals as follows: a real x_A is *computable* if $A \subseteq \mathbb{N}$ is a computable set. In other words, x_A is computable if there is an effective procedure to write down its binary expansion one bit after another. This is essentially the definition of Turing [24] and it is robust because Robinson [20] and others (see [12, 19]) have shown that, computable reals can be defined equivalently by Dedekind cuts, Cauchy sequences and other representations of reals too. For instance, x is computable iff it has a computable Dedekind cut $L_x := \{r \in \mathbb{Q} : r < x\}$; iff there is a computable sequence (x_s) of rational numbers which converges to x *effectively* in the sense that $(\forall n \in \mathbb{N})(|x_n - x_{n+1}| \leq 2^{-n})$, etc. This means that the computability of reals is independent of their representations. The class of computable reals is denoted by **EC** (for **E**ffectively **C**omputable).

Similarly, we can define the notion of Turing reducibility of reals as follows: x_A is *Turing reducible to* x_B (denoted by $x_A \leq_T x_B$) iff $A \leq_T B$. Two reals x, y are *Turing equivalent* (denoted by $x \equiv_T y$) if $x \leq_T y$ and $y \leq_T x$. The Turing degree $\deg_T(x)$ of a real x is defined as the class of all reals which are Turing equivalent

---

[*] This work is supported by DFG (446 CHV 113/240/0-1) and NSFC (10420130638).

A. Beckmann et al. (Eds.): CiE 2006, LNCS 3988, pp. 584–594, 2006.

to x, i.e., $\deg_T(x) := \{y \in \mathbb{R} : y \equiv_T x\}$. Obviously, these are independent of their representations too (see, e.g., [5]). Because of the correspondence between reals and subsets of natural numbers, we can identify the Turing degree $\deg_T(x_A)$ of a real x_A and the Turing degree $\deg_T(A) := \{B \subseteq \mathbb{N} : A \equiv_T B\}$ of the set $A \subseteq \mathbb{N}$. Thus, we can say that a degree of real is c.e. if it contains at least a c.e. set.

This nice story does not work any more if we consider the computability notion which are strictly stronger or weaker than the standard computability. In the following sections, we summarize the results related to these stuffs.

2 Strong Computability

Let F be a class of functions $f : \mathbb{N} \to \mathbb{N}$. By means of functions of F we can define the following classes of reals.

$$\mathcal{C}_1(F) := \{x \in \mathbb{R} : (\exists f, g \in F)(\forall n)(|x - f(n)/g(n)| \leq 2^{-n})$$

$$\mathcal{C}_2^b(F) := \{x \in \mathbb{R} : (\exists f \in F)((\forall n)(0 \leq f(n) < b) \ \& \ x = \sum_{n \in \mathbb{N}} f(n)b^{-n})\}$$

$$\mathcal{C}_2(F) := \{x \in \mathbb{R} : (\exists f \in F)((\forall n)(0 \leq f(n,b) < b) \ \& \ x = \sum_{n \in \mathbb{N}} f(n,b)b^{-n})\}$$

$$\mathcal{C}_3(F) := \{x \in \mathbb{R} : (\exists f \in F)(\forall m, n)(n/(m+1) < x \iff f(n,m) = 0)\}$$

$$\mathcal{C}_4(F) := \{x \in \mathbb{R} : (\exists f \in F)(x = [f(0), f(1), f(2), \ldots])\}$$

$$\text{where } [a_0, a_1, a_2, \ldots] = a_0 + \cfrac{1}{a_1 + \cfrac{1}{a_2 + \cdots}}$$

These classes correspond respectively to the representations of reals by (fast) Cauchy sequences, b-adic expansion, uniform b-adic expansion, Dedekind cuts and continued fractions. If F is the class of computable functions, then all classes defined above are the same. This means that the standard computability of reals is independent of the their representations. For the class PF of polynomial time computable functions, we have

Theorem 2.1 (Ko [8, 9])

1. $\mathcal{C}_4(PF) = \mathcal{C}_3(PF) = \mathcal{C}_2^2(PF) \subsetneq \mathcal{C}_1(PF)$;
2. $\mathcal{C}_1(PF)$ *is a real closed field.*

The primitive recursiveness of the reals was first systematically investigated by Specker [23]. He shows that decimal Dedekind cuts, expansions and Cauchy sequences lead to different versions of primitive recursiveness of reals. Later on, Peter [13], Mostowski [11], Lehman [10] investigated other versions of primitive recursiveness of reals. We summarize some of their results as the following theorem, where PR is the class of primitive recursive functions.

Theorem 2.2 (Specker [23], Peter [13], Mostowski [11], Lehman [10])

1. $\mathcal{C}_4(PR) \subsetneq \mathcal{C}_3(PR) = \mathcal{C}_2(PR) \subsetneq \bigcap_{b>1} \mathcal{C}_2^b(PR) \subsetneq \mathcal{C}_1(PR)$;
2. $x \in \mathcal{C}_3(PR)$ *iff the function* $\lambda n.\lfloor nx \rfloor$ *is primitive recursive;*

3. $x \in \mathcal{C}_4(PR)$ iff $x \in \mathcal{C}_1(PR)$ and x is recursively irrational in the sense that $(\forall n, m)(|x - m/n| \geq 1/g(n))$ holds for a primitive recursive function g.
4. The class $\mathcal{C}_1(PR)$ is a field.

The reals of classes $\mathcal{C}_1(PF)$ and $\mathcal{C}_1(PR)$ are called *polynomial time computable* and *primitive recursive*, respectively.

3   Computably Enumerable Reals

The classes of polynomial time computable, primitive recursive and computable reals have very nice computability as well as mathematical properties. They have also many practical applications. However, there are practical values which can not be characterized by these reals. For example, the length of a curve is defined as the limit of the lengths of polygons which approximates the curve. In this case, we might have a computable increasing sequence of the polygon-lengths which does not converges effectively. To investigate such kind of values, we have

Definition 3.1. *A real x is c.e. (co-c.e.) if there is an increasing (decreasing) computable sequence (x_s) of rational numbers which converges to x. The classes of c.e. and co-c.e. reals are denoted by* **CE** *and* co-**CE**.

C.e. and co-c.e. reals are also called *left* and *right computable* because they can be approximated from the left and right side in the real axis and their classes are denoted by **LC** and **RC**, respectively. Left and right computable reals together are called *semi-computable* and the class of all semi-computable reals is denoted by **SC**. A real x is semi-computable iff there is a computable sequence (x_s) of rational numbers which converges to x *1-monotonically* in the sense that $|x - x_t| \leq |x - x_s|$ for all $t > s$ (see [25, 1]).

The binary expansions of c.e. reals are not necessarily c.e. as observed by Jockusch [21] but they are strongly ω-c.e. as shown in the following theorem.

Theorem 3.2 (Calude, Hertling, Khoussainov and Wang [2]). *A real x is c.e. iff $x = x_A$ and A is a strongly ω-c.e. set in the sense that there is a computable sequence (A_s) of finite sets which converges to A such that, if $n \in A_s - A_{s+1}$, then there exists an $m < n$ with $m \in A_{s+1} - A_s$ for all n and s.*

The binary expansion leads naturally to an infinite hierarchy of c.e. reals. Soare [21] called a c.e. real x *stably c.e.* if its binary expansion is d-c.e. Jockusch's observation gives an example of stably c.e. real. Since there exists a strongly ω-c.e. set which is not d-c.e., the class of all stably c.e. reals is strictly between the classes of binary c.e. and c.e. reals. There is no reason to stop here. In general, a c.e. real is called *h-stably c.e.* for a function h if its binary expansion is an h-c.e. set. Thus, the k-stably c.e., for constant $k \in \mathbb{N}$, and the ω-stably c.e. reals can be defined accordingly [25]. By Theorem 3.2, the classes of h-stably c.e. reals collapse to the level $\lambda n.2^n$-stably c.e. for $h(n) \geq 2^n$ for all n. For lower levels, however, we have a proper hierarchy.

Theorem 3.3 (Weihrauch and Zheng [25]). *For any constant k, there is a $(k+1)$-stably c.e. real which is not k-stably c.e. and there exists an ω-stably c.e. real which is not k-stably c.e. for any $k \in \mathbb{N}$.*

Additionally, Downey [4] calls a real *strongly c.e.* if its binary expansion is c.e. and Wu [26] calls a real k-*strongly c.e.* if it is the sum of up to k strongly c.e. reals and calls k-strongly c.e. reals *regular*. Wu shows that, for any $k \in \mathbb{N}$, a k-strongly c.e. real is $2k$-stably c.e. and there is a $(k+1)$-strongly c.e. real which is not k-stably c.e. This, together with Theorem 3.3, implies that, for any $k \in \mathbb{N}$, there is a $(k+1)$-strongly c.e. real which is not k-strongly c.e. and there exists a c.e. real which is not regular.

The semi-computable reals have a very useful necessary condition as follows.

Theorem 3.4 (Ambos-Spies, Weihrauch and Zheng [1]). *If $A, B \subseteq \mathbb{N}$ are Turing incomparable c.e. sets, then the real $x_{A \oplus \overline{B}}$ is not semi-computable.*

In particular, this theorem implies that any non-computable c.e. degree contains a non-semi-computable real and the class of c.e. reals is not closed under subtraction.

4 Difference of c.e. Reals

The classes of c.e. and semi-computable reals are introduced naturally by the monotonicity of sequences and have a lot of nice computability-theoretical properties. However, neither of them have nice analytical property. For example, they are not closed under subtraction. This motivates us to explore their arithmetical closure and leads to the following definition.

Definition 4.1. *A real x is called* d-c.e. *(difference of c.e.) if $x = y - z$ for c.e. reals y, z. The class of all d-c.e. reals is denoted by* **DCE**.

By Theorem 3.4, the class **DCE** is a proper superset of **CE** and **SC**, because $x_{A \oplus \overline{B}} = x_{2A} - x_{2B+1}$ is d-c.e. but not semi-computable if A and B are Turing incomparable c.e. sets. But the difference hierarchy collapses since **DCE** is obviously closed under addition and subtraction. Moreover, the class **DCE** is also closed under multiplication and division and hence is a field. This follows from another nice characterization of d-c.e. reals as follows.

Theorem 4.2 (Ambos-Spies, Weihrauch and Zheng [1]). *A real x is d-c.e. iff there is a computable sequence (x_s) of rational numbers which converges to x weakly effectively in the sense that the sum $\sum_{s \in \mathbb{N}} |x_s - x_{s+1}|$ is finite.*

Because of Theorem 4.2, d-c.e. reals are also called *weakly computable* in literatures [25, 1, 27]. Now it is easy to see that **DCE** is closed under arithmetical operations $+, -, \times, \div$ and hence it is the arithmetical closure of **CE**. Recently, Raichev [14] shows that **DCE** is actually a real closed field.

There is another very interesting characterization of d-c.e. reals relating to the Solovay reduction which classifies the relative randomness of the real numbers.

A real x is Solovay reducible to y if there exist two computable sequences (x_s) and (y_s) of rational numbers converging to x and y, respectively, such that $|x - x_s| \leq c(|y - y_s| + 2^{-s})$ for some constant c and all s (see [22, 31]). Solovay [22] shows that any c.e. real is Solovay reducible to a c.e. random real. For d-e.c. reals we have the following result.

Theorem 4.3 (Rettinger and Zheng [17]). *A real number is d-c.e. if and only if it is Solovay reducible to a c.e. random real.*

The binary expansion of c.e reals are always $\lambda n.2^n$-c.e. For d-c.e. reals Zheng [29] shows that there exists a d-c.e. real which is even not Turing equivalent to any ω-c.e. set. More generally, about the Turing degrees of d-c.e. reals we have the following results.

Theorem 4.4 (Downey, Wu and Zheng [3])

1. *Any ω-c.e. Turing degree contains a d-c.e. real; and*
2. *There exists a Δ_2^0-Turing degree which does not contain any d-c.e. reals.*

We close our discussion about d-c.e. reals with an interesting necessary condition as follows.

Theorem 4.5 (Ambos-Spies, Weihrauch and Zheng [1]). *For any set A, if x_{2A} is a d-c.e. real, then A is h-c.e. for $h = \lambda n.2^{3n}$.*

By Ershov's hierarchy theorem, there exists a Δ_2^0-set A which is not $\lambda n.2^{3n}$-c.e. and hence x_{2A} is not a d-c.e. real. This implies immediately that the class **DCE** does not exhaust all reals with a Δ_2^0-binary expansions.

5   Divergence Bounded Computable Reals

The class **DCE** is an arithmetical closure of **CE** and is actually a real closed field. It is then natural to ask, if it is closed under total computable real functions. Before answering this question, we introduce another class of reals at first.

Let $h : \mathbb{N} \to \mathbb{N}$ be a function. A real x is called *h-bounded computable* (*h*-bc, for short) if there is a computable sequence (x_s) which converges to x *h-bounded effectively* in the sense that the number of non-overlapping index-pairs (i, j) with $|x_i - x_j| \geq 2^{-n}$ is bounded by $h(n)$ for all n. A real x is called *divergence bounded computable* (dbc for short) if it is *h*-bc for a computable function h ([18]).

Surprisingly, the class **DBC** is just the closure of c.e. reals (and hence of d-c.e. reals) under total computable real functions.

Theorem 5.1 (Rettinger et al [18])

1. $x \in \mathbf{DBC}$ *iff $x = f(y)$ for a c.e. real y and a total computable real function.*
2. *The class **DBC** is a real closed field.*

Notice that, if A is an ω-c.e. but not $\lambda n.2^{3n}$-c.e. set, then x_{2A} is divergence bounded computable but not d-c.e. by Theorem 4.5. This implies that **DBC** properly extends the class **DCE**. On the other hand, by a diagonalization we can show that there is a Δ_2^0-real which is not dbc. The next theorem shows that, even the Turing degrees of dbc reals do not exhaust all Δ_2^0-Turing degrees and this extends Theorem 4.4.

Theorem 5.2 (Zheng and Rettinger [30]). *There is a Δ_2^0-Turing degree which does not contain any divergence bounded computable reals.*

6   Computable Approximable Reals and More

We have seen that there exists real with a Δ_2^0-binary expansion which is not dbc. The reals of Δ_2^0-binary expansion are called **0**'-*computable* because they are the limits of **0**'-computable sequences of rational numbers which converges effectively (Ho [7]). In addition, **0**'-computable reals are simply the limits of computable sequence of rational numbers without any extra condition and hence they are called *computably approximable* (c.a., for short). The class of c.a. reals is denoted by **CA**. The class **CA** is the largest class we discussed so far. Actually, we have the following finite hierarchy:

$$\mathbf{EC} = \mathbf{CE} \cap \mathrm{co\text{-}CE} \subsetneqq \begin{matrix} \mathbf{CE} \\ \mathrm{co\text{-}CE} \end{matrix} \subsetneqq \mathbf{CE} \cup \mathrm{co\text{-}CE} = \mathbf{SC} \subsetneqq \mathbf{DCE} \subsetneqq \mathbf{DBC} \subsetneqq \mathbf{CA}.$$

The class **CA** shares a lot of properties of the computable reals. For instance, it is a real closed field and is closed under computable real functions, and so on.

Of course, the class **CA** can be extended further. To this end, we introduce the following general definition of arithmetical hierarchy of reals. Let $\Gamma_{\mathbb{Q}}$ denote the class of computable functions $f : \mathbb{N}^n \to \mathbb{Q}$ for some n. Θ_{i_n} denotes "$\sup_{i_n}$" if n is odd, and "$\inf_{i_n}$" if n is even, and $\overline{\Theta}_{i_n}$ denotes "$\inf_{i_n}$" if n is odd, and "$\sup_{i_n}$" if n is even.

Definition 6.1 (Zheng and Weihrauch [34])

1. $\Sigma_0 = \Pi_0 = \Delta_0 := \{x \in \mathbb{R} : x \text{ is computable}\}$;
2. *For* $n > 0$,

$$\Sigma_n := \{x \in \mathbb{R} : (\exists f \in \Gamma_{\mathbb{Q}}) x = \sup_{i_1} \inf_{i_2} \sup_{i_3} \cdots \Theta_{i_n} f(i_1, \cdots, i_n)\};$$
$$\Pi_n := \{x \in \mathbb{R} : (\exists f \in \Gamma_{\mathbb{Q}}) x = \inf_{i_1} \sup_{i_2} \inf_{i_3} \cdots \overline{\Theta}_{i_n} f(i_1, \cdots, i_n)\};$$

3. $\Delta_n := \Sigma_n \cap \Pi_n$.

If $x \in \Sigma_n (\Pi_n, \Delta_n)$, *then we also say that* x *is* $\Sigma_n (\Pi_n, \Delta_n)$-*computable.*

Obviously, we have $\Sigma_1 = \mathbf{CE}$, $\Pi_1 = \mathrm{co\text{-}CE}$. The equality $\Delta_2 = \mathbf{CA}$ follows from the following more general results.

Theorem 6.2 (Zheng and Weihrauch [34]). *For any* $n \geq 1$ *and* $x \in \mathbb{R}$

1. $x \in \Sigma_{n+1} \iff (\exists f \in \Gamma_{\mathbb{Q}}^{\emptyset^{(n)}}) (x = \sup_{i \in \mathbb{N}} f(i))$;
2. $x \in \Pi_{n+1} \iff (\exists f \in \Gamma_{\mathbb{Q}}^{\emptyset^{(n)}}) (x = \inf_{i \in \mathbb{N}} f(i))$;

3. $x \in \Delta_{n+1} \iff (\exists f \in \Gamma_{\mathbb{Q}}^{\emptyset^{(n)}})(x = \lim_{i \to \infty} f(i)$ effectively;

4. $x \in \Delta_{n+2} \iff (\exists f \in \Gamma_{\mathbb{Q}}^{\emptyset^{(n)}})(x = \lim_{i \in \mathbb{N}} f(i))$;

where $\Gamma_{\mathbb{Q}}^{A}$ denotes the class of A-computable functions $f : \mathbb{N} \to \mathbb{Q}$.

Theorem 6.3 (Zheng and Weihrauch [34])

1. For any set $A \subseteq \mathbb{N}$, $A \in \Delta_n^0$, if and only if $x_A \in \Delta_n$.
2. For $n \geq 1$, $\Delta_n \subsetneqq \Sigma_n$ and $\Delta_n \subsetneqq \Pi_n$.

7   Ershov's Hierarchy of Δ_2

Let's go back to the class **CA** again and try to introduce the hierarchies similar to Ershov's hierarchy [6] of Δ_2^0-subsets of natural numbers. According to the representations of reals we used, three such kinds of hierarchies can be defined.

Firstly, we consider the binary expansion. Let h be any function. A real x is called h-*binary computable* if $x = x_A$ for an h-c.e. set A ([32]). The k-binary computable for any constant k and ω-binary computable reals are defined accordingly. Let k-b **EC** (for $k \in \mathbb{N}$), ω-b **EC** and h-b **EC** denote the classes of all k-, ω- and h-binary computable reals, respectively. In addition, the class $\bigcup_{k \in \mathbb{N}} k$-b **EC** is denoted by $*$-b **EC**. By Ershov's hierarchy theorem, we have an infinite hierarchy k-b **EC** $\subsetneqq$ $(k+1)$-b **EC** $\subsetneqq$ $*$-b **EC** $\subsetneqq$ ω-b **EC** $\subsetneqq$ **CA** for all constant k. Obviously, 1-b **EC** is the class of strongly c.e. reals and hence 1-b **EC** $\subsetneqq$ **CE**. Furthermore, we have

Theorem 7.1 (Zheng and Rettinger [32])

1. k-b **EC** $\not\subseteq$ **SC** for $k \geq 2$;
2. **CE** $\not\subseteq$ $*$-b **EC** and $*$-b **EC** $\subsetneqq$ **DCE**;
3. ω-b **EC** is incomparable with **DCE**.

Secondly, for Dedekind cuts, we call a real x h-*Dedekind computable* if its Dedekind cut L_x is h-c.e. ([32]). The k- (for $k \in \mathbb{N}$), $*$- and ω-Dedekind computability are defined accordingly. Their classes are denote by h-d**EC**, k-d**EC**, $*$-d**EC** and ω-d**EC**, respectively. By definition, we have 1-d**EC** = **CE** and **SC** $\subseteq$ 2-d**EC**. For other levels, we have

Theorem 7.2 (Zheng and Rettinger [32])

1. k-d**EC** = **SC** for $k \geq 2$;
2. ω-b **EC** = ω-d**EC**.

To introduce an Ershov-style hierarchy of reals by means of Cauchy sequence representation, we consider the number of big jumps of a sequence ([32]). For any function $h : \mathbb{N} \to \mathbb{N}$, a real x is called h-*Cauchy computable* if there is a computable sequence (x_s) of rational numbers which converges to x h-*effectively* in the sense that the number of non-overlapping index pairs (i, j) with $i, j \geq n$ and $2^{-n} < |x_i - x_j| \leq 2^{-n+1}$ is bounded by $h(n)$ for all n. Other types of Cauchy computability can be defined accordingly and the classes of h-, k- ($k \in \mathbb{N}$), $*$- and ω-Cauchy computable reals are denoted, respectively, by h-cEC, k-cEC, $*$-cEC and ω-cEC. By definition, we have 0-cEC = EC and ω-cEC = DBC.

For Cauchy computability we have the following hierarchy theorem.

Theorem 7.3 (Zheng and Rettinger [32]). *If f, g are total computable functions such that $f(n) < g(n)$ for infinitely many n, then g-cEC $\not\subseteq$ f-cEC.*

Thus, we have an Ershov-type hierarchy that k-cEC $\subsetneq$ $(k+1)$-cEC $\subsetneq$ $*$-cEC $\subsetneq$ ω-cEC for any $k \in \mathbb{N}$. Other properties of Cauchy computability are summarized in the following theorem.

Theorem 7.4 (Zheng and Rettinger [32])

1. *The class k-cEC is incomparable with the classes* **CE** *and* **SC** *for any $k > 0$.*
2. *There are $x, y \in$ 1-cEC such that $x - y \notin$ $*$-cEC. Therefore, k-cEC and $*$-cEC are not closed under addition and subtraction for any $k > 0$.*
3. *ω-b EC $\subsetneq$ ω-cEC and k-b EC $\subsetneq$ k-cEC for any $k \geq 1$ or $k = *$.*

8   Hierarchy of DBC Reals

In section 5 we have introduced h-bounded computable reals. More generally, for any class C of functions, we call a real C-*bounded computable* (C-bc) if it is h-bc for an $h \in C$. The classes of h-bc and C-bc reals are denoted by h-**BC** and C-**BC**, respectively. Obviously, if C is the class of all computable functions, then C-**BC** = **DBC**.

On the other hand, if $\liminf h(n) < \infty$, then only rational numbers can be h-bc. Therefore, we cannot anticipate an Ershov-type hierarchy in this case. Moreover, if there is a constant c such that $|f(n) - g(n)| \leq c$ for all n, then f-**BC** = g-**BC**. This means that a hierarchy theorem like Theorem 7.3 does not hold neither. Nevertheless, we have another version of hierarchy theorem.

Theorem 8.1 (Zheng [28]). *If $f, g : \mathbb{N} \to \mathbb{N}$ are computable and satisfy the condition $(\forall c \in \mathbb{N})(\exists m \in \mathbb{N})(c + f(m) < g(m))$, then g-**BC** $\not\subseteq$ f-**BC**.*

The next theorem shows that a lot of classes C-**BC** are fields.

Theorem 8.2 (Zheng [28]). *Let C be a class of functions $f : \mathbb{N} \to \mathbb{N}$. If, for any $f, g \in C$ and $c \in \mathbb{N}$, there is an $h \in C$ such that $f(n + c) + g(n + c) \leq h(n)$ for all n, then C-**BC** is a field.*

The hierarchy of the class **DBC** can be extended to the hierarchy of Turing degrees of dbc-reals as follows.

Theorem 8.3 (Rettinger and Zheng [16]). *Let $f, g : \mathbb{N} \to \mathbb{N}$ be monotonically increasing computable functions such that $g(n + 1) \geq g(n) + 2$ and*

$$(\forall c \in \mathbb{N})(\forall^{\infty} n)(f(\gamma n) + n + c < g(n)) \tag{1}$$

for some constant $\gamma > 1$. Then there exists a g-bc real x which is not Turing equivalent to any f-bc real.

9   Monotone Computability Hierarchy

As a generalization of 1-monotonically convergence mentioned in section 3, we call a sequence (x_s) converges to x *h-monotonically* if $(\forall n, m \in \mathbb{N})(n < m \Longrightarrow h(n)|x - x_n| \geq |x - x_m|)$ for a function $h : \mathbb{N} \to \mathbb{R}$. A real x is called *h-monotonically computable* (*h-mc*) if there is a computable sequence of rational numbers which converges to x *h*-monotonically ([15]). If $h = \lambda n.c$ for a constant $c \in \mathbb{R}$, then any *h*-mc reals are called *c-mc*. We call a real *monotonically computable* (*mc*) if it is *c*-mc for some constant c and *ω-monotonically computable* (*ω-mc*) if it is *h*-mc for a computable function $h : \mathbb{N} \to \mathbb{N}$. The classes of *h*-mc, *c*-mc, mc and *ω*-mc reals are denoted by $h\text{-}\mathbf{MC}, c\text{-}\mathbf{MC}, \mathbf{MC}$ and $\omega\text{-}\mathbf{MC}$, respectively.

By definition, we have obviously $0\text{-}\mathbf{MC} = \mathbb{Q}$, $1\text{-}\mathbf{MC} = \mathbf{SC}$ and $c\text{-}\mathbf{MC} = \mathbf{EC}$ for $0 < c < 1$. For $c \geq 1$ we have the following dense hierarchy theorem.

Theorem 9.1 (Rettinger and Zheng [15]). *For any real constants $c_2 > c_1 \geq 1$, $c_1\text{-}\mathbf{MC} \subsetneqq c_2\text{-}\mathbf{MC}$.*

This implies immediately that $\mathbf{SC} \subsetneqq \mathbf{MC}$. Furthermore, it is shown in [15] that, $\mathbf{MC}$ is a proper subset of $\mathbf{DCE}$. That is, we have $\mathbf{SC} \subsetneqq \mathbf{MC} \subsetneqq \mathbf{DCE}$.

Theorem 9.2 (Zheng, Rettinger and Barmpalias [33]).

1. *If $h : \mathbb{N} \to (0, 1]$ is a computable function, then*
   - *If $\sum_{n \in \mathbb{N}}(1 - h(n)) = \infty$, then $h\text{-}\mathbf{MC} = \mathbf{EC}$;*
   - *If $\sum_{n \in \mathbb{N}}(1 - h(n))$ is computable, then $h\text{-}\mathbf{MC} = \mathbf{SC}$;*
   - *If $\sum_{n \in \mathbb{N}}(1 - h(n))$ is non-computable, then $\mathbf{EC} \subsetneqq h\text{-}\mathbf{MC} \subsetneqq \mathbf{SC}$.*
2. *If h is a monotone and unbounded computable function, then $h\text{-}\mathbf{MC} = \omega\text{-}\mathbf{MC}$. Furthermore, the class $\omega\text{-}\mathbf{MC}$ is incomparable with $\mathbf{DCE}$ and $\mathbf{DBC}$*

References

1. K. Ambos-Spies, K. Weihrauch, and X. Zheng. Weakly computable real numbers. *Journal of Complexity*, 16(4):676–690, 2000.
2. C. Calude, P. Hertling, B. Khoussainov, and Y. Wang. Recursively enumerable reals and Chaitin Ω numbers. *Theoretical Computer Science*, 255:125–149, 2001.
3. R. Downey, G. Wu, and X. Zheng. Degrees of d.c.e. reals. *Mathematical Logic Quarterly*, 50(4/5):345–350, 2004.
4. R. G. Downey. Some computability-theoretic aspects of reals and randomness. In *The Notre Dame lectures*, volume 18 of *Lect. Notes Log.*, pages 97–147. Assoc. Symbol. Logic, Urbana, IL, 2005.
5. A. Dunlop and M. Pour-El. The degree of unsolvability of a real number. *CCA 2000, Swansea, UK, September 2000, LNCS 2064*, pages 16–29.
6. Y. L. Ershov. A certain hierarchy of sets. i, ii, iii. (Russian). *Algebra i Logika*, 7(1):47–73, 1968;7(4):15–47, 1968; 9:34–51, 1970.
7. C.-K. Ho. Relatively recursive reals and real functions. *Theoretical Computer Science*, 210:99–120, 1999.

8. K.-I. Ko. On the definitions of some complexity classes of real numbers. *Math. Systems Theory*, 16:95–109, 1983.

9. K.-I. Ko. *Complexity Theory of Real Functions*. Progress in Theoretical Computer Science. Birkhäuser, Boston, MA, 1991.

10. R. Lehman. On primitive recursive real numbers. *Fundamenta Mathematicae*, 49:105–118, 1960/61.

11. A. Mostowski. On computable sequences. *Fundamenta Mathematicae*, 44:37–51, 1957.

12. J. Myhill. Criteria of constructibility for real numbers. *The Journal of Symbolic Logic*, 18(1):7–10, 1953.

13. R. Péter. *Rekursive Funktionen*. Akademischer Verlag, Budapest, 1951.

14. A. Raichev. D.c.e. reals, relative randomness, and real closed fields. In *CCA 2004, August 16-20, 2004, Lutherstadt Wittenberg, Germany*, 2004.

15. R. Rettinger and X. Zheng. On the hierarchy and extension of monotonically computable real numbers. *J. Complexity*, 19(5):672–691, 2003.

16. R. Rettinger and X. Zheng. A hierarchy of on the Turing degrees for divergence bounded computable reals. In *CCA 2005, August 25-29, Kyoto, Japan*, 2005.

17. R. Rettinger and X. Zheng. Solovay reducibility on d-c.e. real numbers. In *CO-COON 2005, August 16-19, Kunming, China*, LNCS, pages 359–368.

18. R. Rettinger, X. Zheng, R. Gengler, and B. von Braunmühl. Weakly computable real numbers and total computable real functions. *COCOON 2001, Guilin, China, August 20-23,*, LNCS 2108, pages 586–595.

19. H. G. Rice. Recursive real numbers. *Proc. Amer. Math. Soc.*, 5:784–791, 1954.

20. R. M. Robinson. Review of "Peter, R., Rekursive Funktionen". *The Journal of Symbolic Logic*, 16:280–282, 1951.

21. R. I. Soare. Cohesive sets and recursively enumerable Dedekind cuts. *Pacific J. Math.*, 31:215–231, 1969.

22. R. M. Solovay. Draft of a paper (or a series of papers) on chaitin's work .... manuscript, IBM Thomas J. Watson Research Center, Yorktown Heights, NY, p. 215, 1975.

23. E. Specker. Nicht konstruktiv beweisbare Sätze der Analysis. *The Journal of Symbolic Logic*, 14(3):145–158, 1949.

24. A. M. Turing. On computable numbers, with an application to the "Entscheidungsproblem". *Proc. of the London Mathematical Society*, 42(2):230–265, 1936.

25. K. Weihrauch and X. Zheng. A finite hierarchy of the recursively enumerable real numbers. In *Proceedings of MFCS'98, Brno, Czech Republic, August, 1998*, volume 1450 of *LNCS*, pages 798–806. Springer, 1998.

26. G. Wu. Regular reals. In V. Brattka, M. Schröder, K. Weihrauch, and N. Zhong, editors, *CCA 2003, Cincinnati, USA*, volume 302 - 8/2003 of *Informatik Berichte, FernUniversität Hagen*, pages 363 – 374, 2003.

27. X. Zheng. Recursive approximability of real numbers. *Mathematical Logic Quarterly*, 48(Suppl. 1):131–156, 2002.

28. X. Zheng. On the divergence bounded computable real numbers. *COOCON 2003, July 25-28, 2003, Big Sky, MT, USA*, volume 2697 of *LNCS*, pages 102–111.

29. X. Zheng. On the Turing degrees of weakly computable real numbers. *Journal of Logic and Computation*, 13(2):159–172, 2003.

30. X. Zheng and R. Rettinger. A note on the Turing degree of divergence bounded computable real numbers. In *CCA 2004, August 16-20, Lutherstadt Wittenberg, Germany*, 2004.

31. X. Zheng and R. Rettinger. On the extensions of solovay reducibility. In *COOCON 2004, August 17-20, Jeju Island, Korea*, volume 3106 of *LNCS*. Springer-Verlage, 2004.

32. X. Zheng and R. Rettinger. Weak computability and representation of reals. *Mathematical Logic Quarterly*, 50(4/5):431–442, 2004.

33. X. Zheng, R. Rettingre, and G. Barmpalias. *h*-monotonically computable real numbers. *Mathematical Logic Quarterly*, 51(2):1–14, 2005.

34. X. Zheng and K. Weihrauch. The arithmetical hierarchy of real numbers. *Mathematical Logic Quarterly*, 47(1):51–65, 2001.

Primitive Recursive Selection Functions over Abstract Algebras

J.I. Zucker[*]

Department of Computing and Software,
McMaster University, Hamilton, Ontario L8S 4K1, Canada
zucker@mcmaster.ca

Abstract. We generalise to abstract many-sorted algebras the classical proof-theoretic result due to Parsons and Mints that an assertion $\forall \mathbf{x} \, \exists \mathbf{y} \, P(\mathbf{x}, \mathbf{y})$ (where P is Σ_1^0), provable in Peano arithmetic with Σ_1^0 induction, has a primitive recursive selection function. This involves a corresponding generalisation to such algebras of the notion of *primitive recursiveness*. The main difficulty encountered in carrying out this generalisation turns out to be the fact that equality over these algebras may not be computable, and hence atomic formulae in their signatures may not be decidable. The solution given here is to develop an appropriate concept of *realisability* of existential assertions over such algebras, and to work in an intuitionistic proof system. This investigation gives some insight into the relationship between verifiable specifications and computability on topological data types such as the reals, where the atomic formulae, *i.e.*, equations between terms of type real, are not computable.

1 Introduction

We investigate a class of problems concerning the relationship between specifiability and computability for a wide class of abstract data types, modelled as many-sorted algebras A, of the following form. Given a predicate P of a certain syntactic class in the specification language $\mathcal{L}(A)$ for A, and a proof of the assertion

$$\forall \mathbf{x} \, \exists \mathbf{y} \, P(\mathbf{x}, \mathbf{y}) \tag{1.1}$$

in a suitable formal system $\mathcal{F}$ for A, can we construct, from this proof, a computable selection function for P, *i.e.*, a computable function f on A such that

$$\forall \mathbf{x} \, P(\mathbf{x}, \mathsf{f}(\mathbf{x})) \tag{1.2}$$

holds in A? (Here the notion of "computable on A" must also be explicated.)

Specifically, we generalise to such algebras a classical proof-theoretic result, due to Parsons [9,10] and Mints [8], which gives a positive solution to the above problem in the case that $\mathcal{F}$ is Peano arithmetic (PA) with induction restricted to

---

[*] Research supported by a grant from Science and Engineering Research Canada. Thanks to J.V. Tucker and three anonymous referees for helpful comments.

A. Beckmann et al. (Eds.): CiE 2006, LNCS 3988, pp. 595–606, 2006.
© Springer-Verlag Berlin Heidelberg 2006

Σ_1^0 formulae, and P is a Σ_1^0 predicate of PA, in which case a primitive recursive selection function f can then be found. As a corollary, a general recursive function which is provably total in PA with Σ_1^0-induction is (extensionally equivalent to) a primitive recursive function.

In [14] this result was generalised to predicates over many-sorted signatures Σ containing the boolean and natural sorts, with their standard operations, and many-sorted Σ-algebras A. The method used was adapted from Mints's method, involving cut-reduction and an analysis of cut-reduced derivations. The result used a generalisation of primitive recursive schemes to many-sorted signatures and algebras. The generalisation went quite smoothly, on the assumption that *equality in A was computable*, so that the atomic formulae of the first-order language over Σ were computably decidable in A.

The case that equality in A is *not computable* provides a serious difficulty for this generalisation. In such a case, a more delicate analysis of formal derivations of assertions of the form (1.1) is required. In this paper such an analysis is given, using an appropriate concept of *realisability* of existential assertions over Σ.

To clarify these issues with an example, consider the topological total algebra of reals

$$\mathcal{R} = (\mathbb{R}, \mathbb{N}, \mathbb{B}; \ 0, 1, +, -, \times, \ \dots \ ) \tag{1.3}$$

("topological" in the sense that all the carriers have topologies in terms of which the basic operations are continuous; "total" in the sense that the basic operations are total [19]). $\mathcal{R}$ containing the carrier $\mathbb{R}$ of reals with its usual topology and its ring operations, as well as the carriers $\mathbb{N}$ and $\mathbb{B}$ of naturals and booleans, with their discrete topologies and standard operations. Note that there is no division operation on $\mathbb{R}$, since there is no continuous total extension of that operation. Similarly, although there is an equality test (*i.e.*, a boolean valued equality operation) on $\mathbb{N}$, there is none on $\mathbb{R}$, since the (total) equality operation on $\mathbb{R}$ is not continuous.[1]

However the *specification language* $\mathcal{L}(A)$, in which the predicates P (1.1) are expressed, has, as atomic formulae, equations between terms of the same sort, for *all sorts* of A, including, *e.g.*, the sort of reals in the above example. It follows that *the atomic formulae in $\mathcal{L}(\mathcal{R})$ are not computable*. In such a case, solving the problem of finding computable selection functions on $\mathcal{R}$ requires a non-trivial concept of *realisability*, as we will see. It will also require restricting our attention to intuitionistic proof systems.

This investigation gives some insight into the relationship between verifiable specifications and computability on topological data types such as the reals, where the atomic formulae, *i.e.*, equations between terms of type real, are not computable.

In particular, it provides an example, in the context of verifiable specifications on such data types, of the general programme proposed by Kreisel [6] of discovering "what more we know when we have proved a theorem than if we only know that it is true".

---

[1] One can define continuous *partial* division and equality operations on the reals [18]; however in this paper we only consider total algebras. We return to this in §6.4.

2   Many-Sorted Algebras; Computation Schemes

We give a brief introduction to many-sorted signatures and algebras. Details may be found in any of [15,16,18,19]. Given a signature Σ with finitely many *sorts s,...* and *function symbols*

$$\mathsf{F} : u \to s, \tag{2.1}$$

where u is the product type $u = s_1 \times \cdots \times s_m$, a Σ-algebra A consists of a carrier A_s for each Σ-sort s, and a total function

$$\mathsf{F}^A : A^u \to A_s$$

for each Σ-function symbol as in (2.1), where $A^u = A_{s_1} \times \cdots \times A_{s_m}$. We let $s,\dots$ range over Σ-sorts, and $u, v, \dots$ over Σ product types.
   We make two assumptions on our signatures Σ and Σ-algebras A.

Assumption 2.1 (N-standardness). *The signatures Σ and Σ-algebras A are N-standard.* That is, they contain
   (a) the sort bool of *booleans* and the corresponding carrier $A_{\mathsf{bool}} = \mathbb{B} = \{\mathsf{true}, \mathsf{false}\}$, together with the standard boolean and boolean-valued operations, including equality at certain sorts called *equality sorts*; and also
   (b) the sort nat of *natural numbers* and the corresponding carrier $A_{\mathsf{nat}} = \mathbb{N} = \{0, 1, 2, \dots\}$, together with the standard arithmetical operations of zero, successor, equality and order on $\mathbb{N}$.

Assumption 2.2 (Instantiation). *For every sort s of Σ, there is a closed term of sort s, called the default term δ^s of that sort.*

The instantiation assumption is used in the proof of Lemma 5.6.
   We will also consider *array signatures* Σ^* and *array algebras* A^*, which are formed from N-standard signatures Σ and algebras A by adding, for each sort s, an *array sort* s^*, with corresponding carrier A_s^* consisting of all arrays or finite sequences over A_s, together with certain standard array operations.
   Let $NStdAlg(\Sigma)$ denote the class of N-standard algebras over Σ. We will present two systems of computation schemes over Σ: PR and μPR.

2.1   PR(Σ) and PR*(Σ) Computation Schemes

Given an N-standard signature Σ, we define PR *schemes* over Σ which generalise those given in [5] for primitive recursive functions on $\mathbb{N}$.
   (a) *Basic schemes: Initial functions*
      (i) *Primitive Σ-functions:*
$$f(x) = F(x)$$

of type $u \to s$, for all the primitive Σ-function symbols $\mathsf{F} : u \to s$, where $x : u$, *i.e.*, x is a tuple of variables of product type u.

(ii) *Projection:*

$$f(x) = x_i$$

of type $u \to s_i$, where $x = (x_1, \ldots, x_m)$ is of type $u = s_1 \times \cdots \times s_m$.

(b) **Inductive schemes:**

(iii) *Composition:*

$$f(x) = h(g_1(x), \ldots, g_m(x))$$

of type $u \to s$, where $g_i : u \to s_i$ $(i = 1, \ldots, m)$ and $h : s_1 \times \cdots \times s_m \to s$.

(iv) *Definition by cases:*

$$f(b, x, y) = \begin{cases} x & \text{if } b = \text{true} \\ y & \text{if } b = \text{false} \end{cases}$$

of type $\text{bool} \times s^2 \to s$.

(v) *Simultaneous primitive recursion on* $\mathbb{N}$: This defines, on each $A \in \mathbf{NStdAlg}(\Sigma)$, for fixed $m > 0$ (the degree of simultaneity), $n \geq 0$ (the number of parameters), and product types u and $v = s_1 \times \cdots \times s_m$, an m-tuple of functions $f = (f_1, \ldots, f_m)$ with $f_i : \text{nat} \times u \to s_i$, such that for all $x \in A^u$ and $i = 1, \ldots, m$,

$$f_i(0, x) = g_i(x)$$
$$f_i(z + 1, x) = h_i(z, x, f_1(z, x), \ldots, f_m(z, x))$$

where $g_i : u \to s_i$ and $h_i : \text{nat} \times u \times v \to s_1$ $(i = 1, \ldots, m)$.

Scheme (v) uses the N-standardness of the algebras, *i.e.* the carrier $\mathbb{N}$.

A $\mathrm{PR}(\Sigma)$ scheme $\alpha : u \to s$ defines, or rather computes, a function $f_\alpha^A : A^u \to A_s$ on each Σ-algebra A.

Lemma 2.3 (Equational specification for PR functions). *For any* $\mathrm{PR}(\Sigma)$ *scheme* α, *we can construct an equational specification* E_α *for the functions* f_α^A *defined by* α *on all* Σ-*algebras* A.

It turns out that a broader class of functions provides a better generalisation of the notion of primitive recursiveness, namely PR^* *computability*. A function on A is $\mathrm{PR}^*(\Sigma)$ computable if it is defined by a PR scheme over Σ^*, interpreted on A^*, *i.e.*, possibly using starred sorts for the auxiliary functions used in its definition. Note that in the classical setting ($A = \mathcal{N} =$ the naturals with their standard operations) this generalisation is not necessary, since $\mathcal{N}^*$ can effectively be coded in $\mathcal{N}$. In general, however, this is not the case; $\mathcal{R}^*$, for example, cannot be effectively coded in $\mathcal{R}$.

We write $\mathrm{PR}(A)$ for the class of functions PR computable on A, etc.

2.2 $\mu\mathrm{PR}(\Sigma)$ and $\mu\mathrm{PR}^*(\Sigma)$ Computation Schemes

The $\mu\mathrm{PR}$ *schemes* over Σ are formed by adding to the PR schemes:

(vi) *Least number* or μ *operator:*

$$f(x) \simeq \mu z[g(x, z) = \text{true}]$$

of type $u \to \text{nat}$, where $g : u \times \text{nat} \to \text{bool}$ is μPR. The interpretation of this is that $f^A(x) \downarrow z$ if $g^A(x, y) \downarrow$ false for each $y < z$ and $g^A(x, z) \downarrow$ true, and $f^A(x)$ is undefined if there is no such z.

This scheme, like *(v)*, uses the N-standardness of the algebra.

These schemes generalise those given in [5] for partial recursive functions on N. Note that μPR computable functions are, in general, *partial*.

Again, a broader class turns out to be more useful, namely μPR* *computability*. This is just PR* computability with μ.

There are many other models of computability, due to Moschovakis, Friedman, Shepherdson and others, which turn out to be equivalent to μPR* computability [16, §7] All these equivalences have led to the postulation of a *generalised Church-Turing Thesis for deterministic computation of functions*, which can be formulated as follows:

Computability of functions on many-sorted algebras by deterministic algorithms can be formalised by μPR computability.*

2.3   Comparison with Imperative Computational Models

In [16] computation on many-sorted Σ algebras was investigated, using imperative programming models: ***While***(Σ), based on the 'while' loop construct over Σ; ***For***(Σ), based similarly on the 'for' loop; and ***While****$^*(\Sigma)$ and ***For****$^*(\Sigma)$, which use arrays, *i.e.*, auxiliary variables of starred sort over Σ.

Writing ***While***(A) for the class of functions ***While***-computable on A, etc., we can list the equivalences between the "schematic" and "imperative" models:

(1) PR(A) = ***For***(A)
(2) PR*(A) = ***For****(A)
(3) μPR(A) = ***While***(A)
(4) μPR*(A) = ***While****(A),

in all cases, uniformly for $A \in NStdAlg(\Sigma)$.

These results are all stated in [16], and can be proved by the methods of [13].

3 Σ_1^* Formulae; The System Σ_1^*-Ind

Let $Lang(\Sigma)$ be the first order language over Σ, and $Lang^*(\Sigma) = Lang(\Sigma^*)$, the first order language over Σ^*. The *atomic formulae* of $Lang(\Sigma)$ are equations between terms of the same sort, for all Σ-sorts (not just equality sorts).

We use the *intuitionistic sequent calculus*, with sequents $\Gamma \longmapsto P$, where the antecedent Γ is a finite sequence of formulae, and the consequent is a single formula P.

We are interested in a certain sublanguage of $Lang(\Sigma^*)$, namely the class of Σ_1^* formulae over Σ, which we now define.

Definition 3.1 (BU quantifiers, equations and sequents).
(a) A *BU (bounded universal) quantifier* is a quantifier of the form '$\forall z < t$', where $z : \mathsf{nat}$ and $t : \mathsf{nat}$. (The most elegant approach is to take this as a primitive construct, with its own introduction rules.)
(b) A *BU equation* is formed by prefixing an equation by a string of 0 or more BU quantifiers.
(c) A *conditional BU equation* is a formula of the form

$$Q_1 \wedge \ldots \wedge Q_n \to Q \tag{3.1}$$

where $n \geq 0$ and Q_i and Q are BU equations. A *conditional BU equational theory* is a set of such formulae (or their universal closures).
(d) A *BU equational sequent* is a sequent of the form

$$Q_1, \ldots, Q_n \longmapsto Q$$

where $n \geq 0$ and Q_i and Q are BU equations. This sequent *corresponds to* the conditional BU equation (3.1).

Definition 3.2 (Elementary and Σ_1^* formulae). A formula of $\mathbf{Lang}(\Sigma^*)$ is
(a) *elementary* if it is formed from Σ^*-equations by applying *conjunctions*, *disjunctions*, and *BU (bounded universal) quantification* (in any order);
(b) Σ_1^* if it is formed from Σ^*-equations by applying *conjunctions*, *disjunctions*, *BU quantification* and also *existential Σ^*-quantification*, i.e., unbounded existential quantification over any sort in Σ^* (in any order)[2].

Lemma 3.3. *If P is an elementary formula all of whose variables are of equality sort, then the predicate defined by P is PR* computable.*

Let T be a set of formulae in $\mathbf{Lang}^*$, which we can think of as *axioms* for a class of Σ^*-algebras. We make the following assumption about T.

Assumption 3.4. *T consists of conditional BU Σ^*-equations.*

Note that this is a stricter condition than conditional Σ_1^* formulae, since it excludes disjunctions and existential quantification. However, this assumption is not unduly restrictive, as it includes axiomatisations by *conditional equations*, and (hence) *Horn formulae*, which are central to the theory of logic programming and abstract data types [7].
    We will define an intuitionistic sequent calculus $\Sigma_1^*\text{-}\mathsf{Ind}(\Sigma^*, T)$ with the axioms T as extra initial sequents.

Definition 3.5 (The intuitionistic sequent calculus $\Sigma_1^*\text{-}\mathsf{Ind}(\Sigma^*, T)$). This system has the following axioms (initial sequents) and inference rules: rules for the first order predicate calculus with equality over the signature Σ^*, including

---

[2] The notation may be a bit confusing: Σ^* refers to a signature with array sorts, whereas Σ_1^* refers to a particular syntactic class of formulae over Σ^*.

rules for the BU quantifier, and *cut*, as in [3] or [11]; the Σ^*-*equality axioms*; the Peano axioms for the sort nat of naturals, including the primitive recursive equational definitions for the nat operations $+$, $\times$, $=$, $<$, and the sequents

$$\text{true} = \text{false} \;\longmapsto\; t_1 = t_2 \qquad \text{and} \qquad \text{S n} = 0 \;\longmapsto\; t_1 = t_2$$

for a variable n : nat and arbitrary terms t_1, t_2 of the same sort; the equation

$$(\mathsf{x}^{\mathsf{bool}} = \text{true}) \;\vee\; (\mathsf{x}^{\mathsf{bool}} = \text{false}); \tag{3.2}$$

a certain set of conditional BU axioms for arrays [3]; the axioms T in sequent form; and the Σ_1^* *induction rule* (where the induction formula $F(\mathsf{a})$ is Σ_1^*):

$$\frac{\Gamma \;\longmapsto\; F(0) \qquad F(\mathsf{a}), \Gamma \;\longmapsto\; F(\mathsf{Sa})}{\Gamma \;\longmapsto\; F(t)}.$$

It follows from Assumption 3.4 that the initial sequents of the calculus Σ_1^*-Ind(Σ^*, T) are all Σ_1^*. In fact, they are all *BU equational* (except for (3.2), which is a disjunction of equations). This is used in the proof of Theorem 2.

Now let $\mathbb{K} \subseteq NStdAlg(\Sigma)$, and let T be a set of formulae in $Lang^*(\Sigma)$ such that $\mathbb{K} \models T$. (We could suppose that T is a complete N-standard axiomatisation of $\mathbb{K}$, *i.e.*, $\mathbb{K}$ is the class of *all* N-standard Σ-models of T, although this is unnecessary for what follows.) The following soundness result then clearly holds:

Lemma 3.6 (Soundness of Σ_1^*-Ind). Σ_1^*-Ind$(\Sigma^*, T) \vdash P \;\Longrightarrow\; \mathbb{K}^* \models P$.

4 Provable Totality of Schemes; PR* Selection Functions

With each μPR$^*(\Sigma)$ scheme $\alpha : u \to s$, we can effectively associate a Σ_1^* formula $P_\alpha(\mathsf{x}, \mathsf{y})$, the *computation predicate for* α, where $\mathsf{x} : u$ and $\mathsf{y} : s$, which represents the graph of the function defined by α, *i.e.*, for all $A \in NStdAlg(\Sigma)$, and for all $a \in A^u$ and $b \in A_s$,

$$A \models P_\alpha[a, b] \quad\Longleftrightarrow\quad \alpha^A(a) \downarrow b.$$

The construction of P_α is defined by structural induction on α. We omit details.

Note that even if the scheme α is defined over Σ only, the definition of P_α generally involves existential quantification over *starred sorts*.

Definition 4.1. A scheme α is *provably total in* Σ_1^*-Ind(Σ^*, T) iff

$$\Sigma_1^*\text{-Ind}(\Sigma^*, T) \vdash \forall \mathsf{x}\, \exists \mathsf{y}\, P_\alpha(\mathsf{x}, \mathsf{y}).$$

Lemma 4.2. *If α is a* PR* *scheme, then α is provably total in* Σ_1^*-Ind.

The central result of this paper is formulated with reference to a class $\mathbb{K}$ of N-standard Σ-algebras and an axiomatisation T of $\mathbb{K}$.

---

[3] listed in [14, §4] and [17, §3.2].

Theorem 1. *Suppose* $\mathbb{K} \models T$ *where* $\mathbb{K} \subseteq \mathbf{NStdAlg}(\Sigma)$, *and* T *consists of conditional BU* Σ^**-equations. If*

$$\Sigma_1^*\text{-}\mathsf{Ind}(\Sigma^*, T) \;\vdash\; \exists \mathbf{y} P(\mathbf{x}, \mathbf{y})$$

where $P(\mathbf{x}, \mathbf{y})$ *is a* Σ_1^* *formula, with free variables* $\mathbf{x} : u$ *and* $\mathbf{y} : s$, *then there is a* PR* *scheme* $\beta : u \to s$ *such that*

$$\text{for all } A \in \mathbb{K} \text{ and all } x \in A^u, \quad A \;\models\; P[x, \beta^A(x)]. \tag{4.1}$$

The function β^A is called a *selection function, Skolem function, realising function* or *witnessing function* for $\mathbf{y}$ in P.

As a corollary, we have a kind of converse to Lemma 4.2.

Corollary 4.3. *Suppose* $\mathbb{K} \models T$, *where* $\mathbb{K} \subseteq \mathbf{NStdAlg}(\Sigma)$ *and* T *consists of conditional BU* Σ^**-equations. If a* μPR* *scheme* α *is provably total in* $\Sigma_1^*\text{-}\mathsf{Ind}(\Sigma, T)$, *then* α *is extensionally* PR* *on* $\mathbb{K}$, *i.e., there is a* PR* *scheme* β *such that* $\alpha^A = \beta^A$ *for all* $A \in \mathbb{K}$.

A stronger version of Theorem 1 replaces (4.1) by a *provability* condition:

Theorem 2. *Suppose* T *consists of conditional BU* Σ^**-equations. If*

$$\Sigma_1^*\text{-}\mathsf{Ind}(\Sigma^*, T) \;\vdash\; \exists \mathbf{y} P(\mathbf{x}, \mathbf{y})$$

where $P(\mathbf{x}, \mathbf{y})$ *is a* Σ_1^* *formula, with free variables* $\mathbf{x} : u$ *and* $\mathbf{y} : s$, *then there is a* PR* *scheme* $\beta : u \to s$ *such that*

$$\Sigma_1^*\text{-}\mathsf{Ind}(\Sigma_\beta^*, T + E_\beta) \;\vdash\; P(\mathbf{x}, \mathsf{f}_\beta(\mathbf{x}))$$

where Σ_β^* *is the extension of* Σ^* *with symbols for the function* $\mathsf{f}_\beta : u \to s$ *defined by the scheme* β, *together with the auxiliary functions used in its definition, and* E_β *is the equational specification for these functions given by Lemma 2.3.*

Theorem 1 is an immediate consequence of Theorem 2, the proof of which uses the technique of realisability, to which we now turn.

5   Realisability

In preparation for the proof of Theorem 2, we define a *realisability relation* between tuples from A and Σ_1^* formulae. First we define the *type* of a Σ_1^* formula.

Definition 5.1 (Type of a Σ_1^* formula). The *type* $\boldsymbol{tp}(P)$ of a Σ_1^* formula P is a particular Σ^*-product type. It is defined by structural induction on P.

   (*i*) $\boldsymbol{tp}(t_1 = t_2) \;=\; \mathsf{bool}$
   (*ii*) $\boldsymbol{tp}(P_1 \wedge P_2) \;=\; \boldsymbol{tp}(P_1) \times \boldsymbol{tp}(P_2)$
   (*iii*) $\boldsymbol{tp}(P_1 \vee P_2) \;=\; \mathsf{bool} \times \boldsymbol{tp}(P_1) \times \boldsymbol{tp}(P_2)$

(iv) $tp(\forall k < t\, P) = tp(P)^*$

where, for any Σ^*-product type u, u^* is the corresponding component-wise starred type; thus, if (say) $u = s_1 \times s_2 \times s_3^* \times s_4^* \times s_5$ then $u^* = s_1^* \times s_2^* \times s_3^{**} \times s_4^{**} \times s_5^*$.

(v) $tp(\exists y^s P) = s \times tp(P)$ for any Σ^*-sort s.

Remark 5.2. (a) The base case, $tp(t_1 = t_2)$, could be defined as any Σ-sort. (b) The doubly starred sorts s^{**} appearing in clause (iv) are not actually present in the signature Σ^*; the doubly indexed (2-dimensional) arrays which they represent are effectively coded by 1-dimensional arrays in the well-known way.

The central concept of this section is a *realisability* relation between term tuples of a particular Σ^*-product type, and Σ_1^* formulae of the same type.

Definition 5.3 (Realisability of Σ_1^* formulae). Let t be a Σ^*-term tuple, and P a Σ_1^* formula, both of the same product type. We define the expression '$t \rhd P$' ("t realises P") to be a Σ_1^* formula, by structural induction on P:

(i) $t \rhd (t_1 = t_2) \equiv t_1 = t_2$.

(ii) $(t_1, t_2) \rhd (P_1 \wedge P_2) \equiv (t_1 \rhd P_1) \wedge (t_2 \rhd P_2)$.

(iii) $(b, t_1, t_2) \rhd (P_1 \vee P_2) \equiv (b = \text{true} \wedge t_1 \rhd P_1) \vee (b = \text{false} \wedge t_2 \rhd P_2)$.

(iv) $t^* \rhd (\forall z < t_0\, P) \equiv \forall z < t_0(t^*[z] \rhd P)$.

(v) $(t_0, t) \rhd (\exists y P) \equiv t \rhd P\langle y/t_0 \rangle$.

Remark 5.4. (a) If P is a formula built up from *equations* using *conjunction* and *BU quantification* only, then $t \rhd P$ is identical to P (by a simple induction on P). In particular, the realisability of a *BU equation* P is the same as P. (b) However, in cases (iii) and (v), the realising tuple contains extra information: it includes a "witness" to the truth of the disjunction or existential quantification.

The above two remarks together imply that for a Σ_1^* formula P, realisability of P implies P. This is stated precisely in the following lemma.

Lemma 5.5. *For any Σ_1^* formula P and term tuple t of the same type, the sequent*

$$t \rhd P \longmapsto P$$

is provable in intuitionistic predicate logic.

In the opposite direction, we can say more: every provable Σ_1^* formula P has a realiser which is PR in the free variables of P. To prove this, we must actually prove a more general statement, *i.e.*, we must show how to construct a realiser for the succedent formula of a Σ_1^* sequent, which is PR not just in the free variables of the sequent, but also in realisers of the antecedent formulae.

The precise formulation is given as Lemma 5.6. This can be viewed as a strengthening of Theorem 2, which follows easily from it.

Lemma 5.6. *Suppose T consists of conditional BU Σ^*-equations, and the Σ_1^* sequent*

$$Q_1, \ldots, Q_m \longmapsto P \tag{5.1}$$

is provable in Σ_1^*-$\mathsf{Ind}(\Sigma^*, T)$. *Let* $Q_1, \ldots, Q_m, P$ *have types* $v_1, \ldots, v_m, v$ *(resp.)* *and* $\boldsymbol{var}(Q_1, \ldots, Q_m, P) \subseteq \mathbf{x} : u$. *Let* $\mathbf{z}_1, \ldots, \mathbf{z}_m$ *be tuples of variables, pairwise disjoint and disjoint from* $\mathbf{x}$, *with* $\mathbf{z}_i : v_i$ *for* $i = 1, \ldots, m$. *Then for some tuple of PR schemes* $\alpha : u \times v_1 \times \cdots \times v_m \rightarrow v$,

$$\mathbf{z}_1 \rhd Q_1, \ \ldots, \ \mathbf{z}_m \rhd Q_m \ \longmapsto \ \mathsf{f}_\alpha(\mathbf{x}, \mathbf{z}_1, \ldots, \mathbf{z}_m) \rhd P$$

is provable in Σ_1^*-$\mathsf{Ind}(\Sigma_\alpha^*, T + E_\alpha)$, *where* Σ_α^* *is the extension of* Σ^* *with symbols for the function tuple together with their auxiliary functions, and* E_α *is the equational specification for these functions.*

Theorem 2 follows immediately from Lemma 5.6 by taking $m = 0$ in (5.1).

Next, in order to prove Lemma 5.6, we need:

Lemma 5.7 (Cut Reduction Lemma). *Every derivation* $\mathcal{D}$ *in* Σ_1^*-Ind *can be transformed into a derivation* $\mathcal{D}'$ *of the same end-sequent containing only* Σ_1^* *cuts. Moreover, if the end-sequent is* Σ_1^* *then every formula in* $\mathcal{D}'$ *is* Σ_1^*.

The proof of Lemma 5.7 proceeds by a technique similar to that in the proof of Gentzen's *Hauptsatz* (see [3, III, §3] or [11, Ch. 1, §5]).

Lemma 5.6 is then proved by induction on the length of a "cut-reduced" derivation of the sequent.

6   Some Concluding Remarks

6.1   Primitive Recursive Realisability

Realisability, as a technique in proof theory, goes back to [4]. Since then many variants have been developed. For a thorough treatment of various versions of realisability applied to Heyting arithmetic and related systems, with extensive bibliography, see [12, Part III]. Many of these versions are based on (Gödel numbers of) *partial recursive* functions on $\mathbb{N}$. We may then ask (ignoring for now the difference between formalisms over $\mathbb{N}$ and over many-sorted algebras) how it is that only PR* and not μPR* realisability is needed in the present study?

The answer lies in our restricted language Σ_1^*, which has no '$\rightarrow$' or (unbounded) '$\forall$'. The only implication and universal quantification present are *global*, i.e., (implicit) implication from the antecedent to the succedent of a sequent, and (implicit) universal quantification over the free variables of a sequent. There is, however, *no iterated implication, or universal quantification in the antecedent of an implication.* This allows realisation with primitive recursive functionals only.

6.2   Comparison with the Results in [14]

In the cited paper, the analogue of Theorem 1 was proved by a proof-theoretic analysis involving constructing (only) PR* *witnessing functions* for existential statements, but not the (more general) *realisers* of Σ_1^* formulae. That proof works only for theories with *decidable* atomic formulae, *i.e.*, computable equality, since for such theories, elementary formulae are PR* computable (Lemma 3.3) and so the only non-trivial logical operation on Σ_1^* formulae is existential quantification. This allows for a much simpler proof of (the analogue of) Lemma 5.6, in which we only have to show, by induction on the length of derivations, the existence of such PR witnessing functions for existential theorems, without the whole apparatus of realisability (Definition 5.3). Regrettably, this assumption (*i.e.*, computability of atomic formulae) was not made explicit in [14].

6.3   Necessity for the Intuitionistic Sequent Calculus

In this paper we had to assume the provability of

$$\forall \mathbf{x} \, \exists y \, P(\mathbf{x}, y) \tag{6.1}$$

in the *intuitionistic* formal system $\Sigma_1^*\text{-Ind}(\Sigma^*, T)$. This provides another contrast with the development in [14], where the provability of (6.1) was assumed to be in the corresponding *classical* system.

To show the necessity for the intuitionistic calculus in the context of the present paper, consider the algebra $\mathcal{R}$ of reals (1.3) and the formula

$$P(\mathbf{x}, \mathbf{y}) \equiv_{df} (\mathbf{x} \neq 0 \wedge \mathbf{y} = 0) \vee (\mathbf{x} = 0 \wedge \mathbf{y} = 1)$$

where $\mathbf{x}$ and $\mathbf{y}$ are real variables. Then (6.1), with this predicate for P, is a truth of classical logic, and can be easily proved in the classical sequent calculus over $\mathcal{R}$. However the unique selection function for (6.1) is clearly not continuous on $\mathbb{R}$, and hence not PR* computable on $\mathcal{R}$, since for functions on topological algebras, PR* (or μPR*) computability implies continuity [18].

To summarise: *if we drop the assumption of computabile equality, we must strengthen the assumption of provability of (6.1) from classical to intuitionistic.*

Note that in the classical result of Mints and Parsons, provability of (6.1) can be assumed to be in the *classical system* PA with Σ_1^0 induction. The results of this paper therefore do not, strictly speaking, by themselves generalise these classical results. Rather, the present results, *together with* those of [14], provide a generalisation of these classical results, to settings where computability of equality may or may not hold.

6.4   Total vs Partial Algebras

In this paper we have considered only total algebras. This is a real restriction, since partial basic functions occur quite naturally in topological algebras; consider, for example, the algebra $\mathcal{R}$ of reals (1.3) augmented with continuous partial operators of division, equality and order [18]. To extend the current theory to such partial algebras would entail adapting the proof theory used here to a logic of partial terms or definedness (see, *e.g.*, [1, pp. 97–99] and [2]). This is likely to be a major undertaking, but one worth pursuing.

References

1. M. Beeson. *Foundations of Constructive Mathematics*. Springer-Verlag, 1985.
2. S. Feferman. Definedness. *Erkenntnis*, 43:295–320, 1995.
3. G. Gentzen. Investigations into logical deduction. In M.E. Szabo, editor, *The Collected Papers of Gerhard Gentzen*, pages 68–131. North Holland, 1969.
4. S.C. Kleene. On the interpretation of intuitionistic number theory. *Journal of Symbolic Logic*, 10:109–124, 1945.
5. S.C. Kleene. *Introduction to Metamathematics*. North Holland, 1952.
6. G. Kreisel. Some reasons for generalizing recursion theory. In R.O. Gandy and C.M.E. Yates, editors, *Logic Colloquium '69*, pages 139–198. North Holland, 1971.
7. K. Meinke and J.V. Tucker. Universal algebra. In S. Abramsky, D. Gabbay, and T. Maibaum, editors, *Handbook of Logic in Computer Science*, volume 1, pages 189–411. Oxford University Press, 1992.
8. G. Mints. Quantifier-free and one-quantifier systems. *Journal of Soviet Mathematics*, 1:71–84, 1973.
9. C. Parsons. On a number theoretic choice scheme II. *Journal of Symbolic Logic*, 36:587, 1971. (Abstract).
10. C. Parsons. On n-quantifier induction. *Journal of Symbolic Logic*, 37:466–482, 1972.
11. G. Takeuti. *Proof Theory (Second edition)*. North Holland, 1987.
12. A.S. Troelstra, editor. *Metamathematical Investigation of Intuitionistic Arithmetic and Analysis*, volume 344 of *Lecture Notes in Mathematics*. Springer-Verlag, 1973. *Second corrected edition*, Institute for Logic, Language and Computation, Technical Notes X-93-05, Amsterdam, 1993.
13. J.V. Tucker and J.I. Zucker. *Program Correctness over Abstract Data Types, with Error-State Semantics*, volume 6 of *CWI Monographs*. North Holland, 1988.
14. J.V. Tucker and J.I. Zucker. Provable computable selection functions on abstract structures. In P. Aczel, H. Simmons, and S.S. Wainer, editors, *Proof Theory*, pages 277–306. Cambridge University Press, 1993.
15. J.V. Tucker and J.I. Zucker. Computation by 'while' programs on topological partial algebras. *Theoretical Computer Science*, 219:379–420, 1999.
16. J.V. Tucker and J.I. Zucker. Computable functions and semicomputable sets on many-sorted algebras. In S. Abramsky, D. Gabbay, and T. Maibaum, editors, *Handbook of Logic in Computer Science*, volume 5, pages 317–523. Oxford University Press, 2000.
17. J.V. Tucker and J.I. Zucker. Abstract computability and algebraic specification. *ACM Transactions on Computational Logic*, 3:279–333, 2002.
18. J.V. Tucker and J.I. Zucker. Abstract versus concrete computation on metric partial algebras. *ACM Transactions on Computational Logic*, 5:611–668, 2004.
19. J.V. Tucker and J.I. Zucker. Computable total functions, algebraic specifications and dynamical systems. *Journal of Logic and Algebraic Programming*, 62:71–108, 2005.

Author Index

Lecture Notes in Computer Science

For information about Vols. 1–3962

please contact your bookseller or Springer

Vol. 4005: G. Lugosi, H.U. Simon (Eds.), Learning Theory. XI, 656 pages. 2006. (Sublibrary LNAI).

Vol. 4004: S. Vaudenay (Ed.), Advances in Cryptology - EUROCRYPT 2006. XIV, 613 pages. 2006.

Vol. 4003: Y. Koucheryavy, J. Harju, V.B. Iversen (Eds.), Next Generation Teletraffic and Wired/Wireless Advanced Networking. XVI, 582 pages. 2006.

Vol. 4001: E. Dubois, K. Pohl (Eds.), Advanced Information Systems Engineering. XVI, 560 pages. 2006.

Vol. 3999: C. Kop, G. Fliedl, H.C. Mayr, E. Métais (Eds.), Natural Language Processing and Information Systems. XIII, 227 pages. 2006.

Vol. 3998: T. Calamoneri, I. Finocchi, G.F. Italiano (Eds.), Algorithms and Complexity. XII, 394 pages. 2006.

Vol. 3997: W. Grieskamp, C. Weise (Eds.), Formal Approaches to Software Testing. XII, 219 pages. 2006.

Vol. 3996: A. Keller, J.-P. Martin-Flatin (Eds.), Self-Managed Networks, Systems, and Services. X, 185 pages. 2006.

Vol. 3995: G. Müller (Ed.), Emerging Trends in Information and Communication Security. XX, 524 pages. 2006.

Vol. 3994: V.N. Alexandrov, G.D. van Albada, P.M.A. Sloot, J. Dongarra (Eds.), Computational Science – ICCS 2006, Part IV. XXXV, 1096 pages. 2006.

Vol. 3993: V.N. Alexandrov, G.D. van Albada, P.M.A. Sloot, J. Dongarra (Eds.), Computational Science – ICCS 2006, Part III. XXXVI, 1136 pages. 2006.

Vol. 3992: V.N. Alexandrov, G.D. van Albada, P.M.A. Sloot, J. Dongarra (Eds.), Computational Science – ICCS 2006, Part II. XXXV, 1122 pages. 2006.

Vol. 3991: V.N. Alexandrov, G.D. van Albada, P.M.A. Sloot, J. Dongarra (Eds.), Computational Science – ICCS 2006, Part I. LXXXI, 1096 pages. 2006.

Vol. 3990: J. C. Beck, B.M. Smith (Eds.), Integration of AI and OR Techniques in Constraint Programming for Combinatorial Optimization Problems. X, 301 pages. 2006.

Vol. 3989: J. Zhou, M. Yung, F. Bao, Applied Cryptography and Network Security. XIV, 488 pages. 2006.

Vol. 3988: A. Beckmann, U. Berger, B. Löwe, J.V. Tucker (Eds.), Logical Approaches to Computational Barriers. XV, 608 pages. 2006.

Vol. 3987: M. Hazas, J. Krumm, T. Strang (Eds.), Location- and Context-Awareness. X, 289 pages. 2006.

Vol. 3986: K. Stølen, W.H. Winsborough, F. Martinelli, F. Massacci (Eds.), Trust Management. XIV, 474 pages. 2006.

Vol. 3984: M. Gavrilova, O. Gervasi, V. Kumar, C.J. K. Tan, D. Taniar, A. Laganà, Y. Mun, H. Choo (Eds.), Computational Science and Its Applications - ICCSA 2006, Part V. XXV, 1045 pages. 2006.

Vol. 3983: M. Gavrilova, O. Gervasi, V. Kumar, C.J. K. Tan, D. Taniar, A. Laganà, Y. Mun, H. Choo (Eds.), Computational Science and Its Applications - ICCSA 2006, Part IV. XXVI, 1191 pages. 2006.

Vol. 3982: M. Gavrilova, O. Gervasi, V. Kumar, C.J. K. Tan, D. Taniar, A. Laganà, Y. Mun, H. Choo (Eds.), Computational Science and Its Applications - ICCSA 2006, Part III. XXV, 1243 pages. 2006.

Vol. 3981: M. Gavrilova, O. Gervasi, V. Kumar, C.J. K. Tan, D. Taniar, A. Laganà, Y. Mun, H. Choo (Eds.), Computational Science and Its Applications - ICCSA 2006, Part II. XXVI, 1255 pages. 2006.

Vol. 3980: M. Gavrilova, O. Gervasi, V. Kumar, C.J. K. Tan, D. Taniar, A. Laganà, Y. Mun, H. Choo (Eds.), Computational Science and Its Applications - ICCSA 2006, Part I. LXXV, 1199 pages. 2006.

Vol. 3979: T.S. Huang, N. Sebe, M.S. Lew, V. Pavlović, M. Kölsch, A. Galata, B. Kisačanin (Eds.), Computer Vision in Human-Computer Interaction. XII, 121 pages. 2006.

Vol. 3978: B. Hnich, M. Carlsson, F. Fages, F. Rossi (Eds.), Recent Advances in Constraints. VIII, 179 pages. 2006. (Sublibrary LNAI).

Vol. 3977: N. Fuhr, M. Lalmas, S. Malik, G. Kazai (Eds.), Advances in XML Information Retrieval and Evaluation. XII, 556 pages. 2006.

Vol. 3976: F. Boavida, T. Plagemann, B. Stiller, C. Westphal, E. Monteiro (Eds.), Networking 2006. Networking Technologies, Services, and Protocols; Performance of Computer and Communication Networks; Mobile and Wireless Communications Systems. XXVI, 1276 pages. 2006.

Vol. 3975: S. Mehrotra, D.D. Zeng, H. Chen, B. Thuraisingham, F.-Y. Wang (Eds.), Intelligence and Security Informatics. XXII, 772 pages. 2006.

Vol. 3973: J. Wang, Z. Yi, J.M. Zurada, B.-L. Lu, H. Yin (Eds.), Advances in Neural Networks - ISNN 2006, Part III. XXIX, 1402 pages. 2006.

Vol. 3972: J. Wang, Z. Yi, J.M. Zurada, B.-L. Lu, H. Yin (Eds.), Advances in Neural Networks - ISNN 2006, Part II. XXVII, 1444 pages. 2006.

Vol. 3971: J. Wang, Z. Yi, J.M. Zurada, B.-L. Lu, H. Yin (Eds.), Advances in Neural Networks - ISNN 2006, Part I. LXVII, 1442 pages. 2006.

Vol. 3970: T. Braun, G. Carle, S. Fahmy, Y. Koucheryavy (Eds.), Wired/Wireless Internet Communications. XIV, 350 pages. 2006.

Vol. 3969: Ø. Ytrehus (Ed.), Coding and Cryptography. XI, 443 pages. 2006.

Vol. 3968: K.P. Fishkin, B. Schiele, P. Nixon, A. Quigley (Eds.), Pervasive Computing. XV, 402 pages. 2006.

Vol. 3967: D. Grigoriev, J. Harrison, E.A. Hirsch (Eds.), Computer Science – Theory and Applications. XVI, 684 pages. 2006.

Vol. 3966: Q. Wang, D. Pfahl, D.M. Raffo, P. Wernick (Eds.), Software Process Change. XIV, 356 pages. 2006.

Vol. 3965: M. Bernardo, A. Cimatti (Eds.), Formal Methods for Hardware Verification. VII, 243 pages. 2006.

Vol. 3964: M. Ü. Uyar, A.Y. Duale, M.A. Fecko (Eds.), Testing of Communicating Systems. XI, 373 pages. 2006.

Vol. 3963: O. Dikenelli, M.-P. Gleizes, A. Ricci (Eds.), Engineering Societies in the Agents World VI. XII, 303 pages. 2006. (Sublibrary LNAI).